The Evolution and

Ecology of Armadillos, Sloths, and Vermilinguas

G. Gene Montgomery, *Editor*

Smithsonian Institution Press

Washington and London

1985

Library of Congress Cataloging in Publication Data
Main entry under title:

The Evolution and ecology of armadillos, sloths, and vermilinguas.

Bibliography:
Includes index.
Supt. of Docs. no.: SI 1.2:Ev6
1. Edentata. 2. Edentata, Fossil. I. Montgomery, G. Gene (Gerald Gene)
QL737.E2E94 1985 599.3′1 84-600292
ISBN 0-87474-649-3

The paper in this book meets the guidelines for permanence and durability of the Committee on Production Guidelines for Book Longevity of the Council on Library Resources.

Contents

ANATOMY AND PHYSIOLOGY

REPRODUCTION

DISEASES AND PARASITES

ARTHROPOD ASSOCIATES OF SLOTHS

ECOLOGY OF LIVING SLOTHS

ECOLOGY OF LIVING VERMILINGUAS

ECOLOGY OF LIVING ARMADILLOS

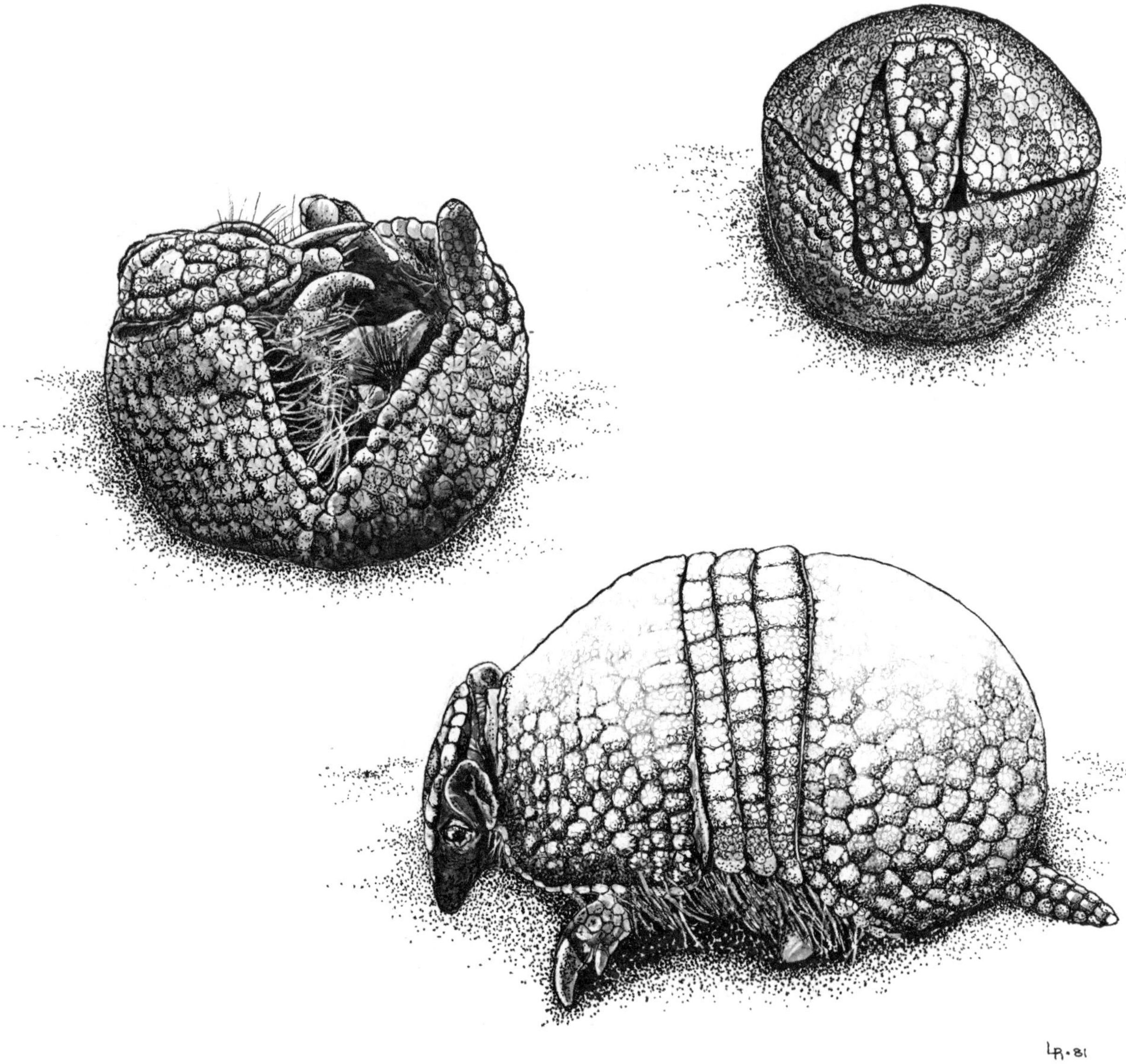
LR·81

INTRODUCTION

The following collection of papers is the result of a symposium, ''The Evolution and Ecology of Xenarthra,'' organized by G. Gene Montgomery, on the occasion of the 59th Annual Meeting of the American Society of Mammalogy held in 1979.

The Xenarthra, the American edentates, are a group of particular interest. The various species play important roles in the ecosystems of tropical America and the south temperate ''Patagonian'' region of South America. One species is invading warm temperate North America. The living forms—the armadillos, treesloths, and anteaters—are diverse; equally specialized, in different ways, are the recently extinct groundsloths and glyptodonts. As an order, the Xenarthra are a classical example of adaptive radiation. They can be diagnosed by certain anatomical features but cannot, any longer, be characterized by common traits of behavior or natural history.

Gene Montgomery not only undertook the tremendous task of editing these proceedings, but he also invited others to add papers to the book to ensure a wider coverage of Xenarthran biology. Unfortunately, a number of obstacles caused long delays in the editing process. Gene compiled the vast majority of the papers, but when it became clear that he was unable to finish the job, others volunteered to complete the few remaining tasks. An introductory chapter and two papers that were unavailable or incomplete could not be included. Despite these absences the book includes forty-three papers on a large variety of aspects of the biology of armadillos, sloths, and vermilinguas and is, without doubt, the most complete work published to date on the biology of these exceedingly interesting animals.

Ira Rubinoff

Director, Smithsonian Tropical
Research Institute

TAXONOMY
AND
DISTRIBUTION

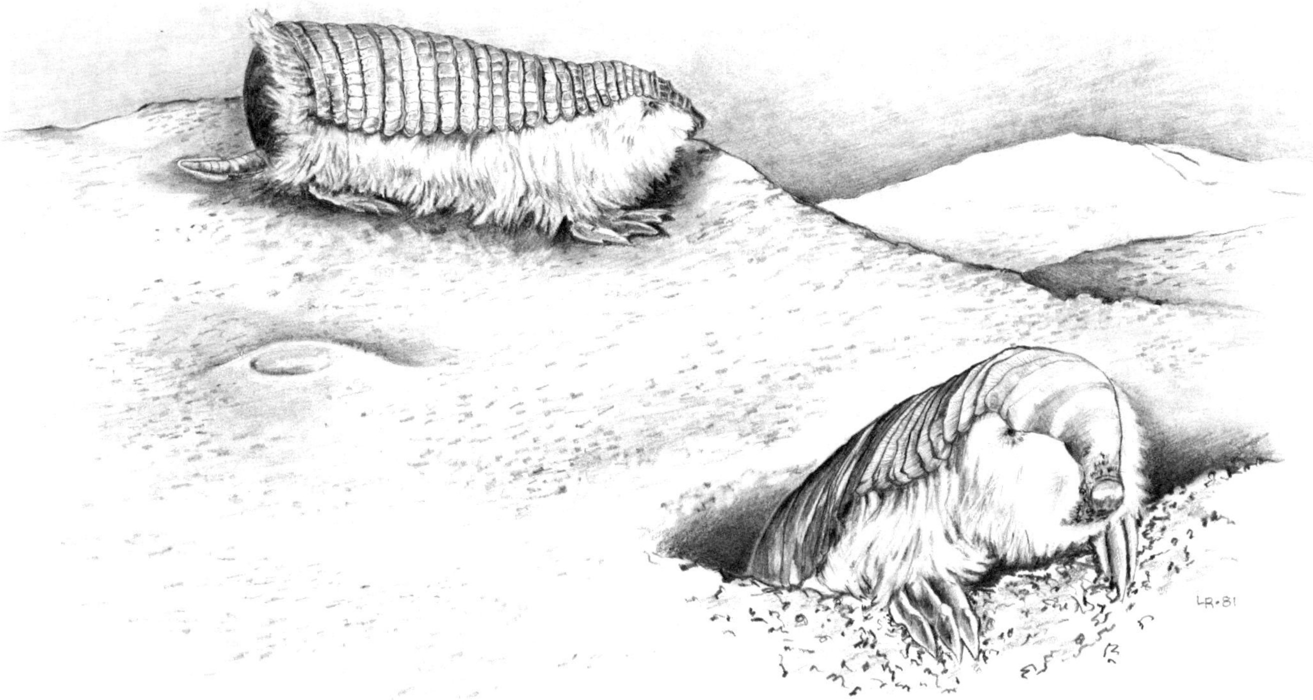

HISTORY OF CLASSIFICATION AND NOMENCLATURE IN XENARTHRA (EDENTATA)

Bryan P. Glass

Oklahoma State University Museum, Stillwater, Oklahoma 74074.

Abstract

Xenarthra is a legitimate descriptive ordinal name for the living neotropical sloths, anteaters, and armadillos, as well as for the extinct ground sloths and glyptodonts. Use of Xenarthra as an ordinal name rather than Edentata for a taxon which includes mostly toothed forms, excepting only the vermilinguas (Myrmecophagidae), recognizes a unique and fundamental skeletal character common to all. In this paper I show that the name Xenarthra is valid, and propose that it replace Edentata as an ordinal name for this neotropical group of mammals.

Abstracto

Xenarthra es un legítimo nombre descriptivo de la Orden que incluye a los perezosos, hormigueros y armidillos neotropicales vivientes, como también a los extintos perezosos terrestres y gliptodontes. El uso de Xenarthra como nombre de una Orden, en vez de Edentata, para un taxon que incluye en su mayoría formas dentadas, con excepción solamente de los vermilinguas (Myrmecophagidae), reconoce una fundamental y singular characterística común a todos los miembros de la Orden. En este artículo yo demuestro la validez del nombre de la Orden que incluye a este grupo neotropical de mamíferos.

Resumo

Xenarthra é um legítiomo nome ordinal, descritivo, para as preguiâs, tamanduās e tatus neotropicais atuais, bem como para as extintas preguiâs terrícolas e gliptodontes. O uso de Xenarthra como um nome ordinal, em vez de Edentata, para um taxon que inclui, na maior parte, formas dentadas, excetuando somente os vermilinguas (Myrmecophagidae), reconhece um carāter esquelético único e fundamental, comum a todos. Neste trabalho, mostro que o nome Xenarthra é válido e proponho que este substitua Edentata como um nome ordinal para este grupo de mamíferos neotropicais.

When Linnaeus (1758) published the tenth edition of *Systema Naturae* he cast the mammals known to him into eight Orders, of which one, Bruta, included elephants, anteaters, manatees, sloths, and pangolins, while another, Bestiae, included armadillos, pigs, some insectivores, and opossums. While Gill (1872), Hay (1902), and other authors used Bruta rather than Edentata, Simpson (1945) concluded that Linnaeus did not recognize the xenarthran edentates as a natural group at all, and in his classification he merely noted Linnaeus' Bruta as a partial synonym of the order Edentata.

The earliest post-Linnaean name that included the South American xenarthrans is Mutici (Wolffer, 1780), probably erroneously credited to Storr by Simpson and others, which came close to embracing the natural group recognized today but included the aardvark, as one of the anteaters, and the pangolins. Simpson (1945), in using the name Edentata as an ordinal name, noted that Mutici was "historically antecedent but now forgotten." Mutica is a Linnaean (1766) name for the Cetacea, though he used Cete, not Mutica, in 1758. Simpson used Mutica as a cohort embracing modern whales and their fossil antecedents, and it would be unfortunate to use so similar a name for such different animals. While Simpson did not pursue the nomenclatural intricacies of these old names, it seems reasonable to consider Mutici as a junior homonym of Mutica, the two spellings probably being based on different understandings of declension, gender, or some other nuance of Latin grammar.

The ordinal name Edentata dates from Cuvier (1798), although the name Edentati (Vicq d´Azyr, 1792) predates it by six years. Simpson (1945) used the former, while listing Vicq d´Azyr´s prior name without comment in his synonymy. Advocating either spelling does not alter basic nomenclature, only authorship.

The classic contents of the Edentata, as used by Cuvier and later authors, were the sloths, anteaters, armadillos, pangolins, and the aardvark. The latter two were removed to separate orders by Huxley (1872), whose Tubulidentata is currently used for the aardvarks but whose Squamata for the pangolins was preoccupied. The name Pholidota for the pangolins dates from Weber (1904).

Other additions to the Edentata occurred with the tremendous increase in knowledge of fossil forms from the late eighteenth century onward. According to Simpson (1945), the ground sloths were added in 1796 and the glyptodonts in 1838, although he cited no references bearing those dates. The Taeniodonta were added to it by Wortmann (1896) but later removed to a separate order (Matthew, 1937), and the Palaeanodonta were added by Matthew (1918). Since 1918 the Order Edentata has usually been recognized as comprising the palaeanodonts, armadillos, glyptodonts, sloths, ground sloths, and anteaters.

Simpson (1945) followed Scott (1937) in dividing the order into two suborders: the Palaeanodonta, for that group only, and the Xenarthra, comprising all others both extinct and recent. Within the Xenarthra, Simpson designated two infraorders: the Pilosa, for those with body hair, and the Cingulata, for the armored forms. Gill (1872) used the word "xenarthral" as an adjective, but Cope (1889) apparently was the first to use Xenarthra as a formal taxon. Simpson's (1945) classification was modified by Romer (1966), who divided the Xenarthra into four superfamilies: the Mylodontoidea containing only the Mylodontidae; the Megalonychoidea comprising the Megalonychidae, Megatheriidae, and Bradypodidae; the Loricata (which he grudgingly indicated was equivalent to the Cingulata), comprising the Glyptodontidae and the Dasypodidae; and the Vermilingua, (Illiger, 1811; Gill, 1872; Osborn, 1910), containing only the Myrmecophagidae. Although Romer (1966, 1968) clearly recognized the Edentata as a distinct order, with priority dating to the eighteenth century, in his chapters entitled "Edentata" he included the pangolins and the aardvark with no more distinction than that accorded the anteaters or other natural members of that Order.

The most recent change in classification is that of Emry (1970), who described an Oligocene pangolin from the White River deposits of Natrona County, Wyoming, and, in an extended review of its relationships, concluded that the Palaeanodonta were ancestral to the pangolins and not to the xenarthrans. Using his scheme, the families Metacheiromyidae, Epoicotheriidae, and Manidae constitute the Pholidota, thus leaving only Romer's seven families (see above) in the Edentata (Xenarthra). Emry's argument that the Palaeanodonta could not be ancestral to the Xenarthra, being of more recent age than the oldest xenarthrans (a view acknowledged by Simpson and Romer) and their similarities are therefore due to convergence, is strong.

According to this latest arrangement all edentates, living and fossil, have "xenarthrous" vertebrae (i.e., they have at least dual and, in many cases, even more intervertebral articulations in addition to the usual zygapophyses), though they are rudimentary in the Bradypodidae (Flower and Lydekker, 1891). They also have ischial articulations with the spinal column, thus incorporating into the sacrum vertebrae that which in other mammals would be caudal elements. In the scapula, the acromion and coracoid processes are extraordinarily developed. Simplification to homodonty or else complete loss of teeth has occurred, and in many but not all xenarthrans the manus flexes inward. These characters are as typical of ground sloths and glyptodonts as they are of living forms.

One could argue that the name Edentata, meaning "toothless" was originally coined to apply to anteaters, pangolins, and aardvarks, and is more properly to be considered as a synonym of the Pholidota and Tubulidentata, which would leave the name Xenarthra clear as the ordinal name for the New World members of this nomenclaturally troubled group.

Given the taxonomic homogeneity of the Xenarthra, their zoogeographic restriction to the Western Hemisphere, their essentially Neotropical habitat, and the availability of this valid name based on a fundamental and unique skeletal character, I propose that the name Xenarthra be adopted as the ordinal name.

LITERATURE CITED

Cope, E. D.
1889. The Edentata of North America. Am. Nat., 23:657-664.

Cuvier, G. L. S. F. D.
1798. Tableau elementaire de l'histoire naturelle des animaux. Paris, J. B. Bailliére, 710 pp.

Emry, R. J.
1970. A North American Oligocene pangolin and other additions to the Philidota. Bull. Amer. Mus. Nat. Hist., 142:455-570.

Flower, W. H. and R. Lydekker.
1891. An introduction to the study of mammals, living and extinct. London. Adam and Charles Black, 763 pp.

Gill, T.
1872. Arrangement of the families of mammals and synoptic tables of characters of the suborders of mammals. Smithsonian Misc. Coll., 230, 98 pp.

Hay, O. P.
1902. Fossil vertebrata of North America. Bull. U.S. Geol. Survey, 179:1-868.

Huxley, T.
1872. A manual of the anatomy of vertebrated animals. New York, D. Appleton and Co., 431 pp.

Illiger, C.
1811. *Prodromus systematis mammalium* etc. Berlin, C. Salfeld., 301 pp.

Linnaeus, C.
1758. *Systema Naturae,* etc. 10th ed. Stockholm, Laurentii Salvii. 824 pp.

1766. *Systema Naturae,* etc. 12th ed. revised. Stockholm, Laurentii Salvii Vol. 1 pt. 1, 532 pp.

Matthew, W. D.
1918. A revision of the lower Eocene Wasatch and Wind River faunas. Part 5 - Insectivora (part.), Glires, Edentata. Bull. Amer. Mus. Nat. Hist., 38:565-657.

1937. Paleocene faunas of the San Juan basin, New Mexico. Trans. Amer. Phil. Soc., N.ser., 30:1-510.

Osborn, H. F.
1910. The age of mammals in Europe, Asia, and North America. New York, MacMillan and Co., 635 pp.

Romer, A. S.
1966. Vertebrate Paleontology. 3rd Ed. Univ. of Chicago Press, Chicago, Ill. 468 pp.

1968. Notes and comments on Vertebrate Paleontology. Univ. of Chicago Press, Chicago, Ill. 304 pp.

Scott, W. B.
1937. A history of land mammals in the western hemisphere. Rev. ed. New York. MacMillan and Co., 786 pp.

Simpson, G. G.
1945. The principles of classification and a classification of mammals. Bull. Amer. Mus. Nat. Hist., 85, 350 pp.

Vicq d'Azyr, M. F.
1792. Systéme anatomique des quadrupedes. Encycl. Methodique. Paris, Vve. Agasse, Vol. 2.

Weber, M.
1904. Die saugetiere. Jena, Gustav Fischer, 866 pp.

Wolffer, F.
1780 *Prodromus methodi Mammalium* Diss. Tubingen (Praeses. Gottl. Konr. Christ Storr). 43 pp.

Wortmann, J. L.
1896. Psittacotherium, a member of a new and primitive order of the Edentata. Bull. Amer. Mus. Nat. Hist., 8:259-262.

THE IDENTIFICATION AND DISTRIBUTION OF RECENT XENARTHRA (= EDENTATA)

Ralph M. Wetzel

Biological Sciences Group, University of Connecticut, Storrs, Connecticut 06268, USA. (Present Address: The Florida State Museum, University of Florida, Gainesville, Florida 32611, USA)

Abstract
An identification key, description of ranges based upon known specimens, and summaries of natural history are presented for Recent species of Xenarthra (= Edentata).

Abstracto
Una clave, descripciones de distribución basadas en especímenes conocidos, y resúmenes de sus historias naturales, se presentan para las especies vivientes de los Xenarthra (= Edentata).

Resumo
Uma chave, descrições, baseadas nos espécimes conhecidos, e sumários da história natural são apresentados para espécies recentes de Xenarthra (= Edentata).

Some genera and species of the order Xenarthra (= Edentata) are well known as types of three extant families, such as *Myrmecophaga tridactyla* L., the giant anteater; *Bradypus tridactylus* L., the three-toed sloth; and *Dasypus novemcinctus* L., the common long-nosed armadillo. The remaining two dozen or so species are not especially controversial (which is, perhaps, too aggressive a word for most Xenarthrans), but do lack uniform support by mammalogists. This is to be expected as the order has received little recent taxonomic attention. The species arrangement and distribution of Xenarthra used in continental faunal catalogues (Cabrera, 1958; Hall, 1981), in encyclopedic presentations of mammals of the world (Moeller, 1975; Walker, 1975), or in reports of more regional or specialized nature are necessarily patchwork compromises of a varied, often less than satisfactory, basic literature.

Against such odds I present a summary of the Recent species of Xenarthra in the form of an identification guide, and elsewhere (Wetzel, 1985) a review of the armadillos, Dasypodidae. Despite being truncate, this key is based upon an examination of most of the original literature, extant holotypes, and other specimens in the major collections of South American mammals.

Except where noted, all statements of distribution are based on specimens identified in this study. All measurements are in millimeters; those for the body were taken from the specimen label, and cranial measurements were made by me. When possible, for each specimen I used the length of the hind foot, which included the longest claw, and the length of the ear taken from the notch. The measurements of both body and skull are presented for only wild-captured specimens whose occipital and occipital-basisphenoid sutures were fused. I believe these data are from the adults and young adults of the populations. These were the sexually mature ones, as suggested by specimens that had supportive reproductive data.

"Vermilinguas" (worm-tongued) is actually a more descriptive vernacular name than "anteaters" for the family Myrmecophagidae, as most members feed heavily upon termites as well as ants.

For vermilinguas and sloths, I give average weights, body measurements, and skull measurements, type localities and pertinent synonyms; these are reported elsewhere for armadillos (Wetzel, 1985).

The mean, if not labeled, can be recognized as the first number listed, followed by ± the standard deviation of the sample and, in parentheses, the observed range and the number in the sample (N). Total length indicates the sum of the lengths of head, body and tail. Where appropriate, the greatest length of the skull is indicated as

CNL (condylonasal length). Rostral length is the distance along the midline from a line through the lacrimal foramina to the tip of the nasal bones (Wetzel, 1985). Tooth counts are given as fractions, maxillary pairs/mandibular pairs. Counts of movable bands of armadillos were made along the middorsal line of adults and subadults and, due to some dorsal fusion of bands in adults, the counts are more conservative than for immatures. Counts of scutes on the fourth movable band include the marginal scutes.

KEY TO THE RECENT XENARTHRA

1a.-- Toothed; tongue not a slender cylinder; tail variable length but never furred except when extremely short; jugal normal size; pterygoids separate and not forming posterior margin of palate: **Bradypodidae**, **Megalonychidae** and **Dasypodidae**; sloths and armadillos..... **5**

1b.-- Toothless; tongue elongate and slender; tail nearly as long or longer than head + body and partially or completely furred; jugal small; pterygoids meeting at midline and forming posterior margin of palate: **Myrmecophagidae**; Vermilinguas (Ant and Termite Eaters)...... **2**

Figure 1. Distribution of two-toed or pigmy anteater *Cyclopes didactylus* **(horizontal lines)** and giant anteater *Myrmecophaga tridactyla* **(vertical lines).** See text for citation of specimens of *C. didactylus* from coastal Pernambuco and Alagoas in extreme eastern Brazil and for most southerly record of *M. tridactyla* in Rio Grande do Sul.

MYRMECOPHAGIDAE; Vermilinguas

2(1)a.-- Rostrum short (less than 35 percent of total length of skull); small, length of head + body less than 500, skull less than 60; arboreal:

Cyclopes didactylus (L.). Two-toed or pygmy anteater; tapacara, (Panama); angelito (Colombia); tamanduá-í.
Myrmecophaga didactyla Linné, 1758.
Cyclopes didactylus Cabrera, 1958; Moeller, 1975; Hall, 1981.

Type locality.-- Surinam.
Distribution.-- Forests of tropical Mexico, Veracruz, into South America west of the Andes (Fig. 1) possibly as far south as northwestern Peru (Grimwood, 1969); east of Andes through forests of Orinoco and Amazon basins; extreme eastern Brazil in coastal Pernambuco and Alagoas (Maceió, MN 17294; Viçosa, Sitio Novo, MN 17295; Usina Sinimbú, MZUSP 7523, 8298); western Brazil (Amazonas) and Amazonian Peru to southeastern Bolivia (Santa Cruz, Buenavista and Santa Cruz).
Comments.-- A small, arboreal, nocturnal (Sunquist and Montgomery, 1973a) specialist on ants which live in twigs and the small stems of lianas; very rarely eats termites (Montgomery, 1979, 1985a, 1985b); hind foot highly modified for grasping small twigs while climbing and feeding, tail prehensile; 2 toes on manus, 4 on pes; color gradient in Central America from golden yellow to yellow with darker middorsal stripe; in South America, silvery gray dorsum with chocolate middorsal stripe, darkest in Amazon basin with loss of prominent stripes but with more yellow on dorsum in peripheral populations.
Size and weight.-- Based on adults from the Amazon basin of Brazil: total length 433 ± 31 (366-500, N 64); length of tail 232 ± 26 (165-295, N 64); length of pes 38.1 ± 3.7 (30-50, N 64); greatest length of skull 50.0 ± 4.1 (47.6-53.3, N 67); ratio of length of rostrum/greatest length of skull 0.28 ± 0.02 (0.24-0.32, N 57). Adults from Mexico to Brazil: length of ear 12.7 ± 1.9 (10-18, N 24); weight 234.7 ± 63.8 g (175-357, N 8).

2(1)b.-- Rostrum elongate (over 40 percent of total length of skull); large, length of head + body more than 600 and skull more than 100; terrestrial, arboreal or both: ***Myrmecophaga*** or ***Tamandua***....................... **3**

3(2)a.-- Digits on manus 4, the fourth inconspicuous; rostrum much longer than rest of skull (65 percent of CNL, N 25); large (total length over 1500); tail with long hair, banner-like; adults entirely terrestrial:

Myrmecophaga tridactyla L. Giant anteater, hormiguero gigante; oso caballo; oso palmera; tamanduá-bandeira.

Myrmecophaga tridactyla Linné, 1758; Cabrera, 1958; Hall, 1981; Moeller, 1975.

Type locality.-- Brazil, Pernambuco.

Distribution.-- Forests and savannas from Belize through South America (Fig. 1) to northern Argentina (Salta, Formosa, Chaco, Misiones) and southern Brazil (Rio Grande do Sul, São Laurenco do Sul, MZUSP 1771).

Comments.-- Large, nocturnal (Montgomery and Lubin, 1977) or diurnal-crepuscular (Shaw et al., 1985), terrestrial, wide-ranging (Montgomery, 1979), acute olfaction (McAdam and Way, 1967); specialist on ants, at least in the llanos of Venezuela (Montgomery, 1979; 1985b) and the Chaco of Paraguay; can be effectively bipedal, as with other members of the family, during moments of inquiry, defense, digging, and feeding; does not burrow, but curls in depression on ground or under shrub, covered with tail. One young, following gestation of 190 days (Asdell,1964); young carried on back of mother, later sometimes travels with mother. Four claws on manus, five on pes, head long, cranium low and narrow, more slender than muscular neck and covered with short, gray hair; black bands edged with white extend from throat and chest over shoulder and back; coarse, long, black hair on midline of back and tail. Endangered by man's activities over much of its range.

Size and weight.-- Based on adults from Brazil, Colombia, Ecuador, Guyana, Peru, Surinam, and Venezuela: total length 2000 ± 241 (1740-2817, N 16); length of tail 734 ± 76 (645-900, N 16); length of pes 165 ± 8 (150-180, N 13); length of ear 46.7 ± 4.7 (35-50, N 9); weight 32.9 ± 7.1 kg (22-39, N 5); CNL 359 ± 19 (313-409, N 52); ratio of rostral length/CNL 0.65 ± 0.02 (0.63-0.73, N 49).

3(2)b.-- Digits on manus 4, fourth digit not inconspicuous; rostrum proportionally intermediate among vermilinguas (43 percent of CNL, N 49); size intermediate (total length 1161, N 16); tail nearly naked for distal three-quarters; terrestrial and arboreal:
Tamandua **4**

4(3)a.-- Always black vested against a paler background; ear shorter (means of geographical samples 40-46); 4 pairs of orbital foramina; posterior border of infraorbital foramen distinctly crescent shaped (Fig. 2):

Tamandua mexicana (Saussure). Northern tamandua; oso hormiguero; tamanduá-mirim.

Myrmecophaga tamandua var. *mexicana* Saussure, 1860.

Tamandua tetradactyla: Cabrera, 1958 and Moeller, 1975, both in part.

Tamandua mexicana: Wetzel, 1975; Hall, 1981.

Type locality.-- Mexico, Tabasco.

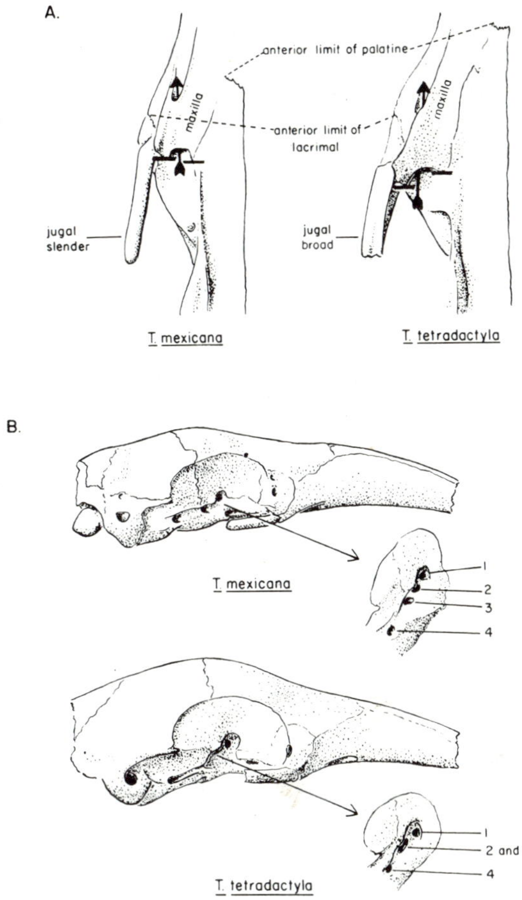

Figure 2. Skulls of *Tamandua*. **A** Right palate. Infraorbital foramen is marked by arrow and medial and lateral margins of the posterior opening, by bars. **B** Right side of skulls and detailed view of orbits, tilted to show foramina: **1** optic foramen; **2** orbital fissure; **3** foramen rotundum; **4** foramen ovale. Foramina **2** and **3** are separate in *T. mexicana* but combined in *T. tetradactyla* or separated by a deeply set, slender septum which is visible from lateral view only at its base (after Wetzel, 1975).

Distribution.-- Southeastern edge of Mexican plateau into South America west of Andes from northwestern Venezuela to northwestern Peru (Fig. 3).

Comments.-- Medium-sized, nocturnal, crepuscular or diurnal, arboreal and/or terrestrial (Montgomery and Lubin, 1977), feeding on a combination of ants and termites (Montgomery, 1985a, 1985b); dens in hole in ground or tree; single young following gestation period of 130-150 days (Silveira, 1968, for Panama); 4 claws on manus, 5 on pes; low cranium smaller than muscular neck, slender rostrum intermediate in proportional and actual length between the short *Cyclopes* and the longer *Myrmecophaga*; forelimb very strongly muscled and modified for tearing at feeding substrate (Taylor, 1985); tail prehensile; black vest on back and sides, tan on head, neck, legs, furred base on tail and middorsal stripe from neck to at least shoulders.

Size and weight.-- Based on adults from Mexico to NW Venezuela: total length 1111 ± 54 (1020-1300, N 60); length of tail 544 ± 47 (400-675, N 59); length of hind foot 98 ± 10 (80-110, N 56); length of ear 44.4 ± 3.6 (37-51, N 42); weight 4.32 ± 0.68 kg (3.2-5.4, N 15); CNL 126.4 ± 5.8 (114.8-138.9, N 72); ratio rostral length/CNL 0.45 ± 0.01 (0.41-0.47, N 71).

4(3)b.-- Pelage variable; uniformly golden, brown, or black, or partially to completely black vested against a paler background; ear longer (means 50-54 nearest range of *T. mexicana*); 3 pairs of orbital foramina in most skulls; posterior border of infraorbital foramen an incomplete crescent, its lateral border distinctly posterior to medial border (Fig. 2):

Tamandua tetradactyla L. Southern tamandua; oso melero; caguaré; tamanduá-mirim.

Myrmecophaga tetradactyla Linné, 1758.

Myrmecophaga longicaudata: Wagner, 1844; as *T. longicaudata*, Cabrera, 1958; Soukup, 1965; Husson, 1978.

Tamandua tetradactyla: Cabrera, 1958, and Moeller, 1975, both in part; Wetzel, 1975; Hall, 1981.

Type locality.-- Brazil, Pernambuco.

Distribution.-- South America east of Andes from Venezuela to Argentina (Sante Fe, Chaco, Salta, Jujuy) and northern Uruguay (Fig.3).

Comments.-- Medium-sized, nocturnal, feeding on a combination of ants and termites with termites predominating, at least in the llanos of Venezuela (Montgomery and Lubin, 1977) to diurnal-crepuscular in winter of Paraguayan Chaco and without predominance of termites over ants. Pelage geographically variable from uniformly pale tan or golden in northwestern to black vested in southeastern part of range. Between these extremes are partially vested, and even black or dark brown phases (Fig. 3).

Size and weight.-- Based on adults from Venezuela to Argentina: total length 1105 ± 71 (955-1300, N 83); length of tail 511 ± 58 (402-672, N 83); length of hind foot 100 ± 8 (80-118, N 71); length of ear 49.4 ± 4.6 (41-57, N 53); weight 4.83 ± 0.90 kg (3.42-7.00, N 41); CNL 129.7 ± 4.6 (120.2-142.2, N 88); ratio of rostral length/CNL 0.43 ± 0.02 (0.38-0.48, N 87).

5(1)a.-- Dermal armor on dorsum of head and body; forelegs much shorter than head + body; tail conspicuous and not furred; zygoma complete: **Dasypodidae**; armadillos.. **10**

5(1)b.-- No dermal armor; forelegs as long or longer than head + body; tail inconspicuous and furred; zygoma

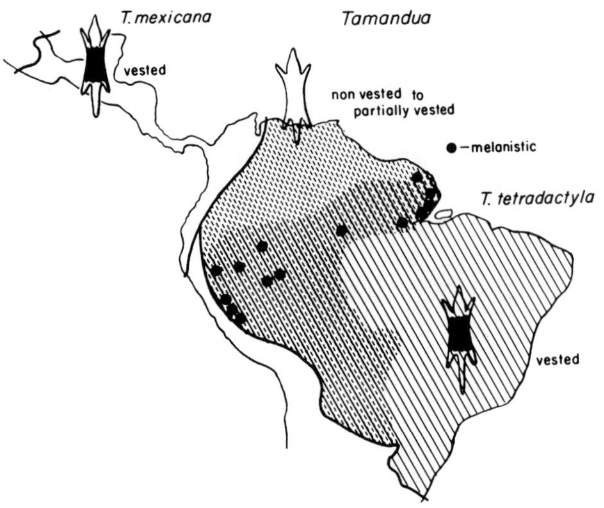

Figure 3. Distribution of *Tamandua* species and geographical variation of pelage (after Wetzel, 1975).

incomplete: **Bradypodidae** and **Megalonychidae**; sloths.. **6**

6(5)a.-- Larger, except for tail; digits 2 on manus, 3 on pes; lacking dense fur layer; forelegs only slightly longer than hind legs; neck shorter and less mobile, cervical vertebrae 5-7, only occasionally 8; teeth with sharp, beveled occlusal surfaces, anterior pair prominent and caninelike, triangular in outline; distinct rostrum, with prenasal bone: **Megalonychidae**; two-toed sloths.... **9**

6(5)b.-- Smaller, except for tail; digits 3 on manus and pes; forelegs 1.5 times as long as hind legs; long, mobile neck with 8 or 9 cervical vertebrae; teeth simple pegs of approximately equal size except for anterior maxillary pair which is always smaller and may sometimes be missing; little rostral development: **Bradypodidae**; three-toed sloths... **7**

BRADYPODIDAE; three-toed sloths

7(6)a.-- Pelage uniformly pale brown over body and head except for pronounced long, black hair on base of neck, projecting over shoulders as two plumes; no male speculum; skull with pterygoid sinuses distinctly inflated:

SUBGENUS ***Scaeopus*** Peters.

Bradypus torquatus Desmarest. Maned sloth; preguiça de coleira.

Bradypus torquatus Illiger, 1811; Cabrera, 1958; Wetzel and Avila-Pires, 1980. *Nomen nudum*.

Bradypus torquatus Desmarest, 1816.

Scaeopus torquatus: Paula Couto, 1979 (errata); Avila-Pires in Wetzel and Avila-Pires, 1980.

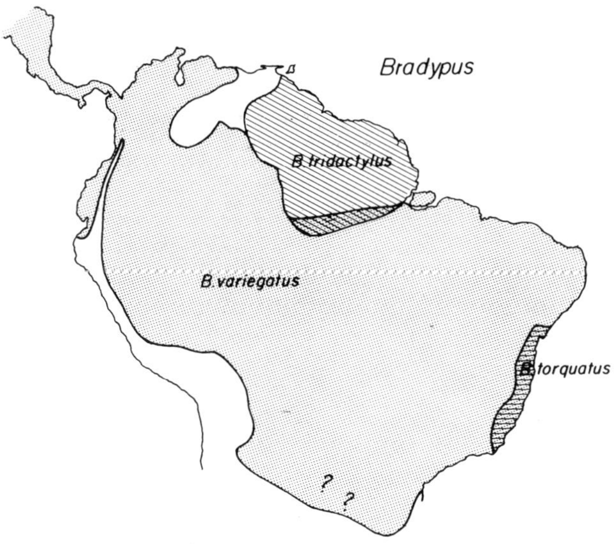

Figure 4. Distribution of three-toed sloths *Bradypus*.

Type locality.-- Brazil, Atlantic drainage of Bahia, Espírito Santo and Rio de Janeiro (Wetzel and Avila-Pires, 1980).
Distribution.-- Populations of this endangered species have dwindled alarmingly with destruction of coastal forests in southeastern Brazil; remnants probably remain in the states of Bahia, Espírito Santo, and Rio de Janeiro (Fig. 4).
Comments.-- Perhaps the most endangered of South American mammals, certainly of the xenarthrans; very little is known about the life history of this sloth. It is considered by T. M. S. Santos (Paula Couto, 1979) to be generically distinct, as *Scaeopus torquatus*.
Size and weight.-- Based on adults from Bahia, Espírito Santo, and Rio de Janeiro: total length 520 (500, 540, N 2); length of tail 49 (48, 50, N 2); length of hind foot 107 (100, 115, N 2); weight 3.88 kg (3.60, 4.15, N 2); greatest length of skull 77.2 ± 3.2 (73.5-84.3, N 10); length of mandibular spout (midline projection of mandible anterior to first tooth) 6.1 ± 0.8 (4.8-7.1, N 8).

7(6)b.-- Pelage with blotches of pale color on darker dorsum; obvious white, dense fur layer; males with middorsal yellow to orange speculum; pterygoids not inflated: SUBGENUS ***Bradypus*** L. (see Wetzel and Koch, 1973; Wetzel and Avila-Pires, 1980).......... 8

8(7)a.-- Throat and sides of face brown; prominent dark brown forehead and suborbital stripe outlining paler color of ocular area of face; no foramina in anterodorsal nasopharynx; skull larger and mandibular spout shorter (Fig. 5):

Bradypus variegatus Schinz. Brown-throated three-toed sloth; ai-ai; preguiça.
Bradypus variegatus Schinz, 1825; Wetzel and Kock, 1973; Wetzel and Avila-Pires, 1980; Hall, 1981.
Bradypus infuscatus Wagler, 1831; Cabrera, 1958; Soukup, 1960; Baker, 1974.
Bradypus tridactylus not Linné, 1758, but Moeller, 1975, in part.
Arctopithecus griseus Gray, 1871a.
Arctopithecus boliviensis Gray, 1871b; as *B. boliviensis* , Cabrera, 1958; Bucher, 1980; Mares, et al., 1981.
Bradypus cuculliger not Wagler, 1831, but Moeller, 1975, in part.

Type locality.-- Brazil, possibly Bahia.
Distribution.-- From Honduras into South America to coastal Ecuador, through Colombia and Venezuela (except for delta of Orinoco, llanos, and Guiana highlands), continuing east of the Andes through the forests of Ecuador, Peru, and Bolivia to northern Argentina (Jujuy, FMNH 21672) and, according to Cabrera (1958), also Formosa, Chaco, Misiones; forested Brazil (except Amapá) to Paraná (Londrina, MZUSP 6639) and Rio Grande do Sul (Cabrera, 1958) (Fig. 4).
Comments.-- Perhaps the most common medium-sized mammal in many neotropical forests; arboreal folivore feeding on leaves of many species of trees and lianas (Montgomery and Sunquist, 1975; 1978); acyclic as regards activity pattern (Sunquist and Montgomery, 1973b); adults are long-lived and females produce one young per year; descend to the forest floor to defecate only about once each week (Montgomery and Sunquist, 1978); live in microsympatry with two-toed sloths *Choloepus*; relatively unaggressive. Larger specimens from Bolivia are part of a gradient of decreasing size northeasterly into the Amazon Basin and are not separable at the species level as *Bradypus boliviensis* Gray (Wetzel, unpublished data; Wetzel and Avila-Pires, 1980).

Size and weight.-- Based on adults from Nicaragua to Brazil: total length 578 ± 59 (420-800, N 108); length of tail 58 ± 14 (38-90, N 101); length of hind foot 122 ± 16 (90-180, N 101); length of ear 13 ± 3 (8-22, N 41); weight 4.34 ± 0.85 kg (2.25-5.50, N 25); greatest length of skull 75.8 ± 4.9 (65.1-87.4, N 107); length of mandibular spout 1.4 ± 1.0 (0.04-3.6, N 108).

8(7)b.-- Throat white or yellowish buff, continuous with pale color of forehead, lacking dark markings on face; foramina in anterodorsal nasopharynx; skull smaller and mandibular spout longer (Fig. 5):

Bradypus tridactylus L. Pale-throated three-toed sloth; ai-ai; preguiça de bentinho.
Bradypus tridactylus Linné, 1758; Cabrera, 1958, in

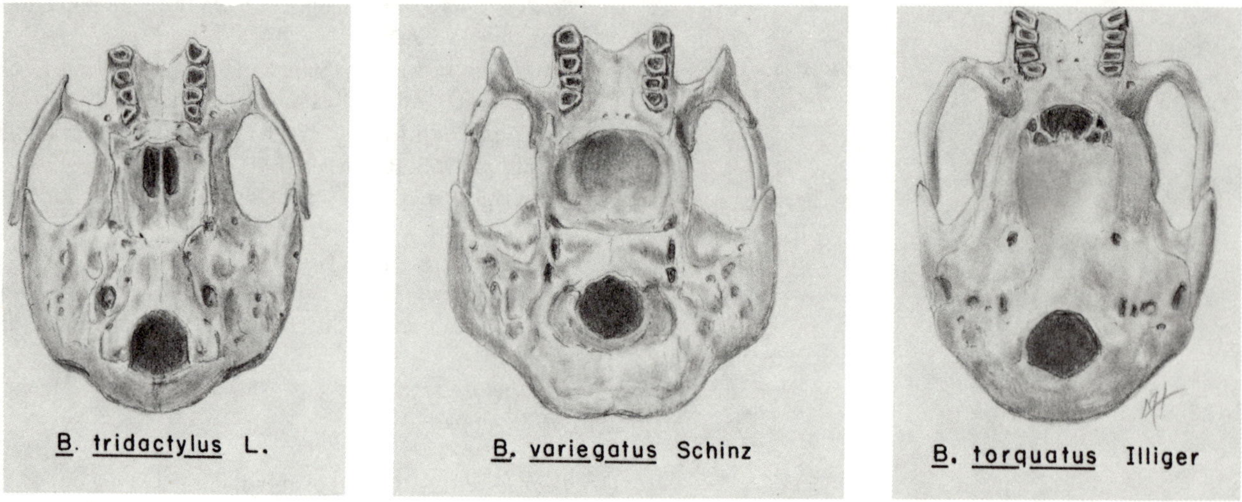

Figure 5. Skulls of three-toed sloths *Bradypus*, posteroventral view. Note foramina in anterior nasopharynx of *B. tridactylus* and *B. torquatus*; the absence of such foramina in *B. variegatus*; expanded pterygoid region and pronounced anteroposterior compression of teeth in *B. torquatus*.

part; Wetzel and Kock, 1973; Wetzel and Avila-Pires, 1980.

Bradypus cucilliger Wagler, 1831; Moeller, 1975, in part.

Arctopithecus flaccidus Gray, 1849; Var. 1 *Dysoni* Gray, 1869 and Cabrera, 1958 (as *Bradypus infuscatus flaccidus*).

Arctopithecus flaccidus Var. 2 *Smithii* Gray, 1869 and Cabrera, 1958 (as junior synonym of *Bradypus infuscatus marmoratus*[Gray, 1849]).

Type locality.-- Surinam.

Distribution.-- In Venezuela the delta of Río Orinoco, state of Bolívar and highlands of T. F. Amazonas, through forested Guyana, Surinam, French Guiana and adjoining northeastern Brazil, south to the Rio Amazonas where it is sympatric with *B. variegatus* along both banks from Rio Negro to mouth (Fig. 4).

Comments.-- See Beebe (1926) for observations on the natural history of this species in Guyana.

Size and weight.-- Based on adults from throughout its range: total length 550 ± 26 (498-600, N 17); length of tail 54 ± 10 (31-75, N 17); length of hind foot 115 ± 15 (95-140, N 15); length of ear 13 (10-15, N 3); weight 4.01 ± 0.28 kg (3.8-4.5, N 7); greatest length of skull 73.8 ± 3.7 (66.1-83.4, N 46); length of mandibular spout 2.6 ± 0.90 (0.9-5.4, N 47).

MEGALONYCHIDAE; two-toed sloths

9(6)a.-- Throat paler and set off by darker pectoral generally finer than guard hairs on rest of body; hair on top of head longer and lighter in color than body hair; 1 pair of small foramina and 2 pairs of large foramina in inter-pterygoid space with the posterior pair opening into pterygoid sinuses (in a few populations shallow fossae replace the anterior pair of foramina); inter-pterygoid width not as proportionally broad anteriorly and narrow posteriorly, maximal width less than twice minimal width; posterior swellings of pterygoid sinuses not as expanded (Fig. 6), widths less than 13.5; maxilla not in broad contact with frontals (Fig. 7); cervical vertebrae usually 6 (5-6, N 12):

Choloepus hoffmanni Peters. Hoffmann's two-toed sloth; unau; preguiça real.

Choloepus hoffmanni Peters, 1859; Cabrera, 1958, in part; Moeller, 1975, in part; Wetzel and Avila-Pires, 1980; Hall, 1981.

Type locality.-- Costa Rica, Herédia, Volcán Barba (Wetzel and Avila-Pires, 1980).

Distribution.-- Lowland to high montane forests from Nicaragua to northwestern South America, through western Colombia to Venezuela west of Cordillera Oriental and south to northwestern Ecuador (Esmeraldas and Pichincha) and east of Andes from Rio Solimões in Peru south to at least 11° S in Peru (Junín, Chanchomayo) and in Bolivia (Beni, Guayaramerín, AMNH 209941) and in western Brazil (southwestern Amazonas and probably Acre) east to northern Mato Grosso (Aripuanã, USMN, uncatalogued) (Fig. 8).

Comments.-- Nocturnal (Sunquist and Montgomery, 1973b) and arboreal; usually less common than the 3-toed sloth in areas of sympatry, still a very common medium-sized arboreal mammal (Montgomery and Sunquist, 1975); broader food habits than the 3-toed sloth, probably feeding on leaves, fruits, buds, and perhaps small vertebrates (Meritt, 1985); adults are long-lived, gestation period approaching 1 year (Eisenberg and Maliniak, 1985); relatively aggressive and

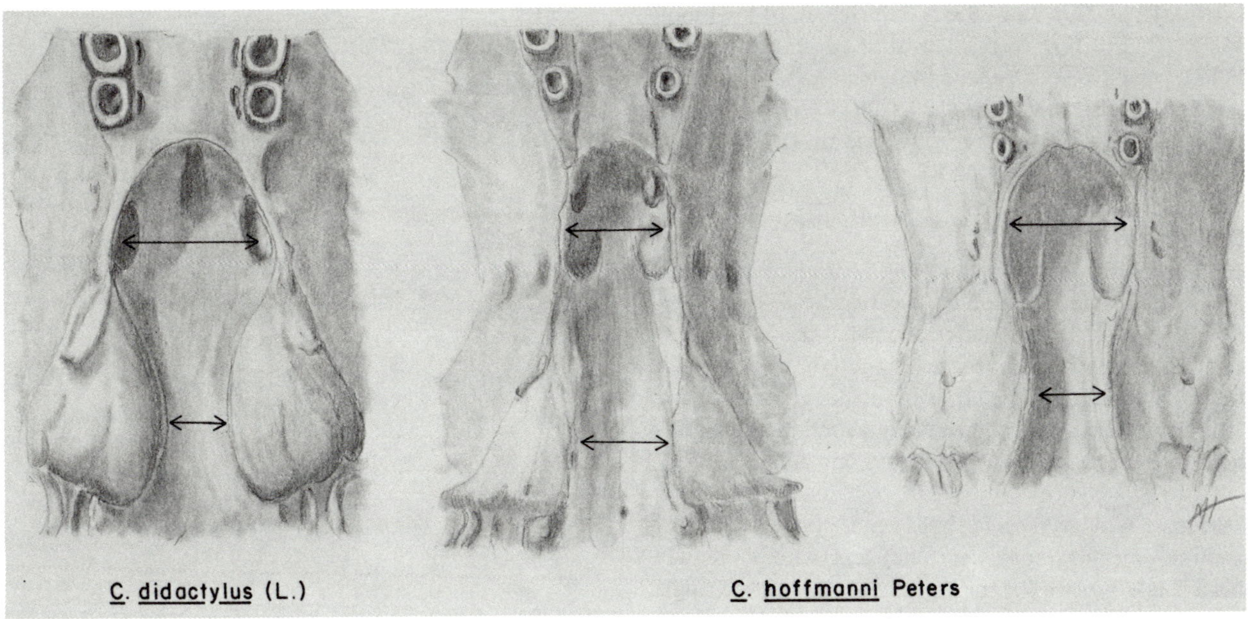

Figure 6. Selected cranial characters of two-toed sloths *Choloepus*, ventral view, **left to right**: *C. didactylus* AMNH 133427, Brazil, Pará, Fordlandia; *C. hoffmanni* AMNH 76655, Peru, alto Ucayali, Lagarto; *C. hoffmanni* AMNH 18896, Panama, Chiriquí, Boquerón. Note in *C. didactylus* the larger anterior and smaller posterior inter-pterygoid space, the greater pterygoid expansions, and absence of the posterior pair of foramina in inter-pterygoid space; in both *C. hoffmanni*, from extremities of the range, proportionally less difference between maximal and minimal inter-pterygoid space, the lesser pterygoid expansions, and two pairs of large foramina in inter-pterygoid space.

capable of inflicting serious wounds with self-sharpening caniniform teeth. See Webb (1985) on the alliance of *Choloepus* with the megalonychid sloths.

Size and weight.-- Based on adults from Costa Rica to Brazil: total length 621 ± 48 (540-720, N 52); length of tail 21 ± 5 (14-30, N 12); length of hind foot 123 ± 27 (100-150, N 45); length of ear 25 ± 4 (20-37, N 33); weight 5.72 ± 0.69 kg (4.54-6.69, N 13); greatest length of skull 106.3 ± 0.6 (97.8-116.5, N 51); ratio of minimal/maximal inter-pterygoid widths 0.59 ± 0.09 (0.31-0.83, N 75).

9(6)b. Throat not distinctly paler and contrasting with pectoral hair; hair on cheeks and throat not distinctly shorter and finer than that of neck and shoulders; 1 pair of small and 1 pair of large foramina in the anterodorsal inter-pterygoid space and usually no posterior foramina that penetrate into the pterygoid sinuses; anterior inter-pterygoid space broad, more than twice as wide as posterior inter-pterygoid space; pterygoid sinuses broadly inflated (Fig. 6), usually wider than 14; maxilla in broad contact with frontals (Fig. 7), not interrupted by lacrimals; cervical vertebrae usually 7 (6-8, N 7):

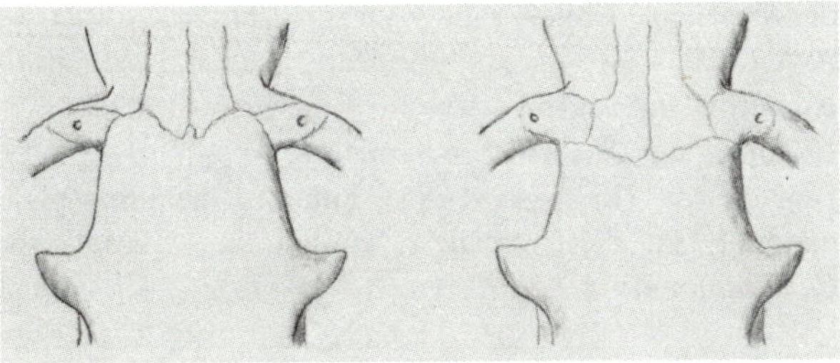

Figure 7. Dorsal view of lacrimal and portions of maxilla, frontal and nasal bones of *Choloepus*. **Left**: note in *C. didactylus* the broad contact of maxilla with frontal bone. **Right**: in *C. hoffmanni* this contact is excluded by interposition of lacrimal and nasal bones.

Choloepus didactylus (L.). Linné's two-toed sloth; unau; preguiça real.

Bradypus didactylus Linné, 1758.

Choloepus didactylus: Cabrera, 1958; Moeller, 1975, in part; Wetzel and Avila-Pires, 1980.

Choloepus florenciae J.A. Allen, 1913; Cabrera, 1958 (as *C. hoffmanni florenciae*).

Type locality.-- Surinam.

Distribution.-- From the delta of Río Orinoco south in Venezuela, west to upper Orinoco drainage in Colombia (Meta), east to Guyana, Surinam, and French Guiana, and in Brazil to the state of Maranhão (Miritiba, MZUSP 2899) on the east and west along both banks of the Rio Amazonas - Solimões to the western Amazon basin of Colombia, Ecuador, and Peru (Fig. 8).

Size and weight.-- Based on adults from Brazil, Colombia, Ecuador, Guyana, Peru, Surinam, and Venezuela: total length 677 ± 55 (630-885, N 41);

length of tail 23 ± 7 (14-33, N 11) ; length of hind foot 148 ± 17 (110-170, N 37); length of ear 28 ± 4 (20-35, N 19); weight 6.07 ± 1.09 kg (4.05-8.45, N 21); greatest length of skull 111.2 ± 5.7 (103.3-119.8, N 42); ratio of minimal/maximal inter-pterygoid widths 0.47 ± 0.08 (0.25-0.62, N 30).

DASYPODIDAE; Armadillos

10(5)a.-- With only tympanic ring instead of bulla and no ossified external auditory meatus; head shield narrow; a row of smaller scutes associated nearer scapular shield than head shield or lacking any scutes between head and scapular shield; without long, white hair: long-nosed ***Dasypus***, domed ***Tolypeutes***, giant ***Priodontes***, and naked-tailed ***Cabassous*** armadillos....... **17**

10(5)b.-- With tympanic bulla and ossified external auditory meatus (bony ear tube) and with either (1) head shield broad (W/L ratios 0.69 to over 1.0), a single row of large nuchal scutes immediately behind head shield and no wider than space between ears, long hair on carapace; or (2) long, white hair on sides and venter: euphractines and pichiciegos...................................... **11**

11(10)a.-- Eyes and pinnae vestigal; long auditory meatus opening posterior side of orbit; sides and venter of body completely covered with dense, long, silky, white hair; posterior margin of body a rounded rump-plate; tiny armadillos (head + body length much less than 200, CNL less than 50); two frontal prominences; teeth 7-8/8: pichiceigos... **16**

11(10)b.-- Eyes and pinnae normal, not vestigal; pinnae not set on rear of orbit; lacking silky hair completely covering sides and venter; body not ending in rump-plate; larger (head + body length more than 200, CNL more than 50); no frontal prominences; teeth 8-9/9-10: euphractines... **12**

12(11)a.-- Large (head + body length more than 400, CNL more than 100); carapace pale yellow or tan; hair white; lacking complete movable band at anterior edge of scapular shield; width of head shield 80 percent or less of length (Fig. 9); zygomatic arch elongate and slender; jugal at symphysis never twice as high as overlying anterior edge of squamosal:

Euphractus sexcinctus (L.). Yellow or white-bristled armadillo; tatú-poyú; tatú-peludo.
Distribution.-- Savannas of southern Surinam and adjacent Pará, southeastern Brazilian highlands to Rio Grande do Sul, Mato Grosso, the Gran Chaco, eastern Paraguay, northeastern Argentina (Corientes, Misiones and Buenos Aires) and Uruguay.
Comments.-- This largest of the euphractine armadillos fights viciously with teeth and claws, and is the only armadillo which tries to bite when handled; food includes a very wide variety of items, from fruits and other plant material to carrion and small vertebrates (Redford, 1985); 3-4 pores for glands in posterior middorsum of carapace.

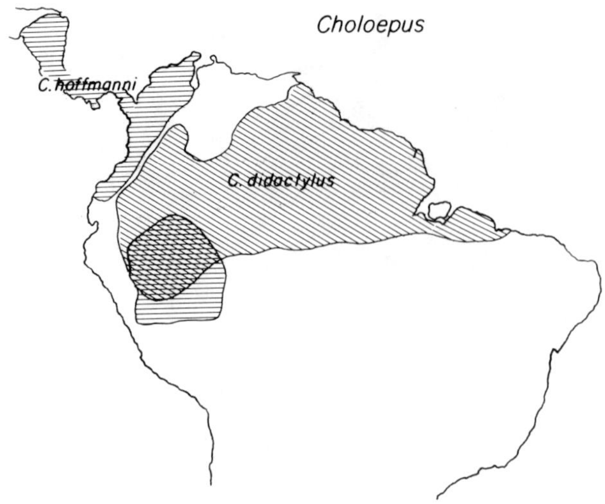

Figure 8. Distribution of two-toed sloths *Choloepus*.

12(11)b.-- Small (head + body length less than 400, CNL less than 100); darker tan to dark brown dorsum and hair; one complete movable band at anterior edge of scapular shield; width of head shield greater than 80 percent of length (Fig. 9); zygomatic arch shorter, jugal at symphysis more than twice as high as overlying anterior end of squamosal: ***Zaedyus*** and ***Chaetophractus*** **13**

13(12)a.-- Ear shorter, length less than 20 (Fig. 9); marginal scutes with apices sharply pointed; nuchal scutes less than 5 in anteroposterior length; teeth 8/9; anterior rostrum slender, width 21 percent of length; height of zygomatic more or less uniform and without distinct notch below orbit:

Zaedyus pichiy (Desmarest). Pichi.
Distribution.-- Argentina from Tucumán south to Strait of Magellan, Chile, and from the Atlantic coast of Argentina west to the Andean grasslands of both Argentina and Chile.
Comments.-- A carnivore-omnivore, apparently preferring carrion (Redford, 1985), but also including plant material in diet; smallest euphractine except for *Chaetophractus vellerosus*.

13(12)b.-- Ear longer than 20 (Fig. 9); apices of marginal scutes more rounded; nuchal scutes longer than 6; teeth 9/10; width of anterior rostrum 25-33 percent of length; zygomatic with distinct notch below orbit: ***Chaetophractus***.. **14**

Figure 9. Anterior views of the smaller euphractine armadillos. **Left to right**: *Chaetophractus villosus* USNM 96515, Argentina, Santa Fe, Río Salado; *C. vellerosus* USNM 200364, Argentina, Mendoza; *Zaedyus pichiy* USNM 4033, Argentina, Patagonia. Note: slender head shield, shorter ears, and lesser anteroposterior length of the incomplete row of mid-nuchal scutes in *Zaedyus pichiy*; broader headshields in *Chaetophractus*; longer ears and intermediate breadth of headshield in *C. vellerosus*. Photograph by R.M. Wetzel.

14(13)a.-- Smaller (head + body length less than 250, CNL less than 80); ear proportionally long, extending to first immobile row of scutes in scapular shield; W/L head shield ratio narrower (0.90, N 17); dorsum variegated brown and pale tan (Figs. 9 and 10).

Chaetophractus vellerosus (Gray). Lesser hairy or crying armadillo; piche llorón or quirquincho chico.
Distribution.-- Southeastern Bolivia and western Paraguay south through Argentina to provinces of La Pampa. Río Negro and Santa Cruz, and possibly the puna of Bolivia, Argentina, and Chile.
Comments.-- Smallest armadillo except for pichiciegos; renal adaptation for desert life (Greegor, 1975); omnivorous, feeding mainly on insects, various vertebrates, and on pods, tubers and roots of plants (Greegor, 1985); loud cries of protest or alarm; lacks middorsal glandular pores.

14(13)b.-- Larger (head + body length more than 250; CNL more than 80); ear shorter; W/L head shield ratio wider (more than 0.90): ***C. nationi*** or ***C. villosus*****........ 15**

15(14)a.-- Dorsum tan, covered with long, pale hair; head shield as wide as long; restricted to high Andean grasslands:

Chaetophractus nationi Thomas. Bolivian hairy armadillo.
Distribution.-- Andean grasslands of Bolivia.
Comments.-- Validity of this species could not be verified (Wetzel, 1985) because of so few specimens available for study. Although this species is retained, additional data are needed for support. My limited observations suggest that it may be a high-altitude subspecies of *C. vellerosus*.

15(14)b.-- Dorsum dark brown with sparse, long, black and brown hairs; W/L head shield ratio 0.95 (N 12):

Chaetophractus villosus (Desmarest). Greater hairy armadillo; tatu-velú or quirquincho común.
Distribution.-- Chaco of Paraguay (and probably also of Bolivia and Argentina); in Argentina from provinces of Mendoza, Córdoba, and Santa Fe south to Santa Cruz

and in adjoining eastern Chile south to at least 52°30'.
Comments.-- Omnivorous, apparently relishes carrion (Redford, 1985); with or without middorsal glandular pores.

16(11)a.-- Sides of carapace free of body, with hair under carapace; 4 or 5 large, rectangular scutes on posterior margin of head shield; smaller (head + body length 120.5, N 2, Frechkop and Yepes, 1949; CNL 40); high frontal prominences, well anterior to orbits:

Chlamyphorus truncatus Harlan. Pink fairy armadillo; pichiciego menore.
Distribution.-- Argentina from southern Catamarca south to Mendoza and east to western Buenos Aires.
Comments.-- Small, fossorial, coming to surface at night; omnivorous (Minoprio, 1945; Rood, 1970).

16(11)b.-- All of carapace attached to body; head shield without posterior row of large scutes; larger (head+ body length 161, CNL 43.8, USNM 283134); lower, broader frontal prominences extending posteriorly to orbits:

Chlamyphorus retusus Burmeister. Greater fairy armadillo; pichiciego mayore.
Distribution.-- Northern Argentina in provinces of Salta, Chaco, and Formosa, through western Paraguay and southeastern Bolivia (probably all of the Gran Chaco).
Comments.-- The life history of this small, fossorial armadillo is virtually unknown, including its food habits.

17(10)a.-- Scapular and pelvic shields flexible (in living animals); movable bands 6 or more; greatest height of zygoma more than 5; no pronounced arch to toothrow and palate in lateral profile; body of mandible slender and more or less uniform in height below toothrow: ***Priodontes, Cabassous*** or ***Dasypus***.......................... **19**

17(10)b.-- Rigid, dome-shaped scapular and pelvic shields; movable bands usually 3 (may be 1 or 2); slender zygoma, greatest height less than 5; distinct dorsal arch to maxillary toothrow and palate in lateral profile; vertical height of body of mandible much greater at middle than at ends of toothrow: ***Tolypeutes***...... **18**

18(17)a.-- Five toes on manus; teeth 8/9 (N 4); except for wider interorbital, skull generally not as wide (mastoidal width/CNL ratios 0.35, N 4):

Tolypeutes tricinctus (L.). Brazilian domed armadillo; tatú-bola.
Distribution.-- Because of few specimens with data and lack of information on present status of this endangered species, only an indefinite range of northeastern highlands of Brazil is known.

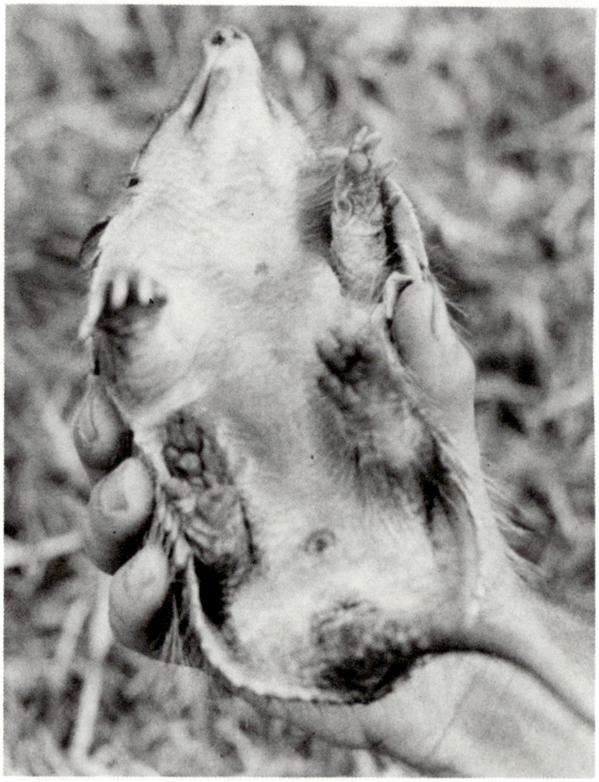

Figure 10. Lesser hairy or crying armadillo *Chaetophractus vellerosus*, Paraguay, Presidente Hayes, Juan de Zalazar. **Above**: lateral view. **Below**: ventral view. Photographs by R.M. Wetzel.

18(17)b.-- Three or four toes on manus; teeth 9/9; wider skull except for interorbital (mastoidal width/CNL ratio 0.39, N 23):

Tolypeutes matacus (Desmarest). Azara's domed armadillo; tatú-bolito; quirquincho bola (Fig. 11).
Distribution.-- Santa Cruz, Bolivia and southern Mato Grosso, Brazil, southward through the Chaco of Paraguay to the province of Santa Cruz, Argentina.
Comments.-- Thought to feed largely, if not entirely, on ants and termites (Redford, 1985), although a more complete study is needed.

19(17)a.-- Less supple carapace, with 6 to 11 movable

bands; margins of scutes of movable bands overlapped by triangular scales; rosettes of rounded scutes on scapular and pelvic shields; foreclaws moderate, not modified as large, scimitar shapes; tail long, over 55 percent of length of head + body and with very slender tip; very long rostrum, 55 percent or more of CNL; height of condyloid process much less than coronoid process: ***Dasypus***; long-nosed armadillos.................. 24

19(17)b.-- Very supple carapace, with 11 to 14 movable bands; margins of scutes on movable bands coincide with margins of overlying scales; rectangular scutes on carapace; large, scimitar-shaped foreclaws; tail shorter than half the length of head + body; rostral length less than half that of skull; height of condyloid process of mandible greater than that of coronoid process: ***Priodontes*** and ***Cabassous***; priodontines......................... 20

20(19)a.-- Tail armored with articulating bony scutes; large size (head + body more than 700, CNL more than 170); longer rostrum; teeth numerous (18/19) and flattened transversely:

Priodontes maximus (Kerr). Giant armadillo, tatú-guasú, tatú-canastra.
Distribution.-- South America east of the Andes from northwestern Venezuela south through Amazon basin to the Gran Chaco and southeastern Brazil.
Comments.-- The largest living xenarthran (Fig. 12), and one of the species most endangered by the activities of man. It is prized as food and hunted throughout its range. The most myrmecophagous of the armadillos, feeding mainly on ants, termites and their larvae and eggs (Redford, 1985).

20(19)b.-- Tail without scutes (but may or may not have visible scales); smaller size (head + body length less than 495 and CNL less than 125); shorter rostrum; fewer teeth (9/8), and any severe transverse compression confined to anterior pairs: ***Cabassous***................... 21

21(20)a.-- Larger (head + body length 410-490, CNL more than 99); scutes on head shield less than 50; palate extending posteriorly to line between bases of zygoma; anteromost teeth flattened transversely:

Cabassous tatouay (Desmarest). Greater naked-tailed armadillo; tatú-ai mayore; tatú de rabo mole grande.
Distribution.-- From southeastern highlands of Brazil through Uruguay, southeastern Paraguay, and Misiones, Argentina.
Comments.-- Although relatively little studied, all species of *Cabassous* probably feed mainly on ants and termites (Redford, 1985).

21(20)b.-- Smaller (head + body length less than 450 and, where sympatric with *C. tatouay*, not exceeding 350; CNL less than 91); more numerous scutes on headshield where sympatric with *C. tatouay* (mean 54.0); palate not extending posteriorly to line between bases of zygoma; anteromost teeth not flattened transversely: ***C. chacoensis, C. centralis*** and ***C. unicinctus***...22

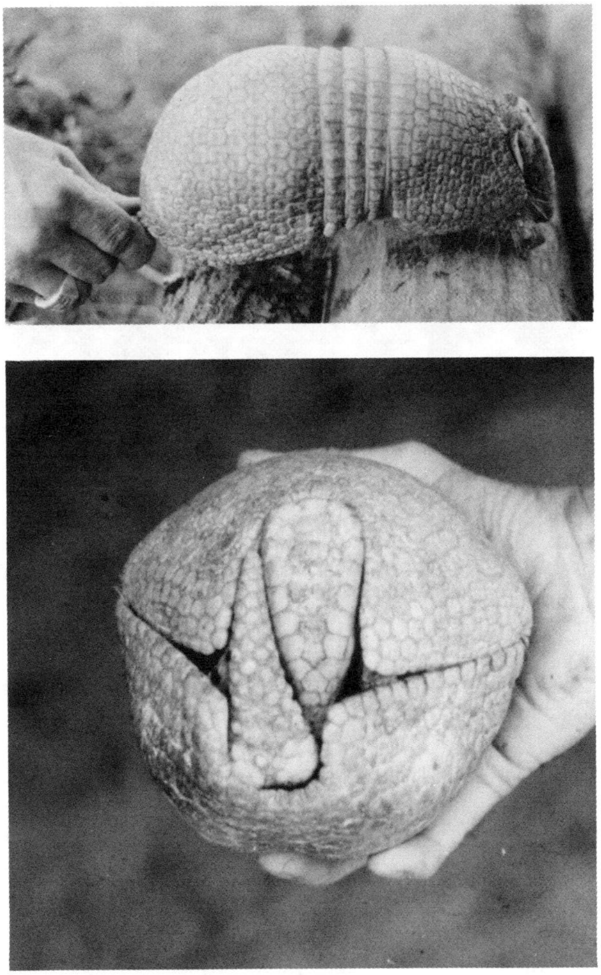

Figure 11. Azara's domed armadillo *Tolypeutes matacus*, Paraguay, Presidente Hayes, Juan de Zalazar. **Above**: lateral view. **Below**: Anteroposterior view with head and tail folded into defensive position. Photographs by R.M. Wetzel.

22(21)a.-- Smallest *Cabassous* (mean length head + body 303, CNL 69.8); ear very short and with fleshy edge, lacking the flared ears of other species; all teeth except for first and last ones flattened anteroposteriorly, not transversely:

Cabassous chacoensis Wetzel. Chacoan naked-tailed armadillo; tatú-ai menore (Fig. 13).
Distribution.-- Restricted to the Gran Chaco (southeastern Bolivia, western Paraguay, northwestern Argentina) except for possibly adjoining Mato Grosso, Brazil.

Figure 12. Giant armadillo *Priodontes maximus*. Photograph by Merlin Tuttle.

22(2l)b.-- Somewhat larger (head + body length more than 320, CNL more than 75); large, flared ears; transverse diameters of teeth usually less than anteroposterior diameters: ***C. centralis*** and ***C. unicinctus***............. 23

23(22)a.-- Skull slender (mean interorbital width 24.3, zygomatic width 40.8); smaller than adjacent populations of *C. unicinctus* (mean head + body length 341, CNL 77.1):

Cabassous centralis (Miller). Northern naked-tailed armadillo; tatú de rabo mole; armado de zopilots. *Distribution*.-- Guatemala to northwestern Venezuela.

23(22)b.-- Skull broader (mean interorbital width 26.6, zygomatic width more than 42); larger than adjoining populations of *C. centralis* (mean head + body length 388, CNL 85.6):

Figure 13. Chacoan naked-tailed armadillo *Cabassous chacoensis* Paraguay, Presidente Hayes, Juan de Zalazar. Note the tiny ear. Photograph by R.M. Wetzel.

Cabassous unicinctus (Linné). Southern naked-tailed armadillo; tatú-ai menore; tatú de rabo mole menore. *Distribution*.-- *C. u. unicinctus* (L.) occurs from Venezuela east of Andes to Amazon-Solimõ es Rivers, there intergrading with *C. u. squamicaudis* (Lund). The latter subspecies extends south to Mato Grosso, Mato Grosso do Sul, Goiás, Sã o Paulo, and Minas Gerais.

24(19)a.-- Dorsum and venter of body densely covered with light brown hair; movable bands 10-12; rostrum long (66-88 percent of CNL) and narrow (anterorostral width only 7 percent of CNL):

SUBGENUS ***Cryptophractus***

Dasypus pilosus (Fitzinger). Hairy long-nosed armadillo; tatú-peludo más (Fig. 14).
Distribution.-- Mountains of Peru (Fig. 15).

24(19)b.-- Hair sparse on venter and inconspicuous on dorsum; movable bands 6-10; rostral length 53-67 percent of CNL and width not less than 9 percent of CNL: remaining ***Dasypus***.. **25**

25(24)a. Large, protruding scales (may be in excess of 17 long) on proximal, posterior surface of hind leg; small, vestigal fifth digit on manus; large (head + body length more than 475, CNL more than 111); posterolateral borders of palate keeled:

SUBGENUS ***Hyperoambon***

Dasypus kappleri Kraus. Greater long-nosed armadillo; tatú-peba mayore.
Distribution.-- Orinoco and Amazon basins (Fig. 15).

25(24)b. Without protruding scales on hind legs; normal fifth digit on manus; smaller size (means of head +body length of remaining species not exceeding 475, CNL less than 111); posterolateral borders of palate rounded, not keeled: subgenus ***Dasypus***... **26**

26(25)a. Scutes on fourth movable band 54-64; tail longer (265-450 and 70 percent or more of length of head + body); more yellow on sides of carapace; longer palate (69 percent of CNL, N 25); larger size (head + body 365-573, CNL 78.7-110.9):

Dasypus novemcinctus L. Common long-nosed armadillo, cachicamo, tatú-hú, tatú-verdadeiro (Fig. 14). *Distribution*.-- Southern United States southward through Central America to South America, on the Pacific side of Andes south to northwestern Peru, Colombia, Venezuela; east of Andes, all of South America to Argentina (Santiago del Estero, Santa Fe, Entre Ríos), and the islands of Grenada, Margarita, Trinidad, and Tobago (Fig. 15). *Note*: Movable bands,

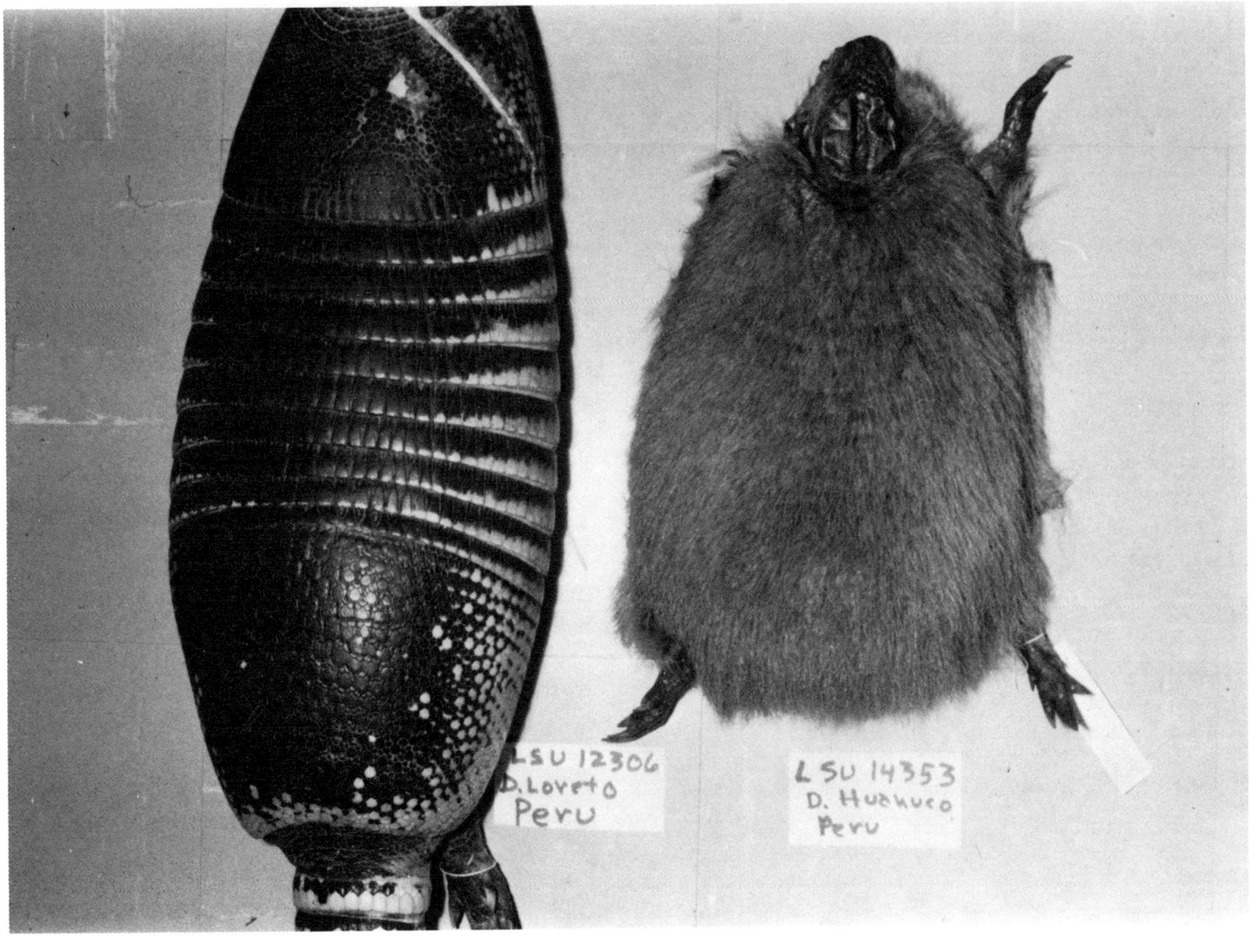

Figure 14. **Left**: Common long-nosed armadillo *Dasypus novemcinctus* LSU 12306, Peru, Loreto, Balta. Note 8 movable bands. **Right**: Hairy long-nosed armadillo *Dasypus pilosus* LSU 14353, Peru, Huánuco, Paso Carpish. Note large ears (here dried and folded) and absence of concealing hair on head. Photograph by R.M. Wetzel.

although varying from 7-10, are more frequently 8 in Argentina and Paraguay, average 8.3 in the highlands of Brazil; 9 in northern South America, and between 8 and 9 in Central America to United States.

Comment.—Currently the most studied of the armadillos in terms of ecology, movements and home ranges (Breece and Dusi, 1985), activity patterns (Layne and Glover, 1985) and food habits (Wirtz et al., 1985; reviewed by Redford, 1985). Primarily insectivorous, tending toward specialization on ants and termites.

26(25)b.-- Scutes on fourth movable band 43-62 (means of samples do not exceed 53); tail shorter (125-205, and less than 70 percent of length of head + body); less yellow on carapace; shorter palate (63 percent, 66 percent, and 66 percent of CNL, N 67, for three remaining species); smaller (head + body length 240-314, CNL 57.9-75.5): ***D. septemcinctus, D. sabanicola*** and ***D. hybridus***.. 27

27(26)a.-- Smaller (head + body length 265, N 8; CNL 63.4, N 15) except for longer ear (31, N 7); fewer movable bands (6.5, N 13); fewer scutes on fourth movable band (46.2, N 13):

Dasypus septemcinctus L. Lesser long-nosed armadillo; tatú-mulita.

Distribution.-- Lower Rio Amazonas south through eastern Brazil to Rio Grande do Sul and west to Mato Grosso, southeastern Bolivia, and northern Argentina (Fig. 15).

Comments.-- *Dasypus mazzai* Yepes is considered a construct of *D. septemcinctus* L. and *D. novemcinctus* L. (Wetzel and Mondolfi, 1979).

27(26)b.-- Larger (means, head + body lengths 290 and 297, CNL 68.0 and 70.2, respectively) except for smaller ears (22-30); more movable bands (means of 8.0 and 6.9, respectively): ***D. sabanicola*** and ***D. hybridus***......... 28

28(27)a.-- More movable bands (8.0, N 27); northern South America:

Dasypus sabanicola Mondolfi. Northern lesser long-nosed armadillo, tatú de sabana.
Distribution.-- Savannas of Colombia and Venezuela (Fig.15).
Comment.-- Insectivorous, tending toward specialization on ants and termites (Pacheco and Naranjo, 1978).

28(27)b.-- Fewer movable bands (6.9, N 20); geographically separated from *D. sabanicola* by *D. septemcinctus*:

Dasypus hybridus (Desmarest). Southern lesser long-nosed armadillo; tatú-mulita, tatú-ira.
Distribution.-- Eastern Paraguay and Argentina, southern Brazil and Uruguay on the northeast to Jujuy province on northwest, south to Mendoza and Río Negro (Fig. 15).

Figure 15. Distribution of long-nosed armadillos *Dasypus*.

ACKNOWLEDGMENTS

Travel and study at the various museums were made possible by funds from Charles O. Handley, Jr.'s Smithsonian Venezuelan Project and grants to me from the American Philosophical Society, the National Geographic Society, the University of Connecticut Research Foundation, and most recently, a grant to Philip Myers and myself from the National Science Foundation. Our fieldwork in Paraguay, which has provided useful series of eight xenarthran species, was supported by the latter three agencies. I am grateful to the numerous institutions and curators who generously permitted study of specimens in their care, including the following whose specimens are cited by the indicated abbreviations:
AMNH--American Museum of Natural History, New York; Sydney Anderson, Richard Van Gelder.
BMNH--British Museum (Natural History), London; G. B. Corbet, I. R. Bishop.
FMNH--Field Museum of Natural History, Chicago; P. W. Freeman, P. Hershkovitz, J. C. Moore.
LSU--Louisiana State University, Museum of Zoology, Baton Rouge; G. H. Lowery, Jr. (deceased).
MN--Museu Nacional, Universidade Federal do Rio de Janeiro; F. D. de Avila-Pires, J. Moojen.
MZUSP--Museu de Zoologia da Universidade de Sã o Paulo; P. E. Vanzolini.
USNM--National Museum of Natural History, Smithsonian Institution, Washington, D.C.; C. O. Handley, Jr., H. W. Setzer, R. W. Thorington, Jr.

LITERATURE CITED

Allen, J.A.
1913. New mammals from Colombia and Ecuador. Bull. Amer. Mus. Nat. Hist., 32: 469-484.

Asdell, S.A.
1964. Patterns of mammalian reproduction, 2nd. ed. Cornell University Press, Ithaca, New York, 670 pp.

Baker, R.H.
1974. Records of mammals from Ecuador. Publ. Mus. Michigan State Univ., Biol. Ser., 5: 129-146.

Beebe, W.
1926. The three-toed sloth *Bradypus cuculliger cuculliger* Wagler. Zoologica 7: 1-67.

Breece, G.A., and J.L. Dusi.
1985. Food habits and home ranges of the common long-nosed armadillo *Dasypus novemcinctus* in Alabama. (this volume).

Bucher, E.H.
1980. Ecología de la fauna chaqueña. Una revisión. Ecosur, Cordoba, 7: 111-159.

Cabrera, A.
1958. Catálogo de los mamíferos de América del Sur. Rev. Mus. Argentino Cienc. Nat. "Bernardino Rivadavia", Zool. (1957), 4(1): 1-307.

Desmarest, A.
1816. *Bradypus*. P. 327, *in* Nouveau dictionnaire d´ histoire naturelle. Second edition, Vol. 4. Deterville, Paris.

Eisenberg, J.F. and E. Maliniak.
1985. Maintenance and reproduction of the two-toed sloth *Choloepus didactylus* in captivity. (this volume).

Frechkop, S. and J. Yepes.
1949. Etude systématique et zoogéographique des Dasypodidés conservés a línstitut. Bull. Inst. Roy. Sci. Nat. Belg., 25: 2-56.

Gray, J.E.
1849. On the genus *Bradypus* of Linnaeus. Proc. Zool. Soc. London, 1849: 65-73.

1869. Catalogue of carnivorous, pachydermatous, and edentate mammalia in the British Museum. London. vii +176 pp.

1871a. On a new species of three-toed sloth from Costa Rica. Ann. Mag. Nat. Hist., Ser. 4, 7: 302.

1871b. Notes on the species of Bradypodidae in the British Museum. Proc. Zool. Soc. London, 1871: 428-449.

Greegor, D.H.
1975. Renal capabilities of an Argentine desert armadillo. J. Mammal., 56: 626-632.

Greegor, D.H.
1985. Ecology of the little hairy armadillo *Chaetophractus vellerosus*. (this volume).

Grimwood, I.R.
1969. Notes on the distribution and status of some Peruvian mammals. Amer. Comm. Internat. Wildlife Protect. and N.Y. Zool. Soc., Spec. Publ. No. 21, 86 pp.

Hall, E.R.
1981. The mammals of North America. 2nd. ed., vol. 1. John Wiley and Sons, New York, xv +600 +90 pp.

Husson, A.M.
1978. The mammals of Suriname. Zöol. Monogr. Rijksmus. Nat. Hist. No. 2. E.J. Brill, Leiden, xxxiv +569 pp.

Illiger, K.
1811. Prodromus Systematis Mammalium et Avium. Salfeld, Berlin, vii +176 pp.

Layne, J.N. and D. Glover.
1985. Activity patterns of the common long-nosed armadillo *Dasypus novemcinctus* in south-central Florida. (this volume).

Linné, C.
1758. Systema naturae. 10th ed., vol.1. Laurentii Salvii, Holmiae, 824 pp.

Mares, M.M., R.A. Ojeda, and M.P. Kosco.
1981. Observations on the distribution and ecology of the mammals of Salta province, Argentina. Ann. Carnegie Mus. Nat. Hist., 50: 151-206.

McAdam, D.W. and J.S. Way.
1967. Olfactory discrimination in the giant anteater. Nature, 214: 316-317.

Meritt, D.A., Jr.
1985. The two-toed Hoffman's sloth, *Choloepus hoffmanni* Peters. (this volume).

Minoprio, J.D.L.
1945. Sobre el *Chlamyphorus truncatus* Harlan. Acta Zool. Lilloana, Tucuman, 3: 5-58 +26 plates.

Moeller, W.
1975. Edentata. 11: 149-181 *in* Grzimekś animal life encyclopedia (B. Grzimek, ed.). Van Nostrand Reinhold Co., New York, 635 pp.

Montgomery, G.G.
1979. El grupo alimentico (feeding guild) del oso hormiguero: convivencia y especialización de las

presas de sustento de los osos hormigueros neotropicales (Edentata, Myrmecophagidae). ConCiencia 6(1): 3-6

Montgomery, G.G.
1985. Impact of vermilinguas (*Cyclopes, Tamandua*: xenarthra = edentata) on arboreal ant populations. (this volume).

Montgomery, G.G.
1985. Movements, foraging and food habits of the four extant species of neotropical vermilinguas (Mammalia; Myrmecophagidae). (this volume).

G.G. Montgomery and Y. D. Lubin.
1977. Prey influences on movements of neotropical anteaters. Pp. 103 - 131, *in* Proceedings of the 1975 predator symposium (R.L. Phillips and C. Jonkel, eds.). Montana Forest and Conservation Experiment Station, Univ. of Montana, Missoula, 268 pp.

G.G. Montgomery and M.E. Sunquist
1975. Impact of sloths on neotropical forest energy flow and nutrient cycling. Pp. 69-98, *in* Tropical ecological systems (F.B. Golley and E. Medina, eds.). Springer Verlag, New York, 398 pp.

1978. Habitat selection and use by two-toed and three-toed sloths Pp. 329-359, *in* The ecology of arboreal folivores (G. G. Montgomery, ed.). Smithsonian Institution Press, Washington, D.C., 574 pp.

Pacheco, J. and C.J. Naranjo.
1978. Field ecology of *Dasypus sabanicola* in the flood savanna of Venezuela. Pp. 13-17, *in* The armadillo as an experimental model in biomedical research. Pan. Amer. Health Org. Sanitary Bur. Sci. Publ. 366, iv +140 pp.

Paula Couto, C. de.
1979. Tratado de paleomastozoologia. Acad. Brasileira Cienc., Rio de Janeiro, 590 pp.

Peters, W.
1859. Hr. Peters theilte der Akademie. Monatsb. K. Preuss. Akad. Wiss. Berlin, 1858: 128.

Redford, K.H.
1985. Food habits of armadillos (Xenarthra: Dasypodidae). (this volume).

Rood, J.P.
1970. Notes on the behavior of the pigmy armadillo. J. Mammal., 51: 179.

Saussure, H. de.
1860. Note sur quelques mammiferes du Mexique. Rev. Mag. Zool., Ser. 2, 12: 3-11.

Schinz, H.R.
1825. Das Thierreich eingetheilt nach dem Bau der Thiere als Grundlage ihrer Naturgeschichte und der verleichenden Anatomie von dem Herrn Ritter von Cuvier. Vol. 4. J.G. Cotta´schen, Stuttgart and Tübingen, xiii +793 pp.

Shaw, J.H., T.S. Carter and J.C. Machado-Neto.
1985. Ecology of the giant anteater *Myrmecophaga tridactyla* in Serra da Canastra, Minas Gerais, Brazil: a pilot study. (this volume).

Silveira, E.K.P. da.
1968. Notas sobre a historia natural do tamandua mirim (*Tamandua tetradactyla chiriquensis* J. A. Allen, 1904, Myrmecophagidae), com referencias a fauna do istmo do Panama. Vellozia, Rio de Janeiro, 6: 9-31.

Soukup, J.
1960. Materiales para el Catálogo de los Mamíferos peruanos. Biota, Lima, 3: 131-161.

Sunquist, M.E. and G.G. Montgomery.
1973a. Activity pattern of a translocated silky anteater (*Cyclopes didactylus*). J. Mammal., 54(3): 782.

1973b. Activity patterns and rates of movement of two-toed and three-toed sloths (*Choloepus hoffmanni* and *Bradypus infuscatus*). J. Mammal. 54: 946-954.

Taylor, B.K.
1985. Functional anatomy of the forelimb in vermilinguas (anteaters). (this volume).

Wagler, J.G.
1831. Mittheilungen über die Gattungen der Sippe *Bradypus*. Isis von Oken, 24: 604-612.

Wagner, J.A.
1844. Die Säugthiere in Abbildungen nach der Natur mit Beschreibungen von Dr. Johann Christian Daniel von Schreber. Suppl., Vol. 4 Erlangen, Leipzig, xii +523 pp.

Walker, E.P.
1975. Mammals of the world, 3rd ed., vol. I (J.L. Paradiso,ed.). The Johns Hopkins Univ. Press, Baltimore, xlviii +644 pp.

Webb, D.E.
1985. The interrelationships of tree sloths and ground sloths. (this volume).

Wetzel, R.M.
1975. The species of *Tamandua* Gray (Edentata, Myrmecophagidae). Proc. Biol. Soc. Washington 88: 95-112.

Wetzel, R.M.
1985. Taxonomy and distribution of armadillos, Dasypodidae. (this volume).

Wetzel, R.M. and F.D. Avila-Pires.
1980. Identification and distribution of Recent sloths of Brazil. Revista Brasileira de Biologia, 40: 831-836.

Wetzel, R.M. and D. Kock.
1973. The identity of *Bradypus variegatus* Schinz (Mammalia, Edentata). Proc. Biol. Soc. Washington 86: 25-34.

Wetzel, R.M. and E. Mondolfi.
1979. The subgenera and species of long-nosed armadillos, genus *Dasypus*. Pp. 43-63, *in* Vertebrate ecology in the northern Neotropics (J.F. Eisenberg, ed.). Smithsonian Institution Press, Washington, D.C., 271 pp.

Wirtz, W.O., D.H. Austin and G.W. Dekle.
1985. Food habits of the common long-nosed armadillo *Dasypus novemcinctus* in Florida, 1960-61. (this volume).

TAXONOMY AND DISTRIBUTION OF ARMADILLOS, DASYPODIDAE

Ralph M. Wetzel

Biological Sciences Group, University of Connecticut, Storrs, Connecticut 06268.

Abstract

The systematics and distribution of the family Dasypodidae (Mammalia; Xenarthra = Edentata) has not been revised or, in the past twenty years, reviewed. Specimens extant in the major collections of the world were examined, including those from recent fieldwork in Venezuela and Paraguay. Consideration of qualitative characters, univariant and multivariant analyses led to a restatement of the genera *Cabassous* and *Dasypus*, the tribe Euphractini, and a comparison of the 8 Recent genera and 20 species in the family. Distribution maps based on the origin of specimens examined were prepared. Taxonomic and distributional problems that cannot be solved without additional fieldwork are pointed out.

Abstracto

La sistemática y distribución de la familia Dasypodidae (Mammalia; Xenarthra = Edentata) no ha sido nunca enmendada, ni siquiera revisada en los últimos veinte años. Para este trabajo, se examinaron los especímenes existentes en las mayores colecciones del mundo, incluyendo los obtenidos en recientes giras de campo en Venezuela y Paraguay. Consideraciones de carcterísticas cualitativas, así como el resultado de análisis univariantes y multivariantes nos ha llevado a un replanteamiento de los géneros *Cabassous* y *Dasypus* de la tribu Euphractini, y a una nueva comparación de los 8 géneros recientes y 20 especies pertenecientes a esta familia. Se prepararon mapas de distribución basados en el sitio de origen de los especímenes examinados. En este estudio, se puntualizaron problemas taxonómicos y de distribución que no pueden ser resueltos sin el resultado de trabajos de campo adicionales.

Resumo

A sistemática e distribuição da família Dasypodidae (Mammalia, Xenarthra = Edentata) não tem sido revisada ou reexaminada no últimos vinte anos. Espécimes existentes nas maiores coleções do mundo foram examinados, inclusive os de trabalhos de campo recentes na Venezuela e Paraguai. Consideração de caracteres qualitativos, análises univariantes ou multivariantes conduziram a uma reafirmação dos gêneros *Cabassous* e *Dasypus*, a tribo Euphractini e uma comparação dos 8 gêneros recentes e 20 espécies na família. Mapas de distribuição baseados na origem dos espécimes examinados, foram preparados. Problemas taxonômicos e de distribuição que não puderam ser solucionados sem trabalhos de campo adicionais são salientados.

Interest in armadillos, Dasypodidae, is heightened by unique features of their anatomy and physiology and their biogeographical position of relative antiquity in South America. Most recently Storrs et al. (1974) have found one species, *Dasypus novemcinctus*, a promising key to problems in leprosy research and are continuing to investigate a broad spectrum of other species of armadillos. Emry's (1970) removal of the palaeanodonts of the early Cenozoic of North America from the order Edentata leaves fossil remains of armadillos of equivalent time in South America as the earliest examples of Xenarthra (= Edentata). Data on armadillos may eventually provide insights to paleoclimatic fluctuations and zoogeographical problems of faunal interchange between South America and North America.

In view of the importance of armadillos, and a probable surge of interest in them as subjects for field and laboratory studies, a review and revision of their taxonomic status is timely. Synthesis and comparisons of the recent species and genera of armadillos cannot, of course, keep pace with the accumulation of specimens and other primary data. In the last three decades the majority of the taxa have been critically compared only by Moeller (1968). Talmage and Buchanan (1954), although centrally concerned with *Dasypus novemcinctus*, reviewed all major aspects of the other 19 species they recognized. Frechkop and Yepes (1949) considered limited numbers of specimens and taxa in the framework of Yepes' (1928) earlier taxonomic opinions. Cabrera (1958) assembled the last definitive listing of species and subspecies, taxonomic synonymies, and distributional summaries.

SOURCES OF SPECIMENS

This comparison of genera and species of armadillos is based upon specimens examined in most of the major collections of South American mammals. Many of these specimens either were not available or, for other reasons, were not examined by earlier writers reporting on armadillos. Space prevents mentioning all collections visited during the study except those represented by specimens cited here. Curators listed were in charge of a collection at the time it was visited. Abbreviations preceding the names of institutions are used in later accounts to identify the location of specimens.

AMNH -- The American Museum of Natural History, New York; R.G. Van Gelder and S. Anderson.

BMNH -- British Museum (Natural History), London; G.B. Corbet, I.R. Bishop.

CONN -- University of Connecticut Museum of Natural History, Storrs; R.E. Dubos.

CM -- Carnegie Museum of Natural History, Pittsburgh; J.K. Doutt, H.H. Genoways, D.A. Schlitter.

ICN -- Instituto de Ciencias Naturales, Universidad de Colombia, Bogotá; E. Barriga B.

IRSNB -- Institut Royal des Sciences Naturelles de Belgique, Brussels; X. Misonne.

LACM -- Los Angeles County Museum of Natural History, Los Angeles; D.R. Patten.

LSU -- Louisiana State University, Museum of Zoology, Baton Rouge; G.H. Lowery, Jr. (deceased).

MACN -- Museo Argentino de Ciencias Naturales "Bernardino Rivadavia", Buenos Aires; J.A. Crespo.

MCZ -- Museum of Comparative Zoology, Harvard University, Cambridge; J.A.W. Kirsch, B. Lawrence, M.E. Rutzmoser.

MEBRG -- Museo de la Estación Biológica de Rancho Grande, Maracay, Venezuela; G. Medina P.

MHN -- Museo de Historia Natural "Javier Prado", Lima; H. de Macedo.

MN -- Museu Nacional, Universidade Federal do Rio de Janeiro; F.D. Avila-Pires., João Moojen de Oliveira.

NMBE -- Naturhistorisches Museum, Bern; P. Lups.

NMV -- Naturhistorisches Museum, Vienna; K. Bauer.

RMNH -- Rijksmuseum van Natuurlijke Histoirie, Leiden; A.M. Husson.

ROM -- Royal Ontario Museum, University of Toronto, Toronto; R.L. Peterson.

SMN -- Staatliches Museum fur Naturkunde, Stuttgart; F. Dieterlen.

USNM -- U. S. National Museum of Natural History, Washington D.C.; C.O. Handley, Jr.

ZMUC -- Universitetets Zoologiske Museum, Copenhagen; F. W. Braestrup, H.F. Baagoe.

METHODS

In the accounts of species, only partial synonymies are presented, but emphasis is given names used in reviews of the family during this century. The holotype is cited if I have examined it.

All linear measurements are in mm. The collector usually recorded the standard external measurements: either the length of head + body or the total length (head + body + tail); length of tail; length of hindfoot including the longest claw; and the length of ear from notch. For most species I have used length of head+ body in order to confine any undetected bias from damaged tails to one parameter. However, the total length method is followed for the three-banded armadillos, *Tolypeutes*, as their small, armored tails only partially extend beyond the pelvic shield. Thus a head+ body length derived by subtraction of tail from total length would be inaccurate. In the accounts, measurements are listed merely as head + body, tail, hindfoot and ear; in all cases, it is the length of that structure which is recorded.

I measured head shields, nuchal scutes, movable bands, and counted numbers of scutes in the fourth movable band. The anteroposterior length of the longest nuchal scute is given merely as a numerical value associated with nuchal scute. Counts of teeth are given as fractions with maxillary pairs as the numerator and mandibular pairs as denominator; bilateral asymmetry was common, and counts are based on means of left and right toothrows.

Only a few of my 15 cranial measurements (Fig. 1) are cited and these are indicated by the following abbreviations: **Head shield w/l**: maximal transverse width and greatest anteroposterior length of the head shield. **CNL, condylonasal length**: distance from most posterior faces of occipital condyles to most anterior tips of nasal bones. **RL, rostral length** : a diagonal length from lacrimal foramen to tip of nasal bones. **RLA, rostral length, adjusted**: distance along midline of rostrum from a line through lacriminal foramina to tip of nasal bones. RLA is derived as the altitude of a triangle in which RL is the hypotenuse and one half the distance

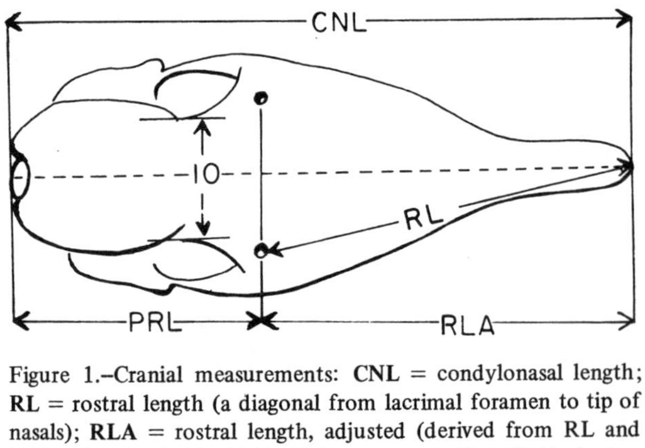

Figure 1.–Cranial measurements: **CNL** = condylonasal length; **RL** = rostral length (a diagonal from lacrimal foramen to tip of nasals); **RLA** = rostral length, adjusted (derived from RL and half of interlacrimal width); **PRL** = postrostral length (derived as CNL-RLA); **IO** = interorbital width.

between lacrimal foramina is the base. **PRL, postrostral length**: distance from a line through the lacrimal foramina to a line across posterior margin of occipital condyles. PRL is derived as the difference between CNL and RLA. **IO, interorbital width**: minimal width between orbits.

Other symbols include: $\bar{y}$ = mean; s = standard deviation of sample; N = number of sample.

DASYPODIDAE
SUBFAMILY **Dasypodinae**
TRIBE **Euphractini**

The euphractine armadillos form a zoogeographically compact unit centered in drier, more open habitats from the high puna to pampas and intervening savanna, steppe, monte, and thorn forest from Bolivia and the Mato Grosso south to Patagonia. Only one species, *Euphractus sexcinctus*, occurs through most of eastern Brazil and, although restricted to a savanna center, north of the Amazon River (Fig. 2).

Euphractines listed from largest to smallest are *Euphractus sexcinctus, Chaetophractus villosus, C. nationi, Zaedyus pichiy,* and *C. vellerosus*. The close relationship of four or more species has been recognized by their inclusion in a single subfamily Euphractinae (Frechkop and Yepes, 1949; Talmage and Buchanan, 1954) a tribe Euphractini (Cabrera, 1958) or the single genus *Euphractus* (Moeller, 1968). Grouping the euphractines within one genus overemphasizes their similarities and would require the erection of subgenera to express the differences and similarities presently indicated at the generic level. These differences are demonstrated by canonical analysis (Fig. 3); by chromosomal differences (Jorge et al. 1977) between *Chaetophractus villosus* (2N = 60, NF = 90), *Zaedyus pichiy* (2N = 62, NF = 94), and *Euphractus sexcinctus* (2N = 58, NF = 102); and in the comparisons that follow.

All the euphractines are stocky armadillos with broad heads, relatively stout teeth, and usually prominent, coarse hair. Hair on the carapace, legs, and venter ranges in color from buffy white or tan to black and in density from sparse to relatively dense. Although hirsuteness is less than in the rare *Dasypus pilosus*, some of the euphractines are known as hairy armadillos. In addition, euphractines have a head shield forming a ledge over the orbits; a single, large nuchal (cervical) band of scutes lying between the ears and no wider than the head shield; 2 small, incomplete rows of scutes on the anteroventral margin of the scapular shield; marginal (most ventral) scutes of carapace rounded anteriorly and curving to wider (dorsoventral diameters) posterior margins that are usually pointed; claws on second and third digits approximately equal, longer than the other claws; tail encased in scutes which are arranged from proximal to distal as (a) rings of single scutes, (b) rings of double rows of scutes to (c) a terminal random pattern of small scutes. Euphractine skulls have a prominent nuchal crest (lambdoidal ridge) curving anteriorly at midline and forming a U- or V-shaped notch; well-developed tympanic bulla and ossified external auditory meatus; palate at midline not extending posteriorly beyond level of toothrow by more than one anteroposterior diameter of longest tooth; 9/10 pairs of teeth in *Euphractus* and *Chaetophractus*, but only rarely so in *Zaedyus*.

GENUS ***Euphractus*** Wagler

Euphractus sexcinctus (Linné). Yellow or six-banded armadillo; gualacate, tatú-peludo.

Dasypus sexcinctus Linné, 1758.

Euphractus sexcinctus: Frechkop and Yepes, 1949; Talmage and Buchanan, 1954; Cabrera, 1958; Moeller, 1968.

Type-locality.-- Brazil, Pará, designated by Thomas (1911) on the basis of similarity in size of head shield of holotype (provenance unknown) with that of specimens from Pará.

Range.-- The Sipaliwini savanna of southern Surinam and its continuation as the Paru savanna in adjacent Pará (Fig. 2); the Brazilian highlands including Mato Grosso and south to Rio Grande do Sul; Uruguay to northeastern third of the province of Buenos Aires; north and west through the northern Argentinian provinces of Misiones, Corrientes, Formosa, Chaco, Salta, Jujuy, and probably northern Santiago del Estero; through both eastern and western Paraguay; and in southeastern Bolivia to Santa Cruz. Tamayo (1973) considered reports of this species in Chile to be assignable to *Chaetophractus villosus*. Within this large range, optimal habitats for *E. sexcinctus* are probably savanna, parkland, forest edge and steppe. More data are needed on its distribution in the eastern Amazon basin.

Measurements.-- $\bar{y}$, s, and N for adult specimens from throughout the range: head + body 446.6 ± 34.2, 23; tail 231.3 ± 14.0, 23; hindfoot 84.2 ± 6.2, 13; ear 38.9 ± 3.5, 14; weight 5.39 kg ± 0.94, 9; ratio of head shield w/l 0.74 ± 0.03, 23; nuchal scute 15.3 ± 1.2, 23; CNL 114.9 ± 4.4, 44.

Comparisons.-- Exceeded in size and weight only by *Priodontes maximus* and *Dasypus kappleri, E. sexcinctus* is the largest of the euphractines. It is pale yellow, tan or reddish tan in color and is sparsely supplied with buffy-white hairs. In addition to size, *E. sexcinctus* differs from other euphractines by its longer ears which extend back to the second or third row of scutes on the scapular shield; in lacking a complete movable band at the anterior edge of the scapular shield; in having a

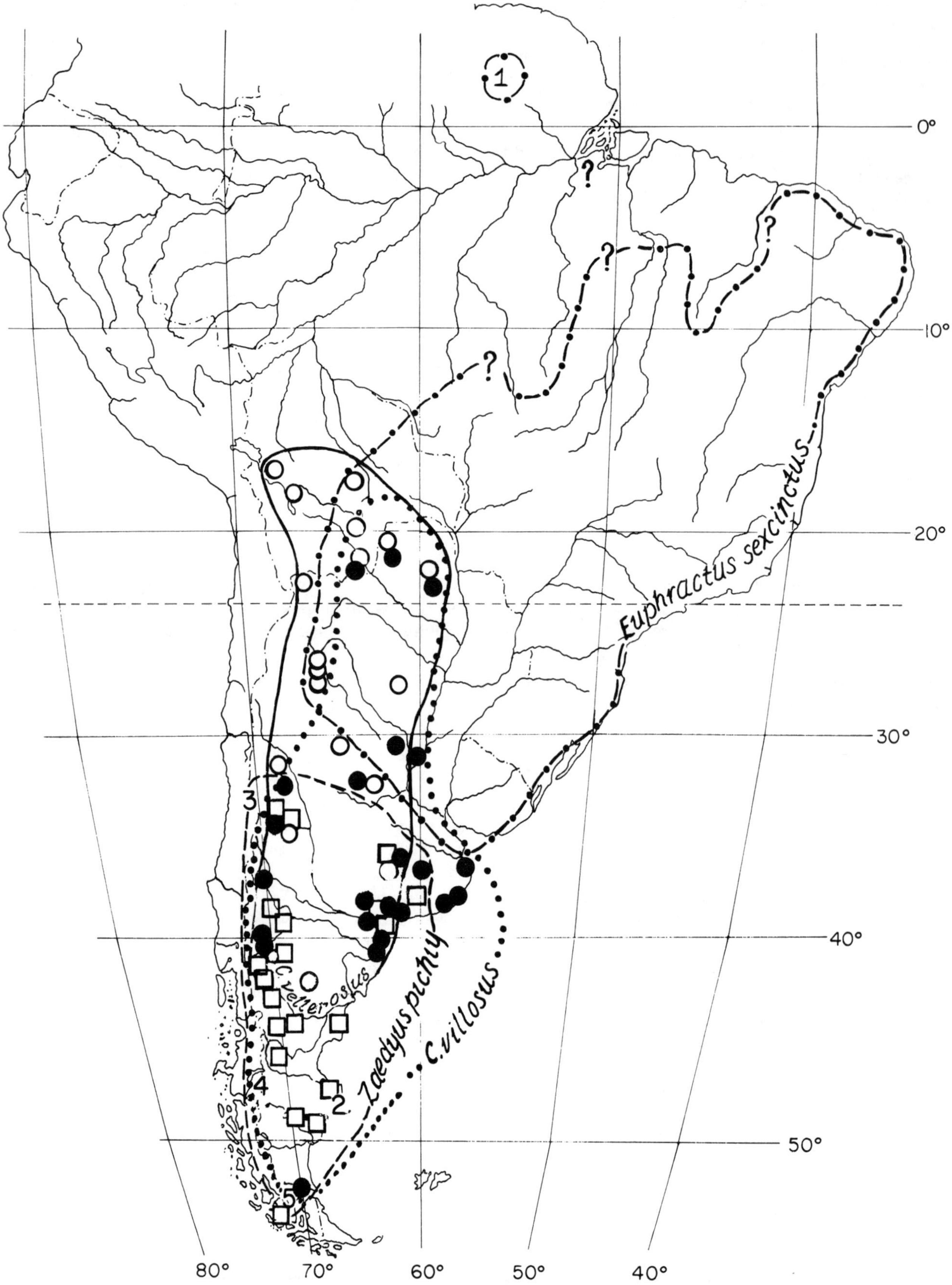

Figure 2.--Euphractine armadillos, distribution and specimens examined. •—•—• *Euphractus sexcinctus* (N = 85); ——— and ○ *Chaetophractus vellerosus* (N = 28) and in puna of Bolivia, *C. nationi* (N = 2); • • • • ◦ and ● *C. villosus* (N = 61); — — — — and □ *Zaedyus pichiy* (N = 36). From literature: **1** *E. sexcinctus*, Sipaliwini Savanna, Surinam, and Paru Savanna, Brazil (Husson, 1973); **2** *C. villosus*, Argentina, Santa Cruz, Pto. Deseado (Crespo, 1974); **3** *C. villosus*, Chile, Prov. Aconcagua, Rio Colorado (Tamayo, 1973); **4** *C. villosus* and *Zaedyus pichiy*, Chile, Prov. Aísen (Tamayo, 1973); **5** *Zaedyus pichiy*, Chile, Strait of Magellan (Texera, 1973).

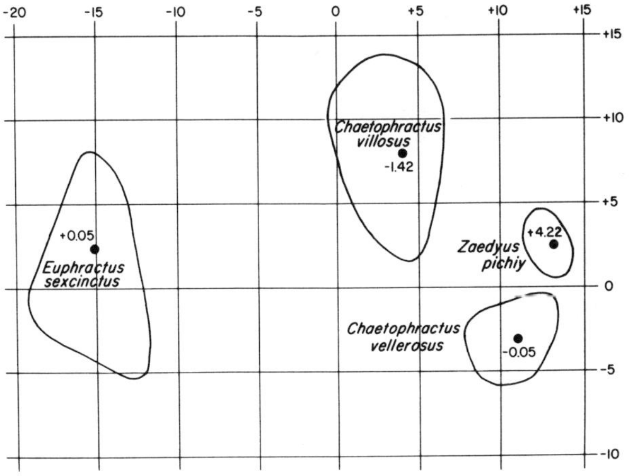

Figure 3.– Euphractini, canonical analysis, 17 variables (13 cranial measurements, 4 of head shield and nuchal scutes). • intercept of group means of two coordinates; numerical value is group mean of third coordinate. All adult specimens: *Euphractus sexcinctus* (N = 22), Paraguay, Uruguay and holotype *E. s. boliviensis; Chaetophractus villosus* (N = 11), Argentina and Paraguay; *C. vellerosus* (N = 17) Bolivia, Paraguay and Argentina (includes holotypes *C. v. vellerosus* and *C. v. pannosus*); *Zaedyus pichiy* (N = 5), Argentina. Note that third coordinate widely separates *Zaedyus* from *Chaetophractus*.

proportionally longer rostrum and a more elongate head shield whose greatest width is only 80 percent or less of its length; and in having two or three openings for glands in the middorsal scutes of the pelvic shield (however, some *C. villosus* as well have glandular pores on the middorsal pelvic shield). *E. sexcinctus* has a more elongate, slender zygomatic arch, and the jugal bone (malar) is never twice as high (dorsoventral height) as the overlying anterior end of the squamosal (Fig. 4).

GENUS ***Chaetophractus*** Fitzinger

Range.-- From the high plateau or puna of Bolivia through the Gran Chaco of Bolivia, Paraguay, and Argentina south through the monte, espinal, and pampas of Argentina to southern Patagonia and in adjoining Chile to the Strait of Magellan (Fig. 2).

Comparisons.-- Differs from *Euphractus* in having a relatively short zygomatic arch, flattened transversely; posterior of jugal bone more than twice as high (dorsoventrally) as the overlying zygomatic process of the squamosal (Fig. 4); head shield proportionally broader; one complete movable band on anterior margin of the scapular shield. Differs from *Zaedyus* in having 9/10 pairs of teeth; pronounced constriction in height of the middle of jugal bone while anterior of jugal is approximately half the height of posterior of that bone; and anteroposterior length of nuchal scutes greater, never shorter than 6.0 in adults.

Chaetophractus vellerosus (Gray). Screaming or small hairy armadillo; piche llorón or quirquincho chico. *Dasypus vellerosus* Gray, 1865. Holotype: BMNH

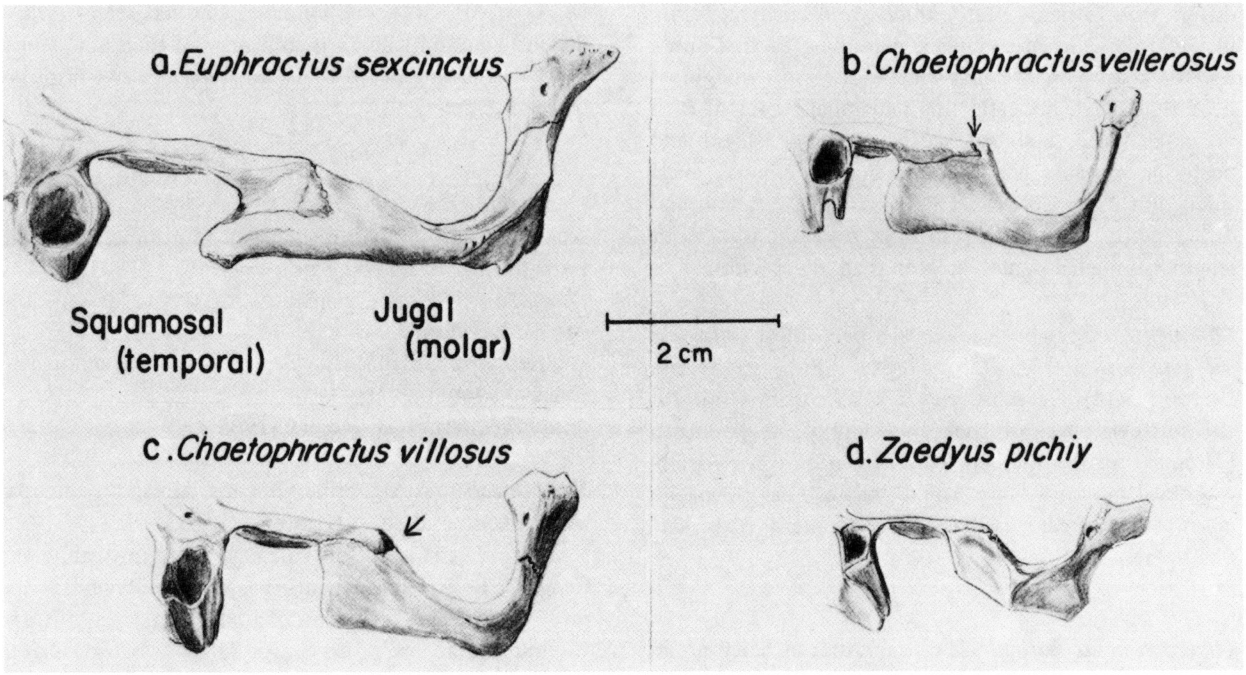

Figure 4.--Euphractini, right zygomas. **a.** *Euphractus sexcinctus*, CONN 15963, Paraguay, Presidente Hayes, Juan de Zalazar (Chaco); **b.** *Chaetophractus vellerosus*, CONN 16593, same locality; **c.** *C. villosus*, CONN 16603, same locality; **d.** *Zaedyus pichiy*, AMNH 25667, Argentina, Santa Cruz, Río Chico. Note: (1) in *Euphractus* vertical height of squamosal (temporal) is greater than jugal (malar) at their suture; in *Chaetophractus* and *Zaedyus* the reverse is true. (2) In *Zaedyus* uniform vertical height of jugal with no constriction below orbits as in other genera. (3) Variation, marked with arrows, in squamosal-jugal suture between specimens of *C. vellerosus* and *C. villosus*. (4) Ossified auditory meatus in all euphractines.

1376, mounted specimen, plus skull 47.11.22.20, coll. W.T. Bridges.
Chaetophractus vellerosus: Yepes, 1928; Frechkop and Yepes, 1949; Talmage and Buchanan, 1954; Cabrera, 1958.
Euphractus villosus desertorum Krumbiegel, 1940.
Euphractus vellerosus: Moeller, 1968.

Type locality.-- Bolivia, Santa Cruz, Santa Cruz (de la Sierra).
Range.-- From the approximate latitude of the city of Santa Cruz, Bolivia, through the Chaco Boreal of that nation and of western Paraguay, south through Argentina to the provinces of Mendoza, San Luis, La Pampa, Río Negro (MACN 31.2, Arroyo Pilcaniyeu), and Santa Cruz (MACN 29.899, no other locality data) (Fig. 2). See comments under *C. nationi* concerning the puna of Bolivia, Argentina, and Chile.
Measurements.-- $\bar{y}$, s, and N for adults from Argentina (N = 15), Bolivia (N = 1, holotype), and Paraguay (N = 4): head + body 230.1 ± 8.3, 19; tail 104.1 ± 9.0, 19; hindfoot 47.0 ± 5.5, 16; ear 30.8 ± 1.8, 14; ratio of head shield w/l 0.90 ± 0.04, 17; nuchal scute 6.9 ± 0.7, 17; CNL 62.6 ± 1.8, 20. Greegor (1974) found the following weights for his sample, no age groups indicated, from Catamarca, Argentina: males, 0.86 kg ± 0.14, 76; females, 0.81 kg ± 0.18, 71.
Comparisons.-- The smallest euphractine, *C. vellerosus* has more abundant hair on the venter and dorsum and differs from other *Chaetophractus* and *Zaedyus pichiy* in the following: ears longer, extending to first fused row of scutes on scapular shield; posterior margin of head shield flat, not reflecting individual curves of each marginal scute; marginal scutes of pelvic shield and posterior movable bands not as sharply pointed; head shield not as broad as in *C. villosus* or *C. nationi* and only slightly broader than in *Z. pichiy*; anteroposterior length of nuchal scutes greater than in *C. villosus* or *Z. pichiy*.
Comments.-- For some ecological aspects of *C. vellerosus* see Greegor (1974, 1975, 1980a, 1980b, 1985) and Crespo (1944). The subspecies *C. v. vellerosus* (Gray) for the northern part of the range and *C. v. pannosus* (Thomas) for the southern part have not been critically examined in recent times. The distress call of this species is a startling scream, as suggested by some of its vernacular names.

Chaetophractus nationi (Thomas). Andean hairy armadillo.
Dasypus nationi Thomas, 1894. Holotype: BMNH 93.9.9.1. Presented by Prof. V. Nation.
Chaetophractus nationi: Yepes, 1928; Talmage and Buchanan, 1954; Cabrera, 1958.
Euphractus nationi: Moeller, 1968, by implication.

Type locality.-- Bolivia, Oruro, Oruro.
Range.-- Puna of Bolivia, departments of Cochabamba (D. A. Meritt, Jr., pers. comm.), Oruro and La Paz. See Comments regarding distribution in Argentina and Chile.
Measurements.-- Holotype, head + body 268; tail incomplete; hindfoot without claws approximately 52; ear 30 (all from Thomas, 1894); head shield w/l 60.8/60.8; nuchal scute 7.7. MCZ 7393, male, Bolivia, La Paz, Viacha, 13,000 ft., head shield w/l 62.8/62.9; nuchal scute 8.0; CNL 76.4.
Comparisons.-- *C. nationi* is intermediate in size between *C. vellerosus* and *C. villosus*; the head shield is proportionally even broader than in most *C. villosus*; hair on the holotype is buff and tan, up to 72 mm long, and chiefly sparse. Hair on MCZ 7393 from Viacha is buffy white and dense, although outlines of movable bands and scutes are visible.
Comments.-- The skull of the young adult holotype is fragmented. Only one intact, adult skull, MCZ 7393, was available as a representative of this species and proved to be distinctly separated in canonical analysis from *C. vellerosus* and *C. villosus*. Specimens from Argentina, Jujuy, Abra Pampa (BMNH 19.8.1.76 and 19.8.1.77) were not separable from *C. vellerosus* except for having broader head shields and are included with *C. vellerosus* in Fig. 2. Tamayo (1973) summarized records of *C. nationi* from the puna of Chile. More specimens are needed to either support the validity of this species or to establish it as a high-altitude subspecies of *C. vellerosus*. D. A. Meritt, Jr. (pers. comm.), who has seen nearly a dozen living individuals, is convinced of their uniqueness as well as having heard their "piercing and awe inspiring bleat or distress call."

Chaetophractus villosus (Desmarest). Larger hairy armadillo; quirquincho mediano or quirquincho común.
Le Tatou Quatrieme ou Tatou Velu Azara, 1801.
Loricatus villosus Desmarest, 1804, and description, 1803. Based on Azara.
Chaetophractus villosus: Yepes, 1928; Frechkop and Yepes, 1949; Cabrera, 1958.
Euphractus villosus: Moeller, 1968.

Type locality.-- Argentina, Buenos Aires, the pampas between 35° and 36° S (Cabrera, 1958).
Range.-- The Chaco of Paraguay (and probably the Chaco of Bolivia and Argentina as well); in Argentina the provinces of Mendoza, Cordoba, and Santa Fe south to Chubut and according to Crespo (1974), as far as Santa Cruz; and in the eastern edge of Chile, (Tamayo, 1973), from at least provinces Bío-Bío south to Magallanes, Dazy Harbour (Inst. Patagonia 213) (Fig. 2).
Measurements.-- $\bar{y}$, s, and N of 9 adults from Argentina and 9 from Paraguay: head + body 330.1 ± 31.3, 18; tail 149.3 ± 12.1, 16; hindfoot 67.1 ± 5.8, 14; ear 28.5

± 2.2, 12; ratio of head shield w/l 0.92 ± 0.04, 22; nuchal scute 6.6 ± 1.3, 12; CNL of 35 adults from Argentina, and 11 from Paraguay 89.7 ± 4.9; weight 2.02 ± 0.59, 7.
Comparisons.-- Second in size to *E. sexcinctus* among the euphractines, *C. villosus* is a darkly colored armadillo with long black hair rather than the tan of *C. vellerosus* or the white of *C. nationi*. It also differs from *C. vellerosus* in the following characters: head shield broader, both in proportions and in actual measurements; scutes on head shield not flat and smooth but bearing designs; posterior margin of head shield not straight but reflecting the outline of individual marginal scutes; nuchal scutes shorter; ears shorter, not extending to the first fused row of scutes on the scapular shield; marginal scutes of pelvic shield not rounded but pointed; some individuals with glandular pores on middorsal pelvic shield. Fig. 4 illustrates the differences between the zygomatic arches of *C. villosus* and *C. vellerosus*.
Comments.-- Crespo (1944) discusses some ecological aspects of this species.

Zaedyus pichiy (Desmarest). Pichi. Le Tatou Septiéme ou Tatou Pichiy Azara, 1801.

Loricatus pichiy Desmarest, 1804, and description, 1803. Based on Azara.
Zaedyus pichiy: Osgood, 1919; Yepes, 1928; Frechkop and Yepes, 1949; Talmage and Buchanan, 1954; Cabrera, 1958.
Euphractus pichiy: Moeller, 1968.

Type locality.-- Argentina, Buenos Aires, Bahia Blanca. Based on the statement "Plata; Bahia-Blan." for *Tatusia minuta*, a junior synonym by Lesson as interpreted by Cabrera (1958).
Range.-- Argentina from the provinces of Mendoza, San Luis and Buenos Aires south to Río Santa Cruz, and from the Atlantic coast west to the Andean grasslands of both Argentina and Chile (Cabrera, 1958; Tamayo, 1973), and south in Chile to the Strait of Magellan (SMN 5688, Punta Arenas; Texera, 1973). See map, Fig. 2.
Measurements.-- $\bar{y}$, s, and N for adult specimens: head + body 273.0 ± 39.0, 5; tail 109.2 ± 8.7, 5; hindfoot 50.4 ± 3.2, 5; ear 13.4 ± 1.5, 5; ratio of head shield w/l 0.88 ± 0.04, 13; nuchal scute 4.5 ± 3.8, 5; CNL 66.7 ± 2.7, 26; teeth 8.1 ± 0.27, 26/8.9 ± 0.49, 25.
Comparisons.-- A small, hairy armadillo with sharply pointed marginal scutes, *Z. pichiy* differs from all other euphractines in lacking teeth on the premaxillary, thus having only 8 maxillary and 9 mandibular pairs. In addition, the jugal (Fig. 4) has a proportionally greater dorsoventral height, and the nuchal scutes are shortest (less than 5.0) in anteroposterior length. Additionally *Zaedyus* differs from the other euphractine to which it is most similar, *C. vellerosus*, in being somewhat larger and in having a head shield proportionally more slender, ears much shorter, and marginal scutes which are sharply pointed.
Comments.-- See Crespo (1944) concerning tunnels.

TRIBE **Priodontini**

This tribe includes two genera, *Priodontes* and *Cabassous*. Characteristics shared by all members include: rounded, blunt snout; relatively large eyes; ovoid, dome-shaped carapace with numerous, narrow transverse bands of which 11 to 14 are movable; 3 short rows of nuchal scutes; sparse hair; large, scimitar-shaped foreclaws, the longest on the third digit; slender mandible with height of condyloid process greater than coronoid process; frontals with distinct dome-like expansions; and, as in *Dasypus* and *Tolypeutes*, a tympanic ring rather than bulla. Frechkop and Yepes (1949) considered the horizontal suture between squamosal and jugal as distinctive; however, this condition occurs also in Euphractini.

GENUS ***Priodontes*** F. Cuvier

Priodontes maximus (Kerr). Giant armadillo; armadillo gigante; tatú-guasú tatú-canastra.

Dasypus maximus Kerr, 1792.
Dasypus giganteus E. Geoffroy St. Hilaire, 1803; Desmarest, 1804.
Priodontes giganteus: Yepes, 1928; Frechkop and Yepes, 1949; Talmage and Buchanan, 1954; Cabrera, 1958.
Priodontes maximus: Thomas, 1880; Allen, 1895; Hershkovitz, 1959; Wetzel and Lovett, 1974; Handley, 1976.

Type locality.-- French Guiana, Cayenne (Kerr, 1792).
Range.-- South America east of the Andes from northwestern Venezuela (state of Yaracuy) to French Guiana south through the Amazon basin and along eastern foothills of the Andes in Colombia, Ecuador, Peru, and Bolivia to northwestern Argentina (provinces of Salta, Formosa, and Chaco), Paraguay, and southeastern Brazil in the state of Rio Grande do Sul (Fig. 5). Moeller's (1968) distribution map included eastern Brazil in the range of *Priodontes*, but I have found no specimens nor other records from Brazil east of the states of Pará, Goiás, and São Paulo.
Measurements.-- $\bar{y}$, s, and N of adults from throughout range: head + body 895.5 ± 63.9, 4; tail 528.2 ± 16.6, 4; hindfoot 188.0 ± 8.5, 4; ear 53.8 ± 6.2, 5; weight, ROM 46260, Guyana, Raa Wau, 29.48 kg; teeth 17.9 ± 1.8, 38 /19.3 ± 2.0, 32; CNL 193.1 ± 6.0, 37.
Comparisons.-- Largest of all Recent species of arma-

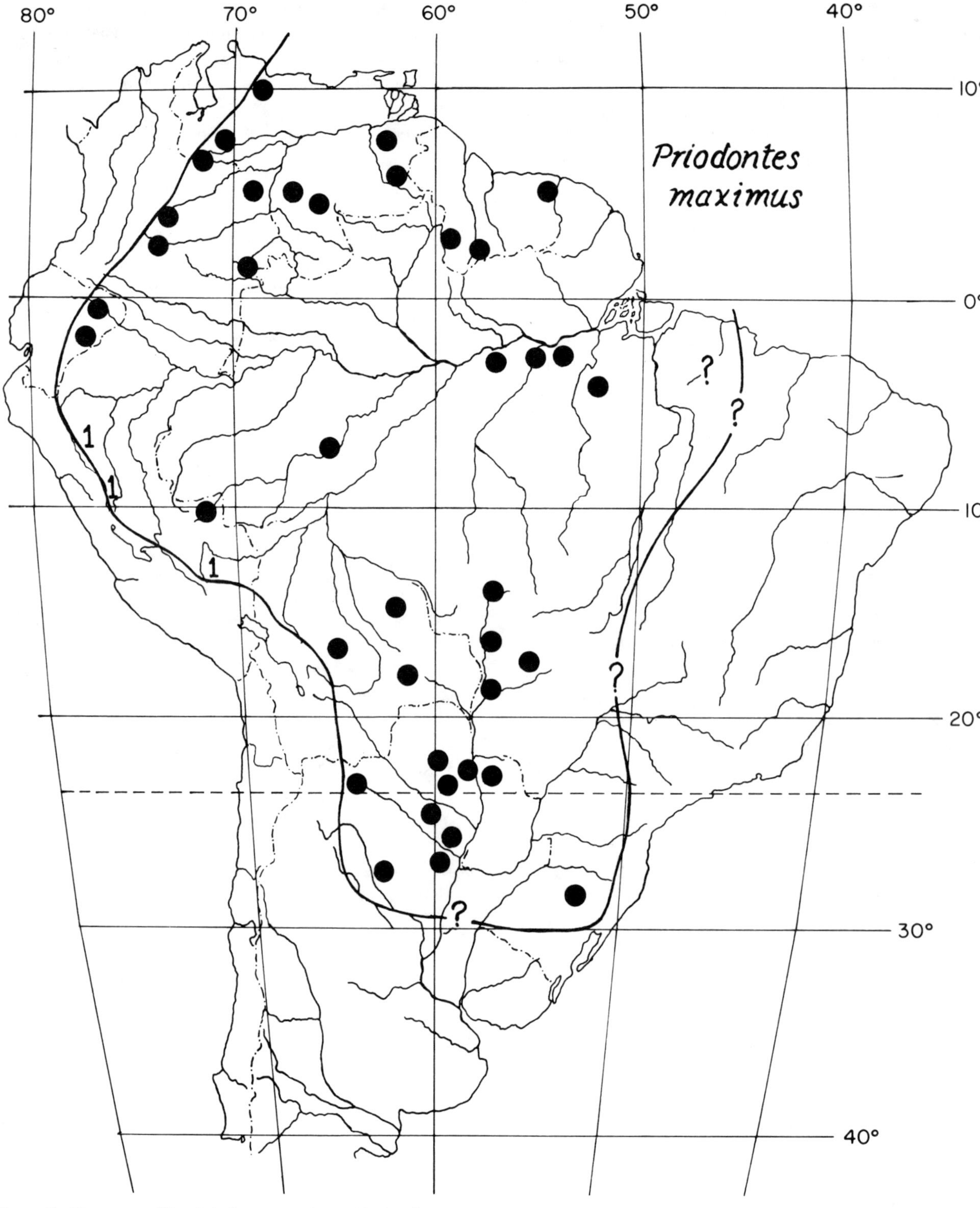

Figure 5.–Giant armadillo, *Priodontes maximus*, distribution and (•) specimens examined. Although the map is based on most of the available specimens with geographical data, only 39 individuals are involved. Note absence of records from northeastern Brazil. 1 citations from Grimwood (1969).

dillos, adult *Priodontes* are easily identified by size alone. *Cabassous* specimens have been mistaken by amateur collectors for very young *Priodontes*, but the soft tail of *Cabassous* is strikingly different from the well-armored tail of *Priodontes*. The carapace of *P. maximus* is dark dorsally and sharply marked laterally by a buffy border. In addition to size, *Priodontes* differs from *Cabassous* in having a proportionally longer

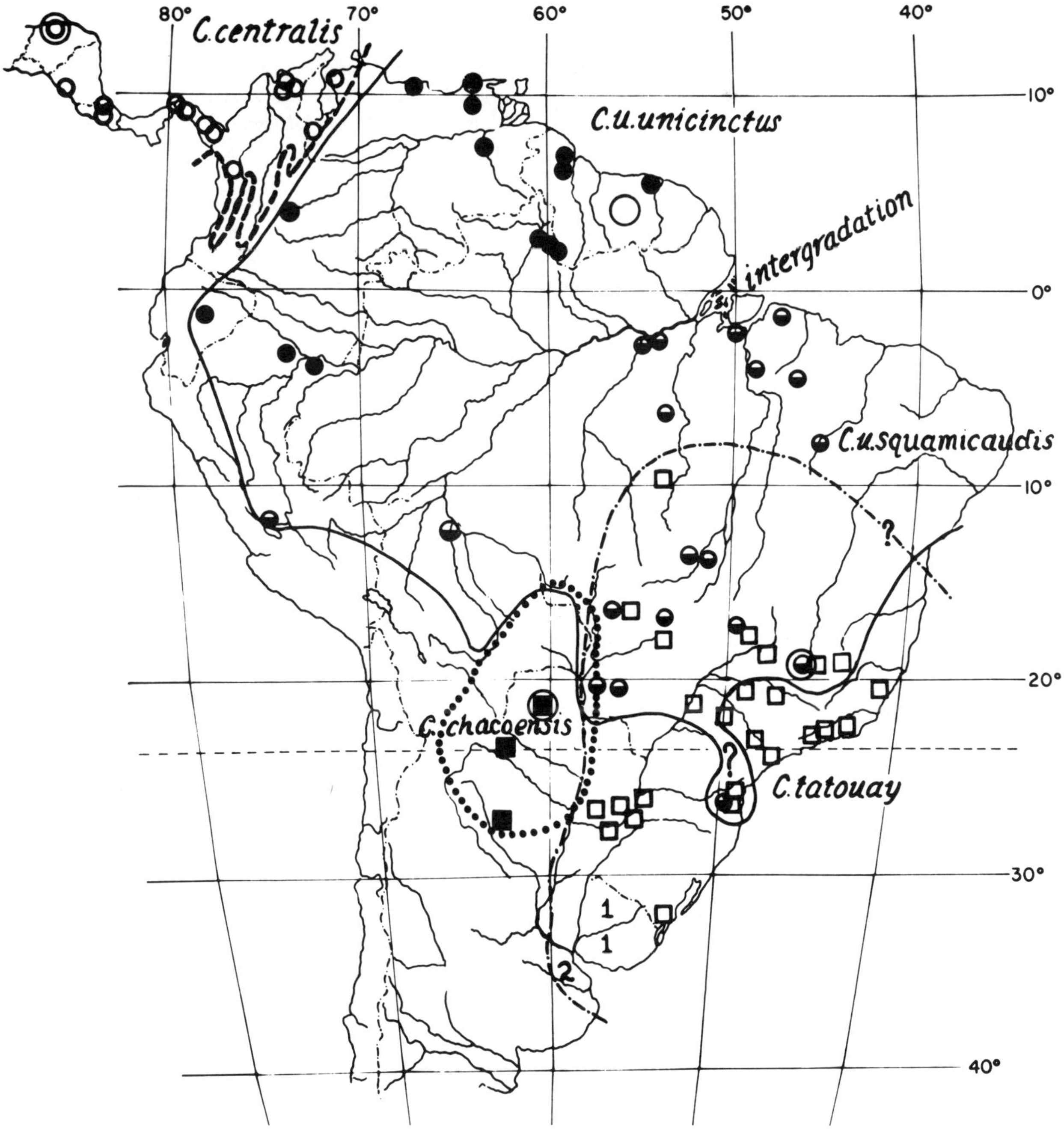

Figure 6.– Naked-tailed armadillos, *Cabassous*, distribution and specimens examined. – – – – and ○ *C. centralis* (N = 26); ——— and ● or ◒ *C. unicinctus* (N = 62); ••••••• and ■ *C. chacoensis* (N = 7); •—•—•—• and □ *C. tatouay* (N = 36). Type-localities are indicated by larger circles. From literature: **1** *C. tatouay* (as *C. gymnurus*, Ximénez, et al., 1972), Uruguay; **2** *C. tatouay* (as *C. unicinctus*, Moeller, 1968). Note: exact location of most northern records of *C. tatouay* is unknown, based on "State of Pará" for BMNH 45.8.25.16 and 46.5.13.15, and NMBE, Goeldi No. 10 (Wetzel, 1980).

rostrum, more numerous teeth, 18/19 pairs, which are very flattened transversely, and chromosomes 2N = 50, NF = 76, as compared to *Cabassous centralis*, 2N = 62, NF = 78 (Benirschke et al., 1969).

GENUS ***Cabassous*** McMurtrie

Range.-- West of the Andes from Guatemala to Venezuela and east of the Andes from Venezuela south to the provinces of Buenos Aires and Santiago del Estero in Argentina (Fig. 6).

Characterization and comparisons.-- Excepting the Chlamyphorinae, *Cabassous* species are the most fossorial of the armadillos. Carapace very flexible, ovoid, somewhat dome-shaped, with a variable number of movable bands, usually 11; 3 short nuchal rows of scales;

tail rounded and without armor, being either without epidermal scales (naked) or scales small and thin; 5 long foreclaws, those on the third, fourth, and fifth digits particularly broad and scimitar-shaped, and first digit bearing a long, very slender claw; 5 blunt claws on hindfoot, the third claw longest; edges of toes and medial side of feet flanked with single line of strong bristles. Skull with short rostrum; teeth peg-shaped and somewhat ovoid, 9 maxillary pairs and 8 mandibular pairs in adults (range 7-10/8-9); body of mandible slender. Differs from *Priodontes* not only in being much smaller but also in having fewer teeth; teeth broader and not so flattened laterally; rostrum much shorter; tail without strong armor.

Comments.-- These armadillos are not commonly seen nor collected. Probably because of their scarcity as specimens in research collections and because various authors were unable to examine most of the holotypes, the taxonomy has become badly tangled. Some light was offered by Thomas (1911) who designated Surinam as the type locality for the type species *C. unicinctus* (Linné) as the name was based upon Seba (1734) and Seba's specimens were usually from Surinam. This opportunity for clarity was, however, frequently ignored or more likely missed, because it was not understood that only a single species of *Cabassous* occurs in Surinam and environs. It will be noted in the following synonymies that many authors reversed the two species that are sympatric in the highlands of Brazil. Qualitative comparisons (Fig. 7), univariant analyses, and canonical analyses (Fig. 8) indicate that there are four species. The large species *C. tatouay*, separable by both size and cranial characters, is sufficiently described (Desmarest, 1803) to be clearly identifiable. The smaller *C. unicinctus* is distributed east of the Andes from Surinam and adjacent areas, through the Amazon basin to the highlands of Brazil. The third species, *C. centralis*, is closely related to *C. unicinctus*, with the Andes forming an approximate, but probably not absolute, barrier between the two species. The fourth species *C. chacoensis* is confined to the Gran Chaco.

Cabassous unicinctus (Linné). Southern naked-tailed armadillo. Synonymy, range and measurements under subspecies.

Comparisons.-- In *C. unicinctus* the short rostrum is proportionally broader than in the other species (Fig. 7); except for the most anterior, teeth are rounded rather than ovoid; palate posterior to the toothrow is short and broad; and the exterior surface of the ear is usually partially scaled. *C. unicinctus* is similar to *C. centralis* in that excepting the most lateral scutes, the majority of scutes on the scapular and pelvic shields are wider than long and the occipital condyle does not bulge posteriorly as in *C. tatouay* but lies near the posterior margin of the

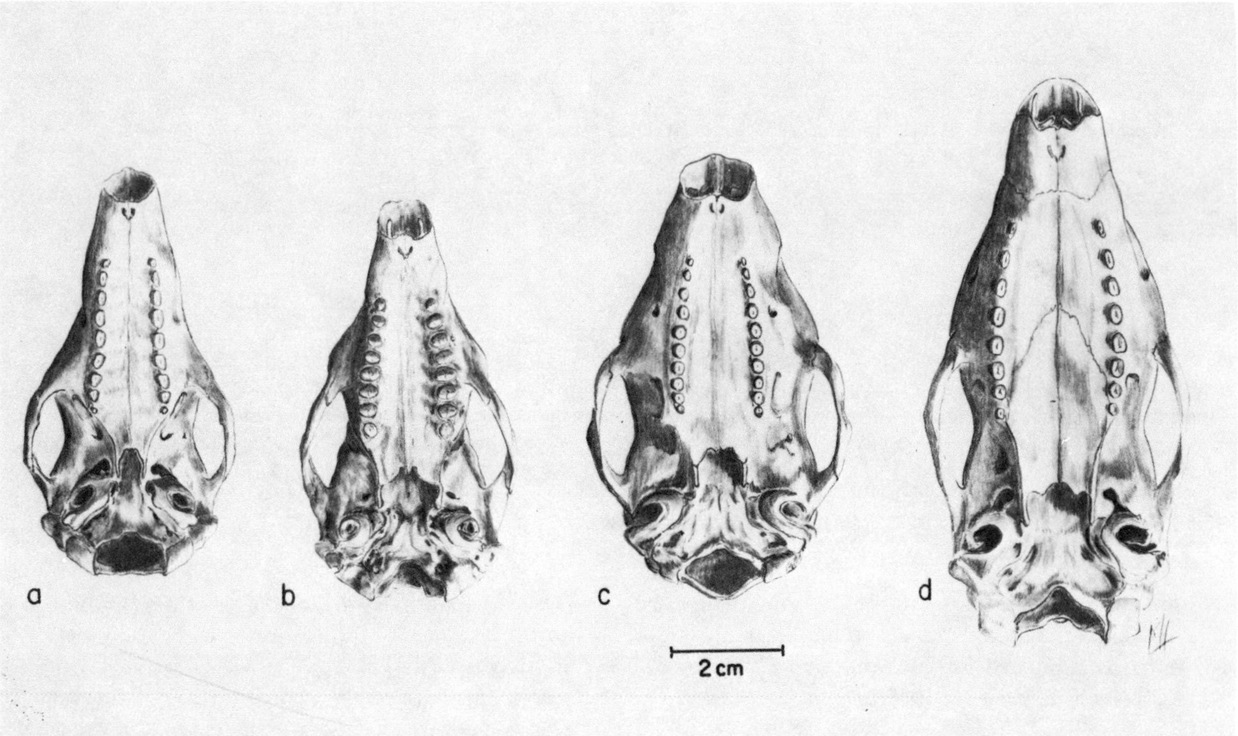

Figure 7.--*Cabassous*. Palatal view of adult skulls. **a.** *C. centralis*, AMNH 23441, Colombia, Magdalena, Bonda; **b.** *C. chacoensis*, holotype CM 67067, Paraguay, Presidente Hayes, Juan de Zalazar. **c.** *C. unicinctus squamicaudis*, AMNH 133318, Brazil, Goiás, Anápolis. **d.** *C. tatouay*, AMNH 133317, Brazil, Goiás, Anápolis (Wetzel, 1980).

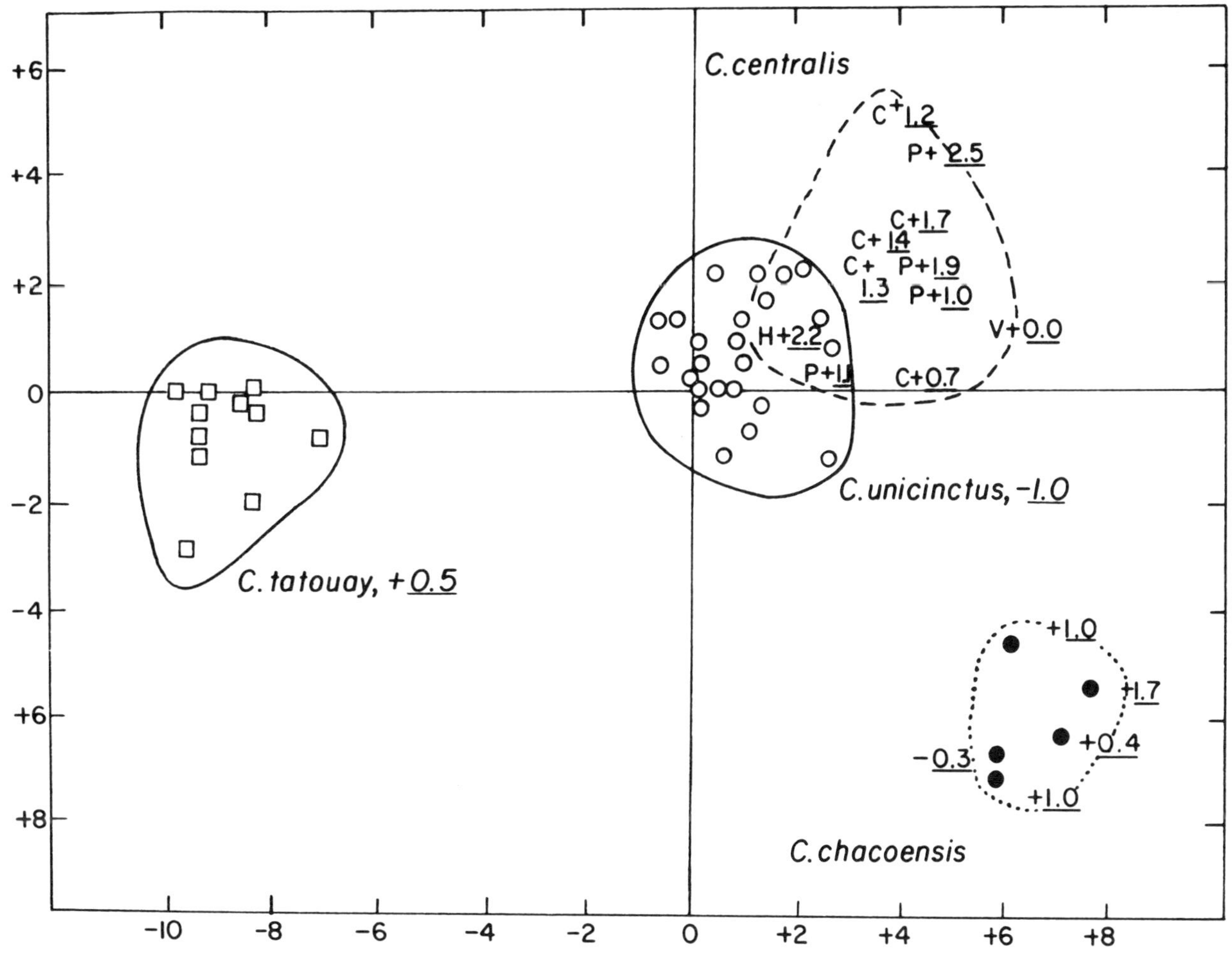

Figure 8.– *Cabassous*, canonical analysis, 11 variables (9 cranial and 2 tooth measurements), N = 51. Interrupted lines enclose specimens not grouped statistically but treated individually. Note overlap of *C. centralis* with *C. unicinctus* is confined to two specimens from Honduras (H) and Panama (P), and does not occur with specimens from Colombia (C) and northwestern Venezuela (V). The third canonical variables of the first two specimens also indicate wide statistical separation from *C. unicinctus* (Wetzel, 1980).

skull. Also note its separation from *C. tatouay* and *C. chacoensis* in Fig. 8 and from *C. centralis* in Figs. 8 and 9.

Cabassous unicinctus unicinctus (Linné). Tatu seu Armodillo, Africanus, mas. Seba, 1734.

Dasypus unicinctus Linné, 1758. Based ultimately, upon Seba.

Ziphila lugubris Gray, 1873. Holotype: BMNH 51.8.25.10, Brazil, "St. Catherine's," state not known. Paratype: BMNH 55.8.28.7, Guyana, Demerara (see Wetzel, 1980).

Cabassous unincinctus, in part: Talmage and Buchanan, 1954; Cabrera, 1958.

Cabassous lugubris, in part: Yepes, 1928 and 1944; Talmage and Buchanan, 1954; Moeller, 1968.

Cabassous unicinctus unicinctus: Wetzel, 1980.

Type locality.-- Surinam, designated by Thomas (1911).

Range.-- From Colombia, Ecuador and Venezuela east of the Andes to Guyana, Surinam, and French Guiana, intergrading in the Amazon basin of Brazil and Peru with *C. unicinctus squamicaudis* (Figs. 6 and 9).

Measurements.-- $\bar{y}$, s, and N of adult specimens from Venezuela and Surinam: head + body 387.5 ± 30.2, 12; tail 184.6 ± 11.9, 12; hindfoot 79.2 ± 3.9, 12; ear 35.2 ± 2.7, 11; weight 2.90 kg ± 0.39, 9; CNL 85.6 ± 2.6, 15; RLA 44.8 ± 1.9, 18; PRL 44.6 ± 1.0, 15. $\bar{y}$ and N of specimens from area of intergradation with *C. u. squamicaudis* (Taperinha and Rio Tapajós near juncture with the Amazon): CNL 81.6, 4; RLA 41.6, 5; PRL 43.2, 4.

Cabassous unicinctus squamicaudis (Lund). Tatu seu Armodillo, Africana, foemina Seba, 1734.

Xenurus squamicaudis Lund, 1845 (1843). Lectotype ZMUC L-33 (Wetzel, 1980).

Dasypus hispidus Burmeister, 1854.

Dasypus gymnurus, var. *loricatus* Wagner, 1855. Holotype: NMV 3043, Brazil, Mato Grosso, Cabeça de Boi.

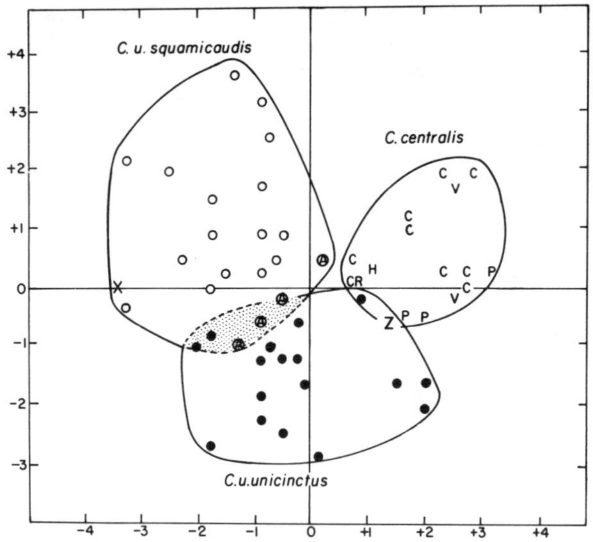

Figure 9.– *Cabassous,* canonical analysis, 9 variables (cranial measurements), N = 54. For *C. centralis,* **C** = Colombia, **CR** = Costa Rica, **H** = Honduras (holotype), **P** = Panama, **V** = northwestern Venezuela. For *C. u. squamicaudis,* open circles enclosing letters A denote specimens from area of intergradation with *C. u. unicinctus,* **X** = *Xenurus latirostris* Gray (holotype). For *C. u. unicinctus,* **Z** = *Ziphila lugubris* Gray (holotype). The greatest displacement of *C. centralis* from *C. u. unicinctus* occurs at points of nearest geographical contiguity (Wetzel, 1980).

Xenurus latirostris Gray, 1873. Holotype: BMNH 51.8.25.9, Brazil, "St. Catherine's," state not known (Wetzel, 1980).
Cabassous loricatus: Yepes, 1928 and 1935; Talmage and Buchanan, 1954; Cabrera, 1958; Moeller, 1968.
Cabassous latirostris: Yepes, 1928; Krumbiegel, 1940.
Cabassous hispidus: Yepes, 1928; Krumbiegel, 1940; Talmage and Buchanan, 1954; Cabrera, 1958; Moeller, 1968.
Cabassous unicintus squamicaudis: Wetzel, 1980.

Type Locality -- Brazil, Minas Gerais, Lagoa Santa.
Range.-- From an area of intergradation with *C. u. unicinctus* along the Amazon-Solimões River from its mouth to the foothills of the Andes in Peru and Bolivia south to the states of Mato Grosso, Mato Grosso do Sul, Goiás, Minas Gerais, and Maranhão (Fig. 6). I have seen no specimens or records for this species from as far southwest as the Argentine province of Mendoza as shown by Moeller's figure (1968: Abb. 9b).
Measurements.-- $\bar{y}$, s, and N of adults from areas south of the Amazon River in Bolivia (Depto. Beni), Brazil (Goiás, Maranhão, Mato Grosso, Mato Grosso do Sul, Minas Gerais, Pará) and Peru (Depto. Junín): head + body 324.0 ± 23.7, 4; tail 115.5 ± 22.6, 4; hindfoot 71.5 ± 5.1, 4; ear 27.0 ± 2.2, 4; CNL 78.1 ± 4.4, 21; RLA 36.2 ± 2.7, 24; PRL 42.1 ± 2.5, 21.

Cabassous centralis (Miller). Northern naked-tailed armadillo.
Tatoua centralis Miller, 1899. Holotype: USNM 19464/35382, coll. Wittkügel.
Cabassous centralis: Yepes, 1928; Talmage and Buchanan, 1954.
Cabassous unicinctus: Moeller, 1968, in part, not of Linné, 1758.

Type locality.-- Honduras, Cortés, Chamelecón.
Range.-- Pacific coastal plain and mountains of Quiché in Guatemala (Handley, 1950); Honduras, Nicaragua, Costa Rica, Panama, to northwestern Venezuela west of the Andes (Fig. 6).
Measurements.-- $\bar{y}$, s, and N of adults from Central America and Colombia: head + body 340.5 ± 23.9, 6; tail 159.5 ± 17.7, 6; hindfoot 65.2 ± 5.4, 6; ear 32.7 ± 2.2, 6; CNL 77.1 ± 3.6, 17; RLA 40.0 ± 1.9, 19; PRL 41.0 ± 2.0, 16.
Comparisons.-- The similarity of *C. unicinctus* is noted under that species. *C. centralis* differs from the other two species in having external surface of pinna without or chiefly without scales; first paired scutes of head shield large and nearly equal; rostrum short but proportionally narrow; palate posterior to toothrow short and narrow (Fig. 7); teeth elongate ovals as in *C. tatouay* but the long axis of midposterior teeth may be diagonal to long axis of the skull. Chromosomes 2N = 62 (Benirschke et al., 1969).
Comments.-- Canonical analyses (Figs. 8 and 9) also indicate the separation of this species from *C. unicinctus.* Character displacement of the population of *C. centralis* geographically nearest the range of *C. unicinctus* is suggested in Fig. 9. Notice that the coordinates for the specimens from Panama, geographically most distant from *C. unicinctus*, are closest to that species.

Cabassous chacoensis Wetzel. Chacoan naked-tailed armadillo.
Cabassous loricatus, not of Wagner, 1855, but of authors: Yepes, 1935, in part for Chaco Austral of Argentina and possibly some unspecified part of Mato Grosso do Sul of Brazil; Minoprio,1945, in part for northern Argentina; Cabrera, 1958, and Moeller, 1968, in parts for Gran Chaco and possibly Mato Grosso do Sul.
Cabassous chacoensis Wetzel, 1980. Holotype: CM 67067.

Type locality.-- Paraguay, Depto. Presidente Hayes, 5-7 km W Estancia Juan de Zalazar.
Range.-- The Gran Chaco (northwestern Argentina, western Paraguay, and southeastern Bolivia) and possibly the adjoining state of Mato Grosso do Sul in Brazil.
Measurements.-- y, range, and N of adults, holotype and

paratypes: head + body 303.0, 300-306, 2; tail 93.0, 90-96, 2; hindfoot 61.0, 61-62, 2; ear 14.5, 14-15, 2; CNL 69.8, 68.5-71.0, 4; RLA 30.7, 29.5-31.5, 5; PRL 38.9, 38.0-39.7, 4.

Comparisons.-- Smaller than any other species of *Cabassous*. Its tiny ear has fleshy expansions of anterior edges, lacking rounded flare of other species. Also differs from all other *Cabassous* in having all but the most anterior and posterior teeth constricted anteroposteriorly rather than transversely.

Cabassous tatouay (Desmarest). Greater naked-tailed armadillo. Tatoú Troisieme, ou Tatou Tatouay Azara, 1801.

Loricatus tatouay Desmarest, 1804, and description, 1803. Based on Azara.

Tatus gymnurus Olfers, 1818. Based on Azara.

Ziphila lugubris Gray, 1865, in part: where holotype cited as "head and body 16 in," but not skull associated with skin BMNH 51.8.25.10 nor paratype from Demerara cited by Gray which are *C. unicinctus*.

Cabassous unicinctus, not of Linné, 1758, but of authors: Yepes,1928, 1935; Krumbiegel, 1940; Frechkop and Yepes, 1949; Talmage and Buchanan, 1954; Moeller, 1968, in part.

Cabassous tatouay: Cabrera, 1958; Wetzel, 1980.

Cabassous gymnurus: Ximénez, et al., 1972.

Type locality.-- Paraguay at 27° S. (Cabrera, 1958).

Range.-- Brazil in states of southern Pará, Mato Grosso, Goiás, Minas Gerais, and Espírito Santo south to Rio Grande do Sul; Uruguay, southern Paraguay east of the Río Paraguay, and Argentina in the province of Misiones (Fig. 6). I have not seen specimens or other data to substantiate Moeller's figure (1968:Abb. 9a) suggesting a record as far west in Argentina as Santiago del Estero nor as far north in Brazil as eastern Pará in the Belém area. Similarly, no specimens were found from Formosa in Argentina, although Cabrera's (1958) speculation that this species probably occurs in the extreme eastern portion of that province seems reasonable. It is also likely that *C. tatouay* may occur or once occurred in the provinces of eastern Chaco, Santa Fe, northern Corrientes and eastern Entre Ríos.

Measurements.-- y, s, and N adults: head +body 457.8 ± 33.2, 5; tail 179.0 ± 19.5, 5; hindfoot 82.2 ± 3.0, 5; weight, AMNH 133314, Brazil, Goiás, Anápolis 6.2 kg; CNL 109.8 ± 4.6, 23; RLA 53.1 ± 2.8, 23; PRL 56.2 ± 2.4, 23.

Comparisons.-- *C. tatouay*, much larger than the other species of *Cabassous*, differs additionally as follows: scutes and scales of scapular and pelvic shield longer than wide; external surface of ear very granular with scales; rostrum much more elongate; teeth elongate ovals, the first pair quite flattened; palate posterior to toothrow elongate, with posterior border extending to a point midway between the basisphenoid-presphenoid suture and the occipital-basisphenoid suture; hamular process elongate and slender.

TRIBE **Tolypeutini**
GENUS ***Tolypeutes*** Illiger

Comparisons.-- *Tolypeutes*, the single genus of the tribe, contains two species, one centered in northeastern Brazil and the other from Bolivia to southern Argentina (Fig. 10). These are the only armadillos capable of rolling into a ball, presenting only armor of the carapace, head shield and tail to a predator. Members of *Tolypeutes* have a hard, very convex carapace, the scapular and pelvic shields composed of rigid ovoid to hexagonal scutes and separated by three, rarely one or two, movable bands; head shield flat, elongated anteriorly, rounded at other margins; pinna round and creased for folding in half when armadillo rolls into a ball; tail short, completely armored with round scutes; penis proportionally elongate; venter with long buff and brown hair; longest curved claw on third fore digit; claws on second, third and fourth digits of hindfoot nearly equal in length, broadly curved or "hoof"-like. Skull with tympanic ring, not bulla; teeth ovoid in cross section. The chromosomes of the one species that has been examined, *T. matacus*, are unique in having "... no truly acrocentric autosomes and the chromosome number (2N = 38) is the lowest in edentates examined so far" (Jorge et al., 1977).

Tolypeutes tricinctus (Linné). Brazilian domed or northern three-banded armadillo; tatú-bola.

Dasypus tricinctus Linné, 1758.

Tolypeutes tricinctus: Sanborn, 1930, in part; Talmage and Buchanan, 1954, in part; Cabrera, 1958; Moeller, 1968.

Type locality.-- Brazil, Pernambuco, designated by Sanborn (1930) who rejected Surinam (Thomas, 1911) on distributional grounds, as Marcgrav's *Historia Naturalis Brasiliae* represented an identifiable regional reference in Linné.

Range.-- The indefinite range of northeastern Brazil (Sanborn, 1930; Cabrera, 1958) reflects the dearth of specimens with data and other distributional information on this rare and endangered species (Fig. 10).

Measurements.-- Measurements of an adult male specimen, MN 4257, from Brazil, Bahia, Bom Jesús de Lapa: total length 270; hindfoot 45; ear 23. Means of 4 adult specimens, MN4257, 4292, 4294 from Bom Jesús de Lapa and MN 1503 from Pernambuco, Serra dos Cariris Velhos: head shield w/l and ratio 34.8/68.6 and 0.51; teeth 8/8.7; CNL 71.1; RLA 35.9; I0 20.1.

Comparisons.-- The extent of variation in three-banded armadillos has contributed to shifting taxonomic treatment ranging from ten specific epithets to a monotypic

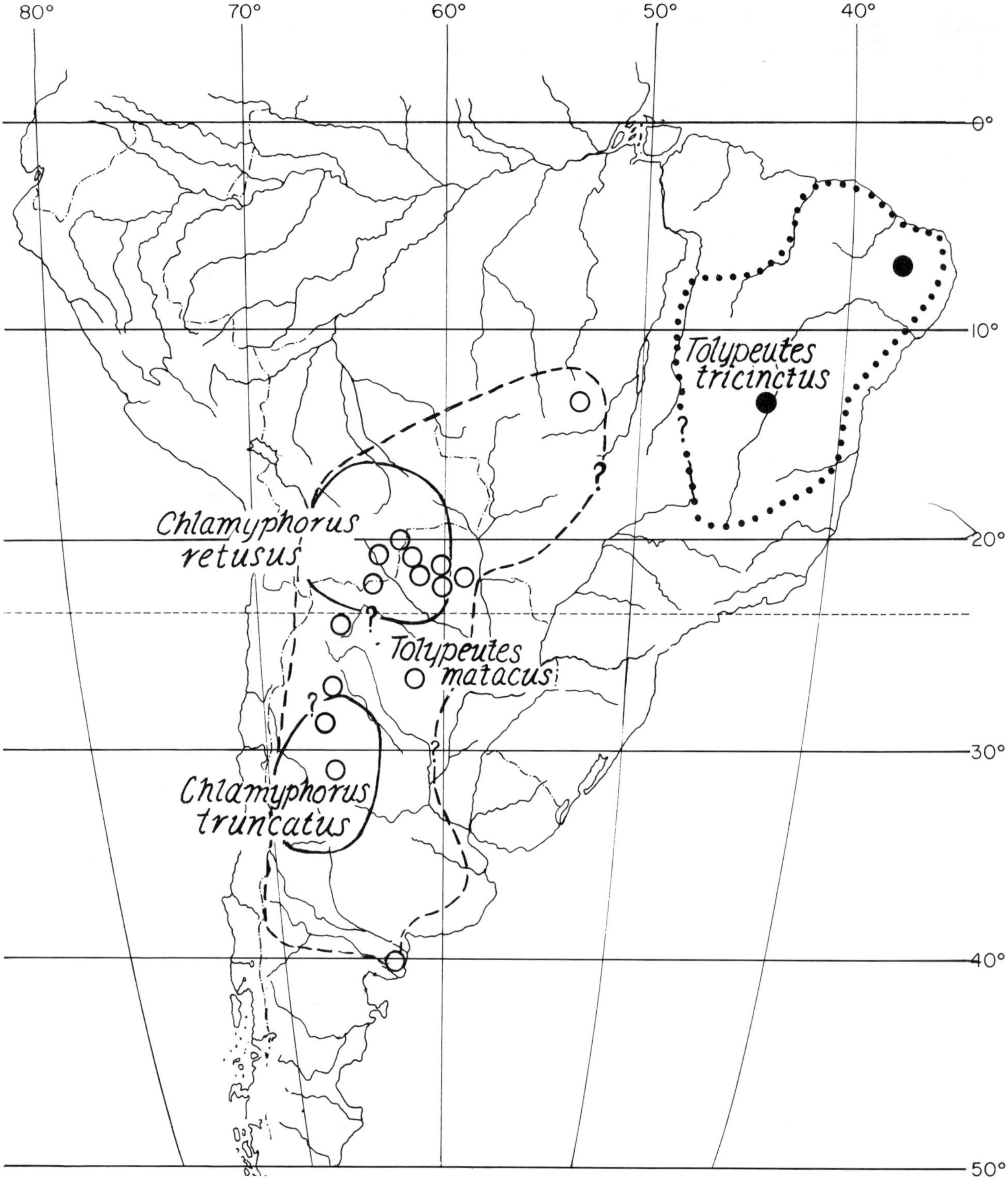

Figure 10.–Distribution of smaller and relatively rare armadillos, based on: (1) ○ and ● = specimens of 3-banded armadillos, *Tolypeutes*, examined (marginal records for *T. matacus* are Argentina, Buenos Aires, Carmen de Patagones, USNM 11822/37864, and Brazil, Mato Grosso, Chavantina das Mortes, MN S-10). (2) Cabrera (1958) for *Chlamyphorus retusus* and *C. truncatus*. The record of *Chlamyphorus truncatus* from the vicinity of Laguna Chasico, southwestern Buenos Aires province (Contreras, 1973) is not shown.

genus (Sanborn, 1930). Canonical analysis (Fig. 11) shows two distinct and widely spaced groups: (1) specimens from near the type-locality of *T. tricinctus* and (2) specimens from the western portion of the range of the genus, Bolivia to Argentina, spanning the type-locality of *T. matacus*. Of the 14 variables in this analysis, those contributing most to separation of the two species are based upon the greater rostral length and smaller interorbital breadth of *T. tricinctus*. Other distinctions of *T. tricinctus* separating it from *T.*

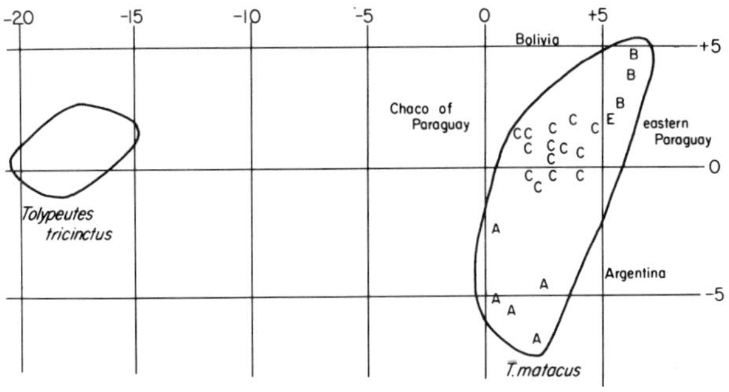

Figure 11.–Three-banded armadillos, *Tolypeutes*, canonical analysis, 14 variables (cranial measurements), N = 27. Individual specimens are not indicated for *T. tricinctus*, N = 4. The following symbols indicate intercepts of coordinates for individual specimens of *T. matacus*: **A**, Argentina; **C**, Chaco of Paraguay; **E**, eastern Paraguay; **B**, Bolivia. Note wide separation of the two species and the geographical gradient displayed by *T. matacus*.

matacus include 8 pairs of maxillary and 9 pairs of mandibular teeth; forefoot with 5 toes rather than 4 or 3; and for the 4 specimens from Bahia and Pernambuco whose measurements are given above, the following pattern of head shield scutes: immediately in front of the posterior marginal scutes is a large median scute, bordered anteriorly by 2 pairs of large scutes which are followed anteriorly by 2 unpaired, median scutes. The scute pattern for my small sample of *T. tricinctus* is not duplicated in the range of variation that Sanborn (1930) presents for *T. matacus* from Mato Grosso.

Tolypeutes matacus (Desmarest). Azara's domed or southern three-banded armadillo; tatú-bolito. Le Tatou Huitieme, ou Tatou Mataco Azara, 1801.

Loricatus matacus Desmarest, 1804 and description, 1803. Based on Azara.

Tolypeutes tricinctus matacus: Sanborn, 1930; Talmage and Buchanan, 1954.

Tolypeutes matacus: Osgood, 1919; Cabrera, 1958; Moeller, 1968.

Type locality.-- Argentina, Tucumán, Tucumán, as selected by Sanborn (1930) from Azara's (1801) range for the species from Tucumán and the pampas of Buenos Aires southward.

Range.-- Santa Cruz, Bolivia, and southern Mato Grosso, Brazil, southward through the Chaco of Paraguay to the province of Buenos Aires in Argentina (Fig. 10).

Measurements.-- $\bar{y}$, s, and N for adults from Bolivia, western Paraguay and Argentina: total length 301.8 ± 17.0, 15; tail 63.7 ± 8.0, 15; hindfoot 42.3 ± 3.2, 15; ear 22.8 ± 3.0, 15; weight, LACM 27349, 27350, Bolivia, Tarija, Capirenda 1.47 kg, 1.59 kg; head shield w/l and ratio 33.9 ± 1.4/66.7 ± 2.2 and 0.51 ± 0.02, N = 25. Teeth 9 ± 0, 32/9 ± 0, 29; CNL 67.7 ± 3.0, 24; RLA 32.7 ± 1.8, 23; IO 16.7 ± 0.7, 23.

Comments.-- See Comparisons, *T. tricinctus*. Concerning Sanborn's (1930) report of a *T. matacus* with 10 pairs of maxillary teeth, the only specimen I have seen with 10 pairs was an immature individual and is not included in the above count. Intercepts of the specimens of *T. matacus* plotted in Fig. 11 are correlated with localities. It therefore seems likely that they present a geographical gradient representing a single species. D. A. Meritt, Jr. (pers. comm.) found a black form of this species isolated in an arid portion of San Juan Province, Argentina.

TRIBE **Dasypodini**
GENUS ***Dasypus*** Linné

Range.-- See map, (Fig. 12) and species accounts.

Comparisons.-- Rostrum long and slender, constituting 55 percent or more of the length of head; carapace smooth and in adults, chiefly dark; 6-11 movable bands; forefoot with 4 claws, longest on second and third digits; hindfoot with 5 claws, longest on third digit. Tail long, over half the length of head and body, terminating in a slender tip. For at least the proximal two-thirds of the tail, armor consists of bands of decreasing diameter, each band formed by 2 or more rows of scutes. Teeth smaller in diameter than in most armadillos; palate extended far posterior to toothrow; tympanic ring as in *Tolypeutes* and Priodontini rather than bulla; mandible longer and more slender than in other armadillos. Four teats; polyembryony reported for *D. novemcinctus* and *D. hybridus* (Galbreath, 1985), quadruplets and possibly polyembryony for *D. sabinicola*, and pairs of young of the same sex for *D. kappleri* (Wetzel and Mondolfi, 1979). Chromosomes, *D. novemcinctus* and *D. hybridus*, 2N = 64, NF = 82-88 (Benirschke et al., 1969).

Dasypus novemcinctus Linné. Common long-nosed armadillo; cachicamo; tatú-hú; tatú verdadeiro.

Dasypus novemcinctus Linné, 1758; Yepes, 1928; Frechkop and Yepes, 1949; Talmage and Buchanan, 1954; Cabrera, 1958; Moeller, 1968.

Dasypus mazzai Yepes, 1933, in part (see Wetzel and Mondolfi, 1979). Holotype: MACN 31-273, carapace, Argentina, Salta, Tabacal, coll. Dr. Mazza.

Type locality .-- Restricted by Cabrera (1958) to Brazil, Pernambuco.

Range.-- From southern United States, Florida, discontinuously west to Texas, southward through Mexico and Central America to South America, west of the Andes to northeastern Peru and east of the Andes, south to the provinces of Santiago del Estero, Santa Fe, and Entre Ríos in Argentina (Fig. 12).

Measurements.-- $\bar{y}$, s, and N of AMNH and MN adult specimens from central highlands of Brazil, Minas Gerais and Goiás, the upper Paranaiba and Grande River

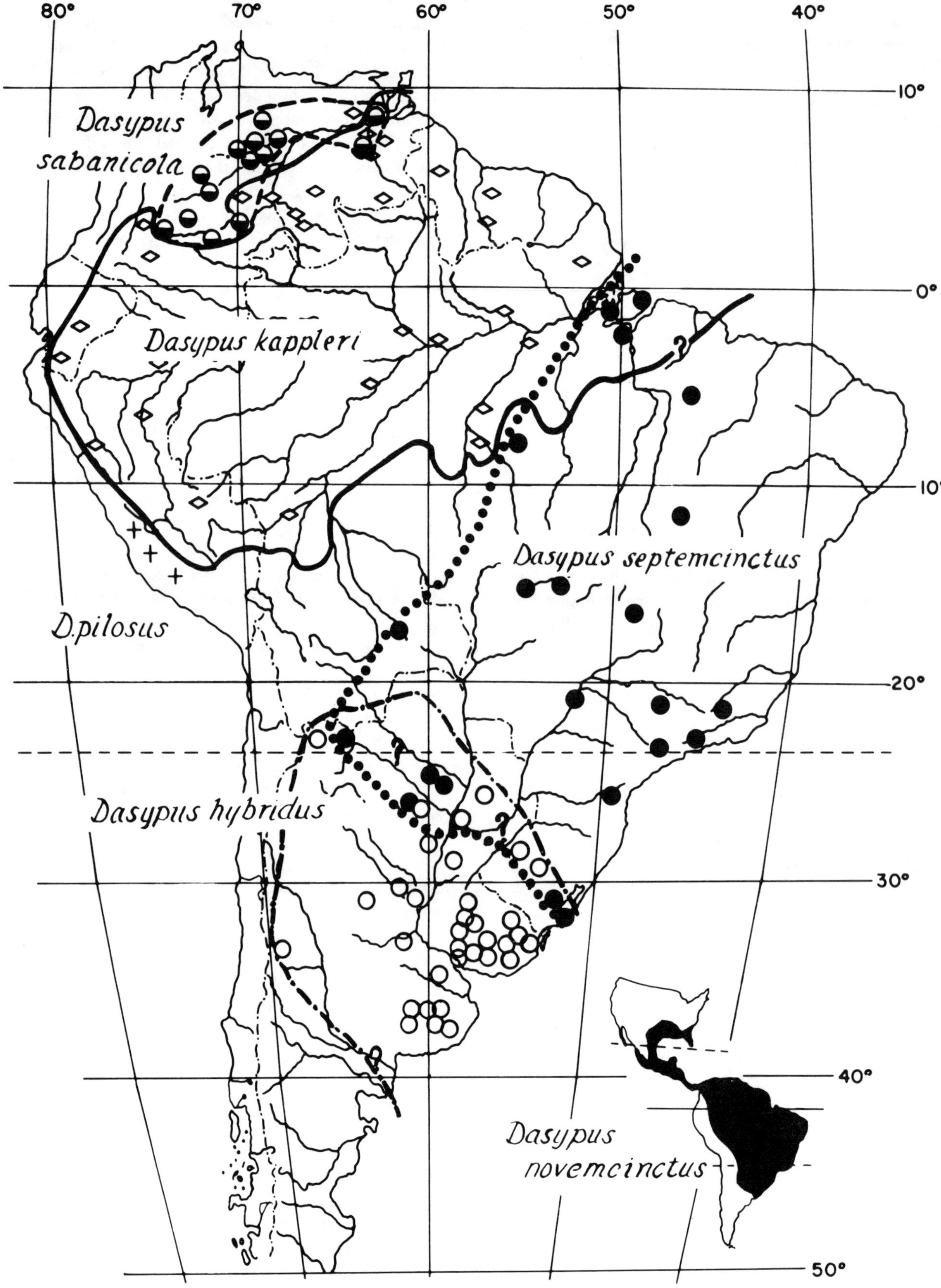

Figure 12.–Long-nosed armadillos, *Dasypus*. Insert, lower right, range of *D. novemcinctus*. Larger map, distribution and specimens examined, remaining 5 species: — — — — and ○ *D. sabanicola* (N = 35); ——— and ◇ *D. kappleri* (N = 54); + *D. pilosus* (N = 4); .... and ● *D. septemcinctus* (N = 32); •—•—•—• and ○ *D. hybridus* (N = 82) (Wetzel and Mondolfi, 1979).

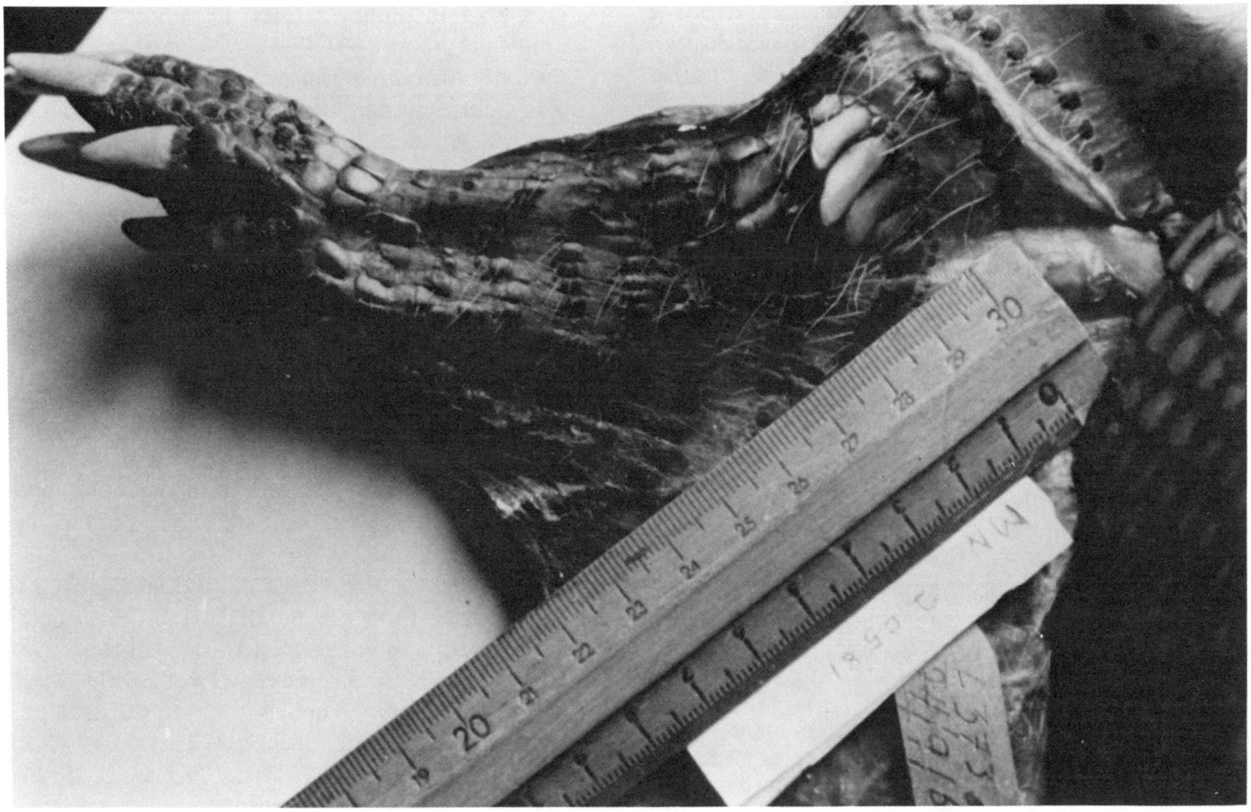

Figure 13.-- "Spurs" on the hindleg of *Dasypus kappleri*. MN 20581, Brazil, Amapá, Rio Tracajatatuba. Scale in mm. Photo. R. M. Wetzel.

drainages: head + body 408.8 ± 23.1, 24; tail 325.0 ± 18.7, 12; hindfoot 92.4 ± 5.5, 23; ear 43.5 ± 3.5, 24; weight 3.3 kg ± 0.56, 16; scutes in fourth movable band 60.9 ± 2.6, 9; teeth 7.4 ± 1.5, 22/8.0 ± 0.4, 22; CNL 90.1 ± 3.5, 27; RLA 53.2 ± 2.4, 28; ratio of RLA to CNL 0.59 ± 0.01, 27.

Comparisons.-- In *D. novemcinctus* the number of movable bands varies from 7 to 10; specimens from the northern and southern portions of the range usually have 8 and those from the middle portion, 9 bands. *D. kappleri* also has 8 movable bands; however, it is not sympatric in the southern portion. *D. novemcinctus* is smaller, has significantly more scutes on the fourth movable band, and lacks *D. kappleri*'s distinctive spurs on the hindleg (Fig. 13) and lateral keels on the palate. In the complex of small, savanna species *D. sabanicola, D. septemcinctus*, and *D. hybridus*, the overall lateral color of adults is dark. *D. novemcinctus* differs from these species in having more numerous yellow, triangular scales on the sides of the movable bands, a greater number of scutes on the fourth movable band, and proportionally smaller ears but a longer rostrum. Where sympatric with one of the smaller armadillos, they are distinguishable because *D. novemcinctus* has a larger number of movable bands and a greater number of scutes on the fourth band.

Dasypus sabanicola Mondolfi. Northern long-nosed armadillo, tatú de sabana.

Dasypus sabanicola Mondolfi, 1968. Holotype: MEBRG 965, coll. J. Ojasti.

Type locality.-- Venezuela, Apure, Achaguas, Hato Macanillal.

Range.-- Savannas of Venezuela from states of Monagas and northeastern Bolívar, west to Apure and savanna of Colombia east of the Andes in departments of Arauca, Meta, and Viachada. The provenance of ICN 1624, skin, as Colombia, Cundinamarca, Tocaima would extend the range of this species to west of the Cordillera Oriental and should be supported by additional data (Fig. 12).

Measurements.-- $\bar{y}$, s, and N of adults from Venezuela (N = 12) and Colombia (N = 15): head + body 290.3 ± 14.9, 26; tail 197.4 ± 13.3, 26; hindfoot 63.0 ± 2.7, 25; ear 26.6 ± 1.6, 25; movable bands 8.0 ± 0.3, 27; scutes in fourth movable band 50.0 ± 2.3, 27; weight 1.49 kg ± 0.33, 7. Teeth 7.5 ± 0.6, 27/7.9 ± 0.5, 27; CNL 68.0 ± 2.9, 27; RLA 37.9 ± 2.0, 27; ratio of RLA to CNL 0.56 ± 0.01, 27.

Comparisons.-- *D. sabanicola, D. septemcinctus* and *D. hybridus* form a unit of small, dark *Dasypus* with an apparent preference for grassland and forest-edge habitats. Each has a comparatively short rostrum, approximately 55 percent of the condylonasal length of the skull. Compared to the sympatric *D. novemcinctus, D. sabanicola* is characterized by smaller size; 8 rather than 9 movable bands; a smaller number of scutes (55 or fewer) on the fourth movable band; a shorter rostrum; a shorter tail with a more pronounced proximal-to-distal decrease in the diameter of the rings; and movable bands that are not as pale laterally, the adults having fewer of the yellow, triangular scales that are prominent on *D. novemcinctus. D. sabanicola* lacks the prominent spurs on the hindlegs, the very long rostrum, and the lateral keels on the posterior palate of the much larger *D. kappleri*.

Dasypus septemcinctus Linné. Brazilian long-nosed armadillo; tatú-mulita.

Dasypus septemcinctus Linné, 1758; Hamlett, 1939; Frechkop and Yepes, 1949, in part; Talmage and Buchanan, 1954; Cabrera, 1958.

Dasypus mazzai Yepes, 1933, in part; Hamlett, 1939; Cabrera, 1958; Olrog, 1976; See Wetzel and Mondolfi, 1979.

Type locality.-- Brazil, Pernambuco.

Range.-- From the mouth of the Amazon south through the highlands of Brazil to Rio Grande do Sul and west to Mato Grosso and the Gran Chaco (Fig. 12).

Measurements.-- $\bar{y}$, s, and N of adults from Argentina, Formosa (N = 3); Bolivia, Santa Cruz (N = 1); Brazil, Mato Grosso (N = 4); Minas Gerais (N = 1); Amazon Basin (N = 4); Rio Grande do Sul (N = 2): head +body 265.0 ± 23.4, 8; tail 147.7 ± 13.5, 7; hindfoot 60, 2; ear 30.9 ± 4.0, 7; weight 1.45 kg, MN 10662, Minas Gerais, Matias Barbosa; movable bands 6.5 ± 0.5, 13; scutes on fourth band 46.2 ± 2.3, 13. Teeth 6.9 ± 0.6, 15/7.2 ± 0.8, 15; CNL 63.4 ± 3.8, 15; RLA 35.6 ± 2.8, 15; ratio of RLA to CNL 0.55 ± 0.01, 15.

Comparisons.-- Because of the similarity of the smaller species of *Dasypus, D. septemcinctus* was considered to include *D. hybridus* by Frechkop and Yepes (1949), Moeller (1968) and others. Hamlett (1939) was able to distinguish between the two species. Our findings (Wetzel and Mondolfi, 1979) indicate three sibling species, with *D. septemcinctus* and *D. hybridus* sympatric along their borders and separable by multivariant analysis (Fig. 14) and univariant comparisons. Except for its longer ears *D. septemcinctus* is smaller in all dimensions than *D. hybridus* or *D. sabanicola* and has fewer scutes on the fourth movable band. Our data on tail length differ from those of Hamlett (1939), but length of the tail is not a satisfactory character because tips of the tail are frequently lost in long-nosed armadillos.

Dasypus hybridus (Desmarest). Southern long-nosed armadillo; tatú mulita; tatuíra. Le Tatou Sixième ou Tatou Mulet Azara, 1801.

Loricatus hybridus Desmarest, 1804. Based on Azara and on description by Desmarest, 1803.

Dasypus hybridus: Yepes, 1928; Hamlett, 1939; Talmage and Buchanan, 1954; Cabrera, 1958.

Dasypus septemcinctus, in part, not of Linné, 1758, but of authors, in part: Frechkop and Yepes, 1949; Moeller, 1968; Ximènez et al., 1972.

Type locality.-- Restricted by Cabrera (1958) to Paraguay, Misiones, San Ignacio.

Range.-- From the Chaco of northern Argentina (Fig. 12) and probably of Paraguay southeast through northeastern Argentina to Uruguay and the highlands of Rio Grande do Sul, Brazil, and south in Argentina to the Río Negro on the east and Mendoza on the west (Cabrera, 1958).

Measurements.-- $\bar{y}$, s, and N of adults from Argentina (N = 10); Uruguay (N = 11); and Brazil, Rio Grande do Sul (N = 5): head +body 297.3 ± 10.6, 11; tail 169.0 ± 9.8, 11; hindfoot 66.3 ± 3.9, 7; ear 27.4 ± 2.6, 11; weight 2.04 kg, 4; movable bands 6.9 ± 0.6, 20; scutes on fourth band 52.6 ± 2.2, 20. Teeth 7.0 ± 0.6, 26/7.8 ± 0.5, 26; CNL 70.2 ± 2.2, 26; RLA 39.1 ± 1.6, 26; ratio of RLA to CNL 0.56 ± 0.01, 26.

Comparisons.-- *D. hybridus* is the largest of the three savanna species. See comparison under *D. septemcinctus*.

Dasypus kappleri Krauss. Greater long-nosed armadillo; tatú-peba grande.

Dasypus kappleri Krauss, 1862; Talmage and Buchanan, 1954; Cabrera, 1958; Moeller, 1968. Lectotype: SMN 285, coll. Kappler.

Type locality.-- Surinam, Marowijne River.

Range.-- From forested areas of Colombia east of the Andes, the delta of the Orinoco River and Venezuela south of the Orinoco to Guyana, Surinam, and French Guiana through most of the Amazon basin (Fig. 12). There are no records for the highlands of Brazil nor for coastal Brazil southeast of the Amazon.

Measurements.-- $\bar{y}$, s, and N of adults from throughout the range: head + body 543.2 ± 22.2, 18; tail 406.1 ± 45.6, 15; hindfoot 119.9 ± 8.4, 15; ear 50.5 ± 3.1, 15; weight 9.68 kg ± 0.7, 11; movable bands 7.8 ± 0.4, 33; scutes in fourth movable band 56.4 ± 2.9, 28. Teeth 7.8 ± 0.7, 45/8.0 ± 0.4, 45. CNL 123.4 ± 4.5, 45; RLA 79.0 ± 3.7, 47; ratio of RLA to CNL 0.64 ± 0.01, 7. Record weight in sample was 10.5 kg, shared by specimens from Brazil, Amapá (MN 20581) and Surinam, Saramacca, Sitagron (RMNH 20962).

Comparisons.-- Second only to *Priodontes maximus* in size, *D. kappleri* is unique in having enlarged, projecting scales or spurs (Fig. 13) arranged in transverse rows on the posterior surface of the hindlegs (the longest scale

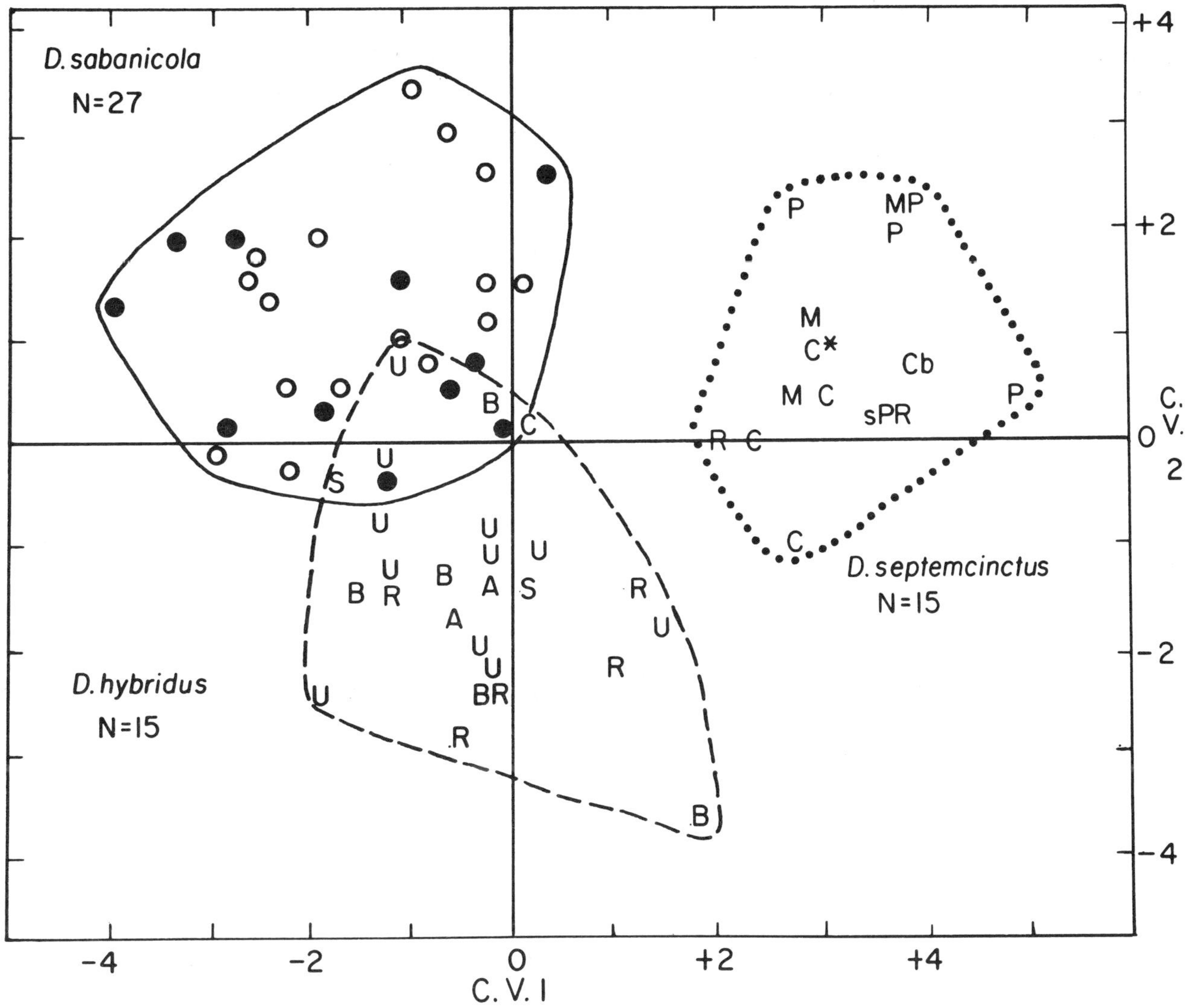

Figure 14.– Canonical analysis, small species of *Dasypus*, 13 cranial measurements. Provenance of specimens: — — — — — *Dasypus sabanicola* ○ = Colombia; ● = Venezuela. *Dasypus hybridus*. Argentina: **A** no other locality data; **B** Buenos Aires province; **C** Chaco province; **S** Sante Fe province. Brazil: **R** Rio Grande do Sul. **U** Uruguay. *Dasypus septemcinctus*. Argentina: **C** Chaco Austral, Formosa; **C*** Salta, Tabacal (type-locality of *D. mazzai*). Bolivia: **Cb** Chaco Boreal, Santa Cruz. Brazil: **M** Mato Grosso (including holotype, *Tatusia megalolepis* Cope); **P** Pará and Maranhão; **R** Rio Grande do Sul; **sP** São Paulo (after Wetzel and Mondolfi, 1979).

may exceed 17 mm in length). The palate differs from that of other members of the genus in having the lateral margins keeled and the posterior margin straight, not distinctly indented at midline. The rostrum is proportionally longer than in other *Dasypus* except for *D. pilosus*. The mean number of movable bands differs from any other *Dasypus* with which it may be sympatric except for the much smaller *D. sabanicola*.

Dasypus pilosus (Fitzinger). Hairy long-nosed armadillo; tatú-peludo más.

Cryptophractus pilosus Fitzinger, 1856. Holotype: NMV 222, museum mount, purchased from animal dealer in London.

Praopus hirsutus Burmeister, 1862. Holotype: MHN 26 or 27, "Guayaquil."

Dasypus pilosus: Frechkop and Yepes, 1949; Talmage and Buchanan, 1954; Cabrera, 1958; Moeller, 1968.

Type locality.-- Montane Peru.

Range.-- Known, in addition to the holotype, from the following specimens, all probably collected in the Andes of Peru: Junín, Acobamba, 8000 ft (2,400 m), BMNH 27.11.1.235; Maraynoic, BMNH 94.10.1.13; Huánuco, Paso Carpish, 7 specimens, LSU (Dan Allen Tallman, corresp.); Huánuco, Acomayo, MHN no number; "Guayaquil," no nation, MHN 26 or 27; depto. ? Santiago, IRSNB 319.

Measurements.-- One specimen (MHN): head + body

440; tail 310; hindfoot 66; ear 50. $\bar{y}$, two specimens (BMNH and MHN): movable bands 11; teeth 5/6; CNL 99.0; RLA 66.4; ratio of RLA to CNL 0.67.
Comparisons.-- The carapace is covered with long, tan hair so dense that the scutes are obscured and the movable bands are barely visible; more hirsute than any other armadillo; greater number of movable bands, 10-12, than in any other species of *Dasypus*. Proportionally, the skull of *D. pilosus* is the most narrow and the rostrum longest of any armadillo.

SUBFAMILY **Chlamyphorinae** Pichiciegos

Comments.-- Small, fossorial armadillos with tiny eyes, probably blind; proportionally large claws on the front digits; carapace thin and flexible, consisting of many transverse rows of narrow scutes, attached anteriorly to the head shield and posteriorly to a rounded plate that caps the rump; sides of the body and head, the venter and legs covered with dense, long, silky, whitish hair; small, armored tail flattened dorsoventrally; ischium fused to the rump-plate; tiny pinna posterior to the eye and immediately ventral to the head shield; large tympanic bulla with a long, ossified external auditory meatus which extends anteriorly along the dorsal side of the zygomatic arch, ending in the posterior side of orbit; skull with a frontal fatty pad which fills a depression between a pair of bony prominences. Although probably derived from the euphractines as suggested by Moeller's (1968) study, I have retained the subfamily separation. To be placed in the Euphractini, separated only at the subtribal level (Moeller, 1968), is to overemphasize the recency of this relationship at the cost of a number of unique, derived characters of pichiciegos.

GENUS ***Chlamyphorus*** Harlan.

Chlamyphorus truncatus Harlan. Pink fairy armadillo; pichiciego menore.
Chlamyphorus truncatus Harlan 1825; Yepes, 1928; Frechkop and Yepes, 1949; Talmage and Buchanan, 1954; Cabrera, 1958; Moeller, 1968.

Type locality.-- Argentina, Mendoza, Río Tunuyán.
Range.-- Argentina (Fig. 10), in the provinces of Mendoza, San Juan, San Luis, La Pampa, eastern La Rioja, southern Catamarca, Cordoba, and western Buenos Aires (Cabrera, 1958; Crespo, 1974).
Measurements.-- Two adults, Argentina, ZMUC 250, sex unknown, La Pampa?, Estación Hucal, and CONN 19956, female, Mendoza: head + body 149, 121; tail 36.5, 37; hindfoot 31; diameter of eye 2.0 (all preceding measured from preserved specimens); teeth 8/8, 7.5/8; CNL 39.7, 35.7.
Comparisons.-- Differs from *C. retusus* as follows: Carapace attached to the body by a thin membrane only along a mid-dorsal line from the head shield to the rump-plate, free of the body along the sides and from the occipital and parietal regions of the head medial to the pinna and orbit; hair extending under the carapace; head shield terminating laterally as a supraorbital and rostral crest, with hair extending dorsally above the eye; pinna minute, surrounded by hair and with base not attached to the carapace; transverse diameter of the carapace decreasing evenly from body to nose, lacking a distinct nuchal constriction; row of three to five large scutes on the head above the eyes, much wider than the scutes anterior and posterior to this row; scutes on the head forming at least one abrupt "step"; tail distinctly spatulate distally; on pes, base of the first toe is more nearly opposite the base of the fifth than the fourth toe; a pair of exaggerated prominences dorsal and anterior to the orbits; cavity between these frontal prominences is filled with a fatty pad and the base of the cavity is not ossified; the first five pairs of teeth are flattened transversely in cross section, followed by no more than two pairs with long axes at angles to the long axis of the toothrow.
Comments.-- For an excellent consideration of many aspects of this species, see Minoprio (1945).

Chlamyphorus retusus Burmeister. Larger fairy armadillo; pichiciego mayore.
Chlamyphorus retusus Burmeister, 1863.
Burmeisteria retusa: Gray, 1865; Yepes, 1928; Talmage and Buchanan, 1954; Cabrera, 1958; Moeller, 1968.
Calyptophractus retusus: Fitzinger, 1871.

Type locality.-- Bolivia, Santa Cruz, Santa Cruz (de la Sierra).
Range.-- From northern Argentina (Fig. 10), in the provinces of Salta, Chaco, and Formosa, through western Paraguay (CONN 16881) and southeastern Bolivia to central Santa Cruz (Cabrera, 1958).
Measurements.-- USNM 283134, topotype, adult male: head + body 161; tail 38; hindfoot 39; CNL 43.8; teeth 8/8. CONN 16881, adult female, Paraguay, Boquerón, 15 km S Filadelfia: Teeth 8/8; CNL 43.5; diameter of eye 2.45.
Comparisons.-- In addition to its larger size, *C. retusus* differs from *C. truncatus* as follows: carapace completely attached to skin of back and head; head shield wider, extending laterally and ventrally to level of eye, thus dense hair lacking dorsal to eye; pinna larger, base attached to posterolateral margin of head shield and with dense hair along only ventral side of base; with distinct nuchal constriction of carapace; head shield more rounded, with an even progression of larger to smaller scutes from posterior to anterior, and without a row of distinctly larger scutes on posterior margin; head shield without distinct "step" in dorsal surface; tail without as distinct a distal widening; on pes, base of first

toe more nearly opposite base of fourth than fifth toe; in skull, fatty pad and frontal prominences not as large, the promincences more posteriorly located, with posterolateral faces bounded by orbits; without un-ossified cleft in depression between prominences; all teeth posterior to first maxillary and first two mandibular pairs distinctly flattened ovals in cross section, with long axes at approximately 45° from long axis of toothrow.

Comments.-- The generic separation of the two pichiciegos can be justified by a number of external and cranial characters. However, as these are chiefly continuous variables, the two species are here considered different evolutionary points on one generic gradient of fossorial adaptation. *C. retusus* has not traveled quite so far along this gradient as evidenced by the following characters: (1) frontal prominences on the skull are smaller, more widely separated, and, hence, (2) no specialized enlarged scutes articulate with these prominences. It seems probable that the frontal prominences and the fatty pad are fossorial adaptations and have some utility in strengthening and protecting the skull as it is thrust forward into soil. (3) Pinnae, although much smaller than in other armadillos, are not as reduced as in *C. truncatus* nor attached to the zygoma. (4) External auditory meatus is not an ossified tube affixed to the zygoma. (5) Nuchal constriction of the carapace is retained, thus providing a posterior boundary, lacking in *C. truncatus*, which defines the head shield. (6) Eyes, although much smaller than in other armadillos, are less reduced than in *C. truncatus*. (7) Temporal musculature is less reduced, with lambdoidal (nuchal) ridge more pronounced and the profile of parietal-occipital less dome-like. Although this feature is not presented as a general fossorial character, it is suggested as part of the *Chlamyphorus* gradient, perhaps correlated with a trend toward smaller-sized food items.

As pointed out to me by Malcolm C. McKenna and George F. Engelmann, the generic name *Burmeisteria* Gray (1865) is not available as it is antedated five months by *Burmeisteria* Salter (1865), described as a subgenus of trilobites. The genus *Calyptophractus* Fitzinger (1871) is next available should a separate genus be later deemed advisable for *Chlamyphorus retusus*.

ACKNOWLEDGMENTS

This paper is a contribution of the Smithsonian Venezuelan Project, supported by a contract (DA-49-193-2788) of the Medical Research and Development Command, Office of the Surgeon General, United States Army. Additional support for travel and fieldwork was received from the National Geographic Society, the University of Connecticut Research Foundation, and the American Philosophical Society. I am indebted to Charles O. Handley, Jr., Smithsonian Venezuelan Project, other curators who generously permitted examination of specimens under their care, Theodore A. Heist who adapted the data for canonical analyses, Drew S. Wetzel for her work with this study and manuscript, Gary J. Galbreath who read the section on Dasypodini, and Dennis A. Meritt, Jr., who read the entire manuscript.

LITERATURE CITED

Allen, J. A.
1895. On the names of mammals given by Kerr in his 'Animal Kingdom,' published in 1792. Bull. Am. Mus. Nat. Hist., 7: 179-192.

Azara, F. de.
1801. Essais sur l'Histoire Naturelle des Quadrupèdes de la Province du Paraguay. Paris, vol. 2, 499 pp.

Benirschke, K., R.J. Low and V.H. Ferm.
1969. Cytogenetic studies of some armadillos. Pp. 330-345, *in* Comparative mammalian cytogenetics (K. Benirschke, ed.). Springer-Verlag, New York, 473 pp.

Burmeister, H.
1854. Systematische Uebersicht der Thiere Brasiliens... Georg Reimer, Berlin, vol. 1, 341 pp.

1862. Beschreibung eines behaarten Gurtelthieres *Praopus hirsutus*, aus dem National-Museum zu Lima. Abhandl. Naturf. Ges. Halle, 6: 147-148.

1863. Ein neuer Chlamyphorus. Abhandl. Naturf. Ges. Halle, 7: 167-171.

Cabrera, A.
1958. Catálogo de los mamíferos de America del Sur. Rev. Mus. Argent. Cienc. Nat. "Bernardino Rivadavia," Zool. (1957), 4: 1-307.

Contreras, J.R.
1973. La mastofauna de la zona La Laguna Chasico, Provincia de Buenos Aires. Physis, Sec. C, Buenos Aires 32, 84: 215-219.

Crespo, J.A.
1944. Contribución al conocimiento de la ecología de algunos Dasipódidos (Edentata) argentinos. Rev. Argent. Zoogeogr., 4: 7-39.

1974. Comentarios sobre neuvas localidades para mamíferos de Argentina y de Bolivia. Rev. Mus.

Argent. Cienc. Nat. "Bernardino Rivadavia," Zool., 11: 1-31.

Desmarest, A.G.
1803. Nouveau Dictionnaire d´ Histoire Naturelle... Paris, 21: 428-435.

1804. Nouveau Dictionnaire d´Histoire Naturelle ... Paris, 24: 28.

Emry, R.J.
1970. A North American Oligocene pangolin and other additions to the Pholidota. Bull. Am. Mus. Nat. Hist., 142: 455-510.

Frechkop, S., and J. Yepes.
1949. Etude systématique et zoogéographique des Dasypodidés conservés a l´Institut. Bull. Inst. Roy. Sci. Nat. Belg., 25: 2-56.

Fitzinger, L.J.
1856. ...ein ausgestopftes Exemplar eines in die Familie der Gurtelthiere. Tagebl. 32, Vers. Deutsch. Naturf. Arzt. Wien, 6: 123.

1871. Die naturliche Familie der Gurtelthiere (Dasypodes). Sitz. Math. Natur., K. Akad. Wiss., Wien, 64: 329-390.

Galbreath, G.J.
1985. The evolution of monozygotic polyembryony in *Dasypus*. (this volume).

Geoffroy Saint-Hilaire, E.
1803. Catalogue des Mamiferes du Museum National d'Histoire Naturelle. Paris, 272 pp.

Gray, J. E.
1865. Revision of the genera and species of entomophagous Edentata....Proc. Zool. Soc. London, 1865: 359-386.

1873. Handlist of the Edentate... Mammals in the British Museum. London, vii + 175 pp.

Greegor, D.H., Jr.
1974. Comparative ecology and distribution of two species of armadillos, *Chaetophractus vellerosus* and *Dasypus novemcinctus*. Ph.D. Thesis. Univ. Arizona, 114 pp.

1975. Renal capabilities of an Argentine desert armadillo. J. Mammal., 56: 626-632.

1980a. Diet of the little hairy armadillo, *Chaetophractus vellerosus*, of northwestern Argentina. J. Mammal., 61: 331-334.

1980b. Preliminary study of movements and home range of the armadillo, *Chaetophractus vellerosus*. J. Mammal., 61: 334-335.

Greegor, D.H.
1985. Ecology of the little hairy armadillo *Chaetophractus vellerosus*. (this volume).

Grimwood, I.R.
1969. Notes on the distribution and status of some Peruvian mammals. Am. Comm. Internat. Wildl. Protect. and N.Y. Zool. Soc. Spec. Publ. No. 21: 1-86.

Hamlett, G.W.D.
1939. Identity of *Dasypus septemcinctus* Linnaeus with notes on some related species. J. Mammal., 20: 328-336.

Handley, C.O., Jr.
1950. Game mammals of Guatemala. Pp. 141 to 162, *in* A Fish and Wildlife Survey of Guatemala (G.B. Saunders, A.D. Holloway, and C.O. Handley, Jr.). US Fish and Wildlife Service, Spec. Sci. Rep. Wildl. No. 5: 162 pp.

1976. Mammals of the Smithsonian Venezuelan Project. Brigham Young Univ. Sci. Bull., Biol. Ser., 20: 1-91.

Harlan, R.
1825. Description of a new genus of mammiferous quadrupeds, of the order Edentata. Ann. Lyc. Nat. Hist. New York., 1: 235-245.

Hershkovitz, P.
1959. Nomenclature and taxonomy of the neotropical mammals described by Olfers, 1818. J. Mammal., 40: 337-353.

Husson, A.M.
1973. Voorlopige lijst van de zoogdieren van Suriname. Zool. Bijdr. Rijksmus. Nat. Hist. Leiden, 14: 1-15.

Jorge, W., D.A. Meritt, Jr., and K. Benirschke.

1977. Chromosome studies in Edentata. Cytobios, 18: 157-172.

Kerr, R.
1792. The Animal Kingdom, a Zoological System of the Celebrated Sir Charles Linnaeus. Edinburgh, 400 pp.

Krumbiegel, I.
1940. Die Saugetiere der Sudamerika-Expeditionen Prof. Dr. Kriegs. I. Gurteltiere. Zool. Anz., 131: 49-73.

Krauss, F.
1862. Ueber ein neues Gurtelthier aus Surinam. Arch. Natur., 28: 19-34.

Linné, C.
1758. Systema Naturae...Tomus I, Ed. 10. Holminae, Laurentii Salvii, pp. 50-51.

Lund, P.W.
1845 (1843).Conspectum Dasypodum. K. Danske Vidensk. Selskabs Nat. Math. Afhandl., 11: lxxxii-lxxxvi.

Miller, G.S.
1899. Notes on the naked-tailed armadillos. Proc. Biol. Soc. Washington, 13: 1-8.

Minoprio, J.D.L.
1945. Sobre el *Chlamyphorus truncatus* Harlan. Acta Zool. Lilloana, Tucuman., 3: 5-58 + 26 plates.

Moeller, W.
1968. Allometrische Analyse der Gurteltierschadel ein Beitrag zur Phylogenie de Dasypodidae Bonaparte, 1838. Zool. Jb. Anat., 85: 411-528.

Mondolfi, E.
1968. Descripcíon de un nuevo armadillo del género *Dasypus* de Venezuela (Mammalia--Edentata). Mem. Soc. Cienc. Nat. La Salle, 27: 149-167.

Olfers, I.
1818. Bermerkungen zu Illiger's Ueberblick der Säugthiere...Pp. 192-237, *in* Neue Bibliothek der wichtigsten Reisebeschreibungen..., (F. T. Bertuch, ed.) Weimar, Vol. 15: heft 2. 304 pp.

Olrog, C.C.
1976. Sobre mamíferos del noroeste argentino. Acta Zool. Lilloana, Tucuman, 32: 5-14.

Osgood, W.H.
1919. Names of some South American mammals. J. Mammal., 1: 33-36.

Salter, J.W.
1865. A monograph of the British trilobites... Palaentographical Society, London, 1864, 216 pp.

Sanborn, C.C.
1930. Distribution and habits of the three-banded armadillo (*Tolypeutes*). J. Mammal., 11: 61-68.

Seba, A.
1734. Locupletissimi rerum Naturalium Thesauri... Amsterdam, 1: 47-48: pl. 30.

Storrs, E.E., G.P. Walsh, H.P. Burchfield and C.H. Binford.
1974. Leprosy in the armadillo: New model for biomedical research. Science, 183: 851-852.

Talmage, R.V. and G.D. Buchanan.
1954. The armadillo (*Dasypus novemcinctus*). Rice Inst. Pam., Monogr. Biol., 41: 1-135.

Tamayo H., M.
1973. Los armadillos en Chile. Situación de *Euphractus sexcinctus* (Linnaeus, 1758)....Not. Mensual, Santiago, 17(203-204): 3-6.

Texera, W.A.
1973. *Zaedyus pichiy* (Edentata, Dasypodidae) nueva especies en La Provincia de Magallanes, Chile. Ans. Inst. Pat., Punta Arenas, 4: 335-337.

Thomas, O.
1880. On mammals from Ecuador. Proc. Zool. Soc. London, 1880: 393-402.

1894. On a new species of armadillo from Bolivia. Ann. Mag. Nat. Hist., ser. 6, 13: 70-72.

1911. The mammals of the Tenth Edition of Linnaeus; an attempt to fix the types of the genera and the exact bases and localities of the species. Proc. Zool. Soc. London, 1911: 120-158.

Wagner, J.A.
1855. Die Saugthiere in Abbildungen nach der Natur mit Beschreibungen von Dr. Schreber. Leipzig, 5: 174.

Wetzel, R.M.
1980. A revision of the naked-tailed armadillos, genus

Cabassous McMurtrie. Ann. Carnegie Mus. Nat. Hist., 49 (20): 323-357.

Wetzel, R.M. and J.W. Lovett.
1974. A collection of mammals from the Chaco of Paraguay. Univ. Conn. Occas. Papers, Biol. Sci. Ser., 2: 203-216.

Wetzel, R.M. and E. Mondolfi.
1979. The subgenera and species of long-nosed armadillos, genus *Dasypus*. Pp. 43-63, *in* Vertebrate ecology in the northern neotropics (J. F. Eisenberg, ed.). Smithsonian Institution Press, Washington, D.C., 271 pp.

Yepes, J.
1928. Los Edentata argentinos: Sistemática y distribución. Rev. Univ. Buenos Aires, Ser. 2ª, 1: 461-515.

1933. Una especie nueva de "mulita" (Dasipodinae) para el norte argentino. Physis, 11: 225-232.

1935. Las especies argentinas del género "Cabassous" (Dasypodidae). Physis, 11: 438-444.

1944. Comentarios sobre cien localidades nuevas para mamíferos sudamericanos. Rev. Argent. Zoogeogr., 4: 59-71.

Ximénez, A., A. Langguth and R. Praderi.
1972. Lista sistemática de los mamíferos del Uruguay. An. Mus. Nac. Hist. Nat. Montevideo, ser. 2, 7: 1-49.

EVOLUTION OF XENARTHRA

THE PHYLOGENY OF THE XENARTHRA

George F. Engelmann

Department of Geology, Marshall University, Huntington, West Virginia. (Present Address: Department of Geography and Geology, University of Nebraska-Omaha, Omaha, Nebraska 68182).

Abstract
The problem of xenarthran (edentate) phylogeny can be resolved by cladistic analysis with varying degrees of success. At the level of relationships within the Theria, the many seemingly primitive characteristics of xenarthrans suggest that, although they are apparently eutherians, they may be the most primitive members of that group and a sister group to all other eutherians. Within the order Xenarthra, relationships among both fossil and living taxa can be resolved with greater precision. The principle dichotomy, between cingulates and pilosans is strongly supported. Within the Cingulata, relationships among well-represented taxa can be presented in some detail. The pampatheres and eutatine armadillos are more closely related to the glyptodonts than they are to other armadillos. The remaining armadillos appear to form distinct groups with both fossil and modern representatives in each group. Within the Pilosa, the vermilinguas (ant-and termite-eaters) and sloths are recognized as distinct groups. The vermilinguas, with little diversity among the available representatives, can be dichotomously subdivided. The sloths are more problematic, and some relationships have not been resolved beyond a multichotomy. Nevertheless distinctive groups are recognizable. These include the mylodonts, the megathreres (including some nothrotheriine sloths) and the megalonychids (in a narrow sense). Relationships among these groups and the two living genera of sloths remain unclear, and the placement of some low-level taxa with respect to these taxa is also uncertain. A consideration of the possibilities of relationships of Xenarthra to certain non-xenarthrans that have been considered edentates, the pangolins and palaenodonts, does not yield conclusive results. The relationships and distribution of the xenarthrans are consistent with a land connection between North and South America that may have existed sometime during the late Cretaceous or Early Tertiary.

Abstracto
El problema de la filogenia de los xenartranos (edentados) puede ser resuelto por medio de análisis cladístico con cierto grado de éxito. Al nivel de sus relaciones dentro de los Teria, las muchas características de los supuestamente primitivos xenartra sugieren que, aunque parezcan ser euteria, ellos pueden ser los miembros mas primitivos de ese grupo y además un grupo hermano de todos los otros euteria. Dentro del orden de los Xenarthra, las relaciones tanto entre taxa fósiles como entre taxa vivientes pueden ser resueltas con mayor precisíon. Se ha apoyado fuertemente el principio de dicotomía entre cingulados y pilosanos. Dentro de los Cingulata, las relaciones entre taxa bien representadas pueden ser expuesta con cierto detalle. Los armadillos, pampateres y eutatinos están más estrechamente relacionados con los gliptodontes que con los demás armadillos. Los armadillos restantes parecen formar grupos distintos con representantes tanto fósiles como modernos en cada grupo. Dentro de la Pilosa, los vermilinguas (osos hormigueros) y los monos perezosos son reconocidos como grupos distintos. Los vermilinguas, con poca diversidad entre los representantes disponibles, pueden ser dividos dicotomicamente. Los monos perezosos son más problemáticos, y algunas relaciones no han sido resueltas más allá de una multicotomía. No obstante se pueden reconocer grupos distintos. Estos incluyen a los milodontes, los megateries (incluyendo a algunos monos perezosos notroteriinos) y los megaloniquidos (en sentido limitado). La relacíon entre estos grupos y los dos géneros vivientes de monos perezosos queda sin aclarar, y la posición de ciertos taxa de bajo nivel con respecto a estos taxa es dudosa. Una consideración sobre las posibilidades de la relación entre los Xenarthra y algunos no-xenartranos que han sido considerados edentados, los pangolines y palaenodontos, no produce resultados concluyentes. Las relaciones y distribución de los xenartranos son consistentes con une conexión terrestre entre Norte y Sur América que pudo haber existido aproximadamente durante el Cretaceo Tardio o a principios del Terciario.

Resumo
O problema da filogenia do Xenarthra (edentata) pode ser solucionado através de análise cladística com variados graus de sucesso. A nível de relações dentro de Theria, as muitas características aparentemente primitivas dos xenarthros sugerem que, embora eles sejam aparentemente eutérios, podem ser os membros mais primitivos desse grupo e um grupo irmão de todos os outros eutérios. Dentro da ordem Xenarthra, as relacões entre a classifição taxonômica de fósseis e a de seres viventes podem ser resolvidos com grande precisão. O princípio dicotomia, entre cingulados e pilosanos é fortemente apoiado. Dentro de Cingulata, as relações entre classificações taxonômicas bem representadas podem ser apresentadas com alguns detalhes. Os tatus pampateros e eutatinos são mais intimamente relacionados aos gliptodontes do que a outros tatus. Os tatus restantes parcem formar grupos distintos com representantes fósseis e modernos em cada grupo. Dentro de Pilosa, os tamanduás (comedores de formigas e térmites) e preguiças são reconhecidos como grupos distintos. Os tamanduás, com pequena diversidade entre os representantes disponíveis, podem ser subdivididos dicotomicamente. Os preguiças são

mais problemáticos, e algumas relações não foram esclarecidas além de uma multicotomia. No entanto grupos distintos são reconhecíveis. Estes incluem os milodontes, os megateteres (incluindo alguns preguiças notroterinos) e os megaloniquides (em um senso reduzido). As relações entre estes grupos e os 2 gêneros viventes de prequiças continuam obscuras, e a colocação de algumas classificações taxonômicas de baixos níveis com respeito a estas classificações taxonômicas é também incerta. Uma análise das possibilidades das relações de Xenarthra com certos não-xenarthros que foram considerados edentatas, os pangolins e palaedontes, não aponta resultados conclsivos. As relações e distribuição dos xenarthros são plausíveis com uma conecção de terra entre a América do Norte e América do Sul, que pode ser existido em alguma época durante o fim do Cretáceo e início do Terciário.

The New World edentates, the xenarthrans, have long been known to biological science and, therefore have occupied a place in classificatory schemes of the Mammalia from the very beginning (cf. Glass, 1985). Consequently, the development of ideas concerning the classification of these animals illustrates the development of systematic philosophies. Early attempts to group organisms in classifications utilized similarity of observable features as the criterion of relationship. In some cases the characters employed were of a rather simple nature and produced such groups as the Edentata, including not only the xenarthrans but also the pangolins and aardvarks of the Old World. As the morphology of organisms was examined more closely and relationships were further refined, the general acceptance of evolutionary theory had a profound effect upon systematic philosophy. Taxonomic relationships were perceived as an expression of phylogeny and an attempt to recognize, or at least speculate on a phylogeny became obligatory for the systematic investigation of any group. The further investigation and elaboration of the relationships of the edentates in the broad sense has led to removal of the aardvarks and pangolins progressively further from the xenarthrans on both morphological and biogeographic grounds. This separation is further supported recently by more modern techniques (cf. de Jong, et al., 1985).

The phylogenetic unity of the Xenarthra, however, has not been seriously questioned and the group remains intact. The price of the strong case for the monophyly of the Xenarthra is the isolation of the group within the Mammalia. Not only is the relationship of the Xenarthra to other mammals obscure, but within the group some taxonomic problems are only beginning to become apparent. The absence of intermediates among the three easily recognized xenarthran adaptive types, sloths, vermilinguas (ant- and termite-eaters) and armadillos, indicates that our already-diverse sample of the group provides only a glimpse of the phylogeny of the group. The continuing addition of new forms, such as *Octomylodon* (Yané, 1977), with strikingly different morphology supports the view that we have much more to learn about xenarthran phylogeny. At the same time, the apparently primitive peculiarities of a form occurring relatively late in the history of the xenarthrans, as does *Octomylodon*, serve to caution us against the naive stringing together of known taxa according to their stratigraphic position in hopes of approximating a true phylogeny.

Inquiry into phylogenetic relationships cannot wait until we have compassed the diversity of fossil forms which may occur in the group. This ideal situation is unlikely ever to be achieved, although, in the case of the xenarthrans, there is little doubt that substantial improvement of the record can and will occur. Rather than abandoning the investigation of phylogeny, I believe that we should continue to formulate hypotheses of relationships and test them with whatever evidence is at hand as it becomes available or as its value is recognized. The most objective method of testing such hypotheses relies on the distribution of characters that can be assumed to be uniquely derived within the group being analyzed. The essential elements of this analytical technique were recognized by many evolutionists but were first explicitly stated and consistently applied by Hennig (1966) as a part of his ideas concerning phylogenetic systematics. The technique has been further elaborated by others (Brundin, 1968; Nelson, 1969; Kavanaugh, 1972; and others), and the method has since come to be known as cladistic analysis.

Phylogenetic hypotheses expressed in the form of a cladogram, and tested by shared derived character distributions can be expected to provide as accurate a picture of relationship as evidence permits, regardless of the completeness or incompleteness of the sample. This is not true of all methods of reconstruction of phylogeny. Furthermore, cladistic analysis permits the piecemeal refinement of an elaborate hypothesis, thus permitting continuing work on an otherwise overwhelming body of knowledge.

I have applied this method to some of the problems of xenarthran phylogeny and have suggested some relationships among fossil and living taxa that support, contradict or further resolve relationships that have been proposed by others. In this paper I will present these hypotheses of relationship with a summary of the corroborating evidence. Many of these ideas have been more extensively developed elsewhere (Engelmann, 1978).

CHARACTERS UNIQUE TO XENARTHRA

Among the reasons for the longstanding acceptance of xenarthran monophyly is the widespread distribution, within the group, of characters unique to the group. The xenarthrous articulations of the vertebrae, which give the group its name, are a well-known feature (Grassé, 1955). The only exception within the group is in the glyptodonts, in which the vertebrae in those regions where the xenarthrous articulations are well developed in other xenarthrans, notably the lumbar, are fused into a bony tube (Scott, 1903-1904). The fusion of the transverse processes of the anterior caudal vertebrae to the ischia to form a single pelvic assembly is a feature not known outside the Xenarthra and, with the exception of *Cyclopes*, the pygmy anteater, is found in all xenarthrans. Another character restricted to the Xenarthra is the presence of dermal ossicles in the skin. Although not known to occur in vermilinguas and known only from mylodonts among the sloths (Sinclair, 1910), the presence of dermal ossicles in some sloths makes them a good xenarthran character.

Other features with less than ideal distributions among the taxa that provide corroboration of xenarthran monophyly include the reduction of the dentition, placement of the infraorbital canal entirely lateral to the body of the maxilla, development of a secondary scapular spine by attachment of the teres musculature on the external surface of the scapula, presence of a *rectus thoracis lateralis* muscle (Windle and Parsons, 1899), a small basicranial muscle running from the pterygoid bone to the tympanic annulus (Lubosch, 1907), extensive *retia mirabilia* in the limbs (Grassé, 1955) and paired post-renal *venae cavae* (Hochstetter, 1898).

THE XENARTHRA AS EUTHERIAN MAMMALS

If the hypothesis of xenarthran monophyly can be considered secure, we certainly cannot be so confident of any hypothesis concerning the placement of the Xenarthra within the Mammalia. In spite of a number of seemingly primitive features such as the presence of sternal ribs, the intraabdominal position of the testes (Wislocki, 1933), the presence of a septomaxilla in the nares (Wegner, 1922), a body temperature that is low relative to that of other mammals (Wislocki, 1933) and a remarkably poor ability to control that temperature (Britton and Atkinson, 1938; McNab, 1985), the placement of the Xenarthra within the Eutheria appears to be the preferred hypothesis of relationship. Corroborating this relationship are such characters as a typically eutherian chorio-allantoic placenta (cf. Benirschke and Powell, 1985) , a reproductive anatomy in which the oviducts fuse without incorporating the ureters (Lillegraven, 1969) and the presence of a *corpus callosum* in the brain (Elliot-Smith, 1898).

Within the Eutheria, the general primitiveness of the Xenarthra suggests a relatively primitive position, and it is not unreasonable to hypothesize that the Xenarthra is the sister group to all other eutherians. This hypothesis receives some support from the fact that the vagina and uterus in xenarthrans are but poorly differentiated (Owen, 1868), lacking a distinct cervix, unlike the condition in other eutherians. The primitive character of xenarthran reproductive anatomy is also suggested by the presence of a longitudinal partition that divides the vagina (Owen, 1868), possibly an indication that the fusion of oviducts is less complete than in other eutherians. This hypothesis receives support as well from analyses of the structure of eye-lens proteins (de Jong et al., 1985) and immunological studies of serum albumins (Sarich, 1985). Both kinds of studies indicate that the xenarthrans diverged from other eutherians nearly at the same time as the marsupial-eutherian split occurred.

RELATIONSHIPS WITHIN THE XENARTHRA

Perhaps of greater interest is the problem of relationships within the Xenarthra. Three distinct morphologies are now recognized among the living xenarthrans, the armadillos, vermilinguas and sloths. Fossil members of the group are also clearly referable to one of these three adaptive types. Recently, students of xenarthran phylogeny have tended to regard this lack of intermediates as significant and have considered the primary division of the group to be trichotomous.

Early classifications made the initial subdivision of the Xenarthra dichotomous by grouping sloths and anteaters in the Pilosa (Fig. 1). The dichotomous division of the Xenarthra is not necessarily supported by immunological comparisons of serum proteins, which indicate considerable evolutionary distance among all three of the xenarthran lines (Sarich, 1985), or by the structure of eye-lens proteins (de Jong, 1985). However, this hypothesis is corroborated by many characters, some of which were noted by Flower (1882) when he proposed the group. The characters include a scapular fenestra immediately above the coracoid process of the scapula, the concave articular surface of the astragalus for the navicular, the posterior displacement of the kidneys to a

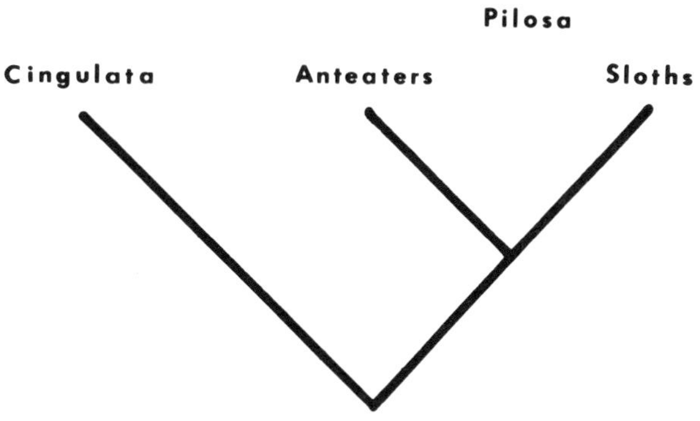

Figure 1: Phylogenetic relationships among major groups within the Xenarthra.

position within the pelvic cavity (Weber, 1928), the location of the testes within the pelvic cavity (Weber, 1928), the expansion of the epitympanic recess dorsally within the squamosal, the reduction or loss of the post-glenoid foramen, the placement of the posterior lacerate foramen, recessed behind a steep structure involving the petrosal, the entotympanic and/or the tympanohyal, the presence of a femoral head of *flexor cruris lateralis* (Windle and Parsons, 1899), interruption of the zygomatic arch and loss of the *pars tuberalis* of the hypophysis (Wislocki, 1938; Hanstrom, 1950).

The sister group of the Pilosa, the Cingulata, includes fossil and living armadillos and the extinct glyptodonts. This group too can be characterized by many shared derived characters such as the modification of dermal ossicles into flattened interlocking plates, fusion of the axis vertebra to one or more cervical vertebrae, fusion of the tibia and fibula, development of a lateral keel on the radial articulation of the humeral trochlea, strong development of the greater trochanter of the femur above the level of the head of the femur, strong development of the third trochanter of the femur, loss of the *foramen rotundum* as a distinguishable cranial foramen, a pronounced post-glenoid fossa and possibly the loss of the *pars intermedia* of the hypophysis (Wislocki, 1938; Hanstrom, 1950).

RELATIONSHIPS WITHIN THE CINGULATA

Let us consider relationships within the Cingulata (Fig. 2). The armadillos and glyptodonts each have a characteristic gross morphology, particularly with regard to the shape of the carapace. These two distinct body plans have been considered to be indicative of a primary dichotomy within the Cingulata (cf. Hoffstetter, 1958). In the armadillos, ossified dermal scutes are fused to one another over the hind quarters of the animal, forming a pelvic buckler. A similar pectoral buckler may be present over the forequarters. The rest of the dorsal surface (between the bucklers or anterior to the pelvic buckler) is covered by scutes arranged into transverse bands that overlap and are moveable over one another. In glyptodonts the entire carapace is solidly sutured together and there are no imbricate mobile bands continuous across the whole structure. However, in some primitive glyptodonts some rows of scutes approximately at the middle of the length of the carapace are not sutured together where they approach the lateral margin of the carapace, providing a definite suggestion of transverse mobile banding (Scott, 1903-4). Patterson and Pascual (1968) have removed the pampatheres from the armadillos and placed them with the glyptodonts, in their interpretation of phylogeny. This implies that Patterson and Pascual regard transverse mobile banding as a primitive cingulate character. I agree with the interpretation of the armadillo type of armor as primitive, but the phylogeny proposed here is somewhat different from that of Patterson and Pascual.

I recognize a primary dichotomy within the Cingulata

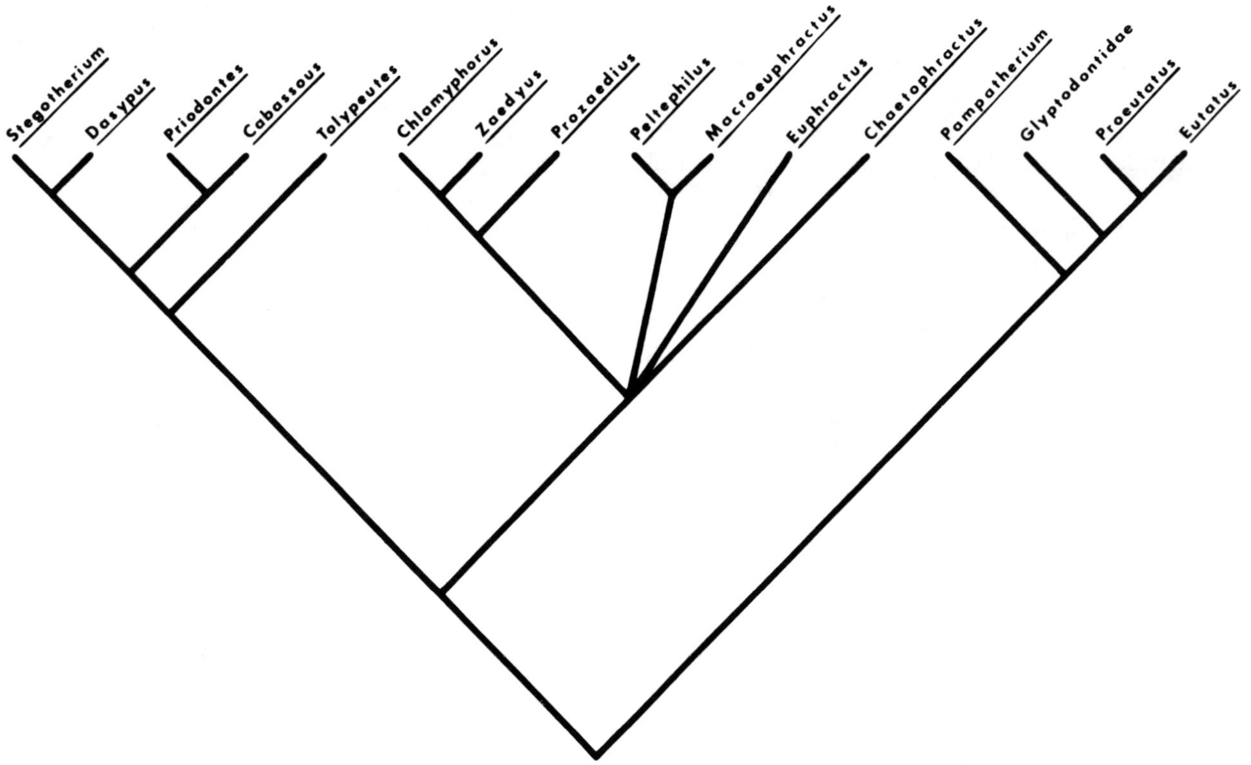

Figure 2. Phylogenetic relationships within the Cingulata.

between a previously unnamed group including the glyptodonts, pampatheres and eutatine armadillos that we may refer to as the Glyptodonta and another previously unnamed group including all of the remaining armadillos, the Dasypoda.

Teeth of the Glyptodonta can be characterized by the presence of a central island of compact dentine that is more resistant to wear than the surrounding dentine that comprises most of the tooth (Hoffstetter, 1958). The Dasypoda, including all living armadillos, is characterized by the simple oval cross-section of the teeth. Lobate dentition of some form, possibly bilobed, is assumed to be primitive for the cingulates.

Glyptodonta-- Within the Glyptodonta the sister groups are the pampatheres on one hand and a group including glyptodonts and eutatines on the other. The pampatheres all have teeth (at least the posterior teeth) that have a reniform bilobed cross-section (Castellanos, 1937). The large size of the pampatheres and some other superficial resemblances to glyptodonts suggest a close relationship between these groups. This hypothesis is also supported by some more specific characters, the reduction of the dental formula to 9/9, the anteroposterior elongation of the teeth, the fusion of three postaxial cervical vertebrae (James, 1957) and the transverse orientation of the glenoid articular surface. However, I believe that the eutatine armadillos are more closely related to glyptodonts than are the pampatheres. This group may be referred to as the Glyptodontoidea, modified from Simpson's (1945) usage. Characters that support this hypothesis include the anteriorly inclined ascending ramus of the mandible, elevation of the basicranial axis relative to the palate, elevation of the manibular condyle relatively high above the tooth row, a narrow, triangular promontorium of the petrosal and an outer layer of compact dentine which is relatively more resistant than the dentine comprising most of the tooth (Hoffstetter, 1958).

The genera *Proeutatus* and *Eutatus* may be united in a group that I call the Eutatini. In addition to those characters that they also share with the glyptodonts, these genera share a square outline of the cranium as seen in dorsal view, an elongate, tubular snout and an elongate premaxilla.

The sister group of the Eutatini, the glyptodonts, have many unique features in common, a continuous globular carapace, trilobate dentition of at least the most posterior teeth, a bony groove delimiting the course of the optic nerve in the lateral cranium and a very short face. The diversity within the Glyptodontidae is fairly substantial but I have not pursued the problem of relationships within this group because of its uniformity of morphology as compared to other xenarthran groups and have treated it as a terminal taxon.

Dasypoda-- The sister group to the Glyptodonta, the Dasypoda, contains two sister taxa that I will refer to as the Dasypodidae, as modified from Simpson (1945), and the Euphracta. The Dasypodidae is most satisfactorily resolved into subordinate groups so we will examine it first. The dasypodids share several derived features, a triangular, anteriorly expanded lacrimal bone, an anteriorly tapering snout, nine or fewer lower teeth, no teeth in the premaxilla, low-crowned teeth, a reduced ventral process of the jugal, the optic foramen close to the sphenorbital fissure, a high-vaulted "globular" carapace (Krieg, 1929) and enrollment behavior as a defensive posture (Walker, 1975).

Within the Dasypodidae, the initial dichotomy is between *Tolypeutes*, the three-banded armadillo, and a group that we will refer to as the Dasypodinae. *Tolypeutes* has numerous unique features including the ability to roll completely into a ball (Walker, 1975). The Dasypodinae can be characterized by two characters, the occurrence of the optic foramen virtually within the sphenorbital fissure and the placement of the infraorbital canal below the zygomatic process of the maxilla.

The Dasypodinae include the sister groups Priodontini and Dasypodini, each containing two well-known genera. The Priodontini links the two living genera *Cabassous* and *Priodontes*, a relationship corroborated by the reduction of the coronoid process of the mandible, the characteristically shaped, enlarged manus claws and the color pattern of the carapace (Walker, 1975). Both genera can be characterized by shared, derived characters but we need not mention them here (cf. Wetzel, 1985a, 1985b).

The Dasypodini includes the modern genus, *Dasypus* and the extinct *Stegotherium*. Characters that support this relationship are the reduction of the dental formula to 8/8 or less, a snout that is very narrow anteriorly, shortening or compression of the postorbital region, a low mandibular condyle and a reduced mandible (Scott, 1903-1904).

The sister taxon of the Dasypodidae, the Euphracta, is a fairly uniform group morphologically. Unfortunately, much of this similarity must be considered primitive for cingulates. There is, however, one feature common to virtually all members of the group, the presence of an ossified bulla with a characteristic configuration (Frechkop and Yepes, 1949).

Within the Euphracta it becomes difficult to recognize convincing shared, derived characters among the probably primitive similarities. Consequently, the first subdivision of this group is an unresolved multichotomy among four taxa. These include the two modern genera, *Euphractus* and *Chaetophractus*, a taxon I will call Peltephilini and a group I shall refer to as Zaedyinae. The two modern genera will not be discussed here (see Wetzel, 1985a). The Peltephilini is a fossil group including the genera *Macroeuphractus* and *Peltephilus*.

This proposed relationship is supported by the common occurrence of a concave lingual surface of the teeth, a reduction of the dental formula to eight upper teeth by loss of the last tooth and the convergence of the anterior teeth on the midline (Lydekker, 1894; Scott, 1903-1904).

The remaining taxon in the multichotomy is the Zaedyinae. In spite of some disturbingly primitive features, the fossil *Prozaedius* is included in this group as the sister group to all other members of the group, the Chlamyphorini. Characters that support this relationship are the presence of a closed bicipital groove, a dental formula reduced to 8/10 and the separation of the pubes medially (Scott, 1903-1904).

The Chllamyphorini, as used here, includes the modern genera *Zaedyus* and *Chlamyphorus*. This relationship is corroborated by the further reduction of the dental formula to no more than 8/9, the nearly vertical anterior margin of the ascending ramus of the mandible, a relatively high mandibular condyle, reduced external ears and a dorsally directed auditory tube (Frechkop and Yepes, 1949).

RELATIONSHIPS WITHIN THE PILOSA

Having reviewed cingulate relationships, I will examine relationships within the Pilosa. Unlike the situation in the Cingulata, the two obvious morphologic groups within the Pilosa, the vermilinguas (ant- and termite-eaters) and the sloths, probably represent a primary dichotomy.

Vermilinguas-- Derived features shared by vermilinguas are the complete absence of teeth, an elongate muzzle, posteriorly extended pterygoids, an ossified auditory bulla and flattened ribs. Within the Myrmecophagidae (Fig. 3) it is possible to distinguish two sister groups. One group that includes the living genus *Cyclopes* and the fossil *Palaeomyrmidon* may be referred to as the Cyclopini and can be distinguished by the strongly concave downward curvature of the basicranial-basifacial axis and the fact that the posteriorly extended pterygoids do not meet along the midline (Rovereto, 1914). The remaining vermilinguas can be placed in the sister group of the Cyclopini, the Myrmecophaginae. All members of this group, for which the feature can be observed, have a distinct groove crossing the promontorium of the petrosal that, according to Guth (1961), bears the internal carotoid.

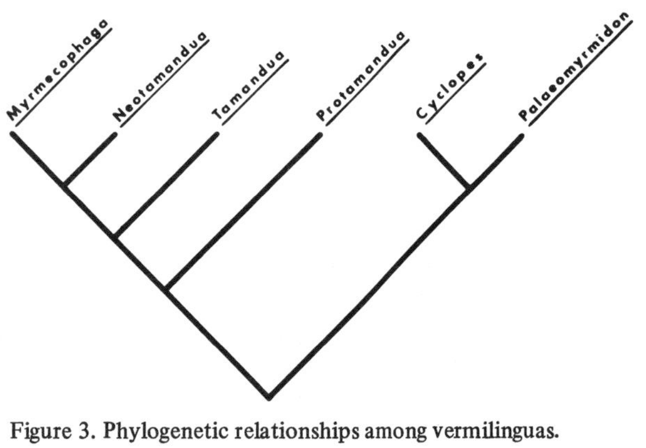

Figure 3. Phylogenetic relationships among vermilinguas.

Relationships within the Myrmecophaginae are difficult to resolve further because of lack of comparable material for all included taxa. However, modifications of the articular facets of the astragalus may be interpreted as corroborating evidence for a hypothesis that *Protamandua* is the sister taxon to remaining myrmecophagines and that among these *Tamandua* is the sister taxon to a group composed of *Neotamandua* and *Myrmecophaga*. This scheme is in essential agreement with that of Hirschfeld (1976).

Phyllophaga-- The only remaining group of xenarthrans, the sister group of the Myrmecophagidae, includes all sloths, fossil and living. The term Phyllophaga is appropriate for this taxon. Characters held in common among sloths include presence in the lumbar vertebrae of paired perforations of the centra (Burlet, 1922), a large asymmetrically developed intravertebral vein (Hoffstetter, 1959), inclusion of the optic foramen within the opening of the sphenorbital fissure, and reduced medial trochlea of the tibial articular surface of the astragalus.

The sloths are perhaps the most diverse group within the Xenarthra, and with only two living genera , perhaps of separate families (Webb, 1985; Wetzel, 1985a), most of that diversity is represented in the fossil record. Therefore, the sample of sloth phylogeny is fairly incomplete both in representation of taxa and the nature of material available for some taxa, and it is difficult to resolve relationships within the Phyllophaga.

Mylodontid Sloths-- Most fossil sloths have been considered to fall within one of three rather large groups, the mylodontids, megalonychids or megatheriids. Of these groups, the mylodontids, as generally accepted, are the most convincingly monophyletic group. In most genera normally referred to the Mylodontoidea there is a characteristic mandibular morphology. The lower margin of the horizontal ramus of the mandible is relatively straight, the angular process is quite deep and the coronoid process of the ascending ramus is relatively low (Stock, 1925).

Within the Mylodontoidea (Fig. 4), relationships can be resolved fairly satisfactorily. The initial dichotomy separates a group including *Orophodon* and *Pseudoprepotherium* from the remaining mylodonts. Both *Orophodon* and *Pseudoprepotherium* are relatively unspecialized in many respects, but both share the

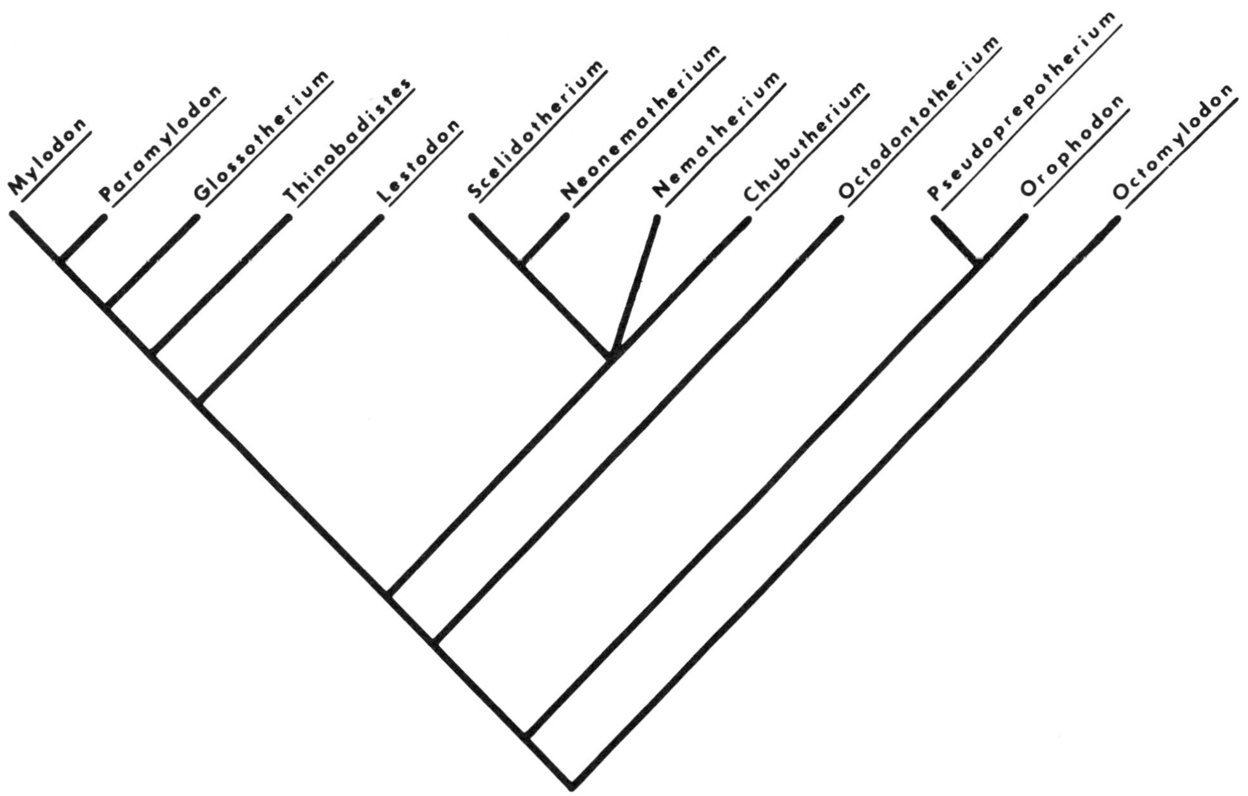

Figure 4. Phylogenetic relationships among mylodontoid sloths.

character of a simple circular cross section of the anterior teeth, a feature which I regard as derived.

The remaining mylodonts all share a distinctive modification of the astragalus. The medial side of the tibial articular surface of the astragalus has been transformed into a knoblike odontoid process that articulates with a corresponding depression in the distal tibia (Hoffstetter, 1958). A similar feature occurs in megatheres, but these two similar features can be distinguished because in the mylodont astragalus there is a considerable separation between the tibial articular surface and the articulation with the calcaneum posteriorly. This group may be subdivided further by separating the primitive genus *Octodontotherium* from those mylodonts that are known to share such features as the confluence of the *Foramen rotundum* with the sphenorbital fissure and the modification of the last lower tooth by anteroposterior elongation and transverse broadening of the anterior lobe. The sloths included in this group are central to the concept of the Mylodontidae.

The Mylodontidae have been divided into the Mylodontinae and the Scelidotheriinae each with a distinctive general morphology. The scelidotheres all share a derived feature in the concave cuboid facet on the distal head of the astragalus (Hoffstetter, 1958). Also, in those scelidotheres in which it can be determined, the last lower tooth has a distinctive T-shaped lobation (Lydekker, 1886). Within the scelidotheres, morphological diversity is not great, resulting mainly from differences in size. The fragmentary nature of *Chubutherium* (Cattoi, 1962) does not allow us to resolve its relationships any more closely than to a multichotomy with *Nematherium* and the remaining scelidotheres, as exemplified by *Scelidotherium*. These scelidotheres can be characterized by their increasingly more triangular teeth arranged in straight parallel rows.

Mylodontine Sloths-- The sister group of the Scelidotheriinae, the Mylodontinae, are united by the loss of the entepicondylar foramen of the distal humerus (Stock, 1925) and possibly by the increased broadening of the snout, although an anteriorly diverging dentition is probably a primitive feature for sloths. Within the Mylodontinae the first dichotomy separates *Lestodon* from the remaining mylodontines which share a characteristic tri-lobed last lower tooth (actually a uniquely expanded anterior lobe of the bilobed tooth). Within this group *Thinobadistes* can be separated from the three remaining genera of mylodonts, all of which exhibit anteriorly inclined occiputs and joining of calcaneal facets of the astragalus (Stock, 1925).

The relationships among these three genera remains problematic because of the questionable nature of the distinction between *Glossotherium* and *Paramylodon* (Hoffstetter, 1952). Inclusion of *Paramylodon* within

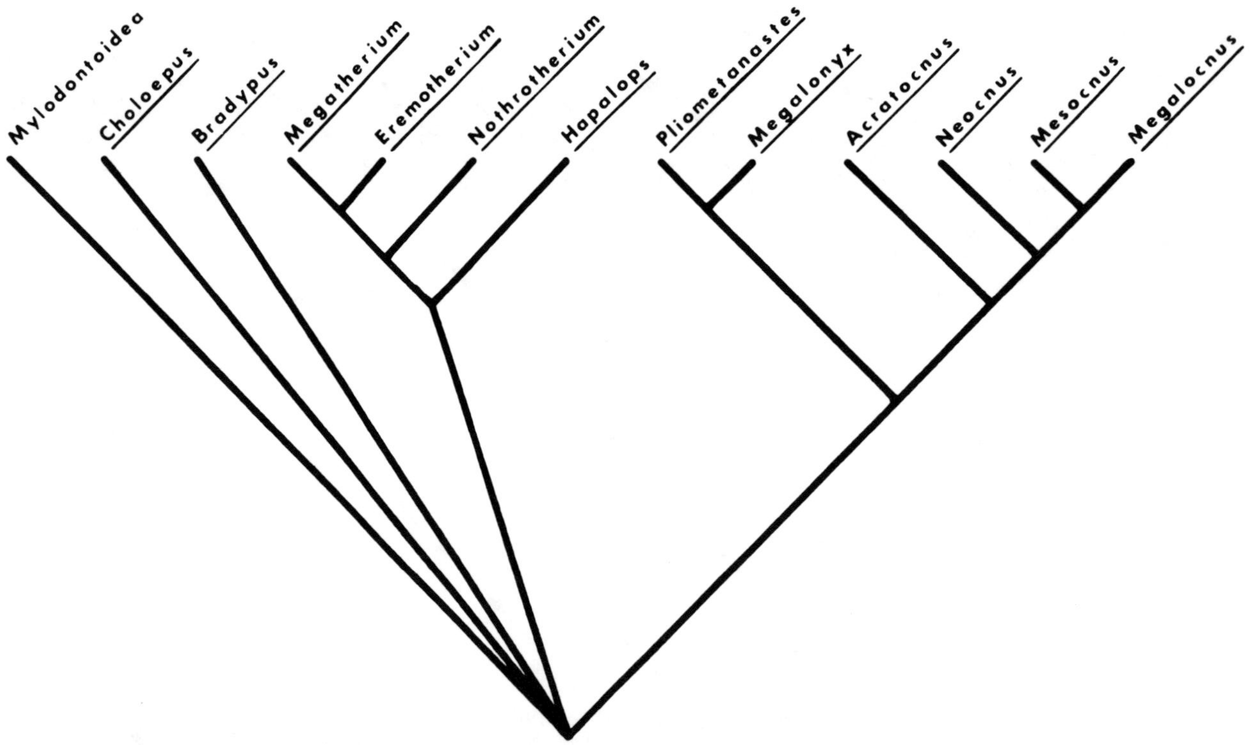

Figure 5. Phylogenetic relationships among sloths with special reference to nonmylodontoid sloths.

Glossotherium makes *Glossotherium* a genus with an exceedingly wide morphological range. The abundant evidence pertinent to this question needs to be thoroughly reviewed. If *Paramylodon* is indeed a distinct and separate taxon, however, the tendency to reduce or lose the first upper tooth may indicate relationship with *Mylodon* in which this tooth is lacking (Stock, 1925; Reinhardt, 1879).

An important genus that has been attributed to the Mylodontoidea but that I have not mentioned yet is *Octomylodon* (Yané, 1977). Interpretation of this taxon as a mylodont may be correct, as the structure of the teeth, with a thick outer layer of compact dentine and a fairly small central region of softer, vascular dentine, is similar to that of mylodonts and unlike other sloths in this respect. A number of features that might provide more convincing tests cannot be determined in the material available. In any case, whether mylodont or not, this sloth has a number of seemingly primitive features and suggests that the primitive dentition in sloths is bilobed in some way.

By discussing the mylodonts first I do not mean to imply that the remaining sloths constitute a monophyletic group. Such a grouping, under the name Megatherioidea (Patterson and Pascual, 1968), seems to be generally acceptable at present. However, I am unable to resolve relationships at this point using good, shared, derived characters. Of these remaining sloths, though, several can be related within convincing monophyletic groups, although relationships among those groups remain unsettled (Fig. 5).

Megatheriid Sloths-- The Megatheriidae includes the megatheres and most, but not necessarily all, of those sloths that have been considered Nothrotheriinae. *Hapalops* and some other Santa Cruzian sloths belong in this group. All of these sloths are characterized by their narrowly elongate premaxilla (Scott, 1903-1904). Within the Megatheriidae, *Nothrotherium* and the megatheres can be considered sister groups, a relationship supported by the common presence of an odontoid tibial articulation of the astragalus similar to that found in the mylodonts, and the tendency of the posterior external mandibular foramen to be located medial to the base of the coronoid process. The large megatheres including *Megatherium* and *Eremotherium* share not only the character of large size but also the pronounced bilophodonty and square cross-section of the teeth.

Megalonychid Sloths-- The other major group of sloths may be called the Megalonychidae although its use here is somewhat restricted over other common usage. I include in the Megalonychidae *Megalonyx* and its close relatives. These sloths all share a marked reduction of any structures extending anterior to the first caniniform teeth. In the skull, the reduction of the premaxilla makes the caniniform teeth effectively terminal and, in the mandible, the symphysial spout anterior to the lower caniniforms is narrow and short (Matthew and de Paula Couto, 1959).

Within the Megalonychidae, *Megalonyx* and *Pliometanastes* appear to be sister taxa sharing a distinctive morphology of the posterior cranium (Hirschfeld and Webb, 1968). These genera together are in turn the sister group of the remaining megalonychids, the antillean sloths (de Paula Couto, 1967).

The antillean sloths are all specialized in having a moderately to strongly reflexed basicranial-basifacial angle. Within this group *Acratocnus* may be considered the sister group to the remaining members. The features that are shared by the antillean sloths exclusive of *Acratocnus* are a greater flexure of the basicranial-basifacial angle and a characteristic meniscoid cross-section of the lower caniniforms.

Finally *Neocnus* can be considered the sister taxon to *Mesocnus* and *Megalocnus*, the last two genera united by the still greater basicranial-basifacial flexure and the extensive development of frontal and parietal sinuses.

The Remaining Sloths-- I have not yet discussed the living sloths. The two genera have been grouped in the Bradypodidae (Simpson, 1945) and there are some features, such as a ball-and-socket articulation between the astragalus and fibula, that support this relationship. More recently, however, such similarities as these genera share, which are undoubtedly selected for an arboreal existence, have been regarded as convergent features and the important differences between *Bradypus* and *Choloepus* have been emphasized. Guth (1961) suggested that *Choloepus* could be more closely related to mylodonts and *Bradypus* more closely related to megatheres. Another, more popular suggestion is that *Bradypus* is closely related to megatheres and *Choloepus* more closely related to megalonychids (Patterson and Pascual, 1968). The relationship of *Choloepus* to the megalonychids is supported by Webb (1985), while the relatively distant relationship of the two living sloth genera is supported by studies of amino acid sequence in eye-lens protein (de Jong, 1985) and by analysis of serum proteins (Sarich, 1985). Although there is a general resemblance of *Bradypus* to the megatheres, including the presence of an ossified bulla, and the strong caniniform teeth of *Choloepus* are reminiscent of *Acratocnus*, I believe that these superficial similarities are not good derived features. Yet convincing derived features in support of any particular arrangement are not apparent. Consequently, any of the relationships mentioned above are possible.

There are, of course, other named genera of sloths not included in this analysis, either because they are based on inadequate material or otherwise unknown to me with regard to useful morphology. However, the relationships as presented may be regarded as a framework to be filled out and modified by further investigation.

RELATIONSHIPS OF XENARTHRANS WITH OTHER MAMMALS

Among modern mammals, one group that shares a number of features with the Xenarthra is the Pholidota. The pangolins resemble the Neotropical vermilinguas in a number of features usually regarded as convergent, especially in view of the absence from pangolins of a reduced dentition, the infraorbital canal lateral to the body of the maxilla, presence of a *rectus thoracis lateralis* muscle (Windle and Parsons, 1899) and a pterygo-tympanic muscle (Lubosch, 1907). Some of these features are found only in pangolins outside of the Xenarthra. In addition to these features, pangolins also share several characters, not with all xenarthrans, but with pilosans in particular. These characters are the concave navicular facet of the astralus, retreat of the kidneys into the pelvic cavity (Tahici, 1966), the relationship of the posterior lacerate foramen to the ear region, expansion of the epitympanic cavity into the squamosal, a femoral head of *m. flexor cruris lateralis* (Windle and Parsons, 1899) and an incomplete zygomatic arch. In spite of the absence of xenarthrous articulations from pangolins, and some other difficulties the possibility of a sister group relationship between pangolins and Xenarthra remains and is not superceded by a clearly superior alternative relationship for the pangolins. It is even possible that the pangolins may be more properly placed within the Xenarthra as the sister group to the Pilosa. The studies of de Jong et al. (1985) and Sarich (1985) argue strongly against such placement.

A fossil group that has been regarded in recent years as ancestral, or at least closely related to the Xenarthra is the Palaeanodonta (Patterson and Pascual, 1968). However, except for the reduced dentition of palaeanodonts, which still retain enamel on their teeth, no convincing derived characters are found in common between the two groups. Emry (1970) has pointed this out and has argued for a close relationship between palaeanodonts and pangolins. It remains possible that the palaeanodonts may be closely related to the Xenarthra through the pangolins. That is, the pangolins and palaeanodonts together may be the sister group of the Xenarthra. However, at the present time I believe that there is no strong evidence for close relationship of Xenarthra to any other particular group of mammals.

XENARTHRAN BIOGEOGRAPHY

There are two principal biogeographic problems concerning the Xenarthra. The first is the problem of the origin of the group and the initial isolation of the Xenarthra and its sister group within South America. The second problem is concerned with those xenarthrans that occur outside South America.

Not only are the xenarthrans morphologically isolated from other mammals, but they are also one of the few groups of such diversity to have such a restricted geographic range. Except for the occurrence of some xenarthrans in North and Central America and the Caribbean islands from the latest Tertiary, the known record of undoubted xenarthrans is almost completely restricted to South America. The conventional explanation of this stratigraphic and geographic distribution has been that xenarthrans originated in South America at a time when that continent was isolated from other major land masses and diversified there throughout the Tertiary until the emergence of the Isthmus of Panama in the latest Tertiary permitted the spread of some suitably adapted forms into other regions of the Americas. However, there are some problems with this model that make it questionable.

For some of these xenarthrans, such as glyptodonts, nothrotheres, megatheres and armadillos, the panamanian route is reasonable, but some sloths, *Megalonyx, Pliometanastes* and *Thinobadistes*, are known from the Hemphillian earlier than the time of emergence of the Isthmus of Panama (Webb, 1978). The phylogenetic hypotheses discussed above affect both of these biogeographic problems.

The earliest record of undoubted xenarthrans is the occurrence in the late Paleocene of South America of the remains of armadillos (Simpson, 1948). However, these early Tertiary armadillos probably fall within the dasypodan branch of my phylogeny of the cingulates and thus imply the existence at that time of the glyptodontan branch and of the Pilosa. This suggests that the first occurrence of xenarthrans is much later than the origin of the group, which probably extends back to the late Cretaceous and possibly further. Evolutionary distances calculated by Sarich (1985) on the basis of rates of substitution in serum proteins indicate as well that xenarthrans are considerably older as a group than was previously supposed.

According to plate tectonic models of paleogeography, the last direct connection of South America to another major land mass prior to the emergence of the Panamanian isthmus was broken when Africa and South America separated about 85 million years ago (Reyment, 1973). As suggested above, it may not be unreasonable to trace the origin of the Xenarthra to this event. If so, one might expect to find the sister group of the Xenarthra in Africa. The modern distribution of the pangolins is African and Southeast Asian and this group is one of the possible sister groups. The fossil record of pangolins, however, extends their range to Europe and North America (Guth, 1958; Emry, 1970). This complicates the picture somewhat and allows the possibility that the event that isolated the Xenarthra from its sister group involved connections between North and South America. If the palaeanodonts, or the Pholidota including the palaeanodonts, are indeed the sister group of the Xenarthra, the deterioration of a connection between North and South America would be strongly suggested as the event responsible for the origin of the Xenarthra.

The second problem of xenarthran biogeography is the occurrence of xenarthrans outside of South America, represented by the northern distribution of some sloths apparently earlier than the establishment of the connection between North and South America as it exists today. Of particular interest is the fact that the entire group of the Megalonychidae as recognized in this paper is restricted to North America and the Antilles. Several South American taxa have been attributed to the Megalonychidae but most are represented only by poor material. Some of these forms, however, notably *Eucholoeops* and *Pliomorphus*, may be closely related to North American megalonychids, but do not exhibit those features that unite the North American megalonychids. Therefore, I believe that the monophyly of that group is secure. Furthermore, within this group, the primary dichotomy separates continental forms from Antillean sloths. Thus two successive breaks between geographic regions could result in the isolation that led to the differentiation of these groups of sloths. The other sloth that occurs in North America earlier than might be expected is the mylodont *Thinobadistes* (Webb, 1978).

In my phylogenetic hypothesis, *Thinobadistes* is considered the sister taxon to the glossotherine sloths. Therefore, at the time of occurrence of that genus, the mylodontoid sloths must have been well differentiated into mylodontids and scelidotheres. Although this may seem to suggest a relatively late lower limit for the time at which the line of *Thinobadistes* first inhabited North America, *Chubutherium* demonstrates that the sloths had already diversified to that point by at least Oligocene time.

Even more out of place according to traditional views is the identification of an edentate from the Paleocene of China (Ding, 1979) and of a myrmecophagid from the Eocene of Europe (Storch, 1981). Although the identification of *Ernanodon* from China is still questionable, the specimen from Europe *Eurotamandua* clearly seems to be a myrmecophagid.

The geographic distribution of the organisms discussed above and their relationships indicate some kind of connection between North and South America prior to the existence of the Isthmus of Panama, possibly in the early Tertiary or the late Cretaceous or both. What is the nature of this connection? One possible route is by way of the Antillean islands. The presence of sloths on some of these islands makes this possibility particularly appealing. In fact, a continuous distribution from South America to North America through the Caribbean Islands that was broken first at the southern end and then at the northern end would be consistent with the distribution and relationships of the Megalonychidae.

Hirschfeld (1971) has suggested this as an early route for dispersal and pointed out that, according to plate tectonic models, the Caribbean Islands may have been much closer throughout the Tertiary than they are now. In fact, if we go back far enough, to the early Tertiary or late Cretaceous, the lands formed over the subduction zone that now produces the islands of the Lesser Antilles may have formed a continuous or nearly continuous connection between North and South America before they were disrupted by the intrusion of the tongue of the Pacific plate that has become the Caribbean plate (Malfait and Dinkelman, 1972). The Greater Antilles are the remains of this ancient archipelago. Rather than adopting Hirschfeld's model that the sloths spread northward by island hopping or waif dispersal in the later Tertiary, I suggest that a more substantial connection was exploited no later than the early Tertiary to produce the presently recognized distribution.

I have presented in this discussion some of my views and hypotheses of xenarthran phylogeny. Although I have not attempted to deal with all of the material that relates to the problem of xenarthran phylogeny, I believe I have set up a framework in which most of the important problems have been considered and, if not resolved, at least exposed. I have made some suggestions concerning biogeographic problems within the Xenarthra and the implications of the proposed phylogeny for those problems. It is clear that we are ignorant of many aspects of the history of the Xenarthra. An improvement of the fossil record, particularly of the late Cretaceous of South America and Africa can be expected to provide important information applicable to these problems.

LITERATURE CITED

Benirschke, K. and H.C. Powell.
1985. On the placentation of sloths. (this volume).

Britton, S.W. and W.E. Atkinson.
1938. Poikilothermism in the sloth. Jour. Mamm. 19: 94-99.

Brundin, L.
1968. Application of phylogenetic principles in systematics and evolutionary theory. Pp. 473-497. *In* Current problems in lower vertebrate phylogeny. (T. Orvig, ed.). Proc. 4th Nobel Symp., Interscience Publishers, John Wiley and Sons, New York. N.Y.

Cattoi, N.
1962. Un nuevo "Xenarthra" del Terciaro de Patagonia. Rev. del Mus Arg. de Cienc. Nat. Bernardo Rivadivia 8: 123-136.

Castellanos, A.
1937. Anataciones sobre la línea filogenética de los Clamiterios. Publ. Univ. Nac. Sta. Fe,8.

de Burlet, H.M.
1922. Ueber durchborte Wirbelkorper fossiler und recenter Edentaten. Morph. Jahrb. 51: 555-584.

de Jong, W.W., A. Zweers, K.A. Joysey, J.T. Gleaves and D. Boulter.
1985. Protein sequence analysis applied to xenarthran and pholidote phylogeny. (this volume).

de Paula Couto, C.
1967. Pleistocene edentates of the West Indies. Amer. Mus. Novitates 2304: 1-55.

Ding Su-yin
1979. A new edentate from the paleocene of Guangdong. Vertebrata Palasiatica 12: 62-64.

Elliot-Smith, G.
1898. The brain of Edentata. Trans. Linn. Soc. London Ser. 2, Zool. 7: 277-394.

Emry, R.J.
1970. A North American Oligocene pangolin and other additions to the Pholidota. Bull. Amer. Mus. Nat. Hist., 142(6): 455-510.

Engelmann, G.F.
1978. The logic of phylogenetic analysis and the phylogeny of the Xenarthra (Mammalia). Unpub. Ph.D. thesis. Columbia University, New York, 329 pp.

Flower, W.H.
1882. On the mutual affinities of the animals composing the order Edentata. Proc. Zool. Soc. London :358-367.

Frechkop, S. and J. Yepes
1949. Etude systematique et zoogéographie des Dasypodidés conservés à línstitut. Bull. Inst. Roy. Sc. nat. Belg. 25, No. 28: 1-12.

Grassé, P.P.
1955. Ordre des Edentates. Pp 1182-1246 *In* Traite de zoologie anatomie, systematique, biologie, Mammiféres. (P.P. Grassé, ed.). L ácademie de Medecíne Edituers, Paris. 17(2).

Guth, C.
1958. Pholidota. Pp 641-647 *In* Traite de paleontologie. (J. Piveteau, ed.). Masson et Cie., Paris. 6 (2).

1961. La région temporale des Edentés. Le Puy, Haute Loire, xviii+ 192 pp.

Hanstrom, B.
1950. The pituitary in some South American and oriental mammals. Acta Univ. Lund. N.S. 46: 1-21.

Hennig, W.
1966. Phylogenetic Systematics. University of Illinois Press, Champaign, Illinois. 263 pp.

Hirschfeld, S.E.
1971. Ground sloths and anteaters (Edentata, Mammalia) from the Tertiery of Colombia, South America. Unpublished Ph.D. thesis, University of California, Berkeley, 257 pp.

1976. A new fossil anteater (Edentata, Mammalia) from Colombia, S.A. and evolution of the Vermilingua. J. Paleo., 50: 419-432.

Hirschfeld, S.E. and S.D. Webb
1968. Plio-pleistocene megalonychid sloths of North America. Bull. Florida State Mus., Biol. Sci., 12: 213-296.

Hochstetter, F.
1898. Ein Beitrag zur vergleichenden. Anatomie de Venensystems der Edentaten. Morph. Jahrb. 25: 362-376.

Hoffstetter, R.
1952. Les mammiferes Pléistocenes de la Republique de l'Equateur. Mem. Soc. Geol. France (n.s.), 31 (66). 391 pp.

1958. Xenarthra. Pp. 535-636, *In* Traité de paléontologie. (J. Piveteau, ed.). Masson et Cie., Paris 6 (2).

1959. Généralité d'une circulation veineuse intrarachidienne chez les Edentés Tardigrades (Paresseux) et Gravigrades. Bull. du Mus. 2° ser. (2): 181-187.

James, G.T.
1957. An edentate from the Pleistocene of Texas. Jour. Paleo., 31(4): 796-808.

Kavanaugh, D. H.
1972. Hennig's principles and methods of phylogenetic systematics. The Biologist, 51(3): 115-127.

Krieg, H.
1929. Biologische Reise-studien in Sudamerika.IX. Gurteltiere. Zeitsch. f. Morph. Oekol. Tiere, 14: 166-190.

Lillegraven J. A.
1969. Latest Cretaceous mammals of upper part of Edmonton Formation of Alberta, Canada, and review of marsupial-placental dichotomy in mammalian evolution. Univ. Kansas Paleont. Cont., 50 (Vert. 12): 1-122.

Lubosch, W.
1907. Das Kiefergelenk der Edentaten und Marsupialer. Jen Denkschr. 7: 521-556.

Lydekker, R.
1886. Description of three species of *Scelidotherium*. Proc. Zool. Soc. Lond.

1894. Contributions to a knowledge of the fossil vertebrates of Argentina. Anales del Museo de la Plata, Paleontologia Argentina, 3: 1-118.

Malfait, B.T. and M.G. Dinkelman
1972. Circum-Caribbean tectonic and igneous activity and the evolution of the Caribbean plate. Geol. Soc. Amer. Bull., 83: 251-272.

Matthew, W.D. and C. de Paula Couto
1959. The Cuban edentates. Bull. Amer. Mus. Nat. Hist., 117: 1-56.

McNab, B.K.
1985. Energetics, population biology, and distribution of xenarthrans, living and extinct. (this volume).

Nelson, G. J.
1969. Origin and diversification of teleostean fishes. Ann. N.Y. Acad. Sci., 167(1): 18-30.

Owen, R.
1868. On the anatomy of vertebrates. Vol. 3, Mammals. Longmans, Green and Co., London. 915 pp.

Patterson, B. and R. Pascual
1968. Evolution of mammals on southern continents. V. The fossil mammal fauna of South America. Quart. Rev. Biol., 43(4): 409-451.

Reinhardt, J.
1879. Beskrivelse af Hovedskallen af et Koempedovendyr, *Grypotherium darwinii*. Kong. Dansk. Vid. Selsk. Skr., nat. mat. Afd.(5), Afd. 12, 4.

Reyment, R.A.
1973. Palaeontological dating of the opening of the South Atlantic Ocean. *In* Implications of continental drift to the earth sciences, Vol. 1, (Tarling and Runcorn, eds.), Academic Press, New York.

Rovereto, C.
1914. Los estratos araucanos y sus fósiles. An. Mus. Nac. Hist. Nat. Buenos Aires, 25.

Sarich, V.M.
1985. Xenarthran systematics: albumin immunological evidence. (this volume).

Scott, W.B.
1903-4. Mammalia of the Santa Cruz beds. I: Edentata. Rep. Princ. Univ. Exp. Patag. 1896-1899, vol. 5.

Simpson, G.G.
1945. Principles of classification and a classification of mammals. Bull. Amer. Mus. Nat. Hist., 85: 1-350.

1948. The beginning of the age of mammals in South America: Part I. Introduction. Systematics: Marsupialia, Edentata, Condylarthra, Litopterna and Notoungulata. Bull. Amer. Mus. Nat. Hist., 91: 5-232.

Sinclair, W.J.
1910. Dermal bones of *Paramylodon* from the asphaltum deposits of Rancho la Brea. Proc. Am. Phil. Soc., 49.

Stock, C.
1925. Cenezoic gravigrade edentates of western North America with special reference to the Pleistocene Megalonychinae and Mylodontidae of Rancho la Brea. Carnegie Inst. Wash. 331, 206 pp.

Storch, G.
1981. *Eurotamandua joresi*, ein Myrmecophagidae aus dem Eozan der "Grube Messel" bei Darmstadt (Mammalia, Xenarthra). Senckenbergiana lethaea, 61: 247-289.

Tahiri, C.
1866. Sur l'anatomie du tractus urogenital de la femelle de *Manis tricuspis*. Bull. Soc. Zool. Fr., 91: 731-734.

Walker, E.P.
1975. Mammals of the world. Third ed., Vol.I, rev. J.L. Paradiso. Johns Hopkins Univ. Press, Baltimore, 644 pp.

Webb, S.D.
1978. A history of savanna vertebrates in the New World. Part II: South America and the great interchange. Ann. Rev. Ecol. Syst., 9: 393-426.

Webb, D.E.
1985. The interrelationships of tree sloths and ground sloths. (this volume).

Weber, M.
1928. Die Saugetiere. Gustav Fischer, Jena, 898 pp.

Wegner, R.N.
1922. Unterschiede der Nasenlochgestaltung und des Os nariale bei den Säugetieren (*Choloepus*) und den Bauriamorphen. Anat. Anz. 97, Ergänzungscheft: 104-111.

1958. Xenarthra. Pp. 535-636, *In* Traité de paléontologie. (J. Piveteau, ed.). Masson et Cie., Paris 6 (2).

1959. Généralité d'une circulation veineuse intrarachidienne chez les Edentés Tardigrades (Paresseux) et Gravigrades. Bull. du Mus. 2° ser. (2): 181-187.

Wetzel, R.M.
1985. The identification and distribution of recent xenarthra (= edentata). (this volume).

Wetzel, R.M.
1985. Taxonomy and distribution of armadillos, Dasypodidae. (this volume).

Windle, B.C.A. and F.G. Parsons
1899. On the myology of the Edentata. Part II. Muscles of the hind limb; and summary of the conclusions respecting the musculature of the order. Proc. Zool. Soc. London: 990-1017.

Wislocki, G.B.
1933. Location of the testes and body temperature in mammals. Quart. Rev. Biol., 8(4): 385-396.

1938. The topography of the hypophysis in the Xenarthra. Anat. Rec., 70: 451-463.

Yané, G. J. S.
1977. Octomylodontinae: nueva subfamilia de Mylodontidae (Edentata, Tardigrada). Descripcion del craneo y mandibula de *Octomylodon robertoscagliai* n. sp., procedentes de la Formacion Arroyo Chasico (edad Chasiquense, Plioceno temprano) del sur del la provincia de Buenos Aires (Argentina). Algunas consideraciones filogeneticas y sistematicas sobre los Mylodontoides. Publ. del Mus. Mun. de Cien. Nat. de Mar del Plata "Lorenzo Scaglia", 2(5): 123-140.

PROTEIN SEQUENCE ANALYSIS APPLIED TO XENARTHRAN AND PHOLIDOTE PHYLOGENY

W. W. de Jong[1], A. Zweers[1], K. A. Joysey[2], J. T. Gleaves[3] and D. Boulter[3]

[1] Department of Biochemistry, University of Nijmegen, 6500 HB Nijmegen, The Netherlands, [2] University Museum of Zoology, Cambridge, U.K., and [3] Deparment of Botany, University of Durham, Durham, U.K.

Abstract

The isolation and amino acid sequence analysis of the eye lens protein α-crystallin A from the xenarthrans *Choloepus hoffmanni, Bradypus variegatus, Tamandua mexicana* and *Myrmecophaga tridactyla* and the pholidotes *Manis javanica* and *Phataginus tricuspis* is described. The amino acid sequence of the major hemoglobin β chain of *Dasypus novemcinctus* is also reported. The proposed sequences are compared with the available sequences of α-crystallin A and Hb β chains from other mammalian species. The α-crystallin A sequences clearly reflect the monophyletic origin of sloths and vermilinguas (anteaters). The most parsimonious solutions for placing the xenarthran sequences in the mammalian α-crystallin A cladogram show this order as one of the oldest offshoots of the eutherian stem, either as a separate branch or on a common base with Hyracoidea and Proboscidea. The pangolin α-A sequence has no shared derived characters with those of the xenarthrans. The sequence can most economically be connected near a common carnivore-ungulate stem. The armadillo Hb β sequence is, in the most parsimonious solutions, joined with the elephant sequence on the same branch.

Abstracto

Se describe el aislamiento y análisis de la secuencia de los amino-ácidos de la proteína α-cristalina A del lente de ojo de los siguientes animales xenartranos: *Choloepus hoffmanni, Bradypus variegatus, Tamandua mexicana* y *Myrmecophaga tridactyla* y de los folidotes *Manis javanica* y *Phataginus tricuspis*. También se informa sobre la secuencia de los amino ácidos de la cadena principal de la hemoglobina β de *Dasypus novemcinctus* . Las secuencias propuestas se comparan con las secuencias disponibles de α-cristalina en cadenas A y Hb β de otras especies de mamíferos. Las secuencias de α-cristalina A claramente reflejan el origen monofilético de los perezosos y osos hormigueros. Las soluciones más parcas para colocar las secuencias xenartranas en el cladograma de la α-cristalina A de los mamíferos, muestran a esta Orden como uno de los ramales más antiguos del tronco euteriano, ya fuera como una rama separada o sobre una base común con los Hyracoidea y los Proboscidea. La secuencia α-A del pangolín no ha compartido características derivadas con los xenartranos. La forma más económica de conectar a esta secuencia es cerca de un tronco común con los carnívoros y ungulados. La secuencia Hb β del armadillo aparece, en las soluciones más parcas, junto con la secuencia del elefante en la misma ramificación.

Resumo

O isolamento e a análise da sequência de aminoácidos das proteínas das lentes dos olhos, α-cristalina A, dos Xenarthra *Choloepus hoffmanni, Bradypus variegatus, Tamandua mexicana* e *Myrmecophaga tridactyla* e o Pholidota *Manis javanica* e *Phataginus tricuspis* são descritas. A sequência de aminoácidos da maior cadeia β da hemoglobina de *Dasypus novemcinctus* é também apresentada. As sequências propostas são comparadas com as sequências da α-cristalina disponíveis nas cadeias A e Hb β de hemoglobina de outras espécies mamíferas. As sequências α-cristalina A refletem claramente a origem monofilética das preguiças e tamanduás. As soluções mais parcimoniosas por colocar as sequências dos Xenarthra no cladograma da α-cristalina A dos mamíferos mostra esta ordem como um dos ramos mais antigos do tronco dos Eutheria, ou como um ramo separado ou como de uma base comum com os Hyracoidea e Proboscidae. A sequência α- A do pangolim não compartilha de nenhuma das características derivadas das dos Xenarthra. A sequência pode, mais economicamente, ser associada próximo ao tronco comum dos carnívoros-ungulados. A sequência da HB β do tatu é, em soluções mais parcimoniosas, associada à sequência do elefante, no mesmo ramo.

The inference of phylogenetic relationships between organisms depends on the evaluation of differences and similarities between fossil or present-day species. Practically any character can be studied for this purpose, although morphological data have been used most extensively. Among the biochemical approaches, comparative immunology and electrophoretical techniques have firmly established themselves as useful aids in taxonomic and phylogenetic studies (Wright, 1974). Studies at the level of DNA, mainly by hybridization techniques, have provided only limited phylogenetic information (Shields and Strauss, 1975).

The most direct information about evolutionary relationships should be derived from comparisons of nucleotide sequences of homologous portions of DNA from different species. Procedures for rapid and relatively simple sequence analysis of mRNA and DNA have recently been developed. A major technical limitation is still the isolation of sufficiently pure mRNAs or DNA-fragments.

Almost as informative as DNA sequences are the amino acid sequences of proteins, which are direct transcripts of the genes. In contrast to immunological and electrophoretic data, which essentially reflect quantitative differences between proteins, sequence analyses enable the precise determination of type and position of amino acid substitutions in homologous proteins. Observed amino acid substitutions can be considered as character states which are extremely suitable for construction of character-cladograms or "phylogenetic trees" (Dayhoff, 1972; Fitch, 1977; Goodman et al., 1979). Although comparative sequence analysis has been applied to only a relatively few proteins and organisms, in the opinion of Simpson (1975) this "body of data already achieved is impressive and highly significant for taxonomy and phylogeny".

From the hundreds of protein sequences determined since the invention of the appropriate techniques in the 1950s, only relatively little phylogenetic information has ensued. This is not because the methods involved are difficult, expensive and time-consuming; the comparative sequence analysis of a set of homologous proteins is at present a rather simple and routine biochemical exercise that can be carried out rapidly. The required equipment is present in many biochemical laboratories, and the cost of using it for this purpose is not particularly great. The main reason for the lack of phylogenetically relevant protein sequence data is probably that few biochemists have been interested in taxonomic problems, and few taxonomists have been in a position to carry out protein sequence analysis. Therefore, most sequences have been determined on proteins from man and easily available domestic or laboratory animals, and very few studies have directly been aimed at specific phylogenetic problems.

Investigations of interest to the study of mammalian phylogeny have mainly been performed on myoglobin (Romero-Herrera et al., 1978), hemoglobin (Goodman et al., 1979), cytochrome c (Moore et al., 1976), pancreatic ribonuclease (Beintema et al., 1977), fibrinopeptides (O'Neil and Doolittle, 1974) and α-crystallin (De Jong et al., 1977). Different proteins undergo evolutionary changes at different rates, depending on their function and conformation (Dickerson, 1971). Fast-evolving proteins such as fibrinopeptides and pancreatic ribonuclease, which can change at a rate of one amino acid substitution per 100 residues in as few as 2 million years (Wilson et al., 1977), are therefore most suitable for the study of intrafamilial or intraordinal relations. Hemoglobin and myoglobin, which evolve at an average rate of 1 percent sequence change in approximately 10 million years, have also provided interesting information about relations within orders, especially those among primates, carnivores, and ungulates. The considerable rate of evolutionary change of these proteins makes them often less suitable for the reliable assessment of relationships among more distant taxa, like the mammalian orders. The occurrence of parallel, back, and superimposed amino acid substitutions hampers the interpretation at this level. A more slowly evolving protein like cytochrome c, although invaluable for the reconstruction of relations among phyla and even kingdoms (Schwartz and Dayhoff, 1978; Lyddiatt and Boulter, 1976) can show too little differences between species from different orders to be useful at that level of analysis.

The eye lens protein α-crystallin, with an intermediate rate of evolution, shows the right degree of difference to be informative at the level of interorder relations in mammals (De Jong et al., 1977). Because no protein sequences had yet been studied from the phylogenetically problematic orders Xenarthra (Edentata), Pholidota, Tubulidentata, Sirenia and Hyracoidea, we began a study of representatives from these taxa. This paper describes the results obtained from the sequence analysis of α-crystallin from sloths, vermilinguas (anteaters) and pangolins, and β hemoglobin from the armadillo.

MATERIALS AND METHODS

Eye lenses were obtained from the following animals: 6 two-toed sloths *Choloepus hoffmanni*, 8 three-toed sloths *Bradypus variegatus*, 1 giant anteater *Myrmecophaga tridactyla*, 1 tamandua *Tamandua mexicana*, 1 common long-nosed armadillo *Dasypus novemcinctus*, 2 Malayan pangolins *Manis javanica* and 2 tree pangolins *Phataginus tricuspis*. Lenses were stored and shipped frozen. Blood was obtained by cardiac puncture from a single adult armadillo *Dasypus novemcinctus*. Solid EDTA was used as an anticoagulatant. The red cells were washed three times with 0.9 percent (w/v) NaCl and stored frozen.

Analysis of α-Crystallin A Chains.--A detailed description of the biochemical procedures involved in the sequence analysis of α-crystallin A chains can be found in De Jong and Terwindt (1976). We describe the successive steps here in more general terms.

The pooled lenses from a species are homogenized in two volumes of buffer, and cell membranes and other insoluble components removed by centrifugation. The supernatant, containing the water-soluble proteins, is brought on a gel filtration column which on elution separates the proteins on the basis of their molecular weight. α-crystallin, being the largest protein aggregate

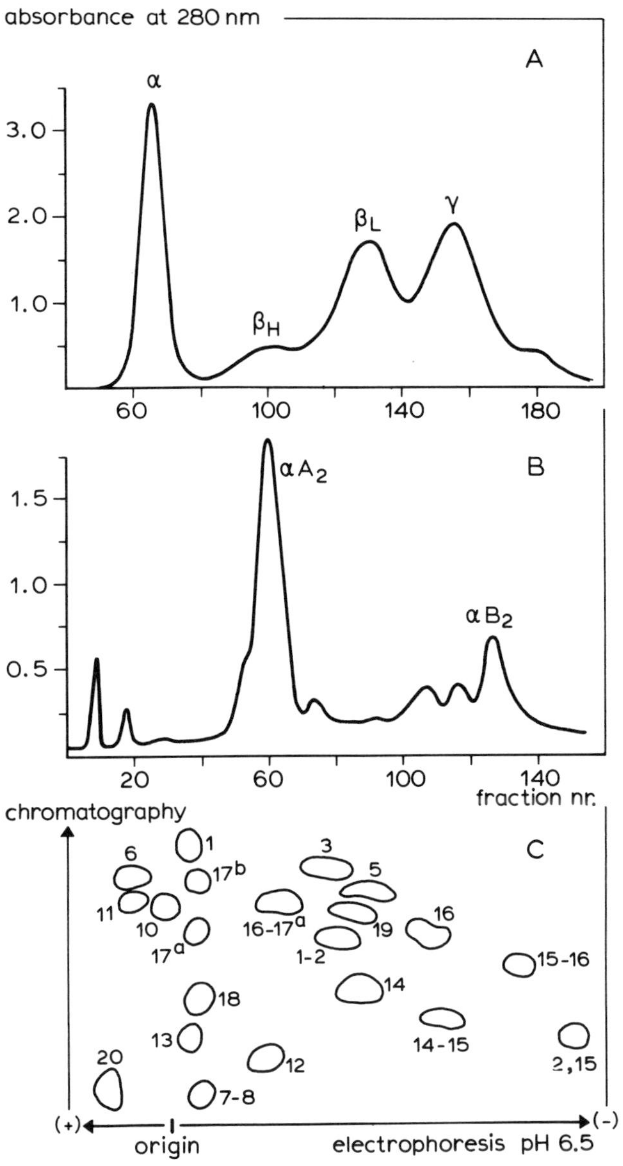

Figure 1. Analytical procedures for the sequence determination of the α-crystallin A chain of the two-toed sloth. **A.** Separation of the water-soluble eye lens protein by gel filtration. **B.** Separation by ion-exchange column chromatography of the composing polypeptide chains of α-crystallin. **C.** Peptide map of a tryptic digest of the α-crystallin A chain. The numbers indicate the tryptic peptides in the order in which they occur in the α A chain (cf. Fig. 3). The hydrophobic peptides T4 and T9 do not occur on the peptide map, but can be isolated by gel filtration.

in the lens cells, elutes as the first peak from such a column, usually well separated from the lower molecular weight β- and γ-crystallins (Fig. 1A). The α-crystallin aggregate, with a molecular weight of about 800,000, is composed of two types of polypeptide chains, α A and α B, occurring in a ratio of approximately 2:1 and each having a molecular weight of about 20,000 (Bloemendal, 1977). The subunit composition of α-crystallin can be made visible by electrophoresis at alkaline pH, in the presence of urea, in order to break hydrogen bonds which otherwise would keep the α A and α B chains together.

Preparative separation of α A and α B chains is performed on the basis of their charge difference, by ion exchange column chromatography, in the presence of urea (Fig. 1B). The isolated α A chain is fragmented by digestion with the proteolytic enzyme trypsin, which normally cleaves the chain at each arginine and lysine residue, and also at cysteine residues when these have been converted chemically into aminoethylcysteine. The resulting mixture of tryptic peptides can be resolved by peptide mapping, where peptides are separated by high-voltage paper electrophoresis followed by paper chromatography in the second dimension (Fig. 1C).

Peptides can be eluted from the paper and used for amino acid analysis and sequence determination by the Edman degradation. Longer tryptic peptides can be further digested with other proteolytic enzymes, like thermolysin, chymotrypsin and protease from *Staphylococcus aureus*. The same enzymes can be used to find overlapping peptides to establish the order of the tryptic peptides in the α A chain.

The proposed sequences of the xenarthran and pangolin α-crystallin α A chains are largely based on amino acid compositions of peptides. When amino acid compositions are found to be identical to those of corresponding peptides from the previously completely determined sequence of bovine α A (Van der Ouderaa et al., 1973) it is assumed that the sequences of these peptides will also be the same. When differences in composition are found with corresponding bovine peptides, the underlying amino acid substitutions are usually localized by Edman degradation or further enzymatic digestions. There is some risk of erroneous sequence inference from amino acid compositions in overlooking reciprocal amino acid substitutions in the same peptide. The chance for such reciprocal substitutions in small peptides is, however, very small (Van Druten et al., 1978).

***Analysis of* Dasypus *Hemoglobin*.**--The hemoglobin in the thawed red cells became largely water-insoluble; thawed cells were therefore extracted with 8 M urea solution, which solubilized not only the hemoglobin but also many other red cell proteins. Heme was removed by dripping the protein solution in acidified acetone, whereupon globin and other protein precipitated. Electrophoresis showed that this precipitate largely consisted of globin and that this globin was composed of a single α-type chain and two β-type chains, in a ratio of approximately 7:3. The globin chains were separated from each other, and from nonglobin protein, by ion-exchange column chromatography in buffers containing 7 M urea. The major β chain was used for further sequence analysis.

The β chain was digested with trypsin, chymotrypsin and thermolysin and the resulting peptides of each digest separated by peptide mapping. The chain was also fragmented chemically by treatment with cyanogen bromide, which cleaves specifically behind methionine residues. These fragments were separated by gel filtration chromatography. The sequence of almost all residues in the Hb β chain was determined by Edman degradation; only a few residues were placed on the basis of amino acid analysis of small peptides by homology with other known Hb β chains. A detailed description of this sequence analysis is given elsewhere (De Jong et al., 1981).

Cladogram Construction.--The sequences we established were used, together with previously published α-crystallin and Hb β sequences, to construct trees according to the principle of parsimony. The four new αA sequences were added to the already available data of αA chains from 17 other mammalian species, the chicken and a frog (De Jong et al., 1977). The method used consisted of two basic parts (J. T. Gleaves, unpub.): 1. A treebuilding procedure similar to that of Dayhoff (1972), in which species are added to an existing tree (starting with a tree of three species) in each possible position. All of the most parsimonious alternatives are kept for the addition of subsequent species. 2. A method for counting the minimum number of substitutions required to account for a given tree, corresponding basically to the preliminary phase of Fitch's (1971) method.

The armadillo Hb β sequence was compared with 43 other vertebrate β-type hemoglobin sequences by Dr. M. Goodman (Wayne State University, Detroit) to construct trees requiring the lowest number of nucleotide replacements, as described by Goodman et al. (1979).

RESULTS

α-crystallin.--The αA chains of two-toed sloth, armadillo, giant anteater, tamandua, and Malayan pangolin migrate in alkaline urea gel electrophoresis in the same position as bovine α A, as do most mammalian α A

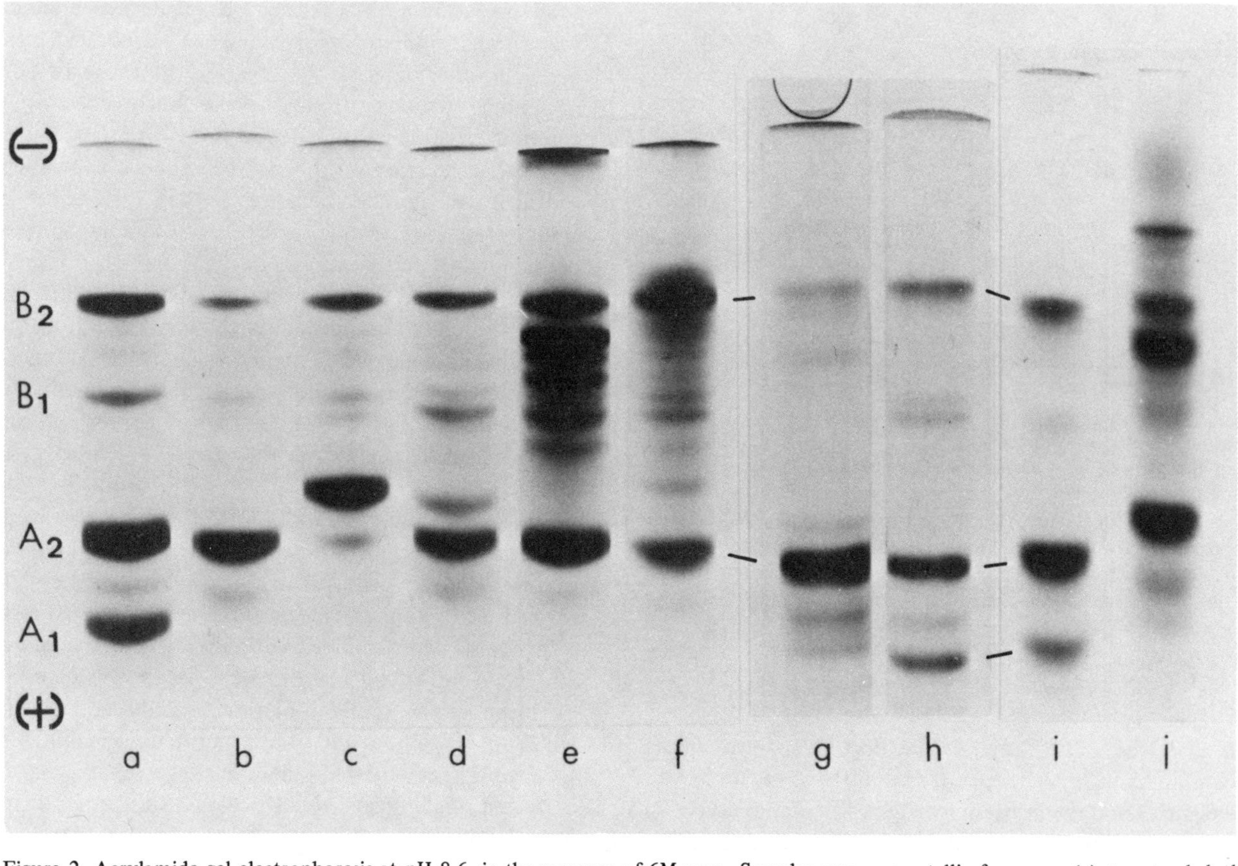

Figure 2. Acrylamide gel electrophoresis at pH 8.6, in the presence of 6M urea. Samples are: α-crystallin from cow (**a**), two-toed sloth (**b**), three-toed sloth (**c**), giant anteater (**d**), total water-soluble lens proteins from tamandua (**e**) and armadillo (**f**), α-crystallin from Malayan pangolin (**g**), calf (**h**), calf (**i**) and lens proteins from tree pangolin (**j**). Samples **a-f**, **g-h** and **i-j** were run on different occasions. Samples **i-j** were run at pH 8.9, instead of pH 8.6. The position of the bovine α-crystallin chains are indicated. In most mammals α A_1 and α B_1 chains are present, which originate *in vivo* by deamidation of the primary gene products α A_2 and α B_2, respectively. This deamidation process apparently hardly occurs in the xenarthran and pangolin α A chains.

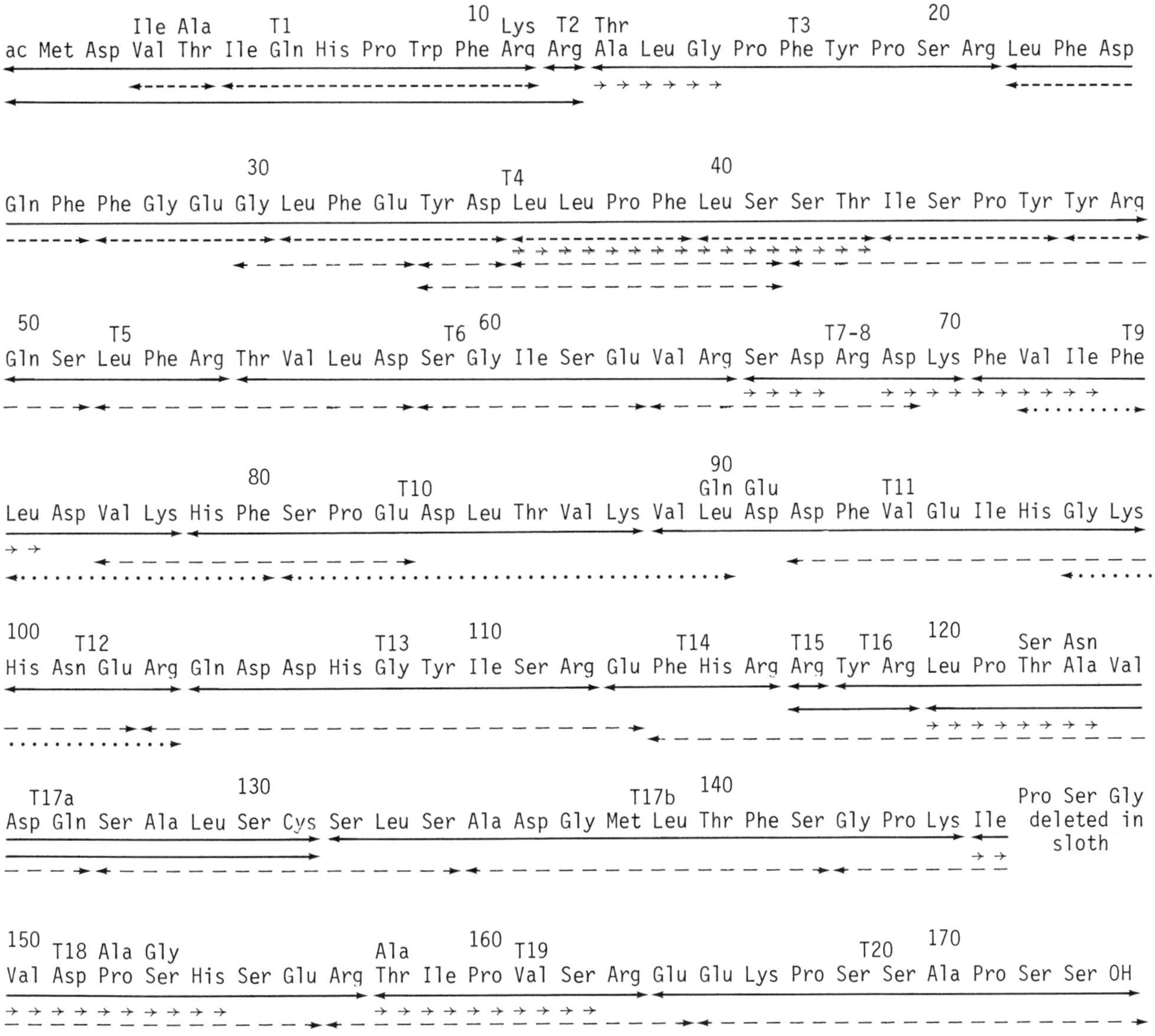

Figure 3. Amino acid sequence of the α-crystallin A chain of the two-toed sloth. Residues that are different in the bovine α A chain are given above the sloth sequence. The tryptic peptides obtained from this chain (cf. Fig. 1 C) are numbered T1, T2, etc. The different types of underlining indicate the peptides that were isolated and analyzed, and by which enzymatic digestion they were obtained: trypsin [——] , chymotrypsin [.......], thermolysin [- - - -] and staphylococcal protease [— — —]. Residues which are placed by dansyl-Edman degradation are indicated by –] –].

chains (Fig. 2). This indicates that no amino acid replacements have occurred which could influence the net charge of the chain. The α A chains of three-toed sloth and tree pangolin are the only mammalian α A chains which have been found to migrate slower than calf α A chain, at a distance indicating an amino acid substitution involving the loss of one negative charge unit. Taking electrophoretic mobility of a protein as a character state, as is often done, this would be considered as a shared derived character of three-toed sloth and tree-pangolin in support of an evolutionary relation between these species. As will be seen, this would be a mistaken conclusion, because the slower electrophoretic mobility of these α A chains is caused by different substitutions in the two chains.

Enough α-crystallin for structural analysis of the total α A chain could be obtained from the sloths, tamandua, and Malayan pangolin. Only a partial analysis could be performed on α A chain from the giant anteater and tree pangolin, whereas the lenses of the armadillo were too small for further analysis. To indicate how the α A sequences were inferred, Figure 3 depicts the proposed sequence of the two-toed sloth α A chain together with the different types of peptides from which this sequence is derived. The sequences of the other α A chains were analyzed in an analogous way. The positions where the

Table 1. Amino acid substitutions in the α-crystallin A chains of xenarthrans and pangolins as compared to the bovine α A chain. *Tamandua* may have a substitution which is not cladistically informative, Ser → Asn in tryptic peptide T17b. In giant anteater and tree pangolin several residues could not be determined (nd) because no satisfactory analysis could be obtained from tryptic peptides T1, T4, T9 and T17b, and T1, T9, T13 and T20, respectively (cf. Fig. 3). Tyrosine in position 74 of Malayan pangolin α A chain could also be in position 71.

Position	*3*	*4*	*7*	*11*	*13*	*34*	*56*	*70*	*74*	*90*	*91*	*92*	*93*	*101*	*122*	*123*	*146*	*147*	*148*	*149*	*152*	*153*	*155*	*158*
Bos	Ile	Ala	His	Lys	Thr	Tyr	Val	Lys	Phe	Gin	Glu	Asp	Phe	Asn	Ser	Asn	Ile	Pro	Ser	Gly	Ala	Gly	Ser	Ala
Choloepus	Val	Thr	His	Arg	Ala	Tyr	Val	Lys	Phe	Leu	Asp	Asp	Phe	Asn	Thr	Ala	Ile		Deleted		Pro	Ser	Ser	Thr
Bradypus	Val	Thr	Gln	Lys	Ala	Ser	Ala	Lys	Phe	Leu	Gly	Asp	Phe	Asn	Thr	Ala	Ile		Deleted		Pro	Ser	Ser	Pro
Tamandua	Val	Thr	His	Lys	Ala	Tyr	Ala	Arg	Leu	Leu	Asp	Asp	Phe	Asn	Thr	Ala	Leu		Deleted		Pro	Ser	Gly	Thr
Myrmecophaga	nd	nd	nd	nd	Ala	nd	Ala	Arg	nd	Leu	Asp	Asp	Phe	Asn	Thr	Ala	Leu		Deleted		Pro	Ser	Gly	Thr
Manis	Ile	Ala	His	Lys	Ala	Tyr	Val	Lys	Tyr	Leu	Glu	Asp	Ser	Ser	Ser	Asn	Val	Pro	Ser	Gly	Thr	Ser	Ser	Ala
Phataginus	nd	nd	nd	nd	Ala	Tyr	Val	Lys	nd	Leu	Glu	Gly	Phe	Asn	Ser	Asn	Val	Pro	Ser	Gly	Ala	Ser	Ser	Ala

xenarthran and pangolin α A chains differ from each other and from bovine α A are given in Table 1. The substitution of an aspartic acid residue by glycine in position 91 and 92 of the α A chains of three-toed sloth and tree pangolin, respectively, explains the lower electrophoretic mobility of these chains.

All substitutions can be explained by single base replacements in the codons for these amino acids, apart from 123 Ala-Asn which requires two base replacements. As usual, most substitutions are chemically conservative, that is, amino acids are replaced by others which have similar chemical properties. Substitutions such as Arg-Lys, Val-Ile or Phe-Tyr will hardly influence the structure and properties of the α A chain. In fact, it is this type of substitutions that cause the neutral or non-Darwinian evolutionary change of proteins that is supposed to play a major role in protein evolution.

The deletion of three residues, 147-149, in the xenarthran α A chain is a unique character. The only other deletion as yet observed is of residue 153 in human and Rhesus monkey α A. Remarkable is the finding of six substitutions between the two species of sloth and at least four differences between the African and Asian pangolins. Considering the generally slow rate of substitution in the α A chain, this is consistent with early evolutionary separations between such species.

The four complete α A sequences obtained were added to the already available data set of α A sequences (De Jong et al., 1977) in order to construct phylogenetic trees (Fig. 4). Although giving a complicated impression, it enables the reader to follow the proposed phylogenetic inferences. Those substitutions which have only been found in a single species are omitted from Figure 4. Such substitutions are autapomorphous for that species, and give no cladistic information. For many positions it is obvious which residue is the ancestral, primitive one, and which residues are derived. For example, position 16 is occupied by proline in all eutherians, chicken and frog, and by serine only in kangaroo and opossum (Fig. 4). Ser-16 is thus assumed to be a shared derived, or synapomorphous character for the marsupials. In other cases, it seems clear that a substitution must have occurred twice or more, like Cys-142 in human, hyrax, and elephant. The substitution Ser-101 in hedgehog is an example of a likely back-substitution. There is, of course, no objective criterion to distinguish between shared derived and parallel, or convergent, substitutions. However, if we see a number of unique substitutions in two or more species that are supposed to be related on the basis of other taxonomic evidence, we can be fairly sure of dealing with genuine synapomorphous characters.

The xenarthrans then clearly posses a number of synapomorphies in their α A chains. The substitutions 56-Ala, 122-Thr, 123-Ala, 152-Pro (also in chicken), 158-Thr and the deletion of residues 147-149 are unique for this order (Fig. 4). The division between Bradypodidae and Myrmecophagidae is not reflected in the α A sequences. Actually the presence of 56-Ala places the three-toed sloth closer to *Tamandua*, a vermilingua (anteater), than to the two-toed sloth, lending support to the contention of Webb (1985), that the two families of sloths are quite distantly related and may have evolved convergently from two kinds of ancestral sloths. Evidence from albumin immunology does not indicate so great a genetic difference between the two kinds of sloth (Sarich, 1985). One has to assume a back-substitution 56 Ala → Val in the two-toed sloth or, less likely, two parallel substitutions 56 Val → Ala in the three-toed sloth and the vermilingua to reconcile the data with the sloth-vermilingua division.

It can easily be seen that the xenarthrans are an old offshoot of the eutherian stem. Especially 91-Glu and 153-Gly appear to be synapomorphous characters of the more recent placental orders, where the xenarthrans still share the apparently primitive residues 91-Asp and 153-Ser with marsupials, chicken, and frog. Less convincing synapomorphies of most placental orders, but lacking in xenarthrans, are 4-Thr, 90-Gln and 152-Ala. Studies of albumin immunology support the view that the xenarthrans are an early form of mammal, indicating that they arose at least 75 to 80 million years ago

POSITION NR	3	4	13	16	17	55	56	58	61	70	74	89	90	91	93	101	122	123	127	129	142	146	147	148	149	150	152	153	155	156	158	168	169
WHALE	I	A	A	P	F	T	V	D	I	K	F	V	Q	E	F	N	S	N	S	L	S	V	P	S	G	M	A	G	S	E	A	S	S
HORSE																			T			I										G	
RHINOCEROS			T			S			V										T			I											
PIG									V																	V							
OX			T																			I				V							
DOG, CAT													L													V							
PANGOLIN											Y		L		S	S										V	T	S					
HEDGEHOG	V	T											L			S	P	S	A				A			L							
RAT, TUPAIA	V	T											L										Q			L							
RABBIT	V	T	T																				Q			L							
HUMAN	V	T	T											D							C	I	Q	T		L		-	T			T	
MONKEY	V	T	T											D								I	Q	T		L		-	T				
ELEPHANT	V	T								Q	L			D							C	I	Q					S					
HYRAX	V	T								Q	L		L	D							C		Q					S					
2-TOED SLOTH	V	T											L	D			T	A				I	-	-	-	V	P	S			T		
3-TOED SLOTH	V	T					A						L	G			T	A				I	-	-	-	V	P	S			P		
TAMANDUA	V	T					A			R	L		L	D			T	A				L	-	-	-	V	P	S	G		T		
KANGAROO		T		S	L			E					L	D		S				I		I	H		D			S		D	S	T	L
OPOSSUM		T		S	L			E		R	Y		L	D	Y	S				I		I	H	T	N		S	S		D	S	T	L
CHICKEN		T			L	S		E			.	I	I	D		S	A			I					D		P	S			P	T	
FROG		T				G	F			R	.	I	L	D		S				I		L	M		S	L	S	S	G		P	T	

Figure 4. Phylogenetically informative amino acid substitutions in 21 mammalian α-crystallin A chains. Chicken and frog α A chains are included for comparison. Vertical lines indicate where residues are identical to the topmost sequence. Only those positions of the total chain of 173 residues are shown at which substitutions have occurred in at least two of the species. Not shown are positions which are identical in all species, or where a substitution has occurred in only a single species. Full species names for previously determined α A sequences can be found in De Jong et al. (1977). The one-letter notation for amino acids is: A, Ala; C, Cys; D, Asp; E, Glu; F, Phe; G, Gly; H, His; I, Ile; K, Lys; L, Leu; M, Met; N, Asn; P, Pro; Q, Gln; R, Arg; S, Ser; T, Thr; V, Val; Y, Tyr; (.) is not determined and (-) indicates a deletion.

(Sarich, 1985).

The situation is less clear with the pangolin α A sequence. It appears to have a number of synapomorphous substitutions in common with carnivores, whales and ungulates: especially 4-Ala and 147-Pro. On the other hand, 3-Ile, 101-Ser (which, however, is Asn in the tree pangolin), and 153-Ser could be taken as primitive characters, placing the pholidotes at the basis of the eutherian tree. In any case, the pangolins and xenarthrans have no synapomorphous substitutions and are not closely related. Sarich (1985) found no particular relationship between xenarthrans and either pangolins or aardvarks on the basis of studies of albumin immunology.

It is obvious that the above-given considerations do not lead to the objectively most probable cladogram of all involved α A chains. Only the strict application of the parsimony principle by elaborate computer programs enables the proposal of the most parsimonious evolutionary tree. The results of such computer analysis provided most parsimonious trees in which the chicken was placed on the marsupial branch. Bringing chicken in a taxonomically more acceptable position, branching off before the marsupial-placental divergence, required only one more substitution and did not change the branching pattern among the mammals. Under the condition that all mammals are more closely related to each other than to chicken or frog, three most parsimonious trees were found, each requiring 119 substitutions. One of these is shown in Figure 5. In one of the alternative trees the positions of *Bradypus* and *Choloepus* are reversed and, in the other, the xenarthran branch has a common base with the hyrax-elephant branch.

Armadillo Hemoglobin β Chain.--The results of the sequence analysis of the major β chain of armadillo hemoglobin are summarized in Figure 6. Because of the much greater variability among hemoglobin chains than among α-crystallins, it is not possible to find remarkable similarities between Hb β chains of armadillo and other mammals by visual comparison of their sequences. Dr. Goodman compared the armadillo Hb β chain sequence with 43 other β-type chains and constructed phylogenetic trees according to different computer-assisted procedures (Goodman et al., 1979). Included among these 43 chains were sequences from representatives of

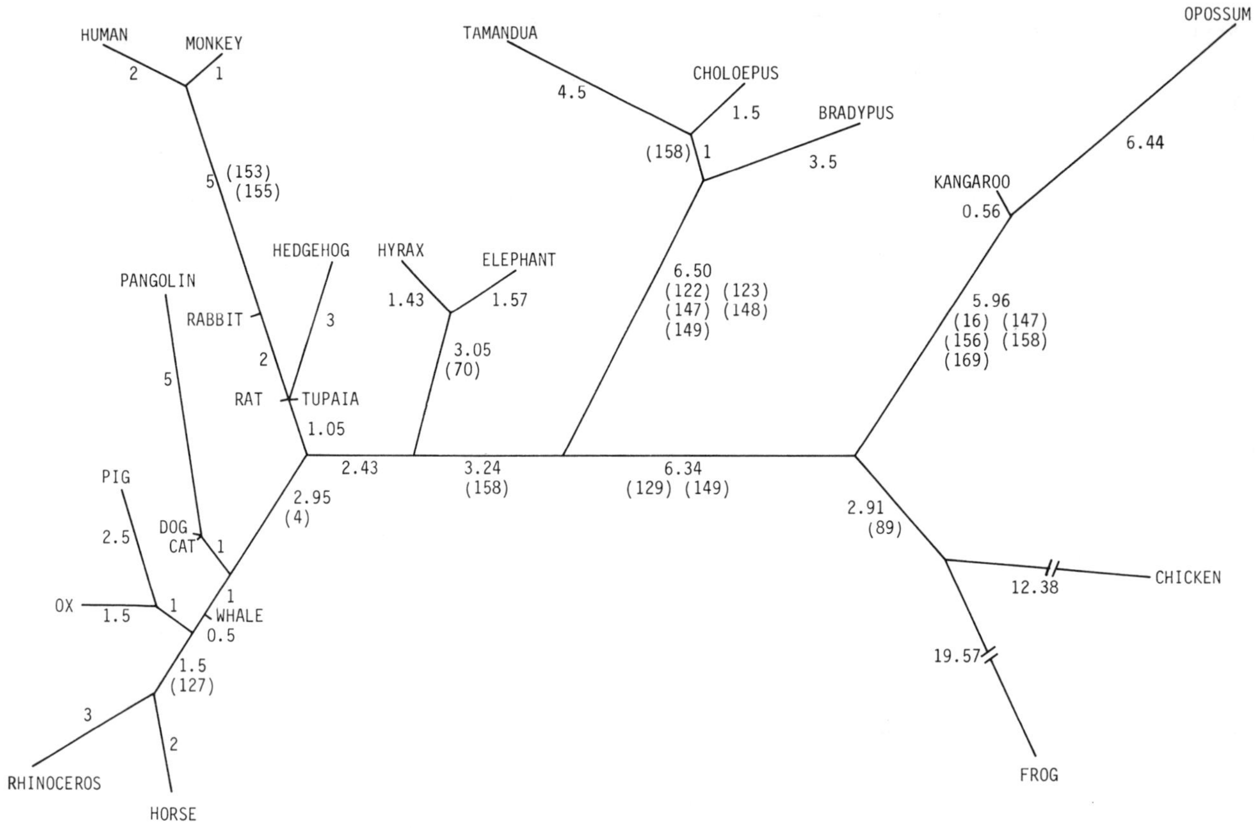

Figure 5. The most parsimonious phylogenetic tree of 23 α-crystallin A chains, under the restriction that all mammals are more closely related to each other than to chicken or frog. Two equally parsimonious alternatives are described in the text. This tree requires 119 substitutions, of which 34 are convergent. Branch lengths are expressed in substitutions. Fractional values arise where there is doubt as to the ancestral sequence at a node. Positions are shown (in brackets) of substitutions other than back, parallel, or trivial (giving no cladistic information).

the following mammalian orders: Primates, Rodentia, Lagomorpha, Scandentia (Tupaia), Carnivora, Proboscidea, Artiodactyla, Perissodactyla, Insectivora, Marsupialia and Monotremata.

Most parsimonious trees were obtained, requiring 886 nucleotide replacements, which strongly supported a joining of the armadillo and elephant branches. To separate the armadillo and elephant Hb β chains would cost an additional 6 nucleotide replacements. The connection of the combined armadillo-elephant branch to the other mammalian taxa is rather unclear, but no particular affinity to the ungulates is obvious. The

Figure 6. Amino acid sequence of the major hemoglobin β chain of the armadillo *Dasypus novemcinctus.*

```
                                   10                                      20
H-Val-Asn-Leu-Thr-Ser-Asp-Glu-Lys-Thr-Ala-Val-Leu-Ala-Leu-Trp-Asn-Lys-Val-Asn-Val-Glu-Asp-His-Gly-Gly-

                 30                                      40                                      50
 -Glu-Ala-Leu-Gly-Arg-Leu-Leu-Val-Val-Tyr-Pro-Trp-Thr-Gln-Arg-Phe-Phe-Glu-Ser-Phe-Gly-Asp-Leu-Ser-Thr-

                                   60                                      70
 -Pro-Ala-Ala-Val-Phe-Ala-Asn-Ala-Lys-Val-Lys-Ala-His-Gly-Lys-Lys-Val-Leu-Thr-Ser-Phe-Gly-Glu-Gly-Met-

                 80                                      90                                      100
 -Asn-His-Leu-Asp-Asn-Leu-Lys-Gly-Thr-Phe-Ala-Lys-Leu-Ser-Glu-Leu-His-Cys-Asp-Lys-Leu-His-Val-Asp-Pro-

                                  110                                     120
 -Glu-Asn-Phe-Arg-Leu-Leu-Gly-Asn-Met-Leu-Val-Val-Val-Met-Ala-Arg-His-Phe-Gly-Lys-Glu-Phe-Asp-His-Trp-

                130                                     140
 -Met-His-Ala-Cys-Phe-Gln-Arg-Val-Val-Ala-Gly-Val-Ala-Asn-Ala-Leu-Ala-His-Lys-Tyr-His-OH
```

phylogenetic trees of mammalian Hb β sequences are not shown because a critical and adequate evaluation of the several alternatives cannot be given within the limited scope of this report.

DISCUSSION

How reliable and how convincing are the phylogenetic inferences from our protein sequence data? It is useful to realize the many problems and limitations connected with the study of phylogenetic relationships based on protein sequences. Misleading conclusions can be drawn when homologous proteins are compared which are coded by genes which have duplicated before the investigated species diverged (Fitch, 1973). An example of such "paralogous" proteins are the hemoglobin α and β chains, for which the genes have originated by an ancient duplication of a single ancestral globin gene. Many vertebrate species, including the armadillo, have multiple Hb α or Hb β chains, and the risk of comparing paralogously related instead of "orthologous" chains is therefore not imaginary. For the α-crystallin A chain there are as yet no indications of gene duplications during mammalian evolution.

A major problem is the apparently frequent occurrence of convergent and back substitutions in the evolution of proteins. It is estimated that up to 50 percent of all substitutions in myoglobin are convergent and 7 percent are back-substitutions (Romero-Herrera et al., 1978). Peacock and Boulter (1975) have shown that correct topologies can be obtained as long as the number of convergent substitutions is smaller than that of the shared derived substitutions. Also disturbing, but not unexpected, is the fact that different tree-construction computer programs can yield different topologies. Quite different branching patterns are often equally, or almost equally, parsimonious and the addition of new sequences can significantly alter a previously proposed topology. It is reassuring that the addition of xenarthran and pangolin α A sequences does not change the apparent relationships among the previously investigated sequences (De Jong et al., 1977).

The "denseness" of a phylogenetic network bears importantly on the accuracy of evolutionary reconstructions from protein sequence data (Holmquist, 1978). The quality of the evolutionary information derived from many closely related sequences is higher than that derived from a few distantly related sequences. A tree is maximally dense when the branch lengths correspond to one nucleotide replacement or less. The α-crystallin A tree approaches this maximal denseness in certain regions. Because of its faster rate of evolution, the Hb β tree has much larger branch lengths and consequently a less optimal denseness and a lower quality of evolutionary information.

The most notable combined evidence obtained from the sequence analysis of xenarthran α A chain and Hb β is the apparent similarity with the corresponding elephant proteins. This may reflect the fact that both groups are ancient, but separate, offshoots of the main placental stem, having conserved a number of shared primitive characters in their proteins. Alternatively, the similarities could be shared derived characters, making the Proboscidea and Xenarthra sister groups. Analysis of additional protein sequences, or other character states, will make it possible researchers to decide which of these two possibilities is the most likely. The pangolin α-crystallin A sequence reveals no shared derived substitutions with the xenarthrans. Although the most parsimonious solution places the pangolin α A sequence near the carnivore stem, this sequence can be connected in a position between the marsupials and xenarthrans at a cost of only two additional substitutions.

How do these protein sequence data relate to current opinions on xenarthran and pangolin phylogeny? Simpson (1945) summarized the history of classification of sloths, anteaters, and pangolins; he left the first two groups in the suborder Xenarthra of the order Edentata and left the pangolins in a separate order, the Pholidota. He stated: "It cannot be denied that the Xenarthra and Pholidota may have had a common ancestry, but this has not been demonstrated as a fact or even as a real probability."

Some authors still consider the pholidotes as the closest relatives of the edentates (Van Valen, 1971; Szalay, 1977), or leave this possibility open (Engelmann, 1985), but others have entirely abandoned this idea (Romer, 1966; Thenius, 1969; McKenna, 1975; Webb, 1985; Sarich, 1985). According to Simpson (1945), the order Edentata also included the fossil suborder Palaeanodonta, known from the Palaeocene to Oligocene of North America. In 1970, Emry described *Patriomanis*, a new North American Oligocene pangolin, and claimed that it could be derived directly from an early unspecialized palaeanodont, such as *Palaeanodon* itself. He interpreted the specialized similarities between the palaeanodonts and Xenarthra as being convergent and he moved the palaeanodonts from the order Edentata to the order Pholidota.

McKenna (1975) subdivided the eutherian mammals into two cohorts, the Edentata and the Epitheria, the former including only armadillos, sloths, and anteaters, and the latter including the pangolins, *incertae sedis*, perhaps within the new magnaorder Preptotheria, which embraced all except five of the living eutherian orders. However, in a recent publication, McKenna (1979) has referred to "the pangolins and their possible edentate allies" from which we deduce that he still leaves open the possibility of their relationship. It is noteworthy that the evidence of α-crystallin A lends support to an early separation of the xenarthrans from the other eutherians.

Most recently, Patterson (1978) has accepted that *Patriomanis* is a pangolin but remains unconvinced of its palaeanodont ancestry. He regards it as possible that early palaeanodonts were ancestral to both Xenarthra and Pholidota, but suggests that these three groups might be better regarded as three separate radiations from an unknown common ancestor and that it is immaterial whether they are regarded as three separate orders or three suborders of Edentata. He notes that the resemblances between some Tubulidentata (aardvarks) and pangolins and xenarthrans "seem clearly to be convergent" and states: " Whatever the relationships of these two groups (pangolins and xenarthrans) and of the extinct Palaeanodonta may eventually prove to be, . . . it is virtually certain that they are not with the aardvarks"; and he concludes that the Tubulidentata are most likely to have ungulate affinities, their ancestry going back to the Condylarthra. This view is supported by the work of Sarich (1985).

Although the pangolin α A sequence is not incompatible with an early and separate origin of this order, the most parsimonious solution places it nearer to the carnivores and ungulates. In view of the general uncertainty about the position of the pangolins, this information from the α A sequence might deserve serious consideration.

Linnaeus initially placed sloths, anteaters, and elephants together with manatees and pangolins in one order, called Bruta (see Glass, 1985), but the finding that Proboscidea and Xenarthra could both be ancient offshoots of the placentals, or even have a sister group relation, finds no further support in the taxonomic and paleontological literature. The elephants, usually grouped together with sea cows (Sirenia) and hyraxes as the subungulates, are without exception more or less explicitly considered as derived from a common condylarth stock with the other ungulate orders. Romer (1966), however, emphasizes that it is far from certain that the ungulate orders form a single natural group. It appears difficult to find in the literature convincing shared derived characters of subungulates and other ungulates. Reinvestigation of available evidence and future molecular studies will certainly reveal whether the separation of the subungulates from the ungulates, bringing them closer to the xenarthrans, indeed corresponds to the actual course of events during mammalian evolution.

ACKNOWLEDGMENTS

Eye lenses and armadillo blood were obtained with the kind cooperation of Dr. G. G. Montgomery and the Smithsonian Tropical Research Institute (Panama), Dr. P. J. H. van Bree (Amersterdam), Mr. E. de Graaf and Dr. P. Zwart (Utrecht), with financial aid from the Netherlands Foundation for the Advancement of Tropical Research (WOTRO). The tree pangolin material was obtained in the course of the Zaire River Expedition (1974), with the help of Dr. J. C. Heymans (Kisangani) and Dr. A. E. Friday (Cambridge), and with financial support from the Royal Society and the University of Cambridge. Dr. M. Goodman (Detroit) kindly made available his tree constructions of vertebrate Hb β chains. Amino acid analyses were kindly performed by Mrs. Marlies Versteeg. This work was supported in part by the Netherlands Foundation for Chemical Research (SON).

LITERATURE CITED

Beintema, J. J., W. Gaastra, J. A. Lenstra, G. W. Welling and W. M. Fitch.
1977. The molecular evolution of pancreatic ribonuclease. J. Mol. Evol., 10: 49-71.

Bloemendal, H.
1977. The vertebrate eye lens. Science, 197: 127-138.

Dayhoff, M.O. (editor)
1972. Atlas of protein sequence and structure, Vol. 5. National Biomedical Research Foundation, Silver Spring, Maryland, xxx +124 pp.

De Jong, W. W., and E. C. Terwindt.
1976. The amino-acid sequences of the α-crystallin A chains of red kangaroo and Virginia opossum. Eur. J. Biochem., 67: 503-510.

De Jong, W. W., J. T. Gleaves and D. Boulter.
1977. Evolutionary changes of α-crystallin and the phylogeny of mammalian orders. J. Mol. Evol., 10: 123-135.

De Jong, W. W., J. A. M. Leunissen, and H. T. Cuijpers.
1981. Primary structure of the major β-chain of armadillo (*Dasypus novemcinctus*) haemoglobin. Biochim. Biophys. Acta, 668:57-62.

Dickerson, R. E.
1971. The structure of cytochrome c and the rates of molecular evolution. J. Mol. Evol., 1: 26-45.

Engelman, G.F.
1985. The phylogeny of the Xenarthra. (this volume).

Emry, R. J.
1970. A North American Oligocene pangolin and other additions to the Pholidota. Bull. Amer. Mus. Nat. Hist., 142: 459-510.

Fitch, W. M.
1971. Toward defining the course of evolution: minimum change for a specific tree topology. Syst. Zool., 20: 406-416.

1973. Aspects of molecular evolution. Ann. Rev. Genet., 7: 343-381.

1977. On the problem of discovering the most parsimonious tree. Amer. Naturalist, 111: 223-257.

Glass, B.P.
1985. History of classification and nomenclature in Xenarthra (Edentata). (this volume).

Goodman, M., J. Czelusniak, G. W. Moore, A. E. Romero-Herrera and G. Matsuda
1979. Fitting the gene lineage into its species lineage, a parsimony strategy illustrated by cladograms constructed from globin sequences. Syst. Zool., 28: 132-163.

Holmquist, R.
1978. A measure of denseness of a phylogenetic network. J. Mol. Evol., 11: 225-231.

Lyddiatt, A. and D. Boulter.
1976. The amino acid sequence of cytochrome *c* from *Asterias rubens* L (common starfish). FEBS Lett., 67: 331-334.

McKenna, M. C.
1975. Toward a phylogenetic classification of the mammalia. Pp. 21-46, *in*: Phylogeny of the primates (W.P. Luckett and F.S. Szalay, editors), Plenum Press, New York and London, XIV +483 pp.

1979. Molecular mammalogy. Syst. Zool., 28: 109-113.

Moore, G. W., M. Goodman, C. Callahan, R. Holmquist and H. Moise.
1976. Stochastic versus augmented maximum parsimony method for estimating superimposed mutations in the divergent evolution of protein sequences. J. Mol. Biol., 105: 15-37.

O'Neil, P. B. and R. F. Doolittle.
1974. Mammalian phylogeny based on fibrinopeptide amino acid sequences. Syst. Zool., 22: 590-595.

Patterson, B.
1978. Pholidota and Tubulidentata. Pp. 268-278. *in* Evolution of African mammals (V.J. Maglio and H.B.S. Cooke, editors). Harvard Univ. Press, Cambridge, Mass. 641 pp.

Peacock, D. and D. Boulter.
1975. Use of amino acid sequence data in phylogeny and evaluation of methods using computer simulation. J. Mol. Biol., 95: 513-527.

Romer, A. S.
1966. Vertebrate paleontology, 3rd ed. University of Chicago Press, viii +468 pp.

Romero-Herrera, A. E., H. Lehmann, K. A. Joysey and A. E. Friday.
1978. On the evolution of myoglobin. Philos. Trans., Royal Soc. London, ser. B (Biol. Sci.), 283: 61-163.

Sarich, V.M.
1985. Xenarthran systematics: albumin immunological evidence. (this volume).

Schwartz, R. M. and M. O. Dayhoff.
1978. Origins of prokaryotes, eukaryotes, mitochondria and chloroplasts. Science, 199: 395-403.

Shields, G. F. and N. A. Straus.
1975. DNA-DNA hybridization studies of birds. Evolution, 29: 159-166.

Simpson, G. G.
1945. The principles of classification and a classification of mammals. Bull. Amer. Mus. Nat. Hist., 85: I-XVI, 1-350.

1975. Recent advances in methods of phylogenetic inference. Pp. 3-19, *in*, Phylogeny of the primates (W.P. Luckett and F.S. Szalay, editors), Plenum Press, New York and London, XIV +483 pp.

Szalay, F. S.
1977. Phylogenetic relationships and a classification of the eutherian mammalia. Pp. 315-374, *in*, Major patterns in vertebrate evolution (M. K. Hecht, P. C. Goody and B.M. Hecht, editors), NATO Advanced Study Institutes Series A, Vol. 14, Plenum, New York, 461 pp.

Thenius, E.
1969. Phylogenie der Mammalia. Walter de Gruyter, Berlin, 722 pp.

Van der Ouderaa, F.J., W.W. de Jong and H. Bloemendal.
1973. The amino-acid sequence of the αA_2 chain of bovine α-crystallin. Eur. J. Biochem., 39: 207-222.

Van Druten, J. A. M., P. G. M. Peer, A. B. H. Bos, and W. W. de Jong.
1978. Reciprocal amino acid substitutions in the evolution of homologous peptides. J. Theor. Biol., 73: 549-561.

Van Valen, L.
1971. Adaptive zones and the orders of mammals. Evolution, 25: 420-428.

Webb, D.E.
1985. The interrelationships of tree sloths and ground sloths. (this volume).

Wilson, A. C., S. S. Carlson and T. J. White.
1977. Biochemical evolution. Ann. Rev. Biochem., 46: 573-639.

Wright, C. A. (ed)
1974. Biochemical and immunological taxonomy of animals. Academic Press, London-New York, XII +490 pp.

XENARTHRAN SYSTEMATICS: ALBUMIN IMMUNOLOGICAL EVIDENCE

Vincent M. Sarich

Departments of Anthropology and Biochemistry, University of California Berkeley, California 94720

Abstract
Immunological distances were measured by the quantitative precipitin technique, using immunological comparisons of sloth, vermilingua (neotropical anteater) and armadillo albumins with one another and with albumins from representatives of other orders of mammals. The albumins of sloths, vermilinguas and armadillos are at least as distinct from one another as are those of carnivores, bats and primates. Albumin immunology indicates that the four xenarthran families separated at least 75-80 million years ago. No special relationships are indicated among the pangolins or the aardvark and any of the xenarthrans.

Abstracto
Se midieron distancias inmunológicas mediante la técnica cuantitativa precipitina utilizando comparaciones inmunológicas de monos perezosos, vermilingües (osos hormigueros neotropicales) y albúminas de armadillos, una con otra, y con albúminas de un individuo de otros órdenes de mamíferos. Las albúminas de monos perezosos, vermilingües y armadillos son tan distintos como aquellos de carnívoros, murciélagos y de primates. La immunología de las albúminas indica que las cuatro familias de los xenartranos se separaron hace 75-80 millones de años. No se ha encontrado ninguna relación especial entre el pangolín o el cerdo hormiguero con ninguno de los xenartranos.

Resumo
Distâncias imunológicas foram medidas através da técnica quantitativa de pricipitação, usando comparações imunológicas entre albuminas de preguiças, vermílinguas (tamanduás) e tatus, e com albuminas de representantes de outras ordens de mamíferos. As albuminas de preguiças, vermílinguas e tatus são pelo menos tão distintas umas das outras como as dos carnívoros, morcegos e primatas. A imunologia da albumina indica que as quatro famílias xenarthras se separaram há, no mínimo, 75-80 milhões de anos. Nenhuma relação especial foi indicada entre os pangolins ou os aardvarks e quaisquer dos xenarthros.

Comparative studies of macromolecular evolution have not included a number of important groups of mammals, and one such group is the Order Xenarthra (Edentata). Beyond an intrinsic interest in the group itself is the important question of relationships to other groups of mammals, in particular Pholodota and Tubulidenta, and the possibility (McKenna, 1975) that the Xenarthra are a sister group to all other placental mammals. Questions of this sort have been approached using both amino acid sequence data, especially on globins but also on eye-lens proteins (de Jong et al., 1985), and immunological comparisons of various rapidly evolving proteins such as serum albumins.

I present data from quantitative immunological comparisons of various sloth, vermilungua (neotropical anteater) and armadillo albumins with one another and with representative albumins from other orders of placental mammals.

MATERIALS AND METHODS

The xenarthran sera used in this study were obtained from Panama through Dr. Robert B. Tesh (*Tamandua tetradactyla, Cyclopes didactylus, Choloepus hoffmanni, Bradypus veriegatus*, and *Dasypus novemcinctus*) and the Los Angeles Zoo (*Myrmecophaga tridactyla*). Dr. Murray Johnson provided sera from *Manis javanica, Orycteropus afer, Soledonon paradoxurus, Scapanus townsendi, Erinaceus europaeus* and *Loxodonta africana.* The sources of other sera used have been reported in previous publications (Sarich, 1969; Sarich, 1970; Sarich and Cronin, 1976).

The albumins of *Tamandua, Cabassous*, and *Bradypus* were purified by polyacrylamide gel electrophoresis (Cronin and Sarich, 1975) and antisera to them prepared in Dutch Belted rabbits (Sarich, 1969). The individual antisera were titered using the microcomplement fix-

Table 1. Quantitative precipitin cross-reactions obtained using antisera to xenarthran albumins, shown as percent similarities between albumins.

Albumen	Bradypus	*Antiserum* Tamandua	Cabassous
Bradypus	100	18	28
Tamandua	20	100	22
Cabassous	25	28	100
Choloepus	70	20	27
Myrmecophaga	22	74	22
Cyclopes	26	50	24
Dasypus	29	19	60
Orycteropus	14	20	32
Manis	10	16	22
Aotus	21	22	28
Homo	10	3	19
Syconycteris	8	14	23
Felis	10	36	35
Procyon	9	14	27
Cavia	18	9	14
Loxodonta	20	20	17
Trichechus	6	9	2
Equus	10	13	25
Bos	10	9	11
Sus	14	10	9
Balaenoptera	10	7	15
Non-xenarthran mean	12	15	20

ation (MCF) procedure and pooled in reciprocal proportion to these titers (Sarich and Wilson, 1966). Because of the large proportion of interordinal comparisons involved, all immunological distances were measured using the quantative precipitin technique (Sarich, 1972), rather than the usual MCF approach. Percent similarities between albumins can be converted to the more familiar MCF immunological distances by multiplying them by 2.1 and subtracting the result from 210.

RESULTS AND DISCUSSION

It is evident that xenarthran albumins can be as different from one another as they are from those of at least some other placental mammals (Table 1). Typical cross-reaction values for interordinal comparisons involving carnivore, bat, and primate albumins fall in the 25-35 percent range (Sarich and Wilson, 1973). Thus, the albumins of sloths, vermilinguas and armadillos are at least as distinct from one another as are those of carnivores, bats, and primates.

Place of the xenarthrans among mammals.-- An objective cladistic analysis of placental mammals would require an appropriate sister group to allow assessments of amounts of change along the various placental lineages. Unfortunately, the only such group available, the marsupials, has albumins too distinct from those of placental mammals to provide reliable cross-reactions. Intraordinal comparisons can show albumins that have clearly accumulated significantly more or less change than the average. For example, the albumin of *Aotus* is much more similar to those of Old World higher primates than is that of *Cebus*, even though there is little doubt on either molecular or morphological grounds that *Aotus* and *Cebus*, both neotropical platyrrhine monkeys, are more closely related to one another than either is to non-platyrrhines (Sarich, 1970). Similarly, the albumin of *Ursus* reacts much better with antisera to non-arctoid carnivore albumins than do those of the other arctoids (Sarich, 1969). These characteristics can be maintained when reference antisera involving interordinal comparisons are used. Thus *Aotus* albumin remains more similar to those of various non-primates such as bats and carnivores than do the albumins of *Cebus* or *Homo* (Sarich and Wilson, 1973). Antisera to artiodactyl, perissodactyl, and proboscidean albumins tend to see all non-ungulate albumins at much the same distance, thus negating the possibility of developing unequivocal cladistic analyses of albumin evolution among nonungulate placentals. The antisera to *Tamandua, Bradypus,* and *Cabassous* albumins do not show this negative feature, possibly indicating a closer relationship of xenarthrans to the various non-ungulate than to ungulates or rabbits (Table 1). I see no logical approach to any precise placement of the xenarthrans among the nonungulate placental mammals using the available albumin data because of the unavailability of suitable outside reference sera.

Relationships Among Sloths, Vermilinguas and Armadillos.-- From the cross reactions based on serum albumin (Table 1), there are no convincing reasons to link any pair of the major xenarthran lineages to the exclusion of the third (Fig. 1). The analysis is predicated on the assumption that the xenarthrans do form a monophyletic unit relative to all other mammals surveyed. The lack of an appropriate outside reference point precludes an objective test of this assumption (see above). The three major xenarthran lineages averaged about 80 units of albumin evolution since the beginning of their adaptive radiation. A similar analysis involving other taxa, would allow about 75 units of change along the average bat or primate lineage since the two orders separated. Yet the average bat-xenarthran or primate-xenarthran distance (using the same species in the comparisons) is about 170 units (Fig. 2). This then leaves only about 15 units, 170 minus 80 (xenarthran evolution) minus 75 (bat-primate evolution) for the internodal segment linking the bat-primate and xenarthran common ancestors. As it is very likely that the bat-primate and xenarthran units are of monophyletic origin relative to one another, the periods of monophyly

	Bradypus	*Tamandua*	*Cabassous*
Bradypus	100		
Tamandua	19	100	
Cabassous	26	25	100
Other Mammals	10	16	20

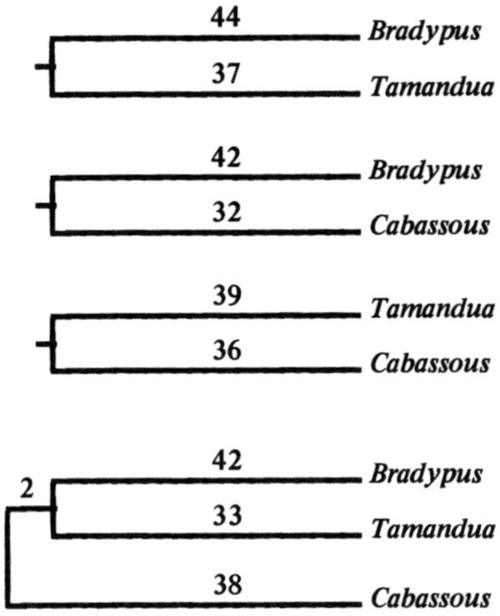

Figure 1: The quantative precipitin data of Tables 1 and 2 are additively apportioned to provide a phylogeny for the three xenarthran lines. The nonxenarthrans serve as outside reference points in determining the amounts of change to be allocated to each of a pair of lineages. Thus the 74 percent difference between the albumins of *Bradypus* and *Cabassous* is apportioned with 42 percent to *Bradypus*, and 32 percent to *Cabassous*, because *Bradypus* albumin is, on the average, 10 percent more distant from the various non-xenarthran albumins.

	Xenarthrans	*Manis*	*Orycteropus*
Xenarthrans	100		
Manis	16	100	
Orycteropus	22	?	100
Other Mammals	15	28	33

Figure 2: Evolutionary distances among the xenarthrans (treated as a unit), *Orycteropus*, and *Manis* are compared. Note that the aardvark and pangolin lines separated from the proto-xenarthran stock at the same time; 48 percent from the present along the xenarthran line. The xenarthrans have averaged only 38 percent change since they separated from one another. This means that they share 10 percent along the common proto-xenarthran line.

must have been rather brief, say 5-10 million years at most.

Albumin Clocks and Xenarthran Evolution.-- Discussion of the temporal implications of the analysis of albumin cross-reactions is in order, although this continues to be the most controversial aspect of studies in protein evolution. Any lineage that can be shown to have existed after separation of the common xenarthran stock from other placental mammals and prior to radiation of the four families with living members, must have been of relatively short duration. Thus separations between the bats and primates on the one hand, and the four extant xenarthran families on the other, were relatively synchronous events. The similar amounts of albumin evolution along the various bat, primate, and xenarthran lines supports this interpretation of synchrony. If lineages leading to the four xenarthran families have existed as long as the bats and primates, they evolved at least 75-80 million years ago.

The genetic distances between two-toed and three-toed sloths are apparently great enough (de Jong et al., 1985) to support the idea that the two extant genera are from different sloth lineages, *Bradypus* being derived from megatherioid sloths and *Choloepus* from megalonychid sloths (Webb, 1985), and may have begun their separate evolution soon after the differences arose in the lineages leading to armadillos, vermilinguas and sloths.

Separation of these lines so early might seem unreasonable in the face of the more traditional ideas of a basic Paleocene radiation linking all placental lines, but it is difficult to see how the molecular evidence can be interpreted within that older hypothetical framework. Indeed, cladistic considerations should have long ago precluded it on a morphological basis as well. For example, a combination of immunological data and the available globin sequence data indicate a series of divergence events going back in time to the beginnings of the adaptive radiation leading to modern nontupaiid primates. These include, in order, those leading to the flying lemur, tree shrews, bats, and finally, the basic nonungulate placental radiation. It is difficult to conceive the most recent common ancestry of forms as diverse as lemurs, galagos, tarsiers, and anthropoid primates as being more recent than about 60-65 million years ago, going back in time to the divergence of the tree shrews and then the bats must involve more millions of years. Whatever my disagreements with Simons on the question of hominid origins, we are in full agreement with the extension of the histories of various placental mammalian orders well back into the Cretaceous. Precisely how far back is a difficult question, because of possible distortions of the immunological distance measurements as we go back in time, but we observe cross-reactions between placental and marsupial albumins at distances of 200-220 units.

Usual usage of the albumin clock (Sarich, 1968; Wilson et al., 1977) allows 55-60 million years for the accumulation of 100 units of immunological distance, thus, we obtain a date of 110-130 million years ago for the marsupial-placental separation. This is entirely in accord with what is known of the history of the two groups (Lillegraven, 1974). When the same clock is applied to

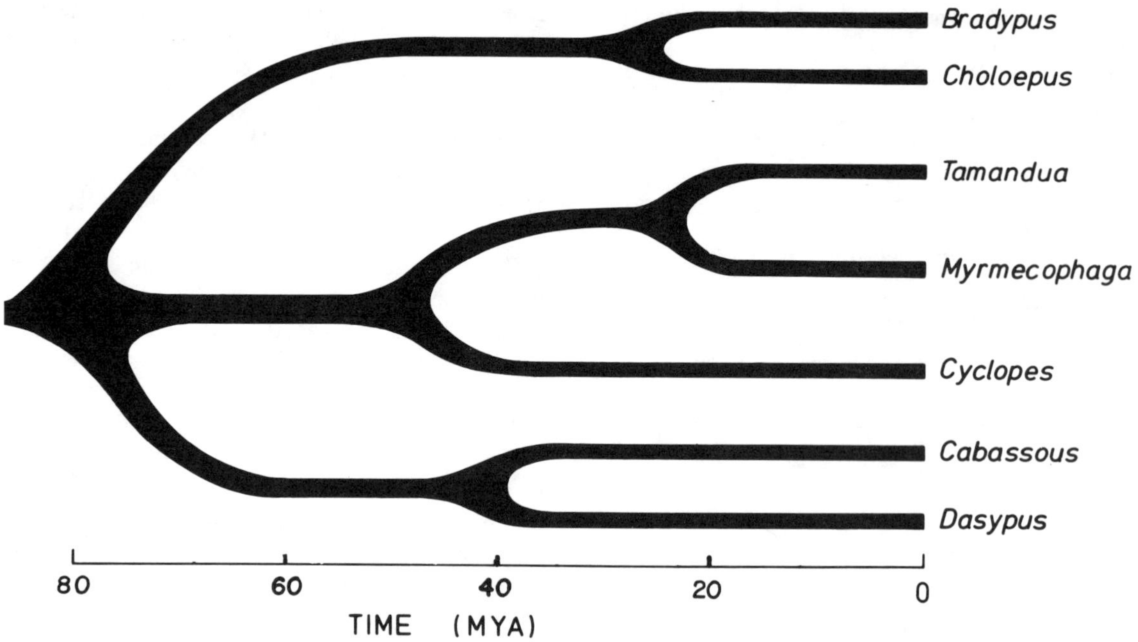

Figure 3: Cladistic and temporal relationships of the seven xenarthran species studied for this article. The tree drawing rules are discussed in Sarich and Cronin (1976). The time scale assumes that 100 units of immunological distance (47 percent difference in the precipitin reaction) between two taxa takes 55 million years to accumulate, on the average.

the Xenarthra, it places the beginning of their radiation at least 80 million years ago.

Possible Relationships of Xenarthrans to Pangolins and Aardvark.-- Our efforts to study these relationships are exploratory, because we have not yet produced antisera to the albumin of a pangolin or aardvark. Thus any immunological distance measurements are unidirectional with a consequent loss of precision. Nonetheless, the comparisons reported here do not indicate special relationships among the pangolins, or the aardvark and any of the xenarthrans. This is evident (Figure 2) because antisera to various carnivore, primate, and insectivore albumins generally show stronger cross-reactions with albumins of both pangolin and aardvark than with those of the xenarthrans, even though the xenarthran-pangolin and xenarthran-aardvark cross-reactions are weaker than those within the Xenarthra (Table 2).

Future Work With Xenarthra.-- I would greatly appreciate receiving sera or tissues from any of the xenarthran species not discussed in this article. Further immunological comparisons will contribute to the understanding of evolutionary relationships within the three major xenarthran clades, but it is most unlikely that the immunological approach will ever produce unequivocal phylogenetic placements among the three or among any other mammalian lines separated by similar genetic distances. To resolve such higher taxon issues, it will almost certainly prove necessary to develop the nucleotide sequences of one or more genes among a variety of mammalian taxa.

Table 2. Quantitative precipitin cross-reactions obtained using antisera to carnivore, insectivore, and primate albumins, shown as percent similarities between albumins. The carnivore antisera were to the albumins of *Felis, Canis,* and *Hyaena*, the primates were *Aotus, Lemur* and *Nycticebus* while the insectivores were *Solendon* and *Scapanus.*

	Antiserum			
Albumin	*Carnivore*	*Primate*	*Insectivore*	*Mean*
Bradypus	11	5	11	9
Tamandua	20	22	9	17
Cabassous	19	25	22	19
Manis	32	25	24	28
Orycteropus	34	44	22	33

ACKNOWLEDGMENTS

I thank all those who have contributed to the growth of our serum and tissue collection. For the studies described here, Drs. Robert Tesh and Murray Johnson were especially helpful. Peter Hornbeck provided superb technical assistance and all the laboratory work was done in the laboratories of Dr. Allan Wilson, Biochemistry Department, University of California, Berkeley. The study was supported by an NSF Systematic Biology grant to the author.

LITERATURE CITED

Cronin, J. E. and V. M. Sarich.
1975. Molecular systematics of the New World monkeys. J. Human Evol., 4: 357-375.

de Jong, W.W., A. Zweers, K.A. Joysey, J.T. Gleaves and D. Boulter.
1985. Protein sequence analysis applied to xenarthran and pholidote phylogeny. (this volume).

Lillegraven, J. A.
1974. Biogeographical considerations of the marsupial-placental dichotomy. Ann. Rev. Ecol. Syst., 5:215.

McKenna, M. C.
1975. Toward a phylogenetic classification of the Mammalia. Pp. 21-46 *In* Phylogeny of the primates. (W. P. Luckett and F. S. Szalay, eds.) Plenum Press, New York.

Sarich, V. M.
1968. The origin of the hominids: an immunological view. Pp. 94-121 *In* Perspectives on human evolution. 1. (P. C. Jay and S. L. Washburn, eds.). Holt, Rinehart, Winston, New York.

1969. Pinniped origins and the rate of evolution of carnivore albumins. Syst. Zool., 18:286-295.

1970. Primate systematics with special reference to the Old World monkeys: a protein perspective. Pp. 175-226 *In* Old World monkeys. (J. R. and P. H. Napier, eds.). Academic Press, New York.

1972. Generation time and albumin evolution. Biochem. Genet., 7:205-212.

Sarich, V.M. and J.E. Cronin.
1976. Molecular systematics of the primates. Pp. 141-170 *In* Molecular anthropology (M. Goodman and R. Tashian, eds.). Plenum Press. New York.

Sarich, V.M. and A.C. Wilson.
1966. Quantitative immunochemistry and the evolution of primate albumins. Science, 154:1563-1566.

1973. Generation time and genomic evolution in primates. Science, 179:1144-1147.

Wilson, A.C., S.S. Carlson, and T.J. White.
1977. Biochemical evolution. Ann. Rev. Biochem., 46:573-639.

THE FOSSIL GIANT ARMADILLOS OF NORTH AMERICA (PAMPATHERIINAE, XENARTHRA = EDENTATA)

Gordon Edmund

Curator of Vertebrate Paleontology, Royal Ontario Museum, Toronto, Canada.

Abstract

The pampatheres form a distinct conservative group which evolved in South America at least since Miocene time. One lineage, represented by *Kraglievichia*, entered North America with the establishment of the Panamanian isthmus and became established in Florida, where it evolved, culminating in the widely distributed Rancholabrean species *Holmesina septentrionalis*. A further invasion of North America occurred in the late Pleistocene, documented by several records from Mexico and Texas tentatively referred to the genera *Pampatherium* and *Plaina.*

Abstracto

Los pampaterios forman un grupo conservativo y differente, que evolucionó en Sur América por lo menos desde tiempos del Mioceno. Una de las líneas de evolucíon, representada por el género *Kraglievichia*, penetró en América de Norte al quedar establecido el Istmo de Panamá, y se estableció en la Florida, donde evolucionó para culminar en la especie Rancolabrea de amplia distribución, *Holmesina septentrionalis.* Durante la última parte del Pleistoceno ocurrió una invasión adicional de América del Norte, documentada a través de varios registros de México y Texas, que han sido tentativamente referidos a los géneros *Pampatherium* y *Plania.*

Resumo

Os Pampatheres formam um grupo distinto e conservador que se desenvolveram na América do Sul, pelo menos desde a era do Mioceno. Uma linhagem, representada pelo *Kraglievichia*, se introduziu na América do Norte com o estabelecimento do istmo panamenho e se estabeleceu na Flórida, onde evoluiu, culminando na amplamente distribuída espécie Rancholabreana, *Holmesina septentrionalis.* Uma posterior invasão da América do Norte ocorreu no final do Pleistoceno, documentada por diversos registros do México e Texas, hesitantemente referida aos gêneros *Pampatherium* e *Plaina.*

The armored xenarthrans (Loricata) include three major fossil groups, armadillos (Dasypodinae), giant armadillos (Pampatheriinae) and glyptodonts (Glyptodontidae). The two armadillo subfamilies are traditionally included in the family Dasypodidae (Romer, 1966; Hoffstetter, 1958), although a recent cladistic analysis (Engleman, 1985) presents strong evidence that the pampatheres are more closely related to the glyptodonts than to the dasypodines.

The early history of the pampatheres is poorly known, although it is obvious that the main center of evolution of the armoured xenarthrans since the Paleocene has been in South America, especially Argentina and Uruguay. The ancestors of the xenarthrans emigrated from North America during the Paleocene, but did not become extinct in North America until the Eocene (Simpson, 1927) where they are represented by a relict form *Metacheiromys* (Fig. 1). The major lineages of armoured xenarthrans appeared quite early in the fossil record (Hoffsteter, 1958), and continued to diversify in isolation for nearly 50 million years, during which time the Panamanian Isthmus was under water.

Fossils from Eocene deposits in Argentina indicate the presence of a variety of armored forms. Most of the currently recognized phyletic lines had become established during the Miocene (Hoffstetter, 1958). Although known only from fragmentary material, *Machlydotherium* (Ameghino, 1902) from the Eocene of Patagonia may represent the earliest recognizable pampathere.

By the time of the re-establishment of the Panamanian isthmus, many genera and species of dasypodines, pampatheres and glyptodonts ranged widely over South America. Representatives of all three groups successfully invaded North America about three million years ago, then became extinct, along with a wide variety of other mammals, only 10 to 15 thousand years ago. *Dasypus novemcinctus* is the sole North American

Figure 1. Life restoration of *Metacheiromys* from the Eocene of North America. This xenarthran was probably a relict descendant of the ancestors of the Loricata.

survivor of this once diverse assemblage.

The term giant armadillo usually refers to the pampatheres, although this name is confusing in view of the very large, living, dasypodines, such as *Priodontes maximus* and the even larger fossil species *Dasypus bellus*. The pampatheres, however, were truly gigantic armadillos, especially the late Pleistocene forms which attained a length approaching 3 m. Their external appearance (Fig. 2) was closer to that of the true armadillos than to that of the glyptodonts, a fact reflected in their osteology (Fig. 3). Pampathere distribution in North America was similar to that of *D. novemcinctus* today (James, 1957) but, like the glyptodonts, they were browsers or grazers, rather than omnivores or insectivores. Because of their large size and heavy armor, their locomotion was also probably similar to that of the glyptodonts, but the presence of moveable bands in the carapace gave the skeleton more flexibility.

This paper will deal only with the North American pampatheres, now adequately known from a good variety of fossil specimens.

ANATOMY

Pampatheres share with dasypodines and glyptodonts a body armor consisting of a large number of osteoderms, commonly referred to as scutes. The glyptodontal armor comprises a single inflexible carapace of tightly sutured scutes. In the pampatheres and dasypodines (s.l.) the armor consists of an anterior and a posterior buckler, each made up of many polygonal plates probably rather loosely sutured together in life. These bucklers are separated by a number of moveable, imbricating bands of specialized scutes, giving the carapace considerable flexibility (Fig. 2). Complete specimens of pampatheres are rare, but all known examples have only three free moveable bands, whereas the dasypodines usually have more. Each imbricating scute has a distinctly raised smooth area which lies beneath the posterior part of the scute ahead of it (Fig. 4). The tail is also armored in a similar manner, with a single ring of slightly overlapping scutes on each vertebral segment. The top of the skull is protected by a casque of polygonal scutes, and other smaller scutes have been found, more or less in place, on the legs and feet. In life, the scutes presumably were covered by a keratinous layer, and coarse hairs or bristles probably grew between the scutes or from pits in the scutes themselves.

The dentition consists of nine teeth per quadrant with no differentiation into molars, incisors, etc. The teeth do not extend to the anterior midline, but form subparallel rows which curve mediad at the anterior end (Fig. 5). Typically, the more posterior teeth are bilobate, while the anterior teeth are reniform to ovate in section. The anterior progression of bilobation in the toothrow has been used as an important character in constructing phyletic sequences (Castellanos, 1937; Robertson, 1976).

Earlier workers described the premaxilla as bearing the two upper anterior teeth (Winge, 1915). Careful preparation and examination by the writer reveals that the premaxillary suture runs generally between teeth one and two. In most cases the premaxilla forms a small part of the anterior wall of the alveolus of the second tooth, but most of the alveolus is clearly formed by the maxilla. This construction was checked on all adequately preserved North American material, as well as on most of the relevant specimens from South America. No specimen was found with two teeth borne on the premaxilla. I thus reject the presence or absence of a second tooth on the premaxilla as a valid taxonomic criterion.

Figure 2. Life restoration of *Holmesina* sp.

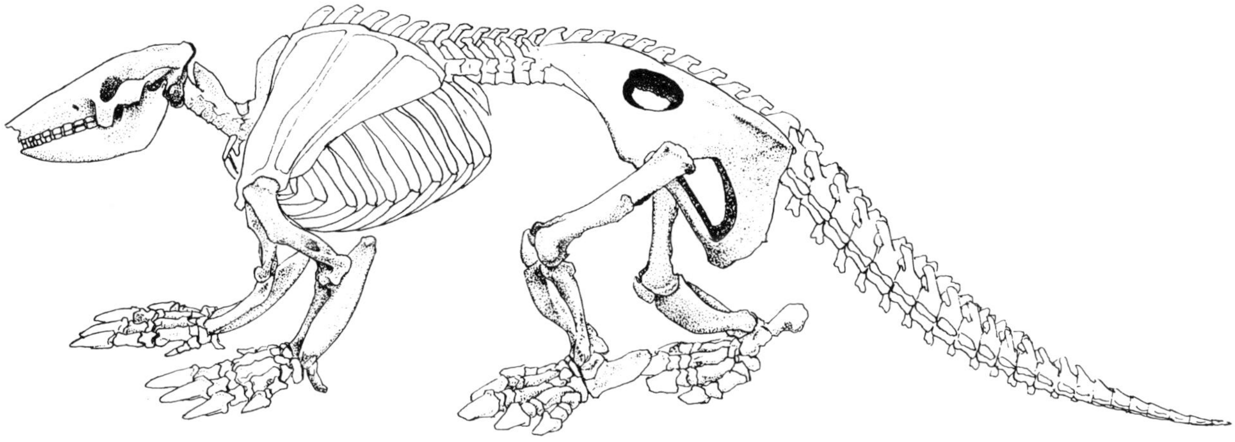

Figure 3. Skeletal reconstruction of *Holmesina* sp. Composite based on late Pleistocene partial specimens.

The skull (Fig. 5) is relatively long and narrow, with a pronounced temporo-orbital constriction. The frontals are inflated by enormous sinuses which extend into the maxillae and reach the parietals. Most of the anterior end of the skull consists of the nasal antrum flanked by the deep alveoli for the open-rooted teeth. The jugal process of the maxilla is robust, as is the same process on the squamosal. They support the well developed jugal arch which has prominent flange-like extensions to increase the area available for attachment of the masseter muscle.

The saggital and nuchal crests are prominent in most individuals, again suggesting powerful jaw muscles. The pterygoid flanges are in line with the tooth row, and are situated well to the rear of the skull. Each flange is composed mainly of the wing of the alisphenoid, with the pterygoid forming only a thin layer on the median wall. The pterygoid also extends forward to meet the palatines in forming the internal nares. In most specimens the pterygoid, apparently poorly sutured, has been lost.

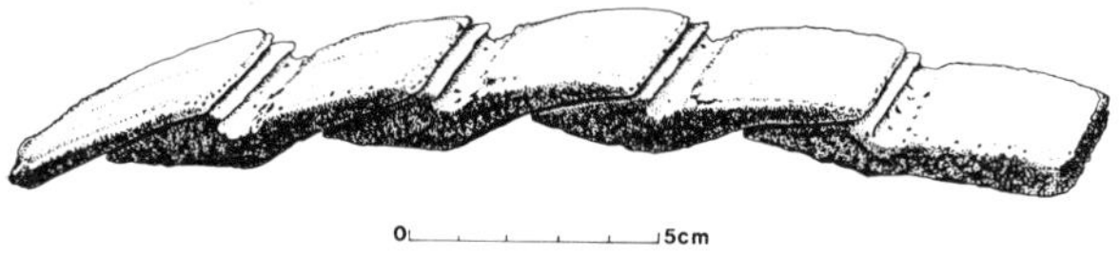

Figure 4. Five articulated scutes of *Holmesina* sp. arranged in their natural imbricating anteroposterior sequence, left to right.

The occipital condyles and foramen magnum are directed more ventrally than posteriorly, suggesting that the head was carried somewhat nose-down, similar to the posture in living armadillos (Fig. 2).

The mandible is also stoutly built, with the ramus exceptionally deep because of the continuously-growing teeth. The region of the angle is broadly rounded and somewhat scalloped in appearance, especially on its inner surface (Fig. 6). The coronoid is near the area of the most posterior tooth, and the flat, rectangular-shaped condyle is raised well above the occlusal plane. The glenoid area of the skull consists of a broad, un-

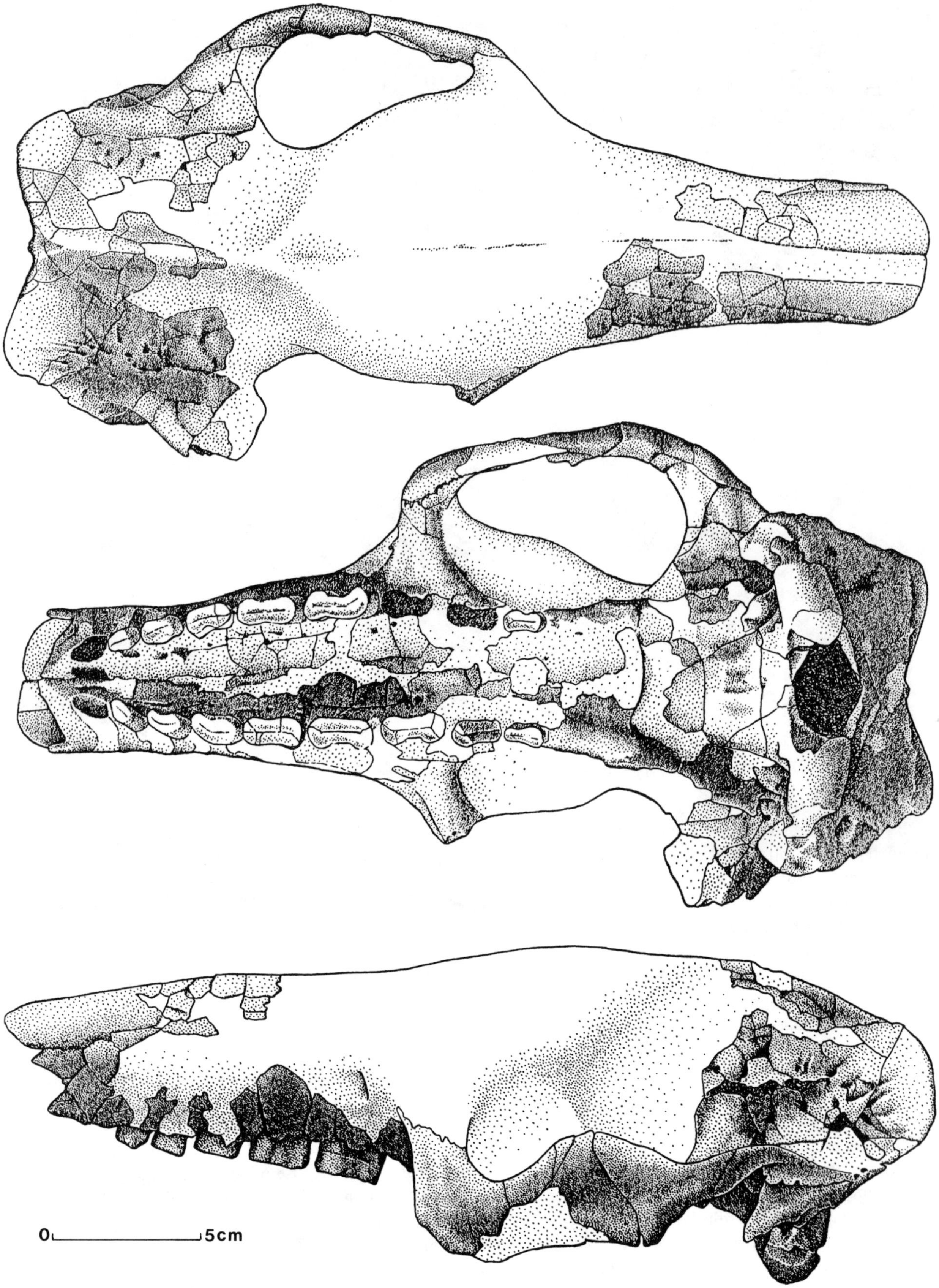

Figure 5. Skull of *Holmesina septentrionalis*, UM 68385, dorsal, ventral and left lateral views.

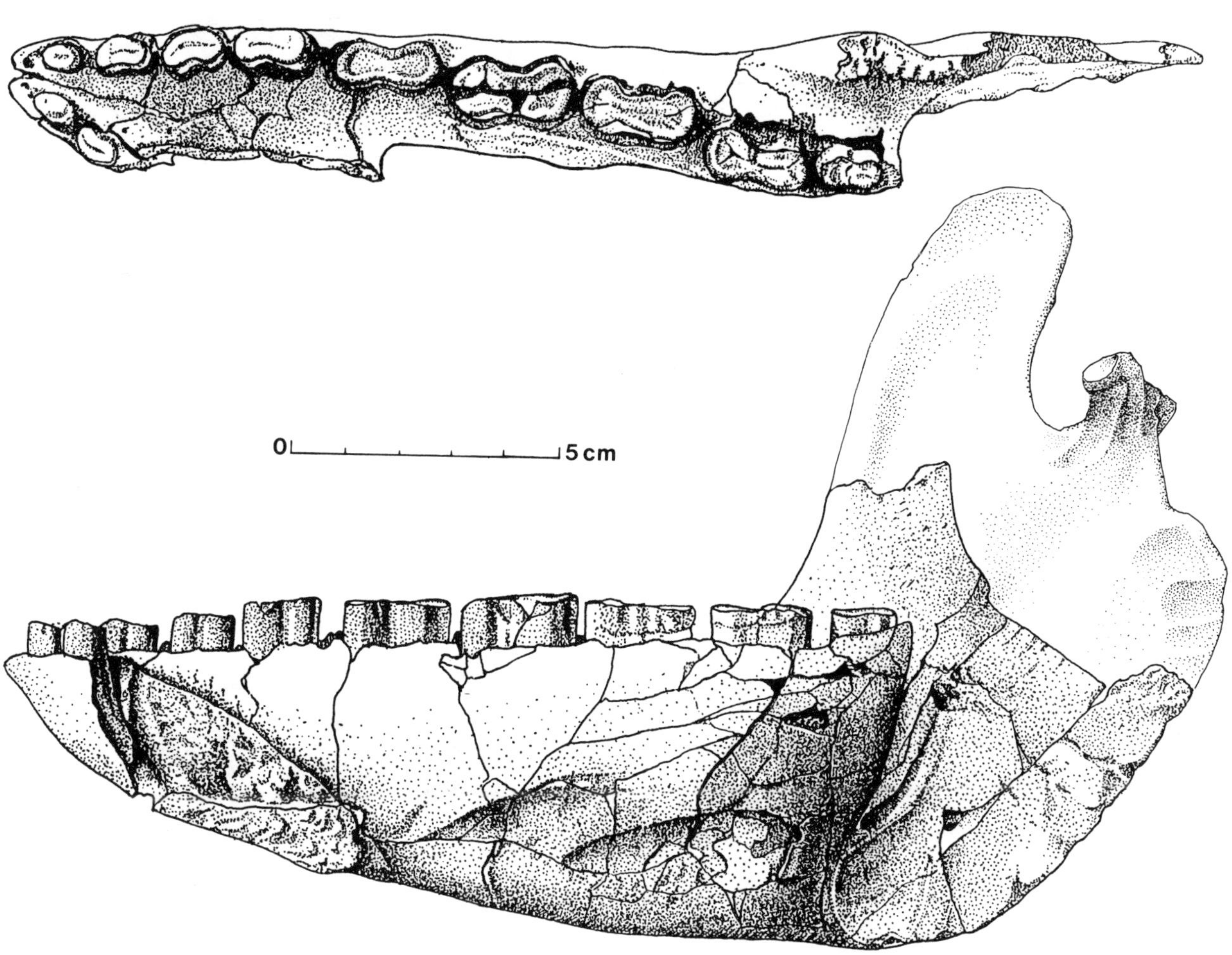

Figure 6. Right hemimandible and symphysis of *Holmesina septentrionalis*, UM 68385, occlusal and medial views. The condyle and coronoid area has been restored after AMNH 13221.

specialized area on the base of the jugal process of each squamosal. Thus, the mandible was not hinged, but free to move to a considerable extent, controlled only by the ligaments and muscle masses. Examination of well-preserved teeth (Fig. 7) reveals clear transverse striations, indicating that the main chewing motion was lateral. In most skulls and mandibles (Fig. 8) the teeth are set in the alveoli in a step-like fashion, so that antero-posterior movement is limited. In some specimens, the posterior (distal) margin of one of the more anterior upper teeth is worn off at about 45°. A matching step-like arrangement of teeth is observed in many mandibles. Obviously the animal chewed with the mandible moving upward and anteriorly until that motion was stopped by the interlocking step-like occlusion. The grinding was then accomplished by lateral movements probably powered largely by the pterygoideus and masseter muscle masses.

The vertebral column resembles that of living armadillos. The atlas is free, with very reduced alar processes. The axis is invariably fused to the next two, or usually three, cervicals. The sixth cervical is always free, but it

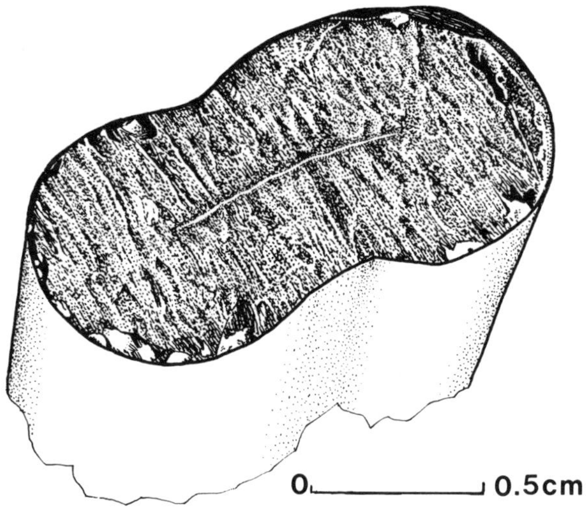

Figure 7. Occlusal surface of an isolated tooth of *Holmesina* from Florida, showing lateral (transverse) striations.

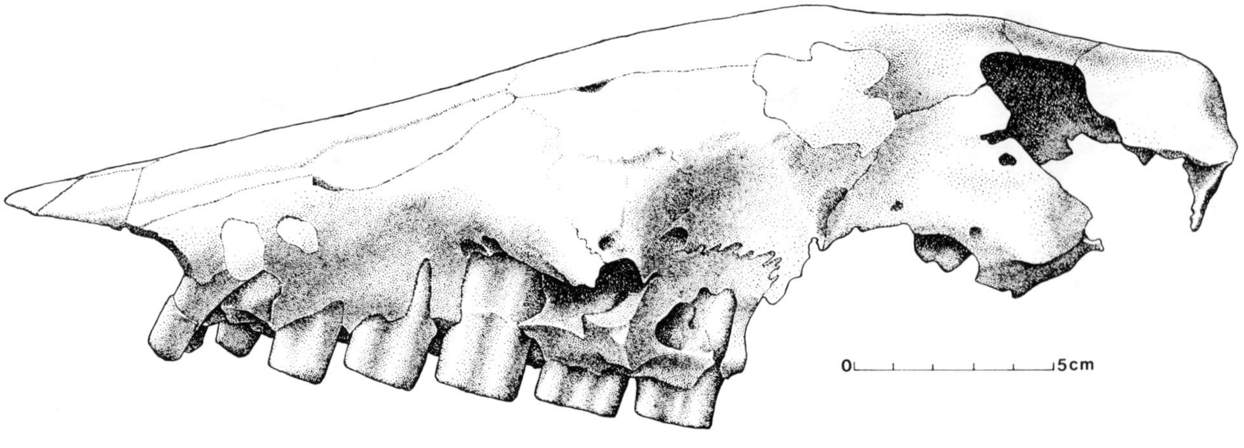

Figure 8. Left lateral view of skull of *Holmesina septentrionalis* UF/FSM 889.

is greatly reduced. Its centrum resembles a transverse rod and the arch is almost vestigial. The seventh (and last) cervical fuses, early in life, with the first, and usually also the second, thoracic vertebra. The remaining thoracic vertebrae are free, although the centra of the anterior ones indicate little opportunity for movement. The additional articulations which characterize the order Xenarthra begin to appear at the anterior end of the thoracic section, and are well developed in the posterior thoracic and lumbar vertebrae (Fig. 9). Some flexibility was possible in this part of the column, correlated with the proximity to the imbricating bands of scutes. The dorsal spines of the lumbar and most thoracic vertebrae terminate in broadened areas, presumably for contact with the carapace.

Apparently two free lumbar vertebrae were present in the Pleistocene forms, but complete vertebral columns are not known. Other lumbar and several caudal vertebrae are incorporated into the pelvis, but their number is uncertain.

The caudal vertebrae are peculiarly constructed to support rings of scutes on the tail. The dorsal spines and transverse processes make contact with the armor, and hemal spines (chevrons), also meeting the armor, are present on at least 12 of the more anterior vertebrae (Fig. 10). The posterior central faces of the more anterior caudal vertebrae are lower on the bone than are the anterior faces. This results in a smooth downward curve, presumably covered in part by the posterior buckler. The tail consisted of at least 24 free vertebrae, exclusive of those incorporated into the synsacrum. The vertebral column thus resembles fairly closely that of a living dasypodine (Winge, 1915). There is some intervertebral fusion in the neck and sacrum, but nothing resembling the extreme fusions seen in the glyptodonts.

The limbs generally are intermediate in proportions and specializations between those of glyptodonts and living *Dasypus*. Because of the bulk of the animal, and the weight and physical constraint of the carapace, the limbs, especially the hind legs, tend to have graviportal adaptations.

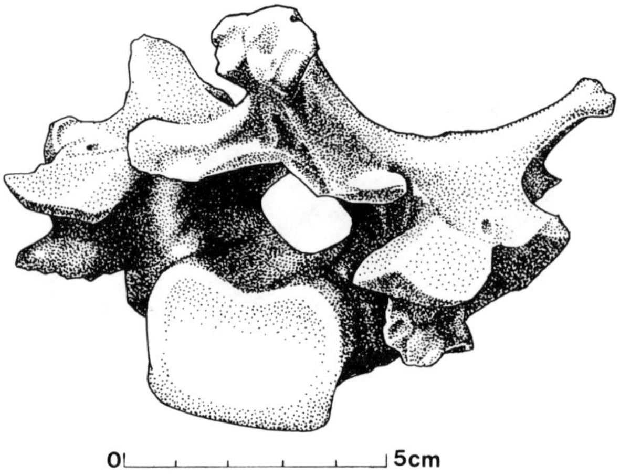

Figure 9. Posterolateral view of a posterior thoracic vertebra of a pampathere (ROM 3854).

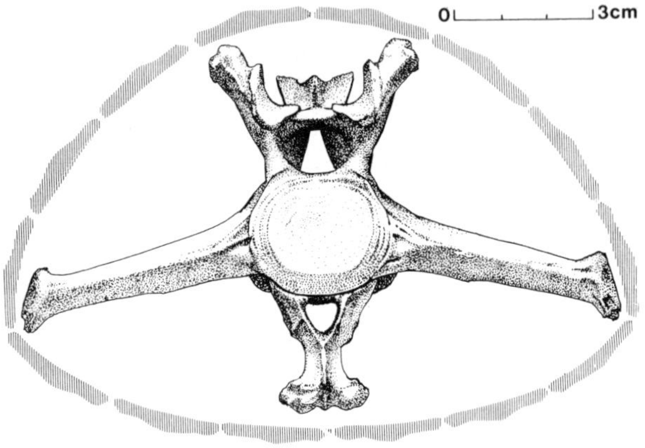

Figure 10. Posterior view of a typical caudal vertebra showing the hemapophysis and probable arrangement of scutes (stippled).

The front limb resembles that of the dasypodines. The

Table 1: Museum acronyms used in this paper.

AMNH	American Museum of Natural History, New York.
FMNH	Field Museum of Natural History, Chicago.
IGM	Instituto de Geologia, Paleontologia, y Mineralogia, Mexico City.
RHB	Private Collection, Roy and Helen Burgess, Venice, Florida.
ROM:B	Brayfield Collection, Royal Ontario Museum.
TMM	Texas Memorial Museum/University of Texas.
UF/FSM	University of Florida/Florida State Museum.
UM	University of Michigan, Museum of Paleontology.
USNM	United States National Museum (Smithsonian Institution).
UZM	University Zoology Museum, Copenhagen, Denmark.

scapula is not completely known, but the acromion process does not unite with the coracoid. There is a well-developed secondary spine bounding the posterior margin of the infraspinous fossa. A long, thin clavicle is known from a partial skeleton from Florida (UF/FSM 9336).

The humerus is well known. As would be expected in this large, heavy animal it is relatively long and robust, with huge muscle scars. The morphology of the head indicates a fore and aft motion, and the distal articulation is greatly expanded medio-laterally to produce a strong elbow joint.

The ulna is long, laterally compressed, and remarkably straight except for the mediad curvature of the olecranon. The relative length of the latter is considerably less than in the fossorial dasypodines. The ulna articulates distally with both the cuneiform and pisiform. The radius is also well developed, especially distally where it expands to articulate broadly with the lunar and scaphotrapezium. The forearm articulations indicate the possibility of a modest degree of rotation of the manus.

The carpus is essentially primitive, with only the scaphoid and trapezium being fused. The remaining carpals are present, supporting five digits. The three central metacarpals are subequal in size and shape, fairly robust, and carry well-developed short toes with strong, straight, pointed unguals. The first and fifth toes on the manus are also complete and bear unguals, but are considerably smaller than the second, third, and fourth toes. No specially developed claws are present in any known pampathere. The manus was almost certainly digitigrade.

The femur is considerably longer than the tibia-fibula, again an indication of graviportal locomotion. The great trochanter reaches only slightly above the height of the head in pampatheres, whereas in the dasypodines it is greatly extended above the head. The third trochanter is enormously developed, borne on a flange projecting about half way down the lateral side of the bone. The tibia and fibula are always fused at both ends, so that it is impossible to distinguish the boundaries of the individual bones, even in subadults. The knee has an ossified meniscus cartilage (lunula) and a cyamella (Pearson and Davin, 1921) in addition to the patella.

The ankle joint is between the double trochlea on the astragalus and the combined tibia-fibula. In contrast with the condition in the dasypodines, the calcaneum does not articulate with the tibia-fibula. The structure of the pes is primitive, with the cuboid, navicular, and all 3 cuneiform bones retained unfused. A well developed sesamoid in the plantar (and also palmar) regions no doubt served the flexor muscles and tendons. All five digits are conserved in the pes, each bearing a broadened hoof-like ungual.

DISTRIBUTION AND TAXONOMY

Because each pampathere skeleton is accompanied by approximately 800 characteristically ornamented osteoderms (scutes), they appear relatively frequently in faunal lists. First described by Lund (1838) from cave deposits in Brazil, several genera and species were soon erected by South American paleontologists. Since the recognition of the first North American pampatheres from Florida (Leidy, 1889b) specimens have been found in several of the southern states, as far north and inland as Kansas, and also in central Mexico. The greatest concentration of records (about 65) is in peninsular Florida, although none have been reported from the Everglades nor the inland area north of Lake Okeechobee. There are several coastal records in Georgia and in North and South Carolina. Only one record is known from the central Gulf States (Louisiana) but there are several along the coast of Texas from Galveston to Corpus Christi. Pampatheres ranged far inland in Texas, especially along the Trinity and Brazos Rivers, and even into adjacent Oklahoma. The most northerly record is Kanapolis, Kansas (Edmund, in Hibbard et al., 1978).

The high number of records from Florida may represent a high population density throughout the Pleistocene, because pampatheres inhabited the area since the beginning of that period. These records also may be attributable to superior opportunities for entrapment, preservation, and discovery. The absence of records from Central America, on the other hand, may be an artifact of less intense collecting effort.

Two phyletic lines of pampatheres had previously been described from North America (Edmund, 1978) and a third is recognized in this paper. One line evolved in Florida from a small form, *Kraglievichia*, which apparently migrated from South America during the latest Pliocene or earliest Pleistocene (Robertson, 1976). A more or less continuous record of its development is now known, with the end form, *Holmesina septent-*

rionalis, becoming widely distributed at the time of its extinction about 10,000 years ago. Two other taxa, represented by specimens from Texas and Mexico, apparently arrived in North America late in the Pleistocene and did not undergo any obvious changes before becoming extinct. One of these taxa is very similar to *Pampatherium* of Brasil and Argentina, and is known from reasonably complete material. The other, so far identified only on the basis of well-preserved scutes, (Fig. 13) is referred to the genus *Plaina*, as described below.

A late Pleistocene pampathere has been known from the southeastern United States since Leidy's 1889 description. Its ancestry remained speculative until Robertson (1976) described a fauna of latest Pliocene age from north-central Florida. He recognized the scutes, bones and skull of the small pampathere as being similar to those of *Kraglievichia paranensis* (Ameghino) from Argentina, and erected the new specific name *K. floridanus* for the Floridean material. At the same time he described a number of pampathere bones and scutes of successively younger ages, and showed that there was a chronocline between *K. floridanus* and the latest Pleistocene species presently known as *Holmesina septentrionalis* (Leidy) Simpson, 1930.

The first specimens to be described were scutes from Florida, and were given the name *Glyptodon septentrionalis* (Leidy, 1889a). The same author subsequently recognized the similarity of his material with that from the Brazilian caves, and accepted Lund's (1838) name *Chlamytherium humboldti* (Leidy, 1889b). Sellards (1915) reexamined the North American material, considered it distinct from the Brazilian, and named it *Chlamytherium septentrionalis* (Leidy). Simpson (1930) decided that it was generically distinct on morphological grounds, and renamed it *Holmesina septentrionalis* (Leidy). I accept this as the valid name for the late Pleistocene pampathere of the southeastern United States and some adjacent areas.

Because both *Chlamytherium* and *Chlamydotherium* were shown to be homonyms, Ameghino (1875) erected *Pampatherium* to include specimens from both Argentina and the Brazilian caves. The latter name was revived and its validity confirmed by Paula Couto (1956) as the type genus of the subfamily (or family of some authors) of giant armadillos, hence the common name pampathere. The names *Chlamytherium, Chlamydotherium*, and chlamythere are still occasionally used, although they are invalid.

Robertson (1976) clearly demonstrated that *Kraglievichia floridanus* of the Blancan Land Mammal Stage of Florida is directly ancestral to *Holmesina septentrionalis*. While it could be argued that since these are species at the ends of a chronocline, only one generic name should apply, I support Robertson's assignment of the Floridian Blancan form to the South American genus *Kraglievichia*. Not only does it reaffirm the close relationship with the Argentinian ancestor, but it serves to indicate that the Blancan form is an early immigrant from which the typically North American form evolved. I restrict the use of the generic name *Kraglievichia* in North America to the small Blancan species (*floridanus*) and also restrict the name *Holmesina septentrionalis* to the largest and latest form derived from *Kraglievichia*. The intermediates in the chronocline are here assigned to *Holmesina* sp. since the erection of other specific names would be meaningless at present.

The geographic and temporal distribution in South America of *Kraglievichia* and other ridge-scuted early pampatheres is not well known. A revision of the pioneering work on pampathere taxonomy by Castellanos (1937, 1946) is long overdue.

Holmesina septentrionalis was apparently well adapted to conditions prevailing towards the end of the Pleistocene, as shown by the rapid extension of its range. *Kraglievichia* and the intermediate-sized *Holmesina* sp. are not known outside of Florida, whereas *H. septentrionalis* appears to have spread rapidly and widely. Pampatheres are generally considered to have been restricted to coastal areas with semitropical or at least Mediterranean climates. This may reflect the areas where civilization has provided best collecting opportunities, at least in South America. The North American *H. septentrionalis* spread up river systems far inland in Kansas, Oklahoma, Texas and Mexico, presumably during the last interglacial period.

In deposits of late Pleistocene age in Texas and Mexico, the range of *H. septentrionalis* overlaps that of the two other pampatheres tentatively referred to the genera *Pampatherium* and *Plaina*.

THE NORTH AMERICAN GENERA

Kraglievichia **and** ***Holmesina***.- As described above, *Holmesina septentrionalis* is clearly derived from *Kraglievichia floridanus*, the evolutionary process taking place in Florida during the span of the Pleistocene. The general description given in the anatomy section applies equally to both species and their intermediates. Robertson (1976) briefly outlined the growth of *K. floridanus* into the larger and more widely distributed *H. septentrionalis*. Except for the much larger size of the latter, the only obvious morphological difference between it and *K. floridanus* is that the third and fourth teeth have progressed toward the morphology of that of the more posterior teeth, i.e. there is an anterior progression of "molarization", if that term can be applied to the virtually isodont teeth of pampatheres.

Adequate series of specimens are now available to document the increasing size of individuals in the *Klaglievichia-Holmesina* chronocline. The size ranges of the three central metatarsals throughout the three

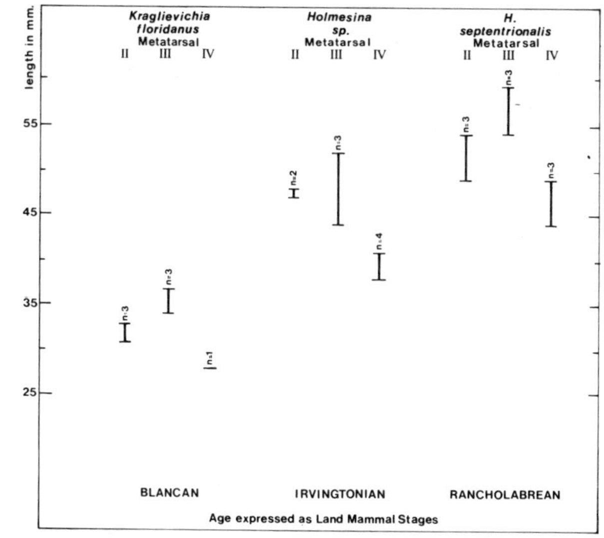

Figure 11. Graph showing increase in size of the three central metatarsals in specimens from the *Kraglievichia-Holmesina* line during the course of the Pleistocene.

Land Mammal Stages is depicted in Fig. 11. While the present number of specimens is inadequate for statistical treatment, the use of such measurements for chronological purposes may eventually be possible. A similar increase through time in the size of the calcaneum is depicted in Fig. 12.

Characteristic of the *Kraglievichia-Holmesina* line is the pattern of ornamentation on the osteoderms, or scutes, as they are commonly called. This is described more fully in the section on *Pampatherium*, but basically, the exposed areas of virtually all scutes have a well defined raised margin and a distinct medial keel or ridge. This pattern is seen in other specimens from South America, and appears to be a useful adaptive taxonomic criterion separating this lineage from that with essentially no ornamentation (Fig. 13).

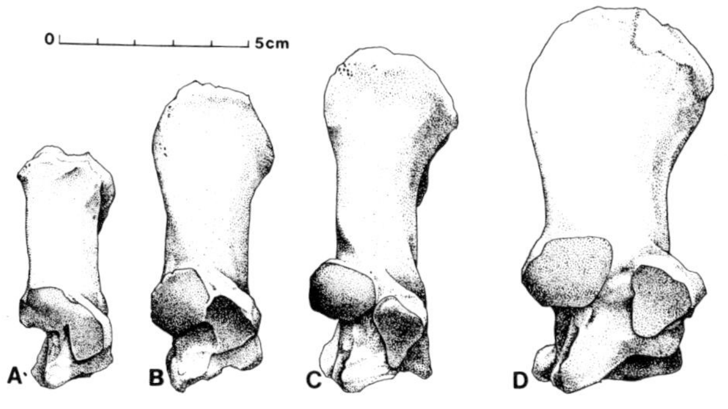

Figure 12. Right calcanei of pampatheres found in Florida, showing gradual increase in size. A. *Kraglievichia floridanus* UF/FSM 10902 (Blancan); B. *Holmesina* sp. UF/FSM 20946 (Early Irvingtonian); C. *Holmesina* sp. UF/FSM 15138 (Late Irvingtonian) reversed from left; D. *Holmesina septentrionalis* RHB 3151 (Rancholabrean).

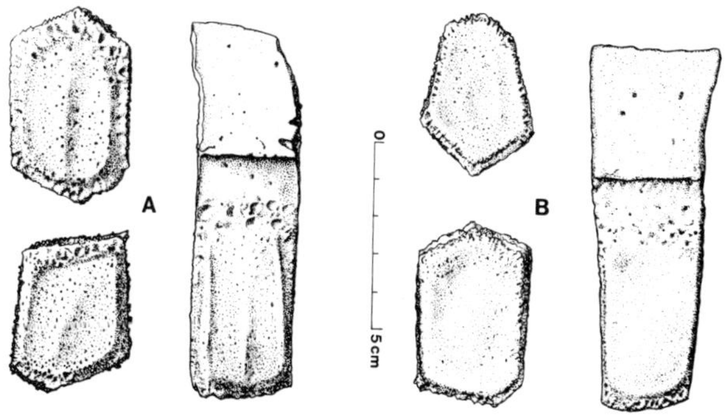

Figure 13. Osteoderms (scutes) of pampatheres. **A.** Two buckler and one imbricating scute of *Holmesina* sp. from Florida (FSM/-UF 15136); **B.** Two buckler and one imbricating scute of *Plaina* sp. from Valsequillo, Puebla, Mexico, uncatalogued, USNM.

For the purpose of this brief review, the general description above, including the drawings (Figs. 5, 6 and 13), are adequate to describe both *Kraglievichia* and *Holmesina.*

Pampatherium.- James (1957) described a reasonably complete but distorted skeleton from Houston, Texas, and assigned it to *H. septentrionalis.* The dimensions, general morphology, and arrangement of scutes agree with those of the more common eastern species, but close comparison with better preserved material from Florida reveals other characters which indicate an independent derivation. In the *Kraglievichi-Holmesina* lineage the scutes retain a constant morphology throughout the Pleistocene. The exposed area of both buckler and imbricating scutes features a raised submarginal band, a fairly prominent median ridge, and a series of deep pits near the anterior margin. In the Houston specimen the margin is only slightly raised, the central ridge is broad and low, and the anterior pits are less well developed.

The dentition is also characteristically different in the specimen from Houston (Fig. 14). In the anterior dentition of *Holmesina*, the second tooth is subreniform in outline, and its long axis is oriented at about 45° to the midline. The third tooth is quite reniform in outline, and is aligned closely with the other alveoli. In the specimen from Houston, both the second and third upper teeth are biconvex and have their long axes in line with the first and fourth teeth. The mandibular dentition is similarly distinctive. Both the dental and the scute characters indicate closer affinity with the material from Brazilian caves which has been redescribed by Winge (1915), and is here referred to *Pampatherium humboldti* (Lund, 1839). Until further studies are made, the specimen from Houston will be referred to *Pampatherium* sp.

Since recognizing that the material from Houston is distinctively different from other North American

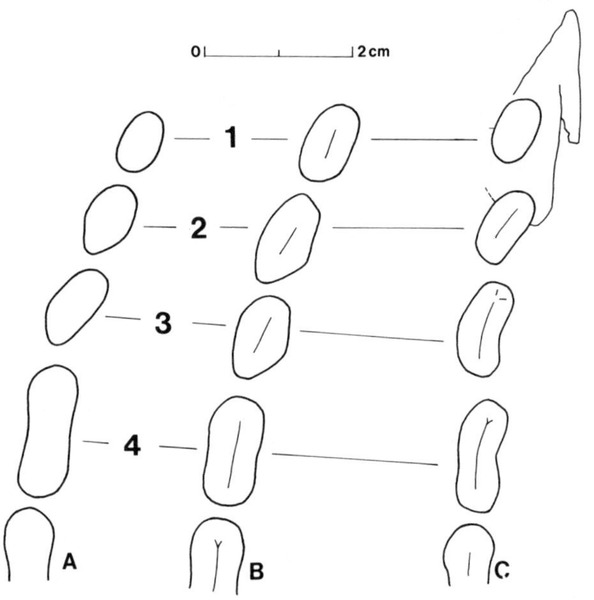

Figure 14. Anterior parts of dentitions of pamaptheres. A. *Pampatherium humbolti*, UZM 2314 (reversed from left): B. *Pampatherium* sp. HCT 4; C. *Holmesina septentrionalis* UM 60385.

pampatheres, I reexamined all available material. Several specimens, all confined to Texas and Mexico, had scutes resembling those of the specimen from Houston. One from Mexico (IGM 56-501) was found with most of the armor in place, along with parts of the skeleton including at least some of the dentition. Most other records are based on scattered and fragmentary material, usually isolated but nevertheless distinctive scutes. A paper describing this animal and its distribution in greater detail is being prepared.

Plaina.- A third taxon of pampatheres, currently based on a large series of scutes from the state of Puebla, Mexico, (Fig. 13B) is tentatively referred to the Pliocene Argentinian genus *Plaina*. No significant differences were noted when they were compared directly with the well preserved but undescribed specimen of *Plaina* sp., FMNH P14414. The size of the scutes is the same, and the ornamentation identical. The raised submarginal band is almost or completely lacking, and the central area is virtually flat and unornamented. A few isolated scutes from localities in Texas may also be referrable to this genus.

Little is known of the Pliocene and Pleistocene pampatheres of northern South America, nor the routes and times of their migration into North America. I anticipate investigating these problems as part of a revision of the subfamily currently underway.

ACKNOWLEDGMENTS

The research was supported in part by National Research Council of Canada Grant A 1408. Much of the material was provided through the generous cooperation of Dr. David Webb, Florida State Museum, University of Florida, and by Field Associates of the Royal Ontario Museum. Most of the illustrations were produced by Anker Odum. I am especially indebted to Dr. Jon Barlow and Dr. Chris McGowan of the Royal Ontario Museum for their critical reviews of this manuscript.

ADDENDUM

Since this paper was submitted, I have examined types and other specimens in South American museums. Several important changes in nomenclature should be noted.

The genus *Kraglievichia* is confined to Argentina, although the name was incorrectly attributed to the small pampathere from the upper Miocene (La Venta) of Colombia. Robertson's (1976) species *floridanus*, assumed to be derived from the latter, should therefore be re-assigned to the genus *Holmesina*.

The genus *Plaina* is based on inadequate material and will probably be relegated to synonymy. The smooth-scuted form from Puebla, Mexico, is now assigned to the *Vassallia-Pampatherium* lineage.

LITERATURE CITED

Ameghino, F.

1875. Notas sobre fósiles nuevos de la formación Pampeana. Ameghino Obras Completas, II: 13-17.

1887. Apuntes preliminares sobre algunos mamíferos extenguidos del yacimiento de Monto Hermoso. Ameghino Obras Completas, V: 339-353.

Castellanos, A.

1927. Breves notas sobre los clamiterios. Publ. Cent. Estud. Ingen. Rosario: 1-8.

1937. Anotaciones sobre la línea filogenética de los clamiterios. Publ. Fac. Cien. Univ. Nacion. Litoral, Rosario, Ser. Tec. - Cien., 8: 1-35.

1946. Una nueva especie de clamiterio *Vassallia maxima* n. sp. Publ. Inst. Fisio. Geol., Univ. Nacion. Litoral, 6(26): 5-47.

Engelman, G.F.

1985. The phylogeny of the Xenarthra. (this volume).

Hibbard, C. W., R. J. Zakrewski, R. E. Eshelman, G. Edmund, C. D. Griggs and C. Griggs.
1978. Mammals from the Kanapolis local fauna, Pleistocene (Yarmouth) of Ellsworth County, Kansas. Contr. Mus. Paleont., Univ. Michigan, 25(2): 11-44.

Hoffstetter, R.
1958. Xenarthra. Pp 533-636 *in* Piveteau, J., Traité de Paléontologie, VI(2) 962 pp.

James, G.
1957. An edentate from the Pleistocene of Texas. J. Paleont., 31 (4): 796-808.

Leidy, J.
1889a. Description of the vertebrate remains from Peace Creek, Florida. Trans. Wagner Free Inst. Sci., 2: 19-32.

1889b. Fossil vertebrates from Florida. Proc. Acad. Nat. Sci. Phila., 1889: 96-7.

Lund, P.W.
1838. Blik paa Brasiliens dyreverden fur sidste jordomvaeltning. K. Danske Vidensk. Selskabs. Naturvid. og Math. Afhand., 8: 61-144.

Patterson, B. and R. Pascual.
1968. Evolution of mammals on southern continents. V. The fossil mammal fauna of South America. Quart. Rev. Biol., 43(4): 409-451.

Paula Couto, Carlos de.
1956. Resumo de memórias de Lund sobre as cavernas de Lagoa Santa e seu conteúdo animal. Univ. do Brazil, Mus. Nac. Publ. Avulsas, 16: 1-14.

Pearson, K. and A. G. Davin
1921. On the sesamoids of the knee joint. Biometrika 13(10): 350-400.

Robertson, J. S.
1976. Latest Pliocene mammals from Haile XVA, Alachua County, Florida, Bull. Florida State Mus., Biol. Sci., 20(3): 111-186.

Romer A. S.
1966. Vertebrate Paleontology. 3rd. Ed. Univ. Chicago Press, 468 pp.

Sellards, E. H.
1915. *Chlamytherium septentrionalis*, an edentate from the Pleistocene of Florida. Amer. J. Sci., 4 (40): 19-31.

Simpson, G. G.
1930. *Holmesina septentrionalis*, an extinct giant armadillo of Florida. Amer. Mus. Novit., 442: 1-10.

1937. *Metacheiromys* and the Edentata. Bull. Amer. Mus. Nat. Hist., 59: 295-381.

Winge, H.
1915. Jordfunde og nulevende Gumlere (Edentata) fra Lagoa Santa, Minas Geraes, Brazilien. E. Museo Lundii. Kjobenhavn., 3 (2): 1-319.

THE SHASTA GROUND SLOTH *NOTHROTHERIOPS SHASTENSIS*(XENARTHRA, MEGATHERIIDAE) IN THE MIDDLE PLEISTOCENE OF FLORIDA

H. Gregory McDonald

Department of Geology, University of Toronto and Department of Vertebrate Paleontology Royal Ontario Museum, Toronto, Canada.

Abstract
The extinct Shasta ground sloth, *Nothrotheriops shastensis*, thought to have been restricted to western North America, is here recorded in Florida, an eastward range extension of 1,750 km. The fauna containing the sloth is late Irvingtonian (early Illinoian) in age, indicating that *Nothrotheriops* was more widely distributed during the earlier part of its history and that the restriction of its range to the southwestern United States was a later, possibly Sangamonian event. *Nothrotheriops shastensis* is one of many western species that reached Florida during the Pleistocene via a Gulf Coast Savanna corridor formed by a lowering of sea level during a glacial advance.

Abstracto
El extinto perezoso terrestre de Shasta, cuya distribución se pensó estuvo restringida al occidente de América del Norte, se informa aquí desde la Florida, una extensión hacia el Este de su conocida distribución de 1.750 km. La fauna que contiene los fósiles de dicho perezoso pertenece a la última parte del período Irvingtoniano, o primera parte del Illinoiano, lo cual indica que el género *Nothrotheriops* tuvo una distribución mucho más amplia durante la primera parte de su historia, y que la restricción de su distribución al suroeste de los Estados Unidos, fue un evento mucho más reciente, posiblemente del período Sangamoniano. *Nothrotheriops shastensis* es una de muchas especies occidentales que llegaron hasta la Florida durante el Pleistoceno a través del corredor de la Sabana de la Costa del Golfo, formada por un descenso del nivel del mar durante el avance de uno de los glaciales.

Resumo
A extinta preguiça terrícola de Shasta, *Nothrotheriops shastensis*, que se considerava restrita ao oeste norte americano, é aqui registrada na Flórida, numa extensão de 1.750 km ao leste. A fauna contendo a preguiça é do final do Irvingtoniano (início do Illinoiano) em idade, indicando que *Nothrotheriops* era mais amplamente distribuído durante a parte primordial de sua história e que a restrição de seus limites ao sudoeste dos Estados Unidos foi um evento posterior, possivelmente Sangamoniano. *Nothrotheriops shastensis* é uma das muitas espécies do oeste que chegaram à Flórida durante o Pleistoceno, via corredor Savana da Costa do Golfo, formado pela descida do nível do mar durante um avançõ glacial.

Attention has recently focused on the Shasta ground sloth, *Nothrotheriops shastensis*, because of its possible association with, and extinction by, early man in North America (Long et al., 1974; Long and Martin, 1974; Martin et al., 1961; Mosimann and Martin, 1975). Thanks to the preservation of plant remains in its dung (Laudermilk and Munz, 1934, 1938; Martin et al., 1961; Hansen, 1978;), as well as mummified specimens of the animal itself (Lull, 1929) in caves of the southwestern United States, our understanding of the animal (McNab, 1985) and its ecology is greater than for any other extinct North American Pleistocene species. However, the late Wisconsinan cave deposits from which this information comes are limited in time and space. The discovery of the remains of *Nothrotheriops shastensis* in Florida in deposits of pre-Rancholabrean age adds a new dimension to our interpretation of the paleozoogeography of the genus.

LOCATION, GEOLOGY, AND ASSOCIATED FAUNA

The locality of the Pool Branch local fauna (Fig. 1, Loc. 25) is the SW 1/4, NW 1/4, sec. 13, T31S, R25E, Homeland Quadrangle, (U.S.G.S. 7 1/2 series, 1952), 3.2 km NE of Ft. Meade, Polk County, Florida (Webb, 1974). The site is marked as number 111 in Figure 2 of Pirkle et al. (1967).

Pirkle et al. (1967) recognized five lithologic units at the locality. The uppermost layer (unit 5) is a loose surface sand. Beneath this layer was the Pleistocene bone bed (unit 4). In the exposure containing the Pool Branch local fauna, unit 5 graded into unit 4 and the

Figure 1. Pleistocene North American records of *Nothrotheriops*: Sites are listed in Table 1.

two could not be separated. Along with the Pleistocene bone, unit 4 contained quartz sand (73.1 percent), clay (5.6 percent), and phosphate (5.1 percent), with reworked Hawthorne limestone from the underlying deposits and silicified pieces of the Miocene colonial coral, *Siderastrea* sp. The sand containing the bone is of multiple origins and was deposited in stream channels or low areas cut into the Miocene marine or estuarine sediments (Pirkle et al., 1967).

The Pool Branch local fauna which includes the *Nothrotheriops shastensis* specimens was collected by Mitchell Hope, John Reynolds, Robert Allen, Jessee Robertson and Dr. S. David Webb of the Florida State Museum. All of the specimens are housed in the Vertebrate Paleontology collections of the Florida State Museum, University of Florida, Gainesville. Specimen numbers are prefixed UF. The Pool Branch sloth material was compared with *Nothrotheriops* from Rancho La Brea and San Josecito Cave housed in the George C. Page and Los Angeles County museums respectively.

Recovered taxa (emended from Webb, 1974) include: *Chrysemys scripta*, *Geochelone* sp., *Trionyx ferox*, *Alligator mississippiensis*, *Holmesina septentrionalis*, *Nothrotheriops shastensis*, unidentified arvicolines, *Ondatra* sp., *Mammut americanus*, *Equus fraternus*, *Tapirus veroensis*, *Platygonus compressus*, *Mylohyus* sp., *Hemiauchenia macrocephala* and *Odocoileus virginianus*. In Webb (1974) the nothrothere was listed as *Megalonyx wheatleyi*. Examination of all of the ground sloth material from the site shows that all specimens are *Nothrotheriops shastensis*. There are two large tooth fragments which may belong to a megathere, but the rest of the sloth material is too small to be referred to that group of giant sloths.

None of the recovered taxa place the fauna precisely

Table 1. Localities of *Nothrotheriops* shown in Figure 1. AMNH = American Museum of Natural History; LACM = Los Angeles County Museum of Natural History; UCMP = Museum of Paleontology, University of California, Berkeley; USNM = United States National Museum.

Locality	*Age*	*Reference*
1. Aden Crater, New Mexico	Wisconsinan	Lull, 1929
2. Brewster Co., Texas	Rancholabrean	Stock, 1925
3. Bustamente Cave, Nuevo Leon	Wisconsinan	Lundelius, (pers. comm. 1978)
4. Cedazo, Aquascalientes	Late Illinoian	Mooser and Dalquest, 1975
5. Clark Co., Nevada	Rancholabrean	AMNH unpublished specimen
6. Coconino Cavern, Arizona	Wisconsinan	Lindsay and Tessman, 1974
7. Conkling Cavern, New Mexico	Wisconsinan	Conkling, 1932
8. Cowboy Cave, Utah	Wisconsinan	Hansen, 1978
9. Curtis, Oklahoma	?	USNM unpublished specimen
10. El Golfo, Sonora	Early Irvingtonian	Shaw, (pers. comm. 1978)
11. Gilliland, Texas	Kansan	Hibbard and Dalquest, 1966
12. Gypsum Cave, Nevada	Wisconsinan	Stock, 1931; Harrington,1933
13. Hawver Cave, California	Wisconsinan	Stock, 1918
14. Irvington, California	Late Irvingtonian	UCMP unpublished specimen
15. Klamath Co., Oregon	?	Packard, 1952
16. Lagoa de Chapala, Jalisco	Rancholabrean	Downs, 1958
17. Lower Sloth Cave, Texas	Wisconsinan	Logan, 1977
18. Manzano Cave, New Mexico	Wisconsinan	Hibben, 1941
19. Mauv Cave, Arizona	Wisconsinan	Laudermilk and Munz, 1938
20. Medicine Hat Fauna 4, Alberta	Late mid-Wisconsinan	Churcher, 1970; Harington, 1978
21. Medicine Hat Fauna 9, Alberta	Kansan	Stalker and Churcher, 1972
22. Newport Bay Mesa, California	Late Sangamonian	Miller, 1971
23. Nichols Site, Arizona	Wisconsinan	Lundin and Royse, 1973
24. Payeatt Cave, Arizona	Wisconsinan	Lindsay and Tessman, 1974
25. Pool Branch, Florida	Early Illinoian	Webb, 1974
26. Potter Creek Cave, California	Wisconsinan	Sinclair, 1905
27. Rampart Cave, Arizona	Wisconsinan	Wilson, 1942
28. Rancho La Brea, California	Wisconsinan	Stock, 1913; 1925
29. Salton Sea, California	?	LACM unpublished specimen
30. Samwel Cave, California	Wisconsinan	Furlong, 1906
31. Sandia Cave, New Mexico	Wisconsinan	Hibben, 1941
32. San Josecito Cave, Nuevo Leon	Wisconsinan	Stock, 1943
33. San Pedro, California	Late Sangamonian	Stock, 1913
34. Shelter Cave, New Mexico	Wisconsinan	Martin et al., 1961
35. Stanton Cave, Arizona	Wisconsinan	Lindsay and Tessman, 1974
36. Tule Springs, Nevada	Wisconsinan	Mawby, 1967
37. Upper Sloth Cave, Texas	Wisconsinan	Logan and Black, 1979
38. Vallecito, California	Kansan	Downs and White, 1968
39. Ventana Cave, Arizona	Wisconsinan	Colbert, 1950
40. Wheeler Co., Texas	Rancholabrean	Hay, 1916
41. Williams Cave, Texas	Wisconsinan	Ayer, 1936

within a glacial or interglacial period. The absence of *Bison*, however, suggests an Irvingtonian age (Savage, 1951). For the purposes of this paper the fauna is considered to be late Irvingtonian (early Illinoian) in age.

THE POOL BRANCH *Nothrotheriops*

Sloth bones from the Pool branch local fauna referred to *Nothrotheriops shastensis* include: UF 11452 left humerus, UF 24665, distal portion of diaphysis of left humerus; UF 24110, right radius; UF 11451, distal end of right radius; UF 24106, right second metacarpal; UF 24107, right fourth metacarpal; UF 11447, right fifth metacarpal; UF 24168, proximal end of right fourth metatarsal; UF 11449; ungual of third digit of manus; UF 24109 and UF 11448, proximal phalanges of manus; UF 11450, two partial neural arches and a centrum of thoracic vertebrae; UF 11444, two partial sternal ribs and two partial costal ribs; and UF 11446, partial sternebrae. Based on left humeri and right radii, at least two adults and a juvenile are represented.

The nearly complete humerus, UF 11452, (Fig. 2-1) is crushed, fragmented and somewhat distorted, especially at the proximal and distal ends. The medial portion of the distal end is missing and the entepicondylar bar is incomplete. The epiphyses of both ends are fused to the shaft and the muscle scars of the diaphysis pro-

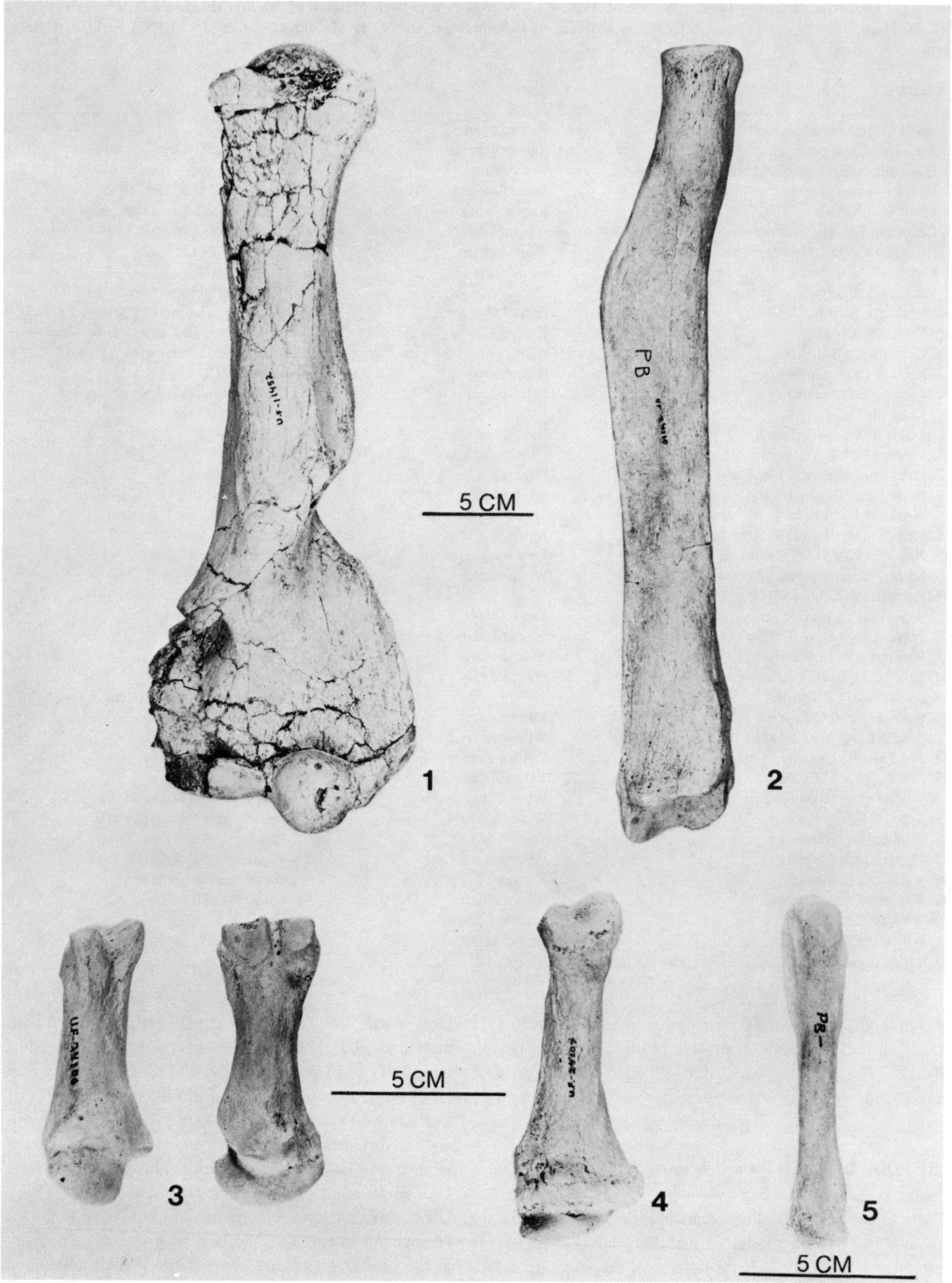

Figure 2. Bones of *Nothrotheriops shastensis*, Pool Branch local fauna, Florida. 1. Left humerus, UF 11452. 2. Right radius, UF 24110. 3. Right metacarpal II, dorsal and medial view, UF 24106. 4. Right metacarpal IV, UF 24107. 5. Right metacarpal V, UF 11447.

nounced and rugose. The deltoid crest is strongly developed and curves posteriorly at its distal end, aiding the supinator crest in strongly defining the musculospiral groove. Comparison of the Pool Branch specimen with a large series of humeri from San Josecito Cave revealed no consistent distinctions in morphology.

The fragment of diaphysis of the second humerus, UF 24665, is smaller, more gracile and less rugose. The deltoid crest is poorly developed and the musculospiral groove not well defined. This bone is from a juvenile animal.

Both right radii, UF 24110 (Fig. 2-2) and UF 11451, are from adult animals with well-fused epiphyses. Both show moderate rugosity but not as pronounced as in the adult humerus. The groove for the M. *extensor radialis* in UF 24110 is short, parallels the long axis of the shaft, and is confined to the anterior surface of the bone; whereas, in UF 11451, the groove is longer, extending proximally along the anterior surface of the shaft and then curving posteriorly. Except for this slight variation in UF 11451, there are no features to distinguish these bones from the radii from San Josecito Cave.

Metacarpal II, UF 24106, (Fig. 2-3) is identical in size and morphology with adult second metacarpals from San Josecito Cave. The articular facet for the first metacarpal is divided into distinct dorsal and palmar facets by a groove. Comparison with a large series of second metacarpals from San Josecito Cave shows a regular progression from specimens having a slight indentation at the distal end of the facet to two completely separated facets. There is only a single facet in most specimens.

Metacarpal IV, UF 24107, (Fig. 2-4) also represents an adult animal. The shaft is strongly rugose but is slender in proportion to its length. The proximal end is not as expanded as in specimens from Rancho La Brea or San Josecito Cave but has about the same mediolateral and dorsopalmar dimensions as the shaft. The articular surface for metacarpal III is a single smooth surface. In some *Nothrotheriops shastensis* this surface is subdivided into dorsal and palmar facets, but the differences are minor and when the Pool Branch specimen was compared to a series from San Josecito Cave no outstanding differences were noted.

Metacarpal V, UF 11447, (Fig. 2-5) is complete, except for a small portion of the distal end. It is lightly built and more gracile than most of the *Nothrotheriops shastensis* fifth metacarpals examined. Despite this, it has a moderately rugose texture and is probably from an adult animal.

The proximal end of a right metatarsal IV, UF 24168, (Fig. 3) exhibits some differences from the other known

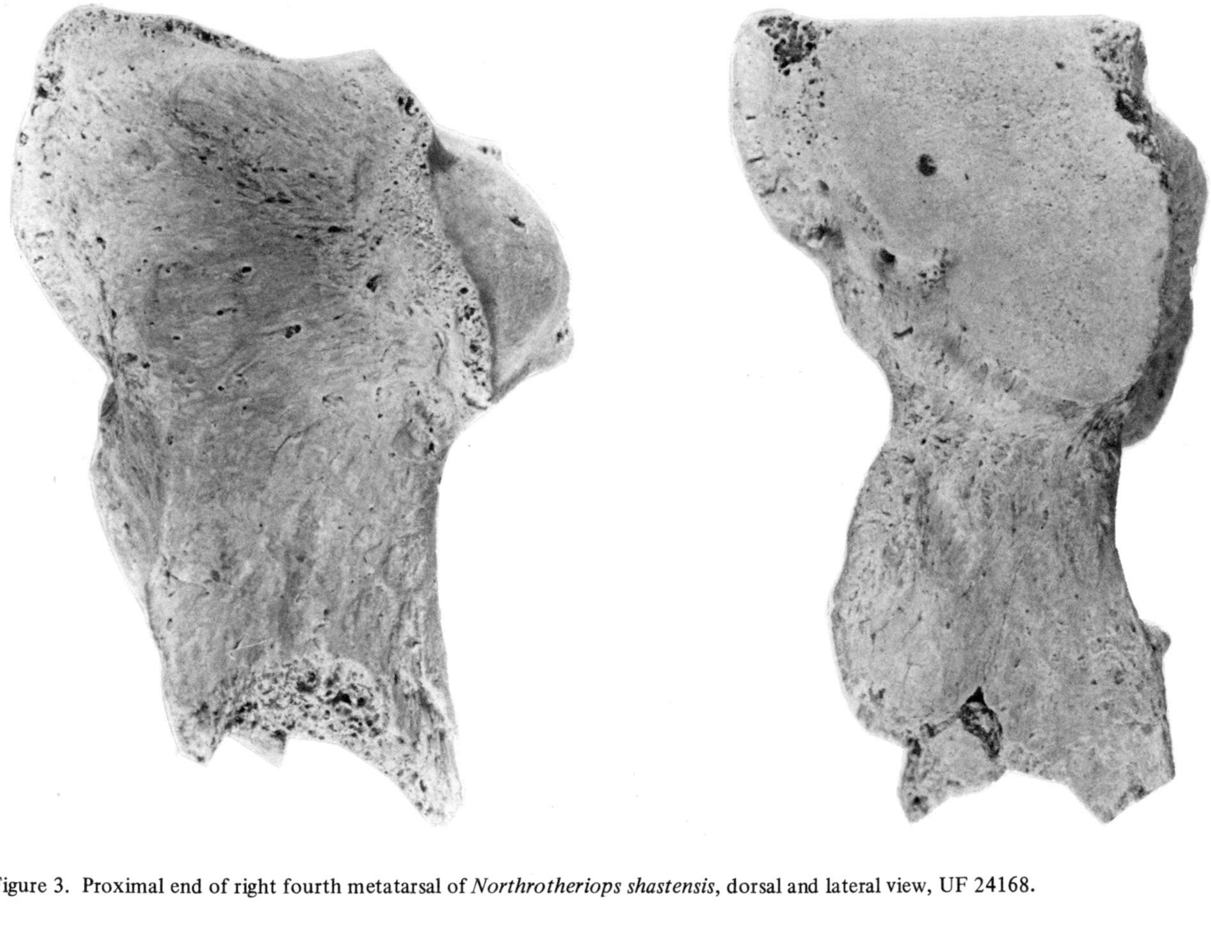

Figure 3. Proximal end of right fourth metatarsal of *Northrotheriops shastensis*, dorsal and lateral view, UF 24168.

specimens. The articular facet for metatarsal III is convex rather than flat. The facet for metatarsal V is elongated proximodistally rather than semicircular as in other *Nothrotheriops.* The shaft of the Pool Branch specimen is markedly compressed dorsoventrally just distal to the edge of the fifth metatarsal facet, a feature not present in other fourth metatarsals of *N. shastensis.* Although smaller in most dimensions than other fourth metatarsals examined, the plantar process of the proximal end is as strongly developed as in the larger specimens.

Comparison of the measurements of bones from the Pool Branch specimens with those from Rancho la Brea (Table 2), indicate few major differences. The morphological similarity between the Pool Branch specimens and those from San Josecito Cave and Rancho la Brea suggests that they belong to the same species, *Nothrotheriops shastensis.*

Table 2. Measurements (mm) of *Nothrotheriops shastensis* from Pool Branch, Florida and Rancho la Brea, California. UF refers to catalog numbers from the Florida State Museum, Gainesville. Catalog numbers for the bones from Rancho la Brea are from Stock (1925).

Measurement	*Pool Branch*	*Rancho la Brea*
Humerus	UF 11452	1874-R-1
Greatest Length	376.0	432.4
Width Across Deltoid Crest	62.5	-
Width of Distal Trochlea	89.0	95.9
Radius	UF 24110	1872-R-1
Greatest Length	257.6	≅ 415
Width at Middle of Shaft	44.1	52.0
Thickness at Middle of Shaft	22.2	30.8
Greatest Width of Distal End	45.1	78.5
Depth of Proximal End	37.0	-
Width of Proximal End	39.4	-
Second Metacarpal	UF 24106	1914-L-1
Greatest Length	75.5	79.5
Depth of Proximal End	30.7	29.0
Width of Proximal End	22.6	24.6
Depth of Distal End	30.4	45.8
Width of Distal End	25.3	28.5
Fourth Metacarpal	UF 24107	-
Greatest Length	126.6	-
Depth of Proximal End	30.3	-
Width of Proximal End	26.4	-
Depth of Distal End	46.2	-
Width of Distal End	28.8	-
Fifth Metacarpal	UF 11447	1917-R-1
Length	≅ 101	≅ 105
Transverse Diam. Proximal End	16.5	24.1
Dorsoventral Diam. Proximal End	19.6	24.9
Fourth Metatarsal	UF 24168	1883-R-2
Depth of Proximal End	42.4	57.5
Width of Proximal End	45.1	58.2

DISCUSSION

Based on the studies of Laudermilk and Munz (1934, 1938), Martin et al. (1961), and Hansen (1978), the ecology of *Nothrotheriops shastensis* has been reconstructed and interpreted as that of an animal adapted to xeric vegetation with food habits comparable to those of the recent desert bighorn sheep *Ovis canadensis.* However, Hansen recognized that the sloth's food habits were opportunistic and noted that the vegetation characteristics of the Shasta ground sloth's diet appeared to be different in other nearby habitats at comparable times. He suggested that *Nothrotheriops* lived under a wider variety of conditions than the limited samples could document.

Virtually all ecological analyses of *Nothrotheriops* have been based on cave sites in the southwestern United States. At Rampart Cave, Arizona, (Fig. 1, Loc. 27) the bulk of the diet (77 percent) of *Nothrotheriops* was desert globemallow *Sphaeralcea ambigua,* Nevada mormontea, *Ephedra nevadensis* and saltbrush *Atriplex* with minor contributors such as sagebrush *Artemisia,* ash *Fraxinus,* tumbleweed *Amaranthus* and sedges (cyperaceae) (Hansen, 1978). The composition of dungballs from nearby Mauv Cave (Fig. 1, Loc. 19) averaged 60 percent common reed *Phragmites,* whereas at Rampart Cave common reed contributed only 0.2 percent to the sloth's diet. Laudermilk and Munz (1934) listed yucca tissue *Yucca brevifolia* and *Agave utahensis,* as forming 80 percent of the plant material present in the dungballs from Gypsum Cave (Fig. 1, Loc. 12).

Neither the Pool Branch local fauna nor the closest comparable one in time from central Florida, Coleman 11A in Sumter County (Martin, 1974), yielded a pollen sample useful to indicate composition of the flora. The oldest published Pleistocene pollen record in Florida is from Mud Lake, Marion County (Watts, 1969), 165 km from the Pool Branch site. Near the base of Watts' column is zone M-6 (the *Pinus-Carya* zone), which is older than 34,500 years B.P.

Zone M-6 is indicative of prairie vegetation with dry oak forest or shrub, and some mesic forest in the vicinity. Desert globemallow, saltbrush and yucca, the major contributors to the Shasta ground sloth's diet at Rampart and Gypsum caves, were not present. However, some of the minor constituents of the Arizona ground sloth's dung were present, including sagebrush, ash, tumbleweed, and sedges. Thus a number of known Shasta ground sloth food species occurred in Florida during the Pleistocene. The recent flora of Florida includes two species of *Atriplex,* six species of *Acacia,* one species each of *Yucca* and *Agave* and *Phragmites communis,* all of which have been identified as food of *Nothrotheriops.* These plants were probably also part of the Pleistocene flora of Florida.

The earliest faunas containing *Nothrotheriops* are

Kansan in age from widely separated localities: Medicine Hat, Alberta, Fauna 9 (Fig. 1, Loc. 21); Vallecito, California (Fig. 1, Loc. 38) and Gilliland, Texas (Fig. 1, Loc. 11). Of these, only Medicine Hat has associated wood and pollen specimens. Mott and Stalker (1972) interpret these as indicating an area of open grassland on upland sites with trees and shrubs occurring along the rivers. Recorded plants include willow *Salix*, Manitoba maple or box elder *Acer negundo*, aspen or cottonwood *Populus*, ash *Fraxinus*, spruce *Picea*, pine *Pinus*, alder *Alnus*, oleaster *Elaeagnus*, Gramineae, Tubliflorae, sagebrush *Artemisia*, Chenopodiinae and some sedges. There are few plants common to the three widely separated localaties during the Pleistocene and the similarity is primarily at the generic level. It is highly probable that those food plants common to all three areas were utilized by *Nothrotheriops*; but, without dung from Florida and Medicine Hat, even that is conjectural.

One broad floral resemblance among all known occurrences of *Nothrotheriops* is the presence of grasslands or open scrubland. It seems clear that *Nothrotheriops* preferred grassland or open savanna habitat. As used by Webb (1977; 1978) this includes, in the broadest sense, all open-country formations with at least a few trees and includes thorn scrub and open deciduous forest.

The semiarid grassland environments of the southwestern United States with which *Nothrotheriops* is commonly associated were apparently in continuous existence throughout the Pleistocene (Martin and Harrell, 1957), and the presence of *Nothrotheriops* and other plant and animal species in Florida implies a savanna corridor from the more arid parts of Mesoamerica or the southwestern United States. We have no direct evidence of these grasslands in Louisiana, southern Mississippi or Alabama, but their occurrence at some time or times during the Pleistocene can be inferred by the presence of species of semiarid grassland origin in Florida. Many of these species are still extant in the state and represent disjunct populations. In addition to the Shasta ground sloth there are: indigo snake *Drymarchon corais*, burrowing owl *Athene cunicularia*, scrub jay *Aphelocoma caerulescens*, sandhill crane *Grus canadensis*, caracara *Caracara cheriway*, mottled duck *Anas fulvigula*, white-tailed jackrabbit *Lepus townsendii*, hognose skunk *Conepatus leuconotus*, southeastern pocket gopher *Geomys pinetus*, an extinct gopher *Thomomys orientalis*, a ground squirrel *Spermophilus* sp., and the extinct armadillo *Dasypus bellus*. Neill (1957) lists a number of plants with a western origin now living in Florida. Altogether, the appearance of *Nothrotheriops shastensis* is not as unusual as it first appears, considering all the other western species which also managed to reach Florida.

The jackrabbit, skunk, southeastern pocket gopher and armadillo have not been found in Florida in sites older than the Nebraskan glaciation (Webb, 1974). The white-tailed jackrabbit is known from two localities, the Nebraskan age Inglis 1A and the Illinoian age Coleman IIA. The hognose skunk is recorded from five localities ranging in age from Nebraskan to early Sangamonian. The southeastern pocket gopher still survives in Florida. It is an endemic species which probably arose from the plains pocket gopher, *Geomys bursarius*. Its occurrence in the state in the Nebraskan as an already distinct species suggests a long history and a connection of grassland between Florida and the Great Plains prior to the first continental glaciation. *Thomomys orientalis* is known only from Sabretooth Cave of Wisconsinan age. The *Spermophilus* is also known only from one locality, Haile XIVA of middle Wisconsinan age. *Dasypus bellus* is known from 21 localities ranging in age from Nebraskan to late Wisconsinan.

The indigo snake is recorded as a fossil at 12 localities, the earliest of which is Sangamonian and it forms part of the recent herpetofauna. The scrub jay is known only from the early Wisconsinan locality, Arredondo (Brodkorb, 1959). The caracara first appears in Florida in the early Sangamonian and is known from five localities as a fossil (Brodkorb, 1964). The mottled duck has only been recovered as a fossil from one locality, the late Wisconsinan Vero Beach (Weigel, 1962). Although no fossil record of the burrowing owl or sandhill crane has been described from Florida, these birds of western affinities are extant in Florida.

The first appearance of these animals in Florida during glacial or early interglacial episodes suggests that their entrance into the state was facilitated by the drop in sea level generated by a glacial advance. The result of such a drop would be formation of a corridor of newly exposed land along the Gulf Coast and this could be the origin of the open woodland corridor postulated by Martin and Harrell (1957). Florida's water table would also be lowered during a glacial episode, thus reducing available water for vegetation, particularly during the dry season of the year. A reduction in available water would limit the development of mesic forests and permit the spread of open scrubland and grassland, suitable habitat for the Shasta ground sloth and the other grassland species.

Based strictly on the fauna at Coleman IIA, Martin (1974) suggests that the late Illinoian was the transitional period from open savanna to subtropical forest in Florida. If a floral transition did occur at that time, the extinction of *Nothrotheriops* in Florida may be due to vegetational changes. The white-tailed jackrabbit and hognose skunk also disappeared from Florida during the late Illinoian and early Sangamonian. Although many of the western immigrants still survive in Florida today, all have a small body size. The large size of the Shasta ground sloth may not have permitted its survival with the loss of its preferred open scrubland and grassland habitat. As pointed out by Guilday (1967), "... in any deteriorating ecosystem, large herbivores are more drastically affected than small herbivores by virtue of their greater demands upon the system for space, food and cover." The disappearance of *Nothrotheriops shastensis* from Florida and the restriction of its range to the southwestern United States during the Sangamonian may have been due to the onset of more mesic conditions.

ACKNOWLEDGMENTS

I thank Dr. S. David Webb, Florida State Museum, for permitting me to borrow the specimens described here and for his many invaluable comments. Ernest L. Lundelius, Christopher A. Shaw and Michael Frazier kindly permitted me to use unpublished data. The following individuals kindly reviewed various drafts of the manuscript and offered suggestions for its improvement, William A. Akersten, Elaine Anderson, A. Gordon Edmund, John E. Guilday, George T. Jefferson, S. David Webb and John A. White. Mrs. Pamela Purves typed the manuscript.

LITERATURE CITED

Ayer, M. Y.,
1963. The archaeological and faunal material from Williams Cave, Guadalupe Mountains, Texas. Proc. Acad. Nat. Sci. Philadelphia, 88: 599-619.

Brodkorb, P.
1959. The Pleistocene avifauna of Arredondo, Florida. Bull. Fla. St. Mus., 4(9): 269-291.

1964. Catalogue of fossil birds Part 2 (Anseriformes through Galliformes). Bull. Fla. St. Mus., 8(3): 335 pp

Churcher, C. S.
1970. The vertebrate faunas of Surprise, Mitchell and Island Bluffs, near Medicine Hat, Alberta. Geol. Surv. Can. Paper 70-1, Part A: 158-160.

Colbert, E. H.
1950. The Fossil Vertebrates. Pp. 126-148 *in* The Stratigraphy and archaelogy of Ventana Cave. (E. W. Haury, ed.) Univ. Arizona Press, Tucson, 599 pp.

Conkling, R. P.
1932. Conkling Cavern: The discoveries in the Bone Cave at Bishop's Gap, New Mexico. West Tex. Hist. and Sci. Soc., 44(4): 6-19.

Downs, T.
1958. Fossil vertebrates from Lago De Chapala Jalisco, Mexico. Cong. Geol. Inter. XX Sesion Cuidad de Mexico 1956 sobretiro de la Seccion Paleontologia, Taxonomy Evolution: 75-77.

Downs, T. and J. A. White
1968. A vertebrate faunal succession in superposed sediments from Late Pliocene to Middle Pleistocene in California. 23rd Internat. Geol. Cong., Prague, 10: 41-47.

Furlong, E. L.
1906. The exploration of Samwel Cave. Amer. J. Sci., ser. 4, 22: 235-247.

Guilday, J. E.
1967. Differential extinction during Late-Pleistocene and Recent times. Pp. 121-140 *in* Pleistocene Extinctions: the Search for a Cause. (P. S. Martin and H. E. Wright Jr. eds.) Yale Univ. Press, New Haven, 453 pp.

Hansen, R. M.
1978. Shasta ground sloth food habits, Rampart Cave, Arizona. Paleobiology, 4(3): 302-319.

Harington, C. R.
1978. Quaternary vertebrate faunas of Canada and Alaska and their suggested chronological sequence. Syllogeus, no. 15. Nat. Mus. Can.: 105 pp.

Harrington, M. R.
1933. Gypsum Cave, Nevada. Southwestern Mus. Pap., no. 8, 197 pp.

Hay, O. P.
1916. Description of two extinct mammals of the Order Xenarthra from the Pleistocene of Texas. Proc. U.S. Nat. Mus., 51: 107-123.

Hibbard, C. W. and W. W. Dalquest.
1966. Fossils from the Seymour Formation of Knox and Baylor Counties, Texas and their bearing on the late Kansas climate of that region. Contrib. Mus. Paleo. Univ. Mich., 21(1): 1-66.

Hibben, F. C.
1941. Evidences of early occupation in Sandia Cave, New Mexico and other sites in the Sandia-Manzano Region. Smithsonian Misc. Coll., 99(23): 1-44.

Laudermilk, J.D. and P.A. Munz.
1934. Plants in the dung of *Nothrotherium* from

Gypsum Cave, Nevada. Carnegie Inst. Wash. Publ., 453: 29-37.

1938. Plants in the dung of *Nothrotherium* from Rampart and Mauv Caves, Arizona. Carnegie Inst. Wash. Publ., 487:271-281.

Lindsay, E. H. and N. T. Tessman.
1974. Cenozoic vertebrate localities and faunas in Arizona. J. Ariz. Acad. Sci., 9(1): 3-24.

Logan, L. E.
1977. The paleoclimatic implications of the avian and mammalian faunas of Lower Sloth Cave, Guadalupe Mountains, Texas. Unpubl. Ms. thesis. Texas Tech. Lubbock, 72 pp.

Logan, L. E. and C. C. Black.
1979. The Quarternary vertebrate fauna of Upper Sloth Cave, Guadalupe Mountains National Park, Texas. Pp. 141-158. *in* Biological Investigations Guadalupe Mountains National Park, (H. H. Genoways and R. J. Baker, eds.) National Park Service Proc. Trans., no. 4, 442 pp.

Long, A. and P. S. Martin.
1974. Death of American ground sloths. Science, 186(4164): 638-640.

Long, A., R. M. Hansen and P. S. Martin.
1974. Extinction of the Shasta ground sloth. Bull. Geol. Soc. Amer., 85: 1843-1848.

Lull, R. S.
1929. A remarkable ground sloth. Mem. Peabody Mus. Yale Univ., 2(2), 39 pp.

Lundin, R. F. and C. F. Royse.
1973. A Late Pleistocene vertebrate fauna from the Nichol's Site, Scottsdale, Arizona. J. Ariz. Acad. Sci., 8(1): 29-33.

Martin, P. S. and B. E. Harrell.
1957. The Pleistocene history of temperate biotas in Mexico and eastern United States. Ecology, 38(3): 468-480.

Martin, P. S., B. E. Sabels and D. Shutler.
1961. Rampart Cave coprolite and ecology of the Shasta ground sloth. Amer. J. Sci., 259(2): 102-127.

Martin, R. A.
1974. Fossil mammals from the Coleman 11A fauna, Sumter County. Pp. 35-99. *in* Pleistocene Mammals of Florida. (S. D. Webb, ed.) Univ. Fla. Press, Gainesville, 270 pp.

Mawby, J. E.
1967. Fossil vertebrates of the Tule Spring Site, Nevada. Nevada St. Mus. Anthro. Papers, 13: 102-128.

McNab, B.K.
1985. Energetics, population biology, and distribution of xenarthrans, living and extinct. (this volume).

Miller, W. E.
1971. Pleistocene vertebrates of the Los Angeles Basin and vicinity (exclusive of Rancho La Brea). Sci. Bull. Los Angeles Co. Mus. Nat. Hist., 10, 124 pp.

Mooser, O. and W. W. Dalquest.
1975. Pleistocene mammals from Aquascalientes, central Mexico. J. Mamm., 56(4): 781-820.

Mosimann, J. E. and P. S. Martin.
1975. Simulating overkill by Palaeoindians. Amer. Sci., 63(3): 304-313.

Mott, R. J. and A. MacS. Stalker.
1972. Palynology of the "Kansan" Carbonaceous Clay Unit near Medicine Hat, Alberta. Geol. Surv. Canada Paper, Part B 72-1: 117-121.

Neill, W. T.
1957. Historical Biogeography of Present-Day Florida. Bull. Fla. St. Mus., 2(7): 175-220.

Packard, E. L.
1952. Fossil edentates of Oregon. Oregon St. Univ. Monog. Stud. Geol., no. 8: 1-12.

Pirkle, E. C., W. H. Yoho and S. D. Webb.
1967. Sediments of the Bone Valley Phosphate District of Florida. Econ. Geol., 62: 237-261.

Savage, D. E.
1951. Late Cenozoic vertebrates of the San Francisco Bay Region. Univ. Calif. Publ. Bull. Dept. Geol. Sci., 28: 215-314.

Sinclair, W. J.
1905. New mammalia from the Quaternary caves of California. Univ. Calif. Publ. Bull. Dept. Geol., 4: 145-161.

Stalker, A. MacS. and C.S. Churcher.
1971. Glacial stratigraphy of the southwestern Canadian prairies; the Laurentide Record. 24th Internl. Geol. Cong., sec. 12: 110-119.

Stock, C.
1913. *Nothrotherium* and *Megalonyx* from the Pleistocene of southern California. Univ. Calif. Publ. Bull. Dept. Geol., 7(17): 344-358.

1918. The Pleistocene fauna of Hawver Cave. Univ. Calif. Publ. Bull. Dept. Geol., 10: 461-515.

1925. Cenozoic gravigrade edentates of western North America. Carnegie Inst. Wash. Publ., 331, 206 pp.

1931. Problems of antiquity presented in Gypsum Cave, Nevada. Scient. Monthly, 32: 22-32.

1943. The cave of San Josecito, Mexico, new discoveries of the vertebrate life of the Ice Age. Calif. Inst. Tech., Balch Grad. School Geol. Sci. Contrib., No. 361, 5 pp.

Watts, W. A.
1969. Pollen diagram from Mud Lake, Marion County, north-central Florida. Bull. Geol. Soc. Amer., 80: 631-642.

Webb, S. D.
1974. Chronology of Florida Pleistocene Mammals. Pp. 5-31, *in* Pleistocene Mammals of Florida. (S. D. Webb, ed.) Univ. Fla. Press, Gainesville, 270 pp.

1977. A history of savanna vertebrates in the New World. Part 1: North America, Ann. Rev. Ecol. Syst., 8: 355-380.

1978. A history of savanna vertebrates in the New World. Part 2: South America and the Great Interchange. Ann. Rev. Ecol. Syst., 9: 393-426.

Weigel, R. D.
1962. Fossil vertebrates of Vero, Florida. Fla. Geol. Surv. Spec. Publ., 10: 59 pp.

Wilson, R. W.
1942. Preliminary study of the fauna of Rampart Cave, Arizona. Carnegie Inst. Wash. Publ., 530: 169-185.

THE INTERRELATIONSHIPS OF TREE SLOTHS AND GROUND SLOTHS

S. David Webb

Florida State Museum, University of Florida, Gainesville, Florida 32611.

Abstract
A review of six major characters supports the Patterson-Pascual hypothesis that *Bradypus* is allied with megatherioid sloths whereas *Choloepus* has close affinities within the megalonychidae. Each living genus thus has separate roots at least as old as the early Miocene (Santacruzean) radiation of sloths in South America. It is suggested that megalonychoid and megatherioid sloths of that age were facultatively arboreal. If so, the term "ground sloth" should be restricted to mylodontoids, which had always been terrestrial, and to the few very large late Cenozoic genera that were independently derived from primitively arboreal groups.

Abstracto
Una revisión de seis características principales, apoya la hipótesis de Patterson-Pascual de *Bradypus* tiene nexos con los perezosos megaterioides, mientras que *Choloepus* muestra afinidades muy cercanas con la familia Megalonychidae. Por lo tanto, cada uno de los géneros vivientes de perezosos tiene raíces separadas, que son tan antiguas como la radiación de los perezosos en Sur América, que ocurrió en el período Santacruzano, de la primera parte del Mioceno. Se sugiere que los perezosos megaloníchidos y megaterioides de esa época eran facultativamente arbóreos. Si ello es así, el término "perezoso terrestre" debe restringirse solamente a los milodontoides, que siempre han sido terrestres y a los pocos y muy grandes géneros de la última parte del Cenozoico que evolucionaron independientemente de grupos que fueron primitivamente arbóreos.

Resumo
Uma revisão dos seis maiores caracteres apóia a hipótese de Patterson-Pascual de que *Bradypus* tem parentesco com as preguiças megaterióides, enquanto que *Choloepus* tem afinidades próximas com os megaloniquídeos. Assim, cada gênero atual tem raízes separadas, pelo menos tão antigas quanto a irradiação de preguiças no início do Mioceno (Santacruzense), na América do Sul. Sugere-se preguiças megaloniquídeas e megaterióides, daquela época, eram arbóreas facultativas. Se assim for, o termo "preguiça terrícola" deveria ser restrito aos milodontóides, que foram sempre terrícolas, e para os poucos gêneros grandes do final do Cenozóico, que foram derivados independentemente dos grupos arbóreos, primitivamente.

The distinction between tree and ground sloths is widely accepted as fundamental. The first fossil sloth to be discovered was the great Pleistocene *Megatherium* described by Cuvier (1799-1800). Discovery of *Mylodon* and *Megalonyx*, also large Pleistocene genera, soon followed. For each of these great genera the term "ground sloth" was clearly appropriate, and the distinctions from the living "tree sloths" *Bradypus* and *Choloepus* were obvious. Sir Richard Owen (1842) coined the term Gravigrada to formalize their distinction from Tardigrada (Brisson, 1756).

The distinction between the two presumed phylogenetic groups, often called tree sloths and ground sloths, has been maintained in all subsequent mammalian classifications. Interestingly, but without explanation, Simpson's (1945) classification further wedged these two sloth groups apart by placing the anteaters between them. No fossil genera have been assigned to the tree sloths, thus all tree sloths are recent and all ground sloths are fossil. This tends further to "clarify" the distinction.

In recent years doubts have arisen regarding the fundamental nature of this sloth dichotomy. The basis for these doubts has been the improved record of fossil sloths. The first expression of a new viewpoint came from the late Bryan Patterson of Harvard's Museum of Comparative Zoology and Rosendo Pascual of the La Plata Museum in Argentina in their broad review of extinct South American mammals (Patterson and Pascual, 1963). In a later contribution (Patterson and Pascual, 1972, p. 270), they presented their viewpoint as follows:

"*Choloepus* shows certain resemblances to the mega-

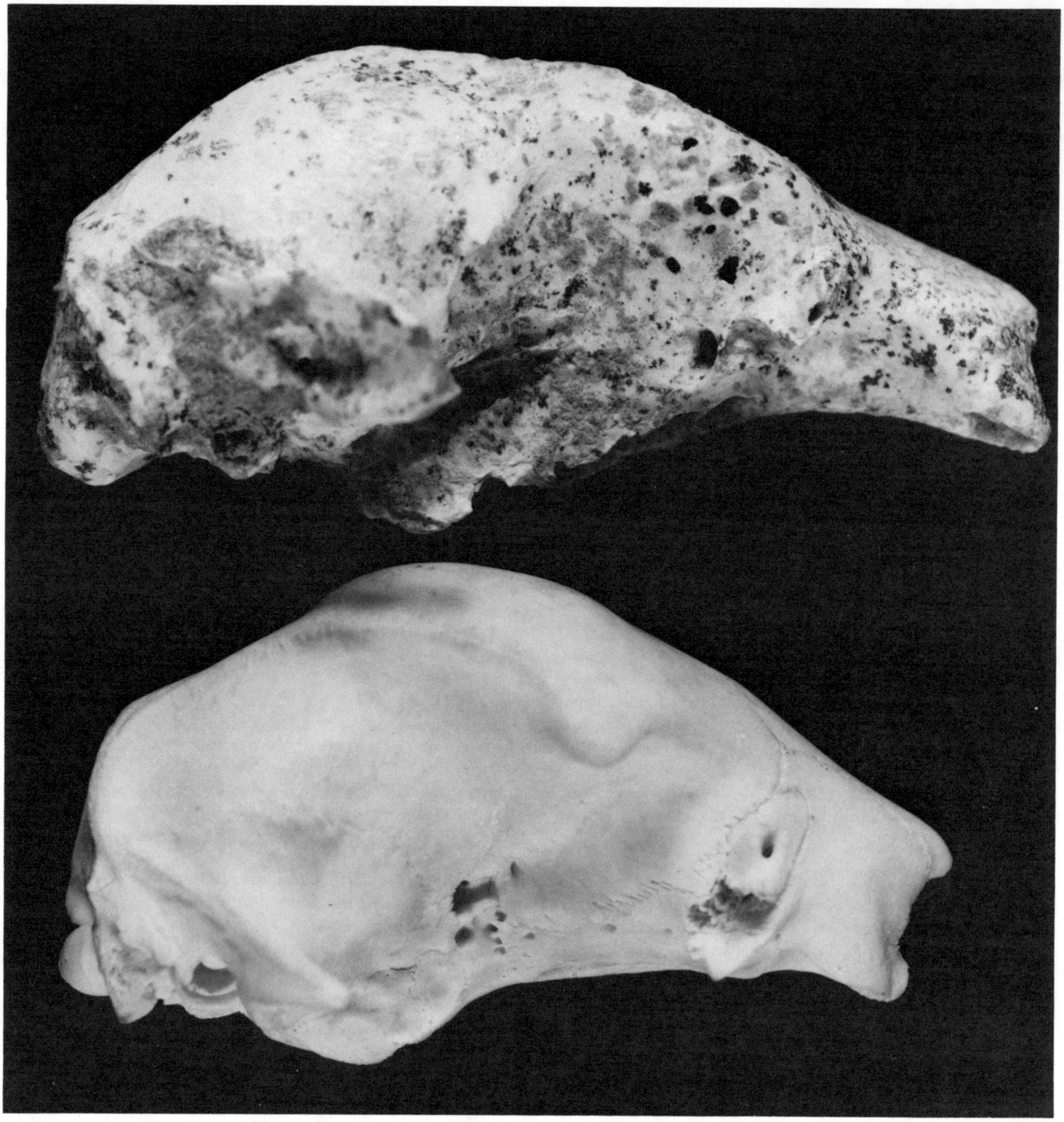

Figure. 1. Right lateral views of the crania of an extinct (*Upper*) and an extant (*Lower*) megalonychoid sloth, to emphasize the many resemblances. ***Upper*** *: Synocnus comes*, UF 25702, late Pleistocene, Dadier Arawak Cave, Miragaone, Haiti, collected by Charles Woods. ***Lower***: *Choloepus hoffmanni*, AMNH 70440, Recent, South America (*via* New York Zoological Society).

lonychids whereas *Bradypus* is somewhat closer to the nothrotheres. Whether or not this is indicative of separate ancestry and subsequent parallelism is uncertain . . . ."

The purpose of this paper is to review the hypothesis of Patterson and Pascual.

ARE ALL FOSSIL SLOTHS GROUND SLOTHS?

Before discussing the phylogenetic question raised here, it is necessary to insist on a clear distinction between phylogenetic and functional categories. As it happens, the living genera of sloths, *Bradypus* and *Choloepus*, are well adapted to an almost wholly arboreal existence. On the other hand, the classical studies of such great extinct sloths as *Megatherium* and *Mylodon* showed them surely to be terrestrial. It has been a common practice, therefore, to refer to all subsequently discovered fossil sloths as ground sloths, although that is not a necessary or logical extension.

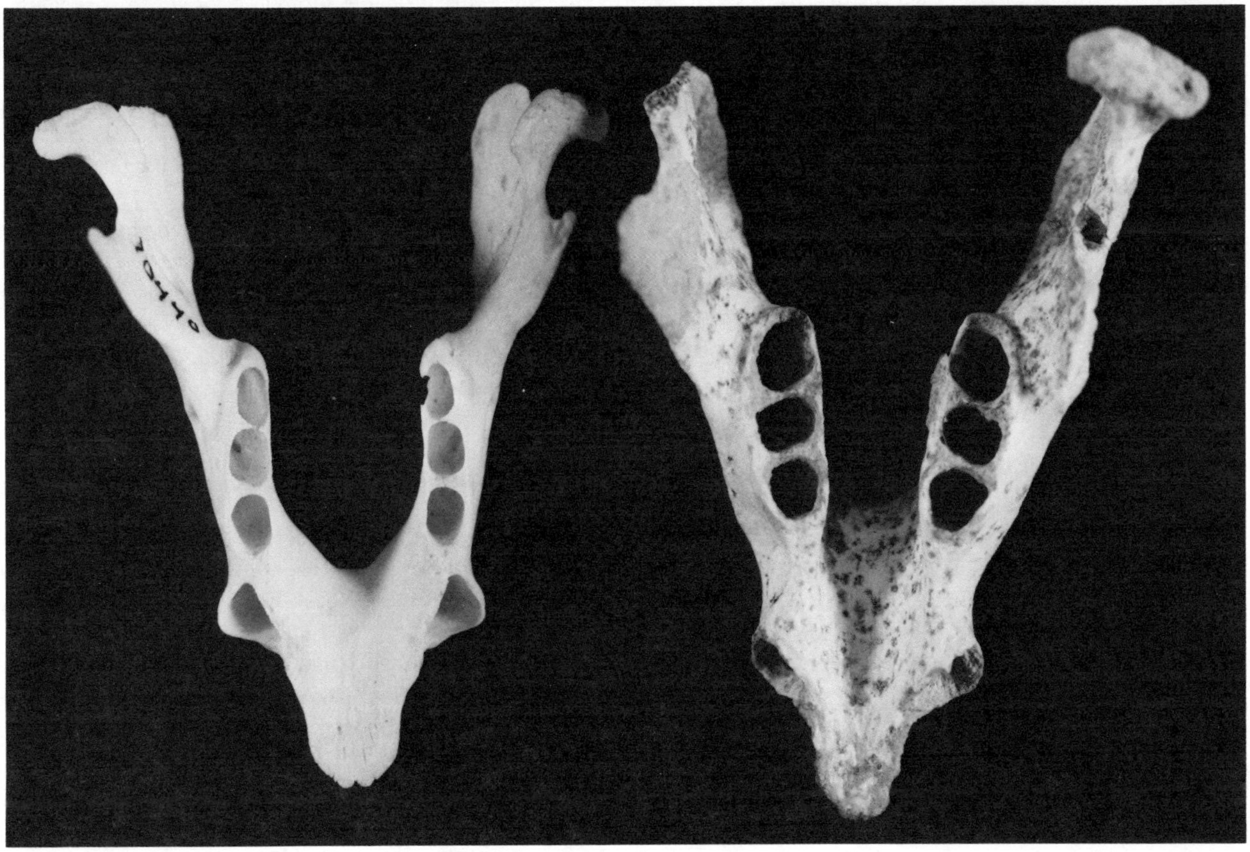

Figure 2. Occlusal views of the mandibles of an extinct (*Left*) and an extant (*Right*) megalonychoid sloth, to emphasize the many resemblances. Specimen data the same as in Figure 1.

It is probably impossible to know whether the Pleistocene sloth *Synocnus* really lived arboreally most or even some of the time. However, in size and many anatomical features, this fossil form is closely comparable to recent *Choloepus* (Figs. 1 and 2). Anthony (1916) suggested that the very similar Puerto Rican genus *Acratocnus* was indeed arboreal. *Neocnus* (= *Microcnus*), an even smaller Antillean fossil sloth, also was regarded as arboreal by Matthew and Paulo Couto (1955).

These rare discussions of possible arboreality of some fossil sloths have had no serious impact on the general assumption that fossil sloths are all ground sloths. In his discussion of arboreal folivores within the mammalia, Eisenberg (1978) recognized these small Antillean sloth genera, along with the larger genera, as "giant ground sloths." It is unfortunate that the question of possible arboreal habits has received so little consideration.

The ultimate ancestors of sloths were probably armadillos, in some broad sense of that word. If so, sloths must have been derived ultimately from terrestrial ancestors. It remains to be determined when some, or many, sloths became arboreal, and in which branches of their family tree.

The mylodontoids left a better record in the late Eocene through Oligocene deposits of South America than did other sloth groups. They exhibit several features which, taken together, suggest that they were particularly well adapted to a terrestrial existence, and indeed may never have abandoned it:

1. No mylodontoids were particularly small; most were the size of a black bear (*Ursus*) or larger.
2. They alone among sloths maintained mosaics of boney plates within their skin, presumably an inheritance from their cingulate ancestors.
3. Their forelimbs were short in proportion to their hind limbs.
4. Their fore- and hind claws had subcircular cross-sections (not compressed) and were nearly straight (not curved).
5. Their cheek teeth became lozenge-shaped for unusually coarse fodder.

The megatherioid and megalonychoid sloths contrast in each of these features with the mylodontoids. Their forelimbs usually equaled or exceeded the hind limbs in length, and their claws were curved and markedly compressed in the transverse dimension. Their cheek teeth retained the primitive peglike arrangement, except for minor alterations, and they evidently had lost the dermal ossifications early in their history. The "heyday" of the nonmylodontoid sloths is the early Miocene (Santacruzean), when some nine genera and several dozen species are recognized from Patagonia (Scott, 1904). In the same deposits mylodontids are extremely rare and have limited diversity, suggesting that they had a different habitus. Most of the megalonychoid and megatherioid sloths fall near the size of *Choloepus*, one of the largest, *Planops magnus*, having a skull length of about 220 mm. These data support Eisenberg's (1978, p. 146) comment that "the South American giant ground sloths (Megatheriidae) clearly derived from an arboreal ancestor. . . ."

If this view is correct, then use of the term ground sloth must be restricted. The late Cenozoic megalonychoids and megatherioids tend to follow Cope's Rule, with at least five genera attaining remarkable size by Pleistocene time. The two most notable are the mammoth-sized megatheriids, *Megatherium* and *Eremotherium*, the latter ranging from the northern Andes into the southeastern United States. Three different megalonychid genera also attained sizes in the range of large bears, namely *Megalocnus* in the Greater Antilles, *Megalonyx* in North America, and a new Central American genus (Webb and Perrigo, 1985). But these are exceptional late derivatives of much smaller sloths from the mid Tertiary. (Similar transitions between arboreality and terrestriality are found in many mammalian families, for example, the Hyracidae, Ursidae, Procyonidae, Sciuridae, and Myrmecophagidae.) The large Pleistocene megatheriids and megalonychids must surely have become ground sloths, but their proximal phylogenetic ancestors were probably tree sloths. Thus, the term ground sloth must be selectively applied and has no broad phylogenetic significance.

DIPHYLETIC RELATIONSHIPS OF *CHOLOEPUS* AND *BRADYPUS*

The Patterson-Pascual hypothesis regarding the diphyletic origin of *Bradypus* and *Choloepus* seems intuitively acceptable to a degree, simply because the two living genera seem remarkably different in many features of their behavior, physiology, and morphology (e.g., Montgomery and Sunquist, 1978). De Jong et al. (1985) feel that the six substitutions of amino-acid sequences in eye-lens proteins of the two sloth genera (α A chain) is consistent with a very early evolutionary separation of the sloth genera; and the serum albumin differences reported by Sarich (1985) indicate that *Choloepus* and *Bradypus* diverged about 35 million years ago, or approximately in the late Oligocene. The Patterson-Pascual hypothesis goes farther than this, however, and states that *Bradypus* has special affinities with the Megatheriidae and that *Choloepus* has its ancestry within the Megalonychidae. Hoffstetter (1969) accepted this probable diphyletic origin, but suggested the erection of distinct families, Bradypodidae and Choloepodidae. He found the proposed paleontological relationships obscure or inconsistent and, to be sure, they were not elaborated.

The following paragraphs cite six osteological features of sloths that support the Patterson-Pascual hypothesis. They are selected because they seem to fulfill the following conditions:

1. They are major characters, the features of which are widely recognizable in fossil and Recent sloths.
2. They differ between *Choloepus* and *Bradypus*.
3. The character state found in *Bradypus* is the same as or closely similar to that in most megatherioids.
4. The character state found in *Choloepus* is the same as or closely similar to that in most megalonychoids.

For each of these characters the primitive and the derived character states are considered. The character states thus provide a basis for inferring phylogenetic relationships.

Several interesting characters are not discussed here, because their distribution was not consistent. For example, *Choloepus* has a large entepicondylar foramen, whereas *Bradypus* lacks it (Lucae, 1884). Some megatherioids also lack the foramen, but others, such as *Nothrotherium*, retain it. In a more thorough study of sloth phylogeny this character ought to be followed, but for this brief survey it is dismissed.

First Tooth.--The most notable feature separating *Choloepus* from *Bradypus* is the nature of the anteriormost tooth in each jaw half. In *Choloepus* the triangular caniniform teeth are the largest teeth of the dentition and are set well ahead of the other (molariform) teeth. Each lower caniniform occludes against the posterior face of an upper caniniform along a nearly vertical face, an arrangement that maintains sharp tips on both upper and lower "fangs." (We mammal paleontologists are notoriously involved with teeth, but this feature in *Choloepus* reaches well beyond the ordinary level of interest, and seems to epitomize differences in their activity levels and agressiveness from *Bradypus*. For example, Brian McNab, Gene Montgomery, and I share similarly scarred thumbs, the result (in wholly separate encounters) of being too slow to escape the bite of an angry *Choloepus hoffmannii*.)

In *Bradypus*, the anteriormost teeth are not caniniform and are not widely separated from the other teeth. The anteriormost upper tooth is a vestigial peg with no occlusal function. The first lower tooth is transversely broad and chisel-shaped. It maintains a strong, oblique, wear surface on its posterior face from shearing against the anterior face of the second upper tooth. When angered, *Choloepus*, but not *Bradypus*, pull objects to their mouth and bite them repeatedly (Montgomery and Sunquist, 1978).

The caniniform self-sharpening front teeth of *Choloepus* closely resemble those found in most Megalonychidae. The principal variations from this same arrangement are the extreme specializations toward rodentlike incisiform teeth in *Megalocnus* and enlarged grinding batteries in *Megalonyx*.

The anteriormost teeth in *Bradypus* share with fossil megatherioids their reduction and their limited separation from the other teeth. No known megatherioid, however, has an identical arrangement, so far as I know. In *Megatherium*, the anteriormost teeth are reduced and are not separated from the others by any diastema, but both the upper and the lower teeth are functional and bilophodont. In still other megatherioids, for example *Notheriops* and *Essonodontherium*, the anteriormost teeth are lost.

The anteriormost teeth are almost surely homologous in all sloths with dental formulae of 5/4. A more difficult question is whether the megatherioid pattern or the megalonychoid pattern represents the derived condition. Probably the more reduced condition is derived and the caniniform arrangement is primitive. The principal basis for this view is the presence of similar caniniform occlusion in most mylodontoid sloths. Also most Santacruzean sloths, even such probable megatherioids as *Planops*, have small caniniform teeth set somewhat in advance of the other teeth.

Jugal Process.--In several respects the sloths were very poorly endowed for a phyllophagous existence by their insectivorous forerunners. One of the most striking inadequacies is the incomplete zygomatic arch and the weakly anchored jugal bone, features which surely limited the area of origin and the force of contraction that the masseter musculature could develop. As each group of sloths elaborated the jugal bone, in order to augment its masticatory ability, the detailed structures diverged.

Sicher (1944) pointed out the considerable differences between the jugal bone of *Choloepus* and that of *Bradypus*. In each genus the bone forms a laterally flattened plate, attached loosely to the maxillary and lacrimal bones. In *Bradypus* the jugal projects a long descending process posteroventrally and a long ascending process posterodorsally. In *Choloepus*, the ascending and descending processes are both shorter and more nearly horizontal.

Similar distinctions appear between most megalonychoids and most megatherioids. In *Megatherium* and *Eremotherium* the jugal is markedly expanded in the vertical plane, with extremely long ascending and descending processes. (In these great Pleistocene genera the jugal contacted the squamosal, thus redeveloping the zygomatic arch.) In megalonychids, the processes are not as well developed, nor do they make such obtuse angles with the horizontal axis of the skull. Even so, *Megalonyx* does finally attain a complete zygomatic arch. The more expanded jugal in *Bradypus* and the megatherioids is clearly the more derived arrangement.

Jaw Articulation.--Sicher (1944) observed an important difference between the mandibular condyles of *Choloepus* and *Bradypus*. In *Choloepus* the mandibular articulating surface is nearly flat, or as Sicher (1944) describes it, "convex in the medial one-third, concave in the lateral two-thirds," whereas in *Bradypus* it is "regularly convex." Moreover, the condyle is much wider (transversely) than long in *Choloepus*, but somewhat longer (anteroposteriorly) than wide in *Bradypus*.

Different modes of jaw articulation are also evident in the glenoid fossae of *Choloepus* and *Bradypus*. In *Choloepus* the articular surface of the squasmosal is short (anteroposteriorly) and wide, whereas in *Bradypus* it is somewhat more elongate but narrow. Sicher (1944) noted in his dissections, besides the absence of an articular disc in both genera, that in *Choloepus* the condyle rests "in contact with the articulating fossa excepting a narrow anterior strip," whereas in *Bradypus* the condyle is "in contact with the anterior part of the articulating fossa only."

A similar set of differences occur in fossil sloths. In megalonychids, as in *Choloepus*, the mandibular condyle is greatly widened with an oblique medial articular facet. The corresponding glenoid fossa is more elongate than in *Choloepus* but with a similar ridge along the medial side of the squamosal. In megatherioids, as in *Bradypus*, the mandibular condyle is hemispherical and sits in a shallow subcircular pocket abruptly truncated posteriorly by the periotic bone.

These articular distinctions are related to different modes of mastication in *Choloepus* and *Bradypus* (Sicher, 1944). *Choloepus* requires a greater repertoire of jaw motions to operate its large caniniform teeth as well as its molariform teeth. Also a very wide gap (up to 33°) is needed in *Choloepus*. Whereas in *Bradypus* a narrow gap is quite sufficient (Virginia Naples, pers. comm.). *Choloepus* uses both retraction and protraction as well as a medial motion during its principal chewing strokes, whereas *Bradypus* uses only protraction and medial motion in the powerful occluding phase. This difference, as Sicher (1944) emphasizes, is reflected in the attrition facets of the cheek teeth.

In fossil megalonichids, as noted by Hirschfeld and Webb (1968), the principal wear facets are, as in *Bradypus*, on the posterior faces of the lower cheek teeth. Nonetheless, a more balanced repertoire, as in *Choloepus*, is evidenced by the presence of strong wear facets on the anterior faces of the lower cheek teeth as well. This is presumably correlated with heavy use of the large caniniforms in megalonychids. The sharp vertical wear faces on the cheek teeth in advanced megatherioids reflects their very powerful oblique mode of mastication. The nothrotheres more nearly resemble *Bradypus* in having shallower wear faces, and the megatheres have simple transverse wear surfaces representing a simple anterior power stroke.

The more versatile masticatory system of *Choloepus* and the megalonychids is presumably the more primitive condition. *Bradypus* and the megatheriods apparently simplified their chewing capacities as they reduced their caniniform teeth. This suggested correlation and historical sequence is supported by the condition in mylodontoid sloths, most of which retain large caniniforms and all of which retain wide mandibular condyles and strong medial ridges along the glenoid fossae. This is, of course, a general statement about the primitive condition and does not address the important secondary specializations in all groups.

Periotic Bone.--The middle ear region of *Bradypus* is enclosed by a nearly hemispherical bulla. In *Choloepus*, that region is largely exposed, only the median side being skirted by the U-shaped ectotympanic bone (Van der Klaauw, 1931). Megatherioids have ossified bullae generally resembling that of *Bradypus*, although some are more flattened and more rugose (Guth, 1962). In possible Santacruzean megatherioids, this condition appears somewhat variable and is often uncertain; nevertheless, some basicrania such as that of *Analcimorphus* also show a relatively complete bulla (Scott, 1904, plate 47).

In megalonychoids no bulla is known. In *Megalonyx* the ectotympanic ring is thicker than in *Choloepus*, but it remains transversely narrow and nonbullate. Presumably the nonbullate condition is the primitive one, as in other primitive mammal groups. It is not surprising, therefore, to discover that mylodontoids share this condition with megalonychoids, a resemblance noted by Guth (1962) and Hoffstetter (1969).

Cranial Flexion.--*Choloepus* has the facial portion of the cranium strongly flexed downward with respect to the basicranial portion. In *Bradypus* these cranial portions lie essentially on the same horizon. This same contrast extends to the megalonychoids with flexed crania and to the megatherioids with flat crania. The contrast becomes extreme when the large Pleistocene forms are compared: *Megalonyx* with its deep short skull and down-dropped palate differs dramatically from *Megatherium* with its long, tubelike cranium. These features carry over to the relationships between the alvolar level and the elevation of the jaw hinge, and to the tendency to develop sagittal and parasagittal crests.

The flat-skulled condition of *Bradypus* and the megatherioids is the more primitive state for this character. Cranial flexion in *Choloepus* and the megalonychoids represents the derived state.

Astragalus.--The proximal surface of the astragalus in *Choloepus* has a nearly flat surface and a greater anteroposterior length than transverse width. In *Bradypus* the astragalar surface has more convex and more concave portions and is about as long as it is wide. The convexity lies toward the medial side and occupies a pocket in the tibia; the concavity is toward the lateral side and is occupied by a process of the fibula. These features arise late in the ontogenetic development of *Bradypus* (Ursing, 1932). In *Choloepus*, the tibia slides broadly over the whole dorsal surface of the astragalus, with less rotation and lateral eversion than in the megatherioids.

The simple astragalus of *Choloepus* has clear counterparts in megalonychoids generally. This is not surprising since that is surely the primitive condition. Similar astragali also appear in the vermilinguas. The peculiar astragalus of *Bradypus* is not exactly recognizable in megatherioids, but the knoblike medial convexity and the shorter anteroposterior dimension are characteristic. Large late Cenozoic megatheriids have greatly exaggerated this feature as their pedolateral feet rotate around the ball-and-socket joint of the astragalus and tibia. *Nothrotheriops* and the Santacruzean genus *Planops* are more nearly comparable to *Bradypus* with gently raised medial convexities of the astragalus (Stock, 1925; Hoffstetter, 1961). Clearly, the derived condition with respect to this character is the contorted ball-like joint in *Bradypus* and the megatherioids, while the nearly flat astragalus of *Choloepus* and of megalonychoids is primitive.

DISCUSSION

Each of the characters presented here tends to support the Patterson-Pascual hypothesis. It should be noted that four of the craniodental characters may be functionally correlated; the enlarged caniniform teeth of *Choloepus* probably required the stronger but more mobile jaw articulation, the more horizontally directed jugal plate, and the strongly flexed cranium. The possibility that such a correlation may exist does not diminish the importance of these characters; indeed, together they appear to represent a very fundamental character complex. The remaining two characters, namely the ectotympanic bone and the proximal surface of the astragalus, tend to support the evidence from the masticatory characters.

It is noteworthy that for most of these characters *Bradypus* and the megatherioids seem to have achieved the derived character states, in comparison with *Choloepus* and the megalonychoids. The exception among characters considered here is cranial flexion. It may be further observed that the enlarged caniniforms in *Choloepus*, while primitive in the broad context of phyllophagous edentates, are specialized beyond the primitive state found in many Santacruzean genera.

More specific analyses of the phylogenetic relationship of *Choloepus* among megalonychoids and of *Bradypus* among megatherioids are needed. The Antillean genera *Synocnus* and *Acratocnus* show remarkable cranial resemblances to *Choloepus* (Figs. 1 and 2). They share with the living genus enlarged caniniforms and short, flexed crania, features that may be derived with respect to known Santacruzean megalonychoids. Webb and Perrigo (1985) have further explored possible megalonychoid relationships and Engelmann (1985) has provided a very useful indication of cladistic relationships of all Phyllophaga. Even so, more detailed studies of the possible relationships between fossil and recent sloth genera are required.

In the rich Santacruzean sloth fauna, megalonychoids and megatherioids were rather subtly distinguishable (Scott, 1904; Hoffstetter, 1969). Possibly these two great groups of sloths diverged in the late or middle Oligocene. According to the Patterson-Pascual hypothesis, as supported by this review, the separation between the *Choloepus* and the *Bradypus* lineages is thus quite ancient. This early separation of the lineages is supported as well by studies involving protein taxonomy (Sarich, 1985; de Jong et al., 1985). This may seem remarkable in view of the many highly adapted arboreal features shared by both modern tree sloth genera. This objection disappears, however, if one accepts the view that Santacruzean and older nonmylodontoid sloths were already at least facultatively arboreal. In that case, most of the arboreal characteristics shared by modern sloths may be primitive features of megatherioid-megalonychoid sloths. In this perspective the giant ground sloths of the late Cenozoic appear as highly specialized end members of the phyllophagous xenarthrans. The recent tree sloths are more typical of (albeit more specialized than) Miocene stages of sloth evolution. We are fortunate to have living relicts of two different Miocene sloth groups.

ACKNOWLEDGMENTS

I am deeply indebted to Donald E. Savage and the late Bryan Patterson for sharing their insights into the mysteries of sloth evolution. I have also received aid and encouragement from my colleages Brian McNab, Charles Woods, and Bruce MacFadden. Gene Montgomery has provided improvements to this paper both as editor and as phyllophagophile. Specimens figured were generously loaned by the University of Florida (UF) and the American Museum of Natural History (AMNH). This work was partly supported by NSF grant DEB 7810716. This is contribution number 174 from the Vertebrate Paleontology Laboratory of the Florida State Museum.

LITERATURE CITED

Anthony, H.E.
1916. Preliminary report of fossil mammals from Porto Rico. Ann. N. Y. Acad. Sci., 27: 193-203.

Brisson, M.J.
1756. La régne animal divisé en IX classes. J. B. Bauche, Paris, 382 pp.

Cuvier, G.
1799-1800. Lecons d'Anatomie comparaée. C. Dumeril, Paris, 521 pp.

de Jong, W.W., A. Zweers, K.A. Joysey, J.T. Gleaves and D. Boulter.
1985. Protein sequence analysis applied to xenarthran and pholidote phylogeny. (this volume).

Eisenberg J.F.
1978. The evolution of arboreal herbivores in the class mammalia. Pp. 135-152 *In*: The Ecology of Arboreal Folivores, (G. G. Montgomery, ed.). Smithsonian Institution Press, Washington, D.C. 574 pp.

Engelman, G.F.
1985. The phylogeny of the Xenarthra. (this volume).

Guth, C.
1962. Apport systématique et phylogénétique de l´ etude de la région temporale chez les Edentés. Coll. Internat. C.N.R.S. no. 104. Evolution des Vertébrés.: 343-358.

Hirschfeld, S.E. and S.D. Webb.
1968. Plio-pleistocene megalonychid sloths of North America. Bull. Fla. Sta. Mus., 12: 213-296.

Hoffstetter, R.
1961. Description d´ une Squelette de *Planops* (Gravigrade du Miocene de Patagonie). Mammalia, 25: 57-96.

1969. Remarques sur la Phylogénie et la classification des Edentés Xénarthres (Mammiferes) actuels et fossiles. Bull. Mus. Nat. d'Hist. Nat., 41: 92-103.

Lucae, J.C.G.
1884. Die Statik und Mechanic der Quadrupeden an dem Skelet und den Muskeln eines *Lemur* und eines *Choloepus*. Abh. Senckenberg Naturf. Gesels ch. Frankfort, 13: 1-92.

Matthew, W.D. and C. de Paula Couto.
1959. The Cuban Edentates. Bull. Amer. Mus. Nat. Hist., 117: 1-56.

Montgomery, G.G. and M.E. Sunquist.
1978. Habitat selection and use by two-toed and three-toed sloths. Pp. 329-359 *In*: The Ecology of Arboreal Folivores (G. G. Montgomery, ed.) Smithsonian Institution Press, Washington, D.C., 574 pp.

Owen, R.
1842. Description of the skeleton of an extinct gigantic sloth, *Mylodon robustus*. R. and J. Taylor, London, 176 pp.

Patterson, B. and R. Pascual.
1963. The extinct land mammals of South America. Proc. XVI Internat. Zool. Congr. Program Vol.: 138-148.

1972. The fossil mammal fauna of South America pp. 247 - 309. *In* Evolution, Mammals and Southern Continents. (A. Keast, F.C. Erk, and B. Glass, eds.) State Univ. of New York Press, Albany.

Sarich, V.M.
1985. Xenarthran systematics: albumin immunological evidence. (this volume).

Scott, W. B.
1904. Edentata. Rept. Princeton Univ. Exped. Patagonia. Vol. V, Paleontology.: 1-364.

Sicher, H.
1944. Masticatory Apparatus of the Sloths. Field Mus. Nat. Hist., Zool. Ser., 19: 161-168.

Simpson, G.G.
1945. The principles of classification and a classification of mammals. Bull. Amer. Mus. Nat. Hist., 85: 1-350.

Stock, C.
1925. Cenozoic gravigrade edentates of western North America with special reference to the Pleistocene Megalonychinae and Mylodontinae of Rancho La Brea. Carnegie Inst. Washington, Publ., 331: 1-206.

Ursing, B.
1932. Uber Entwicklung und Bau des Hand- und Fuss-Skeletts bei *Bradypus tridactylus*. H. Ohlssons, Lund., pp. 1-107.

Van der Klaauw, C.J.
1931. The auditory bulla in some fossil mammals. Bull. Amer. Mus. Nat. Hist., 62: 1-352.

Webb, S.D. and S. Perrigo.
1985. Two new Megalonychid sloths from El Salvador. (this volume).

NEW MEGALONYCHID SLOTHS FROM EL SALVADOR

S. David Webb[1] and Stephen Perrigo[2]

[1]Curator of Fossil Vertebrates, Florida State Museum, Gainesville, Florida 32611. [2]Curator of Paleontology, Museo de Historia Natural de El Salvador, Parque Saburo-Hirao, San Salvador, El Salvador, C. A., and Field Associate, Florida State Museum, Gainesville, Florida 32611.

Abstract

Two sloth mandibles from central El Salvador represent the first substantial sample of megalonychidae in Central America. One jaw represents a new genus and species at least as large as *Megalonyx jeffersoni* with shorter symphysis, more robust cheek teeth, lower coronoid process, and a primitive, pointed caniniform tooth. The other jaw represents a new medium-sized species of *Megalonyx*, most closely comparable to *M. leptostomus*. Together these specimens emphasize that Central America was the probable route of interamerican megalonychid dispersal.

Abstracto

El informe de dos mandíbulas de perezosos del área central de El Salvador representa la primera muestra substancial de la familia Megalonychidae de América Central. Una de las mandíbulas representa un nuevo género y especie, por lo menos tan grande como *Megalonyx jeffersoni*, pero con la sínfisis más corta, los dientes de la mejilla más robustos, el proceso coronoide más bajo y un diente caniniforme puntiagudo y primitivo. La otra mandíbula representa una nueva especie de *Megalonyx* de mediano tamaño, que puede ser comparada más favorablemente con *M. leptostomus*. En conjunto, estos especímenes fortalecen la idea de que América Central fue la probable ruta de dispersión interamericana de los perezosos megaloníchidos.

Resumo

Duas mandíbulas de preiguças da região central de El Salvador representam a primeira amostra substancial de Megalonychidae na América Central. Uma mandíbula representa um novo gênero e espécie pelo menos tão grande quanto *Megalonyx jeffersoni* , com sínfise menor, molares mais robustos, processo coronóide menor e um dente caniniforme pontiagudo e primitivo. A outra mandíbula representa uma nova espécie, de tamanho médio, de *Megalonyx*, mais proximamente comparável a *M. leptostomus*. Juntos, esses espécimes enfatizam que a América Central foi a provável rota para a dispersão interamericana de Megalonychidae.

On 21 November 1978, while prospecting for fossils in the central part of El Salvador, Stephen Perrigo discovered two megalonychid sloth mandibles. Although the two rami were in physical contact they represent different genera, one of which is new. These specimens add significantly to our knowledge of sloth evolution and distribution.

AGE AND OCCURRENCE

Both sloth mandibles were found at the weathered surface of a fine, tuffaceous sandstone. This sandstone unit lies in the upper part of the sedimentary succession of the Barranca del Sisimico, Department of San Vicente, in central El Salvador (Fig. 1). The sandstone unit conformably overlies thinly bedded diatomaceous shales interbedded with thin dark beds of coarse, tuffaceous sands. Fossil vertebrates have been collected at several levels in this section.

The age of the sloth-bearing sandstone is early- to mid-Pleistocene. The sedimentary succession of the Barranca del Sisimico is considered to be part of the Plio-Pleistocene Cuscatlán formation (M. Schmidt-Thomé, 1975). The fossil site is situated in the central graben or central interior valley of El Salvador, a long subsidence zone associated with the active volcanic belt immediately to the south (Stirton and Gealey, 1949). This sequence is certainly younger than the vast ignimbrite flows of Miocene to Pliocene age which occur in the northeastern region of the country and extend into Honduras where Williams and McBirney (1969) obtained Potassium/Argon dates on ignimbrites ranging from 19 to 7 million years ago. In the Barranca del Sisimico section, other fossil vertebrates have been collected and will be reported upon elsewhere. Other fossil taxa found in the upper sandstone near the megalonychids described here include *Cuvieronius* (a short-jawed gomphotheriid proboscidean) and *Eremo-*

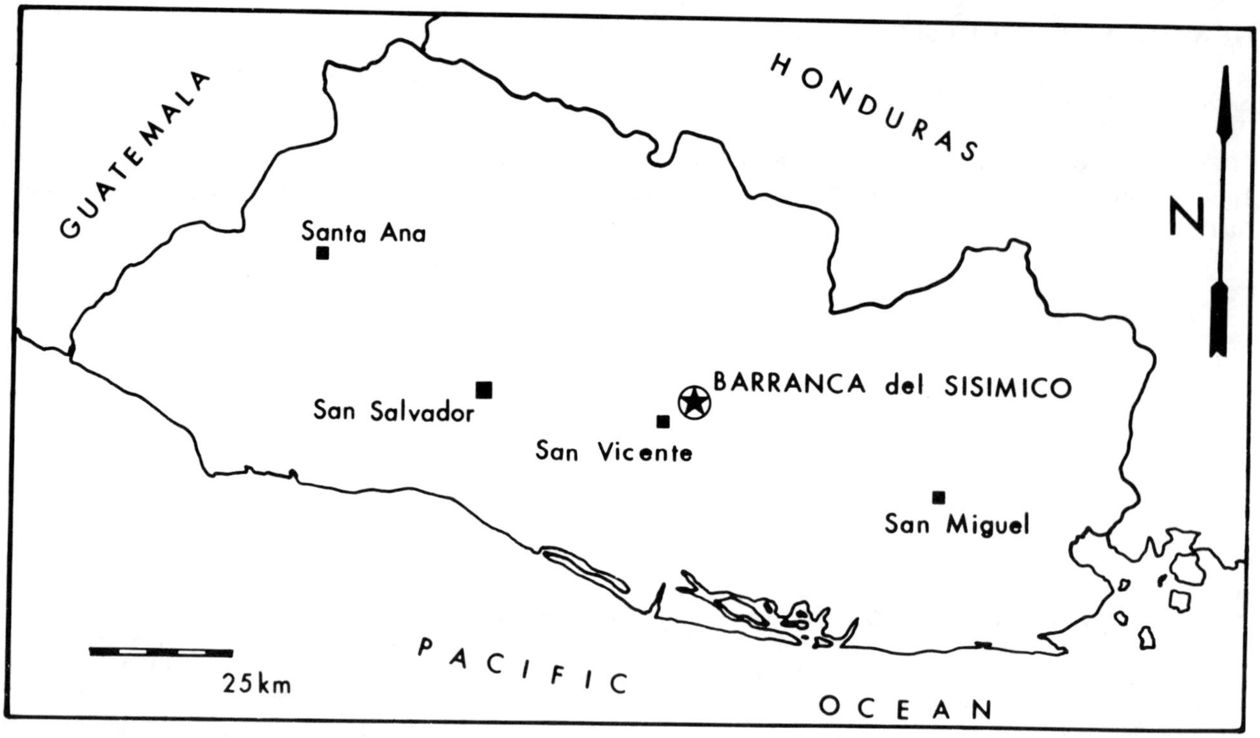

Figure 1. Map of El Salvador and adjacent countries to locate Barranca del Sisimico, site where the sloths were found. Specimens occur in the Cuscatlán Formation within the central graben of El Salvador.

therium (a giant megatheriid ground sloth). We tentatively regard the age of this unit as early to middle Pleistocene.

SYSTEMATICS

Meizonyx, new genus

Etymology.-- *Meizon*, Gr., "greater," comparative of *megas*; and *Onyx*, Gr. "claw"; to suggest comparison with *Megalonyx* of North America.

Type.-- ***Meizonyx salvadorensis***, the only species known.

Diagnosis.-- Size equals or exceeds that of *Megalonyx jeffersoni*. Caniniforms in line with molariforms, unlike Antillean sloths; triangular with vertical, self-sharpening occlusal surface, as in primitive megalonychids. Symphyseal region deep but shorter than in *Megalonyx*; symphyseal spout vestigial. First molariform largest tooth, quadrangular, relatively very wide, with anterolingual corner high and vastly expanded, producing concavities on anterior and lingual faces. Second molariform quadrangular. Third molariform unreduced, subquadrangular to ovate. Widest part of mandible at M2; ascending ramus covering M3 laterally. Articular condyle nearly as high as coronoid process, unlike *Megalonyx*.

Meizonyx salvadorensis, new species

Etymology.-- For the Republic of El Salvador in which this species was discovered.

Type.-- Specimen number 70-168 in the collection of the Museo de Historia Natural de El Salvador, essentially complete left mandibular ramus with caniniform and three cheek teeth (Fig. 2). Collected by Stephen Perrigo at Barranca del Sisimico, Department of San Vicente, El Salvador.

Diagnosis.-- As for genus.

Description of mandible.-- The short, deep massive proportions of the *Meizonyx* mandible are its most striking general features. The depth of the horizontal ramus exceeds that in all but one of the many known mandibles of *Megalonyx* (McDonald, 1977) and the transverse thickness of the ramus vastly exceeds that in even the largest known *Megalonyx*. The length of the symphyseal part of the mandible (which begins anterior to the first cheek tooth) is much shorter than in comparable sized *Megalonyx*. In comparison with *Megalonyx*, the rostrum of *Meizonyx* had a "bulldog" aspect.

The mandibular symphysis is short and deep. The symphyseal "tongue" was vestigial or possibly absent. This point is unsure because the upper 17 mm of the

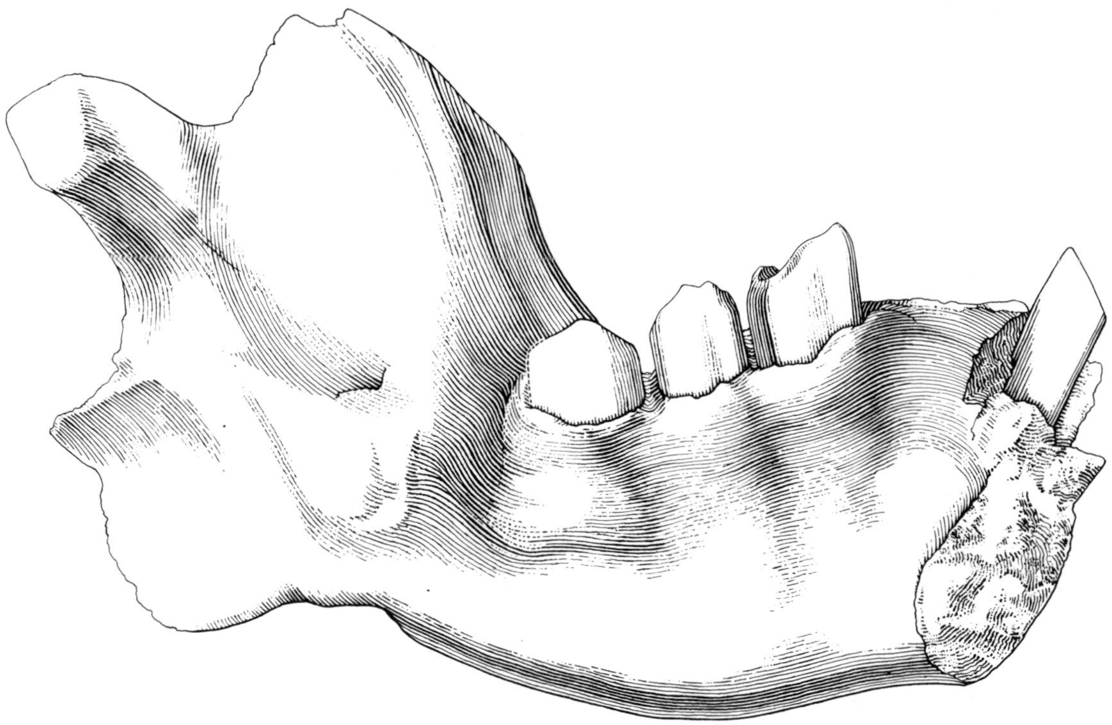

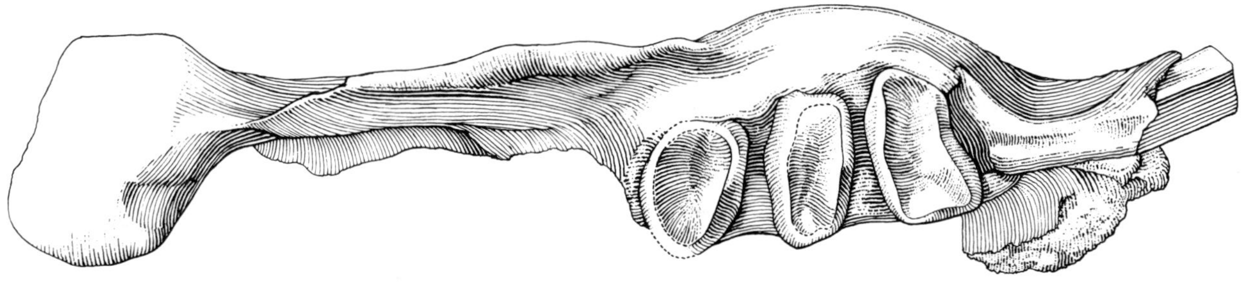

Figure 2. *Meizonyx salvadorensis* type mandible, lingual and occlusal views. Museo de Historia Natural de El Salvador number 70-168. Scale X 0.7.

symphyseal midline is broken. Nonetheless, the broadly rounded contours of the bone in the lower two-thirds of the symphyseal midline and of that anterior to the caninform provide clear limits to the lateral and ventral extent of the symphyseal tongue, and indicate that it was even more restricted than in *Megalonyx jeffersoni.* The anterior face of the symphyseal region is wholly convex in *Meizonyx*, whereas in *Megalonyx* it is concave below the symphyseal tongue. The lingual (posterior) face is flat whereas in *Megalonyx* it is deeply concave in the ventral portion. The sagittal thickness of the symphysis in its ventral portion is much thicker in *Meizonyx* than in *Megalonyx*.

In *Meizonyx* the diastema posterior to the caniniform tooth is about the same length as in large *Megalonyx* mandibles (Table 1). However, since the caniniform of *Megalonyx* is about twice as long anteroposteriorly as that of *Meizonyx*, the distance from the anterior end of the caniniform to the first molariform in *Meizonyx* is only about 60 percent to 75 percent of that in *Megalonyx* of comparable size. No mental foramen is evident in the region surrounding the caniniform tooth of *Meizonyx*, but it may be broken.

The central corpus of the horizontal ramus is remarkable for its great depth and thickness. The maximum width of the ramus is 55.8 mm, as measured across the second molariform tooth.

The ascending ramus is remarkably expansive, in keeping with the massive appearance of the whole mandible. The length of the ascending ramus, from the

Table 1. Measurements of *Meizonyx salvadorensis* mandible (in mm).

Caniniform length	19.0
Caniniform width	17.3
Molariform 1 length	24.7
Molariform 1 width	35.0
Molariform 2 length	22.6
Molariform 2 width	33.9
Molariform 3 length	26.5
Molariform 3 width	30.5
Diastema C1 to M1	42.2
Length M1 to M3 inclusive	87.2
Length C1 to M3 inclusive	147.6
Length of mandible (to condyle)	262.0
Maximum width of mandible (at M2)	55.8
Depth labial to caniniform	90.2
Depth labial to M1	96.5
Depth labial to M2	92.3
Depth labial to M3	88.4
Depth coronoid to angular	160.5

origin of the coronoid process back to the dorsal edge of the angular process, greatly exceeds the length of the horizontal ramus, taken from the coronoid forward. In *Megalonyx* the length of the posterior part of the jaw is equal to or less than that of the anterior part. The anterior origin of the coronoid process largely covers the last cheek tooth from lateral view, unlike *Megalonyx*. The anterior edge of that process is much thicker and turns outward much more broadly than in any known *Megalonyx* mandibles. Along the lateral edge of the coronoid process is a very strong crest for ligaments of insertion from the temporal muscles. The strong crests and the everted coronoid edge indicate that *Meizonyx* possessed a more powerful *M. temporalis* complex than did *Megalonyx*.

Similarly the great lateral expanse of the ascending ramus suggests a more powerful *M. masseter* complex in *Meizonyx* than in *Megalonyx*. That surface is more convex and more heavily crested than in *Megalonyx*, and the crest for *M. zygmomaticomandibularis*, extending from the base of the coronoid process to the anterior end of the angular process, is particularly strong. On the other hand, this crest lies posterior to the last cheek tooth in *Meizonyx*, whereas in *Megalonyx* it usually reaches opposite the anterior end of that tooth.

As in megalonychids generally, a large coronoid foramen opens laterally near the base of the coronoid process. In this case it lies 16 mm below the alveolar border of the last cheek tooth. It provides an anterolateral exit for the dental canal that conducts the mandibular branch of the trigeminal nerve and accompanying vessels.

The angular portion of the mandible is deeper and laterally more convex than in *Megalonyx*. Likewise the concavity in the ventral profile of the mandible is shallower in *Meizonyx* than in *Megalonyx*. The ventral edge of the angular process is turned medially and, on that side, it presents an extensive rugose crest for the deep portion of the internal pterygoid muscle.

The tip of the articular condyle lies over 100 mm above the occlusal plane of the cheek teeth (projected posteriorly). The condyle is only about 5 mm below the tip of the coronoid process. The massive condyle is transversely widened (nearly 50 mm), broadly convex on its dorsal surface, and flat on its ventral surface. In general it is similar to but more robust than the condyle in *Megalonyx*.

Description of the dentition.-- The lower dentition of *Meizonyx* consists of a caniniform and, following a diastema, three molariform teeth, as in other Megalonychidae. The caniniform is in line with the molariforms, quite unlike the laterally placed caniniforms in Antillean sloths. All of these teeth bear a thin outer covering of cement, except on attrition surfaces, and consist of two layers of dentine, the outer layer being more durable than the inner layer (Owen, 1840-45).

The lower caniniform is nearly triangular with a sharp point at its posterior apex. It is straight and rises steeply (at an angle of about 60°) above the occlusal plane. The lingual wall is flat and meets the labial wall at an acute angle, thus producing a sharp dorsal tip. The labial wall has a persistant groove near its upper edge. The tip is at about the same level as, and in anteroposterior alignment with, the cheek teeth. The flat anterior wear face of the caniniform is vertical in a plane that is turned slightly labially from a perfect transverse plane. The wear striations are nearly vertical. The arrangement of the lower caniniform in *Meizonyx* closely resembles that in *Choloepus hoffmannii* (the living two-toed tree sloth), except that it is larger. In the living sloth the self-sharpening caniniforms perhaps serve both for aggressive behavior and for food gathering (Montgomery and Sunquist, 1978), and the same dual function may be postulated for extinct sloths with similar caniniforms.

The first molariform tooth is subtrapezoidal with a great transverse width. The lingual wall is longer than the labial wall. Although broadly comparable to its homologue in *Megalonyx*, the first molariform of *Meizonyx* differs in several respects. Most importantly, it is the largest tooth of the series, contrary to the situation in *Megalonyx*. Its width is 35 mm and its maximum anteroposterior length is 24.7 mm, producing a crown area of nearly 875 mm^2. The highest anterolingual corner of the tooth is broadly expanded anteriorly and lingually, adjoining parts of the anterior wall and lingual wall are concave, whereas in *Megalonyx* these surfaces are convex.

In other respects the first molariform resembles that tooth in *Megalonyx*. The anterior transverse crest slopes ventrally to the labial side, while the posterior transverse crest slopes ventrally to the lingual side. The central basin, consisting of softer inner dentine, is deeply excavated by wear. The posterior slope of the posterior transverse crest forms a steeply dipping attrition facet on which wear produces a dorsal knife edge of durable dentine. A lighter wear face occupies the posterior surface of the anterior transverse crest, and stop faces occur along the anterolingual faces of both transverse crests.

The second molariform tooth is more quadrangular than the corresponding tooth of *Megalonyx*. It also differs in having a minor but persistant concavity on the lingual face. As in *Megalonyx*, the heaviest wear surface occupies the posterior slope of the posterior transverse crest.

The third molariform tooth is more ovate than the preceding tooth, although it too can be described as subquadrate, with the transverse dimension wider than the anterposterior. It resembles the corresponding tooth in *Megalonyx*. The short lingual wall is set more posteriorly than the longer labial wall giving the tooth a skewed orientation.

Relationships.-- Within the context of megalonychid sloths, *Meizonyx salvadorensis* was a giant, weighing at least half a ton. It equalled or surpassed the largest *Megalonyx jeffersoni* specimens from North America in size. Both of these megalonychid species exceeded somewhat the size of *Megalocnus rodens* from the Greater Antilles. In several respects, some of which may be size-related, *Meizonyx* is the most progressive known megalonychid. The short, thick jaw, the massive molariform teeth dominated by the uniquely augmented first molariform, and the vast masticatory muscle insertions on the ascending ramus, all indicate that it was the most powerful herbivore of its family.

Two of the three largest megalonychid sloths had modified caniniform teeth which provided for improved food apprehension and mastication; *Meizonyx* alone retained the primitive self-sharpening caniniforms. In *Megalonyx* the crowns of the caniniforms were expanded to provide grinding surfaces approximately double the area of the largest molariform. In *Megalocnus* the caniniforms were transformed into powerful incisiform teeth, comparable in function to the incisors of a large beaver (Castoridae). The caniniform of *Meizonyx* differs subtly from the primitive meniscoid caniniform outline, seen for example in *Acratocnus*, by its more flattened lingual wall. The closest comparison of the *Meizonyx* caniniform is with that of the smaller sloth, *Pliometanastes* (Hirschfeld and Webb, 1968).

Some characters of the symphyseal region further suggest a special relationship between *Meizonyx* and *Pliometanastes*. In both genera the symphysis is deep and the lower caniniform is aligned with the molariforms and steeply oriented. These features presumably represent a fundamental reorganization of the symphyseal architecture, which contrasts markedly with the pattern in Antillean sloths and in *Choloepus* (Fig. 3).

Meizonyx also shares several derived characters with *Megalonyx*. Many of these common characters, however, may have evolved convergently as consequences of large size. This explanation seems likely, for example, for the quadrangular molariform teeth and the deep mandible. The more vertical canine and deep symphysis are possible common inheritances from *Pliometanastes*. There is stratigraphic continuity in North America between *Pliometanastes* of the early Hemphillian, the primitive species, *Megalonyx mathisi* in the medial Hemphillian, and *M. curvidens* in the late Hemphillian. This chronocline strongly suggests derivation of *Megalonyx* from *Pliometanastes* in North America. The reduced symphysis, evident in *Megalonyx* and especially clear in *Meizonyx* may be the most significant derived character shared by these two genera. Even so, we suspect it is convergent.

Resolution of the relationships among *Meizonyx*, *Pliometanastes* and *Megalonyx* will require a fuller record of *Meizonyx*. At present the simplest hypothesis is to regard *Meizonyx* as the Central American derivative of *Pliometanastes*. Likewise, *Megalonyx* may be the North American derivative of *Pliometanastes*, a lineage that is fairly well documented. By about early Pleistocene time *Megalonyx* lived syntopically, if not sympatrically, with *Meizonyx* in El Salvador.

***Megalonyx obtusidens*, new species**

Etymology.-- *Obtusus*, L. blunt; *Dens*, L. tooth; *obtusidens* thus refers to the more quadrangular aspect of the teeth.

Type.-- Specimen number 70-167 in the collection of the Museo de Historia Natural de El Salvador, left mandibular ramus with all three molariform teeth (Fig. 3). Collected by Stephen Perrigo at Barranca del Sisimico, Department of San Vicente, El Salvador.

Diagnosis.-- Middle-sized species, generally comparable to *Megalonyx leptostomus*. Lingual walls of cheek teeth unusually elongate. First molariform nearly square; length of lingual wall equal to that of labial wall. Third molariform large, ovate and strongly rotated with "posterior" crest posteriolingually oriented; anteroposterior length nearly equal to transverse width.

Description.-- The type mandible of *Megalonyx obtusidens* represents a mature individual with vertical-sided cheek teeth. Its size is comparable with such medium-sized species as *Megalonyx leptostomus* (Hirschfeld and Webb, 1968).

Only the central part of the mandible is preserved. The broken anterolingual edge indicates the posterior edge of

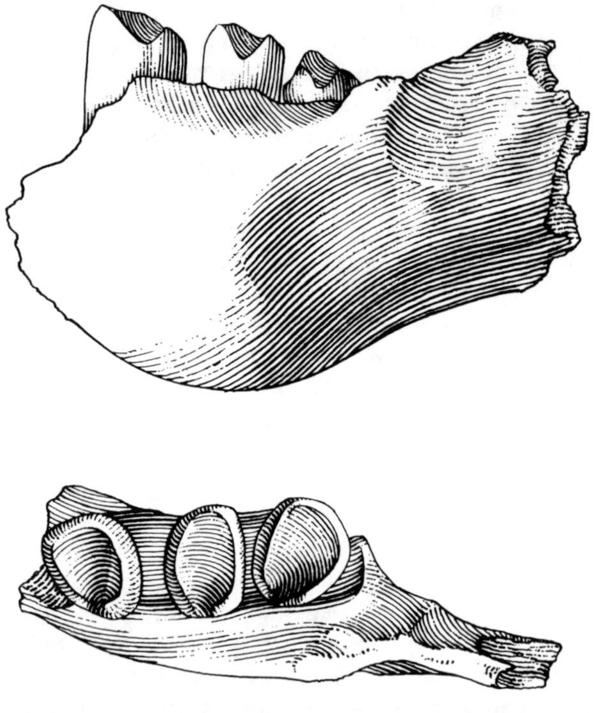

Figure 3. *Megalonyx obtusidens* type mandible, labial and occlusal views. Museo de Historia Natural de El Salvador number 70-167. Scale X 0.7.

the symphysis. The symphysis originates just anterior to the root of the first molariform, the usual position for *Megalonyx*. The horizontal ramus is nearly 67 mm deep below the first molariform and more than 36 mm wide at the thickest point below the third molariform tooth. The ventral border of the mandible rises to a long concave surface posterior to the roots of the cheek teeth. The ascending ramus arises near the middle of the last molariform tooth. The coronoid foramen opens laterally at a point about 34 mm posterior to the last tooth and about 12 mm. dorsal to its alveolar border.

Table 2. Measurements of *Megalonyx obtusidens* mandible (in mm.).

Molariform 1 length	17.9
Molariform 1 width	20.3
Molariform 2 length	17.2
Molariform 2 width	21.5
Molariform 3 length	19.5
Molariform 3 width	22.2
Length M1 to M3 inclusive	63.2
Crown height of molariform 2	70.4
Maximum width of mandible (at M2)	36.5
Depth labial to M1	63.2
Depth labial to M2	66.8
Depth labial to M3	61.6

The diagnostic features of *Megalonyx obtusidens* are found in the cheek teeth. In each tooth the lingual wall is more extensive anteroposteriorly than in any other species of *Megalonyx*. The heights of the mature, ever-growing cheek teeth are about the same as in *Megalonyx leptostomus*.

The first molariform tooth is quadrate with its width only slightly exceeding its length (Table 2). All four walls of the tooth are convex. The lingual wall is at least as long as the labial wall, a feature that distinguishes this species from most other *Megalonyx*. The anterolingual and posterolabial corners are, as usual, the high points. The labial notch is strong as in the other cheek teeth.

The second molariform tooth is subquadrate and less compressed anteroposteriorly than in most *Megalonyx* species. The lingual wall in particular is long and flattened.

The third molariform tooth is large, ovate and strongly twisted. Its anteroposterior length nearly equals its width. The "posterior" transverse crest in fact runs posterolingually, so that the tooth appears to be rotated almost 45° from a "proper" transverse orientation. The posterior wear surface shows the heaviest attrition on

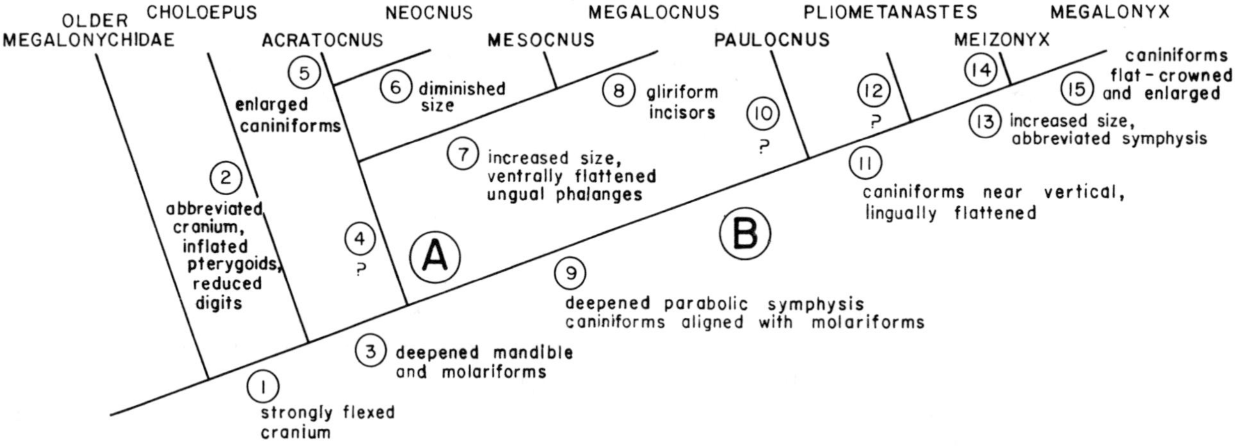

Fig. 4. Cladogram of later Megalonychidae. Numbers indicate shared-derived characters. **A** indicates possible unity of Greater Antillean Megalonychidae; and **B** indicates possible unity of northern South American, Central American and North American genera.

this tooth as on the others. As in other *Megalonyx* species, the slopes of these posterior attrition faces become noticeably shallower from the first molariform back to the third molariform tooth (McDonald, 1977).

Relationships.-- In size and mandibular morphology *Megalonyx obtusidens* is closely comparable to *M. leptostomus*. The cheek teeth have comparable height also. The latter species was widespread in North America during the Blancan (latest Pliocene) (Hirschfeld and Webb, 1968). It is not yet clear whether *Megalonyx obtusidens* is of similar age.

Megalonyx obtusidens exhibits several distinctive features in its lower cheek teeth. The general distinction appears to be the anteroposterior elongation of each tooth especially along the lingual wall. These dental patterns fall well outside the normal range of dental variation in North American *Megalonyx*. Whether populations geographically intervening between El Salvador and the United States may have bridged this morphological gap remains to be discovered.

ZOOGEOGRAPHIC IMPLICATIONS

The occurrence of two kinds of megalonychid sloths in El Salvador focuses attention on the possible role of Central America in megalonychid distribution. The question is whether sloths extended their ranges from South America to North America by way of the mainland or alternatively across the Caribbean by island-hopping. Both Megalonychidae and Mylodontidae appeared in North America about 9 million years ago. In the case of mylodontidae, there has never been serious doubt that they moved over land through Mesoamerica to North America (Webb, 1978). For megalonychidae, however, the diversity of Greater Antillean genera has encouraged a number of zoogeographers to postulate megalonychid distribution to North America ***across***, rather than ***around***, the Caribbean Basin. A stimulating recent example is Rosen (1976) who has "tracked" the megalonychids by way of the Greater Antilles to North America. He implies that they may have been transported passively by geological processes, indeed a slothful mode of transport.

The critical question to help resolve this choice of distribution routes is whether North American megalonychids are more closely related to Antillean or to Central American megalonychids. The evidence presented here favors Central America (Fig. 4). The two North American genera, *Megalonyx* and *Pliometanastes*, have their principle affinities with the two Central American taxa, *Megalonyx* and *Meizonyx*. The two regions share the genus *Megalonyx*; they also share probable sister-group relationships between *Meizonyx* on the one hand and *Pliometanastes* and *Megalonyx* on the other. None of the several Antillean genera share special relationships with either North American genus, although *Acratocnus* and *Pliometanastes* share many primitive features. Thus, the new evidence presented here indicates that the interamerican distribution of megalonychid sloths, like that of mylodontid sloths, probably proceeded by way of Central America rather than by way of the Greater Antilles.

It is notoriously difficult to deduce former distribution routes of organisms when appropriate fossils are not abundantly sampled from relevant regions. Until now the near absence of megalonychids from Central America has prevented any judicious comparison of evidence from the two possible distribution routes.

ACKNOWLEDGMENTS

We are especially grateful to El Museo de Historia Natural de El Salvador for permission to study and describe these new megalonychids. We thank Greg McDonald and Gordon Edmond of the Royal Ontario Museum, and George Engelmann of the American Museum in New York for helpful suggestions regarding sloths and their paleobiology. Howard Converse restored the sloth mandibles from their original weathered condition. Lauren Keswick prepared the illustrations.

LITERATURE CITED

Hirschfeld, S.E. and S.D. Webb.
1968. Pleo-Pleistocene Megalonychid Sloths of North America. Bull. Fla. Sta. Mus., 12:213-296.

McDonald, H.G.
1977. Osteology of the extinct gravigrade edentate, *Megalonyx*, with observstions on its ontogeny, phylogeny, and functional anatomy. M.S. thesis, Univ. Fla., Gainesville. 328 pp.

Montgomery, G.G. and M.E. Sunquist
1978. Habitat selection and use by two-toed and three-toed sloths. Pp. 329-359 *In* The Ecology of Arboreal Folivores (G. G. Montgomery, ed.), Smithsonian Institution Press, Washington, D. C. 574 pp.

Owen, R.
1840-45. Odontography. London, Hippolyte Bailliere. 2 vols., 655 pp.

Rosen, D.E.
1976. A variance model of Caribbean biogeography. Systematic Zool., 24: 431-464.

Schmidt-Thomé, M.
1975. Das Diatomitvorkommen im Tal des Rio

Sisimico (El Salvador, Zentralamerika). Geol. Jb., B 13, S.: 87-96.

Stirton, R.A., and W.K. Gealey
1949. Reconnaissance geology and vertebrate paleontology of El Salvador, Central America. Bull. Geol. Soc. Amer., 60:1731-1754.

Williams, and A.R. McBirney.
1969. Volcanic History of Honduras. Univ. Calif. Pubs. Geol. Sci., 85: 1-101.

THE SOMATIC CHROMOSOMES OF XENARTHRA

Wilham Jorge[1], **Ana Tereza Orsi-Souza**[2], **and Robin Best**[2]

[1]Departamento de Genética, UNESP, Campus de Botucatu, 18.600, Botucatu, São Paulo, Brasil;
[2]Instituto Nacional de Pesquisas da Amazonia (INPA), 69.000, Manaus, Amazonas, Brasil

Abstract
The chromosomes of 26 individuals of the families Bradypodidae (sloths) and Dasypodidae (armadillos) were analyzed. Seven were successfully C- and/or G-banded. The diploid number 64 in *Dasypus novemcinctus* and *D. hybridus* was confirmed. There was great karyotypic similarity between the species. Karyotype of *Chlamyphorus truncatus* was similar to those of the Euphractini. Three-toed sloths *Bradypus variegatus* and *B. tridactylus* have a Robertsonian mechanism of karyotype evolution. The two-toed sloth *Choloepus* is characterized by numerical polymorphism (2n=51, 52, 53 and 64) and by additional chromosomes with no homologous partners, interpreted as supernumerary or B- chromosomes. B_3 was the only presumed X chromosome by its size, shape, and C-banding, suggesting XO females.

Abstracto
Fueron estudiados los cromosomas de 26 individuos de las familias Bradypodidae (monos perezosos) y Dasypodidae (armadillos). Fueron marcados con éxito bandas en 7 individuos usando C- y/o G. El número diploide de 64 fue confirmado en *Dasypus novemcinctus* y *D. hybridus*. Hubo gran similaridad cariotipada entre los especies. El cariotipado de *Chlamyphorus truncatus* fue similar a los de eufractinos. Monos perezosos de tres dedos *Bradypus variegatus* y *B. tridactylus* tienen un mecanismo Robertsoniano de evolucíon cariotipada. El mono perezoso de dos dedos *Choloepus* se caracteriza por polmorfismo numérico (2n = 51, 52, 53, y 64) y por cromosomas adicionales sin homólogos, interpretados como cromosomas supernumerarios o cromosomas B. El cromosoma B_3 fue el único cromosoma X presumible por su tamaño, forma y bandeado de C, sugiriendo hembras XO.

Resumo
Os cromossomos de 26 indivíduos das famílias Bradypodidae (preguiças) e Dasypodidae (tatus) foram analisados. Sete foram analisados com bandas C e G. O número diplóide 64 foi confirmado no *Dasypus novemcinctus* e *D. hybridus* e houve grande similaridade cariotípica entre as duas espécies. O cariótipo do *Chlamyphorus truncatus* foi similar aos do grupo Euphractini. A preguiça de 3 dedos, *Bradypus variegatus* e *B. tridactylus* têm um mechanismo Robertsoniano de evolução do cariótipo. A preguiça de 2 dedos, *Choloepus* caracteriza-se pelo polimorfismo numérico (2n = 51, 52, 53 e 64) e por apresentar cromossomos adicionais sem seus respectivos homólogos, interpretados como supernumerários ou cromossomo B. B_3 foi considerado cromossomo X pelo seu tamanho, forma e marcação C, sugerindo fêmeas XO.

The order Xenarthra is of special interest to geneticists because of the unusual sex chromosomes and unpaired chromosomes in sloths (Corin-Frederic, 1969; Jorge et al., 1977; Jorge, 1981) and the sex ratio in Hoffmann's sloth from Panama of one male to 11 females (Montgomery and Sunquist, 1975; Meritt, 1976). They are also of interest because some armadillos, *Dasypus novemcinctus* and *D. septemcinctus*, are the only vertebrates known to exhibit polyembryony, in which a fertilized ovum regularly gives rise to monozygotic embryos (Hamlett, 1933; Galbreath, 1985).

In this paper, we review literature on sloth and armadillo karyotyping and report our results concerning karyotypes, sex chromosomes, B-chromosomes and taxonomic relationships among some xenarthran species.

MATERIALS AND METHODS

Sources of animals, sex, and number of specimens analyzed are presented in Table 1.

To determine chromosome number, routine analyses of lymphocyte culture and bonemarrow suspension cells were used. Chromosome slides were prepared by the air-dry method and stained with Giemsa. C-banding was carried out following the schedule of Jorge and Benirschke (1977).

RESULTS AND DISCUSSION

Chromosome Number and Sex Chromosomes.-- The xenarthrans have diploid chromosome numbers ranging from 38 to 64 (Table 2). The lowest number is in the armadillo *Tolypeutes matacus* (2n=38). All the others described have chromosome numbers between 50 and 64.

The sex chromosome mechanism is XX:XY in the armadillos, vermilinguas (anteaters) and three-toed sloths. Two-toed sloths *Choloepus* have an odd chromosome number, presence of unpaired chromosomes, and are highly polymorphic.

Dasypus novemcinctus *and* D. hybridus.-- The common long-nosed armadillo is the only species of xenarthran that is so widely distributed; it can be found from Argentina to Mexico and the United States (Wetzel, 1985).

Chromosome studies have demonstrated a constant number in this species of 2n=64. The only difference in the karyotype was found in the morphology of some chromosome pairs. Pairs 11, 18 and 27 in specimens from Texas are acrocentrics, while in individuals from Colombia they are submetacentrics (Benirschke et al., 1969; Jorge et al., 1977). In specimens from Belém, Brazil, the karyotype is similar to the one from Texas, except for the eleventh pair (Assis, et al., 1976). Furthermore, specimens from Botacatu, in southeastern Brazil, had a similiar karyotype to the one from Texas (Fig. 1).

Table 1. Collection localities and number of animals karyotyped. All localities except Mendoza, Argentina were in Brazil. Highway refers to the Belém-Brasília highway.

Species	*Locality*	*Number Karyotyped*	
		Male	*Female*
Dasypus novemcinctus	Botucatu	1	2
Dasypus hybridus	Botucatu	2	1
Chlamyphorus truncatus	Mendoza	-	1
Bradypus variegatus	Belém	3	4
Bradypus variegatus	Curari Island	-	2
Bradypus tridactylus	Manaus	2	1
Choloepus hoffmanni	Belém	-	2
Choloepus hoffmanni	Highway	-	2
Choloepus didactylus	Highway	-	1
Choloepus sp.	Manaus	1	-
Choloepus sp.	Manaus	-	1

Table 2. Chromosome number and sex chromosome constitution of Xenarthra studied to date, except the genus *Choloepus*. 2n=diploid number; M=metacentric, SM=submetacentric, A=acrocentric.

Family Species	*Collection Locality*	*2n*	*Sex Chromosomes*	*Shape of X and Y*		*Banding Pattern Analyzed*	*Reference*
Dasypodidae (armadillos)							
Euphractus sexcinctus	Argentina	58	(XX/XY)	SM	SMA		Benirschke et al., 1969
	Brazil (Botacatu)	58	(XX/XY)	SM	-	G	Jorge, et al., 1977
Chaetophractus villosus	Argentina	60	(XX/XY)	A	A		Benirschke et al., 1969
	Argentina	60	(XX/XY)	A	A	G and C	Jorge et al., 1977
Zaedypus pichiy	Argentina (Mendoza)	62	(XX/XY)	A	SM		Meritt et al., 1973
	Argentina	62	(XX/XY)	A	SM	G	Jorge et al., 1977
Priodontes giganteus	?	50	(XX/XY)	SM	M		Benirschke and Wurster, 1969
Cabassous centralis	Colombia	62	(XX/XY)	SM	M		Benirschke et al., 1969
Tolypeutes matacus	Argentina	38	(XX/XY)	M	A	G	Jorge et al., 1977
Dasypus novemcinctus	Texas	64	(XX/XY)	SM	A		Beath et al., 1962
	Texas	64	(XX/XY)	SM	A	G	Jorge et al., 1977
	Colombia	64	(XX/XY)	SM	A		Benirschke et al., 1969
	Brazil (Belém)	64	(XX/XY)	SM	A		Assis et al., 1976
	Brazil (Botacatu)	64	(XX/XY)	SM	A	G	Jorge, 1981
Dasypus hybridus	Uruguay	64	(XX/XY)	SM	A		Saez et al., 1964
	Argentina	64	(XX/XY)	SM	A	G	Jorge et al., 1977
	Brazil (Botacatu)	64	(XX/XY)	SM	A	G	Jorge, 1981
Chalamyphorus truncatus	Argentina (Mendoza)	58	(XX/ ?)	?	?	G and C	Jorge, 1981
Myrmecophagidae (vermilinguas)							
Myrmecophaga tridactyla	-	60	(XX/XY)	M	SM		Hsu, 1965
Tamandua tetradactyla	-	54	(XX/XY)	SM	SM		Hsu, 1965
Tamandua longicaudata	Colombia	54	(XX/ ?)	M	?	G	Jorge, et al., 1977
Bradypodidae (sloths)							
Bradypus infuscatus	Brazil (Belém)	54	(XX/XY)	SM	M		Jorge, 1981
Bradypus tridactylus	Brazil (Manaus)	52	(XX/XY)	SM	M		Jorge, 1981
Bradypus variegatus	Brazil (Manaus-Curari)	55	(?/XY,XO?)	SM	-		Jorge, 1981

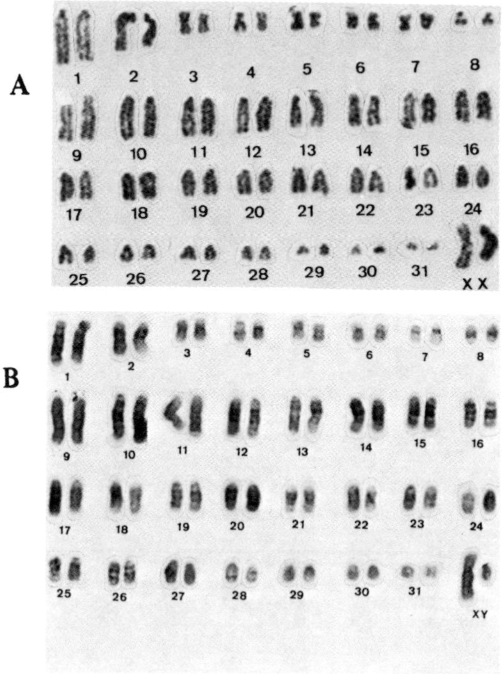

Figure 1. **Top.** Karyotype of female common long-nosed armadillo, *Dasypus novemcinctus* from Botucatu, southeastern Brazil. 2n=64. G-banding. **Bottom.** Karyotype of male seven-banded armadillo, *Dasypus hybridus* from Botucatu, 2n=64. There are no apparent differences between the chromosomes of this species and those of the *D. novemcinctus* from the United States and Botucatu. G-Banding.

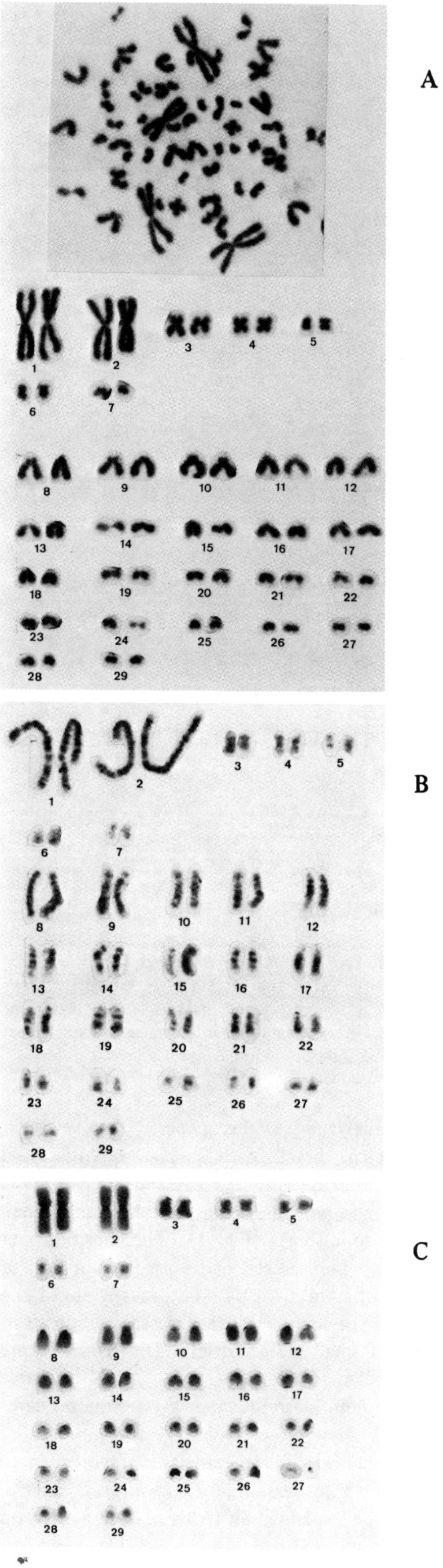

Figure 2. Karyotype of *Chlamyphorus truncatus* Harlan, 2n=58. **A.** Female karyotype of *C. truncatus* using conventional staining. From its size, the number 2 may be the X-chromosome. **B.** Female karyotype of *C. truncatus.* Note that pairs 1 and 2 are unusually long for Xenarthrans. G-Banding. **C.** C-Banding pattern of *C. truncatus.* Note the regular constitutive heterochromatin near centromeres.

Comparing both karyotypes *D. novemcinctus* and *D. hybridus* from Botucatu (Fig. 1), no differences were found, suggesting that there is a close taxonomic relationship between them.

Chlamyphorus truncatus.-- The diploid chromosome number of the female pichiciego from Mendoza Province, Argentina, was 2n=58. Because the specimen was a female, sex chromosome could not be identified. This karyotype is characterized by pairs 1 and 2 that are unusually long for xenarthrans (Fig. 2), and this karyotype shows a great similarity with that of species of Euphractini.

Bradypus.-- The three-toed sloth belongs to the family Bradypodidae, which includes three genera:*Bradypus, Scaepus* and *Choloepus.* Both species of three-toed sloths (Wetzel and Avila-Pires, 1980) from Amazonia are

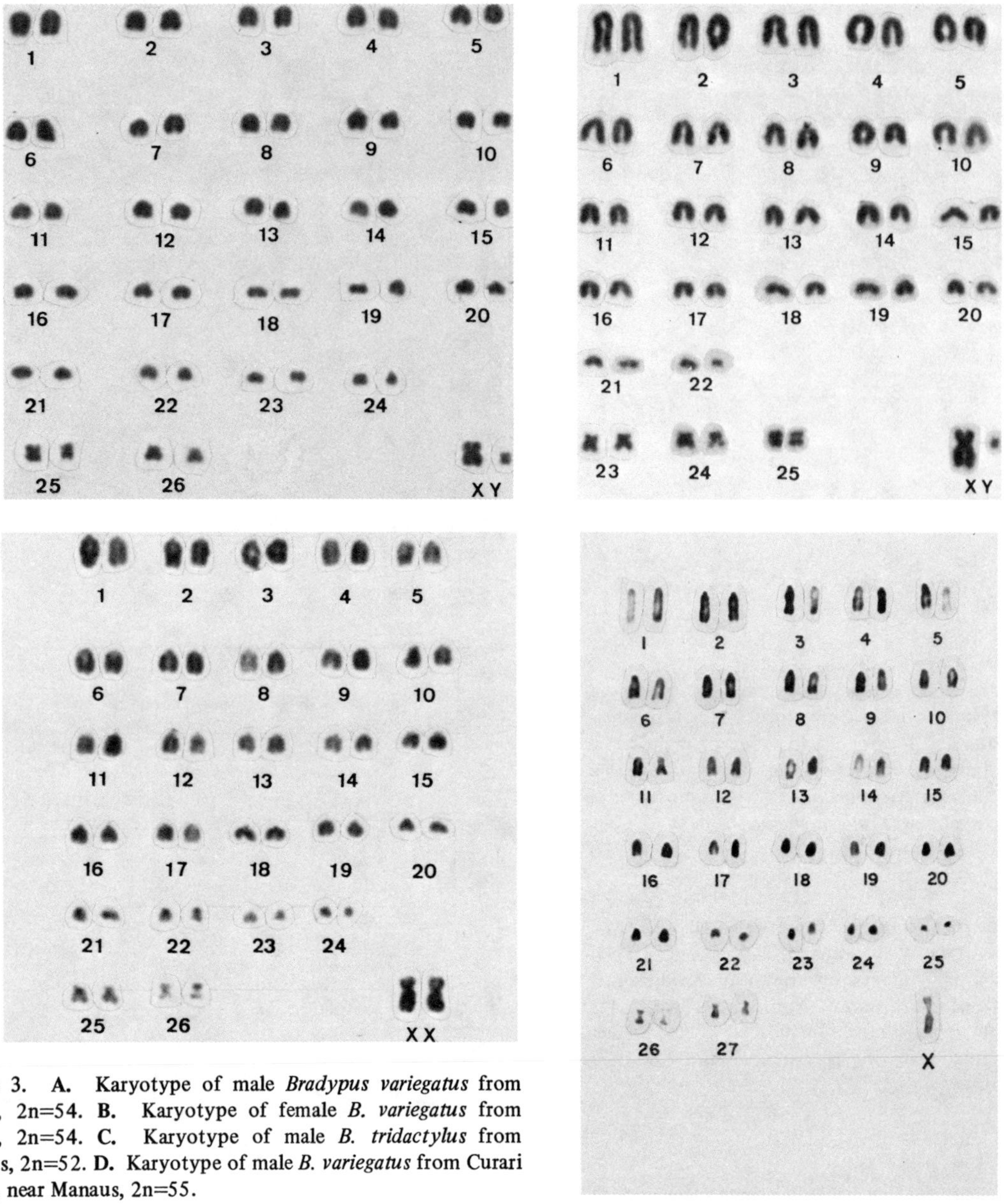

Figure 3. **A.** Karyotype of male *Bradypus variegatus* from Belém, 2n=54. **B.** Karyotype of female *B. variegatus* from Belém, 2n=54. **C.** Karyotype of male *B. tridactylus* from Manaus, 2n=52. **D.** Karyotype of male *B. variegatus* from Curari Island, near Manaus, 2n=55.

cytologically analyzed in this paper: *B. variegatus* collected in Belém, Brazil, with 2n=54; *B. tridactylus* from Manaus, Brazil, with 2n=52; and *B. variegatus* from Curari Island, about 30 km from Manaus, Brazil, with 2n=55 chromosomes (Fig. 3). The chromosomes are grouped into two classes. One class is composed of 22 to 25 pairs of acrocentric chromosomes arranged in a decreasing order of size. The other class is composed of 2 to 3 pairs of small metacentrics. The X and Y are typical mammalian sex chromosomes. The X is a large submetacentric chromosome; the Y, a minute chromosome.

The transformation of acrocentric chromosomes to metacentric or "vice vẹrsa" in *B. tridactylus* and *B. veriegatus* can be explained in terms of fusion ⇔fission of chromosome pairs 23 and 24 of *B. variegatus* and pair 25 of *B. tridactylus*, both having the same number of arms (NF=56) and different diploid numbers.

The sex ratio of *B. tridactylus* was 48 percent female in 85 animals studied, which confirms the expected ratio of 1:1 for the XX:XY mechanism (Table 3).

Table 3. Sex ratios in *Bradypus* and *Choloepus* from Amazonia, Brazil (from Jorge, 1981).

Species	*Collection Locality*	*Percent Males*	*Total*
B. tridactylus	Manaus	48.2	85
B. variegatus	Belém	46.1	13
B. variegatus	Curari Island	43.7	71
Choloepus sp.	Manaus	55.5	18

Table 4. Chromosome number and size relationship of autosomes, unpaired chromosomes, and sex chromosomes of *Choloepus*. 2n= diploid number; SM=submetacentric; A=acrocentric; M=metacentric; P=punctiform; s=small; m=medium; l=large.

Species	*Collection Localities*	*2N*	*SM*	*A*	*M*	*Unpaired*	*Male/ Female*	*Sex Cro-mosomes*	*Reference*
C. hoffmanni	Panama, Costa Rica, and Ecuador	49	5 1	15 s-l	4 s		4/6	XO/XY	Corin-Frederic, 1969
C. hoffmanni	Panama, Colombia	50	5 1	15 s-l	4 s		1/1	XX/XY	Jorge et al., 1977
C. hoffmanni	Brazil (Amazonia)	51	11 s-l	10 s-l	4 s-l	1 SM (X?)	0/2	XO?/ ?	Jorge, 1981
C. didactylus	Brazil (Belém)	53	7 s-l	14 s-l	3 s	5	0/2	XO?/ ?	Jorge, 1981
C. didactylus	Brazil (Amazonia)	53	3 1	18 s-l	2 s	7	1/0	-	Jorge, 1981
C. didactylus	Brazil (Manaus)	64	15 s-m	11 s-m	4 s	4	0/1	-	Jorge, 1981
C. didactylus	Brazil (Manaus)	52		23 s-l	2 s	1 1 SM; 1 P	0/1	XO?/ ?	Jorge, 1981
C. didactylus	Brazil (Manaus)	64	10 s-l	18 s-m	3 s	1 1 SM; 1 s SM	1/0	XY?/ ?	Jorge, 1981

The analyses of the three karyotypes indicates a Robertsonian process of karyotype evolution. The fusion/fission mechanism involves the pairs 23, 24, 25 and 26 in *B. tridactylus* and *B. variegatus* (2n=54), both having the same number of arms (NF=56). In *B. variegatus* (2n:55) the two dot-like elements would be the Y chromosomes and the XYY sex chromosome mechanism could be involved.

Choloepus.-- In the two-toed sloths, there are at least six different numbers of chromosomes: 49, 50, 51, 52, 53 and 64 (Table 4), and at least seven distinct karyotypes (Fig. 4). Two specimens with 2n = 53 chromosomes had the same chromosome number; however, their individual chromosomal elements were remarkably unlike. The specimens with 49 chromosomes were described by Corin-Frederic (1969). In the male the Y-material is translocated on an autosome and the female had an XO sex-chromosome constitution. The specimens with 50 chromosomes (Jorge et al., 1977) were characterized by having the largest Y chromosome found among mammals so far, and heteromorphism of the two X in the female was revealed by C-banding techniques (Fig. 4A). The specimens with 51, 52, 53, and 64 chromosomes are described below.

The karyotype with 51 chromosomes, classified as *C. hoffmanni*, was found in two females from the same region in Amazonas, along the Belém-Brasília Highway. It was characterized by having 3 chromosomes: 25, 26, and 27 (presumed X) without their respective homologues. The heterochromatin arrangement revealed by C-banding technique showed that the number 27 (presumed X) was easily distinguished from the others because it appeared totally euchromatic (Fig. 4C, 4D). Athough males from this area have not been studied, we suppose an XO sex chromosome mechanism in these females. This assumption is supported by C-banding pattern of number 27, which corresponds in size and morphology to the X chromosome of the males with 2n=50 chromosomes. Numbers 25 and 26 were considered as a heteromorphic pair or B chromosomes, depending on their behavior in meiosis.

Karyotypes of two females (mother and daughter) of *C. hoffmanni* with 2n=53 are shown in Fig. 4E, 4F, and 4G. The absence of Barr bodies in cells from brain squashing led us to propose an XO mechanism for females. Under this assumption, the B_3 chromosome would be the presumed X-chromosome by its shape, as well as the heterochromatin arrangement revealed by C-banding technique.

The karyotype of the only male of *C. didactylus* showed an odd diploid number and seven elements without evident partners. The B_3 chromosome was the supposed X-chromosome on the basis of the C-banding pattern (Fig. 4H). The chromosomes of specimens from Manaus are quite different from the others (Fig. 4I, 4J).

The *Choloepus* chromosomes appear to vary geographically. This geographical variation may correspond to chromosome races within the species. The first group came from Central America, Peru, and Colombia, showing a diploid number 49 to 50 with a basic karyotype of 40 acrosubmetacentric chromosomes (1 to 20 pairs) and 8 small metacentric (21 to 24 pairs). There were no chromosomes without partners. Females were XX or XO; males presented polymorphic Y chromosomes (Corin-Frederic, 1969; Jorge et al., 1977).

The specimens from the Belém-Brasilía highway had a diploid number of 51 to 53. This variation was due to different numbers of unpaired chromosomes. The basic karyotype was made up of 42 acro-submetacentric chromosomes (1 to 21 pairs) arranged in a decreasing order of size and 6 small metacentric (22 to 24 pairs), except for *C. didactylus* which had 4 small metacentrics. However, in *C. didactylus*, the Y chromosome was translocated on a small metacentric arising from the B_6 chromosome (Fig. 4B-4H). The third group was made up of specimens whose karyotypes varied greatly among

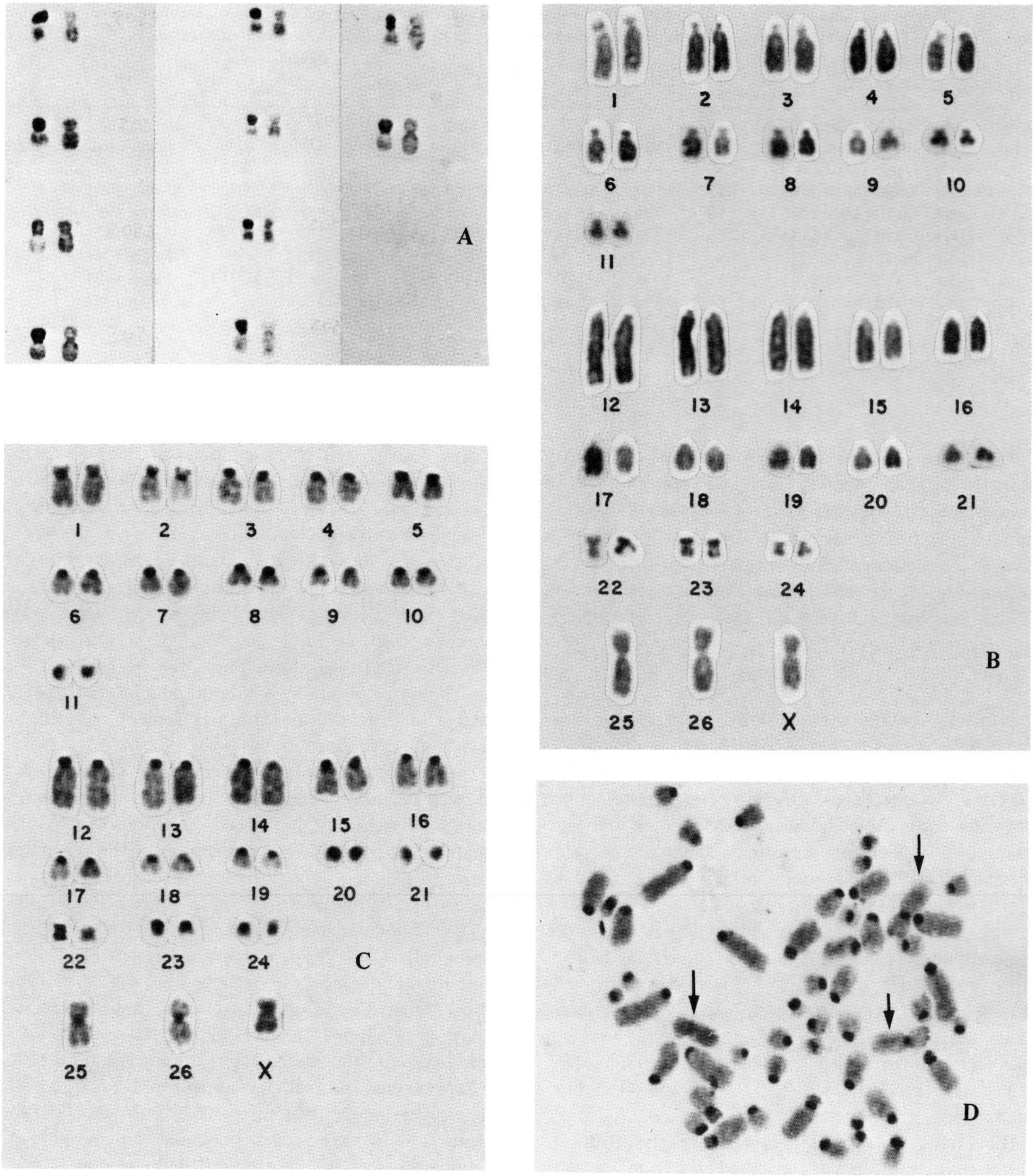

Figure 4. The chromosomes of the genus *Choloepus*. **A.** *C. hoffmanni*, 2n=50 (Jorge et al., 1977). C-band heteromorphism shown in 10 different metaphases. One X-chromosome is heavily stained. The second homologous pair correspond to the X-male chromosome. **B.** Karyotype of *C. hoffmanni*, 2n=53, from Belém, with its five unpaired elements in two females analyzed (mother and daughter). There is a single presumed X-chromosome because the smear from brain was chromatin-negative. **C.** C-banded karyotype of female, 2n=51. D. C-banded metaphase of the female, 2n=51. The unpaired element can be distinguished in all autosomal pairs by its euchromatic pattern. **E.**

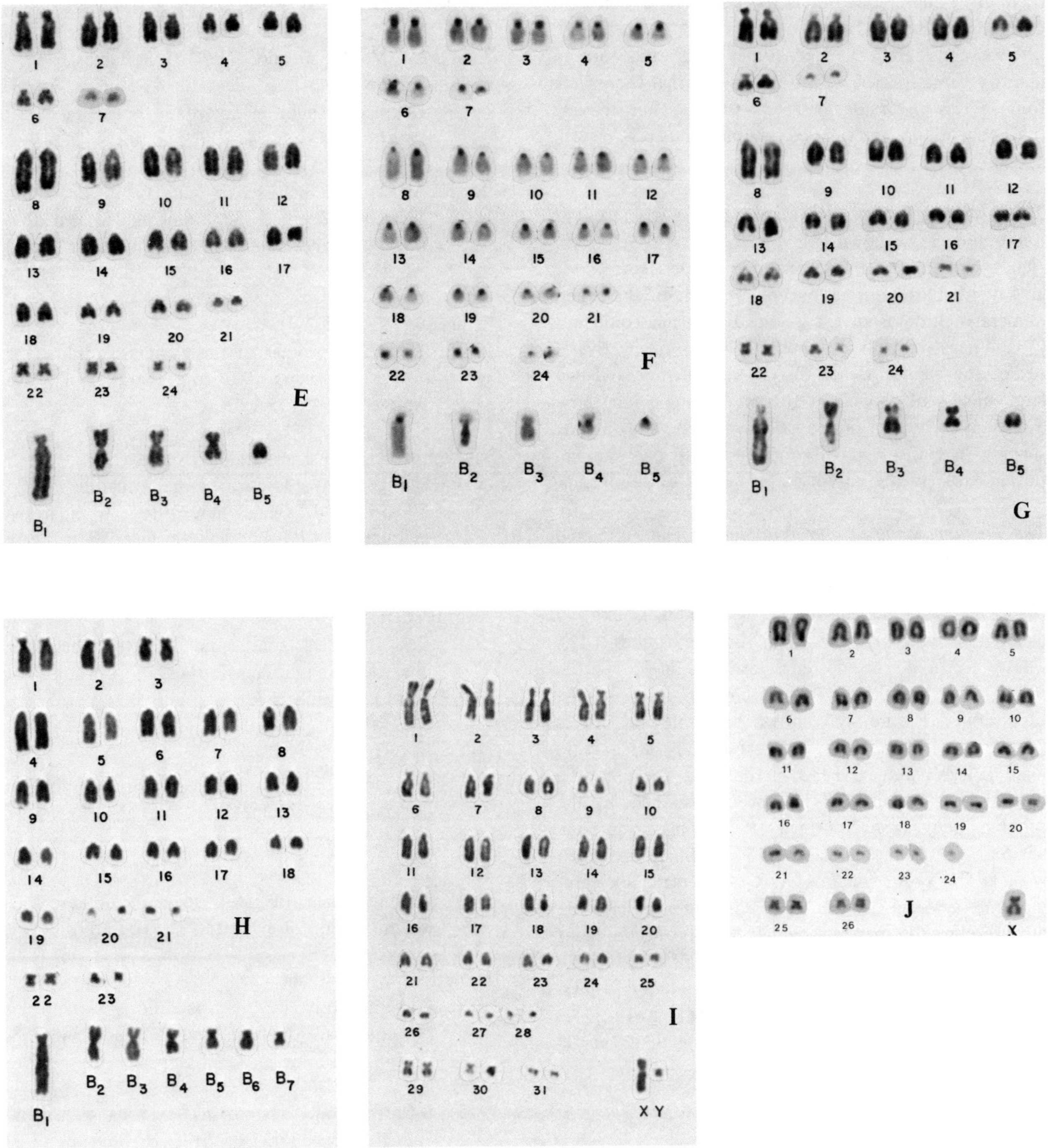

Karyotype of *C. hoffmanni*, 2n=53, from Belém (mother) with its five unpaired elements. There is a single presumed X-chromosome (B3?) because the smear from brain was chromatin negative. **F.** C-banded karyotype of *C. hoffmanni* (mother), 2n=53. **G.** Karyotype of *C. hoffmanni* from Belém (daughter). **H.** Karyotype of a male *Choloepus* sp., 2n=53, with its seven unpaired elements. **I.** Karyotype of a male *Choloepus* sp., 2n=64, from Manaus. The sex chromosomes are speculative. This is the only specimen without unpaired chromosomes. **J.** Karyotype of a female *Choloepus* sp., 2n=52, from Manaus.

themselves and among others from the first and second groups (Fig. 4I and 4J).

If we considered the karyotypes as a valid tool to identify these species, we would assume that there were four species of *Choloepus* instead of two, thus denying the conventional systematics on this genus (cf. Wetzel, 1983).

The C-banding Patterns.-- The C-banding patterns of *Choloepus* are shown in Fig. 4A, 4C, 4D, 4F. The C-band material was located in the pericentromeric area of the chromosomes except for some of the supernumerary chromosomes; e.g., the B_3 chromosome was C-band negative in all specimens analyzed, irrespective of sex of the animal; B_2 of *C. hoffmanni* presented two large blocks of C-band in the chromosome long arm. These two types of constitutive heterochromatine suggest that the heterochromatin might play an important role in the karyotype evolution of this genus.

The supernumerary chromosomes.-- The unpaired chromosomes of two-toed sloths can be divided into 3 types according to size: unpaired small chromosomes (B_5), biarmed medium-sized (B_2 and B_4), and an element which is larger than A-chromosomes (B_1). These unpaired chromosomes can be interpreted as so-called supernumerary or B-chromosomes. The supernumeraries are well known in plants and insects (Muntzing, 1974; Jones, 1975), but they are rare in mammals (Volubujev, 1980). However, they may be heteromorphic pairs, as has been described in the echidna *Tachyglossus* (Murtagh, 1977). This monotreme has 63 chromosomes in male and 64 in female associated with an X_1X_2Y male / $X_1X_1X_2X_2$ female sex chromosome system. Additionally, both sexes have six mitotic chromosomes (a-f) which have no homologous partners. The B-chromosomes in mammals are generally small and heavily stained by C-band technique (Jones, 1975). In sloths, some B's are quite large (Fig. 4E, F, G, and H), and also euchromatic, indicating their genetic activity.

ACKNOWLEDGMENTS

Many persons provided assistance and encouragement. I express my particular gratitude to Drs. Kurt Benirschke and Dennis Meritt, Jr., and to Roberto Martins for technical assistance. This work was supported by Conselho Nacional de Pesquisas (CNPq) Brasil - Proc. 2222.1409/77 and Fundação de Amparo a Pesquisa do Estado de São Paulo (Brasil), Proc. Biol. 79/1797.

LITERATURE CITED

Assis, M.F., M. M. Sampaio, and R. M. S. Barros.
1976. Estudo cariotípico de tatu. Resumos da 28a Reunião Anual da Sociedade Brasileira para o Progresso da Ciência (SBPC) p. 269.

Beath, M. M., K. Benirschke, and L. E. Brownhill.
1962. The chromosomes of the ninebanded armadillo, *Dasypus novemcinctus*. Chromosoma (Berl.), 13: 27-38.

Benirschke, K., and D. H. Wurster.
1969. The chromosomes of the giant armadillo, *Priodontes giganteus* Geoffroy. Acta Zool. Path. Antwerp., 49: 125-130.

Benirschke, K., R. J. Low, and V. H. Ferm.
1969. Cytogenetic studies in some armadillos. Pp. 330-345 *In* K. Benirschke, editor. Comparative Mammalian Ctyogenetics. Springer-Verlag, New York. 645 p.

Corin-Frederic, J.
1969. Les formules gonosomiques dites aberrantes chez les Mammiferes Euthériens. Chromosoma (Berl.), 27: 268-287.

Galbreath, G.J.
1985. The evolution of monozygotic polyembryony in *Dasypus*. (this volume).

Hsu, T. C.
1965. Chromosomes of two species of anteaters. Mamm. Chrom. Newsletter, 15: 108-109.

Hamlett, G. W. D.
1933. Polyembryony in the armadillo: Genetic or physiological? Quart. Rev. Biol., 8: 348-358.

Jones, R. N.
1975. B-Chromosome system in flowering plants and animal species. Pp. 1-85 *In* G. H. Bourne and J. E. Danielli, editors, International Review of Cytology. Academic Press, New York. 420 p.

Jorge, W.
1981. Estudo cromossomico de algumas especies da Ordem Edentata. Tese de Livre-Docência apresentada ao Instituto Básico de Biologia Médica e Agrícola - UNESP - Botucatu - São Paulo, Brasil.

Jorge, W., and K. Benirschke.
1977. Centromeric heterochromatin and G-banding of the Red Brocket Deer, *Mazama americana temama* (Cervoidea, Artiodactyla) with a probable non-Robertsonian translocation. Cytologia, 42: 711-721.

Jorge, W., D. A. Meritt, and K. Benirschke.
1977. Chromosome studies in Edentata. Cytobios, 18: 157-172.

Meritt, D. A. Jr.
1976. Sex ratios of Hoffmann's sloth, *Choloepus hoffmanni* Peters and three-toed sloth, *Bradypus infuscatus* Wagler, in Panama. Am. Midl. Nat., 96: 472-473.

Meritt, D. A., R. J. Low, and K. Benirschke.
1973. The chromosomes of *Zaedyus pichiy*. Mamm. Chrom. Newsletter, 14: 108-109.

Montgomery, G. G., and M. E. Sunquist.
1975. Impact of sloths on neotropical forest energy flow and nutrient cycling. Pp. 89 *In* E. Medina and F. Golly, editors, Trends in Tropical Ecology: Ecological Studies IV. Springer-Verlag, New York. 398 pp.

Muntzing, A.
1974. Accessory chromosomes. Ann. Rev. Genet., 8: 243-266.

Murtagh, C. E.
1977. A unique cytogenetic system in Monotremes. Chromosoma (Berl.), 65: 37-57.

Saez, F. A., M. E. Drets, and N. Brum.
1964. The chromosomes of the mulita (*Dasypus hybridus*): a mammalian edentate of South America. Pp. 163-170 *In* Mammalian cytogenetics and related problems in Radiobiology. Pergamon Press, New York. 370 pp.

Volubujev, V. T.
1980. The B Chromosome system of mammals. Genetica, 52/53: 333-337.

Wetzel, R.M.
1985. The identification and distribution of recent xenarthra (= edentata). (this volume).

Wetzel, R., and F. D. Avila-Pires.
1980. Identification and distribution of the recent sloths of Brazil (Edentata). Rev. Brasil. Biol., 40(4): 831-836.

PROTEIN VARIATION IN POPULATIONS OF *DASYPUS NOVEMCINCTUS* AND COMPARISONS TO *D. HYBRIDUS, D. SABANICOLA* AND *CHAETOPHRACTUS VILLOSUS*

Paul R. Ramsey and Benson A. Grigsby

Department of Zoology, Louisiana Tech University, Ruston, LA 71272.

Abstract
Horizontal starch gel electrophoresis was used to investigate genic variation at 28 loci encoding for 19 proteins in 265 samples of *Dasypus novemcinctus* L. from 13 localities in the United States. Only one polymorphic locus, an isocitrate dehydrogenase (IDH-1), was identified in southwestern populations. Armadillos in Florida, introduced in the 1920s, were monomorphic for all loci studied. A "marker" IDH-1 allele will indicate gene flow from the southwestern population into the Florida group. Average proportion of heterozygotes per locus (H) was about 1 percent to 2 percent for *D. novemcinctus* in Louisiana. Three South American species, *D. hybridus* (Desmarest) (n = 1), *D. sabanicola* Mondolfi (n = 5), and *Chaetophractus villosus* Desmarest (n = 3) share 7, 6 and 1 of 10 allelic products in blood, respectively, with *D. novemcinctus*. Interspecific variation was revealed in 14 of the 28 loci. IDH-1 was polymorphic also in *D. sabanicola*.

Abstracto
Se usó el método horizontal de electroforesis en gel de almidón para investigar las variaciones de los genes en 28 loci en los genes que codifican la síntesis de 19 proteínas distintas, usando para ello 265 muestras de *Dasypus novemcinctus* L. de 13 localidades distintas en los Estados Unidos. En poblaciones del suroeste, solamente se identificó un locus polimórfico, el cual resultó ser de actividad para el isocitrato de dehidrogenasa (IDH-1). Armadillos de la Florida, que fueron introducidos durante la década de 1920, demostraron ser monomórficos para todos los loci de genes estudiados. La existencia de un alelo IDH-1 "marcador" indica un flujo de genes de la población del suroeste hacia el grupo de la Florida. La proporción en promedio de heterocigotos por locus en los genes (H) fue de un 1-2 percent para *D. novemcinctus* de Louisiana. Tres especies suramericanas, *D. hybridus* (Desmarest) (n = 1), *D. sabanicola* Mondolfi (n = 5), y *Chaetophractus villosus* Desmarest (n = 3) comparten con *D. novemcinctus* 7, 6, y 1 respectivamente, de 10 productos alélicos en la sangre. Catorce de 28 loci revelaron variaciones interespecíficas. El locus IDH-1 también resultó ser polimórfico en *D. sabanicola.*

Resumo
A eletroforese horizontal em gel de amido foi utilizada para investigar-se a variação gênica em 28 loci, codificando para 19 proteínas, em 265 amostras de *Dasypus novemcinctus* L., de 13 localidades dos Estados Unidos. Somente um locus polimórfico, um isocitrato de desidrogenase (IDH-1), foi identificado nas populações do sudoeste. Os tatus na Flórida, introduzidos por volta de 1920, eram monomórficos para todos os loci estudados. Um alelo IDH-1 "marcador" indicará o fluxo de gen da população do sudoeste para o grupo da Flórida. Uma proporção média de heterozigotos por locus (H) foi de cerca de 1 percent a 2 percent para *D. novemcinctus* em Louisiana. Três espécies sul americanas, *D. hybridus* Desmarest (n = 1), *D. sabanicola* Mondolfi (n = 5), e *Chaetophractus villosus* Desmarest (n = 3) compartilhavam 6, 6 e 1 dos 10 produtos alélicos no sangue, respectivamente, com *D. novemcinctus*. Uma variação interespecífica foi revelada em 14 dos 28 loci. IDH-1 foi polimórfico também em *D. sabancola*.

Electrophoretically determined allelic products now are used routinely for studies in genetics, systematics, and population biology with a wide variety of animals (Avise, 1974; Lewontin, 1974; Powell, 1975; Selander, 1977). Electrophoretic studies have been particularly useful in testing ecological or evolutionary hypotheses (Gaines and Krebs, 1971; Selander et al., 1971). From the viewpoint of population biology, analysis of protein variation provides the opportunity to examine the genetic consequences of two population phenomena, polyembryony and rapid range expansion, in the common long-nosed armadillo, *Dasypus novemcinctus.*

Along with summaries of the general ecology, taxonomy, and morphology of the species, Talmadge and Buchanan (1954) discussed the reproductive cycle which results in polyembryony. After a delay of three to four months, the blastocyst implants and divides into four embryonic growth centers attached by a common placenta to the uterine mucosa. Monozygotic quadruplets are born approximately four months later. The demographic and genetic implications of such a reproductive pattern are of interest, especially since the common long-nosed armadillo extended its range from Mexico through eastern Texas into Louisiana and Mississippi to as far north as Arkansas during the past century (Humphrey, 1974). Spread of the species was favored by the advance of civilization during the 19th century, which included clearing of woodlands and the removal of large carnivores (Taber, 1939; Fitch et al., 1952).

Further aid to the range extension has been provided by transportation and introduction. Permanent establishment of *D. novemcinctus* in Florida occurred after a pair escaped from a private zoo at Cocoa, Brevard County, in 1924; others escaped from an overturned circus truck at Titusville in 1936 (Sherman, 1936; Fitch et al., 1952; Talmage and Buchanan, 1954). The nucleus of the armadillo population in Florida (and the southeastern United States) was perhaps fewer than 10 animals, probably obtained in eastern Texas. Another introduction at Miami during World War I described by Bailey (1924) apparently did not succeed. Three major changes in distribution have occurred since 1954 (Humphrey, 1974): (1) continued range expansion northward and eastward by the Texas group, (2) expansion northward and westward by the introduced Florida group, and (3) a range contraction in western Texas. Northern winter cold (9 annual freeze-days) and western aridity (380 mm annual precipitation) are thought to determine present distributional limits. *D. novemcinctus* has a continuous range from northern Argentina to Oklahoma and from sea level to 3,000 m (Walker, 1968). Other *Dasypus* species are confined to South America (Wetzel, 1985).

Electrophoretic data can contribute greatly to our knowledge of the systematics and evolution of armadillos. A member of the order Xenarthra (Edentata), which includes vermilinguas (neotropical anteaters) and sloths, the family Dasypodidae contains 13 species in 8 genera plus 6 species of *Dasypus* (Wetzel, 1985; Wetzel and Mondolfi, 1979). Apart from showing the reproductive pattern of regular polyembryony, *D. hybridus* from Argentina and *D. novemcinctus* apparently are closely related. Both species have a diploid number of 64 chromosomes with 46 acrocentric elements (Benirschke et al., 1968).

Recently, *D. novemcinctus* has emerged as an animal model for leprosy and other diseases. Subsequent to Storrs' (1971) summary of the diet, taxonomy, morphology, biochemistry, physiology, and known diseases of the common long-nosed armadillo, the animal has been studied extensively for prospective use in biomedical research. Kirchheimer and Storrs (1971) reported on the development of leprosy in armadillos inoculated with *Mycobacterium leprae*. Purtilo et al. (1975) investigated the immune system of *D. novemcinctus* and found it to be intact and well developed. Walsh et al. (1977) reported on naturally occurring

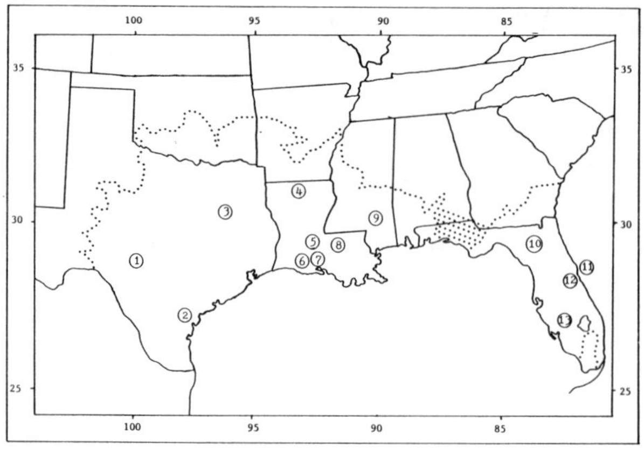

Figure 1. Sample localities within the range of *Dasypus novemcinctus* in the United States. Distribution limits are from Humphrey (1974); stippled area represents zone of contact for southwestern and southeastern populations.

leprosylike disease in *D. novemcinctus*. Storrs et al. (1975) were successful in inducing leprosy in *D. hybridus*. Since genetic influences on disease states are well documented, the degree of genic variation in species which serve as disease models or in other biomedical capacities is of considerable interest.

Although numerous biological aspects of armadillos have been investigated, reports of electrophoretic examination of protein systems are few. Burns et al. (1975) used polyacrylamide disc electrophoresis and densitometer scanning to demonstrate lactate dehydrogenase in *D. novemcinctus*. Dhindsa et al. (1971) presented the banding pattern of hemoglobin in the same species. No studies on biochemical variation as an indicator of genic differentiation have been published heretofore. This lack of information prompted the present studies: (1) to identify polymorphic proteins, optimal tissue for their assay, and appropriate electrophoretic regimes for use in future studies; (2) to establish the degree of genic variation within and among populations and possible differences between the introduced Florida populations and the southwestern United States populations; and (3) to compare genic similarities of four species of armadillos.

MATERIALS AND METHODS

Samples of *D. novemcinctus* (n = 265) were collected during 1976-78 from 13 localities across the range of the species in the southwestern and southeastern United States (Fig. 1, Table 1). Animals from localities 5, 7 and 8 had been housed for as long as one year in a colony at the Gulf South Research Institute (GSRI) at New Iberia, Louisiana. In addition, limited samples of three South American species including *D. sabanicola* (n = 5), *D. hybridus* (n = 1; blood only) and *Chaetophractus villosus* (n = 3; blood only) were available at the GSRI facility (Table 1).

Table 1. Localities and sample sizes for armadillos used in electrophoretic studies. Samples 1-13 are for *Dasypus novemcinctus*.

Sample Number	*Locality*	*Number of Armadillos*
1	Texas: Junction	3
2	Texas: Welder Wildlife Foundation near Sinton	2
3	Texas: Engeling Wildlife Management Area near Palestine	8
4	Louisiana: Ruston	48
5	Louisiana: Ville Platte	13
6	Louisiana: Little Pecan Hunting Reserve near Grand Chenier	12
7	Louisiana: Vicinity of New Iberia	49
8	Louisiana: Grosse Tete	8
9	Mississippi: Hattiesburg	5
10	Florida: Hatchet Creek Hunting Reserve 24 km E. of Gainesville	8
11	Florida: Merritt Island National Wildlife Refuge	45
12	Florida: Platt Ranch near Melbourne	31
13	Florida: Fish-eating Creek Wildlife Management Area near Palmdale	33
14	Venezuela: State of Apure at Rancho El Frio *Dasypus sabanicola*	5
15	Argentina: Province of Buenos Aires near Azul *Dasypus hybridus*	1
16	Argentina: *Chaetophractus villosus*	3

Plasma, hemolysate, and tissue extracts were prepared according to methods outlined by Selander et al. (1971). Blood samples from 255 *D. novemcinctus* and the other species were taken with heparinized, evacuated tubes and stored on wet ice until plasma and red cell fractions were separated and stored at -60°C. Plasma and toluene-treated hemolysate fractions were diluted 1:24 for examination of albumin and 1:14 for examination of hemoglobin, respectively. Plasma samples were treated with Rivanol to confirm the transferrin banding pattern (Chen and Sutton, 1967). For 188 *D. novemcinctus* and the 5 *D. sabanicola*, subsamples of heart, kidney, liver, abdominal muscle, and testis were homogenized 1:1 wt/vol in deionized water to prepare an amalgamated extract. Single-tissue extracts from both species were prepared to examine tissue specificity of isoenzymes and general proteins.

Techniques for horizontal starch gel electrophoresis and protein staining were those described by Selander et al. (1971), Ayala et al. (1972) and Shaw and Prasad (1970). One ml of 13 mM NADP or 15 mM NAD was added to appropriate 11 percent wt/vol electrostarch gels and cathodal trays to enhance staining and resolution of bands on the gels (Table 2). Protein systems were

Table 2. Proteins examined, numbers of scorable loci and buffer systems used for electrophoresis of armadillo tissues.

Protein system (loci)	*Buffer-gel system*
Oxidoreductases	
α-Glycerophosphate dehydrogenase (1)	Tris-citrate pH 8.0 (+ NAD)
Sorbitol dehydrogenase (1)	Tris-citrate pH 8.0 (+ NAD)
Lactate dehydrogenase (2)	Tris-citrate pH 8.0 (+ NAD)
Malate dehydrogenase (2)	Tris-citrate pH 6.7 (+ NADP)
Isocitrate dehydrogenase (2)	Tris-citrate pH 6.7 (+ NADP)
6-Phosphogluconate dehydrogenase (1)	Tris-maleate pH 7.4 (+ NADP)
Glucose-6-phosphate dehydrogenase (1)	Poulik pH 8.7 (+ NADP)
Glutamate dehydrogenase (1)	Tris-citrate pH 8.0 (+ NAD)
Superoxide dismutase (2)	Poulik pH 8.7 (+ NADP)
Transferases	
Glutamate oxaloacetate transaminase (2)	Tris-citrate pH 6.7 (+ ADP)
Phosphoglucomutase (2)	Tris-citrate pH 6.7 (+ NADP)
Hydrolases	
Esterases (3)	Poulik pH 8.7 (+ NADP)
Acid phosphatase (1)	Tris-citrate pH 6.7 (+ NADP)
Leucine amino peptidase (1)	Tris-HCl pH 8.5
Isomerases	
Phosphoglucose isomerase (2)	Tris-citrate pH 6.7 (+ NADP)
Nonenzymatic	
Albumin (1)	Poulik pH 8.7 (+ NADP)
Transferrin (1)	Poulik pH 8.7 (+ NADP)
Protein-3 (1)	Poulik pH 8.7 (+ NADP)
Hemoglobin (1)	Poulik pH 8.7 (+ NADP)

Table 3. Tissue specificity of isoenzymes and general proteins in *Dasypus novemcinctus*. Symbols represented are (+) activity, (–) weak activity and (o) no activity.

	Tissue Extract						
Locus	*Heart*	*Kidney*	*Liver*	*Muscle*	*Testis*	*Plasma*	*Hemolysate*
α-GPDH	+	+	+	+	+	o	o
SDH	–	+	+	–	–	o	o
LDH-1	+	+	–	+	+	–	+
LDH-2	–	–	+	+	o	o	o
MDH-1	+	+	+	+	+	–	–
MDH-2	+	+	+	+	–	o	o
IDH-1	o	+	+	o	o	o	o
IDH-2	+	o	–	–	o	o	o
6-PGDH	+	+	+	+	+	o	+
G-6-PDH	–	+	+	–	+	o	+
GDH	–	+	+	o	o	o	o
SOD-1	+	+	+	–	–	?	+
SOD-2	+	+	+	–	o	o	o
GOT-1	+	+	+	+	–	o	–
GOT-2	+	+	+	+	–	o	o
PGM-1	–	+	+	+	+	o	–
PGM-2	+	+	–	–	o	o	o
ES-1	+	+	+	+	+	+	+
ES-2	–	+	+	–	–	–	o
ES-3	o	o	o	o	o	o	+
ES-4	+	+	+	+	+	+	+
ES-5	o	o	o	o	o	–	–
ASP	o	+	+	–	+	o	o
LAP-1	+	+	+	+	+	–	–
PGI-1	–	–	–	–	–	o	+
PGI-2	+	+	–	–	–	o	o
ALB	–	–	–	–	–	+	o
TRF	o	o	o	o	o	+	o
PT-3	o	o	o	o	o	+	–
HB	o	o	o	o	o	o	+

numbered in increasing fashion by their anode-to-cathode position. Alleles were designated by the relative mobility of their protein products; all *D. novemcinctus* alleles were arbitrarily designated 100. Alleles at homologous loci in other species were assigned a value based on the mobility of their protein product relative to the 100 allele of *D. novemcinctus*.

RESULTS

Five buffer-gel systems were used to assay for 19 proteins encoded by 28 structural loci (Tables 2 and 3). Among the proteins examined, only one polymorphic system, an isocitrate dehydrogenase IDH-1, was found (Table 4). IDH-1 was variable in samples of *D. novemcinctus* taken from the southwestern United States and in *D. sabanicola*. Animals collected from Florida showed a complete lack of variability for the protein systems examined. Comparison of *D. novemcinctus* to *D. sabanicola*, *D. hybridus*, and *C. villosus*, revealed interspecific variation in 14 of the 28 loci (Table 4). A post-albumin general protein has a variable pattern both among species and between samples of *D. novemcinctus*. This variation may have a genetic basis but was not considered here as the product of a single locus.

Tissue specificities (Table 3), genic differences among the species (Table 4), and banding patterns (phenotypes) on gels for individual protein systems are summarized below. The name of the protein is followed by its abbreviation and Enzyme Commission number.

αGlycerophosphate dehydrogenase (α GPDH; 1.1.1.8).-- This anodally migrating tissue enzyme (also named glycerol-3-phosphate dehydrogenase) apparently is encoded by one locus and shows one allele across the range of *D. novemcinctus*. *D. sabanicola* differs from *D. novemcinctus* by possessing another allele, α $GPDH^{119}$.

Sorbitol dehydrogenase (SDH; 1.1.1.14).-- SDH shows a single zone with bands of similar mobility in *D. novemcinctus* and *D. sabanicola*. Activity of this enzyme was absent (or too weak), such that plasma and hemolysate of *D. hybridus* and *C. villosus* yielded no information.

Table 4. Relative mobilities of allozyme bands which differed among four species of armadillos.

Locus	Dasypus novemcinctus	Dasypus sabanicola	Dasypus hybridus	Chaeophractus villosus
α-GPDH	100	119	–	–
LDH-2	origin	origin	origin	+ 3 mm
IDH-1	100/85	85/61	–	–
6-PGDH	100	100	100	76
G-6-PDH	100	94	94	80
GDH	100	80	–	–
SOD-1	100	111	108	76
SOD-2	100	111	–	–
PGM-1	100	100	100	63
ES-3	100	100	100	89
ALB	100	102	100	100
TRF	100	106	100	124
PT-3	100	100	100	92
HB	100	100	110	66

Lactate dehydrogenase (LDH; 1.1.1.27).-- Two LDH systems are present in armadillos. LDH-1 is primarily a heart isoenzyme and is monomorphic. LDH-2 is most prevalent in muscle tissue, and a slightly more anodal band in *C. villosus* accounts for the only variation seen in this system (Table 4). The five-banded pattern of two homotetramers and three heterotetramers seen on gels was demonstrated also by Burns et al. (1975).

Malate dehydrogenase (MDH; 1.1.1.37).-- Two NAD-dependent enzymes were monomorphic in *D. novemcinctus*, and a MDH-1 band of similar mobility was observed in the other species. The two isoenzymes, cytoplasmic MDH-1 and mitochondrial MDH-2, migrate anodally and cathodally, respectively.

Isocitrate dehydrogenase (IDH; 1.1.1.42).-- Two anodally migrating systems were scorable. No data concerning IDH-1 (cytoplasmic) or IDH-2 (mitochondrial) are available for *D. hybridus* and *C. villosus*. IDH-2 appears at the sample origin and is monomorphic in both *D. novemcinctus* and *D. sabanicola*. However, IDH-1 is polymorphic for two alleles in both species. In *D. novemcinctus*, alleles were designated IDH-1^{100} and IDH-1^{85}. Heterozygotes for this dimeric enzyme show a three-banded pattern (Fig. 2). Genotypic frequencies from two regional samples for IDH-1 were in expected Hardy-Weinberg proportions (Table 5); no sex differences were noted when all samples from the southwestern United States were pooled. IDH-1^{100} did not occur in the small sample from Mississippi. *D. sabanicola* has the IDH-1^{85} allele found at high frequency in *D. novemcinctus*. Four *D. sabanicola* were IDH-$1^{85/85}$ homozygotes while one was a heterozygote, IDH-$1^{85/61}$ (Fig. 2).

6-Phosphogluconate dehydrogenase (6-PGDH; 1.1.1-.44).-- Armadillos possess one locus for this enzyme, and the four species showed no polymorphism. From its one hemolysate sample, *C. villosus* showed a slower band, 6-$PGDH^{76}$.

Glucose-6-phosphate dehydrogenase (G-6-PDH; 1.1.1.-49).-- The one product in *D. novemcinctus* was evidenced by a single anodally migrating band. Alleles for *D. sabanicola* and *D. hybridus* were designated G-6-PDH^{94}, while that for *C. villosus* was G-6-PDH^{80}. Another, less distinct band of hexose-6-phosphate dehydrogenase activity which occurred near the sample origin was not consistently scorable. Although the kidney and liver enzyme is better scored from a Tris-borate-EDTA gel (pH 8.0), NADP added to the gel and cathodal tray did not preclude subbanding by red cell G-6-PDH.

Glutamate dehydrogenase (GDH; 1.4.1.2.).-- Primarily a kidney and liver enzyme, GDH showed a single band in armadillos. No polymorphism was noted, but *D. sabanicola* had a more cathodal band, GDH^{80}. Use of techniques described by Nelson et al. (1971) greatly

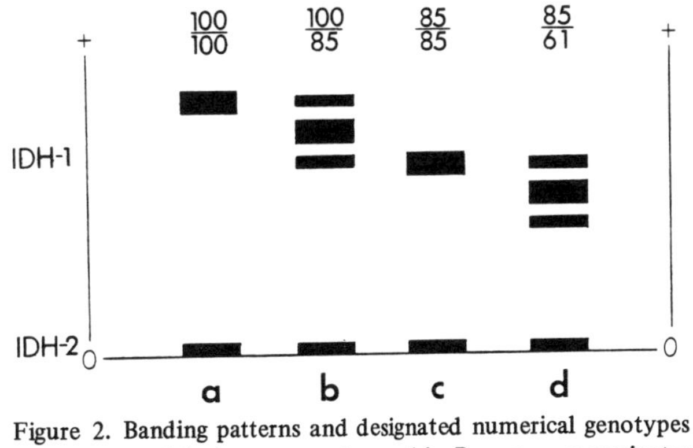

Figure 2. Banding patterns and designated numerical genotypes for isocitrate dehydrogenases observed in *Dasypus novemcinctus* (a, b, c) and *Dasypus sabanicola* (c, d).

improved resolution of this system. A weak, trailing subband, also noted by those investigators, was observed in about one-third of our samples.

Superoxide dismutase (SOD; 1.15.1.1; = tetrazolium oxidase, indolephenol oxidase).-- Two anodally migrating isoenzymes were scorable as negatively stained zones. Neither of the two systems was polymorphic in *D. novemcinctus*, but variation among the four species was evident. The alleles are designated $SOD\text{-}1^{111}$, $SOD\text{-}1^{108}$, and $SOD\text{-}1^{76}$ for *D. sabanicola, D. hybridus*, and *C. villosus*, respectively. SOD-2, not scorable in these blood samples, showed a slightly greater migration for *D. sabanicola* tissue extracts (Fig. 3). Presence of SOD in plasma is questionable (Table 3) because hemoglobin contamination may modify the blue formazan deposited in the gel, causing apparent achromatic zones (Beauchamp and Fridovich, 1971). Misra and Fridovich (1977) developed a procedure involving oxidation of dianisidine in the presence of riboflavin which results in positively stained bands.

Glutamate oxaloacetate transaminase (GOT; 2.6.1.1).-- Two isoenzymes occur in armadillos. Cytoplasmic GOT-1 migrates anodally and mitochondrial GOT-2, not appearing in blood, migrates cathodally. Both systems were nonvariable in *D. novemcinctus*, and no variation was noted among the other species.

Phosphoglucomutase (PGM; 2.7.5.1).-- An anodal PGM-1 and a slightly cathodal PGM-2 were scorable. Hemolysate from *C. villosus* revealed another allelic product, $PGM\text{-}1^{63}$. A faintly staining diffuse zone, frequently observed anodal to PGM-1, may indicate a third PGM locus.

Esterases (ES; 3.1.1.*).-- While at least five esterase systems were revealed, only three of these were consistently scorable (Table 3, Fig. 3). No variation was noted in any of the systems for *D. novemcinctus* An unscorable hemolysate system, ES-5, was distinct in *D. hybridus*, but a diffuse band occurred in the other species. The only interspecific variation occurred in ES-3 where *C. villosus* possessed an allele designated $ES\text{-}3^{89}$ (Fig. 3, Table 4). Esterases are scorable on LiOH(+ NADP) gels and ES-3 is resolved best on Tris-maleate (pH 7.4). However the information reported here is from

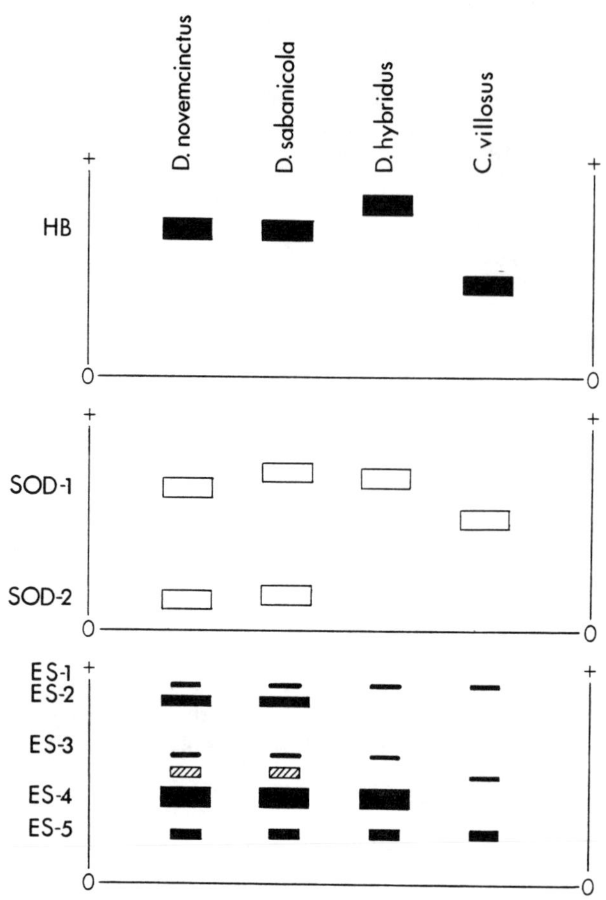

Figure 3. Observed electrophoretic phenotypes for hemoglobin, superoxide dismutases and esterases in four species of armadillos.

Poulik gels (pH 8.7). The band shown between ES-3 and ES-4 in Fig. 3 represents a subband of ES-4, although the possibility of a separate system cannot be excluded. ES-4 was present in hemolysate samples, but diffuse bands prevented accurate representation for *C. villosus*.

Acid phosphatase (ASP; 3.1.3.2).-- A distinctly banding, monomorphic system and a diffuse, more anodal zone were observed in tissue extracts.

Leucine amino peptidase (LAP; 3.4.1.1).-- Two anodally migrating peptidase systems were identified upon staining with L-leucyl-β-naphthylamide-HCl as the substrate and fast garnet as the dye. Red-to-purple bands indicated activity in both systems; the more anodal

Table 5. Genotypic and allelic frequencies for IDH-1 in samples from northern and southern Louisiana. Numbers in parentheses are expected values.

Area	*Genotypes*			*Allele Frequency*	*95 Percent Confidence*
	100/100	*100/85*	*85/85*	*100*	*Limits*
Northern (n = 36)	2 (2.3)	14 (13.5)	20 (20.3)	0.250	0.157-0.363
Southern (n = 44)	1 (1.5)	14 (13.1)	29 (29.4)	0.182	0.101-0.286

LAP-1 was scorable and monomorphic. These amino acyl naphthylamidases are called LAP systems here (but see Manwell and Baker, 1970). General esterase activity increased substantially when a 0.2M Tris-HCl (pH 5.7) stain buffer was used.

Phosphoglucose isomerase (PGI; 5.3.1.9).-- Two loci encode for PGI; anodal and slightly cathodal bands were observed along with a more distinct, intermediate band, the possible heterodimer. This protein did not vary among the species examined. Suggestion of two loci for PGI in armadillos contrasts with the usual reports of a single locus for mammals (Fisher et al., 1980). Although the interpretation is tentative, the two major bands show evidence of tissue specificity (Table 3). A similar three-banded phenotype was described for white-tailed deer (Manlove et al., 1975).

Albumin (ALB).-- This plasma protein forms the most anodal band on gels stained with amido black. *D. novemcinctus, D. hybridus* and *C. villosus* possess identical alleles, while *D. sabanicola* possesses a slightly faster product designated ALB^{102} (Table 4). The locus is also scorable from liver homogenates without dilution.

Transferrin (TRF).-- *D. novemcinctus* and *D. hybridus* possess identical alleles for this protein which is monomorphic in *D. novemcinctus. D. sabanicola* and *C. villosus* possess TRF^{106} and TRF^{124} alleles, respectively (Table 4). Allelic designations are provisional, however, until more samples of these species can be examined. Cathodal subbanding prohibited clear resolution of banding patterns; on similar grounds, a weak polymorphism in *D. novemcinctus* could not be discounted.

Protein-3 (PT-3).-- This monomorphic, most cathodal plasma protein migrates similarly in *D. novemcinctus, D. sabanicola* and *D. hybridus.* Another band, $PT\text{-}3^{92}$, was found in *C. villosus.*

Hemoglobin (HB).-- This erythrocytic protein showed no polymorphism, yet electrophoretic mobility at pH 8.7 differed among three of the four species (Fig. 3). *D. novemcinctus* and *D. sabanicola* have similar HB bands, but *D. hybridus* hemoglobin migrates faster (HB^{110}) and that of *C. villosus* is slower (HB^{66}). The minor, far anodal band shown by Dhindsa et al. (1971) was evident in our undiluted samples but is not shown in Fig. 3. Such minor hemoglobin components have been seen in mice (Selander et al., 1971). The minor and major bands contain products of both alpha and beta loci, but are reported here as a single locus. De Jong et al. (1981) stated that two types of beta chain in a ratio of 3:7 occurred in the hemoglobin of the one *D. novemcinctus* individual from Panama which they investigated. This proposed gene duplication was not detected by our techniques.

DISCUSSION

Several limitations in the data must be considered when the usual electrophoretic technique is used to estimate genic variation. Electrophoretic methods reveal proteins encoded by an unknown percentage of the genome. Loci examined must be assumed to represent a random sample of structural genes only. Additionally, many amino acid changes do not alter the net charge of the protein. Singh et al. (1976) outlined an array of more sensitive techniques such as varying gel conditions, denaturation, immunological methods, enzyme kinetics, and peptide mapping and fingerprinting to aid in detecting "hidden" allelic variation. Combinations of two pH values and two gel concentrations revealed three times the previous number of allelic forms for xanthine dehydrogenase in *Drosophila pseudoobscura.* This design also increased the average population frequency of heterozygotes by 30 percent .

By usual electrophoretic methods used here, the only polymorphism detected in *D. novemcinctus* was found in samples from Texas and Louisiana. The IDH-1 polymorphism probably occurs also in Mississippi, but the small sample precludes confirmation. Obviously, small sample sizes prohibit accurate conclusions on protein variation within the three other species. For example, transferrin apparently migrates differently among the three species, but suitable evaluation of between-sample variation in subbanding was possible only for *D. novemcinctus.* Likewise, no index of similarity or genetic distance may be applied with confidence to such limited data.

Characteristics such as the novel reproductive physiology, a relatively recent range expansion, introduction and population "bottlenecks" suggested that genetic variation might be low in *D. novemcinctus* populations in the United States. This genetic homogeneity is illustrated by calculation of H, the average proportion of heterozygotes per locus or average individual heterozygosity. For the northern and southern Louisiana samples, H was 1.3 percent and 1.1 percent , respectively. An estimated overall heterozygosity of about 1 percent is extremely low when compared to H for other species. Invertebrates and vertebrates show an estimated genic heterozygosity per population of 13 to 15 percent and 5 to 10 percent , respectively, while the average across all animals is about 9 percent (Selander and Kaufman, 1973; Lewontin, 1974; Powell, 1975).

Lewontin (1974) proposed that unique population structures and/or recent evolutionary histories produce some species which are notably less variable than others. Genomes may be relatively monomorphic due to high and consistent inbreeding, expansion from a small pool of variation, or strong selection for uniform genotypes. *D. novemcinctus* reproduces sexually, yet within-litter genetic variance is essentially zero. By producing four isogenic offspring, a female's reproductive effort in terms of genetic diversity is as if she had produced a litter of one. Nevertheless, part of the advantage of sexual reproduction, an array of diverse offspring, is retained because considerable variation in phenotype occurs among armadillo siblings. Cytoplasmic inequality of the four primordial buds is responsible for the median 2.4-fold maximum difference within quadruplet sets in body and organ weights, scute number and biochemical parameters (Storrs and Williams, 1968). Such phenotypic flexibility would enhance the colonizing potential of armadillos as postulated for clones of parthenogenetic cockroaches (Parker et al., 1977).

Inbreeding and the random processes, genetic drift and founder effect, undoubtedly have contributed to the low degree of heterozygosity found in *D. novemcinctus*. The reproductive pattern of *D. novemcinctus* is not likely to decrease heterozygosity in the absence of accompanying factors such as inbreeding. It is possible that *D. novemcinctus* expanded into the United States from a large pool of variation. Genetic bottlenecks resulted from establishment of new populations by relatively few individuals plus inbreeding within these range-front populations. Similarly, these processes may have been repeated in the early 1920s when armadillos were introduced into Florida. Future studies must establish the genic composition of source populations from which the United States populations were derived. The proximate cause of the original expansion is unknown, but genetic factors may have been involved (Humphrey, 1974).

Selection for generalized, relatively homozygous genomes might be a critical determinant for the success of colonizing immigrants. Selander and Kaufman (1973) suggested that genic polymorphism occurs less in mobile organisms which experience their environment as the average of several microhabitats (i.e., fine-grained environment; see Levins, 1968). If extended to colonizers, this hypothesis could be tested only if H is measured prior to a range expansion and if stochastic effects are negligible during migration cycles. Neither condition is likely to occur by design.

Protein variability may be more directly related to the metabolic role of a particular polypeptide than to selection for some level of heterozygosity. Powell (1975) discussed the evidence that allelic diversity for any given enzyme is not confined to a single taxonomic group and that variability of an enzyme may be related to its function. Gillespie and Kojima (1968) previously had divided enzymes into those active in energy metabolism and those of unknown physiological function. They surmised that the glucose-metabolizing enzymes tend to show less variability than those of unknown physiological function. Kojima et al. (1970) postulated that the substrates for the glucose-metabolizing enzymes usually come from a prior reaction and are of a single form which changes in concentration only. Substrates for non-glucose-metabolizing enzymes come from the environment and may differ both quantitatively and qualitatively.

IDH, a glucose-metabolizing enzyme that catalyzes the conversion of isocitrate to α-ketoglutarate in the Krebs cycle, is relatively monomorphic in animals. It shows an average of 0.055 heterozygotes for mammal species and 0.082 for all animals (Powell, 1975). *D. novemcinctus* had an average IDH heterozygosity for all samples of 0.336. Cameron and Vyse (1978) reported an average heterozygosity for IDH of 0.28 for elk. Their samples from Yellowstone National Park revealed an average heterozygosity of 1.2 percent across 24 loci where IDH, as in *D. novemcinctus*, was the only polymorphic locus. The cotton rat, *Sigmodon hispidus*, has extended its range in the United States by northward movement comparable to that by armadillos, opossums and several other species. Johnson et al. (1972) reported genic heterozygosity of 2 percent to 3 percent in populations of *S. hispidus*. Although IDH was polymorphic, other loci, notably GOT-1, made greater contribution to the low value of H.

Thus, low heterozygosity may be associated with certain polypeptides, classes of enzymes, or nonspecific biochemical diversity. In seeking explanations for the low heterozygosity of armadillos, other studies must examine samples from throughout the range of *D. novemcinctus* as well as large samples of other species. One important objective would be to compare species displaying polyembryony to species of *Dasypus* not possessing this trait.

One consequence of the present study concerns its contribution to the accumulating, though circumstantial, evidence for a genetic hypothesis for susceptibility to leprosy. All tested armadillos do not show equivalent response to inoculations with *M. leprae*, and populations of *D. novemcinctus* from Florida and Venezuela differ in their susceptibility from those in Louisiana (E. Storrs, pers. comm.). Further attention must be focused at the zone of contact for the southwestern and southeastern groups of *D. novemcinctus* (Fig. 1). It is reasonable to assume that the "marker" allele IDH-1^{100} will move through the homozygous IDH-$1^{85/85}$ populations in Florida as the groups commingle. When gene flow involving IDH-1 and the postulated susceptibility gene(s) is indicated, observed changes in leprosy susceptibility would lend support to the hypothesis of its inheritance.

Future studies should also focus on the taxonomy of the dasypodids and degrees of intrapopulation variation. From the mobilities of the allelic products (Table 4), the degree of relatedness of the four species in this study can only be suggested. When only the loci encoding for allelic products present in blood are employed in comparing the species, *D. novemcinctus* and *D. hybridus* share 7 of 10 alleles for a 30 percent difference. Likewise, *D. novemcinctus* differs from *D. sabanicola* by 40 percent and from *C. villosus* by 90 percent . If all systems are used to compare *D. novemcinctus* and *D. sabanicola*, a 50 percent difference is seen. Complete taxonomic knowledge of dasypodids, a group to be used as laboratory species or animal models, is vital since it is obvious that species and populations of armadillos have differential susceptibility to disease. However, no accurate conclusions on relationships can be drawn without some appraisal of intraspecific variation.

Addendum

Recent examination of *D. novemcinctus* from the state of Sinaloa, Mexico (n=4) and central Venezuela (n=3) revealed that several additional polymorphic proteins must exist in populations from Central America and/or South America. Among animals from Venezuela, heterozygotes were detected for PGM-1, ES-4, and possibly transferrin. All specimens from Venezuela had uniformly different single bands for α-GPDH, G-6-PDH, GDH, PGI-2 and albumin.

ACKNOWLEDGMENTS

We thank G. A. McFarland, Bill and Betty Jo Johnson, Terry Majors and James Waddill for their assistance in collecting specimens. Dr. Eleanor E. Storrs at the Medical Research Institute of the Florida Institute of Technology provided samples and information while she was at the Gulf South Research Institute, New Iberia, Louisiana. She obtained the *D. hybridus* and *C. villosus* originally from Dr. Rueben Lombardo, CEPANZO, Buenos Aires, Argentina.

LITERATURE CITED

Avise, J.C.
1974. Systematic value of electrophoretic data. Syst. Zool., 23:465-481.

Ayala, F.J., J.R. Powell, M.L. Tracey, C.A. Mourao, and S. Perez-Salas.
1972. Enzyme variability in the *Drosophila willistoni* group. IV. Genic variation in natural populations of *Drosophila willistoni*. Genetics, 70:113-139.

Bailey, H. H.
1924. The armadillo in Florida and how it reached there. J. Mamm., 5:264-265.

Beauchamp, V., and I. Fridovich.
1971. Superoxide dismutase: Improved assays and an assay applicable to polyacrylamide gels. Anal. Biochem., 44:276-287.

Benirschke, K., R.J. Low, and V.H. Ferm.
1968. Cytogenetic studies of some armadillos. Pp. 330-344 *In* Comparative mammalian cytogenetics (K. Benirschke, ed.). Springer-Verlag, New York, 473 pp.

Burns, T.A., G. Chen, and J.C. Lin.
1975. Electrophoretic patterns and kinetics of heart and muscle lactate dehydrogenase of the nine-banded armadillo, *Dasypus novemcinctus*. Comp. Biochem. Physiol., 51:385-387.

Cameron, D.G., and E.R. Vyse.
1978. Heterozygosity in Yellowstone Park elk, *Cervus canadensis*. Biochem. Genet., 16:651-657.

Chen, S., and H.E. Sutton.
1967. Bovine transferrin: sialic acid and the complex phenotype. Genetics, 56:425-430.

De Jong, W.W., J.A. Leunissen and H.T. Cuijpers.
1981. Primary structure of the major β-chain of armadillo (*Dasypus novemcinctus*) haemoglobin. Biochem. et Biophys. Acta, 668: 57-62.

Dhindsa, D.S., A.S. Hoversland, and J. Metcalfe.
1971. Comparative studies of the respiratory functions of mammalian blood. VII. Armadillo (*Dasypus novemcinctus*). Respir. Physiol., 13:198-208.

Fisher, S.E., J.B. Shaklee, S.D. Ferris and G.S. Whitt.
1980. Evolution of five multilocus isoenzyme systems in the chordates. Genetica 52/53: 73-85.

Fitch, H.S., P. Goodrum, and C. Newman.
1952. The armadillo in the southeastern United States. J. Mamm., 33:21-37.

Gaines, M.S., and C.J. Krebs.
1971. Genetic changes in fluctuating vole populations. Evolution, 25:702-723.

Gillespie, J.H., and K. Kojima.
1968. The degree of polymorphism in enzymes involved in energy production compared to that in nonspecific enzymes in two *Drosophila ananassae* populations. Proc. Nat. Acad. Sci., 61:582-585.

Humphrey, S. R.
1974. Zoogeography of the nine-banded armadillo (*Dasypus novemcinctus*) in the United States. BioScience, 24:457-462.

Johnson, W.E., R.K. Selander, M.H. Smith, and Y.J. Kim.
1972. Biochemical genetics of sibling species of the cotton rat (*Sigmodon*). Univ. Texas Publ., 7213:298-305.

Kirchheimer, W.F., and E.E. Storrs.
1971. Attempts to establish the armadillo (*Dasypus novemcinctus* Linn.) as a model for the study of leprosy. I. Report of lepromatoid leprosy in an experimentally infected armadillo. Int. J. Lepr., 39: 692-701.

Kojima, K., J. Gillespie, and Y.N. Tobari.
1970. A profile of *Drosophila* species' enzymes assayed by electrophoresis. I. Number of alleles, heterozygosities, and linkage disequilibrium in glucose-metabolizing systems and some other enzymes. Biochem. Genet., 4:627-637.

Levins, R.
1968. Evolution in changing environments. Monogr. Pop. Biol., 2. Princeton Univ. Press, 120 pp.

Lewontin, R.C.
1974. The genetic basis of evolutionary change. Columbia University Press, New York, 346 pp.

Manlove,M.N., J.C. Avise, H.O. Hillstad, P.R. Ramsey, M.H. Smith, and D.O. Straney.
1975. Starch gel electrophoresis for the study of population genetics in white-tailed deer. Proc. Ann. Conf. SE Assoc. Game and Fish Comm. 29: 392-403.

Manwell, C., and C.M.A. Baker.
1970. Molecular biology and the origin of the species. University of Washington Press, Seattle, 394 pp.

Misra, H.P., and I. Fridovich.
1977. Superoxide dismutase and peroxidase: a positive activity stain applicable to polyacrylamide gel electropherograms. Arch. Biochem. Biophys., 183:511-515.

Nelson, R.L., M.S. Povey, D.A. Hopkinson, and H. Harris.
1977. Electrophoresis of human L-glutamate dehydrogenase: tissue distribution and preliminary population survey. Biochem. Genet., 15:87-91.

Parker, E.D., Jr., R.K. Selander, R.O. Hudson, and L.J. Lester.
1977. Genetic diversity in colonizing parthenogenetic cockroaches. Evolution, 31:836-842.

Powell, J.R.
1975. Protein variation in natural populations of animals. Pp. 79-119, *In* Evolutionary biology, Vol. 8 (T. Dobzhansky, M. K. Hecht, and W. C. Steere, eds.). Plenum Press, New York. 396 pp.

Purtilo, D.T., G.P. Walsh, E.E. Storrs, and C. Gannon.
1975. The immune system of the nine-banded armadillo (*Dasypus novemcinctus* Linn.). Anat. Rec., 181:725-734.

Selander, R.K.
1977. Genic variation in natural populations. Pp. 21-45 *In* Molecular evolution (F. J. Ayala, ed.). Sinauer Associates, Inc., Sunderland, Massachusetts, 277 pp.

Selander, R.K. and D.W. Kaufman.
1973. Genic variability and strategies of adaptation in animals. Proc. Nat. Acad. Sci., 70:1875-1877.

Selander, R.K., M.H. Smith, S.Y. Yang, W.E. Johnson, and J.B. Gentry.
1971. Biochemical polymorphism and systematics in the genus *Peromyscus*. I. Variation in the old-field mouse (*Peromyscus polionotus*). Univ. Texas Publ., 7103:49-90.

Shaw, C.R., and R. Prasad.
1970. Starch gel electrophoresis--a compilation of recipes. Biochem. Genet., 4:297-320.

Sherman, H.B.
1936. List of the recent wild land mammals of Florida. Proc. Fla. Acad. Sci., 1:102-128.

Singh, R.S., R.C. Lewontin, and A.A. Felton.
1976. Genetic heterogeneity within electrophoretic "alleles" of xanthine dehydrogenase in *Drosophila pseudoobscura*. Genetics, 84:609-629.

Storrs, E.E.
1971. The nine-banded armadillo: A new model for leprosy and other biomedical research. Int. J. Lepr., 39:703-714.

Storrs, E.E. and R. J. Williams.
1968. A study of monozygous quadruplet armadillos in relation to mammalian inheritance. Proc. Nat. Acad. Sci., 60:910-914.

Storrs, E.E., G.P. Walsh, and H.P. Burchfield.
1975. Development of leprosy in another species of armadillo *Dasypus hybridus* (L.): genetic and immunologic implications. J. Trop. Med. Hyg., 78:216-218.

Taber, F.W.
1939. Extension of the range of the armadillo. J. Mamm., 20:489-493.

Talmadge, R.V. and G.D. Buchanan.
1954. The armadillo (*Dasypus novemcinctus*); a review of its natural history, ecology, anatomy and reproductive physiology. The Rice Institute Pamphlet Monograph in Biology, 41:1-135.

Walker, E.P.
1968. Mammals of the world. The Johns Hopkins Press, Baltimore, 644 pp.

Walsh, G.P., E.E. Storrs, W. Meyers, and C.H. Binford.
1977. Naturally acquired leprosy-like disease in the nine-banded armadillo (*Dasypus novemcinctus*): recent epizootiologic findings. J. Reticuloendothel. Soc., 22:363-367.

Wetzel, R.M.
1985. Taxonomy and distribution of armadillos, Dasypodidae. (this volume).

Wetzel, R.M. and E. Mondolfi.
1979. The subgenera and species of long-nosed armadillos, genus *Dasypus*. Pp. 43 to 63 *In* Vertebrate ecology in the northern Neotropics (J. F. Eisenberg, ed.), Smithsonian Institution Press, Washington, D.C., 271 pp.

GENETIC COMPARISON OF *DASYPUS NOVEMCINCTUS* AND *D. HYBRIDUS* BASED ON ELECTROPHORETIC STUDIES

Oliver A. Ryder and Robert L. Davis

Research Department, Zoological Society of San Diego, P.O. Box 551, San Diego, California 92112.

Abstract
The extent of genic divergence of the armadillo species *Dasypus novemcinctus* and *D. hybridus* was studied, using starch and polyacrylamide gel electrophoresis in a comparison of liver enzymatic activities, plasma proteins and hemoglobins. These preliminary studies support the genetic distinction of the two armadillos at the species level.

Abstracto
Se estudió la extensión de la divergencia genética de las especies de armadillos *Dasypus novemcinctus* y *D. hybridus*, por medio de la electroforesis en gel de poliacrilamida y en gel de almidón, para comparar las actividades enzimáticas del hígado, de las proteínas del plasma y de las hemoglobinas. Estos estudios preliminares confirman la distinción genética de estos dos armadillos al nivel de especies.

Resumo
O âmbito de divergência gênica das espécies de tatus, *Dasypus novemcinctus* e *D. hybridus* foi estudado pela eletroforese em gel de amido e poliacrilamida numa comparação das actividades enzimáticas do fígado, proteínas do plasma e hemoglobinas. Diferenças gênicas apreciáveis coerentes com a classifição taxonômica vigente ocorrem entre essas duas espécies.

There is little information about genetic similarities of the Dasypodinid armadillos. *Dasypus novemcinctus* Linnaeus, 1758, and *D. hybridus* Desmarest, 1819, have similar or identical karyotypes consisting of 2n = 64 chromosomes, NF = 82 (Jorge et al., 1978; Benirschke et al., 1969). In addition to the considerable external morphological similarity, females of both species have a simple uterus, unlike most other armadillos. These two species also are remarkable for their variable polyembryony (Galbreath, 1985). The range of *D. hybridus* is generally confined to Argentina while *D. novemcinctus* has a wider range including northern Argentina, all countries east of the Andes, Ecuador, Colombia, Panama, Central America, Mexico and the southern United States (Wetzel, 1985).

This study was undertaken to examine biochemical genetic similarities between two armadillo species whose similarities in external morphology, karyotype, and variable polyembryony indicate close evolutionary relatedness. It was considered plausible that the common long-nosed armadillo *D. novemcinctus* and the mulita *D. hybridus* might be so similar as to be sibling species or subspecies (Benirschke, 1969; Jorge et al., 1978).

MATERIALS AND METHODS

Dasypus novemcinctus collected in Florida and *D. hybridus* (kindly provided by Dr. Luis Y. Melendez, Centro Panamericano de Zoonosis, Buenos Aires, Argentina) were kept at the San Diego Zoo. Blood samples were obtained from three *D. novemcinctus* and four *D. hybridus* individuals, separated into plasma and red cell fractions and frozen at -70°C until used. Liver samples from five *D. novemcinctus* individuals, including both sexes, and from a single female *D. hybridus* were obtained at necropsy and frozen at -70°C. Liver homogenates were prepared and horizontal starch gel electrophoresis was performed according to the procedures of Sealander et al. (1971). Enzymes and nonenzymatic proteins were identified using staining methods described by Harris and Hopkinson (1976) and Shaw and Prasad (1970) with minor modifications. Polyacrylamide gel electrophoresis was conducted utilizing 2 mm slab gels, 7.5 percent acrylamide under conditions described by Davis (1964). Staining of proteins in polyacrylamide gels was performed according to Holbrook and Leaven (1976).

RESULTS

Starch Gel Electrophoresis.-- Electrophoretic comparison of the hemoglobins of *D. novemcinctus* and *D. hybridus* demonstrated clear differences between these two species (Fig. 1). Sixteen enzymatic activities or intracellular proteins representing 23 presumptive loci were analyzed by starch gel electrophoresis. Of the 23 presumptive loci, 17 were apparently identical in *D. novemcinctus* and *D. hybridus* (Tables 1 and 2). Without discrimination between loci whose rates of change may vary we may calculate a genetic identity, $I = 17/23 = 0.74$ which yields a value for the average number of electrophoretically detectable substitutions per locus between the two armadillo species of $D = -\ln I = 0.30$.

We identified two and possibly three polymorphic systems (XDH, Es-3 and possibly one alpha GPD locus). We were unable to identify any polymorphic (heterozygous) loci in our single *D. hybridus* liver sample.

Polyacrylamide Gel Electrophoresis.-- Fresh plasma samples of *D. novemcinctus* and *D. hybridus* were subjected to polyacrylamide slab gel electrophoresis with the intent of resolving as many proteins as possible (Fig. 2). The serum albumins of the two armadillo species appear to have similar or identical electrophoretic mobilities. Separate analysis of haptoglobin phenotypes revealed no differences between *D. novemcinctus* and *D. hybridus* (data not shown). We treated the plasma protein gel data as suggested by Sarich (1977), comparing the number of protein bands with the same mobility divided by the number counted in both, and estimate that eight of fourteen bands counted have the same electrophoretic mobility.

Comparison of the plasma proteins of three *D. novemcinctus* and four *D. hybridus* yielded an estimate of their genetic identity, $I = 0.57$ which corresponds to an average substitution per locus of $D = 0.56$. Transferrin polymorphism in *D. hybridus* was apparent, as well as presumptive polymorphism in at least three additional plasma proteins; albumins, transferrins and haptoglobins

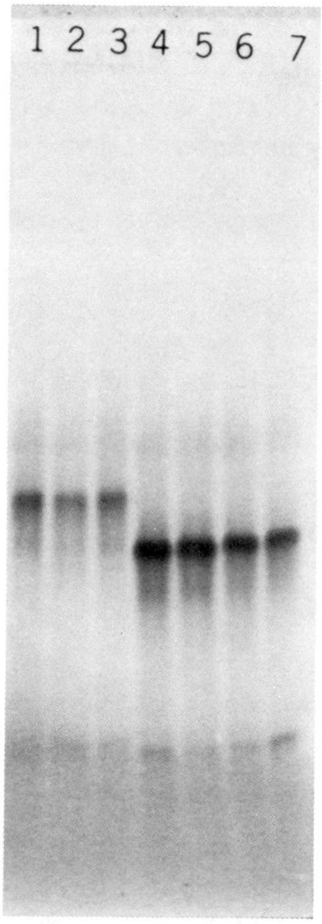

Figure 1. Electrophoresis of hemoglobins from *D. novemcinctus* (slots 1-3) and *D. hybridus* (slots 4-7).

Table 1: Monomorphic enzymatic activities in *Dasypus novemcinctus* and *D. hybridus*.

Enzymatic Activity	*Buffer System*
Alcohol Dehydrogenase	Tris-citrate, pH 6.3
Malic Enzyme	Tris-borate-EDTA, pH 8
Fumerase	Tris-citrate, pH 6.3
Acid Phosphatase-1	LiOH, pH 8.2
Malate Dehydrogenase	Poulik
Peptidase	Poulik
Aldehyde Oxidase	LiOH, pH 8.2
Xanthine Dehydrogenase	Poulik
Gluconate Oxilate Transaminase	Poulik
Adenylate Kinase	Tris-citrate, pH 8
Isocitrate Dehydrogenase	Tris-citrate, pH 8
Lactate Dehydrogenase	Tris-maleate, pH 7.4

Table 2. Enzymatic activities or proteins with varying electromorphs in *Dasypus novemcinctus* and *D. hybridus*. F = fast; I = intermediate; S = slow.

Enzymatic Activity or Protein	*Buffer System*	*Relative Mobility of Electromorphs*	
		D. novemcinctus	***D.hybridis***
Glucose-6-phosphate Dehydrogenase	Tris-borate-EDTA, pH 8	F	S
Acid Phosphatase-2	LiOH, pH 8.2	F	S
Esterase-1	LiOH, pH 8.2	F	S
Esterase-2	LiOH, pH 8.2	F	S
Esterase-3	LiOH, pH 8.2	F	I,S
α -glycerol Phosphate Dehydrogenase-1	Tris-maleate, pH 7.4	S	F
α -glycerol Phosphate Dehydrogenase-2	Tris-maleate, pH 7.4	S	F,I
Hemoglobin	LiOH, pH 8.2	F	S

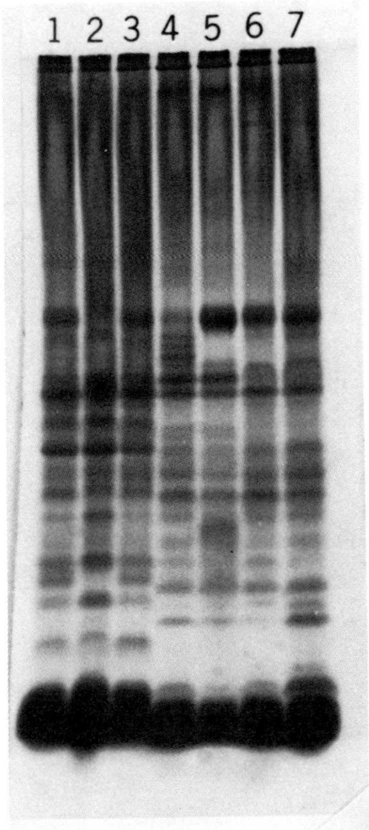

Figure 2. Electrophoresis of plasma proteins of *D. novemcinctus* (slots 1-3) and *D. hybridus* (slots 4-7). Serum albumins are the very darkly stained bands at the bottom of the electropherogram. Haptoglobins are not readily visible in the photograph. Presumptive trasferrins migrated approximately 40 percent as far as albumins.

could be identified with a fair degree of confidence from our gels (Fig. 2). At least two presumptive plasma protein polymorphisms were apparent in *D. novemcinctus*.

DISCUSSION

These studies compared the electrophoretic mobilities of a random assemblage of proteins and enzymatic activities representing many presumptive genetic loci by both starch gel electrophoresis and the less well-recognized band-counting method using polyacrylamide gel electrophorectic fractionation of plasma proteins (Sarich, 1977). The latter technique provides greater resolution than starch gel electrophoresis for two reasons. First, greater uniformity and smaller pore size of the polyacrylamide matrix gel allows a larger number of separate protein moieties to be resolved. Second, blood serum proteins as well as nonspecific esterases of the liver accumulate amino acid substitutions at rates considerably greater than the substitutions rates in the intracellular enzymes normally analyzed by starch gel

Table 3. Summary of genetic distance estimations. For comparison, we list the results both without regard to the proportion of rapidly evolving loci (I) and divided into intracellular and extracellular enzymes or protein species (D).

	I	*D*
Intracellular Proteins and Enzymatic Activities, Including Esterases (Starch Gels)	0.74	0.30
Extracellular Proteins (Polyacrylamide Gels)	0.57	0.56
Polyacrylamide Gels + Esterases	0.47	0.75
Starch Gels Excluding Esterases	0.85	0.16

electrophoresis (King and Wilson, 1975; Sarich and Cronin, 1976; Sarich, 1977). Consequently, comparison of serum proteins and liver esterases may provide greater resolving power for genetic distance estimates involving recently diverged (1-5 million years) lineages (Sarich, 1977).

These preliminary studies support the genetic distinction at the species level of the two armadillos, *D. novemcinctus* and *D. hybridus* (Table 3). The estimated time of divergence from their common ancestor would be in the range of 1.5 to 8 million years ago (Sarich, 1977) depending on whether fast-evolving or slow-evolving loci are employed for the estimate. The estimated time of divergences depends on which loci are defined as fast-evolving or slow-evolving. We obtained lower estimates of genetic distance based on intracellular enzymes and proteins than for the extracellular proteins and the liver esterases than those predicted by Sarich (1977).

ACKNOWLEDGMENTS

We thank Dr. M. Soulé for the use of his electrophoresis laboratory and Drs. K. Benirschke and B. Lasley for providing samples. This work was supported by grants from the NIH (NRSA F32HD05103, GM23073 and HD10190), NSF (1-398) and the Zoological Society of San Diego.

LITERATURE CITED

Benirschke, K., R.J. Low, and V.H. Ferm.
1969. Cytogenetic studies of armadillos. Pp. 330 to 344 *in*, Comparative Mammalian Cytogenetics (K. Benirschke editor). Springer-Verlag, New York. 473 pp.

Davis, B.J.
1964. Disc electrophoresis II. Method and application to human serum proteins. Ann. N.Y. Acad. Sci., 121: 404-427.

Galbreath, G.J.
1985. The evolution of monozygotic polyembryony in *Dasypus*. (this volume).

Harris, H. and D.A. Hopkinson.
1976. Handbook of enzyme electrophoresis in human genetics. American Elsevier. New York. 306 pp.

Holbrook, I.B. and A.G. Leaver.
1976. A procedure to increase the sensitivity of staining by Coomassie brilliant blue G250 - perchloric acid solution. Analyt. Biochem., 75: 634-636.

Jorge, W., D.A. Meritt, and K. Benirschke.
1978. Chromosome studies in Edentata. Cytobios., 18: 157-172.

King, M.C., and A.C. Wilson.
1975. Evolution at two levels in humans and chimpanzees. Science, 188: 107-118.

Nei, M.
1971. Interspecific gene differences and evolutionary time estimated from electrophoretic data on protein identity. Am. Nat., 105: 385-398.

Sarich, V.M.
1977. Rates, sample sizes, and the neutrality hypothesis for electrophoresis in evolutionary studies. Nature, 265: 24-28.

Sarich, V.M. and J.E. Cronin.
1976. Molecular systematics of the primates. Pp. 141-170, *in* Molecular anthropology, (M. Goodman, editor). Plenum Press, New York. 466 pp.

Sealander, R.K., M.H. Smith, S.Y. Yang, W.E. Johnson, and J.B. Gentry.
1971. Biochemical polymorphism and systematics in the genus Peromyscus. I. Variation in the old-field mouse, *Peromyscus polionotus*. Studies in Genetics VI. Univ. Texas Publ. loc, 7103: 49-90.

Shaw, C.R. and R. Prasad.
1970. Starch gel electrophoresis of enzymes - a compilation of recipes. Biochem. Genet., 4: 297-320.

Wetzel, R.M.
1985. Taxonomy and distribution of armadillos, Dasypodidae. (this volume).

ANATOMY AND PHYSIOLOGY

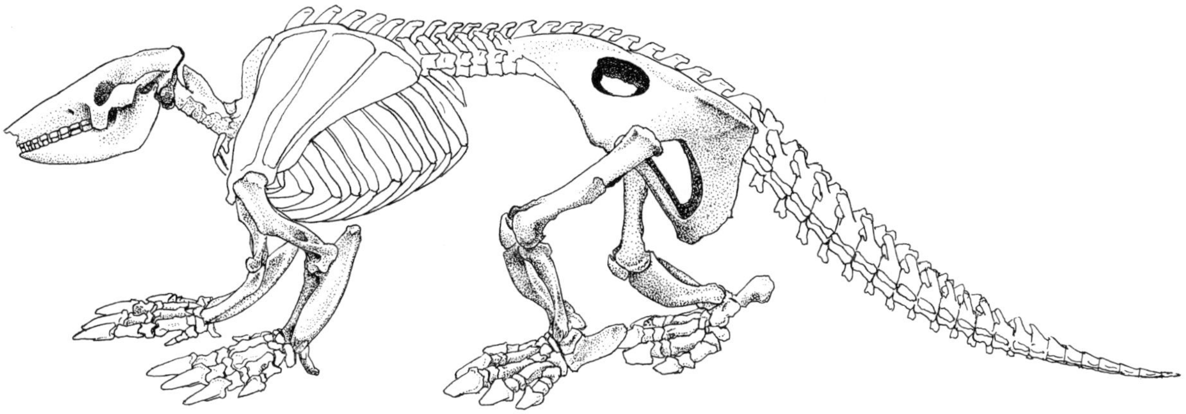

ADAPTATIONS FOR SUSPENSORY BEHAVIOR IN THE LIMBS OF TWO-TOED SLOTHS

Frank C. Mendel

Department of Anatomical Sciences, State University of New York at Buffalo, Buffalo, New York 14214.

Abstract

Positional behavior of free-ranging, two-toed sloths *Choloepus* is nearly unknown, but the few observations of wild and captive sloths suggest that they use suspensory behavior habitually if not exclusively during their active periods. A multitude of anatomical modifications, presumably for this kind of behavior, is readily discernable in the appendages of these animals. The remaining digits of the hands and feet of *Choloepus* have become virtual hooks that can accommodate only a limited range of support sizes. Nearly all joints between limb segments, however, have extremely lax capsules and smooth, rounded mating surfaces that allow considerable ranges of motion. Although the muscle mass is small relative to other mammals of comparable weight, powerful flexion of the limbs is accomplished by increasing the mechanical advantage of the "normal" flexors and supplementing them with a number of unique arrangements of "other than normal" flexor muscles. These traits are interpreted as adaptive for highly acrobatic, energy-conserving, inverted suspensory behavior under suprastrates of diameter less than 10 cm. The potential importance of lianas as preferred supports is discussed.

Abstracto

El comportamiento en posición de los perezosos de dos dedos *Choloepus* que se mueven sin restricciones, es casi desconocido. Sin embargo, las pocas observaciones de perezosos ferales o en cautiverio sugieren que ellos utilizan habitual sino exclusivamente, un comportamiento de suspensión durante sus períodos activos. Existen muchas modificaciones anatómicas, presumiblemente para este tipo de comportamiento, que son fácilmente discernibles en los apéndices de estos animales. Los dedos que aún persisten en las manos y los pies de *Choloepus* se han convertido en virtuales ganchos que solamente pueden acomodar sostenes dentro de un limitado rango de tamaños. Casi todas las coyonturas entre los segmentos de los miembros, sin embargo, tienen cápsulas muy flojas y superficies apareadas lisas y redondeadas, lo cual permite rangos considerables de movimiento en dichos miembros. A pesar de que la masa muscular es pequeña con relación a la de otros mamíferos de peso comparable, la flexión poderosa de los miembros se alcanza aumentando la ventaja mecánica de los flexores "normales" y suplementándolos con un número de disposiciones singulares de otros músculos flexores. Estos se interpretan como rasgos de adaptación para un comportamiento de suspensión invertida altamente acrobático y tendiente a la conservación de energía debajo de superestratos con un diámetro menor de 10 cm. Se discute la importancia potencial de las lianas como sostenes preferenciales de los perezosos.

Resumo

O comportamento posicional das preguiças reais de ocorrência livre *Choloepus* é quase desconhecido mas as poucas observações de preguiças selvagens e de cativeiro sugerem que elas habitualmente usam comportamento suspensório senão durante seus períodos ativos. Uma multidão de modificações anatômicas, presumivelmente por esse tipo de comportamento, é prontamente distinguível nos membros desses animais. Os dígitos remanescentes dos pés e mãos de *Choloepus* tornaram-se ganchos virtuais que podem acomodar somente um certo limite de tamanhos de suporte. Quase todas as junções entre os segmentos dos membros, contudo, têm cãpsulas extremamente soltas, superfícies de união arredondadas que permitem consideável âmbito de movimento. Embora a massa muscular seja pequena, em relação a outros animais de peso comparável, a flexão poderosa dos membros é acompanhada pelo aumento da vantagem mecânica dos flexores "normais" e suplementando-os com um número de arranjos únicos dos músculos flexores "alem dos normais." Essas peculiaridades são interpretadas como adaptações a comportamento altamente acrobático, conservadores de energia e suspensão invertida sob supraestratos de diâmetros menores que 10 cm. A importância potencial de cipóes como suportes preferidos é discutida.

What little is known of the habits and habitats of two-toed sloths *Choloepus* indicates perhaps why our ignorance about them is so profound. They are solitary animals that are active only at night. They are almost entirely arboreal, preferring large-crowned emergent trees laced with abundant lianas. They move little if at all during the day, and seem to deliberately attempt to bury themselves in dense masses of lianas; green algae grows in their fur providing excellent camouflage (Sunquist and Montgomery, 1973; Montgomery and Sunquist, 1975; 1978). It is not surprising then that virtually nothing is known of the positional behavior (postures and locomotor modes) of free-ranging *Choloepus*. The few observations of wild and captive animals (Burton, 1947; Seth-Smith, 1934; Wislocki, 1928; Tevis, 1925; Beebe, 1923, 1926; Bopp, 1954; Britton, 1941; Hagen, 1940; Lundy, 1952; Skutch, 1947; Goldman, 1920; Swanson, 1945; Preble, 1924), suggest that they utilize inverted suspensory behavior habitually if not exclusively during their active periods. Indeed, two-toed sloths are apparently so committed behaviorally and morphologically to this habit that they are incapable of supporting themselves in the usual quadrupedal stance assumed by most tetrapods (Mendel, 1981a). Grand (1978), McNab (1978) and Montgomery and Sunquist (1978) propose that two-toed sloths committed themselves so irrevocably to this kind of behavior because suspensory behavior may be the most energy efficient means of exploiting arboreal environments.

A number of modifications in the musculo-skeletal systems of the appendages of two-toed sloths are readily interpretable as adaptations to highly acrobatic, inverted suspensory behavior and energy conservation. This paper is not meant to provide a detailed anatomical account of appendages of *Choloepus* (see Mackintosh, 1875; Lucae, 1884; Humphry, 1869; Miller, 1935; Windle and Parsons, 1899), but rather a brief review of some of the more obvious anatomical modifications with an attempt to relate these modifications to known or probable biological roles (*sensu* Bock and Von Walhert, 1965).

ANATOMICAL MODIFICATION OF LIMBS

Adaptations of limbs of two-toed sloths can be viewed in terms of three mutually interacting complexes: (1) modification of hands and feet into relatively rigid hooks, (2) maximizing mobility at nearly all joints proximal to the midcarpal and transverse tarsal joints, but with (3) emphasis on powerful flexion of the limbs.

Modification of Hands and Feet Into Relatively Rigid Hooks

Hands and feet (autopodia) of sloths are so modified from the basic pentadactyl configuration that these features are used to distinguish the two genera of tree sloths. Actually, the number of visible toes on each foot is the same in both genera and it is the number of visible fingers on each hand that distinguishes them.

Intact feet and particularly the hands of *Choloepus* look like, and are used very much like hooks. The reduced number of digits, the relatively straight, inflexible and syndactylous digits surmounted by large clawed distal phalanges provide the bases for this configuration. These hooklike appendages are mounted on joints (wrist and ankle and transverse tarsal) that have remarkable ranges of motion (see below) so that the relatively rigid parts of the hand and foot can be manipulated and placed in a wide variety of positions. Digits I and IV of the hand are reduced to metacarpal splints; all remnants of digit V are absent (Fig. 1). It is digits I and V that are reduced to splints in the foot (Fig. 2); these splints are not visible in the intact hand or foot. The remaining digits, II and III in the hand and II-IV in the foot, are bound together throughout their length to the level of the distal interphalangeal joints. The actual hook element of the autopodia is formed by distal phalanges, which are about 3-4 cm in length; each is covered with a

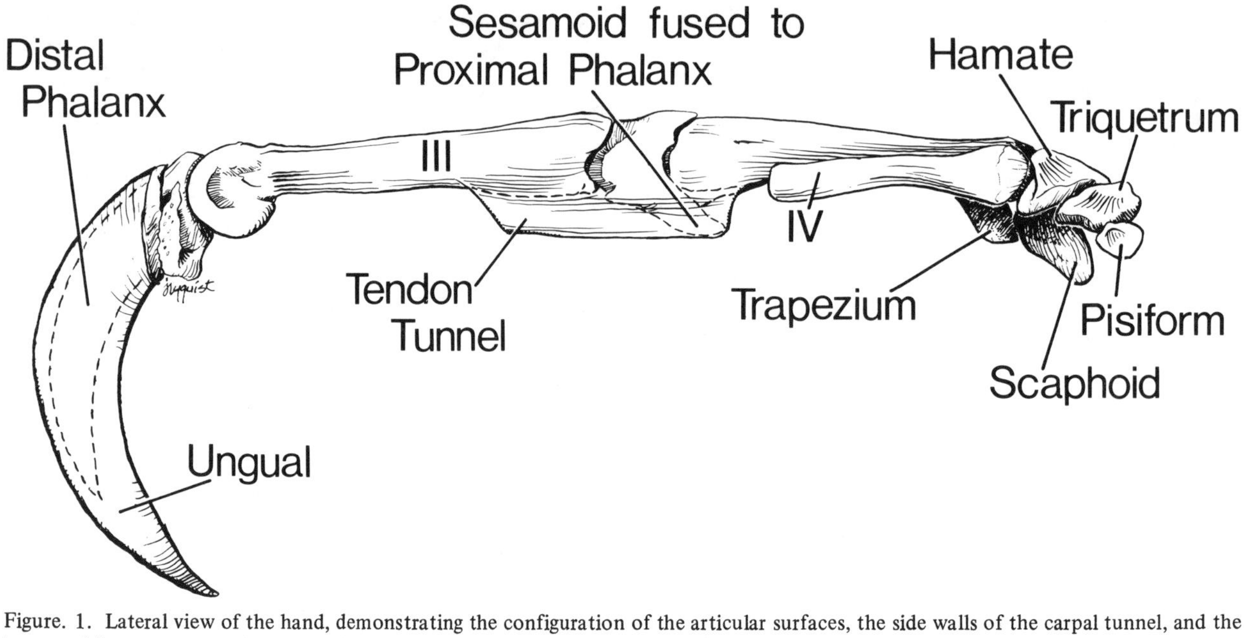

Figure. 1. Lateral view of the hand, demonstrating the configuration of the articular surfaces, the side walls of the carpal tunnel, and the bony and ligamentous components of the tendon tunnel.

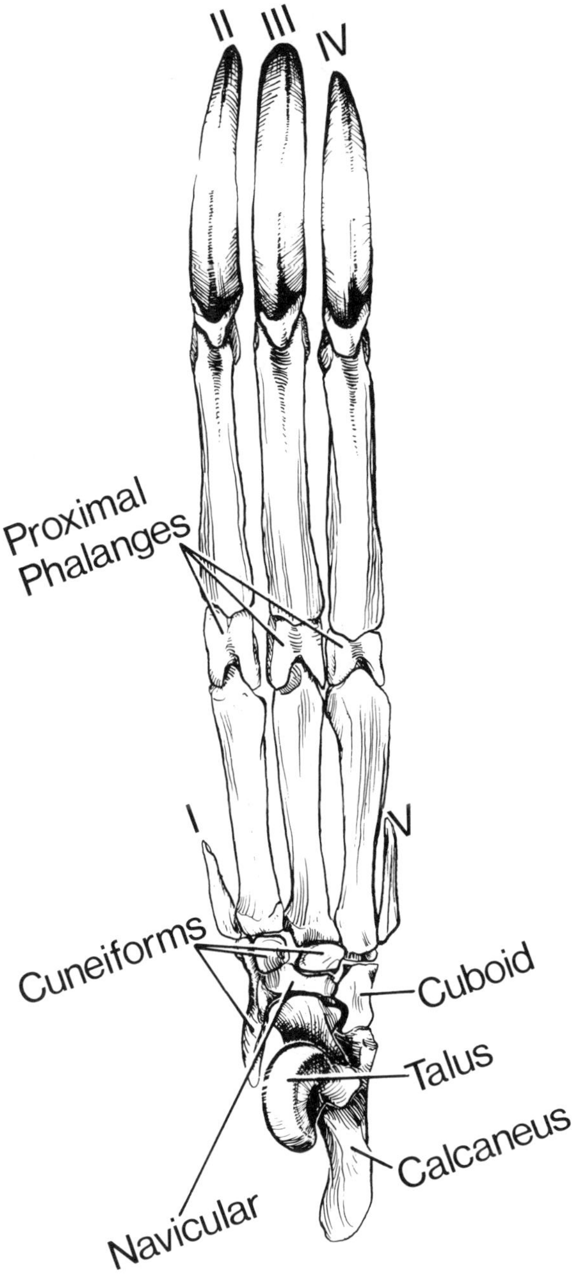

Figure. 2. Dorsal view of the foot, demonstrating the interlocking articulations at the distal and proximal interphalangeal and metacarpophalangeal joints, the development of the first and fifth digits and the relationship among the tarsals.

laterally compressed claw that extends only another 2-3 cm beyond the tip of the distal phalanx, i.e., the claw has good bony support.

The distal phalanges (and claws) are acutely curved, so that when the distal interphalangeal joints are in the straight end-on (neutral) position the tips of the claws are at about 90° to the long axes of the digits (Fig. 1). These phalanges are capable of some 70° of flexion from the neutral position at which time the claw tips contact the volar pads. Extension is arrested at some 90° from the neutral position when a spline on the proximodorsal surface of the distal phalanx engages a fossa on the distodorsal surface of the intermediate phalanx. Full extension of the distal phalanges therefore puts the tips of the claws essentially in line with the long axes of the digits. The articulating surfaces of the distal phalanges are gently concave, but are bisected by vertical splines that fit in grooves in the pulleylike distal ends of the intermediate phalanges (Figs. 1 and 2). This tongue and groove arrangement allows the large range of flexion and extension just mentioned, but prohibits virtually all abduction-adduction and rotation.

The intermediate phalanges of both the hands and feet are relatively robust straight bones that are nearly square in cross section at midshaft. The intermediate phalanges of digit III in either hand or foot are slightly more robust and longer than those of the paraxial digits. The proximal articulating surfaces of these bones are raised into vertical splines that have proximally directed projections on their dorsal and volar extremes. These projections fit into depressions on the dorsal and volar extremes of the articular grooves on the distal ends of the proximal phalanges (Figs. 1 and 2). When these respective male and female surfaces are engaged, they serve as bony stops that limit range of motion. No extension beyond the straight end-on position is permitted at the proximal interphalangeal joints, and only about 25° of flexion, and virtually no abduction-adduction or rotation is possible. The joint capsules are heavily reinforced by ligaments on the laterovolar aspects of these joints.

The proximal phalanges of the hands and feet have been functionally reduced to proximal and distal articular surfaces that are separated from one another only by a thin horizontal crosspiece of bone (Figs. 1 and 2). The mating vertical walls of the proximal phalanges permit no extension at the metapodialphalangeal and proximal interphalangeal joints beyond the straight end-on position. The volar and proximal aspects of the proximal phalanges are the sites of attachment for two spurlike sesamoid bones directed proximoventrally at 30° to the long axes of the digits (Fig. 1). The proximal tips of the sesamoids may or may not be fused together. During flexion, these spurs are jammed against the volar surfaces of their respective metacarpals or metatarsals and thereby prevent flexion of more than 40° at the metacarpophalangeal joints and approximately 50° at the metatarsophalangeal joints. They also channel the massive flexor tendons over the narrow volar surfaces of the metapodials. The channels thus formed are continued on the volar surfaces of the proximal ends of the intermediate phalanges by raised bony lips on the borders of these phalanges. Again abduction-adduction and rotation are prevented by the tongue and groove construction of the joint surfaces and strongly reinforced joint capsules.

The metapodials of the claw-bearing digits are nearly triangular in cross section (apex directed volarly). The metapodials of digit III of the hand and foot are slightly longer and more robust than those of the paraxial claw-bearing digits. As mentioned, the metatarsals of digits I and V and metacarpals of digits I and IV are laterally compressed splints, usually slightly more than one-half the length of the other metapodials, which are tethered by ligaments that run from their distal tips to the laterovolar aspects of the adjacent metapodials (Figs. 1 and 2). The distal ends of the metapodials of claw-bearing digits are also bound together by short transverse ligaments. These bindings preclude independent movements of the metapodials as does the lack of a cleft between more distal elements of the digits, save for the distal phalanges. The intermetapodial joint surfaces and those for the articulations with the carpals or tarsals are short and flat, much like those among the metatarsals and cuneiforms of human feet. These joints admit little abduction, adduction or rotation. This is true except for the articulations of the fourth metacarpal with the third and of the fifth metatarsal with the tarsals. Between the former there is a relatively complex relationship (see Mendel, 1979, for a more complete description) and between the latter there is no articulation at all (Mendel, 1981c).

The distal rows of carpals and tarsals are somewhat unusual in that the trapezium is fused to the base of the pollical metacarpal and the medial cuneiform has a long proximally projecting process that forms a hamate-like process very similar to one found on the scaphoid (Figs. 1 and 2). The scaphoid articulates with the fused trapezium at the level of the other carpometacarpal joints. It is at this juncture, i.e., the midcarpal and transverse tarsal joints, that the structural rigidity and reduction in movement capability seen in the distal portions of the hands and feet give way to highly mobile elements bound loosely by joints that allow extreme ranges of motion.

Maximizing Mobility at Nearly all Joints Proximal to the Midcarpal and Transverse Tarsal Joints

Movement between carpals of the proximal and distal rows is complex, particularly between the triquetrum and hamate; the wrist joint allows wide ranges of movements in nearly all planes. Flexion at the wrist averages about 100°, extension 81°, adduction 99°, and abduction 37° (Mendel, 1979). The proximal articulating surfaces of the proximal carpals, particularly the scaphoid and lunate are convex both dorsoventrally and radio-ulnarly and these mate with a shallow concavity on the radius. The pisiform does not participate in the wrist joint, but articulates solely with the triquetrum and projects volarly into the palm forming a hamatelike process (Fig. 1) the significance of which will be discussed later. The distal end of the ulna has been reduced to a styloidlike tip that participates in the wrist articulation only during extreme ulnar deviation of the hand. The triquetrum articulates with the radius only at a very small facet and normally is separated from the distal end

The ankle and transverse tarsal joints allow the body to turn nearly 180° under a fixed foot. A large portion of this mobility occurs at the transverse joints. The head of the talus has a bulbous kidney-shaped articulating surface along its medioinferior aspect that mates with the proximoinferior aspect of the navicular and a grooved facet on the medioproximal surface of the cuboid (Fig. 2). The rounded perimeter of the head of the talus is separated indistinctly by a fovea from a short flat articulating surface that mates with the proximo-superior surface of the navicular. The fovea receives a conelike projection from the navicular that serves as a pivot for turning of the talus on the relatively fixed navicular and cuboid, i.e., the talus, navicular and cuboid articulate in a configuration similar to the radius, ulna and humerus. The gently concavoconvex calcaneocuboid joint surface allows free motion of the calcaneus on a relatively fixed cuboid.

Motion at the ankle joint is primarily in the plane of flexion and extension although limited ranges of inversion and eversion are also possible. Flexion from a position where the dorsum of the foot is at 90° to the anterior surface of the tibia (neutral position) is about 100°-110°. Further flexion is prevented by contact of the calcaneal tuberosity with the flexor surface of the leg. Extension from the neutral position is about 50°-60°. Flexion and extension of the foot occurs primarily around the fibular malleolus, which has a medially facing rounded axellike projection fitted into a conchelike articulating surface on the lateral aspect of the talus (Figs. 2 and 3). There is no articulation with the tibial malleolus that is some 4-5 mm medial to the trochlea of the talus. This space is, however, encroached upon to some degree by the talar trochlea during eversion of the foot. The ankle joint capsule, like that of the wrist, is very lax and allows the participating bones to be separated by several millimeters.

Pronation (35°- 45°) and supination (90°) of the hand is quite marked. The distal radio-ulnar joint is a highly mobile syndesmosis, whereas the articulating surfaces of the proximal radio-ulnar joint are hominoid-like, having a roughly circular radial head with articulating surface for the ulna over a little more than one half its circumference. The radial notch on the ulna is but a shallow depression and the annular ligament is poorly developed. The entire elbow joint capsule is poorly reinforced by ligaments and is very lax. Indeed, the humerus and ulna can be easily separated by about 5 mm within an intact capsule, so too, can the radius be separated both from the humerus and the ulna. A rounded capitulum, which grades without interruption into a trochlea, fit respect-

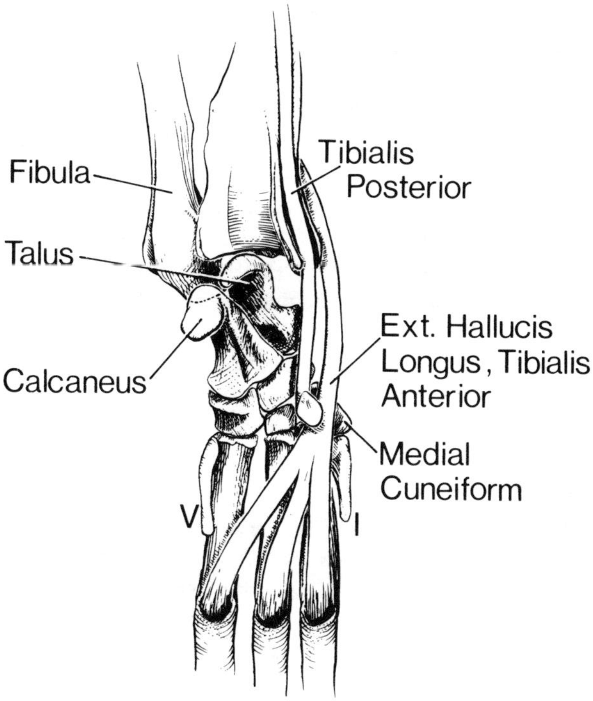

Figure. 3. Volar view of the flexed left foot, demonstrating the disposition of the prime accessory flexors and invertors of the foot. Dotted lines through the calcaneus and talus indicate the outline of the axlelike lateral malleolus.

ively into a broad shallow fovea and smoothly curved semilunar notch. These configurations plus an extremely short olecranon process permit a full 90°of extension of the forearm on the arm and 80°of flexion (measured from a position where the forearm is perpendicular to the arm). In addition, the ulna can be medially rotated 10°- 50°from its resting position on the humerus, which greatly enhances the capacity to supinate the forearm.

The only joints proximal to the metapodials that do not promote wide ranges of motion are the tibiofibular joints. The fibula is quite robust, particularly its distal third; there is no indication of fusion with the tibia, but there is virtually no movement of the fibula on the tibia. These bones articulate at flat intermetapodial-like surfaces and are bound tightly together by strong ligaments.

The capsule of the knee is also a very flimsy structure although not to the degree seen in the elbow; the loose capsule includes the patella and a large sesamoid in its lateral wall (Fig. 4). The large, nearly flat superior tibial articular surfaces articulate with the posteriorly projecting femoral condyles and allow some 30°of medial rotation and 15°of lateral rotation of the leg on the thigh when the leg is flexed 90°on the thigh. The leg can be extended to the full end-on position and flexed through 150°from the fully extended position.

The glenohumeral and hip joints also have very loose and thin walled capsules and allow extreme ranges of motion. The head of the humerus is well above its tuberosities as is the head of the femur above its trochanters (Figs. 4 and 5). The proximal articulating surface of the humerus is somewhat elongated anteroposteriorly and directed dorsally and cranially at an angle of about 40°- 50°to the shaft of the bone. It is set at about 80°(medial) to the axis of the distal articulating surface. It articulates with the scapula in an extremely shallow pear-shaped glenoid fossa. Flexion of the humerus on the scapula ranges about 40°and extension about 80°measured from an arbitrarily defined neutral point at which the spine of the scapula is perpendicular to the long axis of the diaphysis of the humerus and to the transverse axis of the distal articulating surface of the humerus. Humeral abduction from the neutral position ranges near 90°and adduction from the neutral position approaches 40°.

The head of the femur approximates two-thirds of a sphere, which fits into a shallow acetabulum. Flexion of the thigh is about 120°, extension 80°, abduction 100°, lateral rotation 45°- 50°, and medial rotation 10°mea-

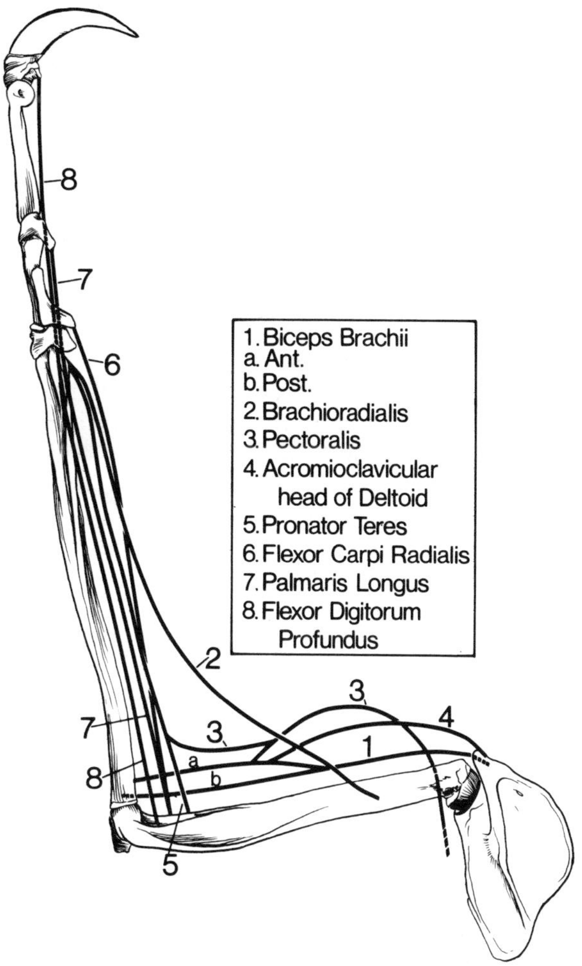

Figure. 4. Lateral view of the forelimb with the forearm fully supinated, demonstrating roughly the lines of action and the approximate attachment sites of the primary flexors of the joints of the forelimb.

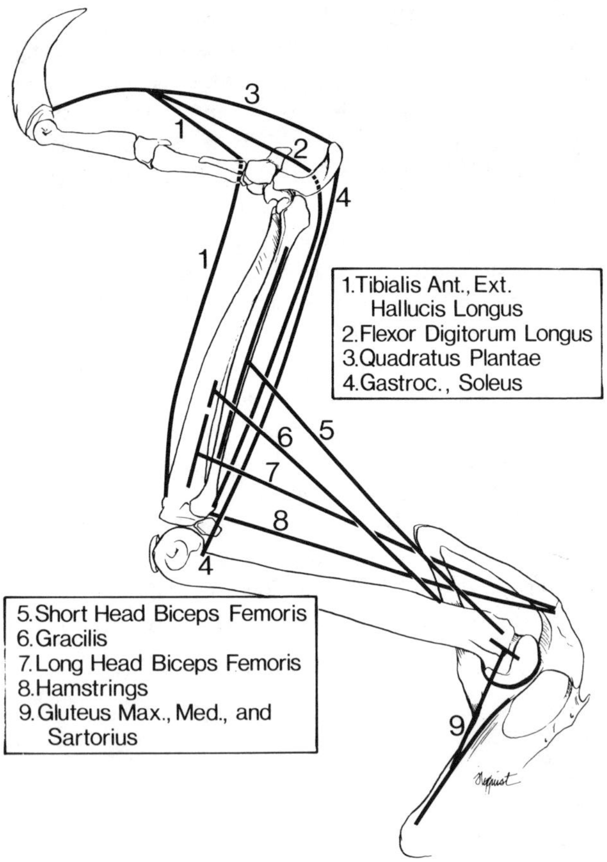

Figure. 5. Lateral view of the hindlimb, demonstrating roughly the lines of action and approximate attachment sites of the primary flexors of the joints of the hindlimb.

sured from a point where the femur is perpendicular to the long axis of the synsacrum.

One of the most interesting elements of sloth morphology is its clavicle and its attachments, particularly to the axial skeleton. The distal end of the clavicle articulates in a small notch on the lateral surface of a tremendously flared acromion process (Fig. 4). This process virtually roofs over the entire head of the humerus. The acromioclavicular joint allows the sternal end of the clavicle to swing dorsally some 40°from a position at right angles to, but in the same plane as, the superior border of the scapula. Such a position is assumed when an animal hangs beneath a support and its scapulae slide ventrally so that its glenoid fossae are considerably anterior to its sternum. The shoulder can be moved cranially until the acromioclavicular articulation abuts the thyroid bone of the animal. In that position, the clavicle is directed caudalward from the acromioclavicular joint at an angle of 50°- 60°. This joint also allows the clavicle nearly 90° of torsion.

The presence of a coracoacromial joint differs among individuals and even between sides of the same individual. When a joint exists, as opposed to a fusion, the distal tip of the acromion process is bound tightly to the lateral tip of the coracoid by dense ligaments that grant little movement at this joint. I speculate that the coracoacromial connection, whether bony or ligamentous, may provide support for the widely flared acromion process.

The sternoclavicular joint is unusual in that the sternum and clavicle do not actually contact one another. Indeed, the sternal end of the clavicle and the manubrium do not even have articulating surfaces for one another, instead being bound together by a fibrous ligament that allows the participating bones to be separated by as much as a centimeter or more. If the clavicle is stripped of muscle attachments and disarticulated from the acromion process, it can be twisted 360°, made to circumscribe complete circles, and positioned at 90°to the sternum in the midsagittal plane. Although this is obviously an unnatural situation, it does illustrate that the capsule is virtually incapable of preventing any movement of the clavicle in any plane. The only circumstance under which it can stabilize the proximal end of the clavicle is while it is under tension. Such tension might occur when the forelimbs of an animal are ventral to its body and its scapulae well cranial and ventral, e.g., when the animal hangs beneath a support. In this position, the clavicles are more or less on end with the sternal ends deep (dorsal) and somewhat medial to the lateral edges of the manubrium, which juts ventrally relative to the remainder of the sternum. This peculiar position of the sternal ends of the clavicles is allowed by the folding of the long and flexible sternoclavicular joint capsules. This folding can continue until the sternal ends of the clavicles reach or slightly surpass the midline (just superficial to the trachea), at which time the capsules are pulled taut, preventing further migration of the clavicles. If both clavicles are in this position simultaneously, they may come into contact with one another deep to the sternum, and in some postures actually abut one another for added support. The capsule then, appears to function much like a stirrup, which must be placed under tension before it provides any support. When an animal lets go of a support with one hand, the non-weight bearing limb tends to fall dorsally, but this may be prevented by the pectoral mass pulling over the rigid but adjustable (with regard to angle) clavicle (i.e., the clavicle may function somewhat like a ginpole over which the forelimb is cantilevered).

Emphasis on Powerful Flexion of Limbs

Because of the loss or reduction in the number of digits and the syndactyly of those remaining, all intrinsic muscles for digits I and V in both hands and feet are absent. The long abductors, flexors and extensors of the pollex and hallux are also absent or have been incorporated into the long flexors of the claw-bearing digits.

The *lumbricals* are reduced in number to just one in the hand and to two or three in the foot. *Interossei* are well developed although they no longer function primarily as adductors and abductors of the digits. They arise from between the metapodials and insert well distally on the dorsal expansions, at the level of the proximal interphalangeal joints. They appear to function primarily as flexors of the metapodialphalangeal and proximal interphalangeal joints.

The long digital flexors of the digits are massive and take origin from nearly the entire available flexor surface. The tendons of these muscles run through deep bony channels as they cross their respective wrist and ankle regions. The carpal tunnel is formed by the concave volar surfaces of the scaphoid and lunate, the pisiform on the ulnar side and a long volar projection of the scaphoid and trapezium on the radial side (Fig. 1). A similar trough is created in the ankle region by a deep groove in the posteromedial aspect of the tibia, the long posterior projection of the calcaneus on the fibular side and the posterovolar projection of the medial cuneiform on the tibial side (Fig. 3). The tendons of the long flexors then course through tunnels on the volar surfaces of the proximal phalanges and proximal thirds of the intermediate phalanges. The roofs of these tunnels are formed by the grooved volar surfaces of the bones, the sidewalls by the sesamoid bones on the volar surfaces of the proximal phalanges, and the floors by dense ligaments (Figs. 1, 4, 5). From the distal mouths of these tunnels, the tendons bowstring across to their attachments on the volar surfaces of the distal phalanges. The points of attachment are separated from the articulating surfaces of the distal interphalangeal joints by some 4-5 mm, which obviously increases the mechanical advantage of the long flexors.

Although *flexor digitorum superficialis* is poorly developed or absent, *palmaris longus* is massive. The latter passes superficial to the flexor retinaculum and inserts on the proximal and volar surfaces of the ligamentous tunnels that house the tendons of *flexor digitorum profundus* (Fig. 4). *Palmaris longus* can be not only a powerful wrist flexor, as is usual, but a powerful flexor of the metacarpophalangeal joints as well, particularly because it runs superficial to the flexor retinactulum (Fig. 4). *Flexor digitorum brevis* is large and inserts at the same places in the foot as *palmaris longus* and *flexor digitorum superficialis,* when it is present, do in the hand (Fig. 5).

The long flexor of the toes is supplemented by three or sometimes four other muscles. The first is a relatively common accessory to the long flexor, *accessorius* or *quadratus plantae.* In two-toed sloths this muscle is very large, originating from the medial, lateral and volar surfaces of an elongate calcaneus, and inserting on the long flexor tendons just proximal to their entry into the ligamentous tunnels (Fig. 5). Another variable muscle that may augment the long digital flexors is *soleus.* This muscle is frequently observed (Humphry, 1869) to send part if not most of its fibers into *quadratus plantae.* Other muscles that contribute to flexors are the combined bellies of *extensor hallucis longus* and *tibialis anterior.* These muscles send their conjoined tendon to the sole of the foot via a notch in the medial surface of the medial cuneiform (Figs. 2, 3, 5). Upon reaching the sole, the tendon divides into three parts each of which inserts on a long flexor tendon just distal to the insertions of *quadratus plantae* and just proximal to the entrance of *flexor digitorum* into the ligamentous tunnels (Figs. 3 and 5). A third muscle *(extensor hallucis brevis)* from the distal end of the *tibialis anterior-extensor hallucis longus* complex (from which it is not always clearly separable), inserts on the medial cuneiform and base of the hallucal metatarsal and augments the ability to invert the foot. *Tibialis posterior* also is a flexor and a particularly powerful invertor of the foot because its tendon is held 5-7 mm medial to its insertion on the medial cuneiform and/or base of metatarsal II by its passage through separate grooves on the posterior aspect of the medially expanded tibial malleolus (Fig. 3).

The long extensors of the digits of both the hand and foot are very small muscles although they are supplemented by relatively large short extensors.

Flexors of the wrist and ankle are far larger than their extensors. *Flexor carpi ulnaris* and *flexor carpi radialis* insert on the volar projections of the pisiform and scaphoid respectively, which provide them additional mechanical advantage (Fig. 4). A similar advantage is accorded both heads of *gastrocnemius* and *soleus* all of which insert on an elongate calcaneal tuberosity (Fig. 5).

Flexion at the elbow and knee is extremely powerful. The lateral and especially the medial supra- and epicondylar regions of the humerus provide greatly expanded attachment areas for the forearm musculature. Nearly all of this musculature, whether carpal or digital flexors or extensors, passes anterior to the elbow and therefore can act to flex that joint. In addition, *brachioradialis* originates high on the humerus and inserts near the wrist; *pronator teres* begins on the supracondylar ridge and has extensive origin on the medial epicondyle, receives a slip from the gladiolar head of *pectoralis major* and inserts on the radius as far distally as the styloid process (Fig. 4). *Brachialis* originates from nearly the entire length of the humerus beginning just distal to the greater tuberosity and inserts, with the posterior head of *biceps brachii,* on the coronoid process of the ulna (Fig. 4). *Biceps brachii* has a single origin from the supraglenoid tubercle but splits into anterior and posterior bellies. The insertion of the posterior belly was just noted; the anterior belly is joined by the superficial gladiolar head of *pectoralis major* and acromioclavicular heads of *deltoid* and inserts via a common tendon on the tuberosity of the radius (Fig. 4). This combination of muscles inserting on the tuberosity of the radius also suggests an emphasis on powerful supination.

Flexion at the knee is powered by the usual *semimembranous* and *semitendinosus,* but is powerfully supplemented by two other muscles. The largest and most important is a huge *biceps femoris,* the long head of which originates as usual from the ischial tuberosity. The long head inserts on the lateral surface of the knee and down one fourth the length of the shaft of the fibula and anterior surface of the tibia. This head is joined at its distal point of insertion by a thin fasicle from *gracilis* , which continues down the fibula to approximately its midpoint (Fig. 5). The short head of *biceps femoris,* from the lateral surface of nearly the entire length of the femur, bifurcates just proximal to the knee to form a sheath over the calf musculature. The lateral insertion begins approximately one third of the way down the fibula and continues nearly to the lateral malleolus; the medial insertion begins several centimeters below the head of the tibia and continues along the medial aspect of that bone to a point several centimeters proximal to the medial malleolus (Fig. 5). The peculiar disposition of *gracilis* and the distal migration of the insertions of both the long and short heads of *biceps femoris* suggests not only an emphasis on flexion of the knee but also on rotation of the leg.

Extensors of both the elbow and knee joints are relatively small compared to the combined flexors of these joints; *triceps brachii* inserts on a very short olecranon process, the mechanics thus dictating weak elbow extension (Fig. 4).

In the relatively omnidirectional joints of the shoulder and particularly the hip, the muscles for rotation, abduction, adduction, flexion and extension are all quite well developed. However, flexion at the shoulder joint is enhanced by the strong development of the acromio-clavicular heads of *deltoid,* particularly because the acromion process extends well anterior to the head of the humerus (Fig. 4). *Supraspinatus* is a huge muscle, occupying more than one half of the dorsal surface of the large scapula and certainly is in a position to aid in flexing the arm as well as medially rotating it.

Flexion of the torso on the thigh appears to be an important movement for two-toed sloths. Captive animals frequently hang head down by their feet alone (i.e., with a fully extended hip) during defensive postures, investigation and locomotion. Recovery of the very long body from this position would appear to require considerable strength. *Iliopsoas, iliacus,* and *rectus femoris* are well developed but are by no means massive. *Sartorius* and both *gluteus maximus* and *gluteus medius* may supplement these flexors because they take origin on the elongated iliac blades and all have fibers that cross the hip joint anterior to its axis and therefore are potential flexors of that joint (Fig. 5).

DISCUSSION

Due to a complete lack of a fossil record, it is not known whether two-toed sloths developed rigid hooklike hands and feet as an adaptation for inverted suspensory behavior or were forced to adopt such behavior because they inherited such structures from an ancestral stock that had evolved them, probably for other purposes. Because all living pilosa have extremely well-developed claws, reduced numbers of digits (at least on the hands) and restricted digital movement (Walker, 1968), it is tempting to speculate that tree sloths probably inherited fairly specialized autopodia, which limited their options in dealing with arboreal supports.

Whatever their origins, the autopodia of extant two-toed sloths are used very much like hooks that fit over the tops of supports. The inability to hyperextend, abduct or adduct the digits or to effectively flex any but the distal phalanges is perhaps a means of maintaining the effectiveness of the long flexors. By keeping the digits relatively straight and parallel to one another, except for the distal phalanges, the pull of the long flexors is concentrated on flexing the distal phalanges. If bony stops were not built in to prevent it, flexion at the metapodialphalangeal and proximal interphalangeal joints could potentially be far ranging as long as the wrist or ankle were straight. But if either were flexed, the ability to keep the distal elements flexed would be greatly diminished. To get an appreciation for this problem, readers should try making their hands into a sloth hand by flexing the fingers at the proximal interphalangeal joints. Even when the wrist is flexed, the intermediate and distal phalanges remain tightly flexed. Now make a tight fist and try flexing the wrist. The grip is broken or at least weakened because the flexor muscles cannot contract further to take up the slack in the tendons created by the increased coiling of the elements of the hand (active insufficiency, c.f. Hamilton, 1976). It may be that the proximal phalanges of two-toed sloths are reduced to little more than proximal and distal articulating surfaces and flexion at the metapodial-phalangeal and proximal interphalangeal joints is reduced to insure that the tendons of the long flexors cannot be rendered ineffective in flexing the distal phalanges, which support the animal. The reduced length of the digits caused by the shortening of the proximal phalanges and the need to maintain a flexor moment at the distal interphalangeal joints adequate to offset the gravitationally induced extensor moment probably inhibit the capacity of the hand to accommodate supports of large diameter (Mendel, 1981b; 1981c).

Adult two-toed sloths have extreme difficulty in maintaining a hold on cylindrical supports with diameters of 9 cm and, in fact, can accommodate supports larger than about 4 cm in diameter only by bringing the tips of the claws (the remainder of the claws does not make contact) well over the top of the support and flexing the wrists and hyperflexing the ankles (Mendel, 1981b; 1981c). The lack of claw contact reduces the security of the hold and the flexed wrist and ankles indicate increased energy expenditure because more muscles (e.g., carpal and tarsal flexors) are probably recruited to maintain the hold. Supports smaller than 4 cm in diameter permit good claw to support contact and probably do not require other than digital flexors to maintain the hold.

This relatively limited capacity to accomodate supports of even moderate size would seem to restrict two-toed sloths to only certain parts of the trees, perhaps to the periphery, where supports of the appropriate size are abundant. Montgomery and Sunquist (1978) noted, however, that two-toed sloths show a decided preference for large crowned emergent trees that have abundant lianas interspersed throughout. These lianas are of relatively uniform and small diameter, crisscross within the crown of a given tree, span from one tree to another and descend to the ground. In other words, lianas may provide for two-toed sloths, with their energy conserving but inflexible hooklike hands and feet, a network of small diameter thoroughfares that allow access to large parts of a given tree, other trees, and even the ground. Use of lianas for movement between trees by three-toed sloths has been described elsewhere (Montgomery and Sunquist, 1975).

The extreme mobility of the limbs of two-toed sloths save for the hook-forming parts of the hands and feet seem to be a means of maximizing the abilities to reach out, support and orient the body in a myriad of positions. Such capacities would seem appropriate for exploitation of an arboreal milieu in which potential supports are likely to run at virtually any angle, and potential food sources such as leaves, fruit and flowers are found in greatest abundance out on small, flexible peripheral branches. Grand (1972) discussed the advantages of suspensory behavior in increasing the feeding sphere of arboreal animals in the small branch milieu (Avis, 1962). The acrobatic abilities and the elongated limbs and trunk of two-toed sloths might be interpreted as means of increasing their feeding spheres as well.

The poorly reinforced and extremely lax joint capsules are highly unusual, and seem to be incapable of exercising more than loose restraint on the various articulating bones. This weak development of the capsules might be a means of further increasing the mobility of the limbs, but such mobility may be secondary to adaptations to suspensory postures in which the limbs are under tensile rather than compressive stress. Under these conditions, it may be more energy efficient to reposition some muscles and adjust the length of others so that they maintain joints in certain positions via passive insufficiency (Hamilton, 1976). For example, the distal phalanges may be maintained at about 90° to the long axes of the digits because the tendons of the long flexors are of insufficient length to allow further extension at the distal interphalangeal joints without some flexion at more proximal joints. But flexion at these joints is resisted by the weight of a pendant animal. Muscles such as *brachioradialis* and *biceps femoris,* which have lines of action well away from the joints they span (Figs. 4 and 5), might maintain their respective joints in some degree of flexion, which requires activity by their antagonists to overcome. Such a configuration would seem to protect the joint capsules from tensile stress, whereas pendant postures appear to reduce shearing forces at joints, which may preclude their need for any development beyond synovial sacs. A low-cost multifunction muscular system may in the long run, be more beneficial to an animal with a low energy budget (McNab, 1985) than developing and maintaining two separate systems (i.e., muscles and ligaments) to secure its joints.

The extreme capacities for ulnar deviation of the hand and inversion in the foot deserve special comment. During normal forward locomotion (pictured in Muybridge, 1957), two-toed sloths adduct (ulnarly deviate) their hands in order to keep the hook element of the leading hand at right angles to the support; this position provides maximum claw surface contact with the support (see Yalden, 1972, and Mendel, 1981a for further discussion). The feet are inverted also as a means of keeping the hook elements of the feet at right angles to the support. This position could also be attained by laterally rotating the thighs, but this would reduce the effective stride length and would splay the knees making them susceptible to impingement on various objects. The ability to invert the feet through nearly 180° provides the animal a secure hold while it lets go of a support with its hands, and turns around. This ability may be important to a nocturnal animal (Bopp, 1953, 1954; Britton, 1941; Goffart, 1971; Menegaux, 1908a, 1980b, 1909a, 1909b; Lundy, 1952; Enders, 1940; McCrane, 1966; Skutch, 1947; Swanson, 1945; Howarth and Toole, 1973; Sunquist and Montgomery, 1973), which is perhaps dependent on olfaction or tactile sensation (Krumbiegel, 1941; Kreig, 1961; Tirler, 1966) to "test" each new support by either smelling it or touching it to determine its weight bearing capacity. This ability to invert the feet probably also increases the range of activity the animal has when hanging by the feet alone to feed on a lower branch or to search for new supports. The unusual, perhaps unique recruitment of the combined *extensor hallucis longus* and *tibialis anterior* as accessory digital flexors may be important in maintaining a secure grasp when the feet are severely inverted because their line of action is more direct in this position than that of the normal digital flexors (Figs. 3 and 5).

Grand (1977, 1978) noted that muscle contributes

only about 25 percent of a sloth's total body weight. McNab (1978) suggests that small muscle mass may be a means of maintaining low metabolic requirements in an animal with inherently low metabolic capacities. Two-toed sloths seem to at least partially compensate for small muscular bulk by increasing the efficiency of the muscles, particularly the flexors. They do so by increasing the mechanical advantage of some muscles, adding one or more fasicles from various muscles to bolster a prime mover and by transferring muscles of one functional group to another. The emphasis seems to be on flexion at nearly all joints save for perhaps the shoulder and hip where the body is pulled and pushed respectively forward on the limbs during locomotion. Two-toed sloths probably lift themselves frequently during climbing, feeding and other activities; they certainly maintain the elbow and knee in flexion during suspensory progression and even while hanging motionless beneath a support. The large size and mechanical advantage accorded *pronator teres* and *brachioradialis* in the forelimb and *biceps femoris* in the hindlimb seem well suited to maintaining this posture. The forelimb muscles just mentioned plus the pectoral mass are probably primarily responsible for pulling the body to higher supports during climbing and *biceps femoris* for recovering the body from positions supported only by the feet or for lowering the body to lower supports.

The reduction of the hands and feet of two-toed sloths to relatively small rigid hooks with concomitant loss of lateral digits and reduced intrinsic musculature, the ability to position the limbs in virtually any relationship to the body, and the enhanced powers of flexion of the limbs in an animal with a relatively small muscle mass can thus be interpreted as adaptations to energy conserving suspensory behavior beneath small diameter supports.

ACKNOWLEDGMENTS

I wish to gratefully acknowledge Mr. Dale Fish and Dr. Frank Kallen for reviewing the manuscript and making many helpful suggestions. My sincerest thanks also to Mr. John Nyquist for preparing the illustrations and to Ms. Shirley Travers for typing the manuscripts. I am grateful to Gene Montgomery for his advice and support and the opportunity to participate in this symposium.

LITERATURE CITED

Avis, V.
1962. Brachiation: The cruicial issue for man's ancestry. Southwest. J. Anthrop., 18:119-148.

Beebe, W.
1923. The leisurely sloth. Zool. Soc. Bull., 26:12-16.

1926. The three-toed sloth. *Bradypus cucullinger cucullinger* Wagler. Zoologica, 7:1-67.

Bock, W. J. and G. Von Wahlert.
1965. Adaptation and the form-function complex. Evol., 19:269-297.

Bopp, P.
1953. Biologische beobachtungen an *Choloepus didactylus* L. in gefangenschaft. Verh. schwiz. naturf. Ges., 133:100-101.

1954. Einige biologische beobachtungen am zweizhenfaulter, *Choloepus didactylus* (Linne, 1766). Saugetierk. Mittelil., Stuttgart, 2:114-118.

Britton, S. W.
1941. Form and function in the sloth. Quart. Rev. Biol., 16:13-34; 190-207.

Burton, M.
1947. Tree sloths: a study in slow motion. Illust. London News, 211:356-357.

Enders, R. K.
1940. Observations on sloths in captivity at higher altitudes in the tropics and in Pennsylvania. J. Mamm., 21:5-7.

Goffart, M.
1968. The problem of slothfulness in the didactyl sloth *Choloepus hoffmanni* Pet. Electromyography, 8: 245-251.

Goldman, E. A.
1920. Mammals of Panama. Smithsonian Miscell. Coll. Publ. No.2498, 69(5):1-309.

Grand T.
1972. A mechanical interpretation of terminal branch feeding. J. Mamm., 53:198-201.

1977. Body weight: its relation to tissue composition, segment distribution, and motor function. I. Interspecific comparisons. Am. J. phys. Anthrop., 47:211-240.

1978. Adaptations of tissue and limb segments to facilitate moving and feeding in arboreal folivores. Pp 231 to 241 *in* The Ecology of Arboreal Folivores (G. G. Montgomery, ed.). Smithsonian Institution Press. Washington, D.C. 574 pp.

Hagen, V. W. Von.
1940. Slowfoot the sloth, a prehistoric throwback,

views his upside-down world. Frontiers, Phila., 4:66-71.

Hamilton, W. J.
1976. Textbook of Human Anatomy (2nd ed.). C. V. Mosby Co., St. Louis, 753 pp.

Howarth, S. T. and J. F. Toole.
1973. Some observations on the circadian rhythm of *Choloepus hoffmanni* the two-toed sloth. Lab. Anim. Sci. 23:377-379.

Humphry, G. M.
1869. The myology of the limbs of the Unau, the Ai, the two-toed Anteater and the Pangolin. J. Anat. Physiol., 4:17-78.

Kreig, H.
1939. Begegnungen mit ameisenbaren und faultieren im freier wildbahn. Z. Tierpsychol. Berlin, 2:282-292.

Krumbiegel, I.
1941. Die Saugetiere der Sudamerica. 14. Faultiere Zool. Anz., 136: 53-62

Lucae, J. C. G.
1884. Statik und mechanic der Quadrupeden an dem Skelett und den Muskeln des *Lemur* und *Choloepus*. Abhandl. Senckenb. Naturf. Gesselsch. Frankfurt, 13:1-92.

Lundy, W. E.
1952. The upside-down animal. Nat. Hist., 61:114-119.

Mackintosh, H. W.
1875. On the muscular anatomy of *Choloeopus didactylus*. Proc. Roy. Irish Acad., 2:66-78.

McCrane, M. P.
1966. Birth, behavior and development of a hand-reared two-toed sloth *Choloepus didactylus*. Intern. Zoo Yearbook, 6:153-163.

McNab, B. K.
1978. Energetics of Arboreal folivores: physiological problems and ecological consequences of feeding on a ubiquitous food supply. Pp 153 - 162 *in* The Ecology of Arboreal Folivores (G. G. Montgomery, ed.). Smithsonian Institution Press, Washington, D.C. 574 pp.

McNab, B.K.
1985. Energetics, population biology, and distribution of xenarthrans, living and extinct. (this volume).

Mendel, F. C.
1979. The wrist joint of two-toed sloths and its relevance to "brachiating" adaptations in the Hominoidea. J. Morph., 162:413-424.

1981a. Use of hands and feet of two-toed sloths *Choloepus hoffmanni* during climbing and terrestrial locomotion. J. Mamm., 62(2): 413-421.

1981b. The hand of two-toed sloths: its anatomy and potential uses relative to size of support. J. Morph., 169(1): 1-19.

1981c. Foot of two-toed sloths: its anatomy and potential uses relative to size of support. J. Morph., 170(3):357-372.

Menegaux, A.
1908a. La marche et la facon de grimper des Paresseux d'apres les observations recentes et notamment celles de M. et Mme Geay, voyageurs du Museum d'Histoire. Bull. Mus. Nat. Hist. Natur., 7:334-337.

1908b. La nourriture des Paresseux. Paris Bull. Soc. Zool. France, 33:159-161 and Bull. Mus. Hist. Nat., 14:337-340.

1909a. Contribution a l'etude des Edentes actuels, famille des Bradypodides. Arch. Zool. Exp. Gen. 5° Ser., 1:277-344.

1909b. Quelques faits Nouveux de la biologie des Paresseux. Rev. Sci. Paris ser., 5 11:523-527.

Miller, R. A.
1935. Functional adaptations in the forelimb of the sloths. J. Mamm., 16:38-51.

Montgomery, G. G. and M. E. Sunquist.
1975. Impact of sloths on neotropical forest energy flow and nutrient cycling. Pp 69 to 98, *in* Tropical Ecological Systems. Trends in Terrestrial and Aquatic Research. (F. B. Golley and E. Medina, eds.). Springer-Verlag, N.Y. 398 pp.

1978. Habitat selection and use by two-toed and three-toed sloths. Pp. 329-359 *in* The Ecology of Arboreal Folivores (G. G. Montgomery, ed.). Smithsonian Institution Press, Washington, D.C. 574 pp.

Muybridge, E.
1957. Animals in Motion. (L. D. Brown, ed) Dover publications, N.Y. 74 pp.

Preble, E. A.
1924. Two-toed Sloth. Nat. Mag., 4:354-382.

Seth-Smith, D.
1934. An upside-down life. The sloth and its very queer relations. Field, 164:1297.

Skutch, A. F.
1947. Adventures with Sloths. Nat. Mag., Wash., 40:521-524.

Sunquist, M. E. and G. G. Montgomery.
1973. Activity patterns and rates of movement of two-toed and three-toed sloths *(Choloepus hoffmanni* and *Bradypus infuscatus*). J. Mamm., 54:946-954.

Swanson, P. L.
1945. Nature's Strap-hanger. Fauna, 7:104-108.

Tevis, M.
1925. An Animal Hanging Basket. Mentor, 13:58.

Tirler, H.
1966. A sloth in the family. The Harvill Press, London. 47 pp.

Walker, E. P.
1968. Mammals of the World (2nd ed.) Vol. I. John Hopkins Press, Baltimore. 646 pp.

Windle, B. C. A. and F. G. Parsons.
1899. On the myology of Edentata. Proc. Zool. Soc. Lond., 314-338; 440-1017.

Wislocki, G. B.
1928. Observations on the gross and microscopic anatomy of the sloths. J. Morphol. Physiol., 46: 317-377.

Yalden, D. W.
1972. The form and function of the carpal bones in some arboreally adapted mammals. Acta Anat., 82:383-406.

FUNCTIONAL ANATOMY OF THE FORELIMB IN VERMILINGUAS (ANTEATERS)

Bruce K. Taylor

Box 984, Southwestern Medical School, University of Texas Health Science Center, 5323 Harry Hines Blvd., Dallas, Texas 75235

Abstract

The three living genera of vermilinguas (anteaters; Myrmecophagidae) are compared in terms of how differences in forelimb morphology affect the mechanical capabilities of the limb, and how these differences correlate with what is known about the field biology of these xenarthrans. *Tamandua* has enlarged powerful central foreclaws, plus capabilities for powerful flexion and axial rotation of more proximal limb segments, important for obtaining food and for defense; *Tamandua* exposes ant and termite prey by ripping open carton nests or infested wood. The claws may be flexed with considerable force even when the remainder of the limb is fully extended, allowing initiation of an attack on an ant or termite nest while keeping the body as far away as possible from the retaliatory responses of the insects. The hand of *Tamandua* has prehensile capabilities, useful for manipulating pieces of prey-nest material and for climbing. *Myrmecophaga* closely resembles *Tamandua* in forelimb morphology; although it is terrestrial and eats mainly ants, many of those ants nest inside wood or in the ground. This similarity is consistent with the hypothesis that *Myrmecophaga* might be descended from an arboreal, termite-eating ancestor and that natural selection has favored retention of its *Tamandua*-like forelimb specializations. *Cyclopes* is also similar to *Tamandua*, but its forelimb differs in several respects: capability for powerful claw flexion is not so extreme, powerful shoulder retraction is enhanced, and capability for elbow retraction is diminished. Differences seem to correlate with *Cyclopes*' small body size, that it does not break open carton nests of ants and termites, and that it commonly occupies very small branches and lianas. Some of *Cyclopes*' unique specializations evolved in response to ecological competition with other arboreal termite- and anteaters.

Abstracto

Se comparan los tres géneros vivientes de osos hormigueros (familia Myrmecophagidae) en términos de la forma como las diferencias en la anatomía del antebrazo afectan la capacidad mecánica del miembro, y de cómo estas diferencias se correlacionan con lo que se conoce de la biología en la naturaleza de estos animales xenartranos. El género *Tamandua* tiene poderosas y abultadas garras centrales anteriores, además de la capacidad para ejercer una poderosa flexión y rotación sobre el axis de los segmentos más próximos del miembro; todo ésto es importante para que el animal obtenga sus alimentos y ejerza su defensa. Los *Tamandua* ponen al descubierto las hormigas y termitas, que constituyen su presa, destrozando sus nidos de cartón y la madera infestada. Las garras pueden doblarse con considerable fuerza, aunque el resto del miembro esté totalmente extendido, permitiéndole al animal así el inicio de un ataque sobre un nido de hormigas o termitas, mientras el resto del cuerpo se mantiene de lo más alejado posible de represalias de parte de los insectos. La mano de *Tamandua* tiene capacidad prensil, lo cual es útil para manipular pedazos de los nidos de su presa y para acciones trepadoras. Los hormigueros del género *Myrmecophaga* se parecen a los *Tamandua* en la morfología del antebrazo, a pesar de que son terrestres y de que se alimentan principalmente de hormigas, muchas de las cuales anidan dentro de madera o en la tierra. Esta semejanza es consistente con la hipótesis de que *Myrmecophaga* puede haber evolucionado de un antecesor arbóreo, comedor de termitas, y que la selección natural ha favorecido la retención de especializaciones del antebrazo como las de *Tamandua*. *Cyclopes* también es similar a *Tamandua*, pero su antebrazo difiere en varios aspectos: la capacidad para una poderosa flexión de la garra no es tan exagerada; la poderosa retracción del hombro es mayor, y la capacidad para retraer el codo es menor. Estas diferencias parecen estar correlacionadas con el pequeño tamaño del cuerpo de *Cyclopes*, como con los hechos adicionales de que para comer no rompe nidos de cartón de hormigas y termitas y que comúnmente ocupa muy pequeñas ramas y lianas. Algunas de las singulares especializaciones de *Cyclopes* han evolucionado como respuesta a la competencia ecológica con otros comedores de hormigas y termitas.

Resumo

Os três gêneros existentes de tamanduás (Myrmecophagidae) são comparados em termos de diferenças na morfologia do membro anterior que atingem as capacidades mecânicas do membro e como essas diferenças se relacionam com o que se conhece sobre a biologia no campo, desses Xenarthra. O *Tamandua* tem as garras centrais do membro anterior aumentadas, mais as capacidades para poderosa flexão e rotação axial dos segmentos mais proximais do membro, importantes para a obtenção de alimento e defesa; *Tamandua* expõe suas presas, formigas e cupins, rompendo formigueiros ou madeiras infestadas. As garras podem ser dobradas com força considerável mesmo quando o restante do membro está totalmente estendido. permitindo iniciar um ataque ao formigueiro ou cupinzeiro, enquanto mantém sue corpo tão afastado quanto possível das respostas em represália dos insetos. A mão do *Tamandua* possui capacidades prênseis, uteis para a manipulação de peças do material do ninho da presa e para trepar. *Myrmecophaga* assemelha-se bastante com

Tamandua na morfologia de seus membros, embora seja terrestre, e alimente-se principalmente de formigas, sendo que muitos desses formigueiros estão dentro de madeiras ou no chão. Essa similaridade é consistente com a hipótese de que *Myrmecophaga* pode ser descendente de um ancestral arbóreo, que se alimentava de cupim e que a seleção natural favoreceu a conservação de suas especializações dos membros similares às de *Tamandua*. *Cyclopes* também é similar a *Tamandua*, mas seu membro difere em diversos aspectos; capacidade para poderosa flexão das garras não é tão extrema a poderosa retração dos ombros é aumenta da e a capacidade para retração do cotovelo é diminuída. A habilidade para apertar objetos de pequeno diâmetro é aumentada neste gênero. As diferenças parecem correlacionar-se com o pequeno tamanho corpóreo de *Cyclopes*, que não se alimenta de formigueiros e cupinzeiros partidos, e comumente ocupa galhos muito pequenos e cipós. Algumas das únicas especializações de *Cyclopes* evoluíram em resposta à competição ecológica com outros animais arbóreos que se alimentam de cupins e formigas.

The mammalian forelimb can serve a number of distinct functions. Locomotor uses of the forelimb may vary greatly, depending on how it is used in such activities as walking, running or climbing. Other possible forelimb functions include use as an organ for obtaining food or otherwise manipulating the environment, and use as a weapon for defense against predators. These different biological roles will place different and sometimes conflicting sets of functional demands on the limb, which may necessitate structural compromises.

In vermilinguas (neotropical anteaters), the roles of the forelimb in such activities as food obtainment and defense are particularly important, and adaptation for these roles has involved changes in morphology which in turn affect the limb's locomotor capabilities. Of the three living genera, *Myrmecophaga* is almost wholly terrestrial and *Cyclopes* is wholly arboreal, while *Tamandua* spends considerable amounts of time both in the trees and on the ground (Montgomery, 1979).

It is my purpose in this paper to review the functional anatomy of the forelimb in vermilinguas and show how the anatomy relates to such factors as habitat and diet. The anatomy of the forelimb and its major functional implications in *Tamandua* have been described in more detail in another paper (Taylor, 1978). A summary of the major features of the *Tamandua* forelimb is included in the following discussion to facilitate comparison among the three genera of vermilinguas.

MATERIALS AND METHODS

The general methodology follows that of Taylor (1976, 1978). The North American opossum *Didelphis marsupialis* was chosen as a representative unspecialized mammal and used as a standard of comparison. The skeletal and muscular morphology of all three genera of vermilinguas was evaluated in terms of: (1) determination of those features of forelimb anatomy which distinguish vermilinguas as a group, as well as those features which distinguish among the three genera; (2) how these features affect the mechanical capabilities of the limb; and (3) how limb morphology relates to what is known about the specialized locomotory and feeding behavior of these animals.

RESULTS AND DISCUSSION

Tamandua

Flexion of digits.-- One of the more striking features of the vermilingua forelimb is the development of extremely large, powerful claws on the central digits, primarily digit III. In *Tamandua*, for example, the clawed distal phalanx of digit III is approximately twice as long as its corresponding metacarpal (Fig. 1). The possession of one claw which is considerably larger than all of the others allows the animal to concentrate all of the force of the limb at a single point. This may be of particular importance in ripping open hard substrates, as well as in defense against predators. Wide ranges of flexion and extension can occur at both the metacarpophalangeal and distal interphalangeal joints. At the proximal interphalangeal joint, however, the configuration of the articulating surfaces is such that little or no movement is possible. In effect, then, each digit

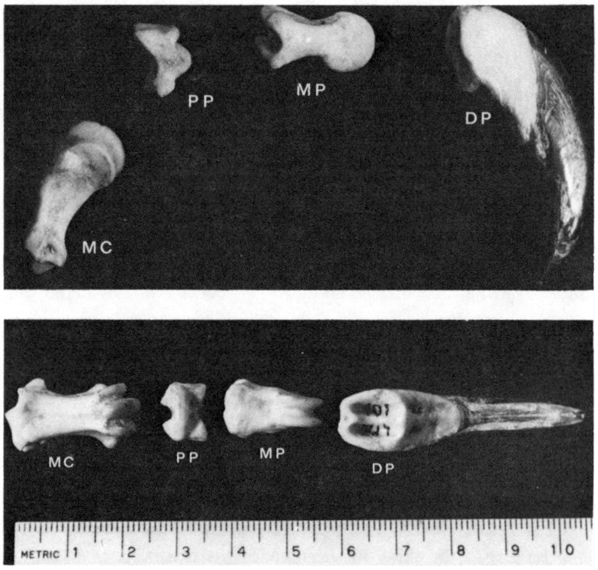

Figure 1. Skeleton of digit III in *Tamandua*, disarticulated, seen in lateral view, above, and in volar view, below. **MC** = metacarpal, **PP** = proximal phalanx, **MP** = middle phalanx, **DP** = distal phalanx. Scale in cm.

possesses only two functional joints, which may increase overall stability under high mechanical loading.

The palm of the hand is equipped with a large, thickened pad consisting primarily of fibrous connective tissue and fat. This pad serves two principal functions. First, the ulnar aspect of the pad provides a weight-bearing surface at the beginning of the stance phase of the walking cycle. As the animal places the hand on the ground, initial contact is made with the pad; the hand subsequently rolls forward so that the weight is then borne on the heads of the metacarpals, the digits being flexed so as to keep the phalanges out of the way (D. W. Yalden, pers. comm.). In arboreal situations, *Tamandua* uses a "pigeon-toed" stance when walking on the surface of limbs about 2 cm diameter and larger; the pad conforms to the curvature of the limb, prevents slippage, and allows the animal to move along the limb without gripping it with the claws. The hind feet contact the limb in a "normal" plantigrade manner, while the prehensile tail trails along the limb, or is held out to one side or the other for balance (Montgomery, unpubl. data).

The second function of the palmar pad is that its volar aspect provides a somewhat resilient surface against which the tips of the claws (principally that of digit III) may be opposed. This feature allows the animal to pinch objects between the claws and the palmar pad, thus providing ample grasping and manipulating capabilities despite the presence of very large claws.

The musculature associated with flexion of the claws is extremely well developed in *Tamandua*. Not only are the usual digital flexors enlarged, but there is also a modification of the *triceps* musculature of the upper arm, whereby a substantial portion of it has become an accessory flexor of the digits. This modification is unique to *Tamandua* and *Myrmecophaga*. In all other mammals, all of the *triceps* musculature inserts on the olecranon process of the ulna (OLEC, Fig. 2) and thus functions as an extensor of the elbow joint. However, in *Tamandua* (and *Myrmecophaga* as well), the medial head of *triceps* (TRM) fails to attach to the olecranon. Instead, it passes medial to it and enters the flexor compartment of the forearm (Fig. 2). It then shares an insertion tendon with *flexor digitorum profundus*, and thus serves as an accessory flexor of the distal phalanges. By far the major portion of the insertion of these two muscles is on digit III.

The significance of this shift in muscle attachment can be further illustrated by comparing the weights of the muscles involved. The effect of changing the attachment of the medial head of *triceps* in the vermilingua is to almost double an already greatly enlarged amount of musculature involved in digital flexion (Table 1). Furthermore, while the forces exerted by the flexor muscles are more or less evenly distributed among all five digits in the opossum, in the vermilingua the insertion is such that the major portion of the muscular force is exerted through digit III. Thus the vermilingua is adapted for extremely powerful claw flexion, and most of this power may be concentrated at a single point of contact with the substrate.

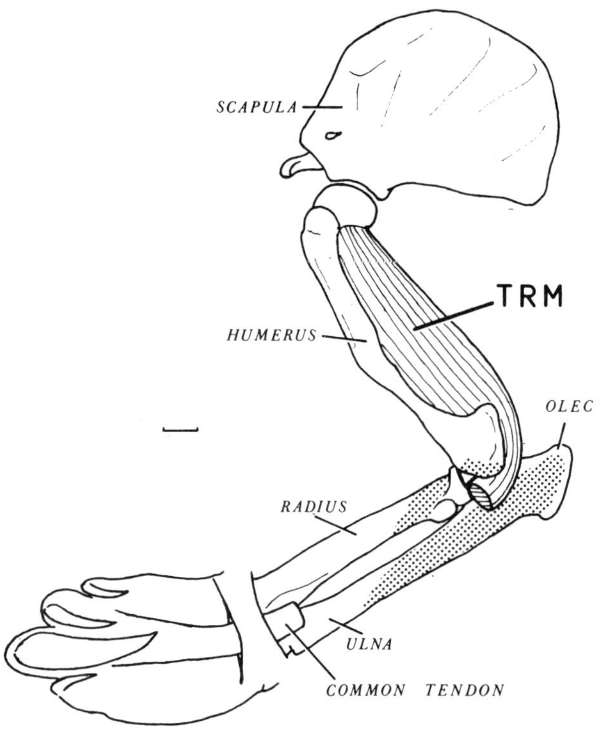

Figure 2. Right forelimb skeleton of *Tamandua*, seen in medial view. Note: (1) medial head of *triceps* muscle (**TRM**), transected as it enters forearm: (2) areas for origin of flexor *digitorium profundus* muscle (shown in stipple); and (3) the common insertion tendon for these two muscles. **OLEC** = olecranon process. Scale: 1 cm.

Another important feature of the modification of the *triceps* musculature in *Tamandua* derives from the fact that as the medial head enters the forearm, it passes directly through the axis of flexion-extension of the

Table 1. Weights of muscles (g) in single specimens of *Didelphis* and *Tamandua*; these two individuals were selected so as to be of virtually identical body size in terms of linear dimensions. (Data from Taylor, 1978).

	Didelphis	*Tamandua*
Musculature flexing proximal interphalangeal joint only:		
Flexor *digitorum sublimis*	0.2	2.0
Musculature flexing distal interphalangeal joint:		
Flexor *digitorum profundus*	3.1	11.7
Medial head of *triceps*	–	10.7
Total	3.3	24.4

elbow joint. As a result, contraction of the medial head has virtually no effect upon movements at the elbow. The animal may thus utilize all of its digital flexors to full effect regardless of the position of the more proximal joints of the limb. The claws may engage the environment with the limb in virtually any posture, and they may be flexed powerfully even when the limb is fully extended.

Elbow and shoulder joints.-- Some features of the musculoskeletal anatomy of the elbow joint in *Tamandua* are directly related to increased capability for powerful flexion of that joint. In all mammals, a portion of the musculature of the dorsal compartment of the forearm takes origin from the humerus and thus crosses the elbow joint on its flexor aspect. In most mammals (as in the opossum, EXT, Fig. 3), this humeral attachment is confined to the supracondyloid ridge. In vermilinguas, however, the ligament connecting this ridge with the deltoid tubercle permits these muscle attachments to be extended proximally more than half the length of the humerus (EXT, Fig. 3). A major skeletal modification is a stout ligament which connects the greatly expanded deltoid tubercle of the humerus (DTB, Fig. 3) to the supracondyloid ridge. This ligament occurs in all three genera of vermilinguas. It is characteristically ossified in *Cyclopes*, but the significance of this ossification is not clear.

The result of the increased area of muscle attachment is a substantial increase in the mechanical advantage for flexion of the elbow of those muscles which arise most proximally on the humerus (compare *Tamandua* and *Didelphis* in Fig. 4).

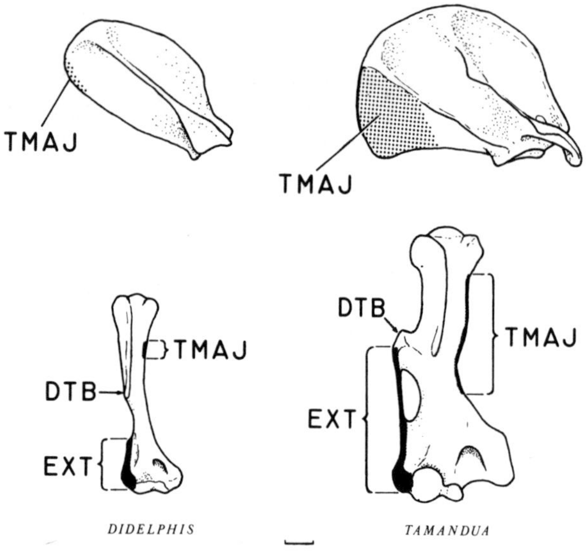

Figure 3. Above: lateral views of scapula in *Didelphis* and *Tamandua*, showing areas for origin of *teres major* muscle (**TMAJ**). Below: anterior views of humerus in *Didelphis* and *Tamandua*, showing areas for insertion of *teres major*, and for origin of dorsal muscles of the forearm (**EXT**). **DTB** = deltoid tubercle. Scale: 1 cm.

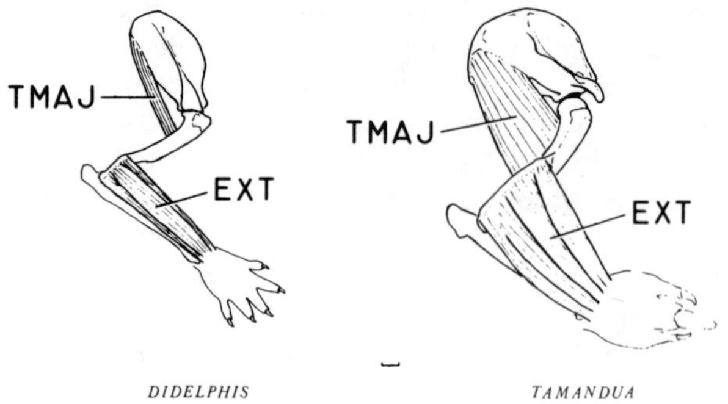

Figure 4. Right forelimb skeletons of *Didelphis* and *Tamandua*, seen in lateral view. Note differences in relative size and mechanical advantage of *teres major* (**TMAJ**) and dorsal forearm muscles (**EXT**). Scale: 1 cm.

In addition to their being arranged so as to have greater leverage across the elbow joint, the muscles are also very large in vermilinguas. For single specimens of *Didelphis* and *Tamandua* of the same linear body size, the following comparisons are noteworthy (data from Taylor, 1978). The combined weights of the muscles of the dorsal compartment of the forearm which arise from the humerus and cross the flexor aspect of the elbow joint (*brachioradialis*, extensors *carpi radialis* and *ulnaris*, extensors *digitorum*, and *supinator*) were 3.0 gm in *Didelphis* and 18.2 gm in *Tamandua*; even from single samples it is clear that this muscle group is substantially larger in the vermilingua. The combined weights of the elbow flexors arising from the scapula and upper humerus (*biceps* and *brachialis*) were approximately equal: 3.2 gm in *Didelphis* and 3.9 gm in *Tamandua*.

Tamandua also exhibits features of its shoulder anatomy which result in enhanced capability for powerfully retracting the humerus (swinging it caudally). One of the major muscles involved in shoulder retraction is *teres major*. This muscle is relatively large in vermilinguas, and there are significant modifications in its attachments which increase its mechanical advantage as compared to the opossum. In all mammals, *teres major* takes origin from the caudal angle of the scapula. The expanded postscapular fossa (characteristic of xenarthrans) provides increased attachment area for this large muscle (TMAJ) in *Tamandua* (Fig. 3). Likewise, the insertion area is extended much further down the shaft of the humerus in *Tamandua* than in *Didelphis*. In addition to accomodating a larger muscle, the increased scapular and humeral attachment areas also increase its mechanical advantage by displacing much of the muscle away from the axis of rotation at the shoulder joint.

The principal muscles responsible for the increased capabilities for forceful elbow flexion and shoulder retraction in *Tamandua* are shown in Figure 4; the homologous muscles in *Didelphis* are also illustrated for comparison. On the basis of this morphology, it may be predicted that *Tamandua* habitually utilizes flexion of the elbow and retraction of the shoulder in order to draw the hand forcefully toward the body. These movements are likely to be important when the animal is tearing apart an object, such as a termite nest or piece of wood, in that flexion of the digits may be supplemented by flexion of the more proximal joints in the forelimb in order to apply more powerful traction to material in which the claws may be hooked.

Axial rotation.-- In addition to features of *Tamandua's* forelimb anatomy which relate to movements of flexion and extension of various joints in the limb, there are a number of features which reflect enhanced capabilities for powerful axial rotation of various limb segments. Factors which increase the capability for pronation and supination of the forearm include expansion of the *supinator, pronator teres,* and *pronator quadratus* muscles, and flaring of the distal part of the shaft of the radius, which increases their moment arms for rotation of the radius. Also, there is a substantial amount of musculature in *Tamandua* (the short head of *biceps*, part of the long head of *biceps*, and *brachioradialis*) which produces concurrent flexion of the elbow and supination of the forearm (see Taylor, 1978, for detailed discussion).

Lateral rotation of the humerus is augmented in *Tamandua* primarily by lateral expansion of the deltoid tubercle (DTB, Fig. 3); this feature is characteristic of xenarthrans and reaches its extreme in vermilinguas. The effect of this lateral expansion is to displace the insertion of the *spinodeltoid* muscle away from the axis of rotation of the humerus, thus increasing its moment arm for lateral rotation of the humerus. Capability for medial rotation of the humerus is also enhanced in *Tamandua* by expansion and increased pennation of *subscapularis*.

That *Tamandua* is morphologically equipped for powerful axial rotation of proximal limb segments and that this type of movement may form an important part of its behavioral repertoire is supported by examination of its digital joint surfaces (Fig. 1). As mentioned above, flexion and extension in the digits can occur only at the metacarpophalangeal and distal interphalangeal joints. The articulating surfaces of both of these joints are deeply grooved and keeled such that no other movements are possible. Axial rotation and side-to-side movements at the digital joints are prevented by an extraordinary degree of bony reinforcement (in mammals with relatively unspecialized forelimbs these movements are prevented only by ligaments).

During the evolution of vermilinguas, the digital joints were apparently habitually subjected to high dislocating stresses in directions other than those of flexion and extension. Such stresses would be produced if axial rotation of more proximal limb segments were used to impart a twisting motion to the hand while the claws were being flexed against resistance; this combination of movements would be particularly useful in loosening pieces of material while the animal is tearing an object apart.

Functional implications.-- The major morphological specializations in the forelimb of *Tamandua* involve enlarged central foreclaws which may be flexed very powerfully, as well as additional capabilities for powerful flexion and axial rotation of more proximal limb segments. All of these features relate to the ability to tear apart hard materials with the claws. This ability is particularly important for *Tamandua*. In addition to ants, *Tamandua* also utilizes termites (genus *Nasutitermes* and others) as food (Montgomery and Lubin, 1977; Montgomery 1979, 1985a, 1985b). The termites spend virtually all of their time within an interconnecting system of nest, covered foraging trails, and gallery systems in dead wood; the nests and foraging trails are constructed of carton, a material which can be very hard. Many of the ants which are used as food have nests inside wood or use hard carton nests constructed by termites (Montgomery and Lubin, 1977; Lubin and Montgomery, 1981; Lubin et al., 1977). Thus, in order to feed on the concentrations of ants or termites in their nests, the vermilingua must be able to tear apart carton and/or wood, both of which are very strong. It is also advantageous for the vermilingua to be able to effect this penetration as rapidly as possible, in order to maximize the number of prey it can take before being driven off by soldier ants or termites (Montgomery and Lubin, 1977; Lubin and Montgomery, 1981).

A major implication of the forelimb anatomy of *Tamandua* is that the claws may be brought to bear with considerable force solely through the mechanism of digital flexion; thus an attack on a prey nest or occupied wood may be carried out, or at least initiated, with the rest of the limb fully extended. This capability would maximize effective reach and thus allow the animal to attack its prey while keeping its body as far away as possible from the retaliatory responses of the insects. This body posture was often observed in the field (G. G. Montgomery, pers. comm.), particularly when *Tamandua* fed on the large nests of *Azteca* ants, which rapidly mobilize thousands of soldiers to the place where the nest is penetrated (Montgomery and Lubin, 1977).

The large palmar pad against which the claws may be opposed allows the hands to be used as a prehensile organ. This provides the animal with the capability for manipulating objects with the hand; for example, pieces of termite nest or infested wood may be grasped, pulled away and held, rather than simply pried apart to fall where they may. Wild *Tamandua* often thus grasped twigs and other small bits of wood in one hand while they licked the ants or termites from them (G. G. Montgomery, pers. comm.). Prehensile hands are also clearly of importance for climbing.

Myrmecophaga

Differences in forelimb anatomy between *Tamandua* and *Myremecophaga* are relatively minor. All of the features of *Tamandua*'s forelimb anatomy discussed above are shared by *Myrmecophaga*. The following differences between these two genera are the most noteworthy, but none of these affects the mechanical capabilities of the limb in any major way.

Cephalohumeralis *muscle*.-- While *Tamandua* possesses a well-developed *cephalohumeralis* muscle, *Myrmecophaga* lacks one entirely. This muscle, which develops from fusion of the *clavotrapezius* and *clavodeltoid* muscle anlage, is a fairly common feature in mammals which lack clavicles (in both *Tamandua* and *Myrmecophaga* the clavicle is vestigial). It is not clear why the *cephalohumeralis* muscle should fail to develop in *Myrmecophaga*, although it may relate to the increased terrestriality of this genus. In *Tamandua*, the *cephalohumeralis* muscle appears to aid in raising the forelimb above and in front of the body during climbing. It may also aid in forcefully striking upward with the limb either at a predator or at a prey nest located above and in front of the animal. *Myrmecophaga*, however, rarely (if ever) climbs, (but see Shaw et al., 1985, for observed exceptions) and it takes its prey on the ground; or, as observed in Venezuela (G. G. Montgomery, pers. comm.) by raising up on its hind legs to feed from the lower parts of small trees while squatting in a bipedal position. The defensive posture of both genera involves rearing up on the hind legs and tail so that they can strike forward and downward with the claws. Thus there appears to be little occasion for *Myrmecophaga* to forcefully elevate the forelimb, and this may correlate with the lack of a *cephalohumeralis* muscle in this genus.

Wrist extension.-- Hyperextension of the carpus, both at the proximal and the middle carpal joints, is somewhat more limited in *Myrmecophaga* than in *Tamandua* (D. W. Yalden, pers. comm.). This difference is probably related to differences in habitat and in body size: the more arboreal *Tamandua* requires a maneuverable wrist, while the heavier *Myrmecophaga* requires a wrist which is more stable in extension, in order to facilitate bearing the body weight while moving on the ground.

Forearm length.-- The length of the forearm relative to the length of the upper arm is greater in *Myrmecophaga* than in *Tamandua* or *Cyclopes* (Table 2). Although the sample sizes of *Myrmecophaga* and *Cyclopes* are too small for statistical analysis, the difference between *Myrmecophaga* and the other two vermilinguas appears to be a substantial one. Elongation of distal limb elements is a commonly-occurring adaptation in terrestrial quadrupeds for increasing stride length. A possible advantage for longer stride length in *Myrmecophaga* is that it may allow the animal to cover large distances while foraging with less expenditure of energy than would be possible with a shorter limb. Montgomery and Lubin (1977) report foraging distances totalling about 10 km per night in Venezuela (approximately 10,000 body lengths, less tail), while arboreal *Cyclopes* and *Tamandua* move about 1,000 and 2,000 body lengths each foraging period, respectively (Montgomery, 1985a).

Table 2. Limb proportions of *Didelphis* and 3 genera of vermilinguas. Shown are the ratio of radius length to humerus length and olecranon length to humerus length for all 4 genera. (Data from Taylor, 1976).

	Didelphis $n=5$	*Cyclopes* $n=1$	*Tamandua* $n=9$	*Myrmecophaga* $n=2$
radius/ humerus	0.95±0.03	0.74	0.75±0.02	1.02
olecranon/ humerus	0.22±0.03	0.18	0.26±0.01	0.30

Digital proportions.-- There are minor differences between *Tamandua* and *Myrmecophaga* in the relative proportions of the bony elements in the digits. In *Myrmecophaga* there is less disparity in size between digits II and III, while the lateral digits are relatively more reduced than in *Tamandua*. These minor differences may relate to differences in use of the hand (*Tamandua* engages in more climbing and tearing, while *Myrmecophaga* engages in more terrestrial walking), but the exact relationships between form and function in this case are not clear.

Functional implications.-- Aside from the above minor differences, the forelimb morphology of *Myrmecophaga* resembles that of *Tamandua* almost exactly. This is puzzling, because the available evidence indicates that forelimb function differs between the two genera to a considerably greater extent. On the one hand, *Myrmecophaga* possesses a suite of forelimb specializations which lead one to conclude that it is nearly as well adapted as *Tamandua* for climbing and for feeding on termites. On the other hand, there is considerable evidence that *Myrmecophaga* rarely, if ever, engages in either of these activities (Montgomery and Lubin, 1977; Montgomery, 1985b; Shaw et al., 1985).

The well-developed foreclaws and powerful grasping

and tearing abilities of *Myrmecophaga* clearly do not play a major role in locomotion and may play only a minor role in obtaining food, except perhaps for digging to find soil-nesting ants or tearing away wood to reach nests of tree-nesting ants like *Camponotus* (Montgomery and Lubin, 1977). The only remaining function in which powerful claws would play a significant biological role is in defense against predators. Two points are clear. First, defensive weapons are important for a large, slow-moving insectivore which often ranges in open country where it can neither hide nor climb to safety. Second, claws are particularly important for a vermilingua; due to its small mouth gape, weak jaws, and lack of teeth, the mouth is useless for defense. It is clear, then, that there is a significant selective advantage in *Mymecophaga*'s possessing large claws and powerful grasping and tearing capabilities for antipredator defense. Shaw et al. (1985) discuss the role of large claws and powerful forelimbs in defense against predators, and conclude that available evidence supports the view that this is their major function.

However, it is not clear why the entire forelimb anatomy of *Myrmecophaga* should be specialized along almost exactly the same lines as that of *Tamandua*. One possible explanation is that the common ancestor of *Myrmecophaga* and *Tamandua* may have been an arboreal termite-eater, and that *Myrmecophaga*'s forelimb anatomy thus reflects prior specialization for climbing and for tearing open termite nests, as well as for defense. If so, then the giant anteater would represent a relatively recent myrmecophagid invasion of a terrestrial, primarily ant-eating niche; predation pressure may have selected for evolution of large body size plus retention of previously acquired forelimb specializations which are now used for defense. It must be emphasized that, while it is consistent with the known facts, this scenario is speculative, and conclusive evidence is lacking.

Cyclopes

The forelimb anatomy of the little silky anteater *Cyclopes* differs from that of the other two genera in a number of significant respects.

Digital flexion.-- The medial head of the *triceps* muscle retains its primitive attachments in *Cyclopes*; that is, it inserts on the olecranon process of the ulna and does not fuse with flexor *digitorum profundus* as in the other two vermilinguas. Thus capability for powerful digital flexion in this genus is somewhat diminished relative to the other vermilinguas, although it is still considerable (flexor *digitorum profundus* is still very well developed in *Cyclopes*). This genus eats only small ants; indeed, there is evidence that it deliberately avoids eating termites (Montgomery, 1979, 1985a, 1985b). Thus its somewhat lessened capacity for claw flexion may relate to the fact that it does not habitually tear open termite nests or termite infested wood. Montgomery (1983) describes the behavior of *Cyclopes* as it feeds on ants living in small stems of lianas and terminal branches of trees. The vermilingua stands on the stem in such a way that its hind feet point in opposite directions, and grasps the stem tightly with the hindfeet, which are highly modified to fit small stems (see below). It maintains torque so that it can stand upright on the twig. It then bends over, inserts the sharp foreclaws into the side of the twig, and opens a slit by prying upward. It removes ants from the interior of the twig with its tongue. The forelimbs are thus adapted more for prying than for tearing away bits of wood or nest.

Shoulder retraction.-- In *Cyclopes*, the *teres major* muscle is relatively larger than in even the other vermilinguas. The large size of this muscle requires large areas for bony attachment, but, (Fig. 5) *Cyclopes*' postscapular fossa is not enlarged to the extent seen in the other two vermilinguas (compare with *Tamandua*, Fig. 3). Instead, a portion of the origin of this muscle has migrated up the vertebral border and anteriorly along the primary spine of the scapula. The significance of this shift in muscle attachment is not clear, although it may be that a large postscapular fossa would impede full retraction of the humerus in *Cyclopes*. The *teres major* muscle also extends its insertion over virtually the entire length of the shaft of the humerus. The muscle is essentially a thick, parallel-fibered sheet which executes a half-twist between origin and insertion. As a result, fibers arising close to the shoulder joint insert far from it (and vice versa), and therefore all of the muscle fibers are relatively long. The greatly increased size of *teres major* (compare with *Didelphis*, Fig. 4) implies specialization for powerful retraction of the humerus; the fact that the muscle architecture permits this size increase without sacrificing fiber length implies that the muscle

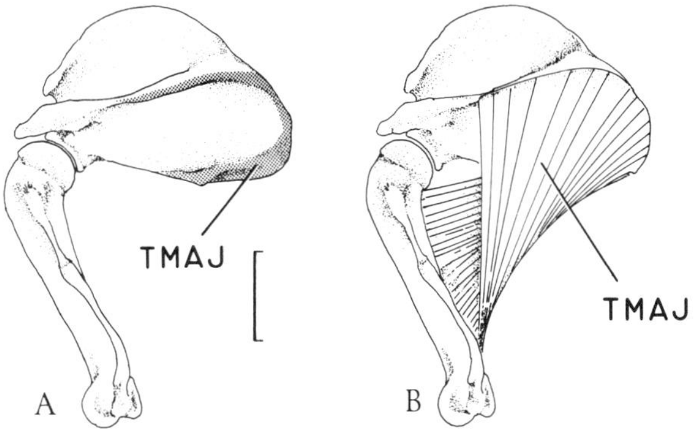

Figure 5. Left scapula and humerus of *Cyclopes*, seen in lateral view. Left: area for origin of *teres major* (**TMAJ**) is shown in stipple. Right: the muscle is shown in place. Scale: 1 cm.

may contract powerfully over a relatively large range of movement of the shoulder joint. It may be predicted, then, that *Cyclopes* is adapted for forceful retraction of the humerus over a wide range of joint positions. It should also be noted that *Cyclopes* possesses a well-developed, fully functional clavicle (lacking in the other two vermilinguas), which may aid in stabilizing the scapula while the humerus is retracted. These are perhaps primarily adaptations for climbing on small vines and branches, in which the foreclaw is used to grasp vertical branches as the animal pulls itself upward.

As in the other vermilinguas, the claws of *Cyclopes* are the only available defensive weapons, but their mode of use in defense is quite different from *Tamandua* and *Myrmecophaga* (Montgomery, 1983). The defensive posture of *Cyclopes* involves holding onto a perch with the hind feet and tail, the hind feet pointed in opposite directions to stabilize the body, erecting the body stiffly, and elevating and protracting the forefeet so that they come to lie alongside the snout (Van Tyne, 1929). From this position, the animal strikes with the foreclaws by bending the entire body forward very quickly, from the waist, and drives the foreclaws into the flesh of the attacker. The *Cyclopes* then works the claws into the flesh by alternately flexing and relaxing the claws. Because of the extreme small size of *Cyclopes*, the primary component of this defensive strike is contraction of the trunk muscles, with some contribution of retraction of the shoulder joint, in order to maximize the sweep of the claws. With a given angular change, utilizing movement at a more proximal joint (such as the shoulder) will result in a greater excursion of the tips of the claws than if a more distal joint (such as a digital joint) is used. In view of the importance of powerful trunk flexion in the defensive behavior of *Cyclopes*, it is of interest to note that *Cyclopes* (and to a lesser extent *Tamandua* and *Myrmecophaga*) possesses extraordinarily well developed ribs and intercostal muscles which lend increased strength and stability to the trunk (Jenkins, 1970).

Elbow extension.-- Extension of the elbow is clearly of diminished importance in *Cyclopes*, even though in this genus the medial head of *triceps* remains an extensor of that joint (which it is not in the other vermilinguas). Most of the elbow extensors (all heads of *triceps*, as well as *anconeus* and *epitrochleo-anconeus*) are relatively smaller in *Cyclopes* than they are even in *Didelphis*. In addition, the mechanical advantage for the elbow extensors, as measured by the relative length of the olecranon process, is also less in *Cyclopes* than in the other two vermilinguas (Table 2).

Cyclopes spends virtually all of its time in the forest canopy, and its diminished capability for elbow extension probably relates both to its extreme arboreality and to its small size. In terrestrial quadrupeds, the elbow extensors play an important role in postural support in that they prevent the elbow from collapsing into flexion under the weight of the body (this consideration is more important for larger animals, since body weight is a function of body volume and thus increases exponentially with a given increase in linear dimensions). The elbow extensor muscles are also important for providing propulsion during quadrupedal walking. Because of its small size, *Cyclopes* would place relatively small demands on its elbow extensors. It probably rarely engages in quadrupedal walking per se, although it does walk upright on small horizontal twigs, grasping them between foreclaw and footpad as it does so (Montgomery, 1983).

Grasping ability.-- *Cyclopes* appears to have an increased capability for delicate grasping movements and manipulations with the hand, as compared to *Tamandua* and *Myrmecophaga*. This increased capability results primarily from the fact that the pisiform bone is elongated and extends into the palmar pad such that it and the enlarged claw of digit III are opposable (Fig. 6). The enlarged opposable pisiform in *Cyclopes* functions as a sort of "thumb" in that it allows the animal to tightly grasp objects of very small diameter. This adaptation may be crucial for resource partitioning between *Cyclopes* and *Tamandua*, since it allows *Cyclopes* to move onto small terminal branches and along small lianas which cannot be utilized by *Tamandua* (Montgomery, 1979, 1985a); these small branches and lianas provide not only a substrate for movement but

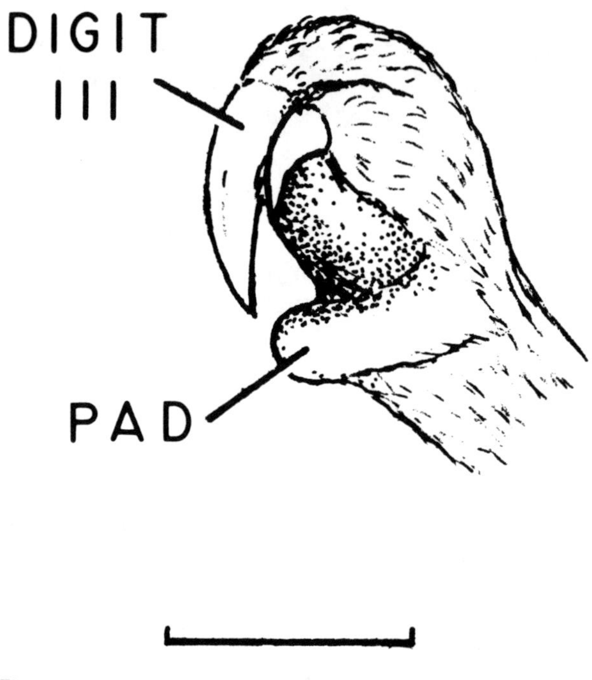

Figure 6. Right manus of *Cyclopes*. Note palmar pad, which is elongated and stiffened by enlarged pisiform bone. Scale: 1 cm.

also the basis of *Cyclopes*' food supply from ants which nest inside the twigs. It is of interest to note that the hind foot of *Cyclopes* possesses an analagous adaptation for increased grasping ability (Jones, 1953).

Functional implications.-- It is likely that these unique specializations of *Cyclopes* evolved in response to ecological competition with other arboreal Myrmecophagidae. The ability to move among the smallest of terminal canopy branches and along the smallest of lianas provides *Cyclopes* with a resource space which is unavailable to *Tamandua*. The fundamental requirements for *Cyclopes* in this regard are greatly diminished body size and increased capability for grasping small-diameter objects. Small body size plus the habitual use of suspensory posturing also diminish the importance of elbow extension in locomotion and place a premium on powerful shoulder retraction.

ACKNOWLEDGMENTS

The following institutions supplied skeletal material and wet specimens for dissection: Field Museum of Natural History, Chicago; Museum of Vertebrate Zoology, Berkeley; National Museum of Natural History, Smithsonian Institution, Washington D.C.; and Lincoln Park Zoo, Chicago. I thank H. R. Barghusen, G. G. Montgomery, and L. B. Radinsky for their helpful comments and discussion. This research was supported by institutional funds from the University of Chicago and by funds from the Field Museum Center for Graduate Studies in Systematic Zoology and Paleontology.

LITERATURE CITED

Jenkins, F. A., Jr.
1970. Anatomy and function of expanded ribs in certain edentates and primates. J. Mamm. 51: 288-301.

Jones, F. W.
1953. Some readaptations of the mammalian pes in response to arboreal habits. Proc. Zool. Soc. Lond., 123:33-41.

Lubin, Y. D. and G. G. Montgomery.
1981. Defences of *Nasutitermes* termites against *Tamandua* anteaters. Biotropica, 13(1):66-76.

Lubin, Y. D., G. G. Montgomery, and O. P. Young.
1977. Food resources of anteaters (Edentata: Myrmecophagidae) I. A year's census of arboreal nests of ants and termites on Barro Colorado Island, Panama Canal Zone. Biotropica, 9:26-34.

Montgomery, G. G.
1979. El grupo alimentico (feeding guild) del oso hormiguero: convivencia y especialización de las presas de sustento de los osos hormigueros neotropicales (Edentata, Myrmecophagidae). ConCiencia 6(1):3-6.

1983. *Cyclopes didactylus* (Tapacara, Serafin de Platanar, Silky Anteater). Pp. 461-463, *In* Costa Rican natural history (D.H. Janzen, editor). Univ. Chicago Press, Chicago, 816 pp.

Montgomery, G.G.
1985. Impact of vermilinguas (*Cyclopes, Tamandua*: xenarthra = edentata) on arboreal ant populations. (this volume).

Montgomery, G. G.
1985. Movements, Foraging and Food Habits of the Four Extant Species of Neotropical Vermilinguas (Mammalia; Myrmecophagidea) (this volume).

G. G. Montgomery and Y. D. Lubin.
1977. Prey influences on movements of neotropical anteaters. Pp. 103-131 *in* Proceedings of the 1975 predator symposium (R. L. Phillips and C. Jonkel, eds.). Montana Forest and Conservation Experiment Station, Univ. of Montana, Missoula, 268 pp.

Shaw, J.H., T.S. Carter and J.C. Machado-Neto.
1985. Ecology of the giant anteater *Myrmecophaga tridactyla* in Serra da Canastra, Minas Gerais, Brazil: a pilot study. (this volume).

Taylor, B. K.
1976. The anatomy of the forelimb in armadillos and anteaters. Ph. D. dissert., University of Chicago. 270 pp.

1978. The anatomy of the forelimb in the anteater (*Tamandua*) and its functional implications. J. Morph., 157:347-368.

Van Tyne, J.
1929. Notes on the habits of *Cyclopes dorsalis*. J. Mamm., 10:314.

THE SUPERFICIAL FACIAL MUSCULATURE IN SLOTHS AND VERMILINGUAS (ANTEATERS)

Virginia L. Naples

Museum of Zoology, University of Massachusetts, Amherst, Massachusetts 01003. (Present address: Department of Biological Sciences, Northern Illinois University, DeKalb, Illinois 60115, USA).

Abstract
The two extant genera of sloths *Bradypus* and *Choloepus* and the three extant genera of vermilinguas *Cyclopes, Tamandua* and *Myrmecophaga* show similar patterns of facial muscle retention. As compared with most placental mammals, facial muscles are reduced in number and complexity, reflecting the relative immobility of the facial area. The muscles are peculiarly differentiated and specialized to permit fine control over mouths and noses. Sloths and vermilinguas depend little on the function of facial vibrissae, which are few in number; musculature associated with vibrissae is reduced and functions primarily to control the anterior rhinarium. There are intergeneric differences in the subdivisions of the facial muscles of sloths. The intergeneric differences among vermilinguas are attributable to allometry and to the extreme lengthening of the rhinarial region in the larger forms. *Cyclopes*, the smallest vermilingua, has a reduced number and complexity of facial muscle elements, even when compared with other vermilinguas, while *Tamandua* and *Myrmecophaga* have increasingly more complex and robust facial muscle elements. A model of masticatory function in vermilinguas is presented, based on field observations of vermilinguas and the structure of their facial musculature. The buccinator, surrounding the mouth and forming the cheek walls, is the most important facial muscle element involved in feeding. The buccinator is uniquely specialized, and could function as a "conveyor belt" mechanism, enabling vermilinguas to swallow prey in a continuous manner. The buccinator complex in sloths is reduced but not specialized. It allows a wide gape and flexible lips and cheek walls. Sloth and vermilingua facial musculature does not resemble that of monotremes, but follows the marsupio-placental plan.

Abstracto
Los dos géneros sobrevivientes de perezosos *Bradypus* y *Choloepus* y los tres géneros sobrevivientes de osos hormigueros *Cyclopes, Tamandua* y *Myrmecophaga* muestran patrones similares de retención de los músculos faciales. Comparados con los de la mayoría de los mamíferos placentarios, sus músculos faciales son de reducida cantidad y poco complejos, reflejando así la relativa inmovilidad del área facial. Los músculos están singularmente diferenciados y especializados para permitir un buen control de la boca y la nariz. Los perezosos y los osos hormigueros dependen poco de la función de las vibrisas faciales, que son pocas en cantidad; la musculatura asociada a las vibrisas es reducida y funciona principalmente para controlar el rinario anterior. Existen diferencias intergenéricas en las subdivisiones de los musculos faciales de los perezosos. Las diferencias intergenéricas entre osos hormigueros son atribuibles a la alometría y al alargamiento extremo de la región rinaria en los ejemplares más grandes. *Cyclopes*, el oso hormiguero más pequeño, tiene una cantidad reducida de elementos musculares faciales poco complejos, aún cuando se le compara con otros osos hormigueros mientras que el *Tamandua* y el *Myrmecophaga* tienen elementos cada vez más complejos y elementos musculares faciales robustos. Se presenta un modelo de la función masticatoria en los osos hormigueros que está basada en observaciones de campo de los mismos y la estructura de su musculatura facial. El buccinador, que rodea la boca y forma las paredes de los cachetes, es el elemento más importante de los músculos faciales empleados para la alimentación. El buccinador es singularmente especializado, y puede funcionar como un mecanismo de "correa transportadora" que le permite a los osos hormigueros ingerir sus presas en forma contínua. El complejo buccinador en los perezosos es reducido pero no especializado. Permite que la boca alcance gran anchura al abrirse y produce flexibilidad en los labios y paredes de los cachetes. La musculatura de los perezosos y los osos hormigueros no se parece a la de los monotremas, pero sí se adapta al plan marsupio-placentario.

Resumo
Os dois gêneros existentes de preguiças *Bradypus* e *Choloepus* e os três gêneros existentes de vermilínguas *Cyclopes, Tamandua* e *Myrmecophaga* mostram padrões similares de retenção de músculos facais. Comparados com a maioria dos mamíferos placentários, os músculos facais são reduzidos em número e complexidade, refletindo a relativa imobilidade da área facial. Os músculos são peculiarmente diferenciados e especializados para permitir um delicado controle sobre boca e nariz. Preguiças e vermilínguas dependem pouco da função das vibrissas faciais, as quais são em pequeno número `a musculatura associada com as vibrissas é reduzida e funciona primariamente controlando a região nasal anterior. Há diferenças inter-genéricas nas sub-divisões dos músculos facias de preguiças. As diferenças inter-genéricas entre vermilínguas são atribuíveis à alometria e ao extremo comprimento da região nasal nas formas maiores. *Cyclopes*, o menor vermilíngua, tem um reduzido número e complexidade de elementos musculares faciais, mesmo quando comparado com outros vermilínguas, enquanto *Tamandua* e *Myrmecophaga* têm elementos musculares facais gradualmente mais complexos e

robustos. Um modelo de função mastigatória em vermilínguas é apresentado, baseado em observações a campo de vermilínguas e na estrutura de sua musculatura facial. O bucinador, circundando a boca e formando as paredes das bochechas é o elemento muscular facial mais importante envolvido na alimentação. O bucinador é particularmente especializado, e poderia funcionar como um mecanismo de "esteira rolante", capacitando vermilínguas a engolir suas presas de maneira contínua. O complexo bucinador em preguiças é reduzido mas não especializado. Ele permite uma ampla abertura de boca e flexibilidade de lábios e pareded da bochecha. A musculatura facial de preguiças e vermilínguas não se parecė com a de monotremados, mas segue o plano marsupio-placentário.

Knowledge of the gross anatomy of the Xenarthra (Edentata) is incomplete, and there are no major interpretive studies of the musculature of any part of the order. The anatomy of the superficial facial musculature of diverse groups of mammals such as rodents (Hill, 1937; Meinertz, 1941; Rinker, 1954; Klingener, 1970) and primates (Huber, 1930; 1931) has been studied extensively, and has been used in studies of the systematics of the lower taxa. This approach may not be practical at the species level, but may be of value in revealing intergeneric relationships (Rinker, 1954). There has been no previous attempt to compare facial musculature of the sloth genera *Bradypus* and *Choloepus*, nor have there been comparative studies of facial muscles of vermilinguas *Cyclopes, Tamandua* and *Myrmecophaga*.

Several workers described facial muscles of sloths and some vermilinguas, but these studies antedate that of Meinertz (1941), in which he sorted out the complex relationship of the different layers of superficial facial musculature in mammals. Research on xenarthran facial musculature in the 19th and early 20th centuries was further hampered by the attempts of several workers to show the derivation of facial musculature from that of the monotremes, to which the xenarthrans were then thought to be closely related (Uekermann, 1912). Other researchers noted the reduced facial musculature in sloths and vermilinguas and decided that it was both primitive and undifferentiated and therefore not worth investigating further (Windle and Parsons, 1889).

MATERIALS AND METHODS

The facial musculature of two adults (sexes unknown) each of *Bradypus infuscatus, Choloepus hoffmanni, Cyclopes didactylus*, and *Tamandua mexicana* and one specimen of *Myrmecophaga tridactyla* was dissected. All specimens were preserved in formalin. The *Myrmecophaga* had been skinned, resulting in the loss of some of the origins and insertions of the facial muscles. Such instances are specifically mentioned in the text. Skeletal material from the University of Massachusetts Museum of Zoology, the American Museum of Natural History and the National Museum of Natural History was also examined.

During dissection, the specimens were allowed to air-dry slightly, making fascial planes and muscle fiber directions more obvious. This caused some shrinkage of the structures, compensated for in the illustrations.

Some variation in size was found in muscles between the left and right sides of the same specimen. Because of this variation and the small sample sizes, differences among genera were not quantified. Qualitative differences or similarities (origin, insertion, topographical relationships and innervations) were generally consistent within species.

The facial musculature is described separately for the sloths and the vermilinguas. For sloths, muscle descriptions are based upon *Choloepus*, which has a more complex arrangement than *Bradypus*. For vermilinguas, muscle descriptions are based on *Tamandua* and variations from this are described for *Cyclopes* and *Myrmecophaga*.

Rinker's (1954) monograph on the comparative morphology of the mammalian genera *Sigmodon, Oryzomys, Neotoma*, and *Peromyscus* (Cricetinae) was used as as a source for muscle terminology. Synonyms used by previous authors for facial muscles of xenarthrans are summarized in Table 1.

SLOTH FACIAL MUSCLES

Platysma and Derivatives.-- *M. platysma myoides* consists of mental, buccal and auricular parts. A deeper layer of the platysma, *M. platysma cervicale* was not observed in either *Choloepus* or *Bradypus*.

M. platysma myoides pars mentalis.-- In *Choloepus*, this muscle originates from the dorsal midline on the posterior part of the cervical region with some fibers coming from the dorsal aspect of the shoulder region (Fig. 1). Muscle fibers pass down around the neck in a band about five mm wide, superficial to the *pars palpebralis* of *M. sphincter coli profundus*. The fibers insert into the lateral aspect of the mental region, and pass deep to those of *M. orbicularis oris* all along the length of their shared insertion. In *Bradypus* this muscle is absent.

M. Platysma myoides pars buccalis.-- This muscle is absent in *Choloepus* (Fig. 1). In *Bradypus* the fibers originate in the fascia of the nuchal region and pass laterally, then turn cranially to pass beneath the ear onto the buccal region. Some fibers appear to originate higher and separately and pass immediately ventral to the ear cartilage and then join to the lower fibers anteriorly. They insert into the corner of the mouth and *M.*

Table 1. Synonyms used for superficial facial muscles of sloths and vermilinguas. Present terminology follows Rinker (1954).

Present Terminology	*Terminology of Uekermann (1912)*	*Terminology of Owen (1857)*
M. platysma myoides	*Platysma*	Not named
pars buccalis	*Platysma*	Not named
pars mentalis	*Platysma*	Not named
pars auricularis	*Orbitoauricularis*	Not named
M. platysma cervicale	*Auricularis posterior*	Not named
M. cervicoauricularis	*Auriculo-occipitalis*	*Levator auriculae anterior* and *posterior*
M. cervicooccipitalis	*Orbito-auricularis*	*Levator auriculae anterior*
M. cervicoaricularis		
medius anterior	Not named	Not named
medius posterior	*Auricularis posterior*	Not named
posterior profundus	*Auricularis postererior*	Not named
M. sphincter colli and derivatives		
pars auris	*Sphincter colli*	Not named
M. auriculooccipitalis	*Auricularis superior*	*Levator auriculae (anterior)*
M. orbicularis oculi	*Orbicularis oculae*	Not named
M. nasolabialis	Not named	Not named
superficialis	Not named	Not named
pars palpebralis	*Sphincter colli*	Not named
M. dilator nasi	*Maxillolabialis*	*Levator nasi*
M. nasolabialis profundus		
pars interna	*Nasalis*	*Orbicularis oris*
pars media superior	*Levator labii superioris alaequi nasi*	Not named
pars anterior	*Levator labii superioris alaequi nasi*	*Orbicularis oris*
pars maxillaris superficialis	*Levator labii inferioris*	*Retractor labii superioris*
pars maxillaris profunda	Not named	Not named
pars internus	*Buccinator internus*	Not named
pars orbicularis	Not named	Not named
pars orbicularis oris	*Orbicularis oris*	*Orbicularis oris*

orbicularis oris at both the upper and lower lips. In one specimen the insertion onto the mandible is completely separated into two parts, one being more ventral than the other.

M. platysma myoides pars auricularis.-- In *Choloepus* the fibers arise from the most anterior region of the base of the ear cartilage (Fig. 1). They pass dorsocranially and fan out onto the parietal region passing external to the fibers of *M. orbicularis oculi superioris*. They disappear in the fascia over the parietal bone before reaching as far anteriorly as the orbit. In *Bradypus*, the condition is similar to that in *Choloepus*, but the muscle has undergone greater reduction.

The retroauricular muscle group is greatly reduced in number and size in both sloth genera. *M. cervico-occipitalis* (Figs. 1 and 2) in *Choloepus* arises from the cranial portion of the ligamentum nuchae and the fibers form a thin sheet which passes craniolaterally, fanning out across the parietal bone, where they disappear into the fascial sheet. This muscle is very thin and was not seen in all specimens. It maintains a continuous sheet of fibers across its width without aponeurotic areas between fibers. In *Bradypus* this muscle is absent.

M. auriculo-occipitalis.-- In *Choloepus* this muscle arises from the dorsal midline, superficial to the fibers of *M. cervico-occipitalis* (Figs. 1 and 2). The muscle inserts on the anterodorsal aspect of the ear pinna, slightly above where the base joins the skull. In *Bradypus* this muscle is less well-developed than in *Choloepus*, but with a similar origin and insertion.

M. cervicoauricularis posterior.-- In *Choloepus* this muscle has two parts (Figs. 1 and 2), both arising from the dorsal midline. The anterior slip inserts onto the middle of the dorsal part of the ear cartilage. The caudal slip inserts on the posterior bulge of the ear cartilage. These muscle slips diverge approximately 10 mm before reaching the base of the pinna. A similar condition is seen in *Bradypus*, although the muscle is less well-developed and less differentiated than in *Choloepus*. *M. cervicoauricularis anterior* and *M. cervicoauricularis posterior profundus* are completely absent in both sloth genera.

M. mandibulo-auricularis.-- The muscle is absent in *Choloepus*, but is well developed in *Bradypus* (Figs. 1 and 4). It arises from the anteroventral portion of the base of the ear cartilage and passes anteriorly and ventrally to insert on the most dorsal and posterior portion of *M. massetericus*.

M. sphincter colli profundus *and derivatives*.-- This group

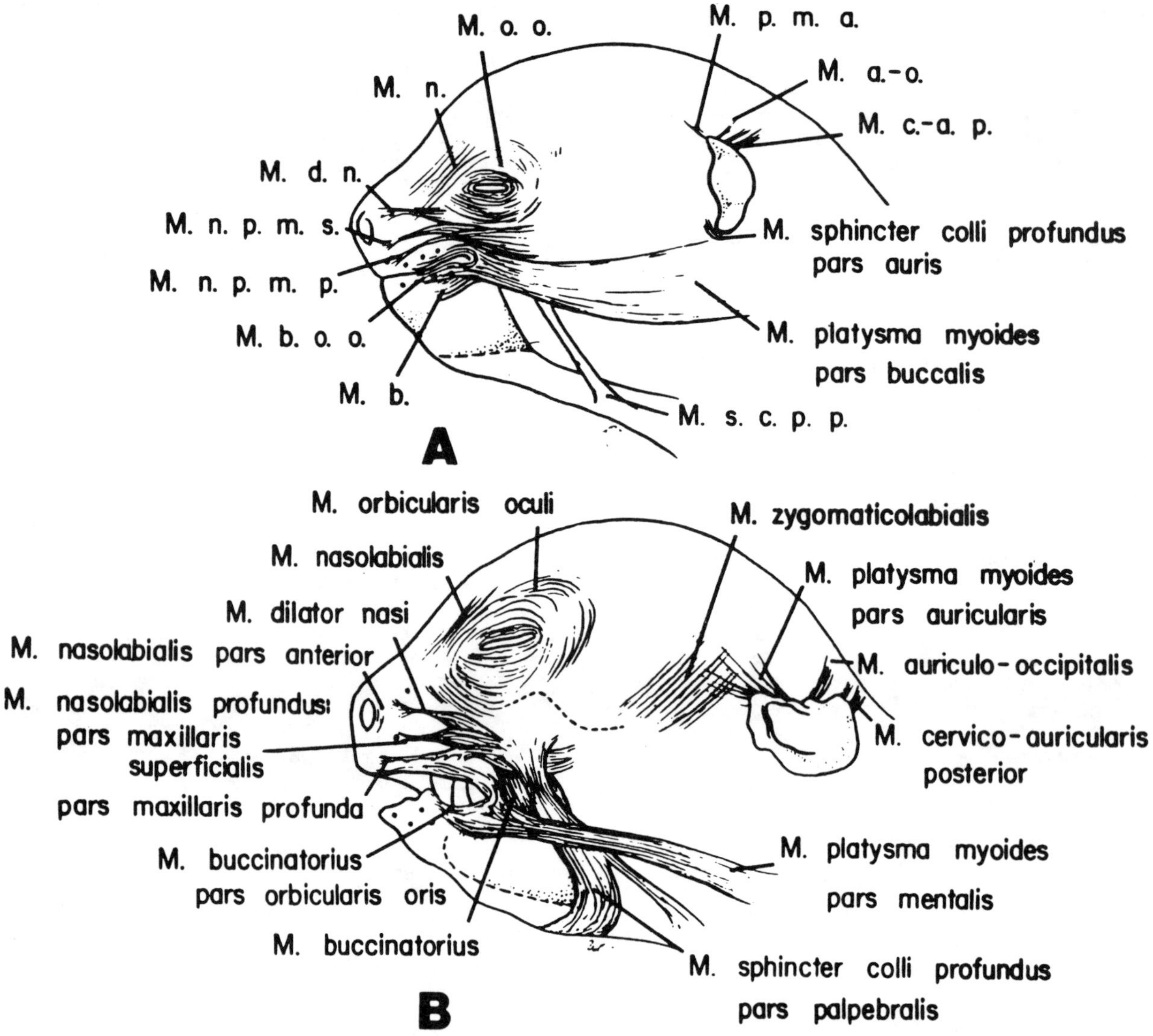

Figure 1. Superficial facial musculature of: A *Bradypus*, and B *Choloepus*.

of muscles is characterized in *Choloepus* and *Bradypus* both by the loss of some major sheet-like components and the simplicity of many of the specialized elements associated with the rhinarium, lips and cheek.

Pars auris.-- No trace of this muscle was found in *Choloepus*. In *Bradypus* this is an independent muscle (Fig. 1), originating on the throat and passing dorsally as a narrow strip to insert in the fascia behind the auricular cartilage. The muscle remains as a vestige near its insertion at the ventral surface of the ear cartilage, but its origin was not discernible.

Pars intermedia ventralis.-- This muscle, which ordinarily lies anterior to *pars auris* is absent in both *Choloepus* and *Bradypus*. The intermediate plate is missing from both sloth genera. *Pars intermedia dorsalis* and derivatives are poorly developed in *Choloepus*. Anteriorly *M. zygomaticolabialis* (Fig. 1), which is derived from an anterior extension of the intermediate plate is reduced to a thin sheet of fibers which are separated by aponeurotic tissue. In *Choloepus*, they arise in the fascia anterior to the ear and pass cranioventrally at a level superficial to *M. platysma myoides pars auricularis*. The fibers insert onto the fascial covering of the posterior projection of the zygomatic arch (jugal portion) superficial to parts of the *pars palpebralis* and the *pars oris* which originate there. All remnants of this muscle are absent in *Bradypus*. *M. frontalis* (auricular part), *M. auricularis anterior superior, M. auricularis anterior inferior* and *M. frontalis* (orbital part) are entirely absent from both sloth genera.

M. orbicularis oculi superior and *inferior*.-- These muscles are well developed anteriorly (Fig. 1). In *Choloepus* the fibers lying in the upper eyelid arise at the anterior

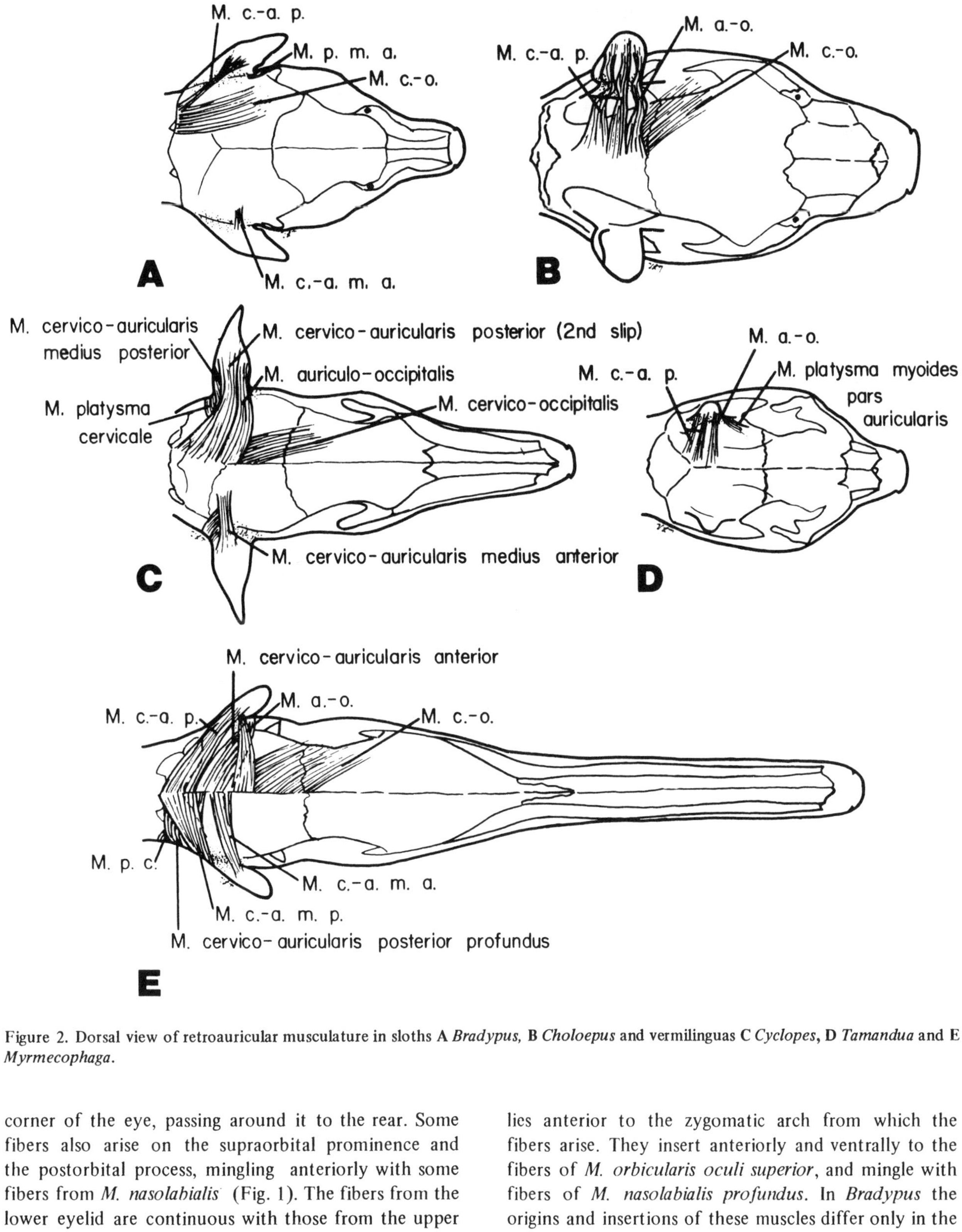

Figure 2. Dorsal view of retroauricular musculature in sloths A *Bradypus,* B *Choloepus* and vermilinguas C *Cyclopes*, D *Tamandua* and E *Myrmecophaga*.

corner of the eye, passing around it to the rear. Some fibers also arise on the supraorbital prominence and the postorbital process, mingling anteriorly with some fibers from *M. nasolabialis* (Fig. 1). The fibers from the lower eyelid are continuous with those from the upper eyelid, and insert on the underside of the anterior corner of the eye; some fibers from the upper eyelid insert on the postorbital prominence. Some fibers from the lower eyelid and the area adjacent and posterior to it insert onto the ascending process of the zygomatic arch. There is another portion of *M. orbicularis oculi inferior* which lies anterior to the zygomatic arch from which the fibers arise. They insert anteriorly and ventrally to the fibers of *M. orbicularis oculi superior*, and mingle with fibers of *M. nasolabialis profundus*. In *Bradypus* the origins and insertions of these muscles differ only in the greater development of a few fibers which arise in the fascia of the frontal prominence and the ascending process of the zygomatic arch. They pass dorsally and grade into the vestiges of *M. sphincter colli profundus intermedia ventralis*.

M. nasolabialis.-- In *Choloepus* this muscle arises from the supraorbital ridge anterior to the postorbital prominence, and inserts into the fibrous pad on the lateral surface of the maxilla; the fibers fan out over the lacrimal bone dorsal to the lacrimal canal (Fig. 1). Here, the sparse fibers terminate in the connective tissue on the lateral surface of the facial region dorsal to the most dorsal fibers of *M. orbicularis oculi inferior* and *M. nasolabialis profundus*. The situation is similar in *Bradypus*, but the muscle is less well-developed and the fibers intermingle more extensively with those of *M. nasolabialis profundus. M. nasolabialis superficialis* is absent in both sloth genera.

Pars palpebralis.-- This is the anterior extension of *pars intermedia*, and it is greatly reduced in sloths (Figs. 1 and 4). In *Choloepus* it arises from the fascia of the midventral surface of the mandible, sweeping dorsally and slightly cranially, and is joined by another slip of fibers originating more posteriorly. They converge approximately at the level of *M. platysma myoides pars mentalis*, under which they pass. The fibers of the muscle insert onto the lateral surface of the zygomatic plate, where they join superficially to the fibers of *M. nasolabialis profundus*, and pass cranially. A different portion of this muscle sheet has been retained in *Bradypus*. The fibers arise in a band three to four mm wide from the fascia of the midventral part of the mandible about halfway posteriorly along the ramus. The fibers pass dorsally beneath *M. platysma myoides pars buccalis* to insert onto the surface of *M. nasolabialis profundus pars maxillaris superficialis* by a thin aponeurosis.

M. dilator nasi.-- In *Choloepus* and *Bradypus*, this muscle arises deep to the origin of *M. nasolabialis profundus pars maxillaris superficialis*. The anterior half is composed of a flat tendon which divides into at least four parts as it approaches its insertion into the dorsolateral border of the rhinarium.

M. nasolabialis profundus.-- In both sloth genera, this muscle consists of three well differentiated parts on either side of the rostrum (Figs. 1 and 3). The muscle slips become tendinous anteriorly. The tendons divide as they pass cranially, with some slips inserting on the nasal cartilage, others into the premaxillary bone, and others into the connective tissue on the sides of the rostrum. The tendons of some of the fibers surround the sparse sinus hairs on the rostrum. In other mammals, this muscle sometimes has additional divisions, which are either missing or extremely poorly developed in sloths. *M. nasolabialis profundus pars interna, pars media superior* and *pars media inferior* are absent in sloths. In *Choloepus* there is a remnant of a *pars anterior* (Fig. 1) which arises from the nasal cartilage at the dorsolateral

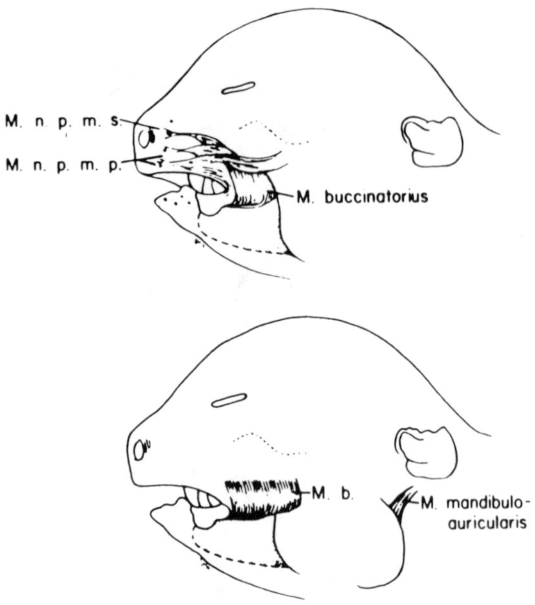

Figure 3. Middle and deep facial musculature layers in *Choloepus*.

margin of the naris, the fibers passing directly ventrally, and inserting into the fat pad and connective tissue of the upper lip. This muscle was not seen in *Bradypus*.

Pars maxillaris superficialis.-- In *Choloepus*, this muscle arises via a short aponeurosis on the ventral portion of the anterior border of the zygomatic plate, but does not extend onto the maxillary bone (Figs. 1 and 3). The muscle inserts by a tendon which gives off a number of branches into the fibers of *M. nasolabialis pars orbicularis oris* on the upper lip. The condition is similar in *Bradypus*.

Pars maxillaris profunda.-- In *Choloepus* this muscle arises from a thin aponeurosis on the anterior side of the zygomatic plate, dorsal to the origin of *pars maxillaris superficialis* (Figs. 1 and 2). There is only one slip, although there are more dorsal fibers which mingle with those of *M. orbicularis oculi inferior* and cannot be clearly distinguished from them. The condition in *Bradypus* is similar. *M. maxillolabialis* and *M. bucconaso labialis* are absent in both sloth genera.

M. buccinatorius.-- In sloths, this muscle, like *M. nasolabialis profundus*, lacks some elements that are present in other mammals (Figs. 1, 3 and 4). *Pars intermaxillaris* and *pars mandibularis* are not differentiated from one another, and it is not possible to distinguish *pars intermaxillaris superficialis* or *pars longitudinalis posterior*. In *Choloepus* the muscle arises from the lateral surface of the maxillary above the row of cheek teeth, terminating directly posterior to the last tooth. The fibers of this muscle form the cheeks, as in other mammals. The muscle inserts into the lateral surface

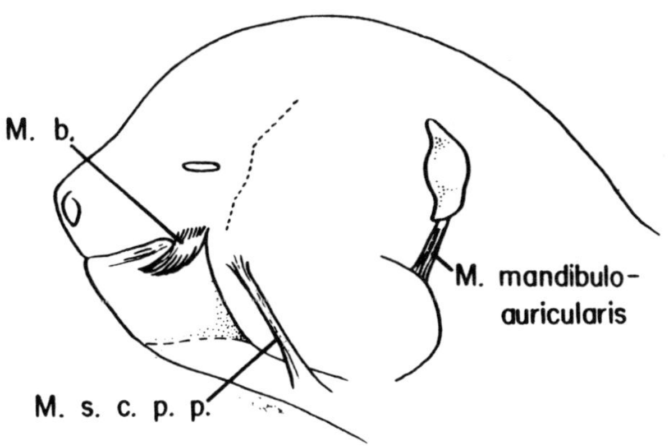

Figure 4. Deep facial musculature layer in *Bradypus*.

of the mandible dorsal to the roots of the cheek teeth. Most of the fibers pass directly ventrally, but the most posterior ones initially travel in a ventral direction toward their insertion. Before reaching it, they pass deep to the more anterior fibers in a cranial direction, along the ventral surface of the muscular insertion. The situation is further complicated by the insertion of fibers of *M. platysma myoides pars mentalis* at the lateral aspect of the lower lip. In *Bradypus* the muscle also originates on the maxilla, but further posteriorly than in *Choloepus*. The muscle insertion is similar to that of *Choloepus*, but there is a greater mingling of the fibers of *M. buccinatorius* with those of *pars orbicularis oris*, and *M. platysma myoides pars mentalis* is entirely absent.

Pars orbicularis oris.-- In *Choloepus* this muscle arises from the skin of the upper lip, principally from the posterior and ventral surface of the connective tissue pad supporting the sinus hairs (Figs. 1 and 4). The muscle fibers pass directly anteriorly- posteriorly, until they reach the posterior end of the buccal opening where they curve ventrally around it. These fibers insert into the lateral aspect of the lower lip, into the skin in this region, and into the subsymphseal region. The fibers are continuous all the way around the buccal opening at the lower lip, and at the rear of the buccal opening are joined by fibers of *M. platysma myoides pars mentalis*. The fibers of the latter muscle pass superficially to those of *M. buccinatorius* and *pars orbicularis*, extending even further anteriorly. In *Bradypus* the origin of *pars orbicularis* is similar, but the insertion differs. The fibers insert into the skin along the lateral surface of the lower lip, and into the subsymphyseal region of denser connective tissue. As in *Choloepus* they are joined by fibers of *M. buccinatorius*. However, instead of fibers from *M. platysma myoides pars mentalis* joining them superficially, they are joined superficially by fibers of the well developed *M. platysma myoides pars buccalis*, which is entirely missing in *Choloepus*. In *Bradypus* this muscle is less well developed, and the fibers are interspersed irregularly with fat and connective tissue.

FACIAL MUSCLES IN VERMILINGUAS

Platysma and Derivatives.-- *M. platysma myoides* consists of mental, buccal and auricular parts. A deeper layer of the platysma *M. platysma cervicole* is also present.

M. platysma myoides pars mentalis.-- In *Tamandua* this muscle arises from the middorsal area of the cervical and shoulder region, with the fibers passing around the neck anteriorly, and remaining distinct from *M. platysma myoides* throughout its course (Fig. 5). The muscle inserts into the corner of the mouth on the ventral surface of the lip, ventral to *M. nasolabialis profundus pars maxillaris superficialis* and deep to *M. orbicularis oris*. In *Cyclopes* the origin has been modified to arise dorsally from the anterior portion of the manubrium sternae, passing anteriorly as a thin strap. The muscle inserts as in *Tamandua*. The origin of this muscle could not be observed in *Myrmecophaga*. It is first observable as a thin slip passing anteriorly, ventral to the ear cartilage. After passing deep to *M. platysma myoides pars palpebralis*, fibers of this muscle are joined by those of *M. platysma myoides pars buccalis*, which insert as in the other two genera.

M. platysma myoides pars buccalis.-- In *Tamandua* this muscle arises with the main fibers of *M. platysma myoides* (Figs. 5 and 6). It inserts ventral to *M. platysma myoides pars mentalis*, and deep to *M. orbicularis* along the lower lip. Throughout its passage the muscle remains distinct from *M. platysma myoides pars mentalis*. In *Cyclopes* the condition is similar to that in *Tamandua*. In *Myrmecophaga* the fibers of *pars buccalis* fuse with those of *pars mentalis* anterior to the eye region, and insert as in the other genera.

M. platysma myoides pars auricularis.-- In *Tamandua* the fibers of this muscle arise from the most anterior region of the base of the ear cartilage (Fig. 5). They pass anteriorly and divide, with some fibers passing above and others below the eye. These fibers assume the position of an *M. frontalis pars auricularis*. The fibers passing ventral to the orbit insert into an aponeurosis ventral to the eye region. In *Cyclopes* the origin of this muscle and the insertion of the upper portion is similar to that in *Tamandua*. The ventral portion of the muscle has a much stronger insertion, curving around the cheek region and passing medially, just posterior to the large lateral buccal salivary gland. In *Myrmecophaga* the condition is similar to that in *Tamandua*.

M. platysma cervicale.-- In *Tamandua* this muscle arises from the fascia near the middorsal line in the region

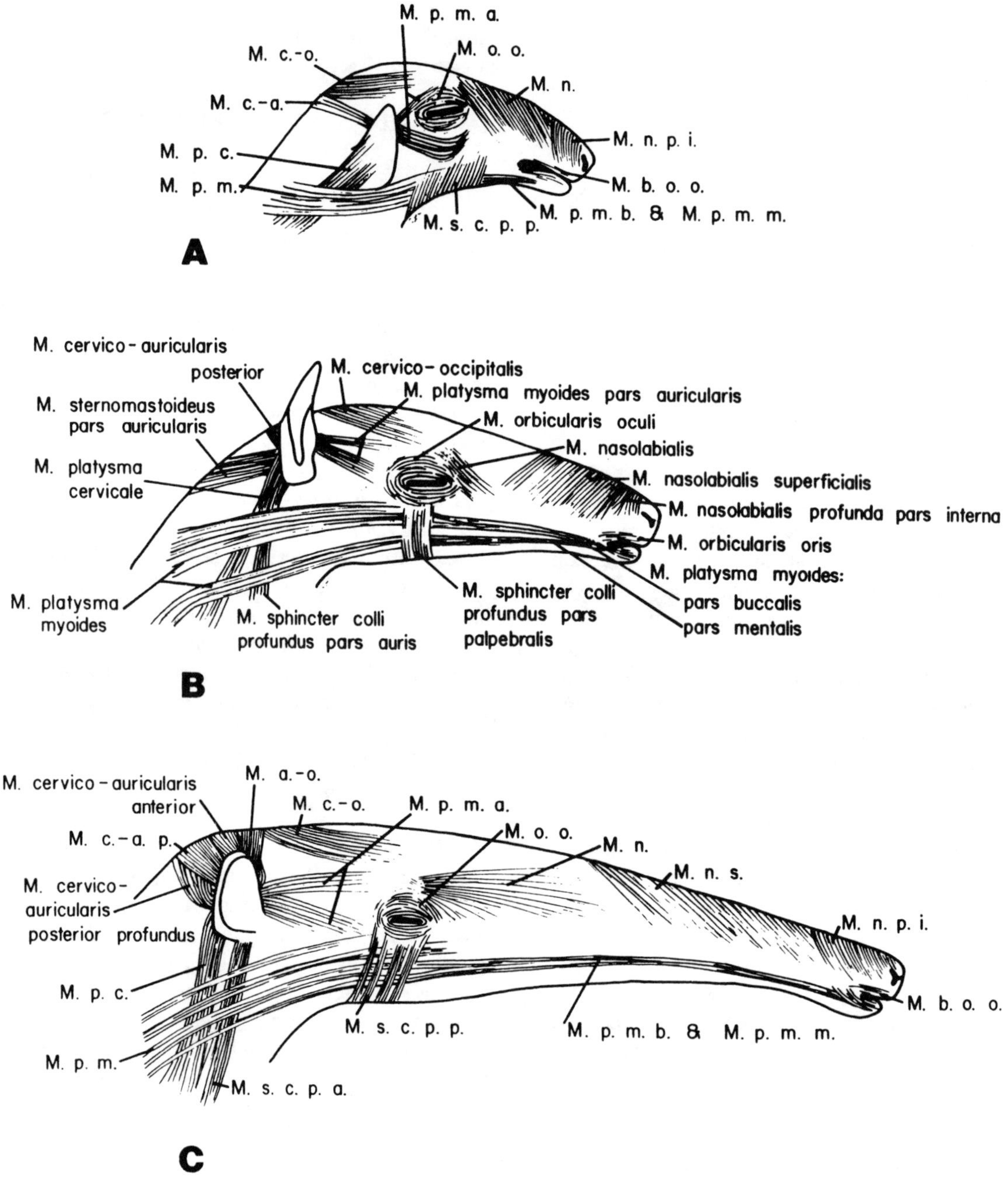

Figure 5. Superficial layer of facial muscles in **A** *Cyclopes*, **B** *Tamandua* and **C** *Myrmecophaga*.

directly caudal to the ear cartilage (Fig. 5). The muscle passes directly posterior to the ear cartilage, deep to *M. platysma myoides* and inserts into the skin of the throat region. In *Cyclopes* this muscle arises from the posterior aspect of the ear cartilage, and inserts as in *Tamandua*. The condition of this muscle in *Myrmecophaga* is similar to that in *Cyclopes*.

Retroauricular Musculature.-- The primitive arrangement of the retroauricular musculature in mammals is in 3 layers. In the 2 smaller vermilingua genera these muscles are reduced in number and development. The outer layer consists of the cervicoauricular muscles which divide into 3 slips. Muscles forming the second layer are the *M. cervicoauricularis medius* group. and the musculature of the deep layer is *M. cervicoauricularis profundus* (Fig. 5).

M. cervicooccipitalis.-- This muscle is the most anterior slip of *M. cervicoauricularis* (Figs. 5 and 6). In *Tamandua* it covers the dorsal area of the skull between the eyes and the ears by a thin layer of muscle fibers originating on the middorsal area over the parietal bones. The insertion of this muscle is very difficult to see, since

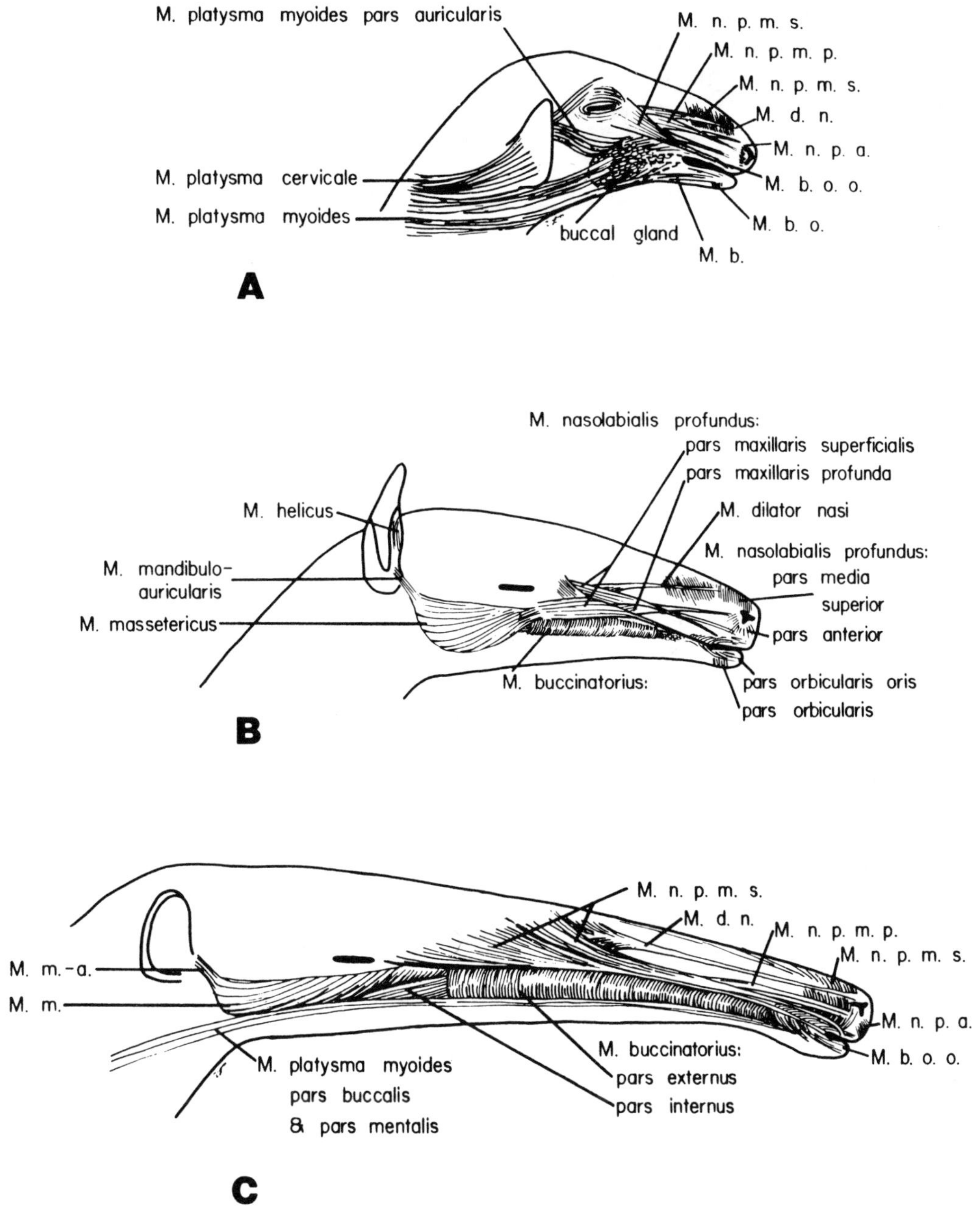

Figure 6. Middle level of facial musculature in **A** *Cyclopes*, **B** *Tamandua* and **C** *Myrmecophaga*.

the fibers become aponeurotic as they pass the skull area dorsal to the eye. In *Cyclopes* the muscle arises further posteriorly than in *Tamandua*, and is less well developed, failing to cover the middorsal area of the skull. In *Myrmecophaga* the muscle is very similar to that in *Tamandua*.

M. cervicoauricularis anterior.-- This muscle was not seen in *Tamandua* or *Cyclopes* (Fig. 5). In *Myrmecophaga* the muscle arises from the dorsal midline of the occiptal bone, with fibers passing anterolaterally immediately caudal to *M. cervicooccipitalis*, inserting on the anterior aspect of the base of the pinna.

M. cervicoauricularis posterior.-- This muscle arises in *Tamandua* from the occipital bone as a fascial sheet, with fibers inserting onto the middorsal portion of the pinna (Fig. 5). Some fibers continue to the most distal part of the pinna. This muscle is quite reduced in *Cyclopes*, arising from a restricted area of the postero-dorsal margin of the occipital bone. It passes onto the pinna as two slips, one onto the most dorsal portion, and the other onto the posterior bulge. In *Myrmecophaga* the muscle arises more posteriorly, from the middorsal line above the cervical musculature, and passes as a single slip onto the base of the pinna. I was unable to determine if the muscles continued along the length of the pinna because the pinnae had been removed when the speciman was skinned.

M. cervicoauricularis medius anterior.-- This muscle in *Tamandua* is moderately well developed, arising as a fascial sheet from the parietal bone, directly deep to *M. cervicoauricularis posterior* (Fig. 5). It inserts on the most dorsal aspect of the base of the pinna, deep to the insertion of *M. cervicoauricularis posterior*. In *Cyclopes* and *Myrmecophaga* the muscle arises more anteriorly. In *Cyclopes* it is extremely reduced while in *Myrmecophaga* it is very well developed.

M. cervicoauricularis medius posterior.-- In *Tamandua*, this muscle arises from the middorsal fascia deep and slightly lateral to the origin of *M. cervicoauricularis medius anterior* (Fig. 5). It inserts on the posterior bulge of the pinna. This muscle is absent in *Cyclopes*. In *Myrmecophaga*, the muscle arises posterior to *M. cervicoauricularis medius anterior*, and is more robust than in *Tamandua*, but its insertion is similar.

M. cervicoauricularis posterior profundus.-- This muscle is absent in *Tamandua* and *Cyclopes* (Fig. 5). In *Myrmecophaga* it is well differentiated, arising beneath the origin of *M. cervicoauricularis medius posterior*, and passing laterally to insert on the base of the ear cartilage ventral to *M. cervicoauricularis medius posterior*.

Intrinsic Ear Musculature.-- These muscles are reduced in number in the vermilinguas, and only two are present.

M. mandibuloauricularis.-- This muscle in *Tamandua* arises from the posterodorsal area of *M. massetericus* (Fig. 6). A rounded robust muscle slip is formed by a group of muscle fibers abruptly changing direction to angle posterodorsally to insert at the anteroventral base of the ear cartilage. The condition of this muscle is similar in *Cyclopes* and in *Myrmecophaga*.

M. helicis.-- This muscle originates in *Tamandua* as a continuation of some fibers of *M. mandibuloauricularis* (Fig. 6). It passes upward along the anterior margin of the pinna, and inserts about one third of the distance up the pinna. This muscle was not found in *Cyclopes* and could not be seen in *Myrmecophaga*.

***M. sphincter colli profundus* and Derivatives**.--This musclulature constitutes the deepest layer of the superficial facial musculature, and is characterized in vermilinguas by both the loss of certain major sheet-like components and by extreme simplicity of many of the specialized elements associated with the rhinarium, lips and cheeks.

M. sphincter colli profundus pars auris.-- This is an independent muscle (Fig. 5). In *Tamandua* it originates in the mid ventral region of the neck, anterior to the origin of *M. sphincter colli profundus*. The fibers pass anterodorsally deep to those of *M. platysma myoides*, and insert onto the ventral and posterior aspect of the pinna. This muscle is absent in *Cyclopes*. In *Myrmecophaga* the muscle arises more posteriorly, from the manubrium of the sternum, and passes anteriorly as a thick rounded band of fibers. The insertion is similar to that in *Tamandua*. *M. sphincter colli profundus pars intermedia ventralis* and *pars intermedia dorsalis* with its derivatives, the "intermediate plate" and its anterior extension, *M. zygomaticolabialis*, *M. frontalis* (auricular and orbital parts), and *M. auricularis anterior* (superior and inferior) are missing in all three genera.

M. auriculooccipitalis.-- In *Tamandua* this is a well developed muscle which arises from the middorsal fascia over the parietal-occipital suture (Figs. 5 and 6). The fibers pass anterolaterally and insert on the anterodorsal aspect of the base of the ear cartilage and pass distally along the length of the pinna. This muscle is absent in *Cyclopes*. The condition in *Myrmecophaga* is similar to that in *Tamandua*.

Orbital and Nasal Musculature.-- *M. orbicularis oculi* is well developed in *Tamandua*, with fibers on the dorsal border of a poorly developed palpebral ligament in the anterior corner of the eye. These pass entirely around the eyelid to insert on the ventral surface of the palpebral ligament (Fig. 5). Some of the fibers are lost in the fascia covering the preorbital process. Other fibers arise caudal to the posterior corner of the eye, and disappear in the fascia beneath it, or intermingle with fibers of *M. sphincter colli profundus pars palpebralis*. The muscle is similarly developed in *Cyclopes* and *Myrmecophaga*.

M. nasolabialis.-- This muscle arises in *Tamandua* from the dorsal and anterior aspect of the palpebral ligament at the anterior corner of the eye (Fig. 5). Fibers pass anteriorly for approximately one half the distance to the nares before fading into the superficial fascia. This musculature is not well developed and the insertion cannot be seen. The muscle is reduced in *Cyclopes* and more strongly developed in *Myrmecophaga*.

M. nasolabialis superficialis.-- This muscle arises in *Tamandua* from the anterior half of the bridge of the rostrum, inserting into the skin and the connective tissue dorsal to the upper lip (Fig. 5). Isolated muscle fibers surround the sparse sinus hairs. In *Cyclopes* this muscle is absent. In *Myrmecophaga* the muscle is similar to that of *Tamandua*.

M. sphincter colli profundus pars palpebralis.-- This muscle arises in *Tamandua* along the midventral line between the mandibular rami (Fig. 5). The fibers pass dorsally superficial to *M. platysma myoides*. They insert in the fascia ventral to the eye, sometimes interlacing with the fibers of *M. orbicularis oculi*. This muscle is similar in *Cyclopes* and *Myrmecophaga*.

All of the remaining muscles in the rostral region are derived from *M.sphincter colli profundus pars oris. M. maxillolabialis* is missing from all three genera.

M. dilator nasi.-- In *Tamandua*, this muscle arises deep to the origin of *M. nasolabialis profundus pars maxillaris superficialis* (Fig. 6). The anterior three-fourths of the muscle become tendinous, and it inserts as a flat tendon into the lateral part of the dorsal border of the rhinarium. In *Cyclopes* and *Myrmecophaga* the muscle has a similar origin and insertion, but is tendinous for the anterior half in *Cyclopes* and the anterior 85 percent in *Myrmecophaga*.

M. nasolabialis profundus--This muscle consists of seven parts. In all three genera parts of this mulculature are reduced and simplified, and *pars media inferior* and *pars anterior profunda* are absent.

M. nasolabialis profundus pars interna.-- In *Tamandua* this muscle arises ventral to the skin over the nasal bones (Fig. 5), extending posteriorly for about 35 mm and inserting into the notch at the anterior corner of the nose. Some fibers pass laterally and ventrally to insert dorsal to the pad of fat above the upper lip. This muscle is vestigial in *Cyclopes* and will not be described. In *Myrmecophaga* the muscle is similar to that of *Tamandua*, but extends posteriorly for approximately 50 mm.

M. nasolabialis pars media superior.-- This muscle arises in *Tamandua* from the bridge of the rostrum with fibers passing ventrally, deep to the tendon of *M. dilator nasi* and inserting into the fibers of *M. orbicularis oris* (Fig. 6). This muscle is similar in *Cyclopes* and *Myrmecophaga* although less well differentiated in *Cyclopes* and better differientiated in *Myrmecophaga*.

M. nasolabialis profundus pars anterior.-- This muscle arises in *Tamandua* from the posterior ventral wall of the narial opening (Fig. 6). The muscle sends transvere fibers between the nostrils, where they insert into the skin of the upper lip. The ventral fibers angle along the lateral surface of the upper lip and mingle with fibers of muscles passing anteriorly through the fat and connective tissue. This muscle is quite complex, but the parts are indistinctly divided. In *Cyclopes* this muscle is very poorly developed, but similar to that in *Tamandua*. In *Myrmecophaga* the muscle is similar to that in *Tamandua*.

M. nasolabialis profundus pars maxillaris superficialis.-- This muscle arises in *Tamandua* from the projection of the lacrimal bone and from the tough connective tissue which replaces the bony zygomatic arch (Figs. 6 and 7). There are two distinct slips, one arising slightly anteriorly and dorsally to the other. They pass anteroventrally, becoming tendinous about half of their way anteriorly. The more dorsal of the two slips inserts into the dorsal aspect of *M. orbicularis oris* of the upper lip. The ventral slip inserts onto the surface of *M. orbicularis oris* directly posterior to the corner of the buccal opening. This muscle is similar in *Cyclopes*, but in *Myrmecophaga* the muscle slips become tendinous about one-third of the distance from their origin.

M. nasolabialis profundus pars maxillaris profunda.-- In *Tamandua*, this muscle arises slightly anteriorly and ventrally, deep to *M. nasolabialis profundus pars maxillaris superficialis*, from the same fascia as the latter muscle (Figs. 6 and 7). It consists of 2 slips, one passing superficial to the other. The outer slip remains fleshy for about 20 mm, approximately half its length, and then becomes a fleshy tendon which inserts immediately dorsal to the tendon of the upper portion of *M. nasolabialis profundus pars maxillaris superficialis*. The tendon gives off one fiber bundle near the posterior corner of the mouth, with the rest of the fibers passing anteriorly to the level of the notch of the naris. There it breaks into three bundles which are then lost in muscle fibers and connective tissue. The deep slip of the muscle passes anteriorly as a fleshy bundle for 18 mm and then becomes tendinous, before passing dorsal to the outer part. It angles upward toward the posterior corner of the naris, but divides and fades into the connective tissue posterior to it. *Cyclopes* and *Myrmecophaga* show similar arrangements of this muscle, with the muscles remaining fleshy for approximately half their length in *Cyclopes* and becoming tendinous after one-fourth their length in *Myrmecophaga*.

M. buccinatorius.-- In many animals this muscle is divided into many parts. The vermilinguas show extreme simplification and some reduction of this muscle group. *M. bucconasolabialis* is absent in all three genera.

In both of the larger genera of vermilinguas, there are well-developed buccal salivary glands situated between the external and internal parts of *M. buccinatorius*. In *Cyclopes* this gland is extremely well developed and expanded posteriorly. With the enlargement of the buccal salivary gland in *Cyclopes, M. buccinatorius* has been greatly modified.

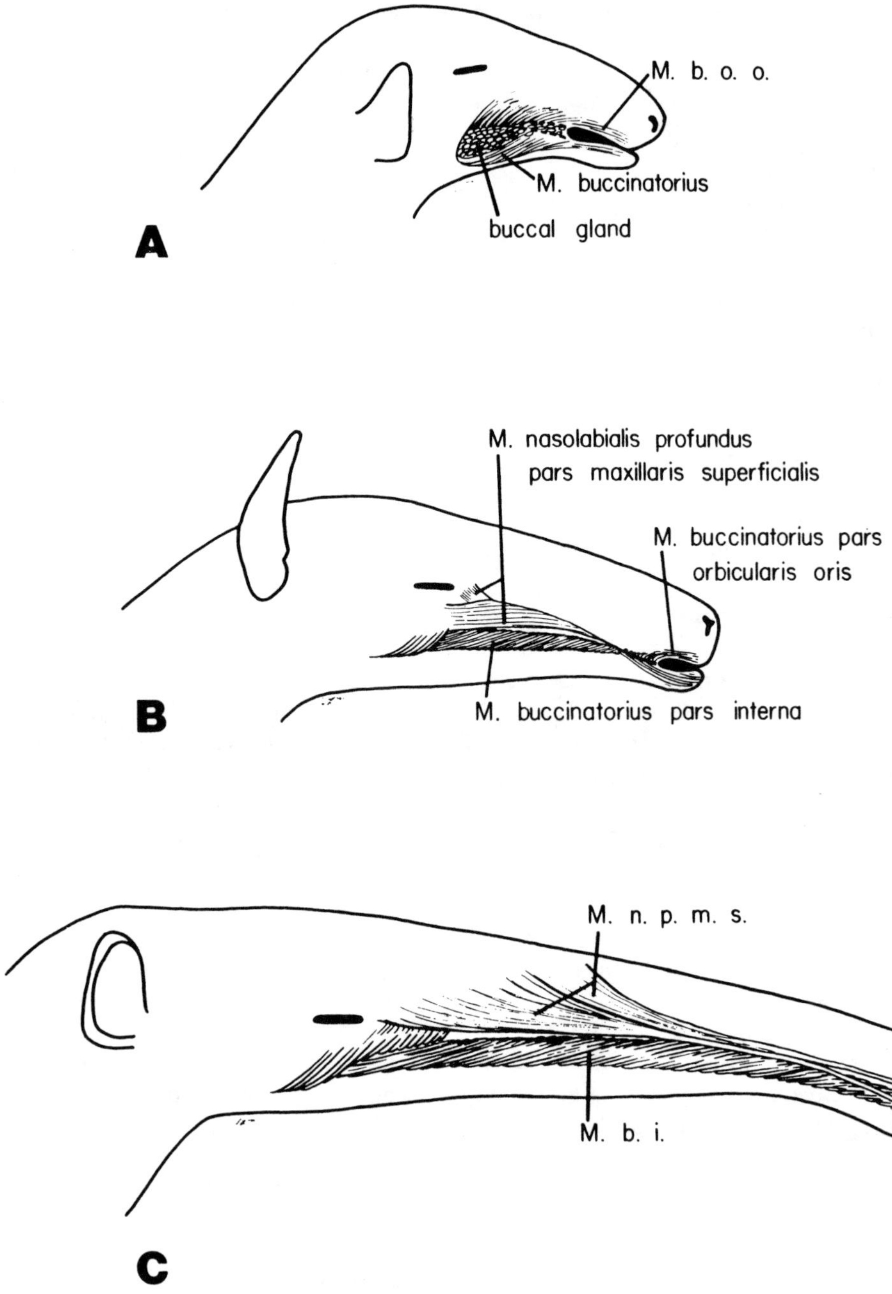

Figure 7. Deep facial musculature of A *Cyclopes*, B *Tamandua* and C *Myrmecophaga*.

M. buccinatorius.-- In *Cyclopes*, this muscle is not divided into an external and an internal portion as in the other two genera (Figs. 6 and 7). It has been modified with the external muscle layer passing superficial to the buccal gland and curving beneath it. Deep to this gland the muscle fibers continue in an anterodorsal direction to join with the fibers of *M. orbicularis oris.*

M. buccinatorius pars externa.-- This muscle in *Tamandua* arises anteriorly from the maxilla, about 12 mm posterior to the corner of the buccal opening (Fig. 6). It continues posteriorly for approximately 33 mm. The muscle inserts directly ventrally onto the edge of the mandible from about 13 mm caudal to the posterior aspect of the buccal opening and extends posteriorly for approximately 33 mm. This muscle sheet is very thin. The muscle in *Myrmecophaga* resembles that in *Tamandua*, with the exception that its anterior origin and insertion are 50 mm posterior to the buccal opening and the muscle continues caudally for approximately 130 mm.

M. buccinatorius pars interna.-- In *Tamandua* this muscle arises from the maxillary bone at the posterior corner of the buccal opening, continuing caudally for approximately 60 mm, with fibers passing almost directly posteriorly (Fig. 7). The muscle inserts along the dorsolateral border of the mandibular ramus. The muscle arrangement is similar in *Myrmecophaga*, although there the muscle extends posteriorly for 210 mm.

M. buccinatorius pars orbicularis oris.-- In *Tamandua* this muscle arises from a dorsoventrally oriented bulge of cartilage just ventral to the anterior edge of the narial opening, giving off fibers which pass caudally (Figs. 6 and 7). Other fibers arise from the maxilla dorsal to the buccal opening, where they are joined by the tendon of *M. nasolabialis profundus pars maxillaris superficialis*. Muscle fibers arising from this tendon join those of *M. orbicularis oris* as the latter pass ventral to the lower lip. The fibers of this muscle insert onto the ventral surface of the mental region of the mandible. In *Cyclopes*, the origins and insertions of the muscle are similar to those of *Tamandua*, but *M. nasolabialis profundus pars maxillaris superficialis* joins *M. orbicularis oris* on the dorsal surface of the upper lip. In *Cyclopes* the posterior border of the buccal opening is lengthened by a deep groove in the surface of the buccal gland, and the fibers pass posteriorly dorsal to it, where they join with *M. buccinatorius* which passes around the buccal gland. On the ventral surface the muscle fibers follow a similar course. In *Myrmecophaga, pars orbicularis oris* is similar to that in *Tamandua*, only more strongly developed.

M. buccinatorius pars orbicularis.-- This muscle arises in *Tamandua* from the lateral border of the mandibles, posterior to the mental symphysis, continuing along the lateral surface for the entire length of the buccal opening (Fig. 6). The muscle fibers continue transversely across the mental region. In *Cyclopes* and *Myrmecophaga* the muscle is similar to that of *Tamandua*, but is not well developed in either genus.

DISCUSSION

Facial muscles of sloths are greatly reduced in number and complexity in comparison with other groups of mammals such as rodents and carnivores (Rinker, 1954; Meinertz, 1941). The primitive condition for those groups is a facial region covered by several layers of sheet-like muscles (Huber, 1930; 1931). One of these muscle sheets, *M. sphincter colli profundus pars superficialis* is completely absent in sloths (Table 2), and *M. platysma myoides* and other portions of the muscle sheet have been lost.

Schulmann (1906) attempted to show that facial muscles of sloths were derived from those of monotremes. The pattern of facial muscle organization in monotremes differs from that common to marsupials and placental mammals (Huber, 1931). Previous studies, particularly before those of Huber (1930; 1931) did not clarify nomenclatural synonomies or homologies of the components of facial musculature (cf. Uekermann, 1912). In spite of these difficulties, published descriptions of many of the components of facial musculature frequently concurred with the descriptions reported here, providing support for the view that facial musculature in sloths and vermilinguas is organized along the marsupio-placental plan.

Although sloth facial muscles are reduced, they are neither degenerate nor undifferentiated in arrangement, as indicated by others (Windle and Parsons, 1899). In both sloth genera, but particularly in *Choloepus*, components of *M. nasolabialis profundus* are well differentiated when compared with other mammals, and permit sloths to flare their nostrils and move the rhinarium. These muscles also control the movement of the few sinus hairs present on the rostral region. Many muscles have been retained in both sloth genera (Table 2). Muscles which move the eyes, nose and lips are similar although they appear to be more strongly developed in *Choloepus*. Both sloth genera have short rostra, and muscular development is sufficient to provide control of fine movements of this part of the face. In contrast, the retroauricular muscles have been lost, except for the most superficial layer, *M. cervicoauriculares* and *M. auriculooccipitalis*. This reflects the small size of the pinna and its relative immobility when compared with other mammals.

Sloths show intergeneric differences in the pattern of the reduction of *M. platysma myoides*. *Pars palpebralis* is present in both genera, but is more strongly developed in *Choloepus*. *Pars mentalis* is present in *Choloepus* but absent in *Bradypus*, while *pars buccalis* is present in *Bradypus* but absent in *Choloepus* (Table 2). Both muscles act to help retract the corner of the mouth, permitting the sloth to grimace or smile; different facial muscles perform the same function in the two genera. Another important difference concerns the relative development of *M. buccinatorius*. In *Choloepus* this muscle is much more strongly developed than in *Bradypus*. This may relate to the larger buccal opening in *Choloepus* and the larger gape. A larger gape allows clearance for the elongate caniniform teeth, which are used in defence and aggressive interactions. The anterior teeth in *Bradypus* do not project far beyond the occlusal plane, permitting ingestion of leaves with a smaller gape; *Bradypus* are far less agressive (Montgomery and Sunquist, 1978).

Facial musculature in vermilinguas is reduced when compared with other mammals, such as rodents, carnivores and primates (Rinker, 1954; Meinertz, 1941). The pattern of muscle reduction is similar among the three genera (Table 2), with the greatest loss of muscular elements in the smaller forms. As in sloths, only portions of the large sheets of facial muscle have been retained.

Table 2. Facial muscles in sloths and vermilinguas.

Muscle	*Sloths* Bradypus	Choloepus	*Vermilinguas* Cyclopes	Tamandua	Myrmecophaga
Platysma and Derivatives					
M. platysma myoides	+	+	+	+	+
Pars buccalis	+	–	–	+	+
Pars mentalis	–	+	+	+	+
Pars auricularis	+	+	+	+	+
M. platysma cervicale	–	–	+	+	+
Retroauricular Musculature					
M. cervicoauricularis					
Pars anterior	–	–	–	–	+
Pars posterior	+	+	+	+	+
M. cervicooccipitalis	–	+	+	+	+
M. cervicoauricularis					
Pars medius anterior	–	–	–	+	+
Pars medius posterior	–	–	–	+	+
Pars posterior profundus	–	–	–	–	+
M. mandibuloauricularis	+	+	–	+	+
M. helicis	–	–	–	+	?
M. sphincter colli profundus and Derivatives					
Pars auris	+	–	–	+	+
Pars intermedia and Derivatives					
Pars intermedia ventralis	–	–	–	–	–
Pars intermedia dorsalis and Derivatives					
"Intermediate Plate"	–	–	–	–	–
M. zygomaticolabialis	–	+	–	–	–
M. auriculooccipitalis	+	+	–	+	+
M. frontalis (auricular part)	–	–	–	–	–
M. frontalis (orbital part)	–	–	–	–	–
M. orbicularis oculi	+	+	+	+	+
M. nasolabialis	+	+	–	+	+
M. nasolabialis superficialis	–	–	–	+	+
Pars palpebralis	+	+	+	+	+
Pars oris and Derivatives					
M. maxillolabialis	–	–	–	–	–
M. dilator nasi	+	+	+	+	+
M. nasolabialis profundus					
Pars interna	–	–	–	+	+
Pars media superior	–	–	+	+	+
Pars media inferior	–	–	–	–	–
Pars anterior	–	+	+	+	–
Pars anterior profunda	–	–	–	+	–
Pars maxillaris superficialis	+	+	+	+	+
Pars maxillaris profunda	+	+	+	+	+
M. bucconasolabialis	–	–	–	–	–
M. buccinatorius	+	+	+	+	+
Pars externus	–	–	–	+	+
Pars internus	–	–	–	+	+
Pars orbicularis	–	–	+	+	+
Pars orbicularis oris	+	+	+	+	+

However, all genera of vermilinguas have more complex and well developed facial muscles than sloths. As in the sloths, *M. sphincter colli profundus pars superficialis* is completely absent, as are *pars intermedia dorsalis, pars intermedia ventralis* and the "intermediate plate".

In contrast to sloths, the pinnae of vermilinguas are relatively large and mobile and the retroauricular muscles which provide pinna movement are well developed. More than one of the three layers of retroauricular muscles is present in all three genera (Table 2). *Myrmecophaga* has a full complement of extrinsic ear muscles from each layer.

Superficial layers of facial muscles are more extensive in vermilinguas than in sloths, because they are better developed, and because additional muscle elements have been retained (Table 2). Parts of all portions of *M. platysma myoides* are present, in most cases as distinct muscles, even though they are not as extensive as in other mammals. *Pars auricularis*, absent in sloths, is particularly well differentiated in vermilinguas.

The complex nasolabialis musculature is strongly developed in vermilinguas, giving them fine control over movements of the rhinarial and buccal regions. For example, in sloths there is only a single slip of *pars maxillaris superficialis* (Table 2). However, all three vermilingua genera have finer control over the width of the mouth than sloths, because a second slip of *pars maxillaris superficialis* is present, inserting onto *M. orbicularis oris* at the corner of the mouth. In *Tamandua* and *Myrmecophaga, M. buccinatorius* has two parts, and provides additional control over the cheek region when the animal is ingesting insects. In *Cyclopes* this muscle has been modified to allow for a greatly expanded buccal salivary gland, and cannot provide as much control over the mouth. Expansion of the salivary gland may allow *Cyclopes* to produce a relatively greater volume of saliva, and help increase efficiency of capture of the ants it uses for food.

The vermilingua genera differ in body size by factors of 1 *Cyclopes*: 3 *Tamandua*: 12 *Myrmecophaga*, and buccal openings are absolutely larger in the larger forms. However, the buccal opening becomes relatively smaller as the body size of the form increases in the series *Cyclopes* to *Tamandua* to *Myrmecophaga*.

As vermilinguas evolved to become more specialized to eat social insects, the tongue became increasingly elongate, and the buccal opening relatively smaller (Winge, 1941), with a concomitant increase in length of the rostrum. The tongue is used to draw insects into the buccal cavity. The elongate nasal region forms a tube with a small anterior opening, and helps channel the tongue more precisely toward a particular target. The tongue is supported by a hyoid apparatus with strongly attached large muscles, allowing manipulation of the tongue while it is fully extended. Pangolins have a comparable arrangement (Doran and Allbrook, 1973).

The elongation of the face required additional specializations in buccal and rhinarial musculature. For example, strong muscular cheek walls help confine prey to the mouth, and a small buccal opening might help prevent escape of living insects from the buccal cavity. *M. buccinatorius* forms the cheek wall in mammals, and is well developed in all three vermilingua genera (Table 2). In *Tamandua* and *Myrmecophaga, M. buccinatorius* is divided into a *pars externus* with fibers oriented dorsoventrally, and a *pars internus* with fibers oriented obliquely in an anterior to posterior direction. This muscle arrangement is not found in other mammals.

In most cases, facial musculature is more strongly developed in *Myrmecophaga*, and less developed in *Tamandua*, with the smallest form *Cyclopes* showing simplification or elimination of facial muscle elements. *Cyclopes* still retains a moderate amount of facial mobility. There is an increase in the relative size of fleshy as opposed to tendinous muscle origins from *Cyclopes* through *Tamandua* to *Myrmecophaga*. The length of the rostrum in the smaller forms is both absolutely and relatively less than in the larger forms. In smaller vermilinguas, facial muscles have a relatively greater surface area in relation to the length of their tendons. In *Myrmecophaga*, the largest form, the tendons are relatively longer and the bellies of the nasal muscles are closer to their origins. This reduces the bulk of the rostrum without sacrificing muscular control of the rhinarium and buccal regions.

***Model for Feeding by Vermilinguas*.**-- Mammals with teeth usually feed by slicing a bite from a large piece of food, then reducing it in size within the oral cavity by chewing, or by taking bite which is small enough so that it can be swallowed whole. The tongue is usually relatively short and highly mobile. It is used to move food within the oral cavity, and to collect and transport food as a bolus to the rear of the cavity where food is swallowed (Hiiemae, 1978). Food usually requires mechanical processing, and several chewing cycles occur between swallows (Thexton et al., 1980).

Ants and termites which form the food of vermilinguas are so small as to preclude the need for biting and mastication; they are swallowed whole (Montgomery, 1985). The rate of ingestion is limited by the rate at which the tongue can transport food, and the rate at which food can be swallowed. The elongate rostrum, which may increase control of the tongue, also increases the distance over which food must be transported before being swallowed.

I propose the following model for the way in which vermilinguas maximize the rate of transport of food within the oral cavity. The rate of transport of insects from the front to the rear of the tube-like oral cavity could be increased by compressing the walls of the tube immediately behind a bolus of insects on a segment of the tongue. As the tongue contracts into the oral cavity, it may become larger in cross section, and may maintain a seal with the walls of the cavity. The expanded buccinator musculature in the cheeks then may contract in synchrony with retraction of the tongue, and a bolus of insects may be propelled toward the throat. As a series of boluses forms on the tongue and are propelled backward, a series of waves of contraction may pass posteriorly along the cheek walls. This manner of moving food to the rear of the mouth would thus act as a "conveyor belt".

Selection for increasing rates of feeding on insects could explain why many mammals which feed on social insects have reduced or absent teeth, and why they have changed to a tongue-based mechanism which provides for continuous food collection and swallowing. Vermilinguas are usually limited to very short feeding bouts because insect prey actively defend their colonies (Montgomery and Lubin, 1977; Lubin and Montgomery, 1981). Selection pressures to maximize food intake rates might thus be strong. Teeth may have been selected against, because they and their support structures would increase bulk of the rostrum and because teeth would interfere with movement of food on the tongue. Armadillo genera such as *Dasypus* and *Priodontes*, which feed extensively on insects, have elongate tongues, weak jaw structure, and small, peg-like teeth. They may thus have followed a similar evolutionary trend as vermilinguas, but have not become so specialized.

Specialization of Facial Musculature.-- In becoming herbivorous, sloths have used a different strategy for obtaining food, and have relatively much larger mouths. They show well developed tongues, which they use in taking leaves into their mouths. Although simple, their teeth are also well developed, and act effectively to cut leaves into pieces small enough for the sloth to swallow. The sloth genera differ from each other in retention of elements of *M. platysma myoides* and *M. sphincter coli profundus*.

There were strong similarities in the pattern of specialization and reduction of the facial muscles between sloths and vermilinguas. However, comparisons of sloth and vermilingua facial musculature to that of other groups fails to indicate special phylogenetic relationships to groups in which facial musculature has been studied (Huber, 1930; 1931: Rinker, 1954). These comparisons indicate, however that extensive development of large muscle sheets is the primitive state for Xenarthra as well as for other groups of mammals, and that a reduced condition in number and complexity of facial muscles is a derived as opposed to an unspecialized condition.

The facial musculature of sloths and of vermilinguas is specialized in unique fashions which reflect their different feeding specializations.

ACKNOWLEDGMENTS

I thank Dr. G. Gene Montgomery and Dr. Richard Thorington for providing preserved sloths and vermilinguas. I thank Dr. Margery Coombs, Dr. Guy G. Musser, and Dr.David Klingener for comments on an early version of this paper, and Dr. Karen M. Hiiemae for comments on the final version. Much of the study was completed as a part of a Ph.D. dissertation at the University of Massachusetts.

LITERATURE CITED

Doran, G.A. and D.B. Allbrook.
1973. The tongue and associated structures in two species of African pangolins, *Manis gigantea* and *Manis tricuspis*. J. Mammal., 54: 887-889.

Hiimae, K.M.
1978. Mammalian mastication: a review of the activity of the jaw muscles and the movements they produce in chewing. Pp. 359-398, *In* Development, function and evolution of teeth (P.M. Butler and K.A. Joysey, editors). Academic Press, London.

Hill, J.E.
1937. Morphology of the pocket gopher mammalian genus *Thomomys*. Univ. Calif. Publ. Zool., 42, no. 2: 81-172.

Huber, E.
1930. Evolution of the facial musculature and cutaneous field of the Trigeminus. Quart. Rev. Biol., 5: 133-188.

1931. Evolution of facial musculature and facial expression. Johns Hopkins Univ. Press, Baltimore. i-xii + 1184 pp.

Klingener, D.
1964. The comparative myology of four Dipodoid rodents (genera *Zapus, Napeozapus, Sicista* and *Jaculus*). Misc. Publ. Mus. Zool. Univ. Mich., No. 124: 1-100.

1970. Superficial facial musculature of *Aplodontia*. J. Mamm., 51: 35-41.

Lubin, Y.D. and G.G. Montgomery.
1981. Defenses of *Nasutitermes* termites (Isoptera, Termitidae) against *Tamandua* anteaters (Edentata, Myrmecophagidae). Biotropica 13: 66-76.

Meinertz, T.
1941. The skin musculature of the Greenland lemming, *Dicrostonyx groenlandicus*. Meddelelser am Gronland, 131: 1-77.

Montgomery, G.G.
1985. Movements, foraging and food habits of the four extant species of neotropical vermilinguas (Mammalia; Myrmecophagidae). (this volume).

Montgomery, G.G. and Y.D. Lubin.
1977. Prey influence on the movements of neotropical anteaters. Pp. 103 to 131, *in* Proceedings of the 1975 predator symposium (R.L. Phillips and C. Jonkel, editors). Montana Forest and Conservation Experiment Station, School of Forestry, University of Montana, Missoula. 385 pp.

Montgomery, G.G.and M.E. Sunquist.
1978. Habitat selection by two-toed and three-toed sloths. Pp. 329-359, *in* The ecology of arboreal folivores (G.G. Montgomery, editor). Smithsonian Institution Press, Washington, D.C. 574 pp.

Owen, R.
1857. On the anatomy of the great anteater (*Myrmecophaga jubata*). Trans. Zool. Soc. London 4: 117-140.

Rinker, G.C.
1954. The comparative myology of the mammalian genera *Sigmodon*, *Oryzomys*, *Neotoma* and *Peromyscus* (Cricteinae), with remarks on their intergeneric relationships. Misc. Publ. Zool. Univ. Mich., No. 83: 1-24.

Schulman, G.
1906. Uber die ventrale Facialmuskulatur einiger Saugetiere, besonders der Monotremen. Festschrift fur Palmen. Nr. 18, Helsingfors: 1-112.

Uekermann, A.
1912. Untersuchungen uber die Gesichtzmuzkulatur der Xenarthra. Zeit. Wis. Zool., 102, no. 3-4: 337-424.

Windle, B.C.A. and F.G. Parsons.
1899. On the myology of the Edentata. Proc. Zool. Soc. Part I, :314-339 and Part II, :990-1017.

Winge, H.
1941. On the interrelationships of the mammalian genera. C.A. Reitzels Forlag, Kobenhavn, Vol. 1: 319-411.

THE EYE OF THE THREE-TOED SLOTH

David Piggins[1] and W.R.A. Muntz[2].

[1] Department of Psychology, University of Guelph, Guelph, Ontario, NIG 2W1, Canada; [2] Department of Biology, University of Stirling, Stirling, FK9 4LA Scotland.

Abstract
The eye of the three-toed sloth *Bradypus tridactylus* Linné is described from various aspects. The refractive findings from nine sloths and the gross ocular anatomy and retinal histology are presented. The sloth eye is capable of optically forming a well-focused retinal image. The retinal compliment of rods and few ganglion cells, together with a thick crystalline lens, indicates that the eye is adapted mainly for lower levels of illumination. However, some useful vision ought to be possible under higher levels of illumination. Notable was the finding of low to minimal refractive error, with moderate variation across individuals.

Abstracto
Se describe el ojo del perezoso de tres dedos *Bradypus tridactylus* Linné desde varios aspectos. Los hallazgos sobre refracción de nueve perezosos, y tanto la anatomía ocular como la histología de la retina son presentados: el ojo del perezoso es capaz de formar ópticamente en la retina una imagen bien enfocada. El complemento retinal de bastoncillos y de unas cuantas células ganglionales, junto con un denso lente cristalino, indica que el ojo está adaptado principalmente para el uso de bajos nivels de iluminación. Sin embargo, debe ser posible la obtención de algo de visión útil a niveles más altos de iluminación. Fue notable el hallazgo en estudios de refracción de, cuando mucho, un bajo grado de miopía en algunos animales y de emetropía y de bajos grados de hiperopía en otros.

Resumo
O olho da preguiça comum, *Bradypus tridactylus* Linné, e descrito em vários aspectos. Os achados refrativos de nove preguiças e a anatomia grosseira e histologia retinal são apresentados. O olho da preguiça e oticamente capaz de formar uma imagem retinal bem focada. O componente retinal do bastão e algumas células ganglionares, juntamente com um cristalino espesso, indicam que o olho está adaptado principalmente para baixos níveis de iluminação. Contudo, alguma visão util deve ser possível sob níveis de iluminação maiores. Notável foi o achado refrativo de, quando muito, uma baixa miopia em alguns animais e emetropia e baixa hiperopia em outros.

In common with the large majority of animal species, little is known about the sensory apparatus and capabilities of the Xenarthrans (Edentates). Most available information is for the two-toed sloth (*Choloepus*; Megalonychidae) and is given in a short chapter by Goffart (1971). There is little available information about the three-toed sloth (*Bradypus*: Bradypodidae). Goffart considers the available evidence for sloth sensory function for the five "classical" senses: touch, smell, taste, sight, and hearing. In the case of sight, much of the evidence for the sloth's visual function stems from the work of Beebe (1926), Watillon and Goffart (1965, 1969), and Goffart (1971), with a number of scattered reports upon various visual aspects by a number of authors over the past century.

Goffart (1971) comments "It has been reported that sloths have bad eyesight, and that they are shortsighted" and concludes "the eye of the sloth with its rod retina but highly contractile pupil is suited for both daylight and nocturnal activity." This is in partial contrast to Walls (1942) earlier comment: "In the lowest orders (Insectivora, Chiroptera, Edentata, Rodentia), the eye is both relatively and absolutely small, in sympathy with the nocturnality of these animals and the unimportance of vision in their lives." Polyak (1957), in his discussion of the importance of arboreal adaptation in primate evolution, comments "The aboreal habit alone, obviously, is not a magic formula guaranteeing phylogenetical progress. The proof of this is in the sloth (*Bradypus*), probably one of the most arboreally adapted animals in existence. By restricting its diet to the foliage of the trees, in which it spends its lazy immobile life, it reduces its senses and intellect to one of the lowest in the entire realm of mammals. The strictly arboreal adaptation in this case was a potent deterrent directly responsible for the low sensory and mental abilities. . . ."

This paper adds variously both support and criticism to these comments in reporting ocular investigations upon the three-toed sloth. The findings compliment our collective knowledge of sloth eyes and vision which has to a great extent previously been gleaned from investigations of the two-toed sloth *Choleopus hoffmanni* Peters.

METHODS

Seven female and one male adult three-toed sloths *Bradypus tridactylus* were captured on an island in the Rio Solimõis near Manaus, Amazonas, Brazil. The small island with its seasonally very dense sloth population is described in Waage and Best (1983). A juvenile female semidomesticated sloth was also used.

The ocular refraction of each eye was determined by retinoscopy. Testing was carried out at 100 cms and at 50 cms upon hand-held animals. Each eye was refracted at least twice. All eyes were tested also under cycloplegia (one drop of 1 percent atropine sulphate solution) at various times following application of the drops. Refraction under cycloplegic conditions commenced when the pupil was moderately dilated from its normal size. Examinations were carried out with low ambient lighting. Ophthalmoscopy was carried out on some sloths to judge pupil size and shape, to examine the fundus and also to estimate the size of the visual fields and optic axes positions.

Two eyes were removed from a *Bradypus tridactylus* obtained from near Manaus. Within two hours of the sloth's natural death, the eyes were placed in Bouins solution, left there for six days, and then kept in 70 percent alcohol. Although the anatomical descriptions and retinal histology are from one specimen, it was felt that the preparation and condition of the eye merited reporting, particularly in view of the scarcity of such eyes.

One eye was sectioned along an anterior-posterior axis passing through the estimated center of the cornea and the optic nerve head and embedded in paraffin. A section of retina taken from the other eye was embedded in paraffin, sectioned at 5 microns and stained with Masson's trichrome stain.

RESULTS

Casual Description of the Eyes in a Living Sloth.-- The eyes of *Bradypus tridactylus* appear set quite wide apart in the animal's head (some 30 mm) and are relatively frontal looking, judging from the relative positions of the center of each cornea (Fig. 1).

Figure 1. Frontal view of a juvenile female three-toed sloth *Bradypus tridactylus* Linné.

Goffart (1971) gives an interocular distance of 31 mm for *Bradypus* and 45-47 mm for *Choloepus didactylus*. Beebe (1926) mentioned that the eyes of the adult *Bradypus* diverge from a frontal body axis by some 25° and by some 34° in *Choloepus didactylus*.

The eyes appear of moderate size with a wide and convex cornea. No eye movements were observed but both slow head-turning and blinking were observed with some frequency. The irides were medium brown in color, exhibiting a moderate range of contraction terminating in a "pinhole" pupil under high levels of illumination (e.g., outside and exposed to daylight). The pupil was circular at all degrees of contraction and placed centrally with respect to the cornea.

The included angle between the diverging optic axis of each eye was estimated at 50° by using the retinoscope beam to obtain a centrally placed corneal reflex and then repeating the procedure with the other eye. The sloth's head was restrained during the determination of reflex positions. This value concurs with that reported for xenarthrans by Beebe (1926) and differs from Prince (1956) and Walls (1942) value of 100°. The monocular and binocular visual fields were also estimated; with the head restrained, the retinoscope beam was moved horizontally, both nasally and temporally until the fundus reflex could no longer be observed. This measure was performed on two sloths whose eyes were under the action of cycloplegia (pupils were larger than normal for the existing light conditions). The horizontal monocular fields were estimated at 70° with a binocular overlap at 35°. A pupil of smaller size would reduce both the size of the monocular fields and the extent of binocular overlap.

"Eye shine" resulting from the presence of a tapetum has previously been reported by Britton (1941), who described it as "bright yellow." However, Watillon and Goffart (1969) could find no anatomical evidence for a tapetum in the eye of *Choloepus*. The current study revealed a strong yellowish fundus reflex in *Bradypus*.

Reports of head movements of the sloth abound in the literature. However, while Enger and Bullock (1965) report spontaneous eye-movements in *Bradypus*, Beebe (1926) reports little movement and virtually no ocular convergence for B. *c. cuculliger* Wagler (*Bradypus tridactylus*, after Wetzel, 1985). Duke-Elder (1958) displays a photograph of *Choloepus* as an example of the "large prominent cyes of nocturnal placentals" and further comments "only a few species have a sufficiently powerful sphincter to effect the mechanically difficult feat of contracting a round pupil to a stenopaeic pinhole." A number of observers have remarked upon the sloth's ability to form a pinhole pupil under conditions of high illumination. Watillon and Goffart (1969) discuss the accommodative and protective role possibly played by a pupil of pinhole dimensions. The influence of pupil size on vision will be discussed below.

***Ocular Refraction of* Bradypus tridactylus**.-- There is a wide variation in refractive error measured across nine sloths (Table 1). Although each pair of eyes were closely similar, in the uncorrected figures three animals shared marked hyperopia, two low hyperopia, two emmetropia and two marked hyperopic astigmatism.

The refractive values (Table 1) were corrected for the distance at which retinoscopy was carried out. The figures are uncorrected for eye size. A consideration of refractive error and eye size is important since it has been reported (Glickstein and Millodot, 1970; Grüsser and Grüsser-Cornhels, 1976; Nuboer and Genderen-Takken, 1978) that with decreasing eye size, eyes refracted retinoscopically exhibit an increasing degree of apparent hyperopia (see review in Grüsser and Grüsser-Cornehls, 1976). Retinoscopic findings are interpreted from reflections at the vitreous humour/inner retina interface; thus, with decreasing eye size, the retinal thickness must be considered in estimations of the eyes' ability to form a clearly focused retinal image within the region of the photoreceptors, which lie at the outer retinal margin, since retinal thickness does not decrease at the same rate as focal length across species.

A correction factor described by Hirschberg (1882) and utilized more recently by Glickstein and Millodot (1970) can be applied to calculate the effective refraction of an eye of small dimension whose refraction has been determined by retinoscopy. Glickstein and Millodot (1970) plotted this correction factor in the function relating eye size and apparent hyperopia for a number of species over a wide range of eye size. The eye size measured for *Bradypus tridactylus* and the apparent hyperopia (taken as the mean of the refractive errors given in Table 1) show remarkably close agreement with Glickstein and Millodot (1970). Using Glickstein and Millodot's (1970) correction (from their plotted

Table 1. Refractive findings for eighteen eyes from nine *Bradypus tridactylus* Linné. Shown are the refractive error, expressed in diopters, which the eye deviated from emmetropia (the optical condition under which a distant object is imaged sharply upon the retina without the benefit of accommodation). The presence of emmetropia is a value of 0. Negative values indicate myopia (colloquially near-sightedness) and positive values hyperopia (colloquially far-sightedness). A difference in refractive error along different meridians of the eye indicates astigmatism. Using the function of Glickstein and Millodot (1970), thc cffective refraction would result in each of the figures being reduced (i.e., rendered more myopic) by some 4.0 diopters. Vt and Hz refer to the vertical and horizontal meridians of the eye, along which these refractive errors were found.

Sloth Number	*Right Eye (Diopters)*		*Left Eye (Diopters)*	
1	+ 7.50		+ 6.50	
2	+ 1.50		+ 1.50	
3	+ 2.00		+ 1.50	
4	+ 7.50		+ 7.50	
5	+ 6.50		+ 6.50	
6	0.00		0.00	
7	0.00		0.00	
8	+ 6.50(Vt)	+ 4.50(Hz)	+ 6.50(Vt)	+ 4.50(Hz)
9	+ 4.00(Vt)	+ 13.00(Hz)	+ 8.00(Vt)	+ 4.00(Hz)
Mean	+ 3.94	+ 4.72	+ 4.20	+ 3.55
Overall Mean		+ 4.1		

function), uncorrected hyperopic values of around 4.00 diopters would convert to emmetropia. Accordingly, values below this would become myopic.

The action of a cycloplegic drug appears not to influence refractive error in any measurable manner. Thus, the sloth apparently does not possess effective accommodation.

Interestingly, sloths have a reputation for being myopic (see Goffart, 1971). Much of the evidence supporting this view appears to rest upon casual behavioral observation and anecdote. However, Watillon and Goffart (1965; 1969) measured the refractive error in six eyes of three *Choloepus hoffmanni* using skiascopy (retinoscopy), finding between 3.0 and 5.0 diopters of myopia which they attributed to the thick crystalline lens and very convex cornea of that species. The eye size and apparent hyperopia relationship appears not to have been considered. The effective refraction, obtained using Glickstein and Millodot's (1970) function, for *Bradypus tridactylus* appears at some variance with these findings. An interspecies difference could account for such a difference in refraction, while an alimentary factor such as drug action could produce increased curvature of the crystalline lens surfaces, rendering it more powerful optically. Both these points were considered by Goffart (pers. comm., 1978). Mendel, Piggins and Fish (in prep.) report low to moderate myopia in nine *Choloepus*, thus indicating a possible interspecies difference, although some variation exists in the findings for each species.

Watillon and Goffart (1965; 1969) infer that *Choloepus hoffmanni* does not accommodate for near objects in the visual field upon the anatomical grounds that they do not possess a ciliary muscle. A similar statement can be made for *Bradypus tridactylus* on optical grounds, since the action of a cycloplegic drug made little difference to refractive error.

Watillon and Goffart (1965; 1969) suggest that accommodation in daylight is unnecessary since a small pupil would provide a stenopaeic aperture giving an appreciable depth of precise vision. While this would be so for pupil sizes of the order of 1 mm, any smaller pupil size would increasingly degrade the retinal image. A theoretical limit is placed upon the resolving power of an optical system by the imaging of point light sources as Airy discs, a diffraction effect which appears as a central bright disc surrounded by a series of lower intensity concentric circles. The size of the disc is a function of aperture size, a smaller aperture producing a larger Airy disc and hence reducing the resolving power (see, for example, Stenström, 1964). Wattillon and Goffart (1969) mention a constricted pupil size of 0.5 mm in daylight which approaches an aperture size at which Airy disc size becomes a consideration. Other than this quoted pupil size, a small pupil in the sloth is often described as a "pinhole," this could help or hinder vision depending upon its actual size. The presence of appreciable myopia in *Choloepus* would render a well-focused retinal image of objects well within forearm length with only moderate pupil closure.

***Gross Ocular Anatomy of* Bradypus tridactylus *Linné*.--** The eye is nearly circular in anterior-posterior section with overall dimensions 8.4 mm axially and 8.3 mm equatorially (Fig. 2). The cornea is quite convex (radius of curvature 3.62 mm) and uniformly thin (0.15 mm) and wide (limbal diameter 6.7 mm). The crystalline lens is thick and rounded with an equatorial:axial length ratio of 1.36 and appears to possess a substantial central core, common in small animals. The radius of curvature of the anterior lens surface is 2.35 mm and that of the posterior surface 1.84 mm. The anterior chamber is deep, with the result that the crystalline lens is set somewhat posteriorally. The irides are thin in cross-section and though there appears to be a ciliary body, its bulk is small.

Watillon and Goffart (1969) describe the eye of *Choloepus hoffmanni* as spherical, as is the case with *Bradypus tridactylus*. However, the eye of *Choloepus* is longer (10.35 mm, though Goffart, 1971, quotes a figure of 10-12.5 mm for both species), possesses a less convex cornea (radius of curvature 6 mm), a thicker cornea (0.35 mm), and a wider cornea (11 mm). The crystalline lens in *Choloepus* is probably less optically

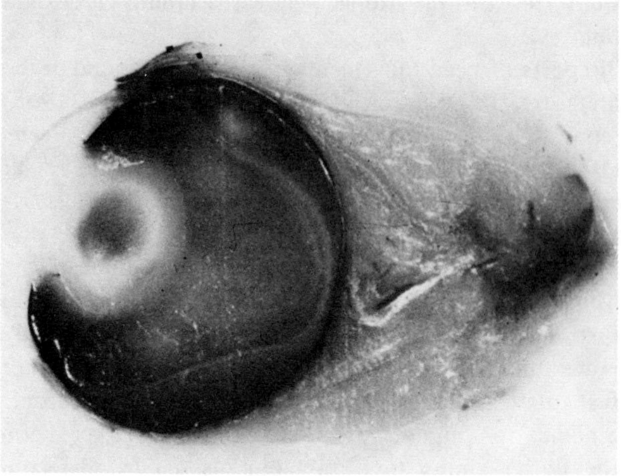

Figure 2. Anterior-posterior (estimated corneal centre-optic nerve head) section of an eye from *Bradypus tridactylus* Linné. The magnification is X 3.75.

powerful than that of *Bradypus* since its equatorial:axial length ratio is higher (1.7); it is not as thick and appears less round. It is not possible to discern the presence of a central core in the section reprinted in Watillon and Goffart (1969), though it appears to be present for *Bradypus*. The presence of this core is usually identified as a central area of higher refractive index than the surrounding lens cortex. Hughes (1972) writes "the effect of a core is to increase the power of the lens relative to a homogenous lens of the same refractive index as the core." A ciliary body is present in *Choloepus*, though a ciliary muscle is lacking. Figure 3 shows the major definable structures, schematically.

***Retinal Histology*.--** The retina has a structure typical of a nocturnal animal (Fig. 4). The receptors are very numerous, with the outer nuclear layer correspondingly thick. Only rods can be seen; cones are rare or absent. While the inner segments of the rods are relatively well preserved, the outer segments are damaged, though their typically cylindrical form can be seen under interference contrast (Fig. 5). Ganglion cells are few in number, indicating a high degree of neuronal convergence.

The retinal histology of this *Bradypus* is quite similar to that reported for *Choloepus* by Watillon and Goffart (1971). Many slender rods with few ganglion cells are identified in both genera. The resulting high neuronal convergence may be identified further in the thin optic nerve of *Choloepus*, a structure which is not easily discernable in this specimen of *Bradypus*. The absence of cones in the sloth retina has been noted by Wislocki (1928) for *Bradypus griseus griseus* Gray (considered by Wetzel, 1985 to be *B. variegatus*) and *Choloepus hoffmanni*.

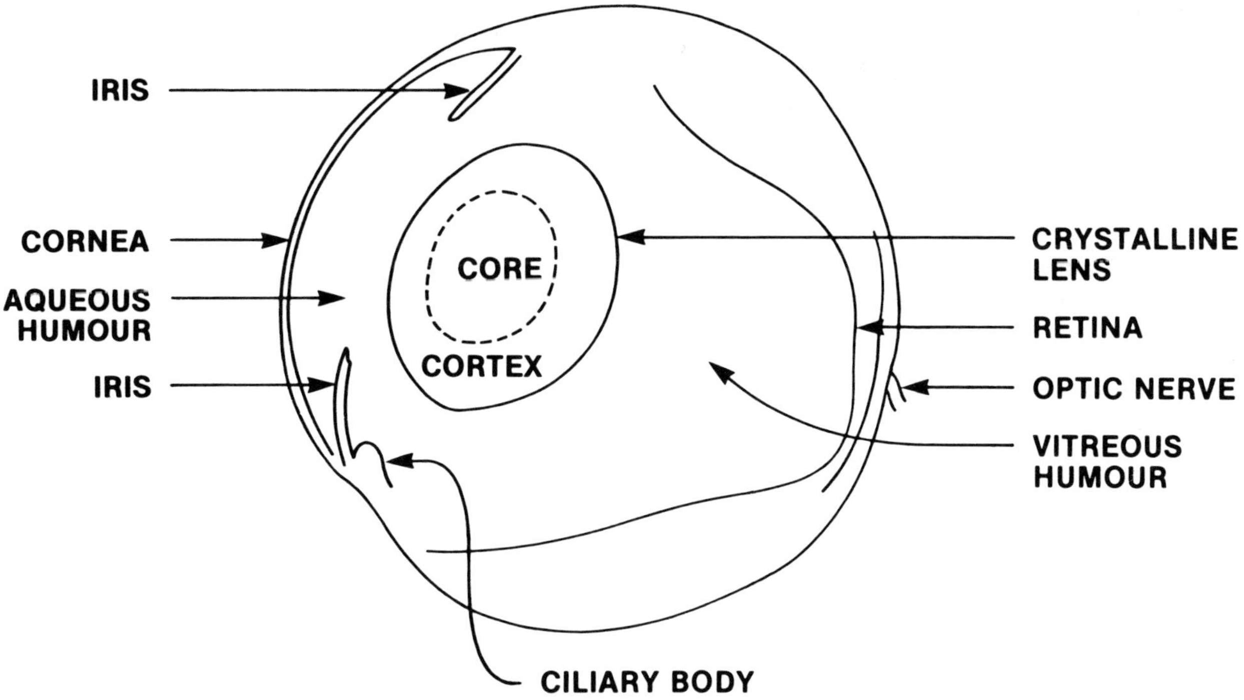

Figure 3. The most readily identifiable anatomical structures in the eye of *Bradypus tridactylus* Linné.

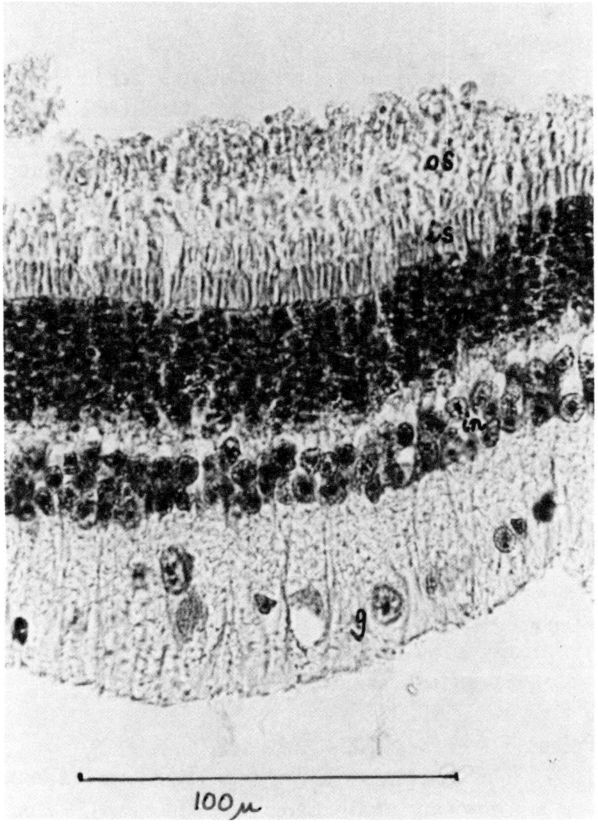

Figure 4. The retina of *Bradypus tridactylus* Linné taken under transmitted light.

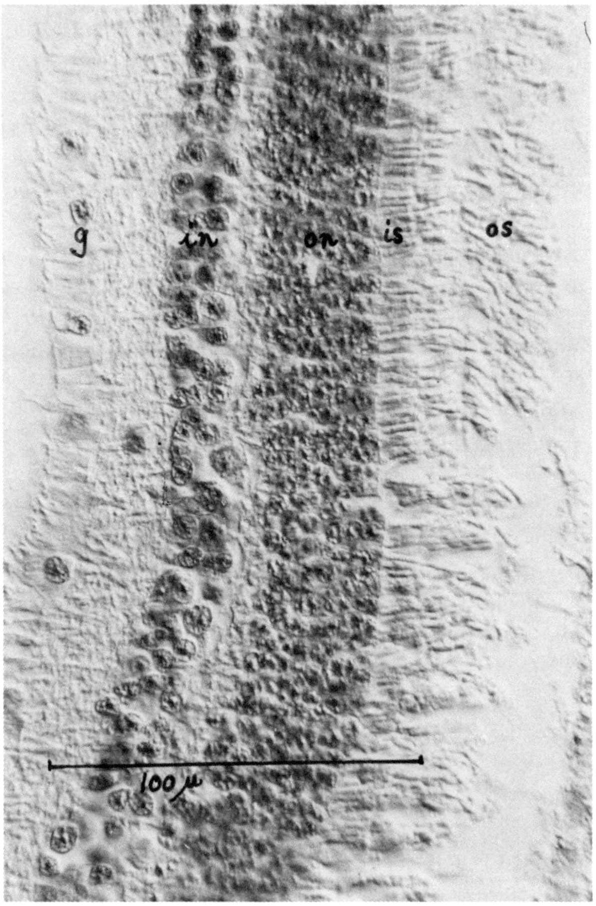

Figure 5. The retina of *Bradypus tridactylus* Linné taken under interference contrast conditions.

DISCUSSION

Many of the previous findings and observations upon the sloth eye, mainly derived from *Choloepus hoffmanni* have been confirmed from the current work on *Bradypus tridactylus*. The results generally support the findings of Beebe (1926), Watillon and Goffart (1965; 1969) and Goffart (1971).

A consensus view would suggest that sloths possess an eye primarily adapted for vision at low light intensities. The eye is of moderate size with a wide cornea and rounded crystalline lens. The retina appears to contain only rods while the low number of retinal ganglion cells and thin optic nerve imply a high degree of neuronal convergence. It is an eye more designed for sensitivity than for acuity. It is possible that sloths have some useful vision in daylight, primarily in the shade. In direct sunlight the presence of a "pinhole" pupil smaller than about 1 mm would more likely hinder than help an eye of already low visual acuity.

In the absence of optical accommodative ability, the myopic *Choloepus* would be well adapted for obtaining a well-focused retinal image of near objects by constricting the pupil. *Bradypus*, though possessing a refraction closer to emmetropia, could also obtain a tolerable measure of retinal image clarity by pupil constriction. However, the absence of a macular (not seen during ophthalmoscopy), the lack of ocular convergence, and the divergent attitude of the optic axes should mean that the three-toed sloth's abilities in near vision are somewhat limited. Although casual observation suggests that some aspects of the sloth's visuomotor behavior for near objects is quite accurate (e.g., claw placement) and might require binocular vision for its execution, the limb movements are sufficiently slow to be monitored utilizing monocular cues to depth. A judiciously placed macular in each eye might more efficiently utilize the presence of the binocular visual field, but such benefit as might accrue would possibly be outweighed by the implied low visual acuity.

ACKNOWLEDGMENTS

We thank Dr. W.E. Kerr, Director of the Instituto Nacional De Pesquisas da Amazonia (INPA) at Manaus, Brazil, for extending an invitation to carry out this work and for providing facilities. We thank Dr. J. Adis and his team for capturing the sloths, and Dr. Domning and Rosemary Bradley for their aid. We particularly thank Robin Best for his invaluable help at all stages of the work.

LITERATURE CITED

Beebe, M.
1926. The three-toed sloth, *Bradypus cuculliger culliger* WAGLER. Zoologica 7: 1-67. Cited in Goffart M. (1971).

Britton, P.O.
1941. Form and function in the sloth. Quart. Rev. Biol. 16: 13-34 and 190-207.

Duke-Elder, Sir S.
1958. The eye in evolution, Vol. 1 System of Ophthalmology. H. Kimpton, London. 843 pp.

Enger, P.S. and T.H. Bullock.
1965. Physiological basis of slothfulness in the sloth. Hvalradets Skriftet 48: 143-160.

Glickstein, M. and M. Millodot.
1970. Retinoscopy and eye size. Science, 168: 605-606.

Goffart, M.
1971. Function and form in the sloth. Pergamon, Oxford. 225 pp.

Grüsser, O.J. and U. Grüsser-Cornhels.
1976. Neurophysiology of the anuran visual system. Pp. 291 to 385, *in* Frog neurobiology: A handbook (R. Llinas and W. Precht, editors). Springer-Verlag, Berlin. 1046 pp.

Hirschberg, J.
1882. Zur dioprik und ophthalmoskopie der Fisch-und Amphibien augen. Arch. Anat (Physiol Abth): 493-526. Cited in O.J. Grüsser and U. Grüsser-Cornhels, 1976. Pp 291 to 385 *in* Neurophysiology of the anuran visual system. Published in Frog neurobiology: a handbook (R. Llinas and W. Precht,editors), Springer-Verlag, Berlin. 1976. 1046 pp.

Hughes, A.
1972. A schematic eye for the rabbit. Vision research, 12: 123-138.

Nuboer, J.F.W. and H. van Genderen-Takken.
1978. The artifact in retinoscopy. Vision research, 18: 1091-1096.

Prince, J.H.
1956. Comparative anatomy of the eye. Thomas, Springfield, Illinois. 418 pp.

Polyak, S.
1957. The vertebrate visual system. (H. Klüver, editor). University of Chicago Press, Chicago. 1390 pp.

Stenström, S.
1964. Optics and the eye. Butterworths, London. 190 pp.

Waage, J.K. and R.C. Best.
1985. Arthropod associates of sloths. (this volume).

Walls, G.L.
1942. The vertebrate eye and its adaptive radiations. Cranbrook Institute of Science, Bloomfield Hills, Michigan. 785 pp.

Watillon, M. and M. Goffart.
1965. Implications physiologiques de la structure de l oeil du Paresseux (*Choloepus hoffmanni* Peters). Archives Internationales de Physiologie et de Biochemie, 73: 163-166.

1969. The eye of the sloth (*Choloepus hoffmanni* Peters). Acta Zoologica et Pathologica Antverpiensia, 49: 107-122.

Wetzel, R.M.
1985. The identification and distribution of recent xenarthra (= edentata). (this volume).

Wislocki, G.B.
1928. Observations on the gross and microscopic anatomy of the sloths, (*Bradypus griseus griseus* Gray and *Choloepus hoffmanni* Peters). J. Morph. and Physiol., 46: 317-397.

FUNCTIONAL COCHLEAR MORPHOLOGY OF THE TWO-TOED SLOTH (*CHOLOEPUS* SP.)

F. Ramprashad[1], J. P. Landolt[2], K. E. Money[2], M. J. Correia[3] and J. Laufer[2]

[1]Department of Zoology, University of Guelph, Guelph, Ontario, Canada N1G 2W1; [2]Defence and Civil Institute of Environmental Medicine, Downsview, Ontario, Canada, M3M 3B9; [3]University of Texas Medical Branch, Department of Otolaryngology, and Physiology and Biophysics, Galveston, Texas, USA 77550.

Abstract

A morphometric study of the cochlea was made from serial sections and graphic reconstructions of the cochlea of two-toed sloths. The cochlea had 2-1/4 spirals and measured approximately 17.4 mm in length. Cochlear morphometric parameters studied were the changes in cross-sectional area of the scalae vestibuli, media, and tympani, thickness of the pars tecta and the pectinata of the basilar membrane, width of the basilar membrane, and density of the bipolar ganglion cells. The change in the cross-sectional area of the scala vestibuli differed from the scala tympani; the former decreased from the oval window and increased to a second maximum value at 12.2 mm from the basal end, while the latter increased from the round window membrane to a maximum value at 4.8 mm and remained unchanged throughout the remainder of the cochlear length. The cross-sectional area of the scala media remained relatively constant throughout the cochlear length. There was a fourfold increase in the change of the width of the basilar membrane from base to apex. There was a gradual increase in width from the base to 12.2 mm followed by a steep increase to a maximum value of 275 μm before it declined to the apex. There was little change in the thickness of the pars tecta and the pars pectinata of the basilar membrane from the base to 8.5 mm before it declined at the apex. These morphometric features indicated a distinct region of morphological discontinuity 10.5 - 13 mm along the length of the basilar membrane. Within this region the stiffness gradient was greatest, and the theoretical position of maximum amplitude of vibration for the frequencies 0.5-1.5 kHz, also coincided. There were 19,900 bipolar-ganglion cells enumerated and the greatest density occurred at 13 mm where 1,600 cells/mm^2 were present. These morphological findings correlate with the known psychoacoustic data on the two-toed sloth.

Abstracto

Se llevó a cabo un estudio morfométrico de la coclea por medio de secciones en serie y de reconstrucciones gráficas de cocleas de perezosos de dos dedos. La coclea tenía 2-1/4 espirales y medía aproximadamente 17.4 mm de largo. Los parámetros morfométricos estudiados fueron los siguientes: los cambios en el área de corte transversal de las escalas de los vestíbulos, de las medias y de los tímpanos; el espesor de las pars tecta y pars pectinata de la membrana basilar; el ancho de la membrana basilar, y la densidad de las células del ganglio bipolar. El cambio en el área transversal de la escala del vestíbulo fue distinto al de la escala del tímpano; el primero de ellos decreció desde la ventana oval y aumentó a un segundo valor máximo, a una distancia de 12.2 mm de la extremidad basal, mientras que el segundo aumentó desde la membrana redonda de la ventana hasta un valor máximo a una distancia de 4.8 mm y permaneció sin cambios por el resto de la longitud coclear. El área de corte transversal de la escala media permaneció relativamente constante a lo largo de toda la longitud coclear. Hubo un aumento total de cuatro veces en el cambio, con el ancho de la membrana basilar desde su base hasta el ápice. Hubo un aumento gradual en el ancho hasta una distancia de 12.2 mm de la base, seguido por un aumento brusco hasta un valor de 275 μm, antes de declinar hasta su ápice. Hubo muy pequeños cambios en el espesor de la pars tecta y de la pars pectinata desde la membrana basilar hasta una distancia de 8.5 mm de su base, para luego decrecer hacia sus ápices. Estos rasgos morfométricos indican que una región distinta de discontinuidad morfológica ocurre entre los 10.5 y 13 mm, a lo largo de la longitud de la membrana basilar. Dentro de esta región, fue mayor la gradiente de rigidez; también coincidió la posición teórica de amplitud máxima de vibración para las frecuencias 0.5 - 1.5 kHz. Se enumeraron 19,900 células de ganglios bipolares y la mayor densidad ocurrió a 13 mm, donde había 1,600 células por mm^2. Estos hallazgos se correlacionan con datos conocidos psicoacústicos del perezoso de dos dedos.

Resumo

Um estudo morfométrico da cóclea foi feito de cortes seriados e reconstruções gráficas da cóclea de duas preguiças reais. As duas voltas e meia da cóclea da preguiça mediam aproximadamente 17,4 mm de comprimento. Parâmetros morfométricos cocleares estudados foram as mudanças; a área do corte transversal da escala vestibular, média e timpânica; na espessura do pars tecta e pectinata da membrana basilar; na largura da membrana basilar; na densidade das células ganglionares bipolares. A mudança na área do corte transversal da escala vestibular diferiu da escala timpânica, o primeiro diminuiu da janela oval e aumentou para um segundo valor máximo de 12,2 mm do final basal, enquanto o último aumentou da membrana da janela redonda para um valor máximo a 4,8 mm e permaneceu inalterado pelo comprimento coclear restante. A área do corte transversal da escala média permaneceu relativamente constante por todo o comprimento coclear. Houve um aumento geral de quatro vezes na mudança com a largura da membrana basilar, da base ao ápice. Houve um aumento gradual na largura da base para 12,2 mm seguido por um abrupto aumento a um valor máximo de 275 μm, antes de declinar para o ápice. Há uma pequena mudança na espessura do pars tecta e do pars pectinata da membrana basilar da base para 8,5 mm, antes de declinar no ápice. Esses aspectos morfométricos indicaram que uma região distinta de descontinuidade morfológica ocorreu entre 10,5 - 13 mm ao longo do comprimento da membrana basilar. Dentro dessa região o gradiente de dureza foi o maior e a posição torácica da amplitude máxima de vibração para as frequências 0,5 - 1,5 KhZ também coincidiram. Houve 19.900 células ganglionares bipolares enumeradas e a mais alta densidade ocorreu a 13 mm, onde havia 1.600 células/mm². Esses achados morfológicos se correlacionam com os dados psicoacústicos conhecidos na preguiça real.

This report is concerned primarily with the morphometrics of the cochlea of the two-toed sloth. Recently there has been a revival of interest in the morphology of the mammalian cochlea. One outcome of this revival has been the attempt to correlate the neuroanatomical data, and psychoacoustic data to the morphometric parameters studied within the cochlea itself (Bruns, 1976a, 1976b; Ehret and Frankenreiter, 1977; Ramprashad et al., 1979). It is within the cochlea that sound is transformed into neural impulses by the stimulation of the mechanoreceptors (sensory hairs) of the organ of Corti.

Precise psychoacoustic data in some Xenarthra (= Edentata) have been determined by Suga (1967), and Peterson and Heaton (1968). Suga (1967) has shown that the frequency range of hearing of the xenarthran audiogram was of the order of 0.3-25 to 30 kHz, with the region of greatest sensitivity occurring at 2-8 kHz at 100 db SPL. Peterson and Heaton (1968) showed that the maximum cochlear-microphonic output of 250μv occurred between frequencies of 0.5-3 kHz. Both studies confirmed that xenarthran hearing was somewhat inferior to that in other terrestrial mammals. Contact-distress calls emitted by young sloths were in the frequency range of 2.0 kHz, i.e., between 1.9-2.6 kHz for the three-toed sloth and 1.4 kHz for the two-toed sloth (Montgomery and Sunquist, 1974). It would be most informative if the morphometric parameters and the neuroanatomical data within the sloth cochlea could be correlated with the psychoacoustic data.

What is known about the structure of the sloth ear indicates that, apart from a few primitive features, it possesses no peculiar structural features, i.e., it is typically mammalian (Weber, 1928; Neubert and Wustenfeld, 1926; Demanez et al., 1968). Primitive features mentioned were: (a) the wall of the osseous external auditory meatus consists of a narrow semicircular annulus tympanicus which is directed upwards and backwards (a feature also found in monotremes, marsupials, insectivores, chiropterans, sirenians and lemurs), (b) the stapes is "columellarlike" (Fig. 2), (c) the lack of insertion of the *tensor tympani* muscle onto the maleus, and (d) the peculiar shape of the tympanic membrane (Goffart, 1971). Half-coil measurements of the cochlear height, number of half turns, cochlear width, and the width and thickness of the basilar membrane of the xenarthran cochlea have been provided in a recent study (Pye, 1979). However, detailed morphometric information is lacking for the xenarthran cochlea. The prime purpose of this report is to ellucidate the morphometric

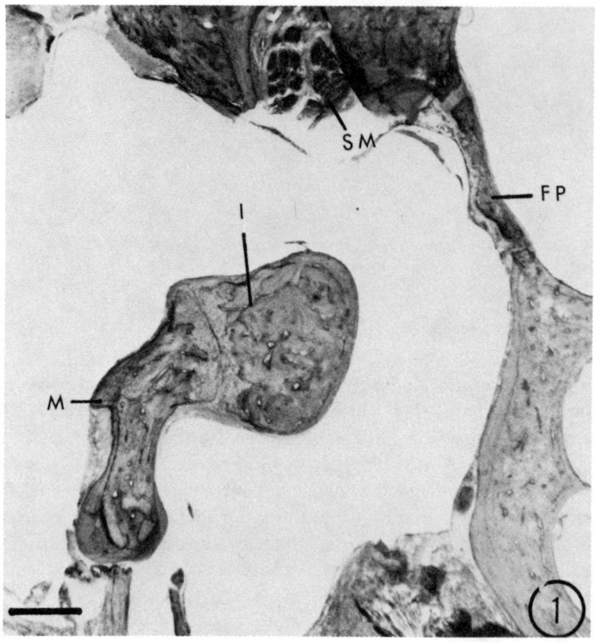

Figure 1. Photomicrograph of a sagittal section through middle ear showing the maleo-incudal joint. (M) - malleus, (I) - incus, (F.P.) - foot plate of the stapes, (S.M.) - stapedius muscle. Scale bar = 255 μm.

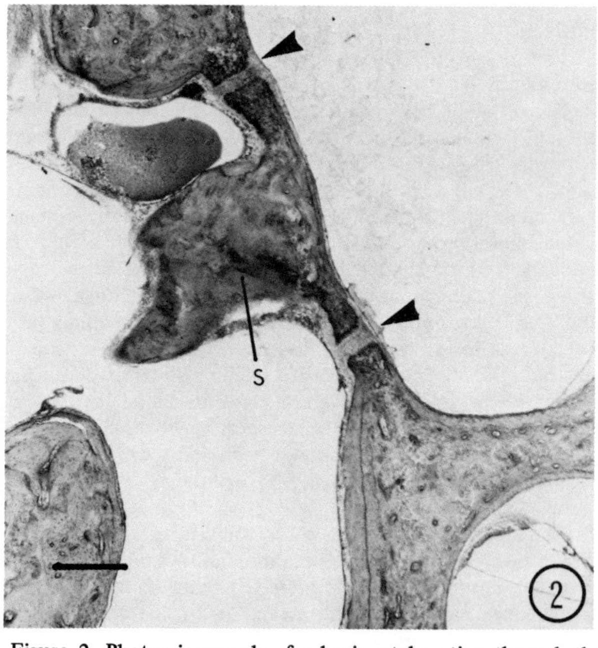

Figure 2. Photomicrograph of a horizontal section through the middle ear showing the primitive "massive" columellar type stapes (S) within the oval window. Note the rather large annular ligament at the arrows, and the absence of an intercrural space. Scale bar = 255 μm.

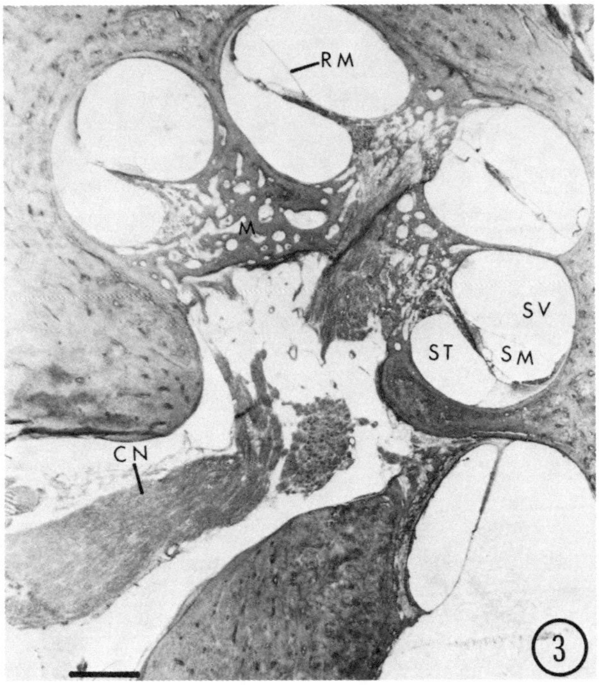

Figure 3. Low power photomicrograph showing the general features of the cochlea of the sloth. Shown are the cochlear nerve (C.N.) entering the bony modiolus, the two and a quarter turn cochlea, (S.V.) - scala vestibuli, (S.M.) - scala media, (S. T.) - scala tympani, and (R.M.) - Reissner's membrane. Scale bar = 255 μm.

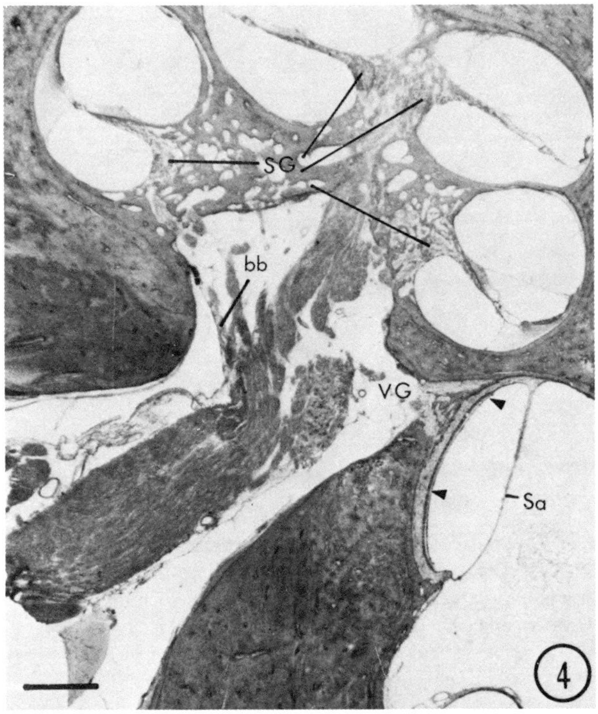

Figure 4. Low power photomicrograph showing the bifurcation of the cochlear nerve into the basal branch (b.b.) which innervates the basal coil only, and the position of the spiral ganglion (S.g.). Illustrated also are (V.G.) - vestibular ganglion, and (Sa) - saccule with its sensory epithelium, the macula sacculi at the arrows. Scale bar = 255 μm.

parameters which could provide some insight into the cochlear mechanics of the two-toed sloth (*Choloepus* sp.).

COCHLEAR MORPHOLOGY

The cochlea of the sloth is typically mammalian. The bony cochlea is a helical tube, the cochlear canal, which forms about two-and-a-quarter turns (Figs. 3, 4). Within the cochlear canal is the membranous organ of Corti. The cochlear canal spirals around the modiolus, a central axis of spongy bone. The modiolus is perforated by numerous bifurcations of the auditory nerve (Fig. 4).

The lumen of the cochlear canal is partially subdivided by a thin plate of bone extending from the modiolus, the osseous spiral lamina, into an upper scala vestibuli and a lower scala tympani; both of which are part of the perilymphatic system (Figs. 3, 4). The scala vestibuli starts from the oval window into which fits the footplate of the stapes, it then spirals upwards and, at the apex of the cochlear canal, joins the scala tympani through a narrow connection, the helicotrema. The term "upwards" refers to an ascending direction in the plane of the cochlear axis. A direction following the course of the cochlear canal is called "spiral" and a radial direction from the axis of the cochlea is called "outwards" (Bredberg, 1968). From the helicotrema, the scala tympani spirals downward and ends posteriorly at the fenestra rotundum.

The fenestra rotundum is closed by a small, thin, fibrous, round window membrane (Fig. 5). The round window membrane functions as a pressure release

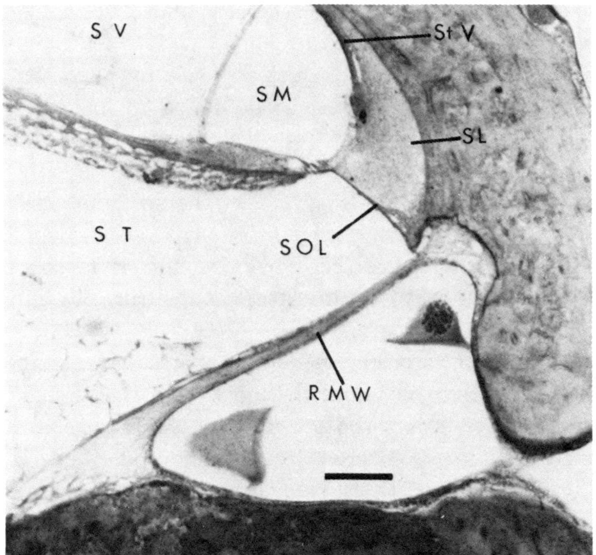

Figure 5. Photomicrograph of a sagittal section at the round window margin of the round window membrane, and the spiral ligament (S.L.) of the organ of Corti. (S.V.) - scala vestibuli, (S.M.) scala media, (S.T.) - scala tympani, (S.O.L.) - secondary spiral osseous lamina. Scale bar = 255 μm.

mechanism for the pressure waves imparted on the cochlear perilymph by the movement of the footplate of the stapes. On the medial floor of the fenestra rotundum is the large ovoid orifice of the cochlear aqueduct, which opens into the scala tympani.

The inner part of the cochlear duct is attached to the osseous spiral lamina. The basilar membrane extends from the osseous spiral lamina outwards to the outer bony cochlear wall. The basilar membrane is firmly attached to the outer bony wall by the thickened periosteum, the spiral ligament (Fig. 5). The spiral ligament is more prominent in the lower basal coil of the cochlea especially near the round window membrane (Figs. 3, 4, 5). The outer wall of the cochlear duct is vascularized with a thick epithelium, the stria vascularis, which ends above the basilar membrane as the promenentia spiralis.

The cochlear duct (the scala media) is triangular when sectioned in the mid-modiolar plane (Figs. 3, 4, 5). Reissner's membrane separates the scala vestibuli from the scala media, and the horizontal basilar membrane separates the scala media from the scala tympani. Above the osseous spiral lamina the periosteum is thickened into the spiral limbus, which extends outwards into the cochlear duct; the tectorial membrane is attached to the vestibular lip of the spiral limbus. In vivo, the tectorial membrane extends over the organ of Corti. The tectorial membrane and the tympanic lip of the spiral limbus enclose the spiral tunnel.

Above the basilar membrane is the specialized organ of Corti (Figs. 6, 7). There are the characteristic inner and outer pillar cells, between which is the inner tunnel (tunnel of Corti). The nucleii of the pillar cells are found in the broad base of the cells above the basilar membrane (Fig. 7).

Medial to the inner pillar cell is a row of inner phalangeal cells which support a single row of inner hair cells (Fig. 7). Outwards from the outer pillar cells are the phalangeal "Deiters" cells which support the outer 3-4 rows of hair cells (Figs. 6, 7). The outer hair cells are cylindrical and the inner hair cells are pear-shaped (Fig. 7). Stereocilia (but no kinocilium as in other mammals) project from the apices of the hair cells into the tectorial membrane.

The hair cells are connected together apically by a thin reticular membrane. It is these hair cells that are the mechanoreceptors which act as transducers for sound energy into nerve impulses. The maximum amplitude of vibration for a particular frequency occurs at a specific location along the length of the membrane (Bekesy, 1941, 1956).

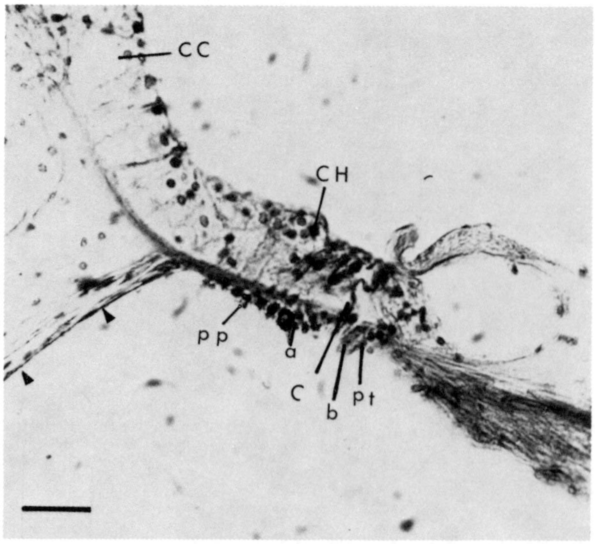

Figure 6. Photomicrograph of a horizontal section of the cochlea at approximately 1.5 mm from the basal end. Note the features of the organ of Corti: the large cells of Hensen (C.H.), and the cells of Claudius (C.C.). Note the absence of a secondary osseous lamina at arrows within the spiral ligament, a feature which is characteristic of low-frequency-hearing mammals. the limits of the pars tecta (p.t.) and pars pectinata (p.p.) are deliniated by the outer pillar cell at C (a and b are shown in greater detail in Fig. 11). Scale bar = 26 μm.

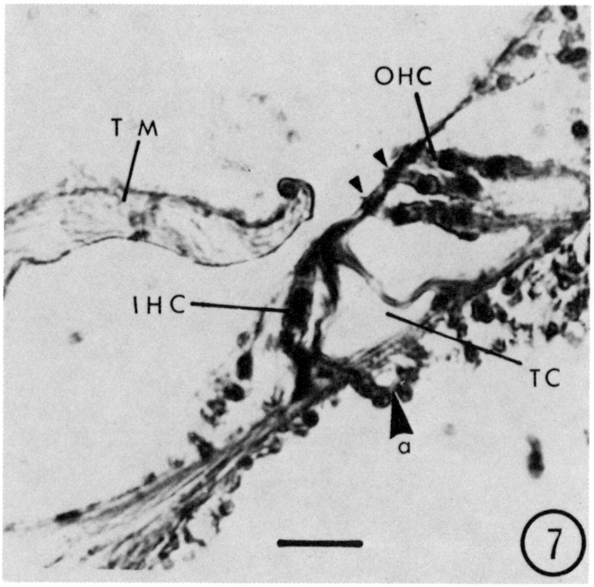

Figure 7. Photomicrograph showing typical mammalian structure of the sensory epithelium of the sloth. Note the appearance of the inner hair cells (I.H.C.), the three rows of outer hair cells (O.H.C.) and the stereocilia at the arrows. Note also the vas spirale at arrow (a) within the pars tecta. (T.M.) - tectorial membrane, (T.C.) -Tunnel of Corti. Scale bar = 11μm.

Changes in the cross-sectional area of the scalae.--Present understanding of hearing in mammals is based on the premise that the acoustical energy, which is transmitted across the ossicular chain, is transformed into hydrodynamic movements of the cochlear perilymphatic and endolymphatic systems thus creating a travelling wave along the length of the basilar membrane (Bekesy, 1941; 1948; 1956). Therefore, measurements of the

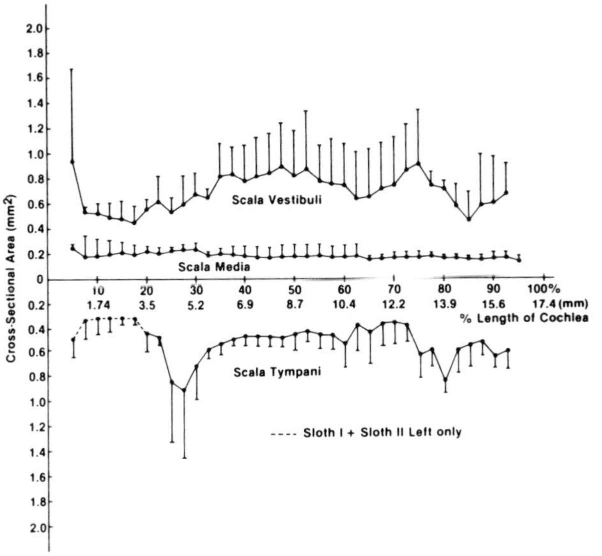

Figure 8. Changes in the cross-sectional area of the scalae vestibuli, tympani and media along the cochlear length. Dots and vertical bars represent means and SD. Note (i) the relatively small change in the scala media along the cochlear length (ii) the rapid decrease in cross-sectional area of the scala vestibuli, (iii) the decreased cross-sectional area between 1-2 mm in the scalae vestibuli and tympani, (iv) the increase in cross-sectional area in the scalae vestibuli and the tympani between 12-13 mm.

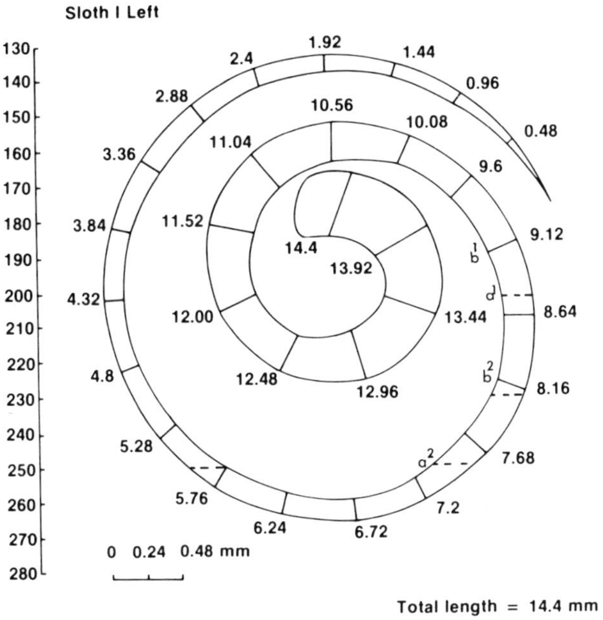

Figure 9. Change in width of the basilar membrane along the cochlear length. Graphic representation of the cochlea follows the procedure of Guild (1921). The widths at b[1], b[2], etc. are taken from a slide in a serial series whose numbers are shown to the left of the figure. Radial lines a[1], a[2] are the true width corrected for the plane of sectioning (Ramprashad, et al., 1979). Note that the cochlea is approximately a two and a quarter turn spiral. Since no measurement was obtainable, the extreme basal end is represented as zero width.

cross-sectional area of the scalae and the width and thickness of the basilar membrane can provide useful information on the cochlear mechanics of the two-toed sloth.

The cross-sectional area of the scala vestibuli is largest at the oval window where it measures 1.6 mm^2, then it decreases to 0.5 mm^2 at 0.5 mm before increasing to a second maximum of 0.9 mm^2 at 12.2 mm. From there, it decreases to the apex of the cochlca (Fig. 8).

The change in cross-sectional area of the scala tympani differs from that of the scala vestibuli (Fig. 8). At the round window membrane, its cross-sectional area measures 0.6 mm^2, then it increases to a maximum of 1.0 mm^2 at 4.8 mm, following which it returns to approximately 0.6 mm^2 for the remainder of the length of the cochlea.

The change in the cross-sectional area of the scala media differs from that of the other two scalae. The cross-sectional area of the scala media measures approximately 0.2 mm^2 and remains relatively constant throughout the cochlear length (Fig. 8). The cross-sectional area of the helicotrema was estimated to measure approximately 0.25 mm^2.

Comparative considerations.--Changes in the cross-sectional area of the scalae have been determined in a number of mammals, for instance in the Heteromyid rodents (Webster and Webster, 1977), in the little brown bat (Ramprashad et al., 1979), in the human, and in the guinea pig (Fernandez, 1952).

In all mammals studied thus far, the scala media remains relatively unchanged throughout the cochlear length. Although the trends in the change of the scalae vestibuli and tympani are similar in all mammals, i.e., a decrease in cross-sectional area from base to apex, there are individual species differences. The curves of the sloth are similar in profile to the Heteromyid rodents (Webster and Webster, 1977) in that there is a second maxima in the apical region of the cochlea. In the human and the guinea pig, the cross-sectional areas of the scalae vestibuli and tympani decrease gradually from base to apex (Fernandez, 1952).

Changes in the width of the basilar membrane.--Although the width of the basilar membrane increases from the base to the apex, the change in width occurs in two distinct stages (Fig. 10). There is a gradual increase in width from 70 μm at the base to 175 μm at 12.2 mm, an increase of 150 percent. Then there is a relatively steep increase, from 175-275 μm between 12.2-14.0 mm, followed by a further increase of 53 percent before it attains a maximum width of 275 μm between 14.0-15.7 mm. The overall change in width from the base to its maximum width is approximately fourfold.

Comparative considerations.--The change in width of the

basilar membrane along the cochlear length in the two-toed sloth is similar to that of the Heteromyid rodents (Webster and Webster, 1977), and of Microchiroptera (Bruns, 1976a; Henson, 1978; Ramprashad et al., 1979). In all these mammals there is an uneven change in width along the cochlear length. However, the change in width is uniform from base to apex in the human, cat, guinea pig and cetacean (Wever, 1938; Fernandez, 1952; Wever et al., 1971; Cabezudo, 1978). Changes in width of the basilar membrane of some mammals suggest that overall changes in width of the basilar membranes appear to have little association with the frequency ranges of their respective audiograms (Tables 1 and 2). For example, the overall change in cochlear length of the two-toed sloth is approximately 3.8; this is of a greater magnitude than that of the lesser horseshoe bat, *Rhinolophus*. Additionally, these two species have similar basal widths of 17 μm (Bruns, 1976a). *Rhinolophus* is an echolocating bat which uses constant frequency modulation and is therefore classed as a high frequency hearing mammal.

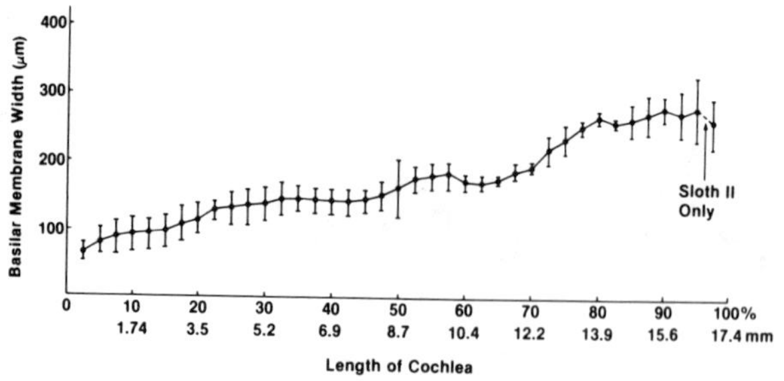

Figure 10. Change in width of the basilar membrane along the cochlear length. Dots and vertical bars represent means and SD. Note that the change in width occurs in two distinct phases; (i) a gradual increase from the base to approximately 10.4 mm, and (ii) a more rapid increase to a maximum of 275 μm.

The hearing range of the sloth is from 0.3 to 30 kHz (Suga, 1967; Peterson and Heaton, 1968). The best

Table 1: Variation in the total cochlear length of mammals with change in the width of the basilar membrane.

Mammal	*Cochlear Length (mm)*	*Basal Width (μm)*	*Maximum Width (μm)*	*Overall Change*	*Reference*
Little brown bat	6.8	57	115	2.0	Ramprashad et al. 1979
House mouse	6.8	40	150	3.7	Ehret and Frankenreiter 1977
Liomys	9	110	150	1.3	Webster and Webster 1977
Dipodomys	10	90	250	2.8	Webster and Webster 1977
Peteronotus	12	38	80	2.1	Henson 1978
Rhinolophus	17	70	155	2.2	Bruns 1978a
Choloepus	17.1	70	275	3.8	This study
Guinea pig	20	70	250	3.6	Fernandez 1952
Cat	22	106	426	4.0	Cabezudo 1978
Human	32	80	500	6.2	Fernandez 1952
Harp seal	33	90	525	5.8	Ramprashad et al. 1979
Pacific White-sided Dolphin	36	30	300	10	Wever et al. 1971
Tursiops	40	25	350	14	Wever et al. 1972

Table 2: A comparison of the total number of ganglion cells and the range of hearing (kHz) in some mammals.

Mammal	*Mean Number of Ganglion Cells*	*Range of Hearing*	*Frequency of Echo-Location*	*Average Body Mass (kg)*	*Author(s)*
Two toed Sloth					
Choloepus	19,900	0.3-30		3.0	Peterson and Heaton 1968
Rabbit	26,700			0.4	Ramprashad et al., 1982
Human	30.500	0.032-20		70	Wever 1949; Schuknecht 1974
Cat	49,962	0.063-52		3.3	Howe 1935; Schuknecht 1974
Harp Seal	52,000	1-32 (air)			Ramprashad 1976;
Pagophilus		0.5-100 (water)		78	Terhune and Ronald 1971, 1972
Little Brown Bat					
Myotis	56,000	0.7-150	25-120	0.005	Ramprashad et al. 1979; Grinnel 1963
Pacific White-sided Dolphin					
Lagenorhynchus	65,000	0.6-80	30-60	95	Wever et al. 1972; Evans 1973
Bottlenose Dolphin					
Tursiops	95,000	1-100	30-69	118	Evans 1973

frequencies for activating the adult sloth ear are in the range 2 to 8 kHz (Suga, 1967). The peripheral auditory function of two-toed sloths is optimized for the range 0.5 to 3 kHz (Peterson and Heaton, 1968). The distress calls of young two-toed sloths are relatively low-pitched, with most of the sound energy concentrated below 1.4 kHz (Montgomery and Sunquist, 1974). It seems likely, that hearing in the adult sloth would be optimized for this range of frequencies, i.e., that these are mammals which hear low frequencies best. Although the sloth's cochlea has typically mammalian features, it lacks a well-developed secondary spiral osseous lamina within the basal coil. This secondary spiral is a characteristic feature found within the cochlea of all high-frequency-hearing mammals, including Microchiroptera (Pye, 1966a, 1966b, 1967; Hall, 1969; Bruns, 1976a, 1976b; Henson, 1978; Ramprashad et al., 1979), Phocida (Mohl, 1968; Ramprashad et al., 1972) and Cetacea (Wever et al., 1971a, b, c, 1972). In the two-toed sloth, the secondary spiral osseous lamina is only present at the round window (Fig. 5). This anatomical feature of the sloth ear is common to other low-frequency-hearing mammals (e.g., the human and the guinea pig.).

Changes in the thickness of the basilar membrane.--The sensory hair cells in the organ of Corti are the transducers of sound energy into nerve impulses. The stimulation of these mechanoreceptors depends entirely on the vibration pattern of the basilar membrane and the subsequent shearing motion of the tectorial membrane and the stereocilia. It follows that morphological discontinuities along the basilar membrane length may alter its physical properties, which would also alter the vibrating properties of the basilar membrane. Therefore, the changes in thickness along the basilar membrane can serve as indicators of the changes of the mechanical and, perhaps, vibrating properties of the basilar membrane.

The mammalian basilar membrane can be arbitrarily divided into two regions, the pars tecta and the pars pectinata (Fig. 6). The outer pillar cell was used as a landmark to delineate the pars tecta from the pars pectinata. The pars tecta and the pectinata are not of uniform thickness throughout the cochlear length (Fig. 11). The pars tecta and the pars pectinata are thickest at 12.6 mm and 9.4 mm, respectively. There is very little change in the overall thickness of the basilar membrane from the base to about 8.0 mm along the cochlear length (Fig. 12). However, it gradually increases and reaches its maximum thickness at 12.3 mm to 12.8 mm before decreasing at the apex. These figures clearly indicate that distinct morphological discontinuities occur along the basilar membrane of the sloth.

Comparative considerations.--Recent morphometric studies on the basilar membrane of bats (Bruns, 1976a, 1976b; Ramprashad et al., 1979), house mice (Ehret and

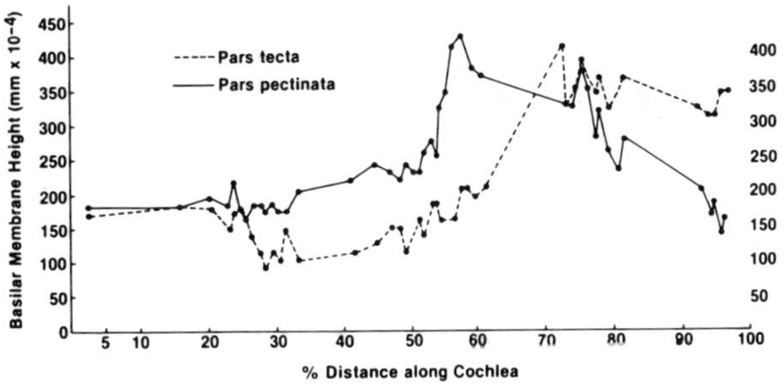

Figure 11. Change in the maximum thickness (height) of the pars tecta and pars pectinata of the basilar membrane. See a, b, in Fig. 6 for reference points which were measured. Note; (i) the similarity of the curves for both pars tecta and pectinata, and (ii) the different position where the maximum thicknesses occur. Dots represent mean values from two temporal bones.

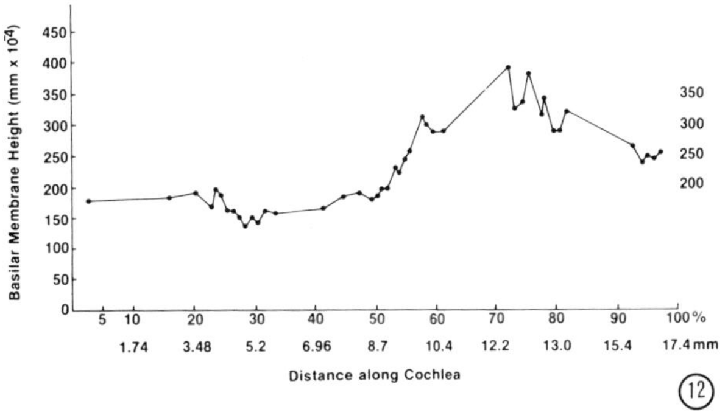

12

Figure 12. Change in the thickness (average height) of the basilar membrane along the cochlear length. Note; (i) the maximum thickness occurs at 12.2 mm, (ii) the thickness of both the pars tecta and pectinata remain constant from the base to approximately 3.5 mm, (iii) the pars tecta and the pectinata are thicker in the apical region than at the basal end, attributable to the thicker mesothelial layer of the basilar membrane (see Cabezudo, 1978).

Frankenreiter, 1977), and cats (Cabezudo, 1978) show that morphological discontinuities occur along the basilar membrane. The region of morphological discontinuity in the sloth was not as pronounced as in the cat or in the bat. In all mammals studied, the changes in the thickness of the pars pectinata influenced the overall change in the thickness of the basilar membrane. In the Microchiroptera (Pye, 1966; 1966b; 1967; Bruns, 1976a; Ramprashad et al., 1979; Henson, 1978), in the Heteromyid rodent (Webster and Webster, 1977), cat (Cabezudo, 1978), and house mouse (Ehret and Frankenreiter, 1977), the basilar membrane was hypertrophied (i.e., hyalinized). However, no such hyalinization was evident within the basilar membrane of the sloth. Instead, the thickening of the pars tecta and the pars pectinata in the sloth were mainly in the mesothelial layer below the basilar membrane. This mesothelial layer is an integral part of the basilar membrane (Cabezudo, 1978). In the cat, both the mesothelial layer and hyalinization of the basilar membrane were present (Cabezudo, 1978).

Ehret and Frankenreiter (1977) suggested that these thickenings would result in subtle changes in the vibrating patterns of the two components of the basilar membrane, and developed a model of the vibration modes of the basilar membrane using the morphological findings. From that model, they were able to account for the electrophysical findings on phase opposition between the inner and outer hair cells (Dallos, 1973; Zwislocki, 1975) by the phase opposition movements of the stereocilia of the inner and outer hair cells during displacement. Ehret and Frankenreiter (1977) concluded that inner and outer hair cells have separate modes of mechanical excitation and separate innervation patterns.

MORPHOLOGICAL CONSIDERATIONS OF HEARING

Stimulation of the mechanoreceptors of the cochlea is dependent on the subtle movements of the basilar membrane. If morphological discontinuities exist along the basilar membrane, then these, in turn, would result in an alteration of the mechanical and subsequently the vibration properties of the basilar membrane.

Morphometric studies of the cochlea in some mammals show that the frequencies (particularly high frequencies) to which the mammal is maximally sensitive coincide within the region of morphological discontinuities (Bruns, 1976a, 1976b; Ehret, 1977; Ehret and Frankenreiter, 1977; Ramprashad et al., 1979). In the two-toed sloth, the region of the basilar membrane between 10.5-13.1 mm should thus represent the region where the frequencies to which the sloth is maximally sensitive are analyzed. This suggestion would be strengthened if it can also be demonstrated that within this region, i.e., 10.5-13.1 mm, the vibrating properties of the basilar membrane are different.

Bruns (1976a, 1976b) demonstrated that the ratio of basilar membrane height/width approximates the stiffness gradient of the basilar membrane. In the sloth, the stiffness gradient decreases from the base to about 4.5 mm, where it increases to its maximum stiffness at 10.5 mm and then remains relatively unchanged to 13.0 mm, before declining to the apex (Fig. 13).

Wherever the plateaux of stiffness occurred in the stiffness curves of other mammals, frequencies to which the mammal was maximally sensitive coincided with that region (Bruns, 1976a, 1976b; Ehret and Frankenreiter, 1977; Ramprashad et al., 1979). Frequencies are also expanded within the region of maximum sensitivity (Bruns, 1976b). Therefore, the plateau between 10.5-13 mm in the sloth may be the region where frequencies to which the sloth is maximally sensitive are represented. This region occurs in the upper middle coil, the region of the cochlea associated with low frequencies.

It would therefore be of particular importance if it could be demonstrated that the position of maximum amplitude of vibration for the sensitive frequencies coincided within this region of the basilar membrane. Greenwood (1961) derived an equation to relate the critical bandwidth to frequency of vibration or the

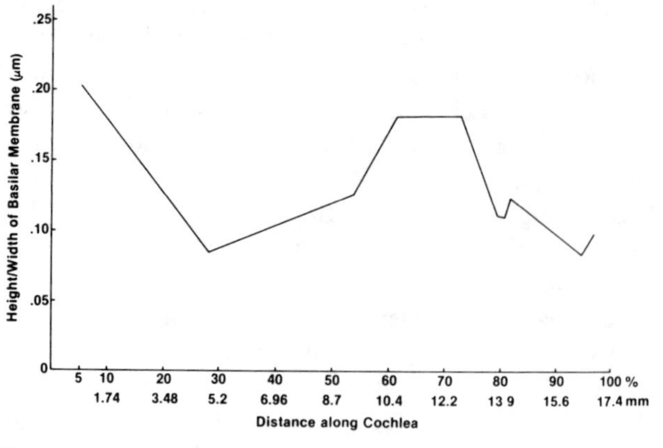

Figure 13. Stiffness gradient along the cochlear length, estimated from the ratio of the height of the basilar membrane to its width. This idealized curve is based on measurements from one sloth. Note; (i) the maximum stiffness occurs at the basal coil, (ii) there is a second area of maximum stiffness within 10.4-13.0 mm from the basal end, (iii) the rapid decline in stiffness at the apex.

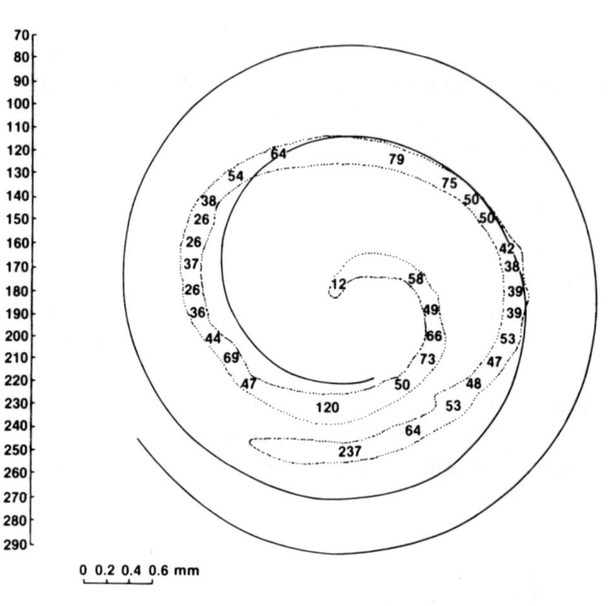

Figure 14. Reconstruction of the coil of the cochlea (procedure of Guild, 1921), with the dotted line showing the course of the spiral ganglion around the cochlea (Ramprashad, 1976). Numbers within the trace of the spiral ganglion are counts of ganglion cells enumerated from a slide of a series that is indicated to the left of the figure. Note; (i) the spiral ganglion includes one and one half turns, (ii) the spiral ganglion starts approximately 0.7 mm from the base of the cochlea and ends approximately 0.3 mm from the apex of the cochlea.

position of the maximum amplitude of vibration to the frequency:

$$f = A(10^{ax}\text{-}1) \qquad (1)$$

where f = frequency in Hz, A and a are constants having values that are species dependent, and x is the distance along the basilar membrane as measured from the helicotrema (apex) in units of "critical bands" (1 critical band – 1 mm length of basilar membrane). Critical bandwidth is defined by Greenwood (1961) as the frequency interval over which the basilar membrane is effective in spatially integrating the energy of vibration. Eldredge (1974) in his review of the cochlear mechanics of the mammalian ear, suggested that existing data closely agree with Greenwood's (1961) initial assumptions. In *Choloepus*, the range of hearing is considered to be between 0.3-30 kHz (Suga, 1967; Peterson and Heaton, 1968).

By substituting f = 30 kHz, the upper limit of the range of hearing in the sloth, into the position/frequency equation and solving for the constant A (with ax = 0.124 x 17 = 2.108) the equation is obtained for the sloth:

$$f = 202.8\ (10^{0.124x} - 1) \qquad (2)$$

It is now possible to suggest the theoretical position along the basilar membrane x, where the maximum amplitude occurs for frequencies within the range of hearing of the two-toed sloth. In particular, for the frequency of 1.4 kHz, the value x = 7.2 mm (or 17.5-7.2 = 10.3 mm from the basal end of the cochlea) is obtained from the equation.

The curve in Fig. 16 demonstrates the theoretical position of maximum amplitude of vibration on the basilar membrane of the frequencies within the range of hearing of the sloth. Frequencies of 0.5-1.5 kHz are represented within 10.2-13.0 mm from the basal end of the cochlea, i.e., the region where morphological discontinuities occur (see Figs. 12 and 13). It is perhaps significant that contact-distress calls of young two-toed sloths were 1.4 kHz (Montgomery and Sunquist, 1974) and that these frequencies are represented within the region of maximum sensitivity of the basilar membrane. These findings suggest the possibility that a fine-tuning mechanism occurs within the basilar membrane itself in the sloth, as has been suggested in bats (Bruns, 1976b; Ramprashad et al., 1979).

Dallos (1970, 1973) has suggested that the larger the size of the helicotrema, the better is the pressure equilibration between the scala vestibuli and scala tympani, since the pressure generated by the travelling waves would fade out earlier as they approach a large helicotrema. The slope of the sensitivity curve became steeper towards the lower frequency limit of hearing as the

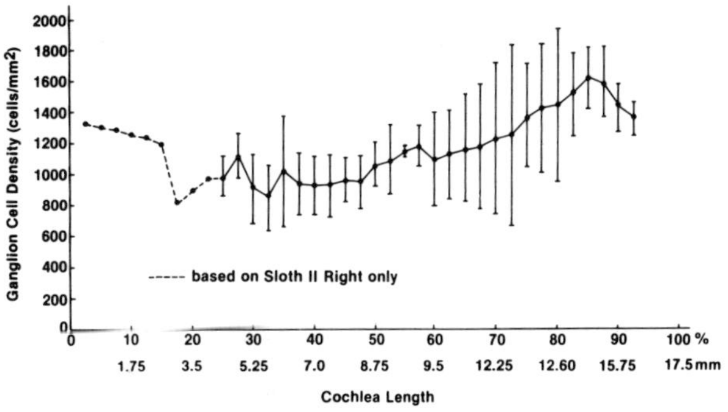

Figure 15. Density of the bipolar ganglion cells along the length of the cochlea. Density is expressed as the mean number of cells per unit cross-sectional area of the spiral ganglion (Ramprashad, 1976; Ramprashad et al., 1979), measured from three temporal bones. Vertical lines show SD. Note; (i) the decrease in density from the base, (ii) the maximum density occurs approximately 15.7 mm from the basal end.

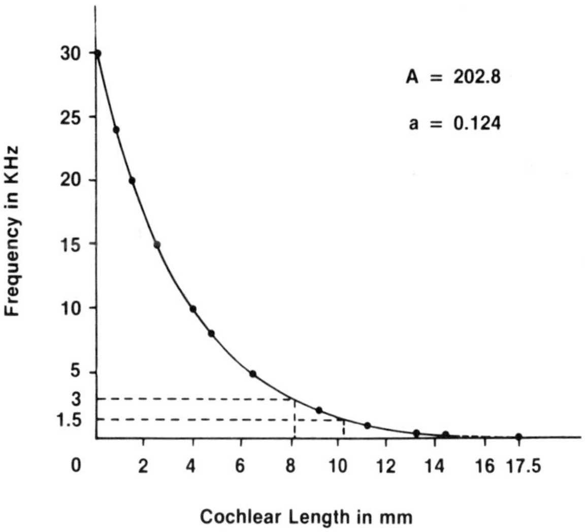

Figure 16. Theoretical frequency map along the cochlear length using Greenwood's (1961) formula for the two-toed sloth. Note; (i) the frequencies below 1.5 kHz are represented between 10.1-13.0 mm from the basal end, (ii) this region coincides with the areas of maximal thickness and cross-sectional area of the pars tecta and pectinata as shown in Fig. 11 and 12, respectively, and within the flat region demonstrating relative stiffness of the basilar membrane in Fig. 13.

helicotrema area increased (Dallos, 1970). In other words, the larger the helicotrema, the lower the low frequency sensitivity of the cochlea. The helicotrema measures 0.325 mm^2 in man (Békésy and Rosenbluth, 1951), 0.152 mm^2 in the cat (Dallos, 1970), and 0.022 mm^2 in the guinea pig (Dallos, 1970). Ehret (1977) showed that the absolute threshold curves in the guinea pig were less steep than in mammals with larger helicotremas (i.e., man and cat). The area of the helicotrema of the sloth was estimated to be 0.25 mm^2, which is as large as in the guinea pig, the cat, and the house mouse, but not as large as that in man. Peterson and Heston

Table 3: A comparison of cell soma widths and lengths of the vestibular and spiral ganglion, and body mass of representative mammals. Applicable entries are from Alexander (1901), Jones and Spells (1963), Ramprashad et al. (1979), and this study.

Order	*Species*	*Mass (kg)*	*Vestibular Ganglion Cell Soma (μm)*	*Spiral Ganglion Cell Soma (μm)*
Homo	Man *Homo sapiens*	70	19 x 27	13 x 15
Artiodactyla	Sheep *Ovis aries*	48	22 x 32	11 x 16
Carnivora	Dog *Canis familiaris*	14	19 x 27	8 x 13
	Cat *Felis domestica*	3.30	16 x 24	9 x 19
Insectivora	Mole *Talpa europaea*	0.51	14 x 16	8 x 11
	Hedgehog *Erinaceus europaeus*	0.03	19 x 27	8 x 14
Lagamorpha	Rabbit *Oryctolagus cuniculus*	0.44	16 x 21	10 x 16
Rodentia	Guinea Pig *Cavia cobaya*	0.25	19 x 27	11 x 16
	House Mouse *Mus musculus*	0.02	11 x 16	6 x 8
	Norway Rat *Rattus rattus*	0.21	16 x 24	8 x 13
Xenarthra	Two toed Sloth *Choloepus* sp.	3		13 x 17
Chiroptera	Horseshoe Bat *Rhinolophus hipposideros*	0.01	8 x 13	5 x 9
	Little Brown Bat *Myotis lucifugus*	0.01	8 x 10	6 x 7

(1968) have shown that the sloth's hearing range is significantly less sensitive than it is in the guinea pig over the entire range of frequencies and requires 10-100 times the sound pressure to produce the standard response. The highest frequency for which the cochlear-microphonic response could be measured was an octave below that of the guinea pig.

The morphometric parameters studied within the sloth's cochlea correlate with the psychoacoustic data. Of significance is the correlation of the frequencies to which the sloth is maximally sensitive with a region of morphological discontinuity along the basilar membrane.

GANGLION CELL DENSITY

The mammalian cochlea is innervated by afferent and efferent nerve systems. The ganglion cell soma of both systems are localized within a bony canal (Rosenthal canal) which spirals around the modiolus (Figs. 3, 4). The afferent ganglion nerve cell is bipolar, with its dendritic end extending outwards to innervate the inner and outer hair cells and its axonal extending centrally to form the cochlear nerve (Bodian, 1962), (Figs. 3, 4). Spoendlin (1968; 1972) has shown that 90 percent to 95 percent of the ganglion cells innervate the inner hair cells while 5 percent to 10 percent of the remaining ganglion cells innervate the outer hair cells in the cat and the guinea pig.

The density of the bipolar ganglion cells in the cochlea of the two-toed sloth decreases from the base where there are 1,300 cells/mm^2 to 3.0 mm from the basal end where there are 800 cells/mm^2. The density then increases gradually, attaining a maximum of 1,600 cells/mm^2 at 13.5 mm from the basal end. The maximum density occurs within the region associated with low frequency vibration, i.e., at the apical region which is also the region where morphological discontinuity of the basilar membrane occurs.

The total numbers of bipolar ganglion cells have been estimated in some mammals; the two-toed sloth has the smallest number of ganglion cells of those studied (Table 2). The total number of ganglion cells is not correlated with body size but rather with the frequency range of hearing in the mammal. Those mammals which utilize high frequencies within their range of hearing have a greater number of ganglion cells as compared to the human. In the sloth, the cell somas measure an average of 13(10-15) μm on its short axis and an average of 17(13-22) μm on its long axis (Table 3). The dimensions of the ganglion-cell soma of the sloth are similar to those in the human despite the apparent disparity in body mass. Cochlear bipolar ganglion cell somas are smaller than those of the vestibular bipolar ganglion cell somas (Table 3).

ACKNOWLEDGMENTS

We thank S. Ramprashad who made the figures, B. Matthews for technical assistance, R. Cardin for editorial comments, and M. Snowdon for typing the manuscript. This work is part of D.C.I.E.M. Research Contract 3278008/2SU78 00209, and represents D.C.I.E.M. Publication 79P30.

LITERATURE CITED

Alexander, G.
1901. Zur Anatomie des Ganglion vestibulare der Saugerthiere. Arch. Ohr. Nas. -Kehlk. Heilk., 51:109-125.

Bodian, D.
1962. The generalized vertebrate neuron. Science 137: 323-326.

Békésy, G. von
1941. Uber die Elastizität der Schnecken trennwand des ohres Akusl. Zeitz., 6:265-278. Translated by E. V. Wever and reprinted. Pp. 469-480 *in* G. Von Békésy, Experiments in Hearing. McGraw Hill, New York, 1960. 745 pp.

1948. On the elasticity of the cochlear parition. J. Acoust. Soc. Am. 20:227-241.

1956. Current status of theories of hearing. Science 123: 779-783.

Békésy, G. von and W. A. Rosenblith
1951. The mechanical properties of the ear. Pp 1085 *in*: Handb. exp. psychol. (S. S. Stevens, ed.), J. Wiley and Sons, New York, 1436 pp.

Bredberg, G.
1968. Cochlear structure and hearing in man, Pp 127-142 *in* Hearing mechanisms in vertebrates (A. V. S. de Reuck and J. Knight, eds.) Churchill, London, 309 pp.

Bruns, V.
1976a. Peripheral auditory tuning for fine frequency analysis by the CF-FM bat *Rhinolophus ferrumequinum*. I. Mechanical specializations of the cochlea. J. Comp. Physiol., 106: 77-86.

1976b. Peripheral auditory tuning for fine frequency analysis by the CF-FM bat *Rhinolophus ferrumequinum*. II. Frequency mapping in the cochlea. J. Comp. Physiol., 106: 87-97.

Cabezudo, L. M.
1978. The ultrastructure of the basilar membrane in the cat. Acta Otolaryngol., 86: 169-175.

Dallos, P.
1970. Low-frequency auditory characteristics: species dependence. J. Acoust. Soc. Amer., 48: 489-499.

Dallos, P.
1973. The auditory periphery. Acad. Press, New York. 548 pp.

Demanez, J. P., M. Goffart and A. Ledoux
1968. La fonction vestibulaire chez l'Unau (*Choloepus hoffmani* Peters). Arch. Int. Physiol. Bioch., 76: 253-267.

Ehret, G.
1977. Quantitative analysis of nerve fibre densities in the cochlea of the house mouse (*Mus musculus*). J. Comp. Neur., 183: 73-88.

Ehret, G., and M. Frankenreiter
1977. Quantitative analysis of cochlear structures in the house mouse in relation to mechanisms of acoustical information processing. J. Comp. Physiol., 122: 65-85.

Eldredge, D. H.
1974. Inner ear cochlear mechanics and cochlear potentials. Pp. 549-584 *in* Handbook of Sensory Physiology. (W. D. Keidel and W. Neff, eds.) Springer-Verlag, New York, 736 pp.

Evans, W. E.
1973. Echolocation by marine delphids and one species of freshwater dolphin. J. Acoust. Soc. Am.,54: 191-199.

Fernandez, C.
1952. Dimensions of the cochlea (guinea pig). J. Acoust. Soc. Am., 24: 519-523.

Goffart, M.
1971. Sensory apparatus, Hearing, Pp 107-109. *in* Function and form in the sloth. Pergamon Press, Toronto, 225 pp.

Greenwood, D.
1961. Critical bandwidth and the frequency coordinates of the basilar membrane. J. Acoust. Soc. Am., 33: 1344-1356.

Grinnell, A. D.
1963. The neurophysiology of audition in bats: Intensity and frequency parameters. J. Physiol., 167: 38-66.

Guild, S. R.
1921. A graphic reconstruction method for the study of the organ of Corti. Anat. Rec., 22: 141-157.

1927. The width of the basilar membrane. Science 65: 67-69.

Hall, J. G.
1969. The cochlea and cochlear nuclei in the bat. Acta Otolaryng., 67: 490-500.

Henson, M.
1978. The basilar membrane of the bat, *Pteronotus p. parnellii*. Am. J. Anat., 153: 143-158.

Howe, H.
1935. The reaction of the cochlear nerve to destruction of its end organs. A study of deaf albino cats. J. Comp. Neurol., 62:73-79.

Igarashi, M.
1966. A standard technique for temporal bone preparation. Monogr. No. 13, Naval Aerospace Medical Institute, Pensacola, Florida, 43 pp.

Jones, G. M. and K. E. Spells.
1963. A theoretical and comparative study of the functional dependence of the semicircular canal upon its physical dimensions. Proc. Roy. Soc., B., 157: 403-419.

Møhl, B.
1968. Hearing in seals. Pp. 172-193, *in* The behaviour and physiology of pinnipeds. (R. J. Harrison, R. C. Hubbard, R. S. Peterson, C. E. Rice and R. S. Schusterman, eds.). Appleton Century Crofts, New York, 411 pp.

Montgomery, G. G. and M. E. Sunquist.
1974. Contact-distress calls of young sloths. J. Mamm. 55: 211-213.

Neubert, K. and E. Wustenfeld.
1926. Morphologie des akustichen organs. Handbuch der zoologie, 8. Lief 29, 1-44.

Peterson, E. A. and W. C. Heaton
1968. Peripheral auditory responses in representative edentates. J. Auditory Res., 8: 171-184.

Pye, A.
1965. The auditory apparatus of the Heteromyidae (Rodentia, Sciuromorpha). J. Anat., 99: 161-174.

1966a. The structure of the cochlea in chiroptera. I. Microchiroptera: Emballonuroidea and Rhinolophoidea. J. Morph., 118: 495-510.

1966b. The Megachiroptera and Vespertillionoidea of the Michrochiroptera. J. Morph., 119: 101-120.

1967. The structure of the cochlea in chiroptera. III. Microchiroptera: Phyllostomatoidea. J. Morph., 121: 241-254.

1979. The structure of the cochlea in some mammals. J. Zool. Lond. 187: 39-53.

Ramprashad, F., S. Corey, and K. Ronald
1972. The anatomy of the seal's ear (*Pagophilus groenlandicus*, Erxleben, 1777). Pp 264-305, *in* Functional anatomy of marine mammals. Vol. I. (R. J. Harrison, ed.) Academic Press, New York. 451 pp.

1976. Population and density of the bipolar ganglion cells in the cochlea of the harp seal *Pagophilus groenlandicus*, Erxleben, 1777. Can. J. Zool., 54: 1918-1926.

Ramprashad, F., K. E. Money, J. P. Landolt, and J. Laufer
1978. A neuroanatomical study of the cochlea of the little brown bat (*Myotis lucifugus*). J. Comp. Neur., 178: 347-363.

Ramprashad, F., J. P. Landolt, K. E. Money, D. Clark and J. Laufer
1979. Morphometric study of the cochlea of the little brown bat (*Myotis lucifugus*). J. Morph., 160: 345-358.

Ramprashad, F., K.E. Money, J.P. Landolt and J. Laufer.
1982. A morphometric study of the rabbit's cochlea. J. Morph. (in press)

Schuknecht, H. G.
1953. Techniques for study of cochlear function and pathology in experimental animals. Arch. Otolaryng., 58: 377-397.

Spoendlin, H.
1968. Innervation patterns in the Organ of Corti of the cat. Acta Otolaryngol., 67: 239-254.

1972. Retrograde degeneration of the cochlea nerve. Acta Otolaryngol., 79: 266-275.

Suga, N.
1967. Hearing in some arboreal edentates in terms of cochlear microphonics and neural activity. J. Auditory Res., 7: 267-270.

Terhune, J. M. and K. Ronald
1971. The harp seal, *Pagophilus groenlandicus* (Erxleben, 1777). X. The air audiogram. Can. J. Zool., 49: 385-390.

1972. The harp seal, *Pagophilus groenlandicus* (Erxleben, 1777). III. The underwater audiogram. Can. J. Zool., 50: 565-569.

Weber, M.
1928. Die Saugetiere. 2nd ed. Gut Fischer, Verlag Gena. 806 pp.

Webster, D. B. and M. Webster
1977. Auditory systems of the Heteromyidae: Cochlear diversity. J. Morph., 152: 153-170.

Wever, E. G.
1938. The width of the basilar membrane in man. Ann. Otol. Rhinol. Laryngol., 47: 37-47.

1949. Theory of hearing. John Wiley and Sons Inc., New York. 484 pp.

Wever, C. G., J. McCormick, J. Palin, and S. H. Ridgway
1971a. Cochlea of the dolphin, *Tursiops truncatus*: The basilar membrane. Proc. Nat. Acad. Sci. (U.S.A.), 68: 2708-2711.

1971b. The cochlea of the dolphin, *Tursiops truncatus*. Hair cells and ganglion cells. Proc. Nat. Acad. Sci. (U.S.A.), 68: 2908-2912.

1971c. The cochlea of the dolphin, *Tursiops truncatus*. General morphology. Proc. Nat. Acad. Sci. (U.S.A.), 68:2381-2385.

1972. Cochlear structure in dolphin, *Lagenorhynchus obliquidens*. Proc. Nat. Acad. Sci. (U.S.A.), 69: 657-661.

Zwislocki, J. J.
1975. Phase opposition between inner and outer hair cells and auditory sound analysis. Audiology, 14: 443-455.

SLOTH HAIR: UNANSWERED QUESTIONS

Annette Aiello

Smithsonian Tropical Research Institute, P.O. Box 2072, Balboa, Panama

Abstract
Hair of tree sloths *Bradypus* and *Choloepus* differs in form and structure from that of all other mammals and may be specialized to accomodate the unicellular algae which grow in it. Conjectures are made concerning the possible relationship(s) between sloths and their algae.

Abstracto
El pelo de los perezosos arbóreos *Bradypus* y *Choloepus* difiere en forma y estructura del de todos los otros mamíferos y puede ser especializado para acomodar la algas unicelulares que crecen en él. Se hacen conjeturas respecto a las posibles relaciones entre los perezosos y sus algas.

Resumo
O pelo de preguiças *Bradypus* e *Choloepus* difere em forma e estrutura do pelo de todos od demais mamíferos e pode ser especializado para acomodar algas unicelulares que nele crescem. Conjecturas são feitas a respeito da possível ou possíveis relações entre preguiças e suas algas.

Among the many odd features of tree sloths *Bradypus* and *Choloepus* perhaps the oddest of all is their hair which, with its algal infestation and peculiar structure, is unlike the hair of any other mammal. Algae have been noted in or on the hairs of several other mammals, most notably inside the hairs of polar bears in captivity (Lewin and Robinson, 1979), but have not been found in wild polar bears, nor do they occur commonly in the few other mammals in which they have been observed.

Green-tinged hair is a consistent feature of 3-toed sloths *Bradypus* living under natural conditions, and the algae responsible for the color are already present on the hairs of animals a few weeks old (Britton, 1941). A number of unrelated alga species have been implicated; Thompson (1972) cultured algae representing four phyla (Chlorophyta, Chrysophyta, Cyanophyta, Rhodophyta) from the hair of *Bradypus*. Although no proof has been provided nor alternatives explored, most biologists assume that the sloth/alga association is of mutual benefit, the alga obtaining shelter and the sloth gaining camouflage. A 3-toed sloth with dry hair appears brownish grey, while a wet animal takes on a green tinge (Beebe, 1926).

HAIR STRUCTURE

Ridewood (1901) described the hair anatomy of eleven living and one extinct species of xenarthrans: *Bradypus tridactylus, Choloepus didactylus, Myrmecophaga tridactyla* (as *M. jubata*), *Tamandua tetradactyla, Cyclopes didactylus* (as *Cyclothurus*), *Chlamyophorus truncatus* (as *Chlamydophorus*), *Euphractus sexcinctus* (as *Dasypus*), *Chaetophractus villosus* (as *Dasypus*), *Tolypeutes matacus* (as *T. conurus*), *Dasypus novemcinctus* (as *Tatusia novemcincta*), *Dasypus pilosus* (as *Tatusia pilosa*), and *Mylodon listai* (extinct).

Typical mammalian hair and that of many xenarthrans, is composed of a central medulla surrounded by a cortex that contains pigment granules, and is enclosed within a thin, fairly uniform, scaly cuticle. The hair structures (cuticle, cortex, and medulla) can be distinguished by selective staining. However, *Bradypus* and *Choloepus* hairs have neither medulla nor pigment granules.

Not only do the hair structures of *Bradypus* and *Choloepus* differ from those of other mammals (other xenarthrans included), but the hairs of the two genera differ as well (Fig. 1). The most obvious external characteristic of *Bradypus* hair is its transversely cracked surface, and in *Choloepus* the striking feature is the fluted appearance due to 8 to 11 longitudinal ribs. *Bradypus* has both over-hair (guard hairs) that harbors algae, and soft underfur which is algae-free. *Choloepus* does not have underfur. In this paper, "hair" will refer to over-hair unless otherwise noted.

Bradypus tridactylus hairs are oval in cross-section and consist of a solid cortex surrounded by a thick extra-cortical layer which harbours algae (Ridewood, 1901). Ridewood interpreted the extra-cortex as "...the cuticle itself, enormously thickened and distinctly cellular, instead of more or less homogeneous and structureless.....a layer not represented, or at least not in this form, in the hair of any other mammal."

The hairs of *Choloepus didactylus* consist mainly of cortex, permeated by irregular branching tubes that may represent a diffuse medulla (Ridewood. 1901). The cuticle enclosing the cortex is thickest at the summits of the above-mentioned ribs, but becomes thinner on the sides and disappears at the bottoms of the grooves except for intermittent cuticular bridges joining rib bases. Within the grooves are strands of extra-cortex in which algae live, and it is Ridewood's view that, as in *Bradypus*, this "extra-cortex" is an elaboration of the cuticle itself.

ALGAE

While many accounts of *Bradypus* hair are vague as to the location of algae, or imply that the colonies are located on the hair surface, Sonntag and Duncan (1922) reported that the algae of *Bradypus tridactylus* are found between the cuticular scales. My observations agree with theirs. In spite of the illusion that the algae rest on the hair surface, they are invisible under the scanning electron microscope, an instrument sensitive to surface structure rather than to color contrast.

Grassé (1955) states that the algae are located in the transverse cracks of *Bradypus* hairs. Hairs which I examined (Table 1) had algae growing right up to the edges of the large cracks, but the tissue within the cracks was pale and appeared to be free of algae. That algae do not inhabit these fissures is not surprising; wetted hairs soak up water as does a sponge, swell slightly, and the cracks close. Algae within a closed crack would have little access to light.

My observations indicate that hairs of *B. variegatus* pass through a distinct age sequence (Fig. 1). Young hairs were white, gray, brownish, or black, and lacked the deep transverse cracks of older hairs. The first traces of algae were seen on these young hairs as tiny dots or very narrow transverse green lines. Older hairs had larger, wider algal colonies and obvious transverse cracks. In apparently older hairs, the cracks were deeper, and the cortical layer could be seen. Upon wetting, even the larger cracks closed considerably. When dry, such hairs gave the effect of beads on a string. The oldest hairs were badly deteriorated; the spongy cuticle had worn off on one side exposing the full length of the cortex. These very old hairs did not appear to have living algae, indeed, only young and middle-aged hairs supported healthy algal colonies. Either algae colonize very narrow cracks in young hairs, or the algae themselves initiate the cracks. Once a crack exceeds a certain depth, algae cease to grow in it. In either case, perhaps the algae hasten the deterioration of *Bradypus* hairs.

The hair of the other *Bradypus* species was similar. *B. tridactylus* hair which I studied was not prominently cracked and may have been from a young or newly molted individual. Hair of *B. torquatus* was essentially like that of *B. variegatus* (Fig. 1). The hair of all three *Bradypus* species readily absorbed water in all parts including the cortex, and the surface of *B. tridactylus* hairs became mucilaginous and sticky when wet.

Choloepus hairs were also wettable, cortex included, but they did not swell when wet. It was not clear whether in *Choloepus* the algae rested upon the surface of the spongy cuticle in the grooves, or were embedded in it. In either case the algae were confined to the

Table 1. Origin of specimens from which hair was obtained. GML refers to Gorgas Memorial Laboratory, Panama. NMNH refers to the National Museum of Natural History, Smithsonian Institution, Washington, D.C.

Species	*Country of Origin*	*Collection Data*
Bradypus variegatus Schinz	Panama, Ancon	29 June 1982; A. Aiello and G. Montgomery.
Bradypus torquatus Desmarest	Brasil, Bahia, nr. Ilheus	NMNH (US 259473).
Bradypus tridactylus L.	Guyana	NMNH (US 256676).
Cabassous centralis (Miller)	Panama	GML
Choloepus didactylus (L.)	Venezuela, T.F. Amazonas, Belen	NMNH (US 388206).
Choloepus hoffmanni Peters	Panama, Arraijan	Dead on Road, Dec. 1982; A. Aiello and R. Cortez.
Cyclopes didactylus (L.)	Panama, Barro Colorado Island	1983; Bonafacio de Leon.
Dasypus novemcinctus L.	Panama	GML
Myrmecophaga tridactyla L.	Panama	GML
Tamandua mexicana (Saussure)	Panama, Cerro Galera	Dead on Road; 1 Dec. 1982; A. Aiello.
Tamandua mexicana (Saussure)	Panama	GML

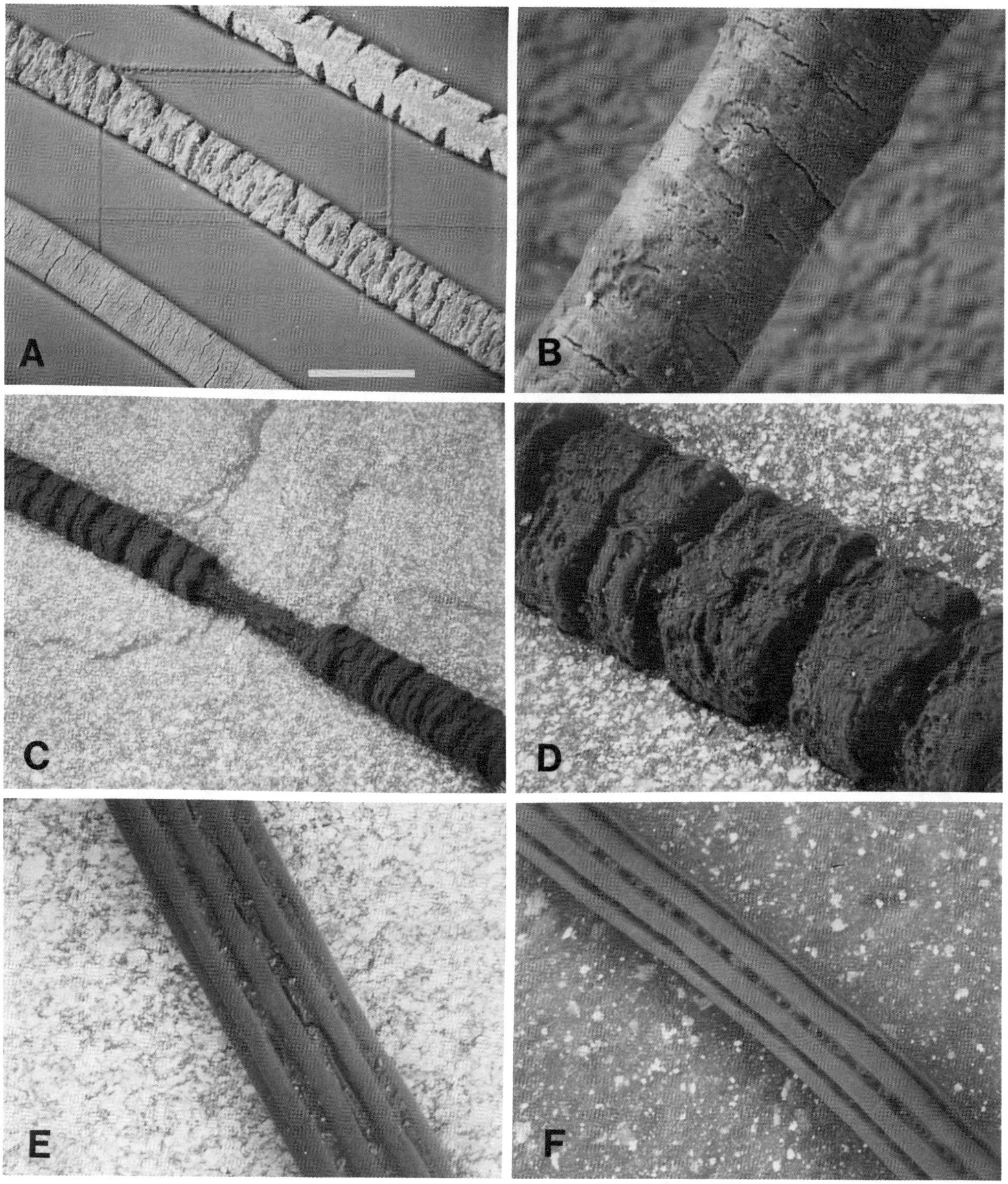

Figure 1. Hair of tree sloths. A. *Bradypus variegatus* Schinz. Three hairs at different stages of deterioration are shown: bottom hair is young and has a few narrow transverse cracks; center hair is older and has larger more numerous cracks; top hair is very old, has deep cracks, and the full length of its cortex is exposed. Scale bar = 0.60 mm. B. *Bradypus tridactylus* L. Hair showing moderate transverse cracking. Scale bar = 0.10mm. C and D. *Bradypus torquatus* Desmarest. Hair has deep transverse cracks and in C a portion of the cortex is partly exposed. Scale bars = 0.40mm and 0.10mm respectively. E. *Choloepus hoffmanni* Peters. Hair with five longitudinal ribs showing. Each rib is enclosed within a scaly cuticle. Scale bar = 0.10mm. F. *Choloepus didactylus* (L.). Hair with four longitudinal ribs showing. Each rib is enclosed within a scaly cuticle which extends across the grooves at frequent intervals and joins the bases of adjacent ribs. Scale bar = 0.10mm.

grooves, and it appears that the hair is better designed than that of *Bradypus* to control the extent of algal colonization.

DISCUSSION

The hair of *Bradypus* and *Choloepus* is apparently specialized to encourage a symbiotic relationship with algae. The hair structures which harbor algae are different, but possibly homologous; no other mammals have either comparable hair structure or algae. Nevertheless, the algae may be accidental guests, or they may be sloth hair parasites which take advantage of the sheltered humidity beneath the hair cuticle and also benefit from the sunbathing activity of their hosts.

If the algae are parasites, then why do sloths have spongy hair that soaks up water and encourages their colonization? Thermal insulation has been suggested (group discussions at STRI). Sloths have difficulty maintaining an even body temperature; exposure to heat raises, and to cold lowers a sloth's temperature (Britton and Atkinson, 1938; Montgomery and Sunquist, 1978; McNab, 1978, 1985).

Certainly, thick fur would be of insulative value to such an animal, but while I agree that the fine fluffy underfur of *Bradypus* probably serves this purpose, I question that spongy, water-absorbing hair would serve as insulation, at least in wet season; more likely it promotes evaporative cooling, and is thus detrimental.

It seems more likely that sloth hair has evolved its unique structure to encourage algal colonization. If so, why?

Camouflage is the reason usually given. But while a green-tinged sloth is perhaps better hidden, the green color is present only during rainy periods, and any sloth is very well concealed by screening vegetation, regardless of its body color.

Would an animal develop such specialized hair for the purpose of half-time improvement of its already excellent camouflage?

The sloth may obtain nutrition or a particular trace element from the algae. Indeed, many leaf-eating animals have difficulty obtaining sufficient nitrogen and it is difficult to explain how the sloth manages to collect enough of this element. It has been suggested (Hoke, 1976) that sloths may prevent certain dietary deficiencies by inadvertently partaking of an occasional ant along with their usual leaves. There are at least two possible methods by which a nutrition transfer from alga to sloth might take place.

Nutrients might be transferred by the sloth licking its hair. While many mammals lick their hair, sloths have not been observed to engage in this activity (Tirler, 1966). Even if they do, it certainly is not a frequent event and nutrition transfer in this way is unlikely.

The nutrients might move into the sloth through the hair itself. Unlike the hair of other mammals, sloth hair (spongy cuticle, and cortex) soaks up water, and even if only one portion is in contact with water, after some minutes the entire length of the hair is moist. Chemicals might diffuse along the hair to the skin surface, and then be absorbed through the skin. The spongy hair structure would promote this process. An analogy can be made to orchid roots, which absorb water and minerals through an outer layer (velamen) of spongy, algae-harbouring epidermal cells. See Dressler (1981) for references about orchid roots.

Dependence upon either algae or an occassional ant or both for trace nutrients might help explain why *Bradypus* does not survive long in captivity; if the algae are not healthy or if, as discussed by Hoke (1976), ants are not available, the sloth would eventually die. *Choloepus*, with its less extensive algal colonies and absence of underfur, is much more active than *Bradypus*. From what can be inferred from its food choices in captivity, it has a more varied diet than *Bradypus*. If sloths do depend upon their algae for proper nutrition *Choloepus* would probably be less dependent on algae than *Bradypus*.

The hair of at least two species of extinct ground sloths was not specialized. Ridewood (1901) examined hairs of *Mylodon listai* from Patagonia and found that they lacked a medullar layer, had a thick scaly cuticle, and showed no trace of a spongy cuticle. They did not at all resemble hairs of tree sloths but instead were very similar to those of modern *Tamandua* and *Daspus*. Ridewood did not mention anything about algae or algal-like bodies. Hausman (1929, 1936) reported that *Nothrotherium shastense* did not have underfur, and that the hairs had medulla (sometimes double), cortex, pigment granules, and scaly cuticle, and "...did not seem to be fundamentally different from those of... modern animals." Alga-like "ovate bodies" were scattered on the surface of these hairs. From the description and illustrations these bodies were solitary, not in algal colonies as in *Bradypus* and *Choloepus*. It is not at all clear that they are the remains of algae, but could be some other form of unicellular life.

SUMMARY

The tree sloths *Bradypus* and *Choloepus* have hair that is strikingly different in form and structure from hair of other xenarthrans (extinct or extant) and other mammals, and is probably specialized to encourage colonization by various unicellular algae. Whether the sloth gains camouflage or nutrition or both from its algae is not yet clear. Certainly this intriguing relationship deserves much more attention than it has received.

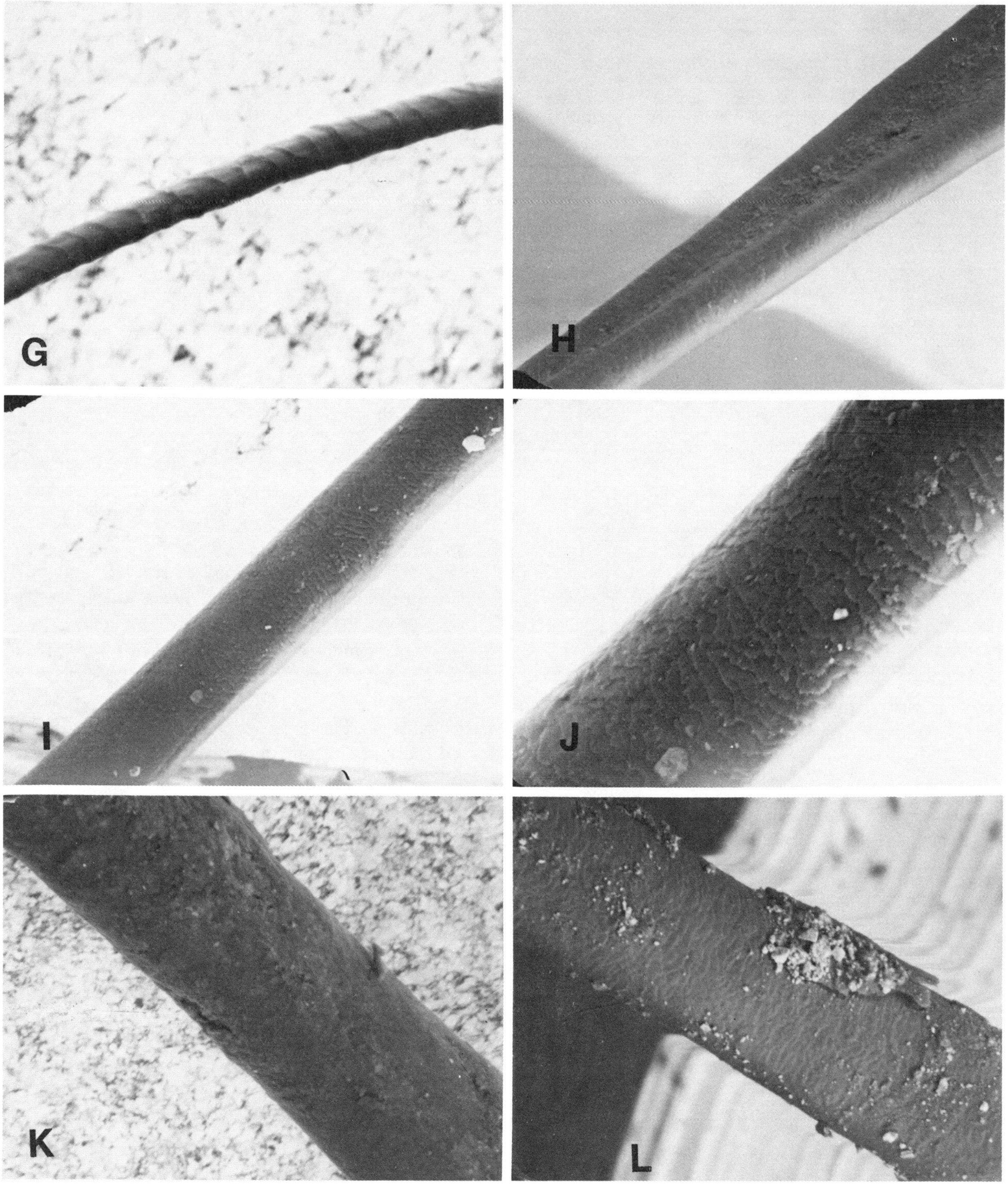

Figure 2. Hair of vermilinguas and armadillos. G. *Cyclopes didactylus* (L.). The hair is much finer than that of other xenarthrans and the cuticle scales are relatively few and large. Scale bar = 0.04mm. H. *Myrmecophaga tridactyla* L. The hairs are compressed and have a broad longitudinal groove on each side. Scale bar = 0.10mm. I and J. *Tamandua mexicana* (Saussure). The scales of the cuticle overlap closely and have ragged edges. Scale bar = 0.10mm and 0.04mm respectively. K. *Cabassous centralis* (Miller). The hairs are covered with debris, but the cuticle scales show through in several places. Scale bar = 0.10mm. L. *Dasypus novemcinctus* L. The hairs have bits of loose debris, but the closely overlapping cuticle scales show clearly. Scale bar = 0.10mm.

ACKNOWLEDGMENTS

I am grateful to G. Gene Montgomery for his encouragement and advice during this project, and to Estorgio Mendez of Gorgas Memorial Laboratory, Panama, Dr. Charles Handley, U.S. National Museum, Washington, D.C., and Patricia and Michael Fogden for their generous donations of hair samples. G. Gene Montgomery and Donald Windsor read the manuscript. Colonel Richard Barrere, U.S. Army, Tropic Test Center, Panama, allowed use of the scanning electron microscope, and the library of the Smithsonian Tropical Research Institute obtained copies of literature not available in Panama.

LITERATURE CITED

Beebe, W.
1926. The three-toed sloth *Bradypus cuculliger cuculliger* Wagler. Zoologica, 7: 1-67.

Britton, S.W.
1941. Form and function in the sloth. Quart. Rev. Biol., 16(1) 13-34.

Britton, S.W. and W.E. Atkinson.
1938. Poikilothermism in the sloth. J. Mamm., 19(1): 94-99.

Dressler, R.L.
1981. The orchids: natural history and classification. Harvard University Press, Cambridge, Massachusetts, 332 pp.

Grassé, P.P.
1955. Ordre des Edentates. Pp. 1182-1246 *In*, Traité de zoologie anatomie, systématique, biologie. Mamminféres. (P.P. Grassé, editor). L'académie de Médecine Edituers, Paris. 17(2): 1171-2300.

Hausman, L.A.
1929. The "ovate bodies" of the hair of *Nothrotherium shastense*. Am. J. Sci., (ser.5)18(106): 331-333.

1936. Further studies of the hair of the fossil ground sloth (*Nothrotherium shastense*), and of its problematical "ovate bodies." Am. J. Sci., (ser.5)31(181):223-228.

Hoke, J.
1976. Discovering the world of the three-toed sloth. Franklin Watts, New York. 88 pp.

Lewin, R.A., and P.T. Robinson.
1979. The greening of polar bears in zoos. Nature, 278(5703): 445-447.

McNab, B.K.
1978. Energetics of arboreal folivores: physiological problems and ecological consequences of feeding on an ubiquitous food supply. Pp. 153-162, *In* The ecology of arboreal folivores. (G. Gene Montgomery, editor). Smithsonian Institution Press, Washington, D.C. 574 pp.

McNab, B.K.
1985. Energetics, population biology, and distribution of xenarthrans, living and extinct. (this volume).

Montgomery, G.G. and M.E. Sunquist.
1978. Habitat selection and use by two-toed and three-toed sloths. Pp. 329 to 359, *In* The ecology of arboreal folivores. (G. Gene Montgomery, editor). Smithsonian Institution Press, Washington, D.C. 574 pp.

Ridewood, W.G.
1901. On the structure of the hairs of *Mylodon listai* and other South American Edentata. Quart. J. Micros. Soc. (n.s.), 44: 393-411.

Sonntag, C.F., and F.M. Duncan.
1922. Contributions to the histology of the three-toed sloth (*Bradypus tridactylus*). J. Royal Mic. Soc., :37-46.

Thompson, R.H.
1972. Algae from the air of the sloth *Bradypus*. Supplement to J. Phycol., 8:8.

Tirler, H.
1966. A sloth in the family. Walker and Co., New York, 47 pp.

ENERGETICS, POPULATION BIOLOGY, AND DISTRIBUTION OF XENARTHRANS, LIVING AND EXTINCT

Brian K. McNab

University of Florida, Gainesville, Florida 32611.

Abstract

Xenarthrans (Edentates) have low body temperatures and low basal rates of metabolism, and usually have high thermal conductances. Low basal rates are associated with burrowing (armadillos) and arboreal habits (tree sloths, some vermilinguas), and with consumption of certain kinds of foods (leaves of trees, ants, and termites). The combination of low basal rates and high conductances produces low body temperatures and high lower limits of thermoneutrality at a given body mass. Low rates of metabolism are important in setting the long gestation periods, low growth rates, extended periods of parental care, and small litter sizes of xenarthrans. These factors combine to make all xenarthrans K-selected, although some armadillos, such as *Dasypus*, may be somewhat more r-selected than other xenarthrans. Low rates of metabolism also have implications for the geographical limits to distribution found in xenarthrans. Among living species, only armadillos venture far into temperate environments. Some armadillos tolerate temperate winters by the use of fat stores, a response that is limited by high thermal conductances; the adaptation that permits the greatest penetration of temperate conditions is found in *Zaedyus* and couples a small mass with seasonal torpor. Anteaters and tree sloths are limited to tropical environments. Extinct xenarthrans can also be divided into those that were limited to tropical and warm-temperate environments (pampatheres and glyptodonts) and those that extended into cold-temperate environments (some ground sloths). The cold temperature tolerance of some ground sloths probably stemmed from a large mass, thick fur coat, and continuously available food. Some megatheriid sloths, however, were restricted to warm-temperate and tropical regions, in spite of large masses. The apparent temperature sensitivity of pampatheres and glyptodonts suggests that the combination of low rates of metabolism and high conductances negated any resistance to low temperatures potentially conferred by a large mass.

Abstracto

Los animales xenartranos (edentados) tienen temperaturas corporales y tasas de metabolismo bajas y, por lo general, poseen una alta conducción termal. Las bajas tasas de metabolismo están asociadas con los hábitos excavadores de los armadillos y con los hábitos arbóreos de perezosos y algunos osos hormigueros, como también con el consumo de algunas clases de alimentos, como hojas de árboles, hormigas y termitas. La combinación de bajas tasas metabólicas con altas conducciones termales producen temperaturas corporales bajas y altos límites inferiores de termoneutralidad por una masa corporal dada. Las bajas tasas metabólicas son importantes en la fijación de los largos períodos de gestación, las bajas tasas de crecimiento, los períodos extendidos de las crías bajo el cuidado paternal y del número reducido de crías típicos de los xenartranos. Estos factores se combinan para que todos los animales xenartranos aparezcan seleccionados hacia el factor K, aunque algunos armadillos, tales como *Dasypus* , están algo más seleccionados hacia el factor r que otros animales xenartranos. Las bajas tasas metabólicas también tienen implicaciones en los límites geográficos de distribución que se encuentran en los animales xenartranos. Entre las especies vivientes, sólo los armadillos se aventuran muy adentro de los ambientes de climas templados. Algunos armadillos toleran los inviernos templados por medio del uso de depósitos de grasa; una respuesta que está limitada por las altas conducciones termales. La adaptación que permite la mayor penetración dentro de condiciones de clima templado, se encuentra en el género *Zaedyus* y se trata de la existencia de una masa corporal pequeña aunada al letargo estacional. Los osos hormigueros y los peresozos arbóreos están restringidos a ambientes de climas templados. Los xenartranos extintos tambíen pueden ser divididos en aquellas cuya distribucíon estuvo limitada a ambientes de clima tropical o templado-caliente, como los pampateros y los gliptodontes, y los que extendieron su distribucíon hacia ambientes de clima templado-frío, como algunos peresozos terrestres. La tolerancia a temperaturas frías de algunos peresozos terrestres probablemente proviene de la existencia de una gran masa corporal, un denso pelaje y alimentacíon disponible en toda época. Algunos peresozos megatéridos, sin embargo, estuvieron restringidos a climas templados-calientes y a regiones tropicales, a pesar de su gran masa corporal. La aparente sensibilidad térmica de los pampateros y gliptodontes sugiere que la combinación de bajas tasas de metabolismo y altas conducciones térmicas en estos animales, se contraponían a cualquiera resistencia a temperaturas bajas que pudiera conferirles su gran masa corporal.

Resumo

Os xenarthros (edentatas) têm baixas temperaturas corpóreas e baixos níveis basais de metabolismo, e geralmente têm altas condutibilidades termais. Os baixos níveis basais são associados com os hábitos de se enterrar (tatus) e arbóreos (preguiças arbóreos e alguns tamanduás), e com o consumo de certos tipos de alimentos (folhas de árvores, formigas e térmites). A combinação de baixos níveis basais e alta condutibilidade produz baixas temperaturas corpóreas e altos limites mínimos de termoneutralidade a uma dada massa corpórea. Baixos níveis de metabolismo são importantes no estabelecimento de longos períodos de gestação, períodos longos de cuidados maternais e tamanhos pequenos de ninhadas de xenarthros. Estes fatores se combinam para tornar todos os xenarthros k-selecionados, embora alguns tatus, como o *Dasypus*, podem ser de alguma maneira mais r-selecionados do que outros xenarthros. Baixos níveis de metabolismo também acarretam implicações para os limites geográficos de distribuição dos xenarthros. Entre as espécies viventes, somente os tatus se aventuram mais longe em ambientes temperados. Alguns tatus toleram invernos de climas temperados através do uso de reservas de gordura, uma reposta que é limitada pela alta condutibilidade à adaptação que permite a maior penetração em condições temperadas é encontrada em *Zaedyus* combina uma massa pequena com um torpor estacional. Tamanduás e preguiças são limitados a ambientes tropicais. Os xenarthros extintos também podem ser divididos entre aqueles que são limitados a ambientes tropicais e temperados quentes (pampateros e gliptodontes) e aqueles que se estendiam a ambientes temperados frios (alguns preguiças terrestres). A tolerância a temperaturas frias de alguns preguiças terrertres provavelmente provinha de uma massa grande, coberturas de pelo espessas, e alimentos continuamente disponíveis. Alguns preguiças megaterídeos, no entanto, eram restritos a regiões tropicais e temperadas quentes, apesar das massas grandes. A aparente sensibilidade a temperaturas dos pampateros e gliptodontes sugere que a combinação de baixos níveis de metabolismo e altas condutibilidades negavam qualquer resistência a baixas temperaturas potencialmente conferidas por uma massa grande.

One of the most distinctive orders of mammals, living or extinct, is the Xenarthra (Edentata), which is limited in distribution to the New World. Few orders contain such a morphologically "strange" assembly as armadillos and pampatheres, glyptodonts, anteaters, and tree and ground sloths. Xenarthrans are equally distinctive in their physiology. The energetics of tree-sloths, armadillos, and anteaters have been described in detail elsewhere (McNab, 1978, 1980b, 1984): all living species are characterized by low body temperatures and low basal rates of metabolism.

The object of this paper is to give a comparative picture of the energetics of living xenarthrans and to place these data in an ecological and evolutionary context. Specifically, I want to raise the question why xenarthrans have such low rates of energy expenditure and what implications these low rates have for the population biology and distribution of xenarthrans, living and extinct.

ENERGETICS

Body Temperature.-- All xenarthrans have low body temperatures (mean 34.1°C, range 32.7° to 35.5°C; Table 1, Fig. 1), whereas most mammals regulate body temperature between 36° and 38°C. Most xenarthrans maintain a constant body temperature at ambient temperatures down to 5°C or 10°C, but the armadillo *Cabassous*, both genera of tree sloths, and the anteater *Cyclopes*, have body temperatures (T_b) that vary with ambient temperature (Fig. 2). *Bradypus, Choloepus*, and *Cabassous* have this dependency in spite of their appreciable masses (3.5 to 6.0 kg). The marked variation of T_b in *Cyclopes* is due in part to its small mass (0.24 kg); the only smaller xenarthran, the armadillo *Chlamyphorus truncatus*, weighs about 0.13 kg and lives underground in a closed burrow system (Minoprio, 1945). The only xenarthran known to enter a deep, reversible torpor (Roig, 1971) and hibernation (Hatcher, 1903) is the armadillo *Zaedyus pichiy*.

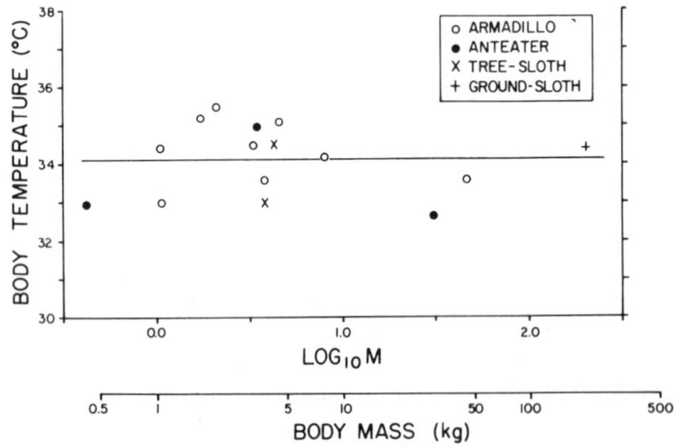

Figure 1. Body temperature of xenarthrans as a function of body mass. The body temperature of the ground sloth *Nothrotheriops* was estimated as indicated in the text. Mean body temperature for all species equals 34.1°C. Data from Table 1.

Basal Rates of Metabolism.-- Most xenarthrans have basal metabolic rates that fall between 40 percent and 60 percent of the rates expected from mass in Kleiber's (1932) relation for placental mammals (Table 1, Fig. 3). *Tolypeutes, Priodontes*, and *Myrmecophaga*, however, have basal rates between 29 percent and 33 percent of expected rates; these genera (2 armadillos and one anteater) are ant- and termite-eaters. Two other myrmecophagous xenarthrans, *Cyclopes* and *Tamandua*, have basal rates between 55 percent and 60 percent of expected rates, although the basal rate of *Tamandua* may fall to 48 percent in individuals free-living in the field (McNab, 1984). Other ant-eating mammals, e.g., echidnas (*Tachyglossus* and *Zaglossus*), have basal rates as low as *Tolypeutes*; others, such as pangolins *Manis* and the aardvark *Orycteropus* have moderate rates of metabolism, similar to those of *Cyclopes* and *Tamandua* (McNab, 1984).

Various factors contribute to the low basal rates of metabolism of xenarthrans, depending upon the species. First, the low basal rates in armadillos appear to be a correlate of their burrowing habits (McNab 1979), and represent a means of reducing heat storage during

Table 1. Parameters of energetics in xenarthrans. $f_m(\%) = 100\ M_b/3.42\ m^{-0.25}$; $f_c(\%) = 100C_m/1.00\ m^{-0.50}$; $F = f_m/f_c$. See text for method used to estimate the values for the extinct ground sloths *Mylodon listai* and *Nothrotheriops shastense*.

Family and Species	*Mass kg*	*Basal Rate $cm^3 O_2$/g.h*	f_m *(%)*	*Minimal Conductance $cm^3 O_2$/g.h°C*	f_c *(%)*	*F*	T_b *(°C)*	T_l *(°C)*	*ΔTl (°C)*	*Source*
Dasypodidae										
Chaetophractus vellerosus	1.11	0.276	47	0.053	177	0.27	34.4	30	4.4	McNab 1980a
Tolypeutes matacus	1.16	0.181	31	0.039	133	0.23	33.0	28	5.0	McNab 1980a
Zaedyus pichiy	1.74	0.226	43	0.027	113	0.38	35.2	28	7.2	McNab 1980a
Chaetophractus nationi	2.15	0.260	52	0.040	185	0.28	35.5	29	6.5	McNab 1980a
Dasypus novemcinctus	3.32	0.243	54	0.039	225	0.24	34.5	28	6.5	McNab 1980a
Cabassous centralis	3.81	0.213	49	0.033	204	0.24	33.6	27.5	6.1	McNab 1980a
Chaetophractus villosus	4.54	0.178	43	0.031	209	0.21	35.1	29.5	5.6	McNab 1980a
Euphractus sexcinctus	8.19	0.151	42	0.019	172	0.24	34.2	26	8.2	McNab 1980a
Priodontes maximus	45.19	0.067	29	0.010	213	0.14	33.6	27	6.6	McNab 1980a
Mylodontidae										
Mylodon listai	1000	0.054	50	0.0019	190	0.26	34.5	6	28.5	This study
Megalonychidae										
Choloepus hoffmanni	4.25	0.188	44	0.018	117	0.38	34.5	24	10.5	McNab 1978
Megatheriidae										
Bradypus griseus	3.79	0.181	42	0.012	74	0.57	33.0	18	15.0	McNab 1978
Nothrotheriops shastense	200	0.081	50	0.0059	264	0.19	34.4	21	13.7	This study
Myrmecophagidae										
Cyclopes didactylus	0.24	0.477	55	0.062	96	0.57	33.0	25	8.0	McNab, 1984
Tamandua tetradactyla	3.50	0.253	58	0.022	130	0.45	35.0	23	12.0	McNab, 1984
Myrmecophaga tridactyla	30.60	0.085	33	0.0054	94	0.35	32.7	15	17.7	McNab, 1984

burrowing activity. Second, the reduction in basal rate in some arboreal mammals is directly related to reduction in muscle mass, when muscle mass constitutes less than 33 percent of total body mass (McNab, 1978): tree sloths (Grand, 1978) have small muscle masses, a correlate of their sedentary habits. It is unclear whether this factor will completely account for the low basal rates of *Bradypus* and *Choloepus*, not to mention

Figure 2. Body temperature of xenarthrans as a function of ambient temperature. Data from McNab (1978, 1980b, 1984).

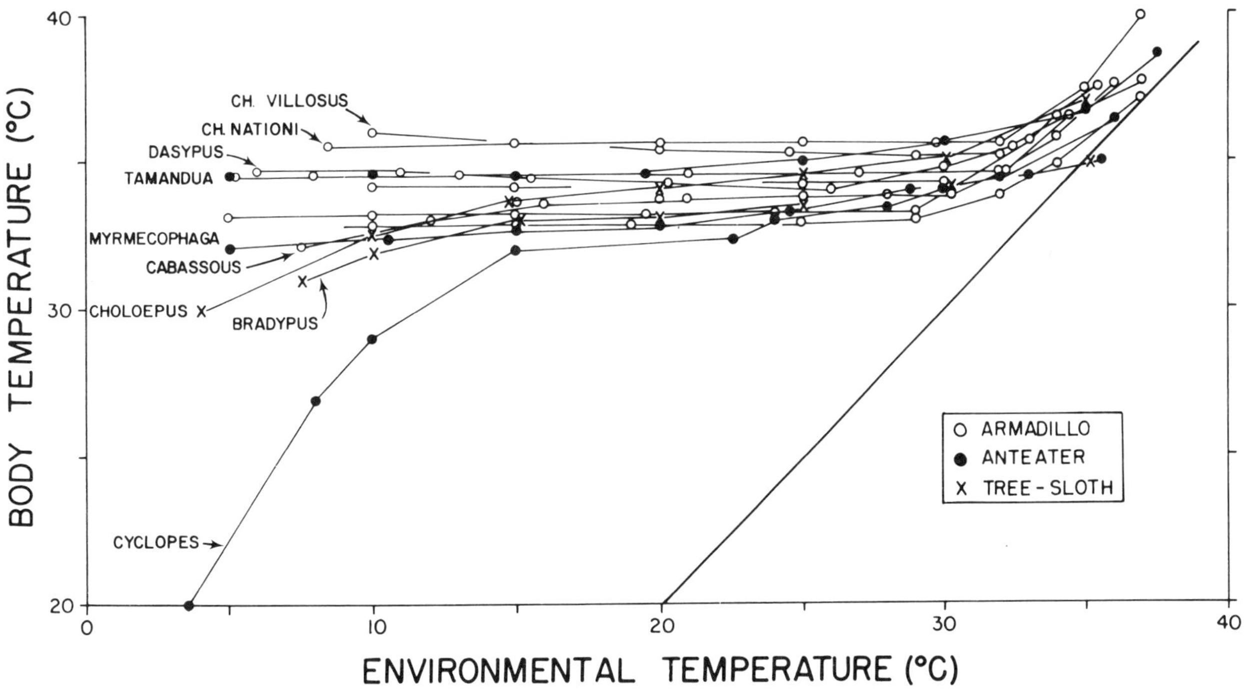

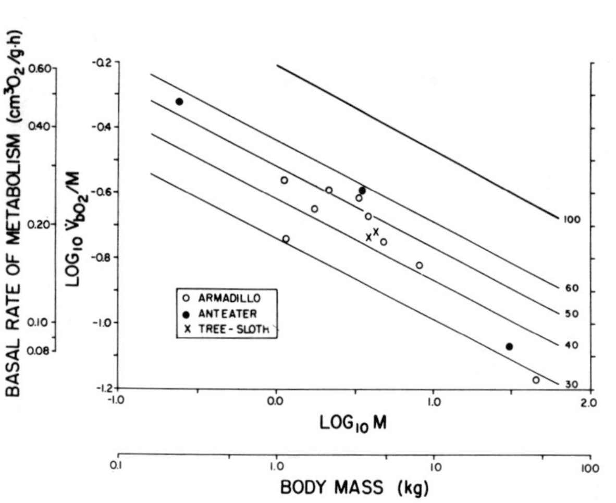

Figure 3. Basal rate of metabolism of xenarthrans as a function of body mass. Parallel curves represent basal rates as a percentage of the values expected from Kleiber's relation (1932). Data from Table 1.

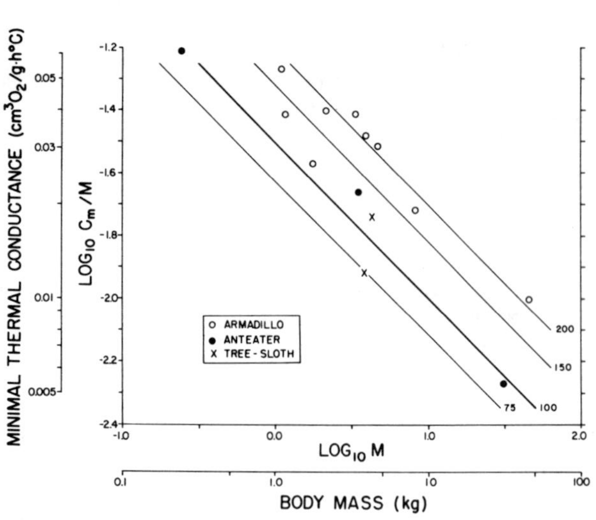

Figure 4. Minimal thermal conductance of xenarthrans as a function of body mass. Parallel curves represent minimal conductances as a percentage of the values expected from the McNab-Morrison relation (1963). Data from Table 1.

Cyclopes, but it surely is not applicable to terrestrial taxa, namely armadillos and the anteater *Myrmecophaga*.

Food habits appear to be influential in setting low basal rates of metabolism in xenarthrans (McNab, 1978, 1980b, 1984). Some foods may be dependably available only during certain seasons (armadillos) or only for a short period of feeding at each feeding stop and require a high investment in energy for movement and foraging (anteaters) (Montgomery, 1979, 1985a, 1985b); the foods may have a low energy content, either because of their molecular makeup (anteaters, tree sloths), or because they are mixed with indigestible matter in the process of being gathered (armadillos, anteaters), or the foods may contain deleterious compounds that require low rates of metabolism to assure adequate rates of chemical detoxification (tree sloths and possibly anteaters) (Montgomery, 1985a, 1985b).

Minimal Thermal Conductance.-- Most xenarthrans have higher minimal thermal conductances than those expected from mass in temperate mammals (McNab and Morrison, 1963), as is expected in tropical mammals (Table 1, Fig. 4). Exceptions are *Bradypus, Cyclopes*, and *Myrmecophaga*, which have low minimal conductances and long hair coats. *Bradypus*, the three-toed sloth, has a lower conductance than *Choloepus* mainly because it has a dense fur undercoat that is lacking in *Choloepus*. Armadillos have especially high minimal conductances, most notably at large masses. High conductances reduce heat storage in burrowing mammals (McNab, 1979). The lowest relative conductances in armadillos are found in the south temperate *Zaedyus* and in *Tolypeutes*. In *Tolypeutes*, at least, the low conductance stems from air trapped in the closed or partly closed carapace. No data are available on the heat exchange of the two smallest armadillos, *Chlamyphorus truncatus* and *C. retusa*, but it should be noted that they are the furriest of the armadillos.

Interaction of Basal Rate, Minimal Conductance, and Body Mass.-- At the lower limit of thermoneutrality, T_l, an endotherm maintains a temperature differential with the environment, ΔT_l, that equals the ratio of basal rate of metabolism to minimal conductance:

$$T_l = \dot{V}$$

$$(1) \qquad = 3.42\,(f_m/f_c)m^{0.25}$$

$$= 3.42\,Fm^{0.25}$$

where f_m and f_c represent the deviations from standard in the basal rate of metabolism and minimal thermal conductance, respectively, and $F = f_m/f_c$ (McNab, 1974). Equation (1) is a maximal estimate of ΔT_l and assumes that an endotherm distinguishes between those temperatures at which physical and chemical means of thermoregulation are used. If an endotherm mixes these means of thermoregulation, ΔT_l will be smaller and T_l will be higher than dictated by minimal conductance. Marked variations in basal rate and minimal conductance independent of body mass, as well as variations in body mass, account for most of the variations in T_l, ΔT_l, and T_b among xenarthrans.

DASYPODIDAE

Armadillos are soil and litter omnivore-insectivores, although some genera (*Tolypeutes, Priodontes* and possibly *Cabassous*) seem to be specialized for termite-eating (see Redford, 1985, for review). Armadillos range

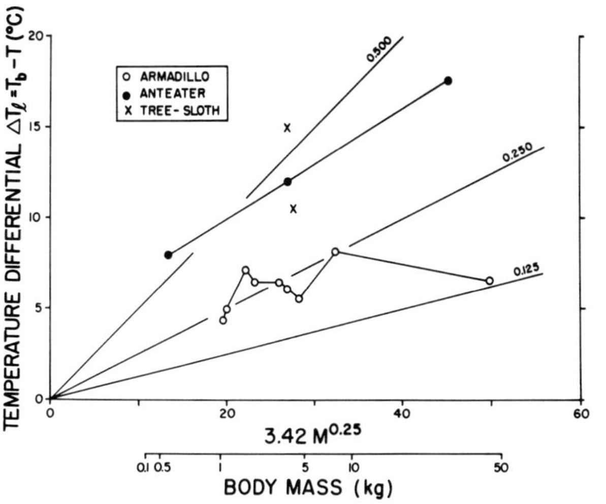

Figure 5. Temperature differential maintained with the environment by xenarthrans at the lower limit of thermoneutrality as a function of body mass. The linear curves labelled 0.500, 0.250, and 0.215 represent differentials that are 50 percent, 25 percent, and 12.5 percent, respectively, of the differentials expected from equation (1). Data taken from Table 1.

in body mass from 0.13 to 60 kg. Because of their low basal rates (ca. 50 percent of expected) and high minimal conductances (ca. 200 percent of expected), armadillos maintain temperature differentials with the environment at the lower limit of thermoneutrality that are only about 25 percent (= 50/200) of the differentials expected from "standard" mammals (Fig. 5). The largest differential found in armadillos, relative to standard mammals is found in the south temperate *Zaedyus*, in which $F = (f_m/f_c) = 0.38$, a value that is about 50 percent greater than that found in most other species. This differential in armadillos tends to be independent of body mass (McNab, 1979; Fig. 5), thereby reducing overheating of large species in burrows; in fact, the largest living species, *Priodontes maximus*, only maintains a differential that is 14 percent of the expected value due to the combination of a very low basal rate (29 percent of expected) and a high minimal conductance (213 percent of expected).

The smallest armadillo, *Chlamyphorus*, seemingly compensates for its small size by a long fur coat. For example, *Chlamyphorus* could maintain ΔT_l at 4.5°C, even if it, like other armadillos, had a basal rate equal to 50 percent of expected rate, if its minimal conductance were about 127 percent of the value expected from mass. Of course, if *Chlamyphorus* maintained a larger ΔT_l, conductance would have to be even lower.

The nearly naked body of armadillos seems to be principally an adaptation to the combination of a large mass and burrowing habits. The further modification of the integument into a series of plates presumably constitutes protection against predation. It is noteworthy that other terrestrial soil and litter omnivore-insectivores (hedgehog, *Erinaceus*; hedgehog tenrecs, *Setifer* and echidnas, *Tachyglossus*) also have a morphological barrier to predation, but in these cases it is a coat of spines. These species also have high conductances and maintain small temperature differentials with the environment (Shkolnik and Schmidt-Nielsen, 1976; McNab, 1980b, 1984).

Whatever the ultimate factors that favored the evolution of the carapace in armadillos, the presence of a carapace has major physiological and ecological consequences. By dictating small temperature differentials with the environment, the carapace must either assure lower body temperatures than found in other xenarthrans, which is not the case (Figs. 1 and 2), or raise the lower limits of thermoneutrality, which is the case (Fig. 6, Table 1). High lower limits mean that armadillos often face environmental temperatures that require an increase in metabolism to maintain body temperature, even in the tropics. As we shall see below, the high thermal conductance associated with a naked carapace is a major factor preventing most armadillos from extensively penetrating temperate climates.

An extinct subfamily of armadillos, the Pampatheriinae was found in South America, but one species, *Holmesina septentrionale*, evolved in North America, its northern limit extending from Texas, Oklahoma, and Kansas east to Florida (Edmund, 1985). Members of this subfamily are of special interest because of their large size (100 kg or more), presumed herbivorous habits, and the likelihood that they early gave rise to glyptodonts (Patterson and Pascual, 1968).

Figure 6. Lower limit of thermoneutrality of xenarthrans and some ungulates as a function of body mass. The curves are derived from equation (1), using the values for body temperature (upper number of each pair) and F (lower number of each pair) that correspond to the mean values for various groups of xenarthrans. Data from Tables 1 and 2.

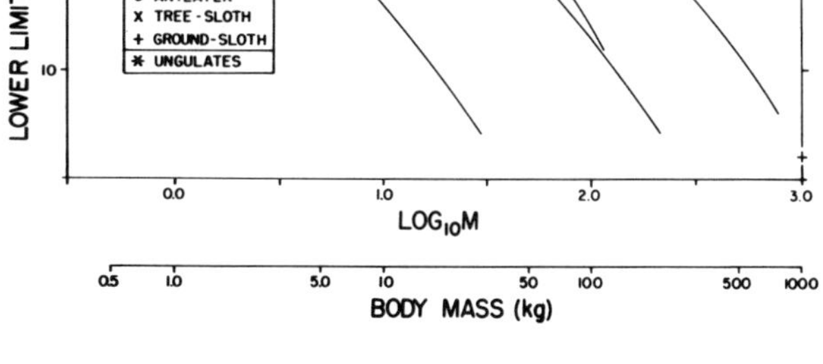

GLYPTODONTIDAE

Glyptodonts were small to huge grazing xenarthrans related to armadillos; they persisted into the Pleistocene and for a time even coexisted with man (Patterson and Pascual, 1968). Glyptodonts were very heavily armored, their shells resembling those of large tortoises. Some species grew to 3 m in length and probably weighed 1,000 kg. Nothing is directly known of the rate of metabolism and life history of glyptodonts, but Gillette (1974), on the basis of tooth structure, suggested that they ". . . were probably browsers, feeding selectively on vegetation along water courses. . . ." Browsing habits in mammals are often associated with low basal rates (McNab, 1978). There is some hope that the collagen fiber technique described below for *Nothrotheriops* may be used to estimate the body temperatures of glyptodonts.

MYLODONTIDAE

Nothing is directly known of the energetics of the ground-sloths belonging to this family, except as implied by their large masses (to ca. 1,000 kg) and thick, shaggy fur coats (Moreno, 1899; Moreno and Woodward, 1899; Woodward, 1900). Both factors theoretically would permit these ground-sloths to occupy seasonally cool to cold climates, although like other ground-sloths, mylodonts were browsing herbivores, which probably means that they too had low basal rates of metabolism.

An estimate of the thermal consequences of a large mass, a thick fur coat, and low basal rate can be made for mylodonts from equation (1). Woodward (1900) measured hair length on a skin of *Mylodon* found in a cave in southern Patagonia (at 50.5° S latitude); hair length varied from 30 to 50 mm on the back of the head, 50 to 65 mm at midback, 70 to 100 mm on the flank, and 150 to 220 mm on the rear limb. If the average length of hair is conservatively assumed to be 80 mm and if the thickness of the fur coat is about one-half of hair length (see Moreno and Woodward, 1899, plate XV, Fig. 1), then an estimate of thermal insulation [= (thermal conductance)$^{-1}$] can be obtained from Scholander et al. (1950, Fig. 3). For example, a coat thickness of 40 mm corresponds to an insulation of 0.045 (kcal/m^{2}day °C)$^{-1}$, which when inverted and converted equals a thermal conductance of 0.093 cal/cm^{2}h °C. If this value is multiplied by the surface-to-volume ratio for a 1-ton mammal (cm^2/g = 10m$^{0.33}$ = 0.10) and the units are converted, thermal conductance is estimated to be 0.0019 cm$^3O_2$/g h °C. The minimal conductance expected for a mammal weighing 1 ton is 0.0010 cm$^3O_2$/g h °C; therefore f_c = 1.90. Most mammals of this mass have very high thermal conductances (e.g., f_c is about 10) to assure adequate heat loss (McNab and Auffenberg, 1976).

Mylodon had such a good fur coat presumably because it had a low rate of metabolism and lived in a seasonally cool environment. If one further assumes that this ground-sloth had a basal rate equal to about one-half of the value expected from mass, the maximal temperature differential maintained at the lower limit of thermoneutrality can be estimated to be:

$$\Delta T_l = 3.42(0.50/1.90)(10^6)^{0.25} \tag{2}$$

$$= 28.5\ °C$$

Thus, if body temperature were about 34.5°C (see *Nothrotheriops* below), T_l would equal about 6°C. If the lower limit of thermoneutrality were not determined by minimal conductance, it would be greater than 6°C. Even this large mammal, then, would often have faced air temperatures below thermoneutrality and would have had to increase its rate of metabolism to keep its body temperature constant.

MEGALONYCHIDAE

Ground sloths of this family often reached a large body size (ca. 500 kg). Some megalonychids, such as *Megalonyx* (originally described by President Thomas Jefferson, 1799), tolerated seasonally cool to cold environments (see below). These sloths presumably were browsers.

Megalonychids chewed their food in a manner different from that used by the (distantly) related family Megatheriidae. These families also differed in the structure of the astragulus. This dichotomy is also found between the living tree sloths *Choloepus* and *Bradypus* (Sicher, 1944, Hirschfield and Webb, 1968; Patterson and Pascual, 1968). It seems likely, therefore, that *Choloepus* and *Bradypus* are, at best, only distantly related and represent convergence between two families of ground-sloths (Webb, 1985). From this view *Choloepus* is a member (or derivative) of the Megalonychidae and *Bradypus* a member (or derivative) of the Megatheriidae (Patterson and Pascual, 1968; Webb, 1985).

The two-toed tree sloth, *Choloepus hoffmanni*, has a low basal rate and a moderate minimal conductance. *Choloepus* has a temperature differential about twice that of armadillos (Fig. 5) as a result of the insulation provided by its long hair coat. Nevertheless, both this sloth and *Bradypus* are among the poorest regulators of body temperature at low ambient temperatures (Fig. 2), mainly because they have very low scopes for metabolism (McNab 1978, 1980a) due to their small muscle masses (Grand, 1978; McNab, 1978). The small muscle masses of tree sloths undoubtedly reflect their arboreal habits. Ground sloths would not be expected to have had such low muscle masses and low scopes; their higher scopes for metabolism, combined with a large

Table 2. **Parameters of energetics in some ungulates and a ground sloth.** $f_m (\%) = 100\ M_b/3.42\ m^{-0.25}$; $f_c(\%) = 100C_m/1.00\ m^{-0.50}$; $F = f_m/f_c$. **See text for method used to estimate the values for the extinct ground sloth** ***Nothrotheriops shastense*** **and the ungulate** ***Bos taurus.***

Species	*Mass* kg	*Basal Rate* $cm^3O_2/g.h$	f_m (%)	*Minimal Conductance* $cm^3O_2/g.h°C$	f_c (%)	*F*	T_b (°C)	T_l (°C)	ΔT_l (°C)	*Source*
Kobus defassa	100	0.267	139	0.010	316	0.44	38.6	15	23.6	Taylor et al., 1969
Taurotragus oryx	150	0.239	138	0.016	620	0.22	36.3	21	15.3	Taylor and Lyman, 1967
Nothrotheriops shastense	200	0.081	50	0.0059	264	0.19	34.4	21	13.7	This study
Bos taurus	225	0.181	115	0.006	275	0.42	38.0	7	31.7	Taylor and Lyman, 1967

mass and long hair coat, would presumably give ground sloths a greater tolerance to low ambient temperatures than is found in tree sloths.

MEGATHERIIDAE

This family includes the largest species of ground sloth, *Eremotherium americanum,* which weighed up to 4 or more metric tons; various smaller ground-sloths; and perhaps the extant three-toed tree sloth, *Bradypus.*

The energetics of the megatheriid ground sloth *Nothrotheriops shastense* can be estimated, in part, from our knowledge of its core body temperature. Body temperature was originally estimated by using its correlation with the proportional abundance of the amino acid hydroxyproline in the collagen fibers of bone (Ho, 1967). Through this technique the body temperature of *Nothrotheriops* was estimated as 34.7°C from a rib (which would likely give core temperature). Ho used a calibration curve of living species that included some erroneous values (e.g., the body temperature of the bat *Eptesicus* is only occasionally 37°C) or inappropriate temperatures (e.g., chicken). When these values are eliminated, the hydroxproline content of the *Nothrotheriops*' rib indicates a recalculated core temperature of 34.4°C.

Nothrotheriops, like living tree sloths, thus had a low body temperature. For a 250 kg mammal, 34.4°C is a very low body temperature, which indicates a low rate of metabolism, a high minimal thermal conductance, or both. Mammals having masses greater than 10 kg generally have much higher thermal conductances than expected from mass (McNab and Auffenberg, 1974). For example, ungulates weighing between 100 and 225 kg have thermal conductances that are 2.8 to 6.2 times the values expected from mass (Table 2). Such high conductances are mainly due to a thin, short hair coat and are a means of minimizing overheating in a body with a small surface-to-volume ratio and a high total rate of metabolism.

The thermal conductance of *Nothrotheriops* can be estimated from the observation of Lull (1929) that hair length in his specimen was about 45 mm. If the thickness of the fur coat was about 23 mm, its insulation would have been about 0.025 $(kcal/m^2\ day\ °C)^{-1}$ (Scholander et al., 1950). Thermal conductance therefore was about 0.167 $cal/cm^2\ h\ °C$, or 0.0055 $cm^3O_2\ g.h\ °C$, assuming that the mass of *Nothrotheriops* was about 250 kg. Such a conductance is 2.75 times greater than the value expected from body mass (McNab and Morrison, 1963). Even though *Mylodon* had about four times the mass of *Nothrotheriops, Mylodon* had a lower thermal conductance relative to standard (i.e., f_c = 1.90 vs. 2.75, respectively). This difference undoubtedly reflects the cooler climate faced by *Mylodon.*

Nothrotheriops fed mainly on woody, desert shrubs (Hanson, 1978), unlike the ungulates that live in the same area today, which mainly feed on grasses (Hanson, 1973). If one assumes that the depression in basal rate is roughly proportional to the fraction of the diet composed of woody plants (McNab, 1978), the basal rate of this ground sloth can be estimated to have been about 50 percent of the value expected from body mass.

The maximal temperature differential maintained by *Nothrotheriops* can be estimated from equation (1) and compared to that estimated from *Mylodon.* If one assumes that f_c equals 2.75 and mass equals 250 kg, then:

$$\text{(3)} \qquad \Delta T_l = 3.42(0.50/2.75)(2.5\text{x}10^5)^{0.25} = 13.9\ °C$$

Thus, the lower limit of thermoneutrality would have been about 20.5°C, or greater if T_l were set by a conductance greater than minimal, in *Nothrotheriops* (Fig. 6). The lower limit of thermoneutrality estimated for *Mylodon* is about 15°C lower, mainly because *Mylodon* was larger and had a thicker coat. The well-developed coat of ground sloths was probably compensatory for low basal rates of metabolism; ground sloths weighing up to a half ton often faced environmental temperatures below thermoneutrality even in the tropics.

The three-toed tree sloth *Bradypus,* like *Choloepus* has a low basal rate of metabolism. *Bradypus,* however, has a much lower minimal conductance than *Choloepus,* mainly because *Bradypus* has a dense, wooly undercoat below the coarse guard hairs (McNab, 1978). *Choloepus* lacks this fur layer. *Bradypus* consequently maintains a larger temperature differential than *Choloepus* (Fig. 5) and a T_l 6°C below that of *Choloepus* (Fig. 6). Like *Choloepus*, *Bradypus* permits body temperature to vary with ambient temperature (Fig. 2), probably because it also has a small muscle mass and thereby a small scope for metabolism (McNab, 1978).

MYRMECOPHAGIDAE

Anteaters have temperature differentials that are two to three times those of armadillos and similar to those of tree sloths (Fig. 5). The differentials of anteaters further contrast with those of armadillos by increasing with mass, whereas those of armadillos tend to be nearly independent of mass.

The large temperature differentials of vermilinguas are due to relatively high basal rates in *Cyclopes* and *Tamandua* (Fig. 3) and to low conductances in *Cyclopes* and *Myrmecophaga* (Fig. 4). *Cyclopes* has a dense fur coat and a low thermal conductance in apparent compensation for its small mass, just as is found in the smallest armadillos. *Myrmecophaga*, in spite of its large mass, has a long hair coat to compensate for its very low basal rate of metabolism. The large mass and apparently high scope for metabolism permit *Myrmecophaga* to maintain body temperature rather independent of ambient temperature, both in the laboratory (McNab, 1982) and in the field (Montgomery, 1985b). *Tamandua* maintains body temperature in the laboratory, but may (often) permit T_b to fall in the field by at least 4 °C (McNab, 1984); unlike *Myrmecophaga* and *Cyclopes*, *Tamandua* often spend periods of inactivity underground or in tree dens (Montgomery, 1985b). The small mass and small scope for metabolism in *Cyclopes* prevent this anteater from maintaining body temperature independent of ambient temperatures below 15 °C (Fig. 2).The presumably small muscle mass and the consequently small scope for metabolism in *Cyclopes* is associated with its obligatory arboreality (McNab 1978, 1980a).

DISCUSSION

Population Biology of Xenarthrans.-- Animals vary greatly in terms of population biology. At one extreme are species that exploit ephemeral resources, have a high fecundity, and readily disperse. Such species are called "r-selected" (MacArthur and Wilson, 1967) to recognize that at times they may increase in abundance at a rate limited only by their maximal intrinsic capacity for population growth (r_m). At the other extreme, there are species that specialize on comparatively stable, limited resources, have a low fecundity, and show little propensity for dispersal; these species were designated "K-selected" (MacArthur and Wilson, 1967) to suggest that their abundance is often limited by the carrying capacity of the environment (K). Pianka (1970) expanded the list of behavioral characteristics that tend to separate r- from K-selected species. Of course, most species fall between these extremes.

In placental mammals there is an intimate relationship between the rate of energy expenditure and the population growth constant r (McNab 1980a). This constant can be described by the relation:

(4) $$r = (\ln R_o)/T$$

where R_o is fecundity and T is generation time; for a particular species r is maximal when R_o is as large and T as small as conditions will permit. R_o increases and T decreases with an increase in rate of metabolism, independent of body mass. Therefore, under given environmental conditions r_m is high in species with high rates of metabolism and low in species of the same mass with low rates of metabolism.

Because variations in rate of metabolism, independent of body mass, are correlated with the kind of food resources used, r_m is also correlated with the food habits of mammals (McNab, 1980a). Mammals that feed on resources that are dependable in space and time and that have a high energy or nutritional content have high rates of metabolism, a large r_m, and show wide fluctuations in abundance. These species tend to be r-selected. But mammals that feed on resources that are patchy in space and time, that have a low energy content, or that have an abundance of antipredator molecules, have low rates of metabolism, a small r_m, and show modest fluctuations in abundance. Such species tend to be K-selected.

When this analysis is applied to xenarthrans, it is clear that all living species are to varying degrees K-selected. They have low rates of metabolism and, based on their small litter sizes, extended parental care, low growth rates (McNab, 1980a), and long gestation periods (Fig. 7), would be expected to have very low intrinsic constants for population growth. Again, these low rates of metabolism result from burrowing habits or with specializing on foods such as the leaves of woody plants or ants and termites.

Armadillos may well have a higher fecundity than other xenarthrans: all xenarthrans other than armadillos have a litter size of one, whereas the only armadillos known to have a litter size of one are the genera *Tolypeutes* and *Priodontes*, which tend to specialize on ants and termites (Redford, 1985); the food habits of all species of armadillo other than *Dasypus novemcinctus* and *D. sabanicola* are very poorly known. The difference in fecundity between armadillos and other xenarthrans is not related to rate of metabolism, which shows that rate of metabolism does not uniquely determine fecundity, even when

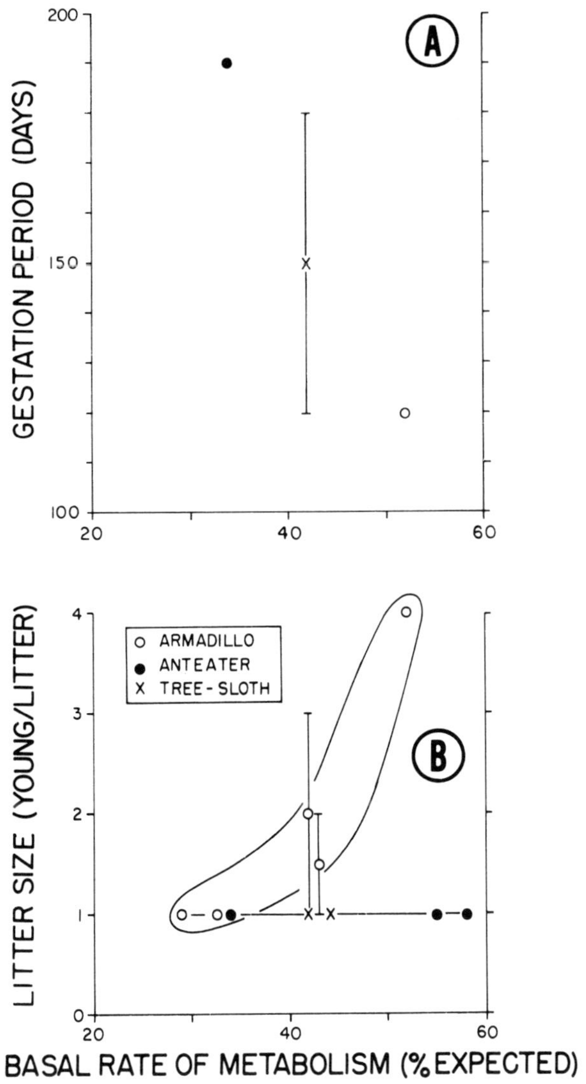

Figure 7. Litter size in xenarthrans as a function of basal rate. Data from Asdell (1964).

body mass is taken into consideration. Among armadillos, however, there is an intriguing correlation of fecundity with basal metabolic rate (Fig. 7).

Finally, a question arises whether there is a relationship between the population biology of prey and that of a predator. For example, variations in rate of metabolism among anteaters and armadillos may be related in part to the ecological properties of their prey. The difference in rate of metabolism between *Orycteropus* and *Myrmecophaga* may reflect the use of colonial termites by the aardvark and the use of ants that belong to small, widely dispersed colonies (at least in Venezuela; Montgomery and Lubin, 1977) by the giant anteater. The very low rate of metabolism of echidnas, from this view, may reflect the ecological properties of their food, rather than representing some intrinsic property of monotremes (McNab 1984).

Energetics and the Distribution of Xenarthrans.-- The patterns of distribution of living and extinct xenarthrans reflect their energetics. For example, the common long-nosed armadillo *Dasypus novemcinctus* appears to face a thermal limit to its distribution in North America (Fig. 8) because it does not seem capable of evading cold winters by entering torpor and because it cannot feed when the ground is frozen or covered with snow (McNab 1980b). Its tolerance to starvation during cold weather increases with body mass and with the amount of fat stored (Layne and Glover, 1983; McNab, 1980b), but is inversely related to thermal conductance. The thermal conductance of *Dasypus* is too high and its young of the year too small and lean to permit penetration into regions with extended cold winters. In contrast, the pichi *Zaedyus pichiy* tolerates a colder climate in southern South America by entering a low temperature torpor and hibernation (Hatcher, 1903; Roig, 1971), a response that requires a small body mass (McNab, 1980b). The western limit to the distribution of *Dasypus* in Texas is produced by its inability to tolerate a dry environment (Humphrey, 1974; Greegor, 1975), whereas *Chaetophractus vellerosus* can tolerate dry environments, in part through an increased renal capacity to produce a highly concentrated urine (Greegor, 1975).

The distributions of pampatheres and glyptodonts in North America (Fig. 8) were similar to that of *Dasypus* at present, even though these herbivorous xenarthrans were much larger, weighing 100 to 500 kg. If these xenarthrans occupied this distribution during one or more interglacial periods, it can be concluded that their temperature sensitivity at a large mass reflected a low rate of metabolism.

The distribution of ground sloths follows several patterns (Fig. 9). *Glossotherium* was found as far north as Washington and southern Alberta in the west and Nebraska, Iowa, and Pennsylvania in the east. *Megalonyx* in eastern North America reached somewhat higher latitudes than *Glossotherium*, and in western North America occurred in British Columbia, Alberta, Yukon Territories, and in Alaska. Some mylodontids and megalonychids were thus found in what were undoubtedly seasonally cool to cold climates. Surely, fur coats and large masses were important contributors to this capacity to tolerate these climates. This pattern contrasts with that found in megatheriid sloths. *Eremotherium*, although the largest ground sloth, was limited in its distribution in North America to the Gulf and Atlantic coastal plains (Fig. 9), and *Nothrotheriops* was restricted to the southern United States, especially to the arid southwest, except for one record from Alberta (MacDonald, 1985). *Eremotherium* does not appear to have tolerated the dry environments used by *Nothrotheriops*.

The difference in distribution between *Glossotherium*

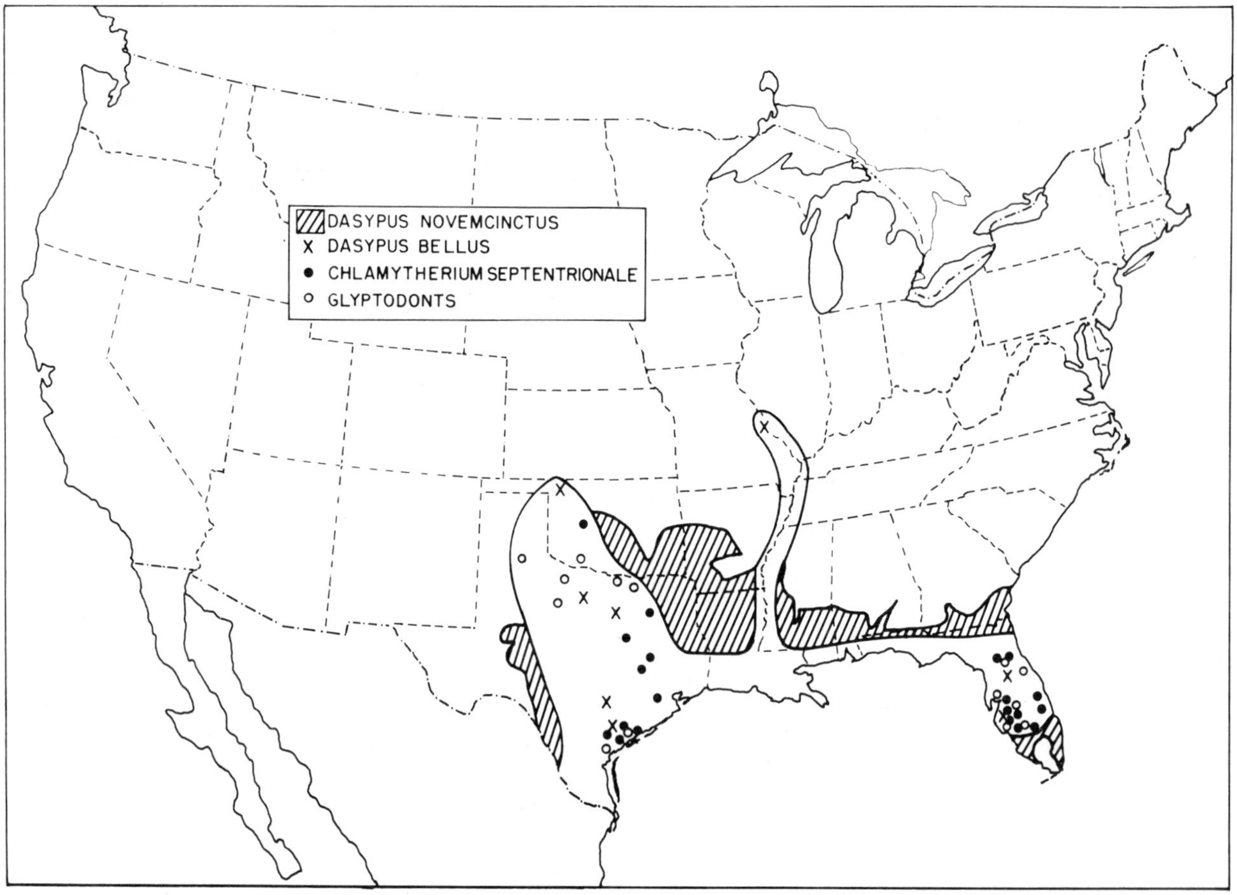

Figure 8. The contemporary distribution of *Dasypus novemcinctus* and the Pleistocene distribution of *D. bellus, Holmesina septentrionale*, and glyptodonts north of Mexico. Data from James (1957), Slaughter (1961), Melton (1964), and Humphrey (1974).

and *Megalonyx* on one hand and *Eremotherium* on the other could represent specialization in food habits, but it more likely reflects a difference in temperature sensitivity, as might be correlated with an unusually low rate of heat production in *Eremotherium*. It is conceivable that *Eremotherium* had a poor fur coat, or even was naked, but there are no known samples of its skin available for study.

Neither *Bradypus* nor *Choloepus* tolerate cool temperate latitudes; this is to be expected in mammals with such low capacities for temperature regulation. *Bradypus* lives at high altitudes (ca. 1500 m) in Panama (Montgomery, pers. comm.), while *Choloepus* lives to at least 2,300 m in northern Panama (Enders, 1940). The hair coat of *Choloepus* living at high altitude was notably thicker than in those living at lower altitude. *Bradypus*, but not *Choloepus*, enters northern Argentina (Yepes 1929), which correlates with the observation that *Bradypus* has a lower T_l (18°C) than is found in *Choloepus* (24°C).

Anteaters also fail to penetrate far into temperate environments. The reasons are twofold. First, they feed on social insects (ants and termites) that are principally tropical in distribution or, at best, have only a seasonal availability in temperate environments. Secondly, anteaters have some difficulty in maintaining a constant body temperature, or do so only with an appreciable expenditure of energy, a response that is incompatible with diminishing food resources. The small size and small scope for metabolism place a more stringent latitudinal limit to the distribution of *Cyclopes* than is the case for *Tamandua* or *Myrmecophaga*. Even these larger genera penetrate higher latitudes in South America only as far as northern Argentina (Yepes, 1929).

Another possible case of termite-eating habits among xenarthrans in warm temperate latitudes is worth considering. The Pleistocene armadillo *Dasypus bellus* was widespread from Texas through Florida (Fig. 8). It differed from *D. novemcinctus* principally by size; it weighed up to 20 or 30 kg. It is unsure whether *D. bellus* had food habits as omnivorous as the common long-nosed armadillo or whether it was more specialized. It is of interest, however, that Light (1930) reported fossil fecal pellets from Florida that contained termites of the mound-building genus *Kalotermes*; these specimens were collected from the same sediments that contained the Seminole Field mammal fauna in which *D. bellus* was found (Simpson, 1929).

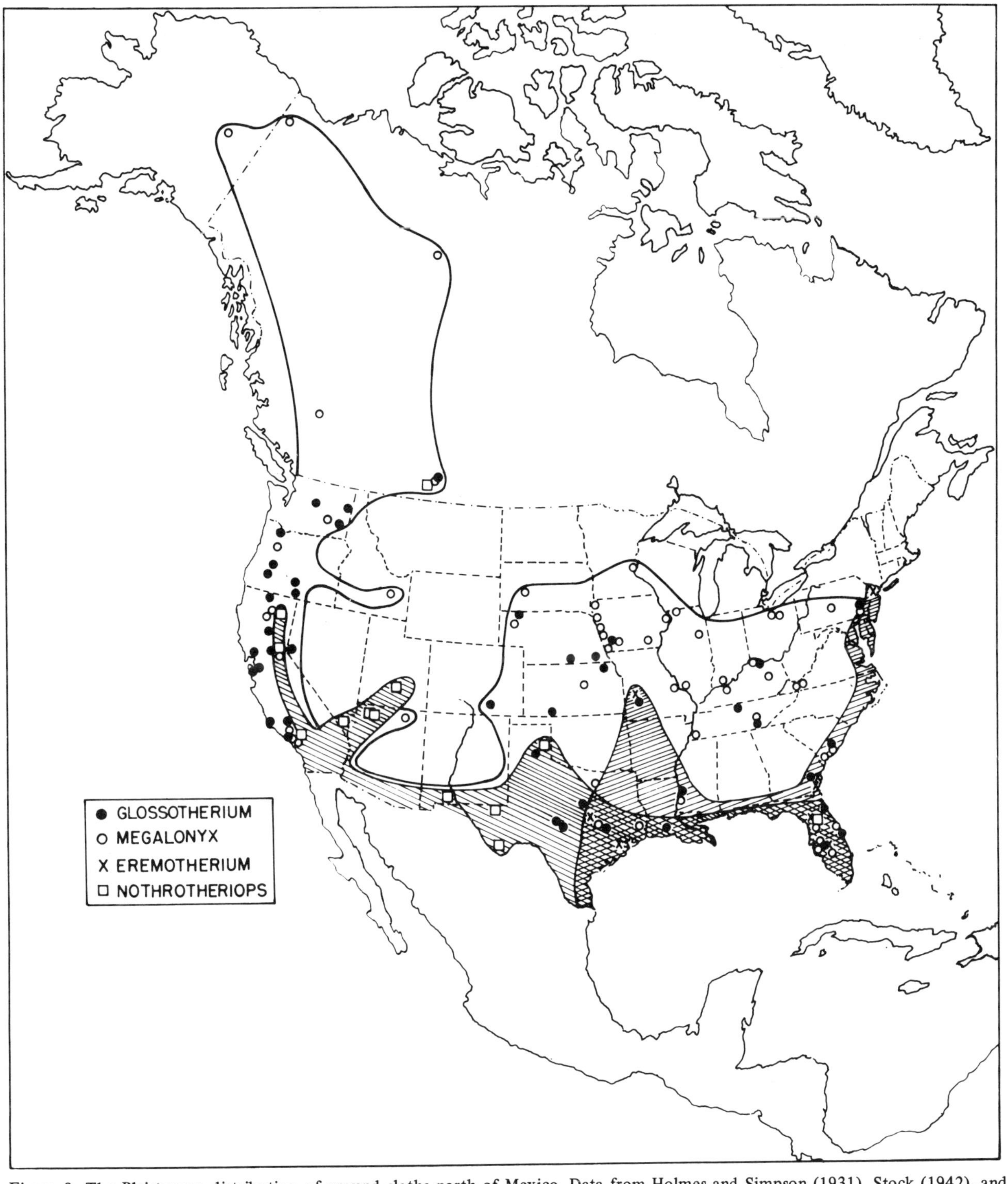

Figure 9. The Pleistocene distribution of ground sloths north of Mexico. Data from Holmes and Simpson (1931), Stock (1942), and Harington (1970).

CONCLUSIONS

Xenarthra is one of the most distinctive orders of mammals. It is essentially neotropical in distribution and its members feed on resources that push their consumers into various evolutionary culs-de-sac. This evolutionary pattern is generally associated with a low reproductive potential, low rate of metabolism, low body temperature, and poor temperature regulation, although a large body size and good fur coat can produce more effective temperature regulation than would otherwise be expected. The general pattern of distribution of xenarthrans in relation to climate are commensurate with the physiological characteristics of living species and suggest aspects of the energetics of fossil species.

It is unfortunate that there was an appreciable loss in diversity of xenarthrans in the late Pleistocene. The major adaptive zones that were lost were large herbivores (ground sloths, glyptodonts, and pampatheres). The study of tree sloths thereby takes on greater significance, because they are the only living xenarthrans that retain these habits.

ACKNOWLEDGMENTS

This study completely depended on the kind cooperation of Dennis Merritt and the Lincoln Park Zoo, Chicago, for the loan of most of the xenarthrans studied; both are gratefully thanked. With the cooperation of John Eisenberg, animals were also borrowed from the National Zoo (Washington, D.C.). Measurements were made on the R/V Alpha Helix expedition to the Rio Negro, Brazil; at the Smithsonian Tropical Research Institute on Barro Colorado Island, Panama; at the Lincoln Park Zoo; and at the University of Florida. S. David Webb and Archie Carr, University of Florida, and G.G. Montgomery, Smithsonian Tropical Research Institute, made many valuable suggestions for this manuscript. This work was aided by NSF grants GB-3477 and GB-20689.

LITERATURE CITED

Asdell, S.A.
1964. Patterns of Mammalian Reproduction. Comstock, Ithaca, New York 670 pp.

Edmund, G.
1985. The fossil giant armadillos of North America (Pampatheriinae, Xenarthra = Edentata). (this volume).

Enders, R.K.
1940. Observations on sloths in captivity at higher altitudes in the tropics and in Pennsylvania. J. Mammal. 21: 5-7.

Gillette, D.D.
1974. A review of North American glyptodonts (Edentata, Mammalia): Osteology, systematics, and paleobiology, unpubl. Ph.D. thesis. Southern Methodist University, Dallas, Texas. 648 pp.

Grand, T.I.
1978. Adaptations of tissue and limb segments to facilitate moving and feeding in arboreal folivores. Pp. 321-341, *in*, The ecology of arboreal folivores. (G.G. Montgomery, editor). Smithsonian Institution Press, Washington, D.C. 574 pp.

Greegor, D.H., Jr.
1975. Renal capabilities of an Argentine desert armadillo. J. Mammal., 56: 626-632.

Hanson, R.M.
1978. Shasta ground sloth food habits, Rampart Cave, Arizona. Paleobiol., 4: 302-319.

Hanson, R.M and P.S. Martin.
1973. Ungulate diets in the lower Grand Canyon. J. Range Mgmt., 26: 380-381.

Harington, C.R.
1970. Ice age mammal research in the Yukon Territory and Alaska. Pp. 35-51, *in* Early man and environments in northwest North America, (R.A. Smith and J.W. Smith, editors), Univ. Calgary Archeol. Assoc. Students Press, Calgary. 136 pp.

Hatcher, J.B.
1903. Narrative of the expeditions. Pp. 1 to 209, *in* Reports of Princeton University expeditions to Patagonia 1896-1899. Vol.1: 1-209. Princeton Univ. Press, Princeton, N.J. 314 pp.

Hirschfeld, S.E. and S.D. Webb.
1968. Plio-Pleistocence megalonychid sloths of North America. Bull. Fla. State Mus., 12: 213-296.

Ho, T.-Y.
1967. Relationship between amino acid contents of mammalian bone collagen and body temperature as a basis for estimation of body temperature of prehistoric mammals. Comp. Biochem. Physiol., 22: 113-119.

Holmes, W.W. and G.G. Simpson.
1931. Pleistocene exploration and fossil edentates in Florida. Bull. Amer. Mus. Nat. Hist., 59: 383-418.

Humphrey, S.R.
1974. Zoogeography of the nine-banded armadillo (*Dasypus novemcinctus*) in the United States. BioScience, 24: 457-462.

James, G.T.
1957. An edentate from the Pleistocene of Texas. J. Paleont., 31: 796-808.

Jefferson, T.
1799. A memoir on the discovery of certain bones of a quadruped of the clawed kind in the western parts of Virginia. Trans. Amer. Phil. Soc., 4: 246-260.

Kleiber, M.
1932. Body size and metabolism. Hilgardia, 6: 315-353.

Layne, J.N. and D. Glover.
1985. Activity patterns of the common long-nosed armadillo *Dasypus novemcinctus* in south-central Florida. (this volume).

Light, S.F.
1930. Fossil termite pellets from the Seminole Pleistocene.Univ. Calif. Publ. Geol. Sci., 19: 75-80.

Lull, R.S.
1929. A remarkable ground sloth. Memoirs Peabody Mus. Yale Univ., 3: 1-39.

MacArthur, R.H. and E.O. Wilson.
1967. The theory of island biogeography. Monographs Popul. Biol. 1. Princeton Univ. Press, Princeton, N.J. 215 pp.

McDonald, H.G.
1985. The Shasta ground sloth *Nothrotheriops shastensis* (Xenarthra, Megatheriidae) in the middle pleistocene of Florida. (this volume).

McNab, B.K.
1974. The energetics of endotherms. Ohio J. Science, 74: 370-380.

1978. Energetics of arboreal folivores: Physiological problems and ecological consequences of feeding on an ubiquitous food supply. Pp 153-162, *in*, The ecology of arboreal folivores, (G.G. Montgomery editor), Smithsonian Institution Press, Washington, D.C. 574 pp.

1979. The influence of body size on the energetics and distribution of fossorial and burrowing mammals. Ecology 60: 1010-1021.

1980a. Food habits, energetics, and the population biology of mammals. Amer. Natural. 116: 106-124.

1980b. Energetics and the limits to a temperate distribution in armadillos. J. Mammal. 61: 606-627.

1984. Physiological convergence among ant-eating and termite- eating mammals. J. Zool, London.

McNab, B.K. and W. Auffenberg.
1976. The effect of large body size on the temperature regulation of the Komodo dragon, *Varanus komodoensis*. Comp. Biochem. Physiol., 55A: 345-350.

McNab, B.K. and P.R. Morrison.
1963. Body temperature and metabolism in subspecies of *Peromyscus* from arid and mesic environments. Ecol. Monographs, 33: 63-82.

Melton, W.G., Jr.
1964. *Glyptodon fredericensis* (Meade) from the Seymour Formation of Knox County, Texas. Mich. Acad. Sci., 49: 129-146.

Minoprio, J.D.L.
1945. Sobre el *Chlamyphorus truncatus* Harlan. Acta Zool. Lilloana, 3: 5-58.

Montgomery, G.G.
1979. El grupo alimentico (feeding guild) del oso hormiguero: convivencia y especialización de las presas de sustento de los osos hormigueros neotropicales (Edentata, Myrmecophagidae). ConCiencia 6: 3-6.

Montgomery, G.G.
1985. Impact of vermilinguas (*Cyclopes, Tamandua*: xenarthra = edentata) on arboreal ant populations. (this volume).

Montgomery, G. G.
1985. Movements, Foraging and Food Habits of the Four Extant Species of Neotropical Vermilinguas (Mammalia; Myrmecophagidea) (this volume).

(G.G. Montgomery, editor), Smithsonian Institution Press, Washington, D.C. xxx pp.

Montgomery, G.G. and Y.D. Lubin.
1977. Prey influences on movements of neotropical anteaters. Pp. 103-131, *in* Proceedings of the 1975 Predator Symposium (R.L. Phillips and C. Jonkel, editors). Montana Forest and Conservation Experiment Station, Univ. of Montana, Missoula. 268 pp.

Moreno, F.P.
1899. Note on the discovery of *Miolania* and of *Glossotherium* (*Neomylodon*) in Patagonia. Geol. Magazine, 6: 385-388.

Moreno, F.P. and A.S. Woodward.
1899. On a portion of mammalian skin, named *Neomylodon listai*, from a cavern near Consuelo Cove, Last Hope Inlet, Patagonia. Proc. Zool. Soc. London, 1899: 144-156.

Patterson, B. and R. Pascual.
1968. Evolution in mammals on southern continents. V. The fossil mammal fauna of South America. Quart. Rev. Biol., 43: 409-451.

Pianka, E.R.
1970. On r- and K-selection. Amer. Natural., 104: 592-597.

Redford, K.H.
1985. Food habits of armadillos (Xenarthra: Dasypodidae). (this volume).

Roig, V.G.
1971. Observaciones sobre la termorregulacion en *Zaedyus pichiy*. Acta Zool. Lilloana, 28: 13-18.

Shkolnik, A., and K. Schmidt-Nielsen.
1976. Temperature regulation in hedgehogs from temperate and desert environments. Physiol. Zool., 49: 56-64.

Scholander, P.F., V. Walters, R. Hock, and L. Irving.
1950. Body insulation of some arctic and tropical mammals and birds. Biol. Bull., 99: 225-236.

Sicher, H.
1944. Masticatory apparatus of the sloths. Field Mus. Nat. Hist., Zool. Ser., 29: 161-168.

Simpson, G.G.
1929. Pleistocene mammalian fauna of the Seminole field, Pinellas County, Florida. Bull. Amer. Mus. Nat. Hist., 56: 561-599.

Slaughter, B.H.
1961. The significance of *Dasypus bellus* (Simpson) in Pleistocene local faunas. Texas J. Sci., 13: 311-315.

Stock, C.
1942. A ground sloth in Alaska. Science, 95: 552-553.

Taylor, C.R. and C.P. Lyman.
1967. A comparative study of the environmental physiology of an East African antelope, the eland, and the Hereford steer. Physiol. Zool., 40: 280-295.

Taylor, C.R., C.A. Spinage, and C.P. Lyman.
1969. Water relations of the waterbuck, an East African antelope. Amer. J. Physiol., 217: 630-634.

Webb, D.E.
1985. The interrelationships of tree sloths and ground sloths. (this volume).

Woodward, A.S.
1900. On some remains of *Grypotherium* (*Neomylodon*) *listai* and associated mammals from a cavern near Consuelo Cove, Last Hope Inlet, Patagonia. Proc. Zool. Soc. London, 1900: 64-79.

Yepes, J.
1929. Los "Edentata" argentinos. Revista Univ. Buenos Aires, 5: 461-515.

REPRODUCTION

ON THE PLACENTATION OF SLOTHS

Kurt Benirschke and Henry C. Powell

San Diego Zoo, San Diego, California, 92112, and Department of Pathology, University of California, at San Diego, California, 92093.

Abstract

Term placentas of *Bradypus tridactylus, Choloepus hoffmanni, Choloepus didactylus* and a mid-term placenta of *Choloepus didactylus* had macroscopic and microscopic features similar to those of *Bradypus griseus*. The first electron microscopic study of two-toed sloth placentas showed that the epithelioid cells in the fetal connective tissue are lysosome-filled macrophages which decrease in number toward maturity. In the mature placenta the microvillous surface of the syncytium abuts the collagenous mantle of the basement membrane that surrounds the prominent maternal vascular endothelial cells. This connective tissue layer is focally interrupted by tongues of syncytial microvilli which reach the endothelium. The sloth placenta is thus correctly identified as endotheliochorial.

Abstracto

Varias placentas terminales de *Bradypus tridactylus, Choloepus hoffmanni, Choloepus didactylus,* y una placenta de mitad de término de *Choloepus didactylus* mostraron rasgos macroscópicos y microscópicos similares a las de *Bradypus griseus*. El primer estudio en microscopio electrónico de placentas de perezosos de dos dedos, mostraron que las células epitelioides en los tejidos conectivos del feto son macrófagos llenos de lisosoma, que disminuyen en números conforme se acerca la maduración de la placenta. En la placenta terminal, la superficie microvellosa del sincicio remata contra el palio colaginoso de la membrana del basamento que rodea las prominentes células endetiliales vasculares de la madre. Esta capa de tejido conectivo aparece interrumpida focalmente por lenguas de microvellos sineiciales que alcanzan hasta el endotelio. La placenta del perezoso es, por lo tanto, correctamente interpretada como una placenta endoteliocorial.

Resumo

Placentas plenamente desenvolvidas de *Bradypus tridactylus, Choloepus hoffmanni, Choloepus didactylus* e uma placenta em desenvolvimento de *Choloepus didactylus* têm características macroscópicas e microscópicas similares ̀as de *Bradypus griseus*. O primeiro estudo de microscopia eletrônica de placentas da preguiçã real *Choloepus* mostrou que células epitelióides no tecido conetivo fetal são macrófagas, cheias de lisossomos, que diminuem em número ̀a medida que amadurece. Na placenta madura, a superfície microvilosa do sincício se encosta ao manto colaginoso da membrana de embasamento que circunda as proeminentes células endoteliais vasculares materna. Essa camada de tecido conetivo é focalmente interrompida por projeções das microvilosidades sinciciais que chegam ao endotélio. Desse modo, a placenta de preguiça é corretamente identificada como endoteliocorial.

Only few placentas of sloths have been studied. Wislocki (1927) found records of seven three-toed sloth *Bradypus* placentas and one of Hoffmann's two-toed sloth *Choloepus hoffmanni*. To these observations he added 28 uteri from pregnant three-toed sloths *Bradypus griseus* in the most complete description of the fetus and membranes of this species to date. Later observations on four young specimens (Heuser and Wislocki, 1935) were the last systematic investigations of reproductive organs of this Xenarthran.

From those studies the sloth placenta can be characterized as deciduate, labyrinthine, endotheliochorial, and nodular-discoid. The presence of a thin, Masson trichrome-staining membrane between the syncytiotrophoblast and maternal endothelium was described by Wislocki (1928). He considered this connective tissue to be a remnant of maternal vessel wall. He strongly rejected the suggestion by de Lange (1926) that the cells considered by Wislocki to be maternal endothelial cells represent trophoblast. For this reason, deLange had suggested that hemochorial placentation was characteristic of the species.

Only one Hoffmann's two-toed sloth placenta had been studied (Turner, 1873) until Wislocki (1928) obtained one other with a fetus 136 mm long which he judged to be at midgestation. With minor exceptions, he considered the placentas from sloths of the two genera to be virtually identical. In *Choloepus hoffmanni*, more fetal connective tissue was found, the trophoblast appeared more reduced, and no amniotic nodules were observed as were found in *Bradypus*.

The placenta of *Choloepus didactylus* has not been described previously. This report briefly summarizes findings on the uteri of two pregnant *Choloepus didactylus* from Brazil, of one *C. hoffmanni* placenta , and one *Bradypus tridactylus* uterus containing a near term fetus. Initial electron microscopic studies were possible on the two placentas of *C. didactylus* even though they were obtained after death.

MATERIALS AND METHODS

Four placentas and fetuses were available, some attached to the uteri of recently dead females. Weights and measurements were taken when feasible. Tissues were fixed in Bouin's solution and in 10 percent formalin. Minute fragments of the two placentas from *C. didactylus* were fixed in buffered glutaraldehyde and processed for electron microscopy. The sections were examined in a Zeiss 10 electron microscope operating at 60 KV.

Case 1. *Bradypus tridactylus*.-- The three-toed sloth was from Belém, Brazil, and it had been dissected by Dr. Z. Silva. The 12 x 5 cm uterus was simplex and pear-shaped; the placenta had a fundal position. The placenta measured 11 x 7 cm in greatest diameter and weighed 60 g (Fig. 1). The fully grown male fetus (Fig. 2) measured 27 cm snout-tail, and had a 15 cm umbilical cord. It weighed 185 g. The fetal adrenals contained a remarkably large fetal cortex similar to that seen in armadillos and higher primates.

Case 2. *Choloepus hoffmanni*.-- This freshly delivered placenta of a Hoffmann's sloth fetus was made available by Dr. Dennis Merrit, Lincoln Park Zoo, Chicago. It measured 20 x 16 x 1 cm and had a marginally attached 9.5 cm umbilical cord (Fig. 3).

Case 3. *Choloepus didactylus*.-- This immature fetus and placenta came from a recently imported 4 kg Brazilian two-toed sloth that died from salmonellosis, gastric ulcer and poor nutrition at the San Diego Zoo. The placenta was attached at the fundus of the unicornuate uterus, weighed 109 g and measured 9 cm in greatest diameter (Fig. 4). The fetus weighed 110 g, and had a 20 cm snout-tail length. The umbilical cord was 12 cm in length. Weights of fetal organs are as follows: liver, 4.3 g; gonads, 0.65 and 0.58 g; adrenals, 0.31 g each; kidneys, 0.55 and 0.58 g; heart, 0.75 g; spleen, 0.07 g; thymus, 0.48 g; lungs, 3.4 g.

Case 4. *Choloepus didactylus*.-- This recently imported two-toed sloth weighed 5 kg at death which was due to

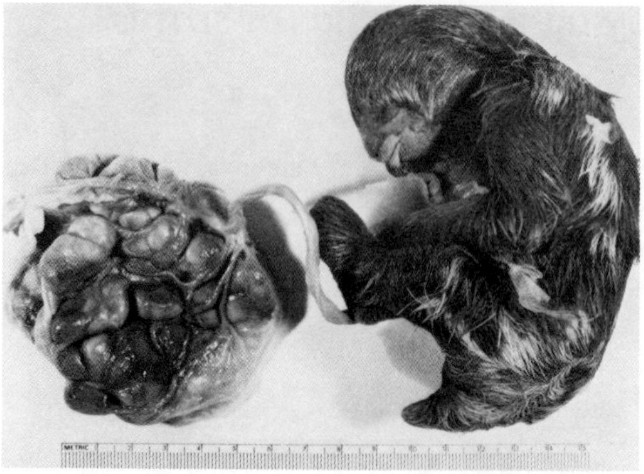

Figure 1. Placenta and membranes of *Bradypus tridactylus* (Case 1) spread out. Note infarction (white) of several lobules or portions thereof in the upper half of photograph.

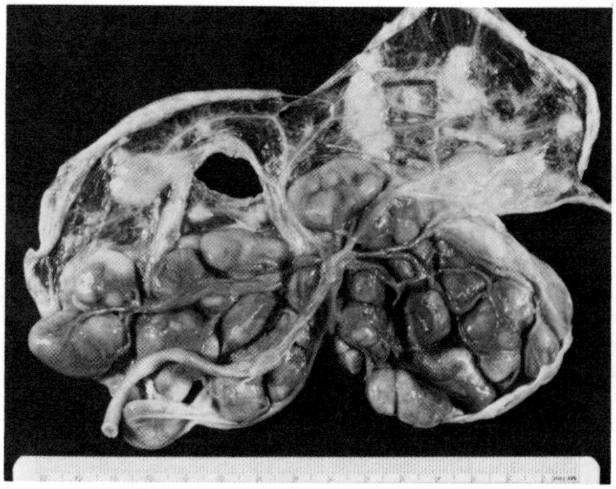

Figure 2. Term *Bradypus tridactylus* (Case 1) showing original position of formalin-fixed fetus and placenta.

salmonellosis and malnutrition. No measurements were taken on the full-term fetus and placenta. The placenta had a similar multilobulated appearance as in case 3 and measured 1 cm in greatest thickness.

FINDINGS

Macroscopic.-- The gross features of the placenta (Figs. 1-4) generally conformed to the description by Wislocki (1927). The placentas were composed of approximately twenty nodular discs held together by membranes and fetal vessels. Generally, the umbilical cords were marginally attached and free membranes made up approximately one-half of the total fetal surface. The amnion showed no nodules and was firmly attached to the chorion. Several placental lobules showed infarctive necrosis, best seen in Fig. 1.

Microscopic.-- Sections of umbilical cords of cases 1 and 4 show them to contain two arteries, one vein, and no

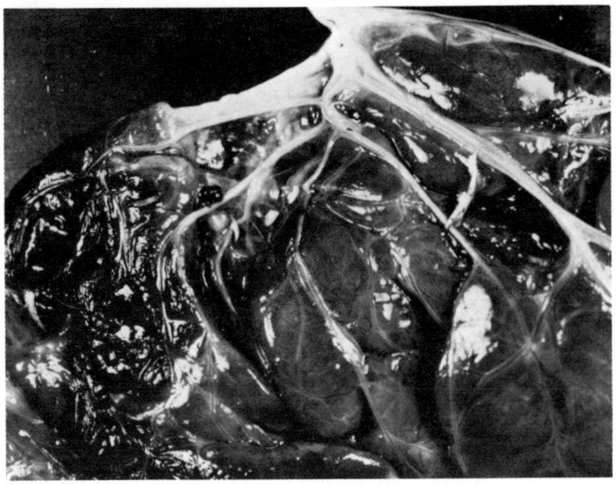

Figure 3. Delivered placenta of *Choloepus hoffmanni* (Case 2). Note marginal insertion of cord and lobular nature of placenta.

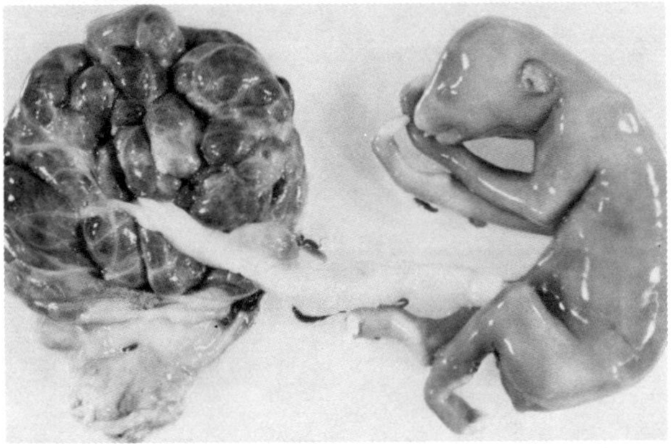

Figure 4. Immature *Choloepus didactylus* and placenta still attached to opened uterus (Case 3). Note the essentially identical appearance of placenta to that in Fig. 1, except for more edematous (immature) umbilical cord.

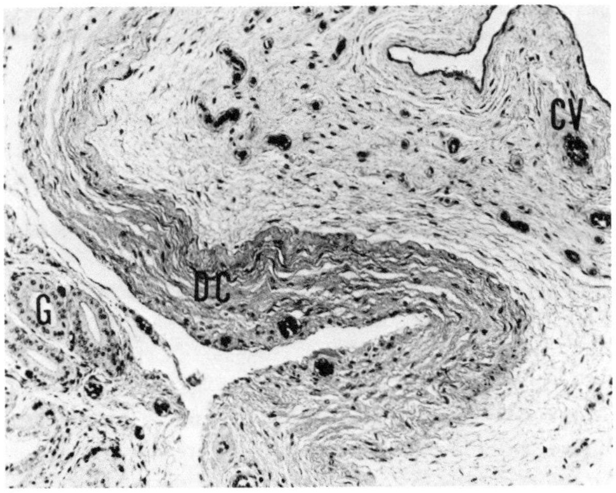

Figure 5. Free membranes of *Choloepus didactylus* (Case 3) attached to the uterus. Maternal endometrial glands of decidua vera are at G, decidua capsularis at DC, and large umbilical vessel branch is at CV. Amnion is at the top right. The large number of small vessels in the thick chorionic connective tissue represent remnants of vessels from a formerly more diffuse placenta. (H and E stain; 100 X magnification).

remnants of ducts. Unlike other observations, the surface of umbilical cord in the mature specimen of *Bradypus tridactylus* (case 1) was diffusely keratinized. Amniotic nodules as described for *Bradypus griseus* by Wislocki (1927) were not found in any specimen. The amnion contained large, pigmented macrophages. The free membranes contained numerous intact fetal vessels in the chorion layer. The larger vessels are within the major chorionic layer as are the major chorionic umbilical branches, while the numerous smaller and more peripheral vessels appear to be the remnants of the former diffuse labyrinth (Fig. 5). Endometrium with glands was found at the base of the placenta and beneath the free membranes in the two specimens observed in the intact uterus (cases 3 and 4). Large bundles of connective tissue made up part of the interlobular septa and undermined the labyrinth at the maternal floor. This connective tissue was separated from the decidua by a thin, basement membrane (Fig. 6). In the other specimens, necrosis was present at the placental base. Trophoblast was not present in the decidua basalis or in the myometrium beneath the implantation site, unlike human placentation. Also, no trophoblast was found within maternal blood vessels.

The histologic features of the labyrinth are confusing because of the large number of "epithelioid" cells (Wislocki, 1927) found within the fetal connective tissue (Fig. 7). They contained large quantities of irregular inclusions of unknown nature (Fig. 8) some of which were slightly brownish. Syncytium occurred in a diffuse manner, around maternal vessels as described previously (Fig. 7) and at the periphery of lobules. Mitoses were not observed and no cytotophoblast was recognized. The endothelial cells of maternal vessels were especially prominent but not invaded by trophoblast. Extra-

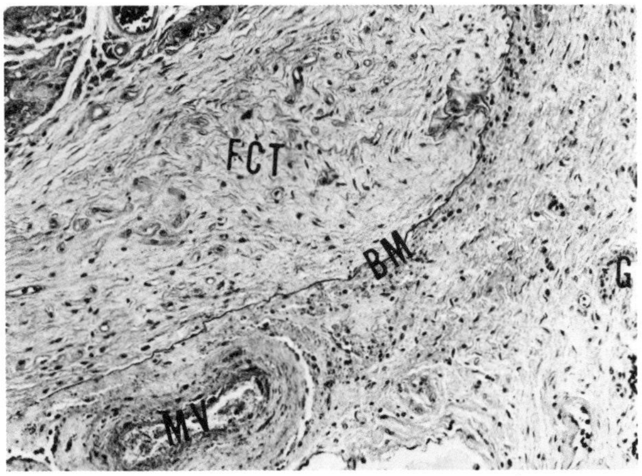

Figure 6. Floor of the placenta from case 3 (Fig. 5). At top left is the edge of the labyrinth, accompanied by a fetal connective tissue septum (FCT) which is delineated by basement membrane (BM). Decidua basalis is below this membrane with maternal vessels (MV) and glands (G). (H and E stain; 100 X magnification).

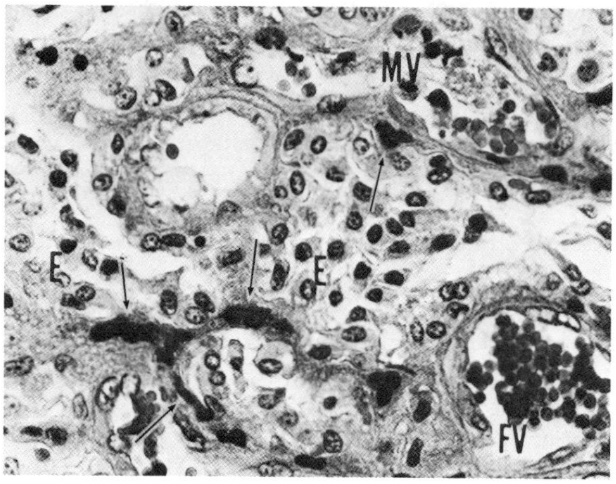

Figure 7. Labyrinth of the placenta from Case 3. There is a large aggregation of epithelioid cells (E) in the fetal connective tissue. Aggregates of syncytial nuclei (arrows) surround the maternal vessels (MV). Fetal vessels (FV) are filled with red cells. (H and E stain; 400 X magnification).

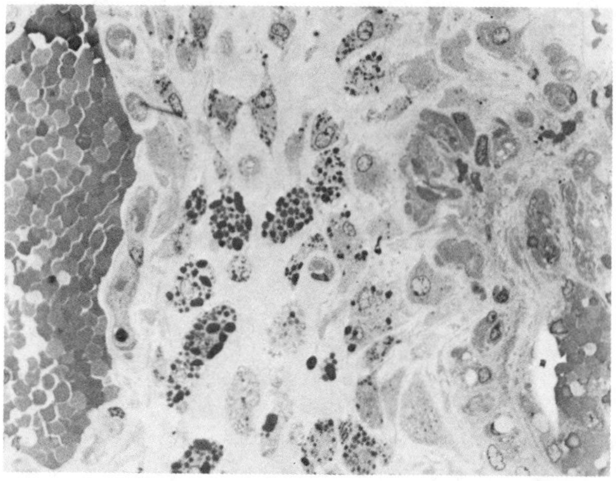

Figure 8. Thin section of the placenta from Case 3 showing epithelioid cells in the center with large, irregular cytoplasmic inclusions. (Toluidine blue, epoxy; 400 X magnification).

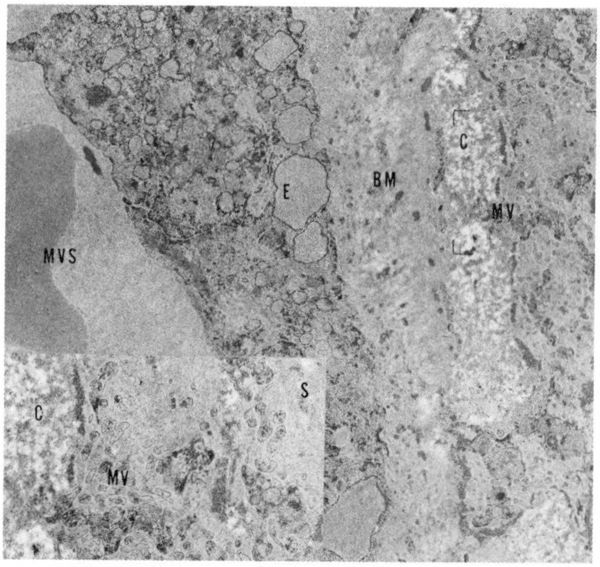

Figure 9. Electronmicrograph of "placental barrier" Case 4, *Choloepus didactylus* at term. Shown are maternal vascular space with red cell (MVS), endothelial cell with dilated endoplasmic reticulum (E), basement membrane (BM), and microvilli (MV) of syncytiotrophoblast (15,000 X magnification). In the inset, the designated area of microvillous syncytial surface is enlarged (22,500 X magnification). Individual microvilli penetrate the collagen and basement membrane of maternal vessel walls. In several areas the collagenous envelope (C) is penetrated by tongues of syncytial trophoblastic cytoplasm (S) which comes to lie adjacent to the basement membrane.

vasation of red cells was seen in some areas of case 3 only and was considered to be of pathologic significance.

Electron microscopic.-- Thin sections of case 3, an immature placenta, revealed considerably more fetal connective tissue than in case 4. The abundant macrophages (epithelioid cells of Wislocki) contained toluidine blue staining inclusions (Fig. 8) and were less prominent at later gestation, as in case 4. The inclusions were largely composed of large electrondense bodies resembling lysosomes.

Despite postmortem autolytic changes, a representative section of the fetal-maternal "barrier" gave sufficient detail for interpretation (Fig. 9). The maternal vessels contained prominent endothelial cells whose cytoplasm was replete with dilated endoplasmic reticulum. Whether this dilatation is of autolytic nature or similar to that shown to exist normally in the cat (Björkman, 1973) is uncertain. Peripheral to this was a basement membrane which, in turn, was enveloped by interrupted collagen bundles. Syncytiotrophoblast abutted the collagen layer and frequently interrupted the bundles with tongues of cytoplasm reaching the vascular basement membrane. Microvilli covered this surface of syncytium and extended in some areas through the basement membrane to the maternal endothelium. No cytotrophoblast was recognized and fetal capillaries were much smaller than the maternal vessel shown (Fig. 6).

DISCUSSION

The examination of three mature and one approximately mid-term sloth placentas from three different species showed essentially similar features, and confirmed the extensive findings on *Bradypus griseus* (Wislocki, 1927). A few minor variations such as keratinization of the umbilical cord surface in case 1 and the absence of amniotic nodules are either individual variations or pathologic circumstances. The histologic appearance of the labyrinth was also found to be as previously described.

This is the first electron microscopic study of the sloth placenta, and although distorted by some postmortem artifacts, it provides the first detailed view of the placental barrier. It supports Wislocki's (1928) rejection of deLange's (1926) contention that this xenarthran has a hemochorial placenta. Syncytium was opposed to a layer of collagen, and basement membrane was peripheral to the prominent maternal endothelial cells. The prominent nature of these endothelial cells is perhaps due to their cytoplasm being filled with dilated endoplasmic reticulum. As Wislocki (1927) correctly observed, a thin, connective tissue barrier separates maternal endothelial cells from syncytium. We found occasional tongues of syncytial cytoplasm and microvilli that penetrate this layer to the endothelial cells.

The placenta compares most closely to that of carnivores. In a review of the differences between epitheliochorial and endotheliochorial placentas Björkman (1973) demonstrated very similar features in dog and cat fetal-maternal areas to those described here. The prominent maternal endothelium in these forms was considered to be possibly the result of response to trophoblast (Enders, 1957). In several of these species, tongues of trophoblastic microvilli can be traced to the maternal endothelial cells, as is the case in the sloths. In most areas, however, a basement membrane and collagen mantle separate endothelium and trophoblast. Whether this connective tissue is of maternal or fetal origin is debated frequently in the literature and is still not resolved.

The only other xenarthran placentas studied in some detail are those of two *Dasypus* species. Despite the fact that these animals also possess a simplex uterus, their placentas differ in macroscopic structure and in their hemochorial relationship to the maternal organism.

ACKNOWLEDGMENTS

We are grateful to Dr. Dennis Meritt, Lincoln Park Zoo, Chicago and Dr. W. Jorge, Botucatu, Brazil, for specimens. Development grants from DRR-NIH (1-S08-RR09002-02), NINCDS (NS 14162) and Basil O'Connor Starter grant NF5-208 are gratefully acknowledged.

LITERATURE CITED

Björkman, N.
1973. Fine structure of the fetal-maternal area of exchange in the epitheliochorial and endotheliochorial types of placentation. Acta Anat., 89 (Suppl. 1): 1-22.

deLange, D.
1926. Quelques remarques sur la placentation de "*Bradypus*".Compt. Rend. Assoc. Anat. Liege, 21: 321-333.

Enders, A. C.
1957. Histologic observations on the chorioallantoic placenta of the mink. Anat. Rec., 127: 231-245.

Heuser, C. H. and G. B. Wislocki
1935. Early development of the sloth (*Bradypus griseus*) and its similarity to that of man. Contrib. Embryol. Carnegie Inst. Wash., 25: 1-13.

Turner,W.
1873. On the placentation of the sloths. Trans. Roy. Soc. Edinburgh, 27: 71-104.

Wislocki, G. B.
1927. On the placentation of the tridactyl sloth (*Bradypus griseus*) with a description of some characteristics of the fetus. Contrib. Embryol. Carnegie Inst. Wash., 19: 209-228.

1928. Further observations upon the minute structure of the labyrinth in the placenta of the sloths. Anat. Rec., 40: 385-395.

THE EVOLUTION OF MONOZYGOTIC POLYEMBRYONY IN *DASYPUS*

Gary J. Galbreath

Committee on Evolutionary Biology, University of Chicago, Chicago Illinois 60637
(Present address: Department of Geology, Field Museum of Natural History, Chicago Illinois 60605; and, Biology Department, Purdue University, Hammond Indiana 46323)

Abstract
Monozygotic polyembryony is unique to *Dasypus* within the Vertebrata. A model is proposed to explain the evolution of this trait, indicating the critical importance of an anatomically preadapted uterus. Relevant anatomical and litter-size data pertaining to several armadillo genera are discussed. It is suggested that models invoking preadaptive constraint deserve more attention from evolutionary biologists.

Abstracto
La poliembrionía monozigótica ocurre solamente en el género *Dasypus*, dentro del grupo de los vertebrados. Este artículo propone un modelo para explicar la evolución de este rasgo, que señala la importancia crítica en dicha evolución de un útero anatómicamente preadaptado. Se discuten datos relevantes, tanto anatómicos como sobre el tamaño de la camada en varios géneros de armadillos. Se sugiere que los modelos que invocan un mayor apremio preadaptivo merecen más atención de parte de biólogos evolucionarios.

Resumo
Poliembrionia monozigótica é exclusiva para *Dasypus*, dentro dos Vertebrata. Um modelo é proposto para explicar a evolução dessa característica, indicando a importância crítica de um útero anatomicamente préadaptado. Dados anatômicos relevantes e tamanho da ninhada, concernentes a vários gêneros de tatus, são discutidos. Sugere-se que hipóteses com referência a préadaptações merecem mais atenção dos biólogos evolucionistas.

As a typical aspect of reproduction, monozygotic polyembryony, first shown to occur in *Dasypus novemcinctus* by Patterson (1913), has been found in no genus of the subphylum Vertebrata other than *Dasypus* (the long-nosed armadillos). A single fertilized egg ultimately gives rise to multiple embryos, following division of the blastocyst's inner cell mass. The process of separate embryo formation and subsequent embryonic development in *D. n. mexicanus* has been discussed by Newman and Patterson (1910), Patterson (1913), and Talmage and Buchanan (1954). Aspects of embryonic development in *D. hybridus* (Fernandez, 1909) indicate the operation of a similar process.

In *D. n. mexicanus* in Texas and Florida, ovulation occurs most often in summer, blastocyst implantation in autumn, and parturition in March and April (Patterson, 1913; Hamlett, 1935; Enders, 1966; Galbreath, unpublished data). It appears likely that many young females, ovulating for the first time, ovulate in the autumn (Enders, 1966). Ovulation probably occurs in multiparous females in response to a reduction in, or cessation of, suckling by their infants. In young females, ovulation presumably occurs when physiological reproductive maturity is attained. In either case, irrespective of the time of year at which fertilization occurs, the blastocyst remains quiescent within the uterine fundus until implantation is triggered, perhaps by some aspect of autumnal climatic change. Mature males typically produce sperm year-round (Galbreath, 1982).

The number of living, monozygotic embryos within a *D. n. mexicanus* uterus is usually four, although occasionally there are three or five (Newman, 1913; Buchanan, 1957; Galbreath, unpublished data). Buchanan (1957) reported a uterus that contained six viable embryos plus an extra amniotic sac. Storrs (1967) recorded the birth in captivity of a litter containing at least five young, and a newly captured female in my possession gave birth to twins. Sets of three viable embryos of *D. n. mexicanus* are sometimes accompanied by a fourth, necrotic, individual. In one set of four embryos from Florida, all were severely skeletally malformed and two were necrotic.

In *D. hybridus*, the number of embryos originally formed reaches a maximum of at least 12 (Fernandez,

1909). However, of 11 embryo sets sufficiently advanced in development to be sexed (Jhering, 1885, 1886; Fernandez, 1909), 10 contained either 8 or 9 viable embryos, and the remaining set contained 7. Reduction of viable embryo number sometimes occurs as a result of intrauterine mortality.

In three cases *D. sabanicola* females gave birth in captivity to same-sex sets of quadruplets, and data from specimens collected in the field are also consistent with the hypothesis that monozygotic polyembryony is typical (E. Storrs, pers. comm.).

Three sets of embryos from *D. kappleri* females were same-sex pairs and a fourth set consisted of two unsexed embryos (R. Wetzel, pers. comm.), suggesting a similar mode of reproduction but a smaller litter size in this species. The presence or absence of monozygotic polyembryony in either *D. septemcinctus* or *D. pilosus* (the two remaining species of the genus) has yet to be investigated.

Two reproductive tracts with bicornuate uteri, said to represent *D. septemcinctus*, were described by Altmann 1924). Their uterine anatomy suggests an absence of monozygotic polyembryony. I have identified these specimens as *Chaetophractus vellerosus* and *Euphractus sexcinctus*. The reproductive tract described in greater detail by Altmann was that of the *C. vellerosus* individual.

In dorsal view the uterus of *Dasypus novemcinctus* is shaped like a kite. The oviducts enter on either side, at the level of the greatest width of the uterus. The most posterior part of the uterus consists of a narrow section termed "cervical." Between the level of the oviducts and the cervical segment lies the uterine "corpus."

The roughly triangular portion of the uterus anterior to the corpus is the "fundus." The fundus is unusual in that it encloses a large section of the uterine cavity. During the period of developmental delay, the blastocyst normally lies in a small endometrial depression at the anterior fundic tip, and implantation (when successful) has invariably been found to occur at this site. This depression lies at the junction of four endometrial furrows, two of which run from the oviduct orifices to the endometrial depression. It is probable that these two furrows guide blastocysts to the fundic tip (Patterson, 1913; Buchanan, 1967).

Buchanan (1967) described a case of two separate conceptuses within one *Dasypus* uterus. At the fundic tip was an implantation site with four embryos. A large, unimplanted blastocyst was also present. Buchanan suggested that the blastocyst originally occupying the more anterior position in the cone-shaped fundic depression had been able to implant sooner than the other blastocyst due to being closer to the sinusoidal layer, and that a decidual reaction had then blocked the potential for implantation in the surrounding endometrial area. Though occasional cases of double ovulation have been found in *D. novemcinctus*, no indication of more than one implantation has been reported. It is probable that the specialized uterus is virtually incapable of allowing the implantation of more than a single blastocyst.

AN EVOLUTIONARY MODEL

The existence of this single known instance of monozygotic polyembryony among the vertebrates has led to speculation regarding possible adaptive advantages that could accrue uniquely to *Dasypus* spp. through the adoption of this trait, as opposed to the normal polyzygotic method of producing multiple young. But I know of no plausible optimization model, and suspect that preadaptive constraints may instead have been critical.

I hypothesize that some common ancestor of all extant armadillo species exhibiting monozygotic polyembryony had a normal litter size of one. A specialized uterus evolved with a large fundic cavity, endometrial grooves, and a fundic apical depression. The adaptive value of permanently positioning the blastocyst in any area conducive to successful implantation might have been high if there was often a period of delay before implantation. This set of adaptations ensured that the blastocyst would be positioned against the endometrium of the anterior section of the uterus, where the sinusoids are largest and are closest to the luminal epithelium in modern *D. novemcinctus*. The overall adaptive advantage was in increased reliability of implantation and/or increased certainty that implantation would occur in an optimal area of the uterus. Such a specialized uterus would have been initially maladaptive in an armadillo species with a polyzygotic litter size of two or more.

Later, some adaptational and/or ecological change produced a situation in which an increase in litter size was both adaptive and energetically possible. But the uterus of this armadillo was now incapable of implanting more than one blastocyst. Monozygotic polyembryony, which occurs as a rare event in many mammalian species, was therefore favored by natural selection as a means of increasing litter size.

It could be argued that the peculiar uterus typical of *D. novemcinctus* (and possibly of some or all other *Dasypus* species) evolved as part of the mechanism of reproduction involving monozygotic polyembryony. But there is no apparent functional or direct causal link between localized implantation and monozygotic polyembryony. My model suggests, instead, that the specialized uterus evolved first and preadapted for monozygotic polyembryony any descendant species that underwent selection for increased litter size.

Should even moderately similar uterine internal morphology, promoting anterior implantation, be found in any species of armadillo with a litter size of one, the model would be strongly supported. The model would be effectively falsified by the proven existence of a *D. novemcinctus*-like uterus in a polyzygotic, polyembryonic armadillo species.

The probable sister group of *Dasypus* (considering only extant armadillo species) consists of a *Cabassous Priodontes* clade, and *Tolypeutes* constitutes the probable sister group of a group containing the above three genera (Engelman, 1978). A group consisting of all four genera then constitutes the probable sister group of all other extant armadillos. All four genera are believed to be holophyletic taxa. Of the nineteen extant armadillo species (cf. Wetzel, 1985), four definitely have bicornuate uteri. These include *Chaetophractus villosus* (Fernandez, 1915), *Chaetophractus vellerosus* (Altmann, 1924), and *Euphractus sexcinctus* (Benirschke et al., 1969). A specimen of *Cabassous centralis* in the Field Museum of Natural History (FMNH 121185) had a bicornuate uterus with horns at right angles to the uterine body. This uterus lacked a fundic cavity, as did the *Tolypeutes tricinctus* uterus mentioned by Garrod (1878). Specializations of the *D. novemcinctus* uterus thus necessarily evolved later than the phylogenetic split between the *Dasypus* and *Cabassous* - *Priodontes* clades. This divergence can be dated to the Miocene or earlier, based on the presence of fossils of a *Dasypus* relative *Stegotherium* in Miocene deposits.

Chlamyphorus truncatus (Minoprio, 1945), *Dasypus hybridus* (Fernandez, 1909), and *D. novemcinctus* have simplex uteri. It is clear that simplex uteri have evolved at least twice within the armadillos. Despite the bicornuate uterus, *Chaetophractus villosus* embryos typically do not occupy the uterine horns (Fernandez, 1915). It is possible that all armadillo species possess "functionally simplex" uteri.

Each of at least eight cases of birth or pregnant uterus examination for captive *Cabassous centralis* disclosed a single neonate or embryo (Meritt, pers. comm.). More than fifty births of *Tolypeutes matacus* in captivity produced a single infant each (Meritt, pers. comm.). Each of three pregnant *T. matacus* collected in the wild contained a single embryo (Krieg, 1929; Sanborn, 1930).

These data on litter size in species of 2 of the 3 extant genera most closely related to *Dasypus* are phylogenetically consistent with the idea that a *Dasypus* ancestor had a typical litter size of one. The existing data concerning armadillo uterine morphology are also consistent with the model.

Evolutionary explanations invoking preadaptive constraint are sometimes treated, a priori, as hypotheses of last resort, worthy of consideration only if all optimization hypotheses are falsified. I believe that two conclusions of general importance are indicated by the present study. (1) Plausible, detailed hypotheses involving preadaptive constraint as a major factor can sometimes be developed. (2) Such hypotheses should be given initial consideration equivalent to that given optimization hypotheses, and should be subjected to attempted falsification whenever possible.

ACKNOWLEDGMENTS

I thank D. Meritt, E. Storrs, and R. Wetzel for providing information, and D. Meritt and the Lincoln Park Zoological Gardens for permission to examine live *Chaetophractus vellerosus* specimens.

LITERATURE CITED

Altmann, F.

1924. Beiträge zur Anatomie des weiblichen Genitales der Dasypodiden. Z. f. Anat. u. Entwicklungsgeschichte, 72: 390-406.

Benirschke, K., R. J. Low, and V. H. Ferm.

1969. Cytogenetic studies of some armadillos. Pp. 330-345, *in* Comparative mammalian cytogenetics (K. Benirschke, ed.). Springer Verlag, New York, 473 pp.

Buchanan, G. D.

1957. Variation in litter size of nine-banded armadillos. J. Mamm., 38: 529.

1967. The presence of two conceptuses in the uterus of a nine-banded armadillo. J. Reprod. Fert., 13: 329-331.

Enders, A. C.

1966. The reproductive cycle of the nine-banded armadillo (*Dasypus novemcinctus*). Pp. 295-310, *in* Comparative biology of reproduction in mammals (I. W. Rowlands, ed.). Academic Press, London, 559 pp.

Engelmann, G. F.

1978. The logic of phylogenetic analysis and the phylogeny of the Xenarthra (Mammalia). Ph. D. Dissertation, Columbia Univ., 329 pp.

Fernandez, M.

1909. Beiträge zur Embryologie der Gürteltiere. 1. Zur Keimblätterinversion und spezifischen Polyembryonie der Mulita. (*Tatusia hybrida* Desm.). Morpholog. Jahrb., 39: 302-333.

1915. Über einige Entwickelungsstadien des Peludo (*Dasypus villosus*) und ihre Beziehung zum Problem der spezifischen Polyembryonie des genus *Tatusia*. Anat. Anz., 48: 305-327.

Galbreath, G. J.
1982. The armadillo, *Dasypus novemcinctus*. Pp. 71-79, *in* Wild mammals of North America: biology, management and economics (J. A. Chapman and G. A. Feldhamer, eds.). Johns Hopkins Univ. Press, Baltimore, 1147 pp.

Garrod, A. H.
1878. Notes on the anatomy of *Tolypeutes tricinctus*, with remarks on other armadillos. Proc. Zool. Soc. London 1878:222-230.

Hamlett, G. W. D.
1935. Delayed implantation and discontinuous development in mammals. Quart. Rev. Biol., 10: 432-447.

Jhering, H. von.
1885. Über die Fortpflanzung der Gürtelthiere. Sitzungsberichte d. Koenigl. Preuss. Akad. d. Wissenschaften zu Berlin 1885: 1051-1053.

1886. Ueber "Generationswechsel" bei Säugethieren. Archiv f. Anat. u. Physiol., Physiol. Abthlg. 1886: 443-450.

Krieg, H.
1929. Biologische Reisestudien in Südamerika. IX. Gürteltiere. Z. f. Morphol. u. Okol. d. Tiere, 14:166-190.

Minoprio, J. D. L.
1945. Sobre el *Chlamyphorus truncatus* Harlan. Acta Zoologica Lilloana, 3: 5-58.

Newman, H. H.
1913. The natural history of the nine-banded armadillo of Texas. Amer. Nat., 47: 513-539.

Newman, H. H., and J. T. Patterson.
1910. The development of the nine-banded armadillo from the primitive streak stage to birth; with especial reference to the question of specific polyembryony. J. Morph., 21: 359-424.

Patterson, J. T.
1913. Polyembryonic development in *Tatusia novemcincta*. J. Morph., 24: 559-684.

Sanborn, C. C.
1930. Distribution and habits of the three-banded armadillo (*Tolypeutes*). J. Mamm.,11:61-68.

Storrs, E. E.
1967. Individuality in monozygotic quadruplets of the armadillo, *Dasypus novemcinctus*. Ph. D. Dissertation, Univ. of Texas, 141 pp.

Talmage, R. V., and G. D. Buchanan.
1954. The armadillo (*Dasypus novemcinctus*). A review of its natural history, ecology, anatomy and reproductive physiology. Rice Inst. Pamph., 41 (2): 1-135.

Wetzel, R.M.
1985. Taxonomy and distribution of armadillos, Dasypodidae. (this volume).

FETAL ENDOCRINOLOGY OF THE COMMON LONG-NOSED ARMADILLO *DASYPUS NOVEMCINCTUS*

Kay C. Nakakura and Bill L. Lasley

Research Department, San Diego Zoo, P.O. Box 551, San Diego, CA 92112 USA.

Abstract
Steroid hormone levels and their production sites were evaluated during the pregnancy of a common long-nosed armadillo. Progesterone was the major steroid produced by the fetal adrenal gland and appeared to be the major contributor to the six- and tenfold concentration from the fetal to the maternal compartment. Dehydroepiandrosterone in the fetal circulation appeared to be a product of the fetal testis. Undetectable levels of bioactive luteinizing hormone or steroidogenic activity was attributed to the placenta. This study reveals more differences than similarities between the feto-placental units of armadillos and primate species.

Abstracto
Se ha evaluado el nivel de hormonas esteroides y sus sitios de produccíon durante el embarazo de un armadillo de trompa larga. La pregesterona fue el esteroide producido en mayor cantidad por la glándula adrenal fetal y parecío ser el mayor contribuyente a la concentracíon de seis y diez pliegues desde el compartimiento fetal al maternal. En la circulación fetal, el dehidroepiandrosterona de la hormona luteinizante bioactiva o la actividad esteroidogénica fue atribiuda a la placenta. Este estudio revela que existen más diferencias que similitudes entre unidades del feto-placental de armadillo y unidades de especies de primates.

Resumo
Os níveis de hormônio esteróide e seus sítios de produção foram avaliados durante a prenhez de uma fêmea de tatu comum de nariz comprido. A progesterona foi principal esteróide produzido pela glàula adrenal fetal e pareceu ser o principal contribuinte para a concentração de seis e dez vezes do compartimento fetal para o maternal. A dihidroepiandrosterona na circulação fetal pareceu ser um produto dos testículos fetais. Níveis indectáveis de hormônio luteinizante bioativo ou atividade esteroigênica foram atribuídas à placenta. Este estudo revela mais diferenças do que similaridades entre as unidades feto-placentárias de tatus e espécies primatas.

In this study we evaluate the steroid hormones of pregnancy in the armadillo for comparison with other animals. The fetal common long-nosed armadillo *Dasypus novemcinctus* has been suggested as a possible experimental model for the primate feto-placental unit because it has a fetal zone in the adrenal gland which involutes following birth, as described for the human and nonhuman primate fetal adrenal gland (Moser and Benirschke, 1962). Although much is known about the hormone profiles associated with pregnancy and the contribution of the fetal adrenal gland in the human, no such profile has been established in the common long-nosed armadillo.

In vitro incubations of adrenal gland slices indicate that the fetal armadillo adrenal is steroidogenically active with conversion of ^{3}H-progesterone to Cortisol and Corticosterone (Bloch and Benirschke, 1964). Fetal armadillo adrenals may provide necessary precursors for Estriol biosynthesis since Dehydroepiandrosterone (DHEA) and 16-hydroxy-DHEA were tentatively identified in fetal plasma (Brinck-Johnsen et al., 1967). Placental aromatase was also found in both early and late gestation. However placental, urinary and plasma Estriol, Estrone, and Estradiol were at very low levels in the pregnant armadillo (Brinck-Johnsen et al., 1967; Brinck-Johnsen, 1970). Thus, despite morphologic similarities with the fetal primate adrenal, the fetal armadillo adrenal may be steriodogenically different.

MATERIALS AND METHODS

The major ^{3}B-hydroxy-Δ^5-steroids and Δ^4 -3-keto-steroids were measured in maternal and fetal serum from five pregnancies. Bioactive Luteinizing Hormone was also measured in maternal and fetal serum and placental homogenates. *In vitro* tissue incubations of maternal and fetal adrenal and placental minces were performed using tritiated Pregneonolone as a precursor. In addition, fetal gonadal tissues were monitored for steroidogenic activity.

Pregnant armadillos were killed by ketamine injection, and maternal blood was drawn from the heart and allowed to clot. The fetuses were exposed, the umbilical cord cut 2 cm from the fetus and 1 - 3 ml of blood was milked from each fetus and allowed to clot. Serum was separated, then stored at -20°C until assayed. Sera from the four fetuses from each pregnancy were pooled. Maternal and fetal adrenal glands, fetal gonads and placentae were removed, weighed and frozen in liquid nitrogen until used. Term fetal weight in common long-nosed armadillos is about 100 g, and mean weights of the four fetuses from each of the five pregnancies were 38.3g, 44.3g, 54.6g, 69.8g and 70.8g.

Maternal and pooled fetal sera were treated with anhydrous diethyl ether to extract the steroids (Mallinckrodt, St. Louis, MO; 10X extraction volume) and the steroids (Sigma Chemical Co., St. Louis, MO) Progesterone, 17α-hydroxyprogesterone, Androstenedione, Testosterone, Pregnenolone, Dehydroepiandrosterone (DHEA) and Androstenediol were separated on celite columns as described previously (Anderson et al., 1976). Cortisol and Corticosterone were extracted with dichloromethane (Mallinckrodt) and separated by celite column chromatography (Manlimos and Abraham, 1975). The following antibodies (AB) were used: 11α-hydroxy-Progesterone-11α hemisuccinate-HSA directed Progesterone-AB (Abraham S7#4), 17OH Progesterone-oxime-BSA directed 17OH Progesterone-AB (Endocrine Science #P17-43), A-3oxime-HSA directed Androstenedione-AB (Abraham S-1557#2), Testosterone-3-oxime-HSA directed Testosterone-AB (Abraham S-741#7), Pregnenolone-3-monohemisuccinate-HSA directed Pregnenolone-AB (Abraham S-873#5), DHEA-3B monohemisuccinate-HSA directed DHEA-AB (Abraham S-1502#7) and Cortisol-21-hemisuccinate-BSA directed Cortisol-AB (Hopper #F39). Adiol was measured using the Testosterone-AB which cross-reacted 80 percent. The Cortisol-AB was also used to measure Corticosterone which had a 20 percent cross-reactivity. The interassay coefficient of variation ranged from 4 percent (Testosterone assay) to 18 percent (Progesterone assay).

Adrenal, gonadal and placental tissues were thawed slowly at 4°C and kept on ice throughout the preparatory procedure. Each tissue was weighed, minced and placed in a plastic vial containing 2 ml radioactive Media 199 (Gibco, Grand Island, NY) and ^{3}H-Pregnenolone (New England Nuclear, Boston, MA; 15,000 cpm/Incubation 1; 29,000 cpm/Incubations 2 and 3). The incubation was carried out in a 37°C shaker bath under 95 percent O_2 and 5 percent CO_2 for 3 hrs, stopped by snap-freezing in methanol:dry ice and kept frozen until extracted three times with 8 ml ether. The steroids were separated and assayed in the same manner as the serum samples. Both radioactive conversion products and immunoreactive steroids were monitored in each fraction.

Ten mg of the steroid was added to the radioactive sample and repeatedly recrystallized until the specific activity in the crystal and mother liquor were identical for two consecutive recrystallizations. The solvent systems used were methanol:H_2O, ethylacetate:isooctane, and ethanol:H_2O.

The celite fraction containing Progesterone was derivatized with MOX (Pierce, Rockford, IL), assessed by mass spectometry following gas-liquid separation and compared to pure crystalline Progesterone derivatized in an identical procedure.

Biologically active Luteinizing Hormone was evaluated in maternal and fetal serum and placental homogenates by the rat interstitial cell Testosterone in vitro bioassay (Dufau et al., 1974; Dufau et al., 1976).

RESULTS

Progesterone was found in high concentrations in the fetus (480 - 920 ng/ml) with lower levels in the mothers (60 -90 ng/ml), providing a 6- to 10-fold concentration gradient from fetus to mother (Fig. 1). 17α-hydroxy-Progesterone concentrations ranged from 1.8 - 13 ng/ml in the mothers and 2.4 - 14.4 ng/ml in the fetuses, demonstrating a small fetal to maternal gradient in two of the five pregnancies. Androstenedione exhibited a small but consistent gradient from fetus (1.1 - 5.5 ng/ml) to mother (0.75 - 3.4 mg/ml). Testosterone concentrations were consistently higher in the three male fetal pools (1.7 - 3.9 ng/ml) compared to the mothers (0.5 - 1.4 ng/ml), whereas the reverse was true in the two female fetal pools (0.4 - 0.7 ng/ml vs. 0.2 - 0.3 ng/ml).

The ^{3}B-hydroxy- Δ^5 steroids, dehydroepiandrosterone (DHEA) and Androstenediol were slightly elevated in the maternal circulation (Fig. 2). DHEA ranged from 4 - 22 ng/ml in the mothers and 3.5 - 13 ng/ml in the fetuses. Androstenediol concentrations were 1 - 2.8 ng/ml in the maternal circulation compared to 0.8 - 1.2 ng/ml in the fetal circulation. Corticosterone values were also higher in the mothers except for armadillo number 21 (30 - 110 ng/ml vs. 8 - 75 ng/ml). Cortisol, the major glucocorticoid produced by the armadillo adrenal was much higher in the mothers as compared to the fetuses (130 - 340 ng/ml vs. 15 - 45 ng/ml).

In order to compare the steroid measurements in the armadillo to values in other species, maternal and fetal serum samples from sheep and horse pregnancies were evaluated in the same assay system. The sheep pregnancy, terminated at 133 days gestation, corresponded to Progesterone concentrations of 11.9 ng/ml in the mother and 0.45 ng/ml in the fetus (Fig. 3). The horse data are mean values of pregnancies ranging from 155 - 280 days gestation, where Progesterone concentrations were $<$ 1 - 3 ng/ml in the pregnant mares and

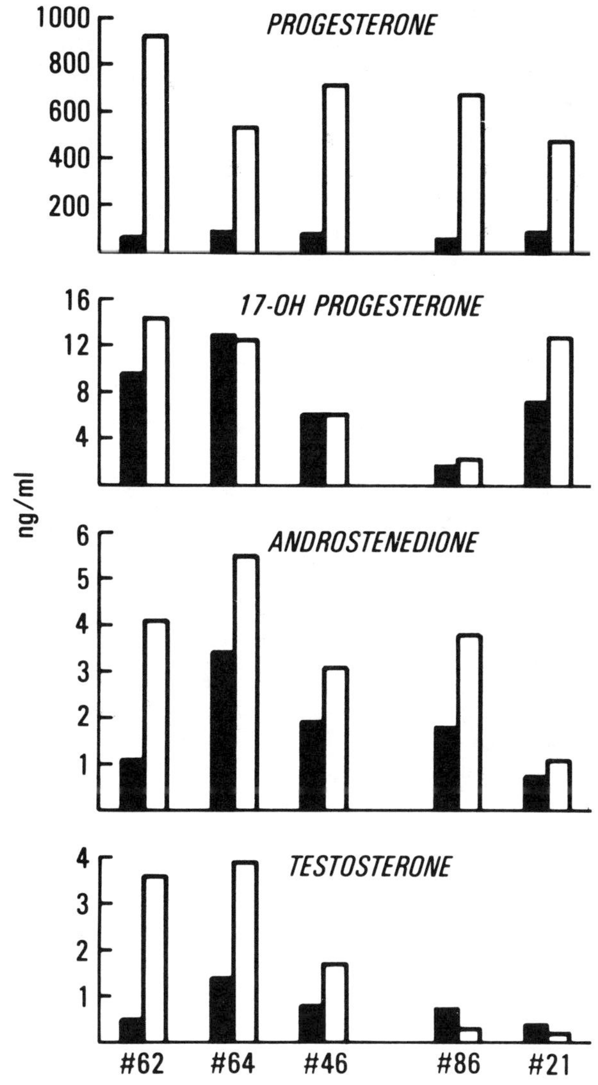

Figure 1. The major Δ^4 and Δ^5 steroids in maternal and fetal serum of five armadillo pregnancies grouped according to fetal sex and size. Circulating steroid concentrations in maternal (solid bar) and fetal (open bar) armadillo serum samples. Male fetuses: armadillo number 62 - mean fetal weight 38.3 g, number 64 - 54.6 g, number 46 - 70.8 g. Female fetuses: number 86 - 44.3 g, number 21 - 69.8 g.

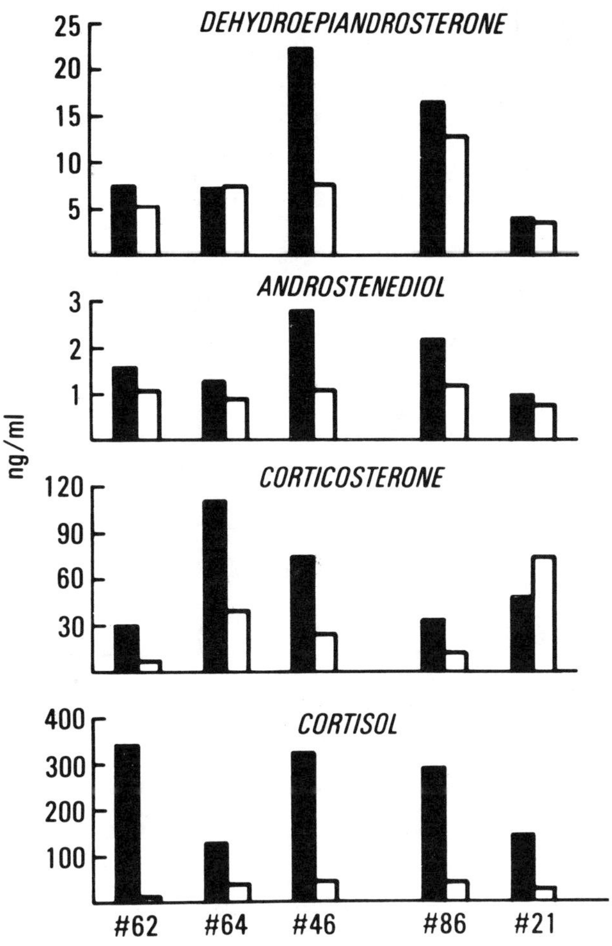

Figure 2. The 3B-hydroxy- Δ^5 steroids in maternal and fetal serum of five armadillo pregnancies grouped according to fetal sex and size. Circulating steroid concentrations in maternal (solid bar) and fetal (open bar) armadillo serum samples. Male fetuses: armadillo number 62 - mean fetal weight 38.3 g, number 64 - 54.6 g, number 46 - 70.8 g. Female fetuses: number 86 - 44.3 g, number 21 - 69.8 g.

1 - 6 ng/ml in the fetuses. DHEA values were much higher in the fetuses (315 ng/ml) compared to the mare (3.3 ng/ml), corresponding to DHEA production by the fetal gonad (Raeside, 1976).

Bioactive Luteinizing Hormone was also quantitated in maternal and fetal armadillo serum and placental homogenates. Luteinizing Hormone activity measured in two placental homogenate samples was at or below the limit of sensitivity of the assay (16 ng/ml protein). Maternal Luteinizing Hormone concentrations from three serum samples were 41 ± 2.5 ng/ml (mean ± SE) whereas bioactive Luteinizing Hormone in serum from four fetal male and four fetal female samples were 131 ± 5 ng/ml and 161 ± 15 ng/ml Luteinizing Hormone, respectively.

Both radioactive conversion products and immunoreactive steroids were monitored from *in vitro* incubations of various tissues of fetuses from three different armadillo pregnancies to observe bulk flow of the steroids. The major radioactive conversion product of ^{3}H-Pregnenolone was ^{3}H-Progesterone in all three fetal armadillo adrenal gland incubations, with 25 percent, 54 percent and 50 percent of the ^{3}H-Pregnenolone being converted to ^{3}H-Progesterone in the incubations, respectively (Figs. 4, 5, 6). The ^{3}H-Progesterone from incubations 2 and 3 was recrystallized to constant specific activity and Progesterone concentration was measured. Progesterone (1.5, 32 and 20 ng/mg tissue, respectively) was also the major immunoreactive steroid as measured by RIA and confirmed by

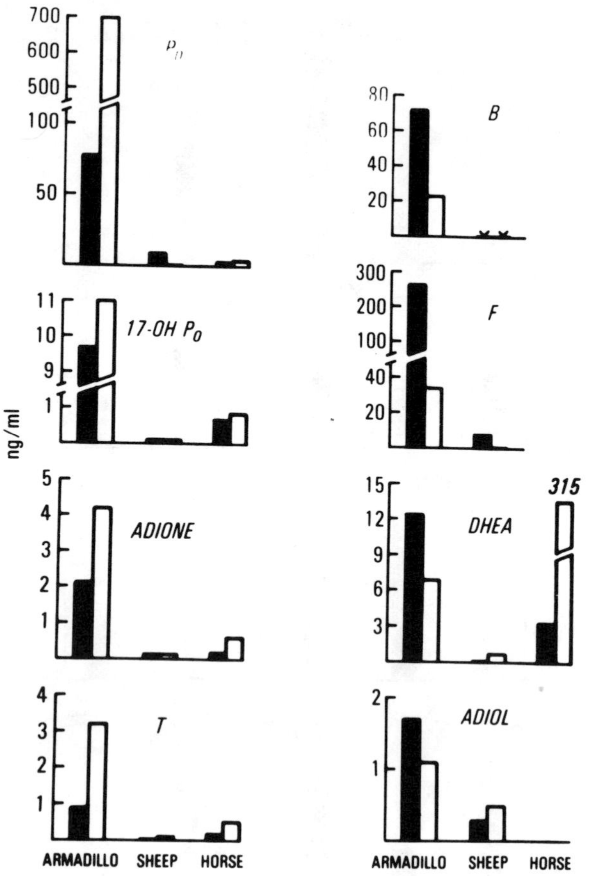

Figure 3. Circulating steroid concentrations in maternal (solid bar) and fetal (open bar) serum samples from armadillo (mean of 5 pregnancies), sheep (133 days gestation) and horse (mean of samples from 155 - 280 days gestation) pregnancies. Po = ^{3}H-Progesterone; B = Corticosterone; 17OH Po = 17a-hydroxy-progesterone ; F = Cortisol; T = Testosterone. Corticosterone, Cortisol, and Androstenediol were not measured in the horse.

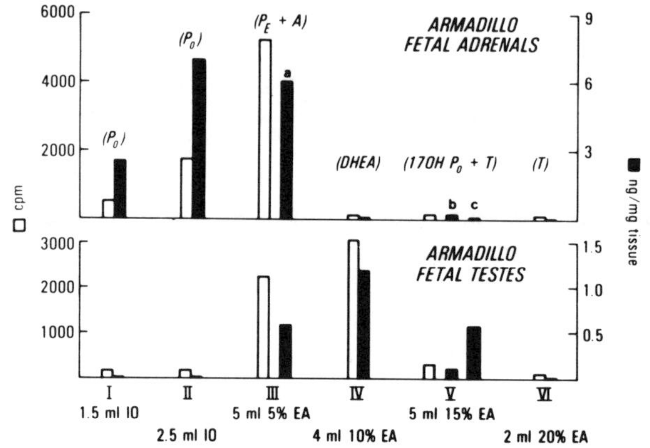

Figure 4. Incubation 1. Radioactive and immunoreactive products from *in vitro* incubation of fetal armadillo adrenal glands and testes. IO = isooctane; EA = ethylacetate; a = immunoreactive Pregnenolone in the fetal adrenals and immunoreactive Androstenedione in the fetal testes; b = immunoreactive 17a-hydroxyprogesterone (17OH Po); c = immunoreactive Testosterone; Po = ^{3}H-Progesterone.

mass spectrometry. Immunoreactive Pregnenolone (6.1 ng/mg tissue) was only evaluated in incubation 1.

The fetal armadillo testes converted 33 percent and 78 percent (incubations 1 and 2, respectively) of the ^{3}H-Pregnenolone to ^{3}H-DHEA which was recrystallized to constant specific activity in incubation 2 (Figs. 4 and 5). The amount of label which remained at Pregnenolone and the amount that was converted to Androstenediol, 17OH Progesterone and Testosterone could not be assessed since these steroids were inseparable in the column system used. Immunoreactive DHEA (1.5 and 14.4 ng/mg tissue), Andorostenedione (0.58 and 0.7 ng/mg tissue), Testosterone (0.58 and 2.2 ng/mg tissue) and 17OH Progesterone (0.1 and 0.2 ng/mg tissue) were found in both incubations of testicular tissue.

The fetal ovary (Fig. 6) and placenta (Figs. 5 and 6) appeared steroidogenically inactive with the label remaining mainly at ^{3}H-Pregnenolone (83 percent and 79 percent, respectively). Traces of immunoreactive Progesterone were found in both the fetal ovaries (0.4 ng/mg tissue) and the placenta (0.3 - 1.3 ng/mg tissue) but other steroids were undetectable. The placenta was virtually devoid of sulfatase activity using DHEA-sulphate as substrate.

Maternal adrenal glands in incubations 2 and 3 converted the ^{3}H-Pregnenolone to ^{3}H-Progesterone (45 percent and 53 percent, respectively), which was recrystallized to constant specific activity (Figs. 5 and 6). Immunoreactive Progesterone (11.7 and 15 ng/mg tissue) and 17OH Progesterone (6.5 and 7.3 ng/mg tissue) were found in substantial concentrations.

Immunoreactive and tritiated Corticosterone and Cortisol were monitored in tissues from incubation 3 (Fig. 6). Although radioactive conversion products were undetectable, high concentrations of immunoreactive Corticosterone (31.1 ng/mg tissue) and Cortisol (30.6 ng/mg tissue) were found in the maternal adrenal glands. Immunoreactive Corticosterone and Cortisol were also detectable in the fetal adrenal glands (2 ng/mg tissue and 0.8 ng/mg tissue, respectively).

DISCUSSION

The primary fetal armadillo adrenal gland steroid is ^{3}H-Progesterone as evidenced by the six- to tenfold concentration gradient from the fetal to the maternal circulation, the high concentrations in the fetal adrenals, and the *in vitro* conversion of ^{3}H-Pregnenolone to ^{3}H-Progesterone. Chemical identity of the immunoreactive Progesterone present in the fetal adrenal was established by mass spectrometry and recrystallization of the ^{3}H-Progesterone to constant specific activity. Circulating Progesterone values in sheep and horse measured in the same assay system were similar to those previously reported in the literature (Bassett et al., 1969; Fylling, 1970; Stabenfeldt et al., 1972; Strott et

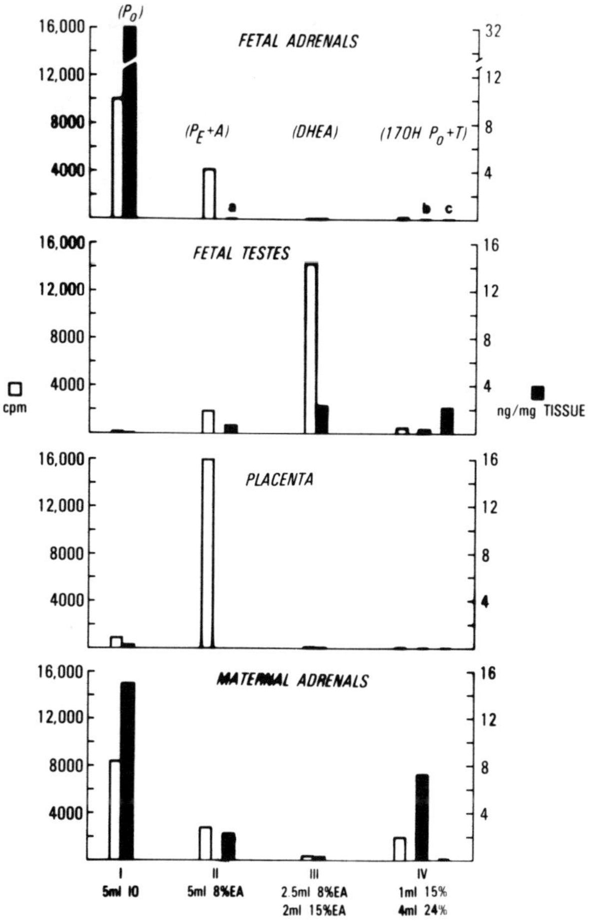

Figure 5. Incubation 2. Radioactive and immunoreactive products from *in vitro* incubation of fetal armadillo ardrenal glands and testes, of placenta, and maternal adrenal glands. IO = isooctane; EA = ethylacetate; a = immunoreactive Androstenedione, Pregnenolone was not measured; b = immunoreactive 17a-hydroxyprogesterone (17OH Po); c = immunoreactive Testosterone (T).

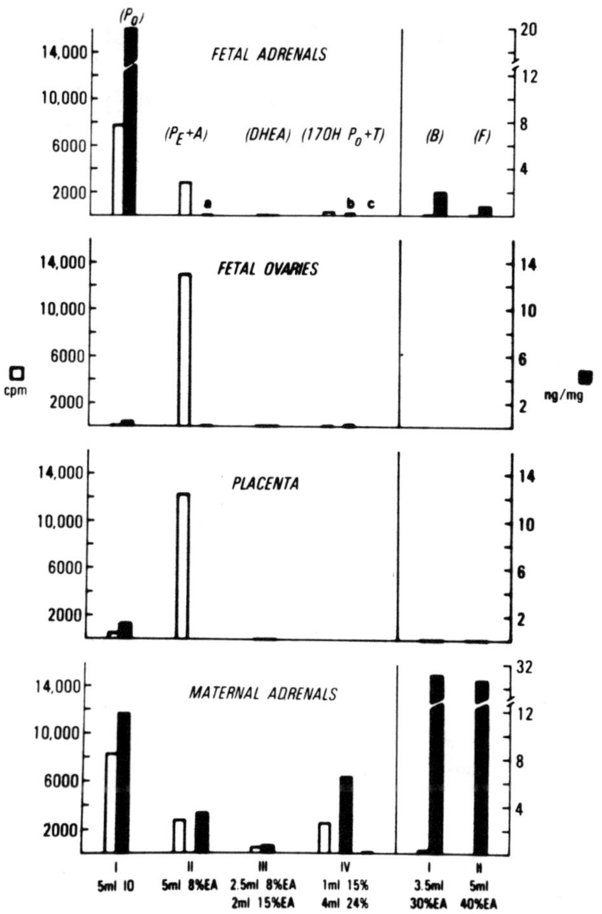

Figure 6. Incubation 3. Radioactive and immunoreactive products from incubation 3. IO = isooctane; EA = ethylacetate; a = immunoreactive Androstenedione (A), Pregnenolone was not measured; b = immunoreactive 17a-hydroxyprogesterone (17OH Po); c = immunoreactive Testosterone (T).

al., 1974; Holtan et al., 1975; Ganjam et al., 1975) and therefore lend credence to the high Progesterone values found in the armadillo.

The maternal adrenals also contained a substantial amount of ^{3}H-Progesterone although Corticosterone and Cortisol levels were approximately twice as high. The lack of labelled Corticosterone and Cortisol in the adrenal incubation is somewhat surprising but is likely due to the high concentration of endogenous Progesterone and its dilution effect on the labelled precursor.

The presence of DHEA in the fetal circulation substantiates the report that DHEA may be a secretory product of the fetal armadillo adrenal gland (Brinck-Johnsen, 1967). However, the higher concentration of DHEA in the maternal circulation as compared to the fetus and the low concentration in fetal adrenal tissue indicate that DHEA is unlikely as a major fetal adrenal steroid in this species. Rather, it appears that in the male armadillo fetus DHEA may be a gonadal product, as found in the horse fetus (Raeside, 1976) since the ^{3}H-Pregnenolone was converted to ^{3}H-DHEA by the fetal testis. There was also a substantial endogenous gonadal pool of DHEA. The additional finding of higher circulating levels of testosterone in the male fetus than in the female fetus and the histologic evidence of Leydig cell hypertrophy indicate that the armadillo fetal testis is steroidogenically active. Unlike the fetal horse, the fetal armadillo ovary is steroidogenically inactive, showing very little conversion of the ^{3} H-Pregnenolone and histologically consisting mostly of oocytes and very little interstitial tissue.

The undetectable levels of placental Luteinizing Hormone and the higher levels in fetal serum relative to maternal serum suggest that the fetal pituitary is the source of Luteinizing Hormone which in turn stimulates the fetal testis to produce androgens. The lack of ovarian stimulation may be attributed to a lack of Follicle Stimulating Hormone, although activity of this hormone was not measured.

The relative steroidogenic inactivity of the placenta in the tissue incubations and the lower concentration of ^{3}H-Progesterone when compared to the fetal adrenal suggest that the armadillo has an unusual endocrine system in which the fetal adrenal may preserve the pregnancy directly through the production of relatively large quantities of Progesterone. Additionally, the undetectable levels of bioactive Luteinizing Hormone, the low steroidogenic capacity and lack of sulfatase activity in the placenta supports the view that little if any endocrine role is provided by the trophoblast. The fetal rhesus monkey adrenal gland also appears to secrete a significant amount of Progesterone early in gestation (Serón-Ferré et al., 1979).

The common long-nosed armadillo is not an appropriate model for the primate feto-placental unit, but the apparent simplicity of the endocrine mechanisms may be advantageous in more basic studies. The lack of placental steroid conversion and Luteinizing Hormone activity point directly to the fetus as providing its own control during the pregnancy. Therefore, the common long-nosed armadillo may provide an excellent model for studying the fetal pituitary tropic factors which likely control both fetal adrenal and gonadal steroid production at mid through late gestation.

ACKNOWLEDGMENTS

We thank Dr. Larry J. Shapiro, Division of Medical Genetics, UCLA School of Medicine, for the placental sulfatase analysis and Dr. Cecil Hornbeck, Veterans Administration Hospital, San Diego, California, for the mass spectrometry analysis. The work was supported by grants HD 10190 and NF 398. Figures 1, 2, 5 and 6 are reproduced with permission from the Journal of Reproduction and Fertility.

LITERATURE CITED

Anderson, D.C., B.R. Hopper, B.L. Lasley, and S.S.C. Yen.
1976. A simple method for the assay of eight steroids in small volumes of plasma. Steroids, 28: 179-196.

Bassett, J.M., T.J. Oxborrow, J.D. Smith and G.D. Thorburn.
1969. The concentration of progesterone in the peripheral plasma of the pregnant ewe. J. Endo., 45: 449-457.

Bloch, E. and K.B. Benirschke.
1964. *In vitro* steroid synthesis by fetal, newborn and adult armadillo adrenals and by fetal armadillo testes. J. Endo., 76: 43-51.

Brinck-Johnsen, T., K. Benirschke and K. Brinck-Johnsen.
1967. Hormonal steroids in the armadillo, *Dasypus novemcinctus.* 1. Oestriol in pregnancy and its *in vitro* biosynthesis by the placenta. Acta Endo., 56: 657-690.

Brinck-Johnsen, T.
1970. Hormonal steroids in the armadillo, *Dasypus novemcinctus.* 2. Oestrone and 17B-oestradiol in pregnancy and their *in vitro* formation by preparation of placentae, early and late in development. Acta Endo., 63: 696-704.

Dufau, M.L., C.R. Mendelson and K.J. Catt.
1974. A highly sensitive *in vitro* bioassay for luteinizing hormone and chorionic gonadotropin: Testosterone production by dispersed Leydig cells. J. Clin. Endo. Metab., 39: 610-613.

Dufau, M.L., R. Pock, A. Neubauer and K.H. Catt.
1976. *In vitro* bioassay of LH in human serum: The rat interstitial cell testosterone (RICT) assay. J. Clin. Endo. Metab., 42: 958-969.

Fylling, P.
1970. The effect of pregnancy, ovariectomy and parturition on plasma progesterone level in sheep. Acta Endo., 65: 273-283.

Ganjam, V.K., R.M. Kenney and G. Flickinger.
1975. Plasma progestagens in cyclic, pregnant and post-partum mares. J. Reprod. Fert., Suppl., 23: 441-447.

Holtan, D.W., T.M. Nett and V.L. Estergreen.
1975. Plasma progestagens in pregnant mares. J. Reprod. Fert., Suppl. 23: 419-424.

Manlimos, F.S. and G.E. Abraham.
1975. Chromatographic purification of tritiated steroids prior to use in radioimmunossay. Analyt. Lett., 8: 403-410.

Moser, H.G. and K. Benirschke.
1962. Fetal zone of the adrenal gland in the nine-banded armadillo, *Dasypus novemcinctus.* Anat. Rec., 143: 47-59.

Raeside, J.E.
1976. Dehydroepiandrosterone in the fetal gonads of the horse. J. Reprod. Fert., 46: 423-425.

Serón-Ferré, M., C. Lawrence, M. Bamamoto, B. Mitchell, P. Siiteria and R. Jaffe.
1979. Production of Δ^4-3 ketosteroids by the Rhesus monkey fetal adrenal gland *in vitro*. Abstract presented at the 26th Annual Meeting of the Society for Gynecologic Investigation, March 21-24, 1979. No. 160.

Stabenfeldt, G.H., M. Drost and C.E. Franti.
1972. Peripheral plasma progesterone levels in the ewe during pregnancy and parturition. Endo., 90: 144-150.

Strott, C.A., H. Sundel and M.T. Stahlman.
1974. Maternal and fetal plasma progesterone, cortisol, testosterone and 17B-estradiol in preparturient sheep: Response to fetal ACTH infusion. Endo., 95: 1327-1339.

TESTICULAR CYCLES OF THE COMMON LONG-NOSED ARMADILLO *DASYPUS NOVEMCINCTUS*, IN NORTH CENTRAL TEXAS

J. Stephen McCusker

Washington Park Zoo, Portland, Oregon, 97221

Abstract

Male armadillos were collected in each quarter of the year. Lengths, widths, and weights of testes were recorded. Sections were studied for spermatogenic activity, and seminiferous tubule diameters were noted. All measurements indicated a definite testicular cycle which correlated with the breeding season in July-September. Equal quantities of spermatids were found in all sections, regardless of season or climatic conditions, indicating continuous spermatogenesis.

Abstracto

Se colectaron armadillos machos durante cada trimestre de un año. Se registraron el largo, ancho y peso de los testículos de estos animales. Se estudió la actividad espermatogénica y se anotó el diámetro de los túbulos seminíferos por medio de cortes para el microscopio. Todas las medidas registradas indicaron la existencia de un definitivo ciclo testicular correlacionado con la época de procreación de julio a septiembre. Se encontraron cantidades iguales de espermátidos en todas las secciones, a pesar de diferencias en las condiciones climáticas y de distintas estaciones, indicando la existencia de un estado de ininterrumpida espermatogénesis.

Resumo

Tatus machos foram coletados em cada trimestre do ano. Comprimentos, larguras e pesos dos testículos foram registrados. Cortes foram estudados para atividade espermatogênica e diâmetros dos túbulos seminíferos foram anotados. Todas as medidas indicaram um ciclo definido que estava correlacionado com a estação reprodutiva em julho-setembro. Quantidades iguais de espermátides foram encontradas em todos os cortes, sem levar em consideração a estação ou condições climáticas. Isso indica espermatogênese contínua.

The primary purpose of this study was to investigate the reproductive cycle of *Dasypus novemcinctus mexicanus* in north central Texas. Males were emphasized; but, because animals were available, skull measurements, body size, body weight, gonadal measurements, and activity cycles for both sexes were recorded (McCusker, 1976).

MATERIALS AND METHODS

From 8 June 1974 to 31 October 1975, common long-nosed armadillos were collected in northcentral Texas from a circular area, with a 70-mile radius, centered in Mineral Wells, Palo Pinto County. The specimens were weighed to the nearest 0.05 kg. The four basic external body measurements were taken from each animal and recorded to the nearest 1.0 mm. Eight skull measurements were taken from each animal.

Gonads were excised in the field prior to dehydration and decomposition. The testes were trimmed of fat, dissected free of tunica vaginalis, epididymis and vas deferens, then weighed to the nearest 0.05 gm. Greatest length and greatest perpendicular width were recorded to the nearest 0.1 mm. Testes were cut in half laterally and immersed in Bouin's fixative. In the laboratory, testes were dehydrated, embedded in paraffin and cut into 5-μ sections. Tissue was stained with hemotoxylin-eosin for light microscopy.

For analysis, data were pooled into seasonal quarters as follows: I, January-March; II, April-June; III, July-September; IV, October-December. Quantitative differences in various parameters were tested for significance using one-way analysis of variance (Sokal and Rohlf, 1969).

RESULTS

Activity Cycles.—Diet and seasonal activity of the common long-nosed armadillo is primarily determined by ambient temperature (Taber, 1945; Talmage and Buchanan, 1954; Breece and Dusi, 1985; Layne and Glover, 1985). In quarters II and III, temperatures were warm,

Table 1. Skull measurements of mature male *D. novemcinctus* from northcentral Texas. The following measurements were taken to the nearest 0.1 mm from each animal: zygomatic breadth (ZB), maxillary tooth row length (MTR), mandibular tooth row length (MdTR), outer occipital condyle span (OOCC), horizontal diameter of the foramen mangum (HFM), verticle diameter of the foramen magnum (VFM), occipitonasal length (ONL), and nasal length (NL). n = sample size, $\bar{x}$ = mean ± two SE, range is in parentheses.

Measure-ment	*Quarter* I	II	III	IV	*All specimens*
ZB	n = 13 $\bar{x}$ = 41.59 ± 0.9 (37.4-43.4)	n = 13 $\bar{x}$ = 42.12 ± 1.3 (38.9-47.4)	n = 5 $\bar{x}$ = 42.88 ± 1.2 (40.9-44.5)	n = 5 $\bar{x}$ = 41.00 ± 2.7 (37.5-44.5)	n = 36 $\bar{x}$ = 41.88 ± 0.68
MTR	n = 13 $\bar{x}$ = 25.23 ± 0.6 (22.7-27.0)	n = 16 $\bar{x}$ = 24.48 ± 0.92 (21.5-27.2)	n = 5 $\bar{x}$ = 25.60 ± 0.88 (24.4-26.9)	n = 5 $\bar{x}$ = 25.04 ± 1.6 (22.9-27.5)	n = 39 $\bar{x}$ = 24.94 ± 0.68
MdTR	n = 13 $\bar{x}$ = 25.25 ± 1.1 (21.8-27.9)	n = 16 $\bar{x}$ = 25.18 ± 1.0 (19.8-28.3)	n = 5 $\bar{x}$ = 26.78 ± 0.72 (25.8-27.9)	n = 5 $\bar{x}$ = 25.74 ± 0.92 (24.5-27.0)	n = 39 $\bar{x}$ = 25.48 ± 0.60
OOCC	n = 13 $\bar{x}$ = 25.23 ± 0.40 (23.9-26.5)	n = 14 $\bar{x}$ = 25.44 ± 0.48 (24.0-26.7)	n = 5 $\bar{x}$ = 25.82 ± 1.1 (23.9-26.9)	n = 4 $\bar{x}$ = 24.75 ± 1.2 (23.5-26.0)	n = 36 $\bar{x}$ = 25.34 ± 0.30
HFM	n = 13 $\bar{x}$ = 11.97 ± 0.40 (10.2-12.9)	n = 14 $\bar{x}$ = 12.31 ± 0.32 (11.1-13.2)	n = 5 $\bar{x}$ = 12.88 ± 0.94 (11.7-14.5)	n = 4 $\bar{x}$ = 12.18 ± 0.20 (12.0-12.4)	n = 36 $\bar{x}$ = 12.25 ± 0.24
VFM	n = 13 $\bar{x}$ = 10.36 ± 0.38 (9.5-11.6)	n = 14 $\bar{x}$ = 10.38 ± 0.22 (9.8-11.2)	n = 5 $\bar{x}$ = 10.28 ± 0.40 (9.6-10.8)	n = 4 $\bar{x}$ = 10.40 ± 0.60 (9.6-11.0)	n = 36 $\bar{x}$ = 10.36 ± 0.18
ONL	n = 13 $\bar{x}$ = 100.38 ± 0.38 (96.0-104.7)	n = 16 $\bar{x}$ = 99.03 ± 1.5 (94.3-105.9)	n = 5 $\bar{x}$ = 101.72 ± 2.9 (98.4-106.3)	n = 4 $\bar{x}$ = 101.35 ± 4.3 (96.8-106.9)	n = 38 $\bar{x}$ = 100.09 ± 1.0
NL	n = 13 $\bar{x}$ = 34.82 ± 0.72 (32.9-37.5)	n = 16 $\bar{x}$ = 33.92 ± 1.3 (25.9-37.4)	n = 5 $\bar{x}$ = 34.46 ± 1.9 (30.9-36.3)	n = 4 $\bar{x}$ = 35.0 ± 3.2 (31.0-37.8)	n = 38 $\bar{x}$ = 34.41 ± 0.70

food was plentiful, and *D. novemcinctus* were active. Quarters I and IV were less optimal; temperatures were lower, food was at a minimum, and the armadillos reduced their activity. During this 17-month study, which included over 200 hours of field work, no armadillos were seen at temperatures below 10°C. During the warmer months (April to September), the armadillos were nocturnal. During the cooler months (October to March), the animals were most active through the warmest part of the day (McCusker, 1976). During exceptionally cold winters, *D. novemcinctus* continued to venture out for feeding, but the weather took its toll on local populations (Kalmbach, 1943).

Three armadillos were collected, two in December and one in January, while basking. In each case, the animal was lying on its ventral surface, alert with its head up and one side of its body close to or in contact with a large rock.

Size and Weight. The occipitonasal length (ONL; Table 1) was used to categorize mature and immature animals (McCusker, 1976). All males with ONL $\geqslant$ 94 mm had mature testes and were included in the analyses of mature animals. Mature males did not vary significantly in occipitonasal length between quarters of the year ($P > 0.1$). Body measurements of mature specimens (Table 2) agreed with those of Burt and Grossenhieder (1952), Hall and Kelson (1959), and Walker (1968). Males showed no significant variance in total length between quarters ($P > 0.1$).

From these analyses, it was evident that variations in gonad size and weight were not due simply to differences in body size.

In contrast to skull and body sizes, animal weights (Table 2) varied significantly between quarters ($P < 0.01$). Animals of similar sizes weighed less during the harsh seasons (quarters I and IV). During these cold

Table 2. Size and weight of male *D. novemcinctus* from northcentral Texas. External body measurements in mm are shown for all male armadillos and total weight in kg is shown for all mature specimens. The following measurements are shown: total length (TL), tail length (T), length of hind foot (HF), length of ear (E), and body weight (W). n = sample size. $\bar{x}$ = mean ± two SE. Range is in parentheses.

Measurement	*Quarter* *I*	*II*	*III*	*IV*	*All Specimens*
TL	n – 13 $\bar{x}$ = 734.5 ± 17.4 (676-784)	n = 16 $\bar{x}$ = 732.9 ± 16.4 (667-801)	n = 6 $\bar{x}$ = 733.8 ± 39.1 (666-793)	n = 6 $\bar{x}$ = 727.0 ± 4.2 (640-771)	n = 41 $\bar{x}$ = 732.7 ± 11.3
T	n = 13 $\bar{x}$ = 333.5 ± 15.3 (265-362)	n = 18 $\bar{x}$ = 321.9 ± 14.4 (225-367)	n = 7 $\bar{x}$ = 315.1 ± 30.3 (254-355)	n = 6 $\bar{x}$ = 309.2 ± 27.5 (251-341)	n = 44 $\bar{x}$ = 322.0 ± 9.6
HF	n = 13 $\bar{x}$ = 91.8 ± 3.6 (75-100)	n = 18 $\bar{x}$ = 91.2 ± 1.6 (86-96)	n = 9 $\bar{x}$ = 90.9 ± 3.4 (81-97)	n = 6 $\bar{x}$ = 93.0 ± 8.1 (74-101)	n = 46 $\bar{x}$ = 91.5 ± 1.6
E	n = 13 $\bar{x}$ = 33.6 ± 1.3 (29-38)	n = 18 $\bar{x}$ = 34.9 ± 1.3 (31-40)	n = 10 $\bar{x}$ = 36.2 ± 0.7 (34-38)	n = 5 $\bar{x}$ = 35.6 ± 1.2 (35-38)	n = 46 $\bar{x}$ = 34.9 ± 0.7
W	n = 12 $\bar{x}$ = 4.1 ± 0.18 (3.4-4.6)	n = 14 $\bar{x}$ = 4.7 ± 0.30 (3.5-5.7)	n = 3 $\bar{x}$ = 5.3 ± 0.28 5.0-5.4)	n = 2 $\bar{x}$ = 3.9 ± 2.6 (2.6-5.2)	

months, food was scarce, foraging time was limited and metabolic drain was probably greatest as the animals reacted to maintain body temperatures above ambient.

Reproduction - Immature Males. Slides were prepared from the testes of six animals of immature skull and body size: one in quarter I, two in quarter II, and three in quarter IV. These testes were smaller and whiter in color than those of mature animals; however, they all had spermatids in the seminiferous tubules. This early sexual development was also found in studies of blood plasma testosterone levels of captive armadillos (Czekala et al., 1980).

Judging from body and skull size of the six animals above, and assuming birth between March and May, the animals collected in quarter I (ONL 91.3 mm) and II (ONL 93.4 and 89.8 mm) were approximately one year old. Those from quarter IV (ONL 88.2, 89.3, 88.7 mm) were too small to have completed two summers. They were born the preceding spring and were 7 to 9 months old. Talmage and Buchanan (1954) reported that physical maturity is attained in about six months. Although the above animals had developed spermatids, and potentially could have donated to the gene pool, they had not reached adult size.

Reproduction - Mature Males. The larger, heavier testes occurred in quarters II (April-June) and III (July-September). Analysis of variance of two testes measurements (Figs. 1, 2) revealed in each case a significant variation between quarters ($P < 0.001$). Quarter II is the prebreeding season and quarter III is the breeding season (Asdell, 1964; Buchanan, 1967; Enders, 1966; Enders and Buchanan, 1959; Kalmbach, 1943; Taber, 1945; Talmage and Buchanan, 1954).

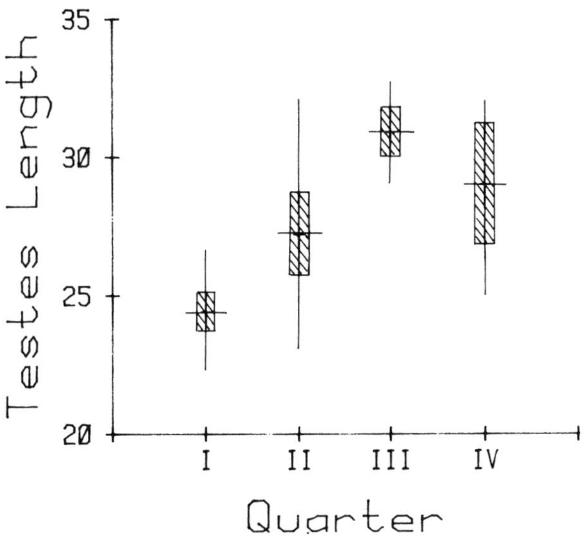

Figure 1. Seasonal changes in lengths of *Dasypus novemcinctus* testes. Horizontal line shows the mean. Shaded area is 2 SE. Vertical line is the range. There were 13 testes in quarter I, 12 in quarter II, 7 in quarter III, and 5 in quarter IV.

Figure 2. Seasonal changes in weights of *Dasypus novemcinctus* testes. Presentation of statistics as in Fig. 1. There were 13 testes in quarters I and II, 7 in quarter III and 5 in quarter IV.

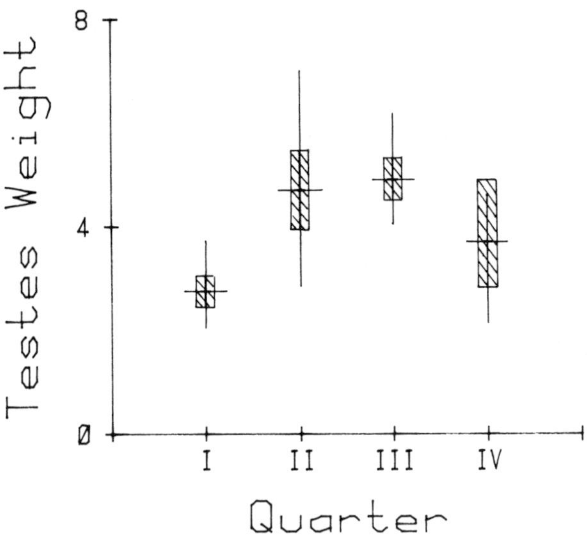

The breeding season for many species of mammals is relatively short and highly regulated. Males are fecund only during the breeding season and their testes increase considerably in size during the fertile period (Aldous, 1937; Asdell, 1964; Bramley and Neaves, 1972; Crowe, 1975; Marshall, 1911-12; Neaves, 1973; Pearson et al., 1952; Pfeiffer, 1956; Wright, 1947; Wright and Rausch, 1955). After the breeding season, the testes may show reduced production of spermatozoa or may become aspermic. In either case, the testes are generally reduced in size and weight (Amann, 1970).

Size and weight differences in the testes of *D. novemcinctus*, however, did not necessarily indicate the level of spermatogenic activity. Even though testes size and weight changed with season (Figs. 1, 2), microscopic examination showed spermatogenic activity throughout the year (Fig. 3). Every testis examined, regardless of the time of year or climatic conditions, contained numerous spermatids. Changes in the size of seminiferous tubules were primarily responsible for the seasonal differences in testis weight and size (Fig. 4). Measurements were significantly different ($P < 0.001$) between quarters, with the largest tubules in quarter II,

Figure 3. Typical sections of *D. novemcinctus* seminiferous tubules showing spermatids and active spermatogenesis. **A** from quarter I (Jan.-Mar.); **B** from Quarter II (Apr.-June); **C** from Quarter III (July-Sept.); **D** from quarter IV (Oct.-Dec). 400X.

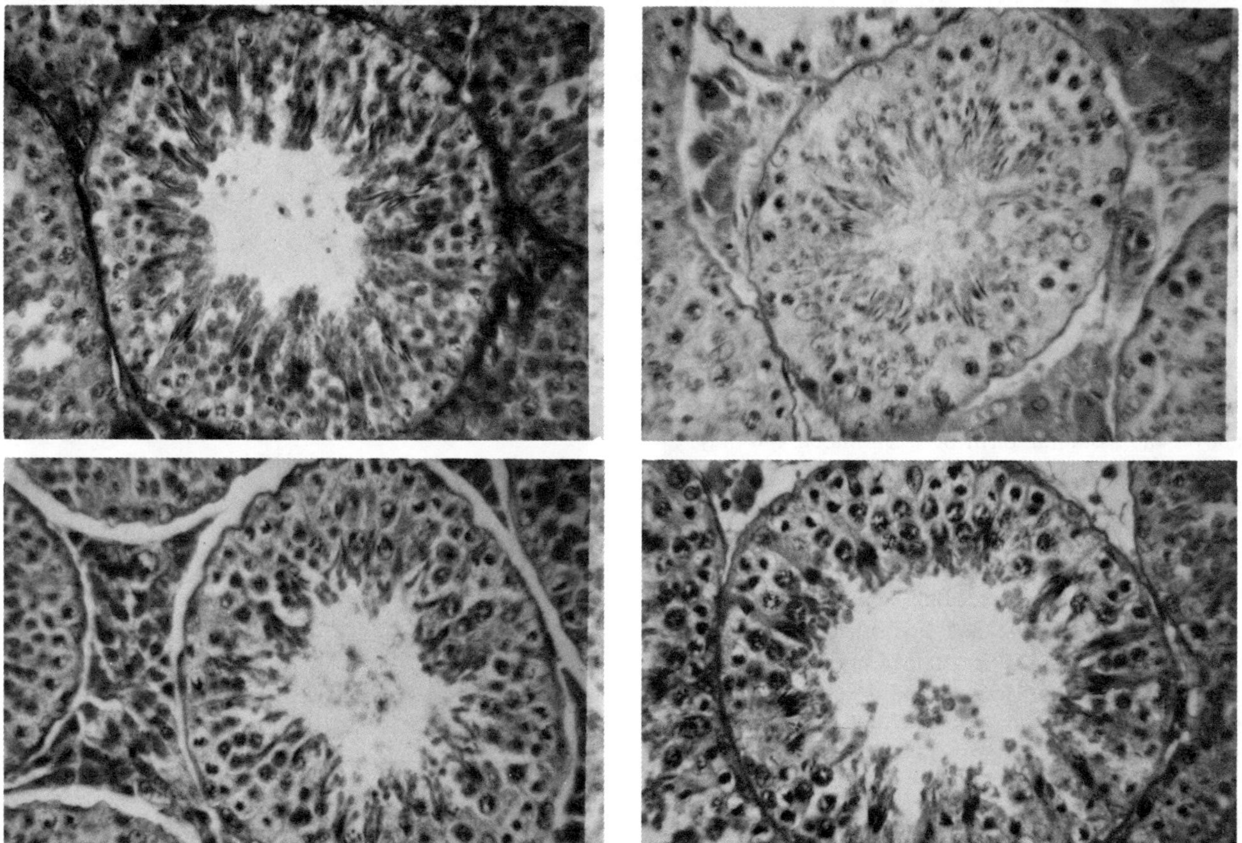

Figure 4. Seasonal changes in the diameter of seminiferous tubules from *Dasypus novemcinctus*. Measurements were taken from 34 males. There were 66 measurements in quarter I, 49 in quarter II, 48 in quarter III and 36 in quarter IV. Presentation of statistics as in Fig. 1.

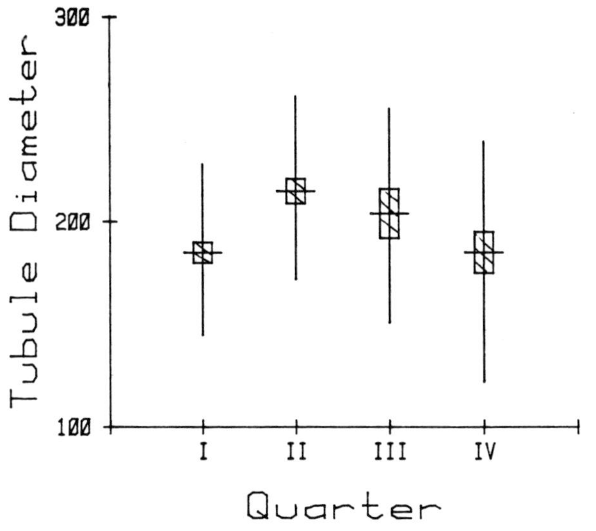

the prebreeding season, and the smallest in quarter IV, the postbreeding season.

DISCUSSION

The activity and geographic range of the common long-nosed armadillo are seasonally restricted by two varibles: ambient temperature and food availability. From October to March in north central Texas, temperatures are less than optimal for *D. novemcinctus*. During these cold months, nocturnal foraging is all but eliminated and the number of active days is greatly reduced (McCusker, 1976). The food supply of the entomophagous armadillo (Kalmbach, 1943; Taber, 1945; Talmage and Buchanan, 1954; Breece and Dusi, 1985; Wirtz et al., 1985) is at a minimum. Thus, breeding in late summer and delayed embryonic implantation (Enders, 1966; Enders and Buchanan, 1959; Kalmbach, 1943; Taber, 1945; Talmage and Buchanan, 1954) are highly adaptive traits in Texas, near the northern limit of its range (Humphrey, 1974). Food and warmth are available at the critical times of breeding, parturition and weaning.

In males, testicular growth and increased seminiferous tubule diameter occur at appropriate times. Spermatogenisis, on the other hand, does not seem to be seasonally restricted even though testes size diminishes in the colder months.

Is it that the common long-nosed armadillo breeds throughout the year? Were this the case, implantation might be delayed for varying periods to allow for a spring or early summer birth. Because of the volume of research which points to specific breeding periods for the female *D. novemcinctus* (Buchanan, 1967; Enders, 1966; Enders and Buchanan, 1959; Kalmbach, 1943; Labhsetwar and Enders, 1968, 1969; Taber, 1945; Talmage and Buchanan, 1954), this is unlikely in populations in the United States. However, it remains a possibility in the unstudied Central and South American populations of *D. novemcinctus* which may be subject to less seasonal influences on food supply and energy

Year-round spermatogenesis might be a secondary or byproduct effect of a constantly high gonadotropic hormone level, which in turn is related to aggressive behavior and territoriality. Although blood testosterone levels in both wild and captive armadillos on expenditures. Data on xenarthran reproduction indicate concomitantly adaptive seasonal breeding and seasonal parturition throughout the order (Asdell, 1964; G. G. Montgomery, pers. comm.), even in populations which are wholly tropical in distribution.

Perhaps the seasonal increase in testes size and weight reflects an energy storage mechanism; that is, as food and water become plentiful in the warm season, a portion of the increased ingestion results in testes enlargement. Testicular lipid storage, however, would be evident in cross section as a marked increase in Sertoli cell volume; this increase was not seen in this study.

Most animals are seasonal breeders and appear to undergo cycles of testicular development and regression. Although reduced production of spermatozoa during the nonbreeding season of the common long-nosed armadillo would seem likely, marked seasonal variability in spermatid number was not seen in this study.

natural light cycles are cyclic, the testosterone levels are relatively high year round (Czekala et al., 1980). High plasma testosterone levels are thought to cause aggressive, territorial behaviors in some mammalian species (Bramley and Neaves, 1972; Neaves, 1973). While high gonadotropic hormone levels are generally linked with reproduction, active reproduction may not always occur.

ACKNOWLEDGMENTS

I thank Drs. Wayne Barcellona, Gary Ferguson, and Art Cleveland for their constant help and encouragement throughout the project.

LITERATURE CITED

Aldous, C. M.
1937. Notes on the life history of the snowshoe hare. J. Mamm., 18: 47-57.

Amann, R. P.
1970. Sperm production rates. Pp. 433 to 482, *in* The testis, (A. D. Johnson, W. R. Gomes and N. L. VandeMark, eds.). Academic Press, New York and London, Vol. 1: 684 pp.

Asdell, S. A.
1964. Patterns of mammalian reproduction. Cornell Univ. Press, Ithaca, New York, 670 pp.

Bramley, P. S. and W. B. Neaves.
1972. The relationship between social status and reproductive activity in male impala, *Aepyeeros melampus*. J. Reprod. Fert., 31:77-81.

Breece, G.A., and J.L. Dusi.
1985. Food habits and home ranges of the common long-nosed armadillo *Dasypus novemcinctus* in Alabama. (this volume).

Buchanan, G. D.
1967. The presence of two conceptuses in the uterus of the nine-banded armadillo. J. Reprod. Fert., 13:329-331.

Burt, W. H. and R. P. Grossenhieder.
1952. A field guide to the mammals. Houghton Mifflin Co., Boston, 240 pp.

Crowe, D. M.
1975. Aspects of aging, growth and reproduction of bobcats from Wyoming. J. Mamm., 56: 177-198.

Czekala, N. M., J. K. Hodges, G. E. Gause, and B. L. Lasley.
1980. Annual circulating testosterone levels in captive and free-ranging male armadillos, *Dasypus novemcinctus*. J. Reprod. Fert., 59: 199.

Enders, A. C.
1966. The reproductive cycle of the nine-banded armadillo (*Dasypus novemcinctus*). Pp. 295-310 *in* Comparative Biology of Reproduction in Mammals, (I. W. Rowlands, ed.), Academic Press, New York and London, 559 pp.

Enders, A. C. and G. D. Buchanan.
1959. The reproductive tract of the female nine-banded armadillo. Tex. Rep. Biol. and Med., 17: 323-340.

Hall, E. R. and K. R. Kelson.
1959. The mammals of North America. The Ronald Press Co., New York, 244 pp.

Humphrey, S. R.
1974. Zoogeography of the nine-banded armadillo (*Dasypus novemcinctus*) in the United States. BioScience, 24: 457-462.

Kalmbach, E. R.
1943. The armadillo: Its relation to agriculture and game. Game, Fish and Oyster Commission, Austin, Texas, 61 pp.

Labhsetwar, A. P. and A. C. Enders.
1968. Progesterone in the corpus luteum and placenta of the armadillo, *Dasypus novemcinctus*. J. Reprod. Fert., 16:381-387.

1969. Pituitary LH content during delayed and post-implantation periods in the armadillo (*Dasypus novemcinctus*). J. Reprod. Fert., 18:383-389.

Layne, J.N. and D. Glover.
1985. Activity patterns of the common long-nosed armadillo *Dasypus novemcinctus* in south-central Florida. (this volume).

Marshall, F. H. A.
1911-12. The male generative cycle in the hedgehog; with experiments on the functional correlation between the essential and accessory sexual organs. J. Physio., XLIII: 247-259.

McCusker, J. S.
1976. The nine-banded armadillo, *Dasypus novemcinctus*, in north central Texas with emphasis on the reproductive biology of the male. Unpubl. M.S. thesis, Texas Christian Univ. 35 pp.

Neaves, W. B.
1973. Changes in testicular Leydig cells and in plasma testosterone levels among seasonally breeding rock hyrax. Biol. of Reprod. 8:451-466.

Pearson, O. P., M. R. Koford and A. K. Pearson
1952. Reproduction of the lump-nosed bat (*Corynorhinus rafinesque*) in California. J. Mamm., 33: 273-320.

Pfeiffer, E. W.
1956. The male reproductive tract of a primitive rodent, *Aplodontia rufa*. Anat. Rec., 84:99-115.

Sokal, R. R. and F. J. Rohlf.
1969. Biometry. W. H. Freeman and Company, San Francisco, 776 pp.

Taber, F. W.
1945. Contribution on the life history and ecology of the nine-banded armadillo. J. Mamm., 26:211.

Talmage, R. V. and G. D. Buchanan.
1954. The armadillo (*Dasypus novemcinctus*). A review of its natural history, ecology, anatomy and reproductive physiology. The Rice Inst. Pamph., 41: 1-35.

Walker, E. P.
1968. Mammals of the world, Vol. I, John Hopkins Press, Baltimore Md., 500 pp.

Wirtz, W.O., D.H. Austin and G.W. Dekle.
1985. Food habits of the common long-nosed armadillo *Dasypus novemcinctus* in Florida, 1960-61. (this volume).

Wright, P. L.
1947. The sexual cycle of the male long-tailed weasel. (*Mustela frenata*). J. Mamm., 28: 343-352.

Wright, P. L. and R. Rausch.
1955. Reproduction in the wolverine, *Gulogulo*. J. Mamm., 36: 346-355.

DISEASES AND PARASITES

LEPROSY IN WILD COMMON LONG-NOSED ARMADILLOS *DASYPUS NOVEMCINCTUS*

E. E. Storrs and H. P. Burchfield

Florida Institute of Technology, Melbourne, Florida 32901.

Abstract

Leprosy has been found in approximately 10 percent of individuals of a common long-nosed armadillo population in Louisiana, in one armadillo from Mississippi, and in three from Texas. Leprosy-like organisms have also been found in one armadillo from Mexico. In continued sampling of armadillos from Florida, no evidence of the disease has been found.

Abstracto

Se ha encontrado lepra en, aproximadamente, el 10 percento de una población de armadillos comunes de nariz larga de Louisiana, lo mismo que en un armidillo de Mississippi y en dos de Texas. Organismos que semejan a los de la lepra se ha encontrado en un armadillo de México. No se encontró evidencia de la enfermidad en un muestreo contínuo de armadillos de Florida.

Resumo

Tem sido encontrada hanseníase em aproximadamente 10 percentagem de indivíduos em uma população de tatus de nove faixas na Louisana, 1 tatu do Mississippi, 1 do Texas, e 1 do México. Em amostragem contínua de tatus da Florida, nenhuma evidência da doença foi encontrada.

Leprosy is a major public health problem today. Individual humans exhibit different responses to the infection, ranging from the most severe lepromatous form with overwhelming numbers of bacilli present to the tuberculoid form where bacilli are rarely found; many people are immune. Resistance to leprosy appears to be primarily regulated by cell-mediated immunity, employing the same immunologic machinery that is thought to control other infectious agents (viruses, fungi, and so forth), neoplastic growth and the rejection of foreign tissue transplants.

Until the 1970's its study was seriously impeded because the causative agent, *Mycobacterium leprae*, could not be grown in artificial media and only a mild infection with microscopic lesions could be obtained in immunologically intact laboratory animals. This left many unanswered questions relating to the epidemiology of the disease, including modes of transmission, effects of genetic and nutritional factors, and so forth, and has slowed progress in the search for drugs to control the disease.

LEPROSY IN ARMADILLOS

It was reported in 1971 that leprosy developed in a common long-nosed armadillo *Dasypus novemcinctus* L. 17 months after inoculation with leprosy bacilli obtained from a human patient (Kirchheimer and Storrs, 1971; Storrs, 1971). Postmortem examination and subsequent histopathologic evaluations confirmed advanced, disseminated disease in this animal (Kirchheimer et al., 1972). Since then, *D. sabanicola* from Venezuela (Convit and Pinardi, 1974) and *D. hybridus* from Argentina (Storrs et al., 1975) also developed leprosy after inoculation with leprosy bacilli, by criteria established at the Workshop on Cultivation of *M. leprae* (Long and Tepper, 1968). *D. novemcinctus* is now well established as a model for leprosy studies (Storrs et al., 1978).

The first report of leprosy-like disease occurring naturally in wild armadillos appeared in 1975 (Walsh et al., 1975). Seven armadillos, 10 percent of the animals collected, had the disease. Further studies corroborated an overall incidence of leprosy of about 10 percent in armadillos from Louisiana (Walsh et al., 1977; Storrs, 1977). A total of 50 armadillos with leprosy was found, 49 from Louisiana and one from Texas. The disease was identical to *M. leprae* by all criteria used for identification of the leprosy bacillus (Meyers et al., 1977; Binford et al., 1977). Other reports of wild armadillos with naturally occurring leprosy-like disease from Mississippi, Louisiana and Texas confirmed the presence of this disease (Fox et al., 1977; Smith et al., 1978; Anderson, 1978; Kirchheimer and Sanchez, 1978).

An exhaustive epidemiologic examination of wild armadillos for the presence of leprosy has not been performed. However, the incidence of disease apparently varies with location. In Louisiana, wild armadillos from the southwestern coastal area of Vermillion Parish, as well as those from the more northern Jackson and Bienville parishes, appear to have a high incidence of disease. On the other hand, 177 armadillos from Hancock County, Mississippi, in the southwestern coastal region adjacent to Louisiana, showed no evidence of leprosy. One animal captured near Natchez, Mississippi, was reported by the Center for Disease Control, Atlanta, Georgia, to have been infected with the disease. We found one animal from northeast Texas which appeared to have leprosy, and two other armadillos from Texas have been found with evidence of the disease (Anderson, 1978; Benirschke, 1980, pers. comm.).

Recent surveys of armadillos in the states of Mexico and Sinaloa in Mexico have produced evidence of leprosy-like disease in one animal (Amezcua, 1981, pers. comm.), but the results have not yet been confirmed.

Armadillos from central Florida appear to be free of leprosy, based on over 400 animals that we have examined. The Florida armadillo population originated from a few imported animals which escaped in the Brevard County area of east central Florida in 1924 and 1936 (see also Wirtz et al., 1985; Ramsey and Grigsby, 1985). These animals must therefore be closely related genetically. They have been isolated from armadillos in the states of Texas, Louisiana, and Mississippi (Humphrey, 1974). The rapid migration of this latter population eastward, and the Florida population westward, should soon result in comingling of the two groups, with possible introduction of leprosy into the Florida wild armadillo population.

DISCUSSION

One can present three tentative hypotheses for the apparent differences in disease incidence between armadillo populations: (a) the results are misleading because of the large numbers of armadillos sampled in Louisiana; (b) the higher incidence of indigenous disease in armadillos in Louisiana is related to the fact that human disease has long been endemic there, and/or the habitat may be more conducive to transmission; and (c) the armadillos in Louisiana have a heritably higher susceptability to infection with *M. leprae* than armadillos from other areas.

With the information currently available, none of the three hypotheses can be ruled out. An epidemiologic study might help in testing hypothesis (a) and might also help in establishing whether hypothesis (b) is valid. This last hypothesis would be very difficult to confirm. Human leprosy has been endemic in Louisiana for at least 150 years, while the armadillos migrated to this state within the last 50 years. The armadillo may have come into direct contact with human leprosy through contact with infected individuals (before or after human death), fomites, or contaminated debris. The mode of transmission of the disease from armadillo to armadillo is not known, but leprosy bacilli are shed from nasal secretions of heavily infected animals and have been found in the milk of a lactating female that had the disease when captured (Walsh et al., 1977).

The mode of transmission of leprosy in humans is not known. It is commonly assumed that prolonged direct contact with other humans with leprosy is necessary for transmission. However, many cases of human leprosy develop with no known prior human contact. Enna et al. (1978) reported that of 1,432 leprosy cases in the continental United States between 1967 and 1976, only 25.8 percent had known direct contact with leprosy patients prior to developing the disease.

Perhaps human-human contact is not necessary for transmission of leprosy, and there are nonhuman reservoirs of *M. leprae*. The armadillo is one possibility. However, in the United States no correlation has been found between development of human leprosy and handling of armadillos. In Venezuela, where human leprosy is endemic, we were told by the Public Health Officer of the state of Apure of a local saying: "people who eat armadillo develop leprosy". Even if an animal was infected, well-cooked meat would not be infective. However, people handling live infected animals in preparing them for the table might run a certain risk. No wild animals with leprosy have been found among the small numbers of armadillos collected in South America.

The possibility that *M. leprae* is harbored in some types of soils could help explain the occurrence of leprosy in people with no direct contact with active human leprosy. It could also explain the infection of wild armadillos, since they are a susceptible species and live in direct contact with the soil, both when feeding and burrowing.

Many species of mycobacteria are found in soil. Some, such as *M. vaccae* and *M. phlei*, are saprophytes. Others sometimes found in soil can be pathogenic to man or animals, including *M. tuberculosis, M. bovis, M. avium* and *M. intracellulare. Lepra bubalorum* in water buffalo *Bubalus bubalis* shows a distribution of nodules in skin and nasal membranes suggesting infection from the mud in which the buffalo wallow (Lobel, 1936). Van Ness (1966) related mycobacterial infections in cattle to alkaline soils, showing that alkaline wetlands either enhance the acid-fast organisms or lower the resistance of the host cattle to infection.

Van Ness (pers. comm.) speculated that human leprosy occurs in areas where the soil is alkaline or calcareous; areas in Louisiana where wild armadillos with leprosy have been found in general have Lake Charles-Crowley soils, which are alkaline swamplands with high humus content. These are areas where human leprosy is also endemic. Coastal soils of Mississippi (including Hancock County), Alabama, and Florida are sandy and human leprosy is not endemic there. None of 177 wild armadillos collected in Hancock County showed evidence of leprosy-like disease.

Kazda et al. (1980; pers. comm.) found acid-fast rods of noncultivable mycobacteria which resemble *M. leprae* in sphagnum bog areas of Norway where leprosy was once prevalent. Sphagnum bogs are acidic with a pH of less than five and their waters contain no more minerals than rainwater. Sphagnum moss also contains small amounts of sphagnol, a phenolic compound, and probably because of this and the acidity of the water, it possesses mild antiseptic properties. During World War I, sphagnum was used as a dressing for wounds. Because of its antibacterial properties, water from sphagnum bogs was used for drinking on sailing ships because it did not become foul on long standing. If mycobacteria do indeed survive in sphagnum bogs for long periods of time, they must be more hardy than most other bacteria.

The Van Ness and Kazda hypotheses of possible *M. leprae* survival in the soil involve extremes of pH-but one (Van Ness) postulates high pH for survival, while the other (Kazda) postulates survival under acidic conditions. It would seem unlikely, unless the organism survives in different forms in two different environments, that both are correct.

Transmission of leprosy by sucking or biting insects has been postulated (Narayanan et al., 1972). The number of bacilli carried by an arthropod in a blood meal from a lepromatous leprosy patient would be much less than the numbers found in nasal secretions of the same patient (Rees, 1972), and passive transmission by flies from environmental contamination by nasal shedding from humans or infected armadillos might pose a more serious threat. Viable *M. leprae* can be collected 2 to 7 days after shedding in human nasal secretions, even in a dessicated state (Davey and Rees, 1974). Flies can carry contaminating mycobacteria either from contaminated secretions to humans, or to other areas of the environment (Geater, 1975).

More epidemiologic information on the prevalence of *M. leprae* in wild armadillos from the United States, and particularly from Latin America, is urgently needed to help determine routes of transmission of the disease from armadillo to armadillo, the possibility of transmission from man to armadillo to man, and to determine possible reservoirs of *M. leprae*.

LITERATURE CITED

Anderson, M.
1978. Leprosy in an armadillo from Texas. Leprosy Scientific Memoranda, Memo L-967.

Binford, C. H., W. M. Meyers, G. P. Walsh, E. E. Storrs, and H. L. Brown.
1977. Naturally-acquired leprosy in the nine-banded armadillo (*Dasypus novemcinctus*): histopathologic and microbiologic studies of tissues from armadillos with naturally-acquired leprosy. J. Reticuloendothel. Soc., 22: 377-388.

Convit, J. and M. E. Pinardi.
1974. Inoculacion de *M. leprae* en dos especies de armadillo: *D. sabanicola* y *D. novemcinctus*. Acta Cient. Venezolana, 25: 51.

Davey, T. G. and R. J. W. Rees.
1974. The nasal discharge in leprosy: clinical and bacteriological aspects. Lepr. Rev., 45: 121-134.

Enna, C. D., R. R. Jacobson, J. R. Trautman, and M. Sturdivant.
1978. Leprosy in the United States (1967-1976). Public Health Reports, 93: 468-473.

Fox, M. D., D. C. Anderson, and A. F. Kaufman.
1977. Leprosy-like disease in a wild armadillo from Mississippi. Leprosy Scientific Memoranda, Memo, L-865.

Geater, J. G.
1975. The fly as potential vector in the transmission of leprosy. Lepr. Rev., 46: 279-286.

Humphrey, S. R.
1974. Zoogeography of the nine-banded armadillo (*Dasypus novemcinctus*) in the United States. Bioscience, 24: 457-462.

Kazda, J., L.M. Irgens and K. Muller.
1980. The isolation of non-cultiviable acid-fast bacilli in sphagnum and moss vegetation by footpad technique in mice. Int. J. Lep. 48: 1-6.

Kirchheimer, W. F. and R. M. Sanchez.
1978. Leprosy in the wild. Leprosy Scientific Memoranda, Memo L-966.

Kircheimer, W. F. and E. E. Storrs.
1971. Attempts to establish the armadillo (*Dasypus novemcinctus* Linn) as a model for the study of leprosy. I. Reports of lepromatoid leprosy in an

experimentally infected armadillo. Int. J. Lep., 39: 693-707.

Kirchheimer, W. F., E. E. Storrs and C. H. Binford.
1972. Attempts to establish the armadillo (*Dasypus novemcinctus* Linn.) as a model for the study of leprosy. II. Histopathologic and bacteriologic post-mortem findings in lepromatoid leprosy in the armadillo. Int. J. Lep., 40: 229-242.

Lobel, L. W. M.
1936. *Lepra bubalorum*. Int. J. Lep., 4: 79.

Long, E. R. and B. S. Tepper.
1968. Workshop on cultivation of *M. leprae* (IX International Leprosy Congress, London.) Int. J. Lep., 36: 559.

Meyers, W. M., G. P. Walsh, H. L. Brown, R. J. W. Rees, and J. Convit.
1977. Naturally acquired leprosy in the nine-banded armadillo (*Dasypus novemcinctus*): reactions in leprosy patients to lepromins prepared from infected armadillos. J. Reticuloendothel. Soc., 22: 369-375.

Narayanan E., K. Shankara Manja, B. M. S. Bedi, W. F. Kirchheimer and M. Balusrebrahmanyan.
1972. Arthropod feeding experiments in lepromatous leprosy. Lepr. Rev., 43: 181-187.

Ramsey, P.R. and B.A. Grigsby.
1985. Protein variation in populations of *Dasypus novemcinctus* and comparisons to *D. hybridus, D. sabanicola* and *Chaetophractus villosus.* (this volume).

Rees, R. J. W.
1972. Can arthropods transmit leprosy? Lepr. Rev., 43: 165-166.

1975. Do flies transmit leprosy? Lepr. Rev., 46: 255-256.

Smith, J. H., S. K. File, B. A. Nagy, D. S. Folse, J. A. Buckner, L. J. Webb, and A. M. Beverding.
1978. Leprosy-like disease of wild armadillos in French Acadiana, Louisiana. J. Reticuloendothel. Soc., 24: 705-719.

Storrs, E. E.
1971. The nine-banded armadillo: a model for leprosy and other biomedical research. Int. J. Lep., 39: 703-714.

1977. Leprosy in the feral armadillo, *Dasypus novemcinctus*. Invitational paper presented at Am. Vet. Med. Assoc. Annual Meeting, Atlanta, Georgia.

Storrs, E. E., C. H. Binford and G. Migaki.
1978. Animal model of human disease. lepromatous leprosy. Animal model: experimental lepromatous leprosy in nine-banded armadillos (*Dasypus novemcinctus* Linn.). Am. J. of Path., 92: 813-816.

Storrs, E. E., G. P. Walsh, and H. P. Burchfield.
1975. Development of leprosy in another species of armadillo (*Dasypus hybridus* L.): genetic and immunologic implications. J. Trop. Med. and Hyg., 78: 216-218.

Van Ness, G. B.
1966. The role of meat inspection in a disease reporting system. Proc. U. S. Livestock Sanitary Assoc. Meeting: 396-402.

Walsh, G. P., W. M. Meyers, C. H. Binford, L. Harris, M. DeRouen, and R. Borne.
1977. Naturally acquired leprosy in the armadillo: the presence of organisms in the milk and mammary glands of a lactating female. Leprosy Scientific Memoranda, Memo L-931.

Walsh, G. P., E. E. Storrs, H. P. Burchfield, E. H Cottrell, M. F. Vidrine, and C. H. Binford.
1975. Leprosy-like disease occurring naturally in armadillos. J. Reticuloendothel. Soc., 18: 347-351.

Walsh, G. P., E. E. Storrs, W. M. Meyers, and C. H. Binford.
1977. Naturally acquired leprosy in the nine-banded armadillo (*Dasypus novemcinctus*): recent epidemiologic findings. J. Reticuloendothel. Soc., 22: 363-367.

Wirtz, W.O., D.H. Austin and G.W. Dekle.
1985. Food habits of the common long-nosed armadillo *Dasypus novemcinctus* in Florida, 1960-61. (this volume).

SLOTHS AS HOSTS OF ARBOVIRUSES

Charles Seymour

Division of Microbiology, St. Francis Hospital and Medical Center, 114 Woodland St., Hartford Connecticut, USA 06105

Abstract

Field surveys of sloths in Panama and Brazil yielded isolates of Saint Louis encephalitis, Murutucu, Oropouche, Anhanga, Bradypus-4, Utive and Utinga viruses, and viruses from the Changuinola and Punta Toro complexes. Serologic evidence from these surveys showed that sloths are also naturally infected by viruses closely related or identical to Venezuelan encephalitis, Mayaro, Ilheus, and Itaporanga viruses. Experimental inoculation of *Bradypus variegatus* with yellow fever, Venezuelan encephalitis and Saint Louis encephalitis viruses produce remarkably long viremias, detectable for up to 23 days, with no apparent illness. Concentrations of circulating virus are high enough to qualify *Bradypus* as potential amplifying hosts for these viruses and for the Oropouche virus. *Choloepus* sloths are more refractory than *Bradypus* to yellow fever and Venezuelan encephalitis viruses, but *Choloepus hoffmanni* are highly susceptible to Saint Louis encephalitis virus. Sloths circulating Saint Louis encephalitis virus are infectious for at least one species of mosquito. Simultaneous Changuinola and Saint Louis encephalitis viremias were seen in one experimental *Choloepus* infected with both viruses, and a chronic or recrudescent Changuinola virus infection was observed in an experimental *Bradypus*. Sloth-specific virus complexes are discussed in relation to their potential for genetic reassortment and chronic infection.

Abstracto

Encuestas en el campo entre perezosos de Brasil y de Panamá rindieron aislamiento de los siguientes virus: Encefalitis de Saint Louis, Murutucu, Oropouche, Anhanga, Bradypus-4, Utive y Utinga, como también de miembros de los complejos virales de Changuinola y Punta Toro. Los perezosos también aparecen naturalmente infectados por virus cercanos o idénticos a la encefalitis venezolana, y a Mayaro, Ilheus e Itaporanga. La inoculación experimental de *Bradypus variegatus* con los virus de fiebre amarilla, encefalitis venezolana y encefalitis de Saint Louis produce viremias notablemente prolongadas, que pueden ser detectadas hasta 23 días después de la inoculación, sin causar ninguna enfermedad aparente. Las concentraciones de virus en la circulación son suficientemente altas como para calificar a los *Bradypus* con hospederos amplificadores en potencia de estos virus. Los perezosos del género *Choloepus* son más refractarios a los virus de fiebre amarilla y de encefalitis Venezolana que los *Bradypus*, pero los *Choloepus hoffmanni* son altamente susceptibles al virus de la encefalitis de Saint Louis. Aquellos perezosos que circulan este virus son infecciosos para por lo menos una especie de mosquitos. Viremias simultáneas con los virus de Changuinola y de la encefalitis de Saint Louis fueron observados en un *Choloepus* experimental infectado con ambos virus, y una infección crónica o recrudecente del virus Changuinola fue observada en un *Bradypus* experimental. La existencia de complejos de virus específicos para los perezosos es discutida en relación a su potencial para la redistribución genética y en las infecciones crónicas.

Resumo

Levantamentos no campo, das preguiças no Panamá e Brasil, permitiram isolar vírus de encefalite de Saint Louis, vírus de Murucutu, Oropouche, Anhanga, Bradypus-4, Utive e Utinga, e vírus dos complexos Changuinola e Punta Toro. As preguiças são naturalmente infetadas por vírus muito proximamente relacionados ou idênticos aos vírus de encefalite venezuelana, Mayaro, Ilhéus e Itaporanga. Inoculação experimental em *Bradypus variegatus* de vírus de febre amarela, encefalite venezuelana e encefalite de Saint Louis produziram viremias notavelmente longas, detectáveis por mais de 23 dias, sem nenhuma doença aparente. As concentrações de vírus circulante eram altas, o suficiente para qualificar *Bradypus* como hospedeiros amplificadores potenciais para esses vírus. As preguiças reais, *Choloepus*, são mais imunes aos vírus da febre amarela e encefalite venezuelana do que *Bradypus*, mas *Choloepus hoffmanni* são altamente susceptíveis aos vírus da encefalite de Saint Louis. Os vírus circulantes da encefalite de Saint Louis das preguiças são contagiosos a, pelo menos, uma espécie de mosquito. Viremias simultâneas da encefalite de Saint Louis e Changuinola foram observadas em um *Choloepus* experimental, infectado com ambos os vírus e uma infecção crônica ou recrudescente por vírus Changuinola foi observada num *Bradypus* experimental. Complexos de vírus específicos de preguiças são discutidos em relação ao seu potencial para reagrupamento genético e infecção crônica.

Field and laboratory studies reviewed here concentrated on arthropod-borne viruses (arboviruses). Although arboviruses fall into any of several morphological taxa (togaviruses, bunyaviruses, orbiviruses, rhabdoviruses), as well as into different antigenic groups, they are united and defined by their ecology. Arbovirus natural cycles depend on biological virus transmission from vertebrate to vertebrate through blood-feeding vectors, which include mosquitoes, ticks, and sandflies. The efficiency of vector infection correlates with virus concentration in the blood of the vertebrate host and with the duration of virus circulation. Therefore, any vertebrate species which can consistently circulate high concentrations of virus for a long period is a potentially efficient arbovirus host. If such a species is frequently infected in nature, and if it is common, it may be epidemiologically important.

The abundance of sloths (Montgomery and Sunquist, 1975) and their frequent, intense, long-lasting and harmless virus infections are strong evidence that sloths are important links in the transmission cycles of some tropical arboviruses. In this paper I review the current information on natural and experimental arbovirus infections of sloths, virus by virus. I conclude with a discussion of the significance of sloths for tropical medicine and basic virology.

FIELD AND LABORATORY STUDIES

Most work on virus infection in sloths has been done in Brazil and Panama. In Brazil, the Belem Virus Laboratory (Instituto Evandro Chagas, formerly the Rockefeller Foundation field laboratory) routinely surveyed forest vertebrates including sloths for arboviruses and arbovirus antibodies (Woodall, 1967; Theiler and Downs, 1971; McLean and Bowen, 1980). These studies began in the 1950s and are continuing.

In Panama, workers at the Gorgas Memorial Laboratory conducted field surveys of wild sloths from 1974 and 1977, and have experimentally infected sloths with Venezuelan encephalitis, Saint Louis encephalitis, and yellow fever viruses (Grayson and Galindo, 1969; Gorgas Memorial Laboratory, 1975; Seymour et al., 1983a). The field projects included a one-year study from September 1974 to August 1975 of radio-marked sloths in the Cerro Azul area. In this study, free-ranging sloths were periodically recaptured for both virus and antibody testing of serial blood samples. Sloths from the Maje area of the Bayano River basin and from other areas of central Panama were also tested for antibodies and occasionally for current virus infection. The Gorgas Memorial Laboratory collaborated with the Smithsonian Tropical Research Insititute on much of the field work (Seymour et al., 1983b, 1983c).

Togaviruses: Antigenic group A (alphaviruses)

Venezuelan encephalitis virus.-- In epidemic form this agent is a devastating, often fatal cause of horse encephalitis. In endemic form it is less virulent and is maintained in swampy forests by a mosquito-terrestrial rodent cycle, with little involvement of animals which live in the forest canopy. However, hemagglutination-inhibition (HI) tests at the Gorgas Memorial Laboratory detected antibody in Panamanian *Bradypus variegatus* sloths against an endemic virus variety. The Belem Virus Laboratory also found HI antibody in sera of central Panamanian sloths; two of 21 *Bradypus* and six of 29 *Choloepus hoffmanni* were positive against Mucambo virus, a South American subtype of Venezuelan encephalitis virus (J. P. Woodall, pers. comm.). The Mucambo HI positives are best interpreted as cross reactions with natural antibody against local Panamanian endemic varieties of Venezuelan encephalitis virus. Neither the Gorgas Laboratory nor the Belem Virus Laboratory confirmed the HI positives by neutralization tests.

Grayson and Galindo (1969) experimentally infected *Choloepus hoffmanni* and *Bradypus variegatus* with an endemic Panamanian strain of Venezuelian encephalitis virus. *Bradypus* sloths developed higher concentrations of virus in the blood (higher viremias) detectable for longer periods of time than did any of 11 other forest species tested except squirrels *Sciurus granatensis*. Viremia peak titers in *Bradypus* were as high as 6.1 $\log_{10}$ weanling mouse intraperitoneal median lethal doses per ml of blood, and viremias were detected for one to two weeks. Virus was detected in kidney and spleen after 19 days (Grayson, 1971). For comparison, the classic rodent reservoir hosts normally develop viremias of comparable intensity, but usually lasting only three to five days.

Bradypus sloths are therefore potential sources of infection for vector mosquitos because of their high and exceptionally long Venezuelan encephalitis viremias. Their forest canopy habitat has removed them from the forest floor focus of intense transmission, although HI antibody suggests at least occasional natural infection. In contrast, two-toed sloths *Choloepus* differed from *Bradypus* in having low Venezuelan encephalitis viremias (3.5 logs per ml) of unreported duration.

Mayaro virus.-- This mosquito-borne agent causes sporadic outbreaks of human febrile illness in forest areas of South America and Trinidad. Sloths may be infected only rarely in nature. One of seven Brazilian *Bradypus tridactylus* tested for HI antibody against Mayaro antigen was positive (Woodall, 1967), but the result was not confirmed by neutralization tests. None of 21 *Bradypus* and 29 *Choloepus*, all from central Panama, was positive (J. P. Woodall, personal communication).

Togaviruses: Antigenic group B (flaviviruses)

Yellow fever virus.-- This notorious mosquito-borne agent can be lethal not only for humans but also for many of the monkey species which are its vertebrate hosts in jungle canopy cycles. Other animals, including sloths, have been considered as possible reservoir hosts of yellow fever virus because of its lethality to monkeys and because it occurs in certain New World areas where monkeys are scarce.

Searching for alternate hosts, Bugher (1951) found that a single Colombian *Choloepus didactylus* "circulated small amounts of virus" following subcutaneous inoculation of low doses of laboratory-passaged yellow fever virus.

In 1974, an outbreak of jungle yellow fever in Panama prompted further sloth inoculations by P. A. Webb and K. M. Johnson (Gorgas Memorial Laboratory, 1975). These experiments used low-passage-level virus to confirm that yellow fever viremias of two-toed sloths *Choloepus* are low. The infection patterns of both *Choloepus* and *Bradypus* were erratic; several animals did not become infected at all, and others developed antibody as late as three to four months after inoculation. However, Webb and Johnson discovered that some *Bradypus* developed extremely high and long yellow fever viremias. Circulating virus was detectable for up to 10 days, with maximum titers of up to 6.5 $\log_{10}$ tissue culture median infectious doses. The long duration is exceptional in any animal; the high titer is typical of the classic monkey hosts. No sloths showed signs of illness attributable to yellow fever virus.

This experimental potential of *Bradypus* sloths as yellow fever hosts prompted field work in two Panamanian forest areas where yellow fever virus had been recently active (Seymour et al., 1983b). None of 70 *Bradypus* (including 21 juveniles) and none of 57 *Choloepus* (9 juveniles) had detectable neutralizing antibody against yellow fever virus. In contrast, monkeys of three species, including 18 of 29 adult howler monkeys *Alouatta villosus* tested, were positive. Since experiments had shown that viremic sloths eventually develop antibody, the field studies are strong evidence that sloths played no important role in the 1974 Panamanian yellow fever activity.

Saint Louis encephalitis (SLE) virus.-- In North America, this New World mosquito-borne virus cycles principally among birds. It periodically causes dramatic widespread epidemics of occasionally fatal human encephalitis. In Central and South America it is known to cause only sporadic human febrile illness and encephalitis. Tropical SLE virus seems to be mosquito-borne also. But although a few strains have been isolated from forest birds, the vertebrate hosts in the cycle are not well understood.

As an antigenic relative of yellow fever virus, SLE virus was included in the original 1974 Gorgas Memorial Laboratory survey for sloth neutralizing antibodies against flaviviruses. With Smithsonian Tropical Research Institute collaboration, the survey was expanded to include a total of 167 *Bradypus* and 99 *Choloepus* from central Panama (Seymour et al., 1983). Eight percent of *Bradypus* and 39 percent of *Choloepus* were specifically positive for SLE virus antibody. Specific SLE virus neutralization means that neutralization titers against SLE virus were at least four times higher than those against yellow fever, Ilheus, and Bussuquara viruses, the only other flaviviruses known to occur in Panamanian forests. In earlier HI tests by the Belem Virus Laboratory on central Panamanian sloth sera, four of 21 *Bradypus* and 12 of 29 *Choloepus* reacted positively against SLE virus antigen (J. P. Woodall, pers. comm.). Though not confirmed by neutralization test, the HI titers were higher against SLE virus antigen than against any other related Panamanian forest viruses. SLE virus was not isolated from Panamanian sloths. However, an isolate from *Bradypus tridactylus* has been reported from Brazil (pers. comm. from F. Pinheiro, cited by McLean and Bowen, 1980).

The high prevalence of antibody in sloths, particularly *Choloepus*, suggested that they are important hosts for SLE virus in Panama. For comparison, a wide variety of other forest vertebrates was sampled at Maje, an area where sloth immunity against SLE virus was found relatively frequently. The results of this survey show that specific SLE virus neutralizing antibody is more prevalent in sloths than in any other group sampled (Table 1), though some groups, especially birds and bats, were inadequately sampled (Seymour, et al., 1983b).

Two nonexclusive hypotheses can explain the high prevalence of previous infection in sloths. First, infected SLE virus vector mosquitoes may infect sloths more frequently than other vertebrates; a sloth population

Table 1. Prevalence of specific neutralizing antibody against Saint Louis encephalitis virus in birds and mammals collected near Maje, Panama, 1974-1976.

Group (no. of species)	*No. positive/ no. tested*	*Percent positive*
Resident birds (39)	1/75	1
Opossums (3)	0/19	0
Bats (6)	0/20	0
Monkeys (4)	0/83	0
Rodents (6)	0/54	0
Rabbits (1)	0/15	0
Armadillos (2)	0/5	0
Tamandua (1)	2/14	14
Sloths (2)		
Bradypus variegatus	9/58	16
Choloepus hoffmanni	25/54	46

would then be a very active virus reservoir. Second, sloths may be infected only as frequently as other vertebrates, but may live longer, with slower population turnover, allowing a gradual but extensive buildup of immune animals.

There are few data by which to test the first hypothesis. The Gorgas-Smithsonian study in Cerro Azul is the only work in which a fairly large number of individual sloths were studied over time. For one year, 28 radio-marked *Bradypus* and two mother-baby *Choloepus* pairs were sequentially bled. Individual *Bradypus* were followed for a mean of 15.9 weeks each; the *Choloepus* pairs averaged 8.9 weeks. One *Bradypus* and one *Choloepus* had SLE virus antibody when first captured. However, both mother and baby of one of the two *Choloepus* pairs developed specific SLE virus antibody between September and November, 1974. This unique study is too limited for quantitative interpretation.

The hypothesis of gradual accumulation of SLE antibody in long-lived sloth populations was evaluated in Panama by testing sloths of different age classes for SLE virus antibody. Juvenile sloths, which were independent of their mothers but not fully grown, were uniformly negative (16 *Bradypus*, 6 *Choloepus*). Infant sloths still rode on their mothers; all of 10 *Bradypus* and 6 of 7 *Choloepus* infants were negative for SLE virus antibody. In contrast, 44 percent of adult Panamanian *Choloepus* and 10 percent of adult *Bradypus* were specifically immune to SLE virus.

If we assume that sloths do not become independent of their mother for at least six months and are not fully grown until at least two years old (Montgomery and Sunquist, 1978; Eisenberg and Maliniak, 1985; Meritt, 1985), these antibody data suggest that sloths are infrequently infected by SLE virus. *Braydpus* are known to live for more than 11 years in the wild (G. G. Montgomery, pers. comm.), and data from zoos show that *Choloepus* can live for decades (Eisenberg and Maliniak, 1985; Meritt, 1985). These observations further support the hypothesis of gradual SLE virus antibody accumulation in long-lived wild sloths. The infection of the two *Choloepus* on Cerro Azul during only a few weeks of observation may be simply fortuitous or may indicate rapid infection, particularly if "jungle" SLE virus occurs in epidemic waves as does its antigenic cousin, yellow fever virus. Such an epidemic transience of tropical SLE is suggested by data from Trinidad (Trinidad Regional Virus Laboratory, 1958).

Experimental infection of sloths (Seymour et al., 1983a) showed that SLE virus infection caused no apparent disease in individuals of either species. After inoculation by the bite of infected mosquitoes, or by needle and syringe, both Panamanian species circulated high concentrations of SLE virus, and their viremia durations were remarkably long. All animals without preexisting antibody became viremic. The geometric mean peak titers in six *Choloepus* was 4.8 $\log_{10}$ suckling mouse intracranial median lethal doses per ml of blood, and 5.3 logs in five *Bradypus*. The median duration of detectable viremia was 11 days in *Choloepus* and 18 days in *Bradypus*. The longest individual viremias were 27 days in one *Choloepus* and 23 days in one *Bradypus*. In birds, the classic North American SLE virus hosts, maximum viremia titers are comparable with those of sloths. However, detectable bird viremias are of much shorter duration, seldom if ever exceeding one week.

The high and extraordinarily long-lasting SLE viremias make these sloths candidates for efficient mosquito infection. SLE virus transmission experiments by L. D. Kramer at Gorgas Memorial Laboratory used colonized *Culex pipiens quinquefasciatus*, an urban mosquito species known as a vector in North America but not in Panama, and *Haemagogus equinus*, a jungle mosquito from which SLE virus has never been isolated. Only sloths circulating very high concentrations of SLE virus (5.1-6.3 logs per ml) could infect almost half the *C. p. quinquefasciatus* feeding on them. Most of these infected mosquitoes were subsequently able to transmit SLE virus to laboratory animals. However, *H. equinus* were not infected by sloths circulating 4.1-5.3 logs of virus per ml. At any given viremia level, chicks and mice infected substantially more mosquitoes of either species than did sloths. The transmission experiments show that a laboratory mosquito-sloth-mosquito-chick cycle is possible; but, in spite of high viremias of long duration, sloths are not as efficient infection sources as are chicks and mice for nonvector mosquito species.

Mosquitoes may not be necessary for SLE virus transmission among sloths under one special circumstance. Seymour et al. (1983a) showed by antibody studies that an experimentally infected mother *Bradypus* transmitted SLE virus to her nutritionally but not socially weaned baby. This result is reminiscent of the nearly simultaneous development of antibody by the free-ranging *Choloepus* mother-baby pair at Cerro Azul (Seymour et al., 1983b).

No experimentally infected sloths in Panama suffered apparent disease attributable to SLE virus. Detectable antibody was slow to develop, perhaps accounting for the long viremias. However, antibody eventually reached high titers, and immune sloths challenged with SLE virus developed no viremia.

Ilheus virus.-- Antigenically related to yellow fever and Saint Louis encephalitis viruses, this New World tropical flavivirus occasionally causes human febrile illness and encephalitis. Neutrilizing antibody against Ilheus virus was found in only one of 167 Panamanian *Bradypus* tested, and in one of 99 *Choloepus*. The positive sera did not cross-react with any other known Panamanian forest flaviviruses. HI antibody unconfirmed by neutralization test has been found in 4 percent of Brazilian *Bradypus* (Woodall, 1967).

Bunyaviruses: Antigenic group C

Murutucu virus.-- Rodents and opossums are the principal vertebrate hosts of this South American virus borne by the *Culex* mosquito, although it has been occasionally isolated from febrile humans (Berge, 1975). Woodall (1967) has reported a single isolation from *Bradypus tridactylus* near Belem, Brazil.

Bunyaviruses: Antigenic group Simbu

Oropouche virus.-- First isolated from a sick human in Trinidad, this agent has caused periodic epidemic febrile illness in Para, Brazil (Pinheiro et al., 1981). Although Oropouche virus is probably transmitted by biting *Culicoides* midges, its epidemiology is still unclear. The only isolates from wild vertebrates have been from a Brazilian *Bradypus tridactylus* captured on the Belem-Brasilia highway in 1960 (Woodall, 1967) and from three other *Bradypus tridactylus* captured in 1971 on the east coast of Para (Pinheiro et al., 1981).

Experimental infection (Pinheiro et al., 1981) of two *Bradypus tridactylus* sloths with Oropouche virus resulted in viremias of 2-6 days duration and peak titers with 5.2-5.4 $\log_{10}$ suckling mouse lethal doses per ml. One of the infected *Bradypus* died. The viremia profiles were similar to those of four species of experimentally infected monkeys.

Utinga virus.-- Related to Oropouche virus but distinct from it, Utinga virus was first isolated from a *Bradypus tridactylus* sloth in the Utinga forest, near Belem, Brazil (Woodall, 1967). P. H. Peralta, of Gorgas Memorial Laboratory, found 25 Utinga virus isolates from *Culicoides* midges and one isolate from *Anopheles* mosquitoes (strain Bayano 531), all collected at Maje, Panama, in 1974 and 1975. Seymour et al. (1983c) found that 61 percent of 23 *Bradypus variegatus* and 52 percent of 23 *Choloepus hoffmanni* from Panama were positive against Utinga virus. In HI tests done at the Belem Virus Laboratory, 3 of 21 *Bradypus* and 9 of 29 *Choloepus* from central Panama were positive (J. P. Woodall, pers. comm.).

These antibody results from wild sloths are confused by possible cross-reactions with Utive virus (see below) and Pintupo virus, another Simbu group virus from Panama; these three viruses form a closely related antigenic complex. Serologic surveys at Maje indicate that these viruses are sloth-specific. None of 75 birds of 39 species and none of 224 mammals of 26 species other than sloths had Utinga-complex virus neutralizing antibody except a single owl monkey *Aotus trivirgatus*.

Utive virus.-- This virus was isolated at Gorgas Memorial Laboratory from a radio-marked adult male *Bradypus variegatus* bled in July, 1975 in the Cerro Azul area of central Panama (Seymour et al., 1983c). The host sloth died before it could be recaptured and bled for seroconversion studies. The validity of the isolation is based on its being a new serotype, which had never been in the Gorgas Memorial Laboratory previously.

Complement fixation (CF) tests show that Utive virus is a member of the Simbu serogroup (Kinney and Calisher, 1981; Seymour et al., 1983c). Within this serogroup, it belongs to a complex of closely related viruses, including Utinga, Pintupo, Oropouche and Facey's Paddock viruses. All members of this complex have been isolated from *Culicoides* midges. Gorgas Memorial Laboratory CF tests of Utive virus, prototype Brazilian Utinga virus, and both Pintupo virus and topotype Panamanian Utinga strain Bayano 531 from Maje *Culicoides* midges show all strains to be identical by CF (Table 2, and Seymour et al., 1983c).

By neutralization and HI test at the Gorgas Memorial Laboratory, however, Utive and Utinga viruses are clearly distinct from each other and from Pintupo virus (Seymour et al., 1983c). Bayano 531 from Panama *Anopheles* is indistinguishable from Utinga (Tables 3 and 4).

Table 2. Complement-fixation reaction relationship between Utinga and Utive viruses. Shown are the reciprocals of the antibody titer/antigen titer.

	Antibody		
Antigen	*Utive*	*Utinga*	*Bayano 531*
Utive	256/256	512/128	512/256
Utinga	256/256	512/128	512/256
Bayano 531	128/512	512/512	256/1024

Table 3. Plaque-reduction neutralization relationship between Utinga and Utive viruses. Shown are the reciprocals of antiserum dilution inhibiting 90 percent of 40-50 plaque-forming units. Incubation overnight at 4°C. The combination which was not tested is indicated by nt.

	Antibody			
Virus	*Utinga prototype*	*Bayano 531 purified*	*Utive purified*	*Utive unpurified*
Utinga prototype	128	128	16	16
Bayano 531 purified	256	256	16	nt
Utive purified	32	16	1024	512
Utive unpurified	16-32	16	1024	1024

Table 4. Hemagglutination-inhibition (HI) relationship between Utinga and Utive viruses. Shown are reciprocals of antibody dilution completely inhibiting 4-8 viral hemagglutinating units.

	Antibody		
Antigen	*Utinga prototype*	*Utive purified*	*Utive unpurified*
Utinga prototype	40	20	10
Utive purified	10	160	20
Utive unpurified	10	80	20

The only other known strain of Utive virus was isolated from a pool of 127 *Culicoides diabolicus* collected at Maje on June 2, 1976.

Serologic surveys of sloths and other Panamanian wild forest vertebrates for Utive virus neutralizing antibody gave results similar to those for Utinga antibody. At Cerro Azul, 3 of 35 *Bradypus* were seropositive, in addition to the *Bradypus* number 121 which yielded the original isolate. One of the seropositive radio-marked animals developed Utive virus antibody during December, 1974, presumably following a recent infection. Also, one of five *Choloepus* from Cerro Azul was seropositive. At Maje, three of 73 *Bradypus* and 22 of 58 *Choloepus* had neutralizing antibody against Utive virus. Negative results for other Maje vertebrates are included in the discussion of Utinga virus.

Phleboviruses: Antigenic Group Phlebotomus Fever

Anhanga virus.-- This agent has been isolated only once, from a *Choloepus didactylus* captured in 1962 in the Castanhal forest, Para, Brazil (Woodall, 1967). None of 29 Panamanian *Choloepus* and 21 *Bradypus* sloths had HI antibody against this virus (Berge, 1975).

Itaporanga virus.-- This agent has been isolated in South America and Trinidad, where it occurs in mosquitoes and primarily arboreal mammals and birds (Berge, 1975). One of the 29 Panamanian *Choloepus* sera was positive against Itaporanga antigen in HI tests at the Belem Virus Laboratory (Woodall, pers. comm.).

Punta Toro virus.-- This virus is found in *Lutzomyia* sandflies; it frequently infects Panamanians in rural areas and can cause febrile illness. It is closely related to several other sandfly viruses from Panama and Colombia (Tesh et al., 1974). These agents form an incompletely characterized antigenic complex with varying degrees of cross-reactivity. Little is known of the natural vertebrate hosts of the viruses of this complex.

A strain designated Pan An 53038 was isolated and reisolated at Gorgas Memorial Laboratory from a free-ranging radio-marked female *Bradypus variegatus* bled December 1, 1974, at Cerro Azul, Panama (Seymour et al., 1983c). Plasma taken 38 days later from the same animal neutralized not only the homologous strain but also prototype Punta Toro and another member of the complex, VP-419 T virus. Plasma taken 40 days before the date of detected viremia was negative against all three viruses. This evidence that the strain Pan An 53038 is a member of the Punta Toro complex was confirmed by CF tests using hamster immune sera.

A Gorgas Memorial Laboratory survey showed that sera from a wide variety of forest animals from central Panama neutralized strain Pan An 53038. Eight of 106 *Bradypus* and 11 of 66 *Choloepus* were positive. The original virus isolate and two other *Bradypus* seroconversions indicate that this and perhaps other Punta Toro-complex viruses were transmitted to radio-marked sloths during November and December, at the end of the rainy season.

Other animals with sera neutralizing strain Pan An 53038 included three different monkey species, *Saguinus geoffroyi, Alouatta villosus* and *Aotus trivirgatus*; an agouti, *Dasyprocta punctata*; an arboreal porcupine, *Coendu rothschildi*; a coatamundi, *Nasua nasua*; and 4 of 7 kinkajous, *Potos flavus*, tested. These positives may reflect infection with cross-reacting Punta Toro-complex viruses other than strain Pan An 53038.

Orbiviruses: Antigenic Group Changuinola

Changuinola virus:-- This and related agents have been isolated over 100 times from phlebotomine sandflies of Panama and Brazil (Tesh et al., 1974; Woodall, 1967). The Changuinola virus strains form an antigenic complex in which viruses are very closely related or identical by CF test but distinct by neutralization test. Until 1975, the only strains from vertebrates were a Changuinola isolate from a febrile human in Panama, and the related Irituia virus from a Brazilian *Oryzomys* rice rat.

The natural vertebrate hosts of this group of viruses were thus unknown until the 1974-1977 Gorgas Memorial Laboratory sloth studies yielded four Changuinola-complex viruses from whole blood of Panamanian sloths (Seymour et al., 1983c). All reacted identically to titer against reference Changuinola antiserum by CF test. Two isolates were from radio-marked *Bradypus* from Cerro Azul, and two were from a *Bradypus* and a *Choloepus* from Aguacate village. All four host animals developed neutralizing antibodies against their homologous isolates.

Neutralization test surveys using two similar Changuinola-complex isolates from sloths showed high antibody prevalence in sloths and low prevalence in other forest vertebrates. Five of 96 Panamanian

Bradypus and 25 of 59 *Choloepus* had neutralizing antibody, but only one howler monkey and one night monkey were positive out of 61 birds and 226 mammals of 26 species other than sloths. Virus isolations and seroconversions from radio-marked *Bradypus* in Cerro Azul showed that they became infected during the December-to-May dry season (two animals), and during the May-to-July early rainy season (three animals).

The Changuinola virus isolates from sloths are remarkable for several reasons. First, they are convincing evidence that this complex of viruses is sloth-specific. Although commonly found in *Lutzomyia* sandflies in both Brazil and Panama, a Changuinola group virus has previously been isolated only once from a wild vertebrate, in spite of thousands of isolation attempts from other neotropical vertebrates. The antibody prevalence pattern supports the virus isolation evidence of sloth specificity.

Second, one Changuinola virus strain (Pan An 307566) was isolated from a *Bradypus* by P. A. Webb six weeks after the sloth was put in insect-proof animal holding quarters at Gorgas Memorial Laboratory. The animal had been inoculated with yellow fever virus and tested for viremia several times with negative results prior to the isolation of strain Pan An 307566. These observations suggest a latent Changuinola virus infection, possibly activated by the stress of captivity, handling, and yellow fever virus inoculation. The experimental yellow fever virus infection itself may have been latent and later activated, because no yellow fever viremia was detected and the sloth developed yellow fever antibody only at four months after inoculation.

Third, a *Choloepus* yielded Changuinola strain (Pan An 341275) three days after arrival in Gorgas Memorial Laboratory, immediately prior to inoculation with Saint Louis encephalitis virus. It was simultaneously viremic for 15 days with naturally acquired Changuinola virus and experimental Saint Louis encephalitis virus. It circulated the Changuinola virus for at least 28 days. High levels of preexisting neutralizing antibody against two other sloth Changuinola virus strains preceeded the appearance of strain Pan An 341275 viremia and strain-specific antibody. This observation suggests that antibody from a previous Changuinola-complex virus infection was not protective against challenge with a new serotype. It is unclear whether the second infection was independently acquired from nature or whether it represents a recrudescence of an earlier latent virus in altered form.

Unassigned Viruses: No Known Antigenic Group

Bradypus-4 virus.-- This apparently unique agent was isolated in February, 1976 from a male *Bradypus variegatus* captured in Chilibre, Panama (Seymour et al., 1983c). The host was experimentally inoculated with Saint Louis encephalitis virus. Bradypus-4 virus, was isolated and reisolated in Vero cells from blood samples taken four days before and three days after Saint Louis encephalitis virus inoculation. The animal developed both CF and neutralizing antibody against Bradypus-4 virus, which is probably an RNA virus with a lipid-containing envelope. The onset of experimental SLE viremia was delayed for two weeks, until after Bradypus-4 virus had been cleared.

Bradypus-4 virus did not react in Gorgas Memorial Laboratory CF tests with National Institutes of Health grouping immune reagents against arbovirus groups A, B, C, California, Simbu, Bunyamwera, Capim, Turlock, Anopheles A, and Anopheles B, nor with specific immune sera against 21 viruses of the phlebotomus fever group. In CF tests at the Yale Arbovirus Research Unit, Bradypus-4 virus immune reagent was negative against 216 worldwide arboviruses and other selected viruses (R. Shope, pers. comm.). These results are strong evidence that Bradypus-4 virus is a new, previously undescribed agent.

DISCUSSION

The wide variety of arboviruses isolated from sloths (Table 5) may be characterized according to antibody

Table 5. Viruses isolated from sloths. Oropouche virus represents 4 isolates, while the Changuinola complex represents 3 isolates.

Virus	*Antigenic Group*	*Sloth Species*	*Locality*	*Reference*
Saint Louis encephalitis	B	*Bradypus tridactylus*	Brazil	F. Pinheiro, cited in McLean and Bowen, 1980
Murutucu	C	*Bradypus tridactylus*	Brazil	Woodall, 1967
Oropouche	Simbu	*Bradypus tridactylus*	Brazil	Woodall, 1967; Pinheiro et al., 1981
Utinga	Simbu	*Bradypus tridactylus*	Brazil	Woodall, 1967
Utive	Simbu	*Bradypus variegatus*	Panama	Seymour et al., 1983c
Punta Toro complex	Phlebotomus fever	*Bradypus variegatus*	Panama	Seymour et al., 1983c
Anhanga	Phlebotomus fever	*Choloepus didactylus*	Brazil	Woodall, 1967
Changuinola complex	Changuinola	*Bradypus variegatus* *Choloepus hoffmanni*	Panama	Seymour et al., 1983c
Bradypus-4	Ungrouped	*Bradypus variegatus*	Panama	Seymour et al., 1983c

and virus isolation data as: sloth-specific (Utinga, Utive, Changuinola), incidental in sloths (Murutucu), or viruses for which the role of sloths in natural cycles is uncertain (all other tabulated viruses).

Viruses specific for sloths.-- These viruses are remarkable because they belong to antigenic complexes, within which viruses are identical by CF test but different by neutralization tests. Both complexes belong to virus taxa distinguished by segmented genomes; both bunyaviruses and orbiviruses are capable of genetic reassortment, in which parent genome segments are repackaged in different combinations during mixed infections (Gentsch et al., 1977; Gorman et al., 1978). Reassortment is one possible explanation for the variety of surface, neutralization-test antigens found in these otherwise identical viruses.

The sloth-specificity and the reassortment potential of these viruses are possibly related to their occurrence as antigenic complexes. Sloths are long-lived sedentary animals with a slow population turnover rate. A monotypic sloth-specific virus should quickly immunize a population, and, in the absence of alternate host species, should die out. Genetic reassortment is a simple method of introducing variation into genotypes and resulting antigenic phenotypes. Since functional immunity relates to surface antigens detected in the neutralization test, immunity would select for changes in these antigens, but internal CF antigens would be conserved unchanged.

Reassortment depends on simultaneous mixed infections, and the chances of such mixed infections are optimal in sloths. They have remarkably extended productive infections, as shown by the long viremias in sloths inoculated with Saint Louis encephalitis, Venezuelan encephalitis, and yellow fever viruses. The simultaneous Saint Louis encephalitis and Changuinola viremias in an experimental *Choloepus* show that in sloths simultaneous productive infections are possible. Sloths may also have long-term latent or recrudescent infections, possibly including that of Changuinola strain YF-149. A latent infection would provide a permanent reservoir of a reassorting virus.

Latency and recrudescence alone may account for the continuation of sloth-specific virus complexes in nature. A young animal infected with a temporarily prevalent serotype may produce recrudescent virus years later, when even a slow population turnover has created enough susceptible young for a renewed infection cycle. Latency and reassortment are not mutually exclusive mechanisms for the continuation of species specific virus complexes. However, latency alone cannot account for variation of antigens crucial to immunity in otherwise identical viruses.

Viruses incidental in sloths.-- Agents such as Murutucu and Venezuelan encephalitis viruses are usually found in vertebrate hosts other than sloths. They clearly do not depend on sloths for continued existence in nature. However, virus isolation and antibody detection prove that sloths are tangentially involved in these virus cycles; when infected, sloths may occasionally be amplifying hosts because of long, intense viremias.

Viruses for which sloths are of uncertain importance.-- Of these agents, Saint Louis encephalitis virus is noteworthy because it can cause epidemic fatal illness in North America, although tropical disease potential is still unclear. High prevalence of natural antibody indicates that many sloths are infected by Saint Louis encephalitis virus at some time in their lives. Experimental infection produces high long viremias infectious for mosquitoes. The importance of sloths to Saint Louis encephalitis virus probably depends on the nature of the cycle. "Jungle" Saint Louis encephalitis virus may naturally occur in "wandering foci," or epidemic waves of transmission, like its cousin "jungle" yellow fever virus. If so, then dense populations of sloths would be almost ubiquitously available for virus amplification; the observed high antibody prevalence would reflect frequent sloth infection during jungle epidemics. On the other hand, jungle Saint Louis encephalitis virus may cycle constantly in stationary, permanent foci of endemic transmission. Then the observed antibody prevalence in sloths would probably reflect only occasional, tangential infections in a population of long-lived animals.

This group of agents also includes Oropouche virus, which is notable as the agent of epidemic, nonfatal, but prostrating, human febrile illness in the tropics. Oropouche virus is similar to the Utinga and Utive sloth viruses in three respects. First, they are all probably transmitted mainly by *Culicoides* midges. Second, sloths are the only wild vertebrates which have yielded virus isolates. Third, the viruses belong to the same antigenic subgroup within the large group Simbu, although Oropouche is not close enough to the sloth viruses to be a member of the Utinga antigenic complex. A corresponding Oropouche antigenic complex would be a logical extension of the parallel between the viruses. However, no variant Oropouche strains have been reported. Tropical medical workers might well be alert for new variants, capable of causing disease even in humans already immune to one antigenic type.

Whether or not sloths are essential or incidental to a natural cycle of an arbovirus, their long experimental viremias are remarkable. Probably the most logical working hypothesis to explain these long viremias is the low metabolic rate of the host (Montgomery and Sunquist, 1978; McNab, 1978; 1985). This hypothesis could be tested by maintaining sloths under conditions which would promote constant body temperatures near 37 °C and measuring viremia peaks and durations. The proximal cause of long-lasting viremia may be a slow or defective immune response; if so, then sloths may prove to be useful systems for immunological studies. Whatever the mechanism and use of the phenomenon, the long and intense viremias of sloths, particularly *Bradypus*, qualify them as potentially important in many poorly understood forest arbovirus cycles.

ACKNOWLEDGMENTS

The following individuals collaborated on field research on Panamanian sloths: C. Seymour, P. H. Peralta, P. A. Webb, K. M. Johnson, L. D. Kramer, and P. Galindo of the Gorgas Memorial Laboratory; G. Gene Montgomery and M. Ojeda of the Smithsonian Tropical Research Institute; and J. Walsh of the International Society for the Protection of Animals. The above-mentioned Gorgas Memorial Laboratory staff studied experimental flavivirus infections in sloths. L. D. Kramer investigated arbovirus transmission to and from sloths by mosquitoes. Smithsonian Institution involvement in the research was supported in part by the Martin Hansen fund and by the Exxon foundation.

Literature Cited

Berge, T. O. ed.
1975. International catalogue of arboviruses including certain other viruses of vertebrates, 2nd ed. USDHEW Publication No. (CDC) 75-8301. 789 pp.

Bugher, J. C.
1951. The mammalian host. Pp 299 to 384, *in* Yellow fever (G. K. Strode, ed.). McGraw-Hill, New York, 710 pp.

Clarke, D. H., and J. Casals
1958. Techniques for hemagglutination and hemagglutination-inhibition with arthropod-borne viruses. Am. J. Trop. Med. Hyg.,7: 756-737.

Eisenberg, J.F. and E. Maliniak.
1985. Maintenance and reproduction of the two-toed sloth *Choloepus didactylus* in captivity. (this volume).

Gentsch, L. Wynne, J. Clewley, R. Shope, and D. H. L. Bishop
1977. The formation of recombinants between snowshoe hare and LaCrosse bunyaviruses. J. Virol., 24:893-902.

Gorgas Memorial Laboratory
1975. 46 Annual Report, Fiscal Year 1974. 94th Congress, 1st Session, House Document No. 94-36. U.S. Government Printing Office.

Gorman, B. M., J. Taylor, P. J. Walker, and P. R. Young
1978. The isolation of recombinants between related orbiviruses. J. Gen. Virol., 41:333-342.

Grayson, M. A.
1972. Discussion on mammalian hosts of VEE. Pp 276 to 277 *in* Venezuelan encephalitis. Pan Am. Health Org. Scientific Publication No. 243., 416 pp.

Grayson, M. A., and P. Galindo
1969. Ecology of Venezuelan equine encephalitis virus in Panama. Am. J. Epidemiol., 88:80-96.

Kinney, R.M. and C.H. Calisher.
1981. Antigenic relationships among Simbu serogroup (Bunyaviridae) viruses. Am. J. Trop. Med. Hyg., 30: 1307-1318.

McLean, R. G., and G. S. Bowen
1980. Vertebrate hosts. Pp. 381 to 450, *in* Saint Louis encephalitis. (T. P. Monath, ed.) Am. Pub. Health Assn., Washington D.C. 680 pp.

McNab, B. K.
1978. Energetics of arboreal folivores: physiological problems and ecological consequences of feeding on an ubiquitous food supply. Pp. 153 to 162 *in* The ecology of arboreal folivores (G. G. Montgomery, ed.). Smithsonian Institution Press, Washington D.C. 574 pp.

McNab, B.K.
1985. Energetics, population biology, and distribution of xenarthrans, living and extinct. (this volume).

Meritt, D.A., Jr.
1985. The two-toed Hoffman's sloth, *Choloepus hoffmanni* Peters. (this volume).

Montgomery, G. G. and M. E. Sunquist
1975. Impact of sloths on neotropical forest energy flow and nutrient cycleing. Pp 69 to 98, in

Trends in tropical ecology (E. Medina and F. Golley, eds.). Ecological Studies 4. Springer-Verlag, New York 398 pp.

1978. Habitat selection by two-toed and three-toed sloths. Pp. 329-359, *in* The ecology of arboreal folivores (G. G. Montgomery, ed.). Smithsonian Institution Press, Washington, D.C. 574 pp.

Pinheiro, F. P., A. P. A. Travassos da Rosa, J. F. S. Travassos da Rosa, R. Ishak, R. B. Freitas, M. L. Gomes, J. W. LeDuc, and O. F. P. Oliva
1981. Oropouche virus. I. A review of clinical, epidemiological, and ecological findings. Am. J. Trop. Med. Hyg., 30: 149-160.

Seymour, C., L. D. Kramer and P. H. Peralta.
1983a Experimental St. Louis encephalitis virus infection of sloths and cormorants. Am. J. Trop. Med. Hyg., 32: 854-861.

Seymour, C., P. H. Paralta and G. G. Montgomery.
1983b. Serological evidence of natural togavirus infections in Panamanian sloths and other vertebrates. Am. J. Trop. Med. Hyg., 32: 844-853.

fections in Panamanian sloths and other vertebrates. Am. J. Trop. Med. Hyg., 32: 844-853.

1983c. Viruses isolated from Panamanian sloths. Am. J. Trop. Med. Hyg., 32 1435-1444.

Tesh, R. B., B. N. Chaniotis, P. H. Peralta and K. M. Johnson
1974. Ecology of viruses isolated from Panamanian sandflies. Am. J. Trop. Med. Hyg., 23:258-259.

Theiler, M., and W. G. Downs
1973. The arthropod-borne viruses of vertebrates. Yale University Press, London. 578 pp.

Trinidad Regional Virus Laboratory
1958. Annual Report, Port of Spain, Trinidad, 147 pp.

Woodall, J. P.
1967. Virus research in Amazonia. Atas do Simposia sobre a biota Amazonica, 6 (Patalogia): 31-63.

THE HEMOFLAGELLATES OF SLOTHS, VERMILINGUAS (ANTEATERS), AND ARMADILLOS

Jeffrey J. Shaw

The Wellcome Parasitology Unit, Instituto Evandro Chagas, Fundação Serviços de Sáude Pública, 66.00 Belem, Para, Brazil

Abstract

Hemoflagellates found in sloths, vermilinguas (neotropical anteaters), and armadillos are listed and their taxonomic status, vectors, distribution and importance to man are discussed. Sloths, particularly two-toed sloths *Choloepus*, harbor the greatest variety of hemoflagellates and in Panama the two-toed sloth is the major vertebrate host of *Leishmania braziliensis panamensis* which causes cutaneous leishmaniasis in man. In the Amazon region the common long-nosed armadillo *Dasypus novemcinctus* is infected with a *Leishmania* not yet isolated from man. Two-toed sloths appear to be the major reservoir of *L. b. guyanensis* in northern Pará, Brazil, and in French Guiana. In an endemic "pian-bois" region of Pará, *Tamandua* anteaters are important secondary reservoirs. All the *Leishmania* of sloths, anteaters and armadillos that have been studied in sandflies are peripylarians, a type of *Leishmania* found in New World mammals, and it is suggested that this group may have evolved in xenarthrans. The genus *Endotrypanum* is intermediate between *Leishmania* and *Trypanosoma* and has retained certain features of its leishmanial ancestors in the vector, but in its vertebrate host it develops in the epimastigotic form and in rare situations the trypanomorphic form. Phlebotomine sandflies are the most probable vectors of xenarthran *Leishmania* and *Endotrypanum*. In drier, more open, areas of Panama, sloths are commonly infected with *Trypanosoma leuwenhoeki*, which may be responsible for some of the *rangeli*-like infections in man. Common long-nosed armadillos are commonly infected with *T. cruzi*, but Brazilian stocks of this parasite belong to a zymodeme that is only rarely found in man. Separate lines of *T. cruzi* may have evolved in xenarthrans and marsupials from a New World insect leptomonad stock. There are enormous gaps in our knowledge of xenarthran hemoflagellates but it is recommended that particular priority be given to the isolation and characterization of stocks from the numerous hosts found in different geographical regions.

Abstracto

Se han enumerado los hemoflagelados que se encuentran en monos perezosos, osos hormigueros neotropicales, y armadillos y se explica su status taxonómico, presencia de vectores, distribución y importancia para el hombre. Monos perezosos, particularmente el mono perezoso de dos dedos *Choloepus*, albergan la mayor variedad de hemoflagelados y en Panamá *Choloepus* es el mayor hospedero entre los vertebrados de *Leishmania brasiliensis panamensis* el cual es la cuasa de leishmaniasis cutánea en el hombre. *Choloepus* es el que lleva la mayor cantidad de *L. b. guyanensis* en el norte de Pará, Brasil y en la Guyana Francesa. En una región endémica "pian bois" de Pará, los osos hormigueros *Tamandua* son un importante hospedero secundario. Todas las *Leishmania* de monos perezosos, osos hormigueros, y armadillos que han sido estudiadas en flebótomus son peripilarios, un tipo de *Leishmania* que se encuentra en mamíferos del Nuevo Mundo, y se sugiere que este grupo ha podido evolutionar en xenartranos. El género *Endotripanum* es intermedio entre *Leishmania* y *Tripanosoma* y ha conservado ciertas características de sus antecesores leishmanianos en el vector, sin embargo en su hospedero vertebrado se desarrolla en la forma epimastigótica y raras veces en la forma tripanomórfica. Flebótomus son los vectores mas probables de *Leishmania* y *Endotripanum* en los xenartranos. En las áreas más secas y abiertas de Panamá, los monos perezosos están infectados por *Tripanosoma leuwenhoeki*, el cual puede ser la causa algunas de las infecciones tipo *rangeli* en el hombre. Los armadillos comunes de trompa larga son usualmente infectados por *T. cruzi*; sin embargo las cepas brasileñas de este parásito pertenecen a un cimodema que solo en raras ocasiones se encuentra en el hombre. Líneas separadas de *T. cruzi* han podido evolucionar en los xenartranos y en marsupiales a partir de un cepa leptomonade de un insecto del Nuevo Mundo. Hay enormes brechas en nuestro conicimiento de los hemoflagelados de xenartranos pero se recomienda que se le dé prioridad particular al aislamiento y caracterización de cepas de los abundantes huéspeded que se encuentran en las diferentes regiones geográficas.

Resumo

No presente trabalho todos os hemoflagelados de preguiças, tamanduás e tatus, são citados com suas posições taxonômicas, vetores, distribuições geográficas e suas importâncias como fontes de infecção ao homem. Preguiças, especialmente as do gênero *Choloepus*, têm o maior número de hemoflagelados e, no Panamá, a preguiça real é um dos principais hospedeiros de *Leishmania braziliensis panamensis*, que é responsável pela leishmaniose cutânea no homem, neste país. Na região amazônica, o tatu galinha *Dasypus novemcinctus*, é comumente infectado com uma espécie de *Leishmania* que, por enquanto, não tem sido encontrada no homem. A preguiça real é o mais importante reservatório silvestre de *L. b. guyanensis* no norte do estado do Pará, Brasil e na vizinha Guiana Francesa. Nas regiões endêmicas de "pian bois" do Pará, tamanduás do gênero *Tamandua* são importantes reservatórios secundários de *L. b. guyanensis*. Todas as leishmanias de preguiças, tamanduás e tatus, cujo desenvolvimento em flebótomos tem sido estudado, pertencem à peripilaria, que até agora só foi encontrado em mamíferos do novo mundo. O autor sugere que essas leishmanias provavelmente evolem em xenarthros. O gênero *Endotrypanum* é intermediáro entre a *Leishmania* e *Trypanosoma* e, no vetor, seu desenvolvimento é mais parecido com *Leishmania* mas, na preguiça, se aproxima aos membros do gênero *Trypanosoma* quanto à sua morfologia. É bem provável que os vetores de *Leishmania* e *Endotrypanum* de xenarthros sejam flebotomíneos. Nas regiões mais secas com menos árvores do Panamá, preguiças de vários gêneros são infectadas com *Trypanosoma leuwenhoeki*, que pode ser responsável por algumas das infecções do tipo *rangeli*, encontrados no homem, naquele país. Infecções por *T. cruzi* são frequentement encontradas no tatu galinha nas Américas, mas as cepas desse parasito no Brasil, pertencem a um zimodema que raramente infecta o homem. É possível que linhagens diferentes de *T. cruzi* evoluam em xenarthros e marsupiais, dos flagelados ancestrais. Nosso conhecimento sobre hemoflagelados dos xenarthros é extremamente escasso e o autor sugere que o isolámento e caracterização de amostras de vários hospedeiros de áreas geográficas diferentes seriam uma das linhas prioritárias de pesquisa.

The modern xenarthra are all that remains of a group of mammals that appeared in the fossil record some 60 million years ago during the Upper Paleocene period in North America (Herschkovitz, 1971), although the xenarthrans may have become genetically distinct as a lineage more than 80 million years ago (Sarich, 1985). All living Xenarthra are, therefore, survivors of a mammalian stock that is older than such groups as the rodents, that appeared in the fossil record some 22 million years later in the early Oligocene period. Their only rivals for being the oldest mammalian group in the New World are the marsupials that go back as far as the Cretaceous period. Unlike the marsupials, who have living members in both Australia and the Americas, the xenarthrans are, and apparently have always been, limited to the New World. To parasitologists such mammals represent a fascinating group, for they are hosts in which typically New World parasites may have evolved and survived.

Protozoologists are presently studying interspecific and intraspecific variations of many groups in great depth and in particular those economically important protozoa such as the leishmanias and trypansomes. New biochemical methods are becoming available (see reviews by Newton and Burnett, 1972; Chance, 1979; Godfrey, 1979) and such techniques in conjunction with studies on morphology, physiology and immunology and *in vivo* and *in vitro* behavior show a considerable degree of diversity, even among organisms that are morphologically very similar. Godfrey (1979) drew attention to the stability that certain isoenzymes showed in different trypanosomatid stocks and felt that it is the insect vectors that are primarily responsible for the maintenance of the stable groups.

In this article I discuss recent advances in the study of hemoflagellates found in xenarthra and relate those studies to possible future lines of research. In the space available it is impossible to describe each parasite in detail and to cite the authors responsible for the various records in the different countries, as given in reviews by Shaw (1969), Hoare (1972), and Lainson and Shaw (1979). For convenience, each parasite is listed separately with its known hosts and geographical distribution. If the record was published after the above-mentioned reviews or illustrates a particularly important point, the source is cited.

PARASITES AND THEIR HOSTS

The hemoflagellates are protozoa of the subphylum Sarcomastigophora (flagellar and/or pseudopodial locomotion): class Zoomastigophorea (no chlorophyll): order Kinetoplastida (possessing a kinetoplast): family Trypanosomatidae (parasites with one flagellum). Members of the genera *Leishmania* and *Trypanosoma* are found in sloths, anteaters, and armadillos, while flagellates of the genus *Endotrypanum* occur in sloths. These three genera are all digenetic in that their life cycle involves both an invertebrate and a vertebrate host.

Genus *Leishmania* Ross 1903

Members of this genus develop as amastigotes in cells of the reticuloendothelial system of their vertebrate host and as promastigotes and paramastigotes in the intestine of their invertebrate host. The only known vectors are phlebotomine sandflies (Diptera: Psychodidae). Based on their development in sandflies, Lainson and Shaw (1979) divided the genus into the sections Hypopylaria, Peripylaria and Suprapylaria. Until now the *Leishmania* found in xenarthra all belong to the section Peripylaria and no members of this section have so far been found in mammals outside the neotropics.

Leishmania braziliensis guyanensis Floch 1954

Choloepus didactylus French Guiana, Brazil (Gentile et al., 1981; Lainson et al., 1981b); ***Tamandua tetradactyla*** Brazil (Lainson et al., 1981a, 1981b).

In the Monte Dourado region of Pará, Brazil infections of *L. b. guyanensis* were found in 27 of 59 two-toed sloths. Of these, 25 isolates were from viscera (liver and spleen), 1 was from viscera and skin and the other was from skin only (Lainson et al., 1981b). These sloths are important reservoirs of *L. b. guyanensis* in French Guiana, where 7 of 15 have been found infected (Gentile et al., 1981).

Infections of *L. b. guyanensis* have also been found in 6

of 27 *Tamandua tetradactyla* anteaters captured in the same region as the infected sloths (Lainson et al., 1981a, 1981b). The stocks from *T. tetradactyla* were biologically and biochemically the same as those isolated from man, sandflies (*Lutzomyia (Nyssomyia) umbratalis* and *L. (N.) whitmani)* and two-toed sloths from the same area. No leishmanial infections were observed in 21 *T. tetradactyla* captured in nonendemic "pian-bois" areas of Pará, south of the Amazon River (Lainson et al., 1979a).

***Leishmania braziliensis panamensis* Lainson and Shaw 1972**

Bradypus veriegatus Costa Rica (Zeledon et al., 1975a), Panama; ***Choloepus hoffmanni*** Costa Rica (Zeledon et al., 1975a), Panama.

This subspecies of *L. braziliensis* has been extensively studied at the Gorgas Memorial Laboratory in Panama for some 25 years. In man it produces cutaneous lesions and is contracted by people who live or work in areas where there is mature forest. Many infected species of mammals and sandflies have been found but the major reservoir/vector cycle seems to involve two-toed sloths and the sandfly *L. (N.) trapidoi*, although four other species have been indicated as possible vectors (Christensen and Herrer, 1973). The extensive survey of Panamanian mammals by Drs. Aristides Herrer and Howard Christensen from 1965 to 1972 revealed infections of *L. b. panamensis* in 67 of 396 two-toed sloths, but in only 2 of 163 three-toed sloths. The incidence of infection in two-toed sloths from Achiote was 37.1 percent and in those from Aguacate 26.1 percent. In both areas mature forest was the dominant vegetation. However, in two-toed sloths captured in La Chorrera, where there was mostly secondary forest, the infection rate was only 4.7 percent

No mention is made of amastigotes in smears from sloths that were infected with *L. b. panamensis* and the sloths were all in good health. Herrer and Telford (1969) noted that in 8 of 13 infections, parasites appeared to be limited to the skin while in other animals they occurred only in the viscera or in both the viscera and skin. In three-toed sloths from Costa Rica, one isolate was from liver and another was from skin. In two-toed sloths one isolate was from blood and the other from liver (Zeledon et al., 1979). In the absence of information on the course of natural or experimental infections it is very difficult to interpret the presence of parasites in different tissues. Possibly animals in which parasites were found in the viscera were in the acute phase while those in which they were limited to the skin were in the chronic phase of the infection.

***Leishmania braziliensis* subspecies**

Choloepus didactylus Brazil.

A leishmania was first seen in a two-toed sloth from Amazonia in 1948 by Professor Leonidas Deane. The parasites were in stained impression smears of liver of sloths, captured in the Abaetetuba region of Pará, Brazil. These sloths were also infected with *Endotrypanum*. At the time he was not sure whether the amastigotes were a leishmanial parasite or if they were a developmental phase of *Endotrypanum*. More recently we isolated *Leishmania* from the skin and viscera of two-toed sloths and there were abundant amastigotes in smears from the viscera of some of the positive animals. This is apparently very different from infections of *L. b. panamensis* in *C. hoffmanni*. Stocks of *Leishmania* from *C. didactylus* are presently being studied and from its behavior in experimentally infected sandflies *L. longipalpis* and hamsters and from its morphology it is clearly closely related to *L. b. guyanensis*.

***Leishmania herreri* Zeledon, Ponce and Murillo 1979**

Choloepus hoffmanni Costa Rica; ***Bradypus veriegatus*** Costa Rica.

This species was described from infections found in 3 of 86 three-toed sloths and 7 of 63 two-toed sloths. The positive three-toed sloths were all infected with *Endotrypanum* and two were also infected with *T. leuwenhoeki*. One infection in a two-toed sloth was a pure one of *L. herreri* while the others were mixed with *Endotrypanum*, *T. leuwenkoeki* and *L. braziliensis*. *L. herreri* was isolated from the spleen and blood of three-toed sloths and from the skin, blood, liver and spleen of two-toed sloths.

Unlike *L. b. panamensis* this species does not produce a visible lesion in hamsters and is differentiated from *Endotrypanum* by its growth as amastigotes in hamster cells grown in tissue culture. It has been isolated from wild-caught sandflies *L. (N.) trapidoi*, *L. (N.) ylephiletor* and *L. shannoni*, which must all be considered as potential vectors. In the original description of *L. herreri* it was stated that it had distinct nuclear and kinetoplast DNA buoyant densities and that the electrophoretic mobilities of its glucosphosphate isomerase (GPI) and 6-phosphogluconate dehydrogenase (6-PGDH) were different from other species of *Leishmania*. No details of these differences were given in the original paper and it is hoped that this most important information will soon become available. The presence of sphaeromastigotes in smears from the skin of inoculated hamsters is more typical of the genus *Trypanosoma* than *Leishmania*.

Some of the promastigote stocks isolated from Panamanian sloths and sandflies by other workers which failed to infect hamsters might have been *L. herreri*.

However, care must be taken in the use of infectivity to hamsters as a taxonomic criteria, especially if the stocks in question have been passaged many times. Both *L. mexicana* and *L. braziliensis* lose their infectivity to hamsters if kept in culture for long periods. Some subspecies of *L. braziliensis* may lose their infectivity after only three or four subpassages.

Leishmania species (not yet described)

Dasypus novemcinctus Brazil (Lainson et al., 1979b)

Stocks isolated from common long-nosed armadillos captured on both the northern and southern sides of the Amazon River are being studied at the Evandro Chagas Institute. During 1978, cultures were made from 14 common long-nosed armadillos and promastigotes were isolated from the blood, liver, and spleen of one, from the blood and liver of another, and from the spleen of a third. All three isolates grew luxuriantly in culture but when inoculated intradermally into hamsters none produced visible lesions. In smears made from the inoculation sites a few days later we found a few intracellular amastigotes. Laboratory-reared sandflies, *L. longipalpis*, were fed promastigotes of the various stocks and they developed as typical peripylarians. Since 1979 (Lainson et al., 1979b) we have isolated more stocks from armadillos and preliminary observations suggest some degree of heterogeneity among the leishmanial stocks from *D. novemcinctus*. None of the *Leishmania* so far found in man appear to be similar to that from armadillos.

The vector of the leishmanial parasite found in common long-nosed armadillos in Brazil is unknown. The apparent absence of infections of this parasite in man could be either because it is not infective to man or because the vector rarely or never bites him.

Genus *Endotrypanum* Mesnil and Brimont 1908.

Parasites of this genus develop as intraerythrocytic epimastigotes in sloths and as promastigotes in cultures and sandflies. The uniqueness of members of this genus is that they are the only trypanosomatids that parasitize red blood cells, a habitat that has been so successfully exploited by the hemosporidia.

Endotrypanum schaudinni Mesnil and Brimont 1908

Choloepus didactylus French Guiana, Brazil; ***C. hoffmanni*** Panama; ***Bradypus veriagatus*** Panama (Herrer and Telford, 1969) Costa Rica; ***B. tridactylus*** Brazil (Shaw, Lainson and Miles, unpublished).

The following infection rates of *E. schaudinni* have been recorded in sloths from Panama: 37 of 110 two-toed sloths and 0 of 39 three-toed sloths (Shaw, 1969), 10 of 85 two-toed sloths and 1 of 77 three-toed sloths (Herrer and Telford, 1969), and 39 of 72 two-toed sloths (Christensen and Herrer, 1979). Fifty percent of the two-toed sloths captured in forested areas were positive for *E. schaudinni* while the infection rate was only 3 percent in those captured in more open areas.

In sloths from Brazil, the following infection rates have been found: 23 of 45 two-toed sloths and 0 of 51 three-toed sloths (Deane, 1961; Deane and Damasceno, 1961) and 27 of 33 two-toed sloths and 1 of 21 three-toed sloths (Shaw and Lainson, unpublished). The above records refer to animals that were caught in the municipalities of Abaetetuba, Ananindeua, Belém, Breves, Maracanã , Moju and Ponte de Pedras of Pará state. Infections of *E. schaudinni* have also been found in 33 of 59 two-toed sloths from the Monte Dourado region of Pará. Of these 16 were mixed infections of *L. b. guyanensis* and *Endotrypanum* (Shaw and Lainson, unpublished). As in Panama the positive animals were associated with undisturbed "terra firme" forest.

The first recorded infection of *E. schaudinni* in Costa Rica was in a three-toed sloth captured in Cartago Province. Subsequent records (Zeledon et al., 1979) do not mention the morphology of the bloodstream forms. It is possible that the two-toed and three-toed sloths from the provinces of Cartago and Limón were infected with *E. schaudinni* while the infections in those from San José Province could have been either *E. schaudinni* or *E. monterogeii*. In all, *Endotrypanum* was found in 7 of 86 three-toed sloths and 23 of 63 two-toed sloths as well as in the wild-caught sandflies *L. (N.) trapidoi, L. (N.) ylephilator* and *L. shannoni*. In some of these sandflies the *Endotrypanum* infections were mixed with *L. braziliensis* and *L. herreri*.

In forested areas of Panama, where *E. schaudinni* is common, 96 percent of the hematophagous insects feeding on two-toed sloths were phlebotomine sandflies. Dissections of wild-caught and laboratory-reared sandflies that had fed on two-toed sloths infected with *E. schaudinni* showed that this parasite could develop in these insects. My efforts and those of other workers to infect a range of haemotophagous insects including mosquitoes and reduviids were completely unsuccesful. These negative results in conjunction with the epizootiological data and the fact that sandflies supported the development of *E. schaudinni* led me to suggest that sandflies could be the natural vector of *Endotrypanum* (Shaw, 1964). Christensen and Herrer (1976) fed 1665 laboratory-reared sandflies on wild-caught sloths and they considered that 136 infections had been derived from sloths with pure *E. schaudinni* infections. In most flies the parasite developed as rounded or oval forms that were attached to various regions of the hind gut, distributed in the gut as follows: 14 percent in the cardia, 52 percent in the midgut, 63 percent in the malpighian tubules, 60 percent in the ileum and 29 percent in the rectal ampulla. In paired feedings they noted that significantly more *L. gomezi* sandflies (70 of 97) were infected than *L. sanguinarius* sandflies (29 of 168). Christensen and Herrer (1979) again noted that *L. sanguinarius* sandflies were less susceptable to infection with *E. schaudinni* than *L. gomezi.* Christensen and Herrer (1976) supported my hypothesis of sandflies being the vector of *Endotrypanum* and their observations of infections in the foregut suggests that transmission is by bite and not, as I had suggested, by fecal contamination.

Electronmicrographs of infected sloth red cells showed that *E. schaudinni* is actually inside the erythrocyte and not merely adhered to its surface. Its basic cellular structure was similar to that of other trypanosomatids but in its posterior end there were numerous round to ovoid electron translucent granules and a few irregularly shaped electron dense granules (Shaw and Bird, 1969). The former appear to coincide with basophilic inclusions seen in Giemsa stained preparations. Croft et al., (1979, 1980) noted clusters of discrete particles in the cytoplasm of the culture forms of certain *Endotrypanum* stocks. These granules were smaller than those seen by Shaw and Bird (1969) in the bloodstream forms and had translucent centers and electron dense walls that were 14 to 18 nm thick. Croft and his colleagues noted that the stocks that had granules had different kinetoplast DNA buoyant densities and different malate dehydrogenase (MDH) and isocitrate dehydrogenase (ICDH) electrophoretic mobilities from the stocks without granules. A possible interpretation of these observations could be a basic difference between Central and South American stocks. This would not seem to be the case, however, since a stock from a two-toed sloth from Panama fell into the same group as two Brazilian stocks (Croft et al., 1980). Preliminary studies in Brazil, using 14 different enzymes, have, however, revealed clear differences between Central and South American stocks of *Endotrypanum* (Shaw, Lainson Povoa and Miles , unpublished). There also appears to be a lesser degree of regional heterogeneity for certain enzymes. For instance, among 16 stocks isolated from sloths captured in the lower Amazon region of Brazil, we identified 8 zymodemes that were only different from each other in a few enzymes. Mannosephosphate isomerase (MPI) was the most variable and the range of forms was reminiscent of those associated with a monomeric enzyme with 3 alleles.

The blood forms but not the culture forms of *E. schaudinni* were infective to laboratory-reared sandflies (Shaw, 1981). The flagellates that developed in the fly from the blood forms were morphologically similar to those seen in culture. It appears that the infectivity in culture is lost very rapidly because even first subpassages of an original isolate failed to infect sandflies. There were significantly more infections of the malphighian tubules and rectal tube and significantly fewer of the pyloric and ileal regions of the sandflies *L.furcata* and *L. flaviscutellata.* In the former species the tips of the malpighian tubules were sometimes so heavily parasitized that they became enlarged. The significance of infections in malpighian tubules is obscure, but it could merely be an aberrant development in an abnormal host. When fed on the same sloth significantly more *L. flaviscutellata* sandflies became infected than *L. longipalpis* sandflies.

***Endotrypanum monterogeii* Shaw 1969**

Choloepus hoffmanni Costa Rica.

This species of *Endotrypanum* is differentiated from *E. schaudinni* by the trypanomorphic morphology of the intraerythrocytic form and serologically by the Nogouchi-Adler test performed on the culture forms. This species has only been found in sloths from the temperate forests of the Tablazo Mountains of San José province. Nothing is known of the vectors.

Genus *Trypanosoma* Gruby 1843.

Flagellates of this genus occur as trypomastigotes in the blood of their vertebrate host and predominantly as epimastigotes in their invertebrate host. The infective stage in the vector is considered to be the small metatrypomastigotes. Trypanosomes of the subgenera *Herpetosoma* Dolfein, 1901, *Megatrypanum* Hoare, 1964 and *Schizotrypanum* Chagas, 1909 have been recorded in xenarthra. The relevant literature is extensive and the interested reader is referred to Hoare (1972) and reviews on the biology of *T. (H.) rangeli* (D'Alessandro, 1976) and heterogeneity within the subgenus *Schizotrypanum* (Miles, 1979).

***Trypanosoma (Herpetosoma) rangeli* Tejera 1920**

Tamandua mexicana Panama.

This trypanosome was described from the reduviid bug *Rhodnius prolixus* of Venezuela and was first found in a vertebrate host, man, some years later. This is the only member of the subgenus that invades the hemolymph and salivary glands. Transmission is by bite and not fecal contamination (Rey-Matiz, 1941). Since that date any *T. lewisi*-like trypanosome that invades the salivary glands of reduviids is generally considered to be *T. rangeli*.

The trypanosomes found in *Tamandua* from Panama (Walton and Sousa, 1967), identified as *T. rangeli*, grew well in blood-agar-base (Difco) media made up with 20 percent rabbit blood but poorly in the same base made up with sheep or human blood. The culture forms were infective to mice and gave rise to trypanosomes that were said to be the same as *T. rangeli*. They were reported to develop in bugs as *T. rangeli* but the species of bug is not given nor are any details about the development. I think that these infections should be referred to as *T. rangeli*-like and perhaps more correctly as an unidentified *Herpetosoma* species.

***Trypanosoma (Herpetosoma) myrmecophagae* Floch and Abonnenc 1948**

Myrmecophaga tridactyla French Guiana.

This trypanosome of the giant anteater develops in the reduviid bug *Rhodnius prolixus* and because of this some workers consider it to be a synonym of *T. rangeli*. The bloodstream trypomastigotes are larger than those of *T. rangeli* and until we know more about the biology and biochemistry of the trypanosomatids of giant anteaters, the name *T. myrmecophagae* must be retained.

***Trypanosoma (Herpetosoma) mesnilbrimonti* Deane 1961**

Choloepus didactylus French Guiana, Brazil.

I recently compared stained preparations of *T. mesnilbrimonti* and *T. preguici* which confirm that the two species are morphologically distinct. In Brazil, Deane found 2 of 25 two-toed sloths were infected, one from the Utinga forest near Belém and the other from Pirituba near Abaetetuba. Trypomastigotes of *T. mesnilbrimonti* are more numerous in the blood than those of *T. preguici*, are smaller than those of the latter species and the position of the nucleus is similar to that of *T. legeri. T. mesnilbrimonti* has only been found in sloths that are also infected with *Endotrypanum*. The vector is unknown and no development has been associated with this species in either insects or culture.

***Trypanosoma (Herpetosoma) leuwenhoeki* Shaw 1969**

Choloepus hoffmanni Panama, Costa Rica (Zeledon et al., 1975b); ***Bradypus variegatus*** Panama, Costa Rica (Zeledon et al, 1975b)

This trypanosome was isolated from the blood of 12 of 39 three-toed sloths and 17 of 110 two-toed sloths. Of these 29 infections, 25 were in animals that had come from areas of predominantly secondary forest and the other 4, represented by two of each species of sloth, were captured in an area where the predominant vegetation was mature forest with clearings of secondary growth. More recent records (Christensen and Herrer, 1979) refer to infections of *T. rangeli* in 15 of 72 two-toed sloths from Panama.

One of the major reasons for naming *T. leuwenhoeki* was that it would not develop in the reduviid bug *Rhodnius prolixus. T. rangeli* was originally described as a parasite of this reduviid, and a trypanosome that will not develop in the original host cannot be the same. Thus I cannot accept the use of the name *T. rangeli* for parasites that will not develop in the bug *Rhodnius prolixus*, be they isolated from sloths or any other mammal. Workers in the field of Chagas' disease have, in general, been a little more prudent and often refer to the presence of a *T. cruzi*-like trypanosome in a given wild mammal.

Christensen and Herrer (1979) successfully transmitted *T. leuwenhoeki* to guinea pigs by the bite of the reduviid bug *Rhodnius pallescens* that had been innoculated intracoelomically with culture forms, strongly suggesting that a reduviid could be the natural vector.

Trypanosoma rangeli was recorded from 3 of 63 two-toed sloths and 4 of 86 three-toed sloths by Zeledon et al. (1979), but I think that these infections in sloths from Costa Rica could also be considered as being *T. leuwenhoeki*.

***Trypanosoma (Herpetosoma) preguici* Shaw 1969**

Choloepus hoffmanni Costa Rica, Panama; ***Bradypus veriegatus*** Costa Rica

While Hoare (1972) put this species into the subgenus *Herpetosoma*, I have difficulties in assigning this species to a subgenus. Its large vacuole in the nuclear region and its general morphology are similar to some species of *Megatrypanum*, such as *T. megadermae*, and some of the more primitive trypanosomes found in birds and reptiles. *T. preguici* infections in both species of sloths have been associated with *E. schaudinni*.

***Trypanosoma (Megatrypanum) legeri* Mesnil and Brimont 1910**

Tamandua tetradactyla Venezuela, French Guiana ,Brazil; ***T. mexicana*** Belize, Panama.

From the recorded infection rates it would seem that *T. legeri* is quite common in *Tamandua* anteaters of both species: *T. tetradactyla*, Brazil 5 of 20 (Deane, 1961). *T. mexicana* Belize 5 of 12 (Lainson, 1965), Panama 5 of 17 (Walton and Sousa, 1967). In these reports, no mention is made of the sex or age of the positive animals nor are details given of the capture sites. Deane (1961) described both short and long forms, whose average lengths were 31.6 and 41.1 μm, which he considered as all being *T. legeri*. Only one basic morphological type was observed by Lainson (1965) in animals from Belize, it was similar to the smaller trypomastigotes described by Deane. Walton and Sousa (1967), however, noted that in the trypanosomal infections of *T. mexicana* from Panama there were smaller broad forms measuring 31.5 μm and longer thin forms measuring 41.8 μm. Because of the behavior of the culture forms isolated from these infections, they considered that the longer thinner trypamastigotes were *T. rangeli* and the smaller shorter ones were *T. legeri*. On morphological grounds this separation into two species is not fully justified as there was no clear separation between the two types.

Because trypomastigotes seen in blood films and cultures are positive does not necessarily mean that the culture forms belong to the trypomastigotes seen in the blood. In 29 infections in sloths from Panama, I only saw trypanosomes in the blood of two animals. Thus it seems possible that Walton and Sousa (1967) could have cultured an occult trypanosome that was mixed with typical *T. legeri*. The only published results on the cultivation of *T. legeri* are those of Walton and Sousa (1967), who noted that it grew poorly in blood-agar base (Difco) and that it could only be maintained in Senekjie's medium containing 30 percent rabbit blood; even then it grew poorly. We have isolated what we consider to be *T. legeri* in nutrient agar medium (Oxoid CM3) made up with 20 percent rabbit blood (Shaw and Lainson, unpublished). Initial isolates grew poorly, but stocks were successfully maintained in this medium although growth has never been luxuriant. The vector of this species is unknown.

Trypanosoma (Schizotrypanum) cruzi **Chagas 1909**

Choloepus hoffmanni Panama; ***Bradypus variegatus*** Panama.
Tamandua tetradactyla Brazil, Venezuela; ***T. mexicana*** Panama; ***Cyclopes didactylus*** Brazil (Miles et al., 1981). ***Cabassous centralis*** Panama; ***C. unicinctus*** Venezuela, French Guiana, Brazil, Argentina; ***Chaetophractus vellerosus*** Argentina; ***Dasypus novemcinctus*** United States, Mexico, Guatemala, Costa Rica, Colombia, Venezuela, French Guiana, Brazil, Argentina; ***D. kapleri*** Venezuela, Colombia; ***Euphractus sexcinctus*** Brazil, Venezuela; ***Tolypeutes matacus*** Argentina; ***Zaedyus pichyii*** Argentina.

Infections of *T. cruzi* have been recorded from many species of xenarthrans. For many years workers in the field of Chagas' disease have suspected heterogeneity among the morphologically similar organisms that are identified as *T. cruzi*. However, studies on isoenzymes show differences that strongly suggest a number of distinct groups (Miles, 1979). Three zymodemes (ZI, ZII and ZIII) have been associated with the acute disease in man (Miles et al., 1978, 1979) and more will undoubtedly be identified.

Infections of *T. cruzi* are uncommon in two-toed sloths and anteaters. In Panama, I found no infections in 149 sloths, Christensen and Herrer (1979) found only one infection in 72 two-toed sloths and Pipkin (1968) recorded one infection from 7 three-toed sloths. Zeledon et al. (1979) did not find *T. cruzi* in 86 three-toed and 63 two-toed sloths captured in Costa Rica. No infections of *T. cruzi* were found in 129 Brazilian sloths captured in the lower Amazon region (Deane, 1961; Lainson et al., 1979). *T. cruzi* has been seen in 3 of 31 *T. mexicana* anteaters from Panama (Pipkin, 1968; Walton and Sousa, 1967) but none were observed in 41 *T. tetradactyla* from the lower Amazon region (Deane, 1961; Lainson et al., 1979). Rodrigues and Mello (1942) recorded infections of *T. cruzi* in 3 of 12 *T. tetradactyla* that had been captured in the forests near Belém, Brazil. Parasites were seen in the blood of the animals and in reduviids that had fed on them, but these authors gave no details of the morphology of the bloodstream trypomastigotes or the flagellates from the bugs. The low infection rates in sloths and anteaters suggest that they are accidental hosts and not the reservoirs. In Panama there may be ecological niches (forest canopy) in which the vector(s) come into sporadic contact with both sloths and anteaters but that no such contact occurs in the lower Amazon region of Brazil. The species of bugs involved and the type of *T. cruzi* are not known.

Of all the xenarthra perhaps only the common long-nosed armadillo, *D. novemcinctus*, can be considered as a reservoir of *T. cruzi*. Chagas (1912) recorded an infection rate of 45 percent in common long-nosed armadillos from Minas Gerais while Freitas' (1950) examinations of specimens of this species from São Paulo were negative. In Pará, Deane (1961) found *T. cruzi* in 6 of 22 *D. novemcinctus*, while we also observed infections in 6 of 22 (Lainson et al., 1979a). Barretto and Ribeiro (1979) quote infection rates of 16 percent in *D. novemcinctus* from Argentina, 5.6 percent from French Guiana, 55.3 percent from Venezuela, and 5.6 percent from Costa Rica. In Panama, Clark and Dunn (1932) noted that 6 of 25 common long-nosed armadillos were infected with *T. cruzi*, while in a survey performed in Panama by the Gorgas Memorial Laboratory in 1965, only one infection was found in 58 animals, Pipkin (1968) failed to find infections in 43 common long-nosed armadillos from Panama.

How can the very different infection rates at different times in the same countries (e.g., Brazil and Panama) be explained? Either some ecological change resulted in different infection rates among armadillos or the disease was present but not detected for technical reasons. In Brazil, some 35 years passed between the positive and negative observations. During this time, tremendous modifications in the habitat almost certainly took place and the vector may have died out while the armadillos were able to adapt to the new conditions. The major factor, however, is perhaps variations in the methods of examination. Chagas (1912) detected infections by seeing parasites in stained blood smears while Freitas (1950) used xenodiagnosis. In Panama Clark and Dunn (1932) found trypanosomes in stained material while the subsequent workers used culture techniques to detect infections. The methods used in the later studies performed in both Brazil and Panama were therefore not strictly comparable although one would suspect that both xenodiagnosis and cultures would be the more sensitive techniques, with infections less likely to be detected by direct searches for parasites in the blood.

Our experience (Lainson et al., 1980) with stocks of *T. cruzi* from common long-nosed armadillos is relevant to the possibility that infections were not detected because of technical problems. We found that these stocks developed very slowly in culture, taking a long time to show positive test results. On occasion, cultures remained negative even though bloodstream trypomastigotes were seen. The armadillo stocks developed poorly in the reduviid bugs *Tritoma infestans, T. brasiliensis, T. sordida, Rhodnius prolixus* and *Dipetalogaster maximus*. Our recent awareness that such stocks belong to zymodeme III (Miles et al., 1981) helps explain difficulties in handling stocks of *T. cruzi* from armadillos. It also suggests that failures to detect infection could arise from unsuitable bugs being used, the use of inappropriate culture media, or because cultures were examined too early. Clearly in the light of the recent observations on the biology of isolates of *T. cruzi* from armadillos, more work is needed in areas where these animals were found to be negative.

In the text books of tropical medicine and parasitology (cf. Wilcocks and Manson-Bahr, 1977), common long-nosed armadillos are the classical reservoir of Chagas' disease, almost certainly because they were the first wild animals to be found infected with a trypanosome that was morphologically similar to the one found in man. Further observations are needed on infections in armadillos from other regions of Brazil and the Americas, especially the identification of the parasites from the different areas. Chagas noted that *Panstrongylus geniculatus* was associated with common long-nosed armadillos and suggested them as the most probable vector among armadillos. This idea was supported by the observations of Barrett et al. (1980), who noted that in Bahia, Brazil, a stock from an armadillo and one from a reduviid bug *P. geniculatus*, found in an armadillo's burrow were both zymodeme III.

Unidentified *Trypanosoma* species

Cyclopes didactylus Brazil (Miles, personal communication).

This is only the second record of a trypanosomatid from the silky anteater. The infection was detected by hemoculture and growth was luxuriant. No parasites were seen in stained blood smears from the positive animal. Many silky anteaters have been examined by hemoculture but none were found infected (Lainson et al., 1979). Since the flagellate found by Miles grows luxuriantly in culture and he only found a single infected animal, it seems possible that it was an accidental infection of a trypanosome of another mammal. Attempts are being made to identify the silky anteater stock.

THE RELATIONSHIP OF INFECTIONS TO HOST BEHAVIOR AND HABITAT

If a vector has a narrower habitat preference than the vertebrate host there will obviously be uninfected populations of the host. In addition, host behavior can facilitate vector-host contact which could increase the infection rate in two equally susceptable hosts that live in the same area. For instance, if two mammals live in the same place but one is herbivorous and the other insectivorous, then the latter will stand a greater chance of being infected with a parasite whose infective stages occur in the insect's feces. Other factors, such as daily range of movement, home range size, and seasonal migration, can also influence the degree of vector-reservior contact. The end product of these fascinating complex relationships is often mixed hemoflagellate infections.

Infection rates of *T. leuwenhoeki* are almost equal in both species of sloths in Panama, while the incidence of *E. schaudinni* is high in two-toed sloths and very low in three-toed sloths. Mixed infections of *T. leuwenhoeki* and *E. schaudinni* are rare but concommitant infections of *L. b. panamensis* and *E. schaudinni* are quite common although there are more infections of the latter. This mosaic of infections has been interpreted as reflecting a wider habitat range of the two-toed sloth as compared to the three-toed sloth (Shaw, 1969). While two-toed and three-toed sloths live in microsympatry, using the same trees on occasion (Montgomery and Sunquist, 1978), two-toed sloths have larger home range spaces, and range more widely during each night's activity (Sunquist and Montgomery, 1973).

The pattern of mixed infections in sloths from Costa Rica is different from those seen in sloths from Panama. The incidence of *L. b. panamensis* appears to be more or less equal in both species of sloth yet there are far more infections of *Endotrypanum* in two-toed than in three-toed sloths. *T. leuwenhoeki* is uncommon in both species but when found, it is in association with either *Endotrypanum* or a *Leishmania* species (Zeledon et al.,1979). It would appear that in Costa Rica, unlike Panama, both species of sloth have equal contact with the habitat of the vector of both *L. braziliensis* and *T. leuwenhoeki*, but that the two-toed sloth has a greater degree of contact with the vectors of *Endotrypanum*.

In Brazil the relationship of infections of *Endotrypanum* and *Leishmania* are more similar to those seen in sloths from Panama. There are proportionally more infections of *E. schaudinni* than *L. braziliensis* but to date no counterpart to *T. leuwenhoeki* has been found in either species of sloth from the lower Amazon region.

There is no satisfactory explanation for the absence of

T. rangeli-like trypanosomes in armadillos from Panama, especially when both sloths and anteaters have been found to be infected with trypanosomes of this complex. It could either be that they are not susceptable or that there is some ecological barrier between them and a suitable vector.

The differences among relative infection rates of *T. cruzi* in sloths, anteaters, and armadillos is probably related to differences in both the habits and habitats of these animals. Sloths are vegetarians, predominantly arboreal, and do not live in any form of hole or burrow: their chances of ingesting a bug contaminated with *T. cruzi* or of being in close contact with a vector species are slight. *Tamandua* anteaters are insectivorous and occasionally pass the day in a hollow tree and, while foraging for ants and termites, they could inadvertantly eat the odd reduviid. However, the only insects found in more than 150 *Tamandua* stomach and fecal samples which were not ants or termites were homopteran mimics of ants. Armadillos are, however, terrestrial, live in burrows and are omnivorous, and clearly stand a greater chance of becoming infected with *T. cruzi* than sloths or anteaters. *T. cruzi* infections are commonest in armadillos, relatively rare in *Tamandua* anteaters and very rare in sloths, suggesting that there is a correlation between food and behavioral habits and infection with this hemoflagellate.

Our present fragmentary knowledge of incidences of mixed infections indicates that the habitat plays a major role in determining the type of parasite, especially when transmission is by bite. It seems, however, that such factors as food habits can assume special importance in the case of trypanosomatids whose transmissions are by passive contamination with contaminated insect fecal material.

VECTORS OF XENARTHRAN HEMOFLAGELLATES

Many insects have been incriminated as vectors of hemoflagellates simply because they have been found to be naturally infected in enzootic or endemic regions. Such information is obviously very important for it supplies primary leads, but these species may represent dead end infections that, for various reasons, are never transmitted to a susceptable host. Epizootiological and experimental evidence conclusively indicates that the sandfly *L. (N.) trapidoi* is the major vector of *L. b. panamensis* among sloths but no actual transmissions have yet been performed. The vectors of the other leishmanias of sloths and armadillos are unknown, but it is likely that they are phlebotomine sandflies. There seems little doubt that the sandfly *L. (N.) umbratalis* is the vector of the *L. b. guyanensis* infections found in *C. didactylus* sloths and *T. tetradactyla* anteaters, but it seems likely that *L. (N.) whitmani* could be an important secondary vector in certain situations because it also occurs on tree trunks. There is increasing evidence to support the hypothesis that sandflies are the vectors of *Endotrypanum*. No species has been singled out as the vector, but recent studies on the susceptability of a limited number of sandfly species to infection indicates that only one is likely to be the major vector in a given area. This view is supported by the epizootiological observations that suggest that different sandflies are responsible for the transmission of *L. b. panamensis* and *E. schaudinni*. There have been no attempts to experimentally transmit *Endotrypanum*.

Experimental transmissions of *Leishmania* by the bite of infected sandflies are very difficult to perform. Some workers failed even when they used sandfly species that were undoubted vectors, on epidemiological grounds. Conversely it may be possible to transmit a *Leishmania* with a sandfly that in nature never comes into contact with the reservoir. This may give a false indication to the role of a particular group of insects as vectors because it does not indicate the vector in a real epizootiological situation. The definition of a vector of a hemoflagellate depends upon a combination of both field and laboratory studies.

None of the vectors of the trypanosomes found in xenarthrans are known but almost certainly those of the different *T. cruzi*-like infections and *T. leuwenhoeki* are reduviid bugs. There is good evidence that the reduviid bug *Panstrongylus geniculatus* is the vector of zymodeme III that is found in common long-nosed armadillos; however, more observations are needed before this bug can be considered as the actual vector. More information is needed on the susceptability of different sandfly and bug species to the now identifiable strains and subspecies of the different hemoflagellates, especially their cycles of development in the different species. Further studies on the hematophagus insects that feed on sloths, anteaters, and armadillos in the wild could give important leads to possible vectors.

Godfrey (1979) suggested that the different isoenzymes seen in the various zymodemes of *T. cruzi* might be adaptions to optimal development and transmission in certain bugs. There is therefore an increasing amount of support for the idea that each trypanosomatid has one major vector and presence or absence of the vector is the limiting factor in distribution of the trypanosomatid rather than distribution of the reservoir host.

XENARTHRA AS RESERVOIRS OF LEISHMANIAL AND TRYPANOSOMAL INFECTIONS OF MAN

To incriminate an animal as a reservoir of a human disease it is first necessary to show that the organism found in it is identical to the disease organism found in man, and then to show that it is the major host rather than an accidental one. A single record of a particular parasite in a wild animal species does not mean that it is a reservoir, especially if no other infections were found in other individuals caught in the same habitat.

In Panama, two-toed sloths *C. hoffmanni* are the major reservoirs of *L. b. panamensis*, the causitive agent of cutaneous leishmaniasis of man. Leishmanial stocks isolated from sloths appeared to be biologically and biochemically the same as those isolated from man. In extensive studies on the infection rates of *L. b. panamensis* in wild animals, Herrer et al. (1973) found that in endemic areas up to 37 percent of the two-toed sloths were infected with this parasite. The importance of two-toed sloths as reservoirs of *L. braziliensis* subspecies in other Latin American countries has yet to be determined. Infections have been found in two-toed sloths near the Amazon river but the parasite has not yet been sufficiently well characterized to associate it with human infections. Common long-nosed armadillos have been found infected with a *Leishmania* by Lainson et al. (1979a) in the lower Amazon Basin, but so far this parasite has not been found in man. The finding of infections of *L. b. guyanensis* in 46 percent of the two-toed sloths *C. didactylus* from an endemic "pian-bois" region of northern Pará clearly indicates them to be the major reservoir. An infection rate of 22 percent in anteaters *T. tetradactyla* means that these must also be considered as important secondary reservoirs (Lainson et al., 1981a, 1981b). The relative importance of different reservoirs can only be assessed by comparing infection rates in laboratory-bred natural vectors fed on wild-infected vertebrate hosts. This is technically difficult, and normally the infection rate in different wild animals is used as an indication of their importance as reservoirs.

Brazilian stocks of *T. cruzi* from common long-nosed armadillos of Bahia and Pará belong to zymodeme III (Barrett et al., 1980; Miles et al., 1981), but we know nothing about stocks from these animals in other parts of the continent. Zymodeme III is rare in man and has only been found in 4 acute cases of Chagas' disease (Miles et al., 1979a) that all became infected at apparently the same time in the capital of Pará state (Shaw et al., 1969). In endemic Chagas' disease regions where human isolates have been studied (Miles et al., 1977; Barrett et al., 1980), none belonged to zymodeme III.

The available evidence, therefore, suggests that the zymodeme of *T. cruzi* associated with common long-nosed armadillos seldom infects man. However, there was no association of armadillos or bugs, such as the reduviid *P. geniculatus*, with the patients from Pará. The position of common long-nosed armadillos as reservoirs in the complex epidemiology of Chagas' disease clearly needs to be investigated in greater detail, especially the enzyme profiles of *T. cruzi* isolates from this species of armadillo and man in other countries. It is difficult to evaluate the importance, if any, of *T. cruzi* infections in sloths and anteaters to human Chagas' disease. To begin with, comparisons need to be made with stocks isolated from these animals and man.

In Panama, man is infected with a *T. rangeli*-like trypanosome that is very similar to *T. leuwenhoeki* of sloths. The high incidence of infection of this trypanosome in both species of sloth from the more open areas leaves no doubt as to the importance of these animals as reservoirs. If some of the *T. rangeli*-like infections in man prove to be the same as *T. leuwenhoeki*, then sloths would be incriminated as a potential source of infection for man. Whether or not sloths could serve as a direct source for the bugs that bite man or as a source of infection for bugs that transmit the infection to a domestic animal that then serves as a reservoir for some domestic bug is an important facet that needs to be investigated in relation to both *T. cruzi* and *T. leuwenhoeki*. In certain regions armadillos, anteaters and sloths are eaten by man. There is thus a direct risk of infection, principally with *T. cruzi*, associated with the preparation and consumption of the animal.

TRYPANOSOMATID EVOLUTION IN RELATION TO XENARTHRAN HEMOFLAGELLATES

The lack of fossil evidence for most of the Protozoa makes it difficult to construct a phylogenetic scheme that can be confidently supported. Most schemes are therefore by necessity based on living organisms. In an excellent review on the radiation of the Eukaryote Protista (Protozoa, Algae, and some Fungi) Sleigh (1979) concluded that the primitive nucleus of kinetoplastids suggests that they evolved directly from the basic primitive flagellate stock.

There are no reasons for rejecting leptomonads as the basic stock of the kinetoplastids and invertebrates as their original hosts (Baker, 1965). It should be noted that *Leishmania* develop as promastigotes (leptomonads) in their vectors, phlebotomine sandflies, and that these insects are considered as perhaps the most primitive of the blood-sucking Diptera, originating possibly as long ago as the Triassic (Edwards, 1926). For the above reasons, parasites of the genus *Leishmania*, especially the leishmania of lizards, are considered to be the most primitive hemoflagellates. Baker (1965) dealt at length with the various theories on the evolution of the trypanosomatids and concluded that the genus *Trypanosoma* is polyphyletic, having evolved along three separate lines that diverged from a primitive hypothetical leptomonad stock of annelids. Based on the biology of certain hemoflagellates of xenarthra, in particular *Endotrypanum*, I suggested (Shaw, 1966) that the subgenus *Schizotrypanum* in the New World and the genus *Endotrypanum* could have both evolved from the insect leptomonad stock.

In the past 15 years, much more information has become available on the trypanosomatids of xenarthrans which supports the line of evolution that I suggested in 1966, and supports the idea that there may have been a line that was particularly associated with xenarthrans. The *Leishmania* found in sloths and armadillos are peripylarians, a group found only in the Americas and considered to be intermediate between the primitive hypopylaria of lizards and the more advanced suprapylaria of rodents (Lainson and Shaw, 1979).

Our present knowledge of the hosts and geographical distribution of peripylarian *Leishmania* strongly supports the idea that this group evolved in xenarthrans. There is at present considerable controversy about the origin of some mammalian *Leishmania* found in the Neotropics, particularly those responsible for the visceral form of the disease in man. Cameron (1956) has suggested that the genus *Leishmania* was introduced by man from the New World to the Old World after Columbus' voyage, while others consider that certain parasites were introduced from the Old World to the New World. This is not the place to discuss the effect man has had on the continental dispersion of leishmaniasis, but I think that our present knowledge of the biology of the peripylarian leishmania of xenarthrans poses a fundamental question. Did the suprapylaria evolve from a peripylarian stock or do they represent a separate line of evolution? Although the evidence is scant, I think that we should seriously consider the possibility that the suprapylaria evolved from the more primitive peripylarians and that rodents could have been responsible for their spread from the New World to the Old World.

Another line of evolution in xenarthrans has resulted in flagellates of the genus *Endotrypanum* that have characters intermediate between those of the leishmania and trypanosomes. In insects they develop as do parasites of the genus *Leishmania*, while in the blood of the sloth they are in the form of epimastigotes and, in one rare species, they have a trypanomorphic form. The genus *Endotrypanum* is perhaps a living example of the ancestoral form of some trypanosomes that has survived because of the antiquity of both its vectors and reservoir hosts.

The idea that trypanosomes of the subgenus *Schizotrypanum* represent a third line that could have developed in xenarthrans needs to be examined in the light of recent biochemical observations on members of this subgenus. Godfrey (1979) pointed out that eight zymodemes of *Schizotrypanum* from bats (Baker et al., 1978) bore no resemblance to *T. (S.) cruzi* from man (Miles, 1979). He suggests that "the heretical possibility should be contemplated that these trypanosomes (referring to *T. (S.) dionisii* and *T. (S.) vespertilionis* of European bats) are not closely related to each other or to *T. (S.) cruzi,* their identical morphology in the blood-stream phase being a result of convergent evolution." Zymodemes I, II, and III of *T. (S.) cruzi* from man may be readily differentiated on at least 5 enzymes (Miles et al., 1980), suggesting a difference that in the future may warrant taxonomic status. The above observations add weight to Godfrey's "heretical possibility". The observations suggest that there may be different lines of *Schizotrypanum* that are infective to man and that they could have evolved independently in some xenarthran stock, such as armadillos, and in other primitive mammals, such as marsupials. Evidence that evolutionary pressures on a leishmanial stock of extinct xenarthrans could have selected the trypomastigotic form comes from the present-day bloodstream forms of *Endotrypanum*.

ACKNOWLEDGMENTS

This work was generously supported by the Wellcome Trust, London, and the Fundaçao SESP of the Brazilian Ministry of Health, Rio de Janeiro. I am particularly indebted to my friends and colleagues for allowing me to cite results from manuscripts that are at present in press and for their permission to quote unpublished observations. Special thanks are due to Drs. Ralph Lainson and Michael A. Miles for their stimulating and very valuable discussions.

LITERATURE CITED

Baker, J. R.
1965. The evolution of parasitic protozoa. Pp 1-27 *in* Third symposium of the British Society for Parasitology, Evolution of Parasites (A.E.R. Taylor ed.). Blackwell Scientific Publications, Oxford, 133 pp.

Baker, J. R., M. A. Miles, D. G. Godfredy, and T.V. Barrett.
1978. Biochemical characterization of some species of *Trypanosoma (Schizotrypanum)* from bats (Microchiroptera). Am. J. Trop. Med. Hy., 27: 483-491.

Barrett, T. V., R. F. Hoff, K. E. Mott, M. A. Miles, D. G. Godfrey, R. Texeira, J. A. Almeida de Sourza, and I.A. Sherlock.
1980. Epidemiological aspects of three *Trypanosoma cruzi* zymodemes in Bahia. Trans. R. Soc. Trop. Med. Hyg., 74: 84-90.

Baretto, M. P., and R. D. Ribeiro.
1979. Reservatórios silvestres do *Trypanosoma (Schizotrypanum) cruzi* Chagas, 1909. Rev. Inst. Adolfo Lutz., 39: 25-36.

Cameron, T. W. M.
1956. Parasites and parasitism. Methuen and Co., Ltd, London, 322 pp.

Chagas, C.
1912. Sobre um trypanosoma do tatú, *Tatusia novemcincta*, transmitido pela *Triatoma geniculata* Latr. (1811). Possibilidade de ser o tatú um depositário do *Trypanosoma cruzi* no mundo exterior. Brasil méd., 26: 305-306.

Chance, M. L.
1979. The identifcation of Leishmania. Pp 53-74 *in* Symposia of the British Society for Parasitology, Volume 17, Problems in the Identification of Parasites and Their Vectors (A. E. R. Taylor and R. Muller, eds.). Blackwell Scientific Publications, Oxford, 221 pp.

Christensen, H. A., and A. Herrer.
1973. Attractiveness of sentinel animals to vectors of leishmaniasis in Panama. Am. J. Trop. Med. Hyg., 22: 578-584.

1976. Neotropical sandflies (Diptera: Psychodidae), invertebrate hosts of *Endotrypanum schaudinni* (Kinetoplastida: Trypanosomatidae). J. Med. Entomol., 13: 299-303.

1979. Susceptability of sandflies (Diptera:Psychodidae) to trypanosomatidae from two-toed sloths (Edentata: Bradypodidae). J. Med. Entomol., 16: 424-427.

Clark, H. C., and L. H. Dunn.
1932. Experimental studies on Chagas' disease in Panama. Am. J. Trop. Med., 12: 49-77.

Croft, S. L., M. L. Chance, and P. J. Gardener.
1979. Ultrastructural and biochemical characterization of strains of *Endotrypanum*. Trans. R. Soc. Trop. Med. Hyg., 73: 322.

1980. Ultrastructure and biochemical characterization of stocks of *Endotrypanum*. Ann. Trop. Med. Parasitol., 74, 585-589.

Deane, L. M.
1961. Tripanosomìdeos de mamìferos da região Amazonica. I. Alguns flagelados encontrados no sangue de mamìferos silvestres do Estado do Parà. Rev. Inst. Med. Trop. São Paulo, 3: 15-28.

Deane, L. M., and R. G. Damsceno.
1961. Tripanosomìdeos de mamìferos da região Amazonica. II. Tripanosomas de macacos da Zona do Salgado, Estado do Parà. Rev. Inst. Med. Trop. São Paulo, 3: 61-70.

D'Alessandro, A.
1976. Biology of *Trypanosoma (Herpetosoma) rangeli* Tejera, 1920. Pp. 327-403 *in* Biology of the Kinetoplastida. Vol I (W.H.R. Lumsden and D.A. Evans eds). Academic Press, London, New York and San Francisco, 563 pp.

Edwards, F. W.
1926. The phylogeny of nematocerous Diptera: a critical review of some recent suggestions. Verhandl. III Internat. Kong. Ent., 2: 111.

Freitas, J. L. P.
1950. Observações sobre xenodiagnòsticos practicados em reservatòrios domèsticos silvestres do *Trypanosoma cruzi* em uma localidade endemica da molèstia de Chagas no Estado São Paulo. O Hospital (Rio de Janeiro), 38: 521-529.

Gentile, B., F. Le Pont, F.X. Pajot, and R. Besnard.
1981. Dermal leishmaniasis in French Guiana: the sloth (*Choloepus didactylus*) as a reservoir host. Trans. R. Soc. Trop. Med. Hyg., 75(4): 612-613.

Godfrey, D. G.
1979. The Zymodemes of Trypanosomes. Pp. 31-53, *In* Symposia of the British Society for Parasitology, Volume 17, Problems in the Identification of Parasites and Their Vectors (A.E.R. Taylor and R. Muller eds.). Blackwell Scientific Publications, Oxford, 221 pp.

Herrer, A., and S. R. Telford.
1969. *Leishmania braziliensis* isolated from sloths in Panama. Science, 164: 1419-1420.

Herrer, A., H. A. Christensen, and R. J. Beumer.
1973. Reservoir hosts of cutaneous leishmaniasis among Panamanian forest mammals. Am. J. Trop. Med. Hyg., 22: 585-591.

Hershkovitz, P.
1971. Edentata. Encyclopedia Britannica, 7: 965-968.

Hoare, C. A.
1972. The Trypanosomes of Mammals. A Zoological Monograph. Blackwell, Oxford and Edinburgh, 749 pp.

Lainson, R.
1965. Parasitological studies in British Honduras. I. A parasite resembling *Trypanosoma (Schizotry-*

panum) cruzi in the coati, *Nasua narica* (Carnivora, Procyonidae), and a note on *Trypanosoma legeri* from the anteater, *Tamandua tetradactyla* (Edentata). Ann. Trop. Med. Parasit., 59: 37-42.

Lainson, R., and J. J. Shaw.
1979. The role of animals in the epidemiology of South American leishmaniasis. Pp. 1-116 *in* Biology of the Kinetoplastida. Vol. II (W. H. R. Lumsden and D. A. Evans, eds.). Academic Press, London, New York and San Francisco, 738 pp.

Lainson, R., J. J. Shaw, and R. D. Naiff.
1980. Chagas' disease in the Amazon Basin: suggestions on transmission *per os*. Rev. Inst. Med. Trop. São Paulo, 22: 294-297.

Lainson, R., J. J. Shaw, and M. Povoa.
1981. The importance of edentates (sloths, anteaters) as primary reservoirs of *Leishmania braziliensis guyanensis*. Trans. R. Soc. Trop. Med. Hyg., 75(4): 611-612.

Lainson, R., J. J. Shaw, H. Fraiha, M. A. Miles, and C. C. Draper.
1979. Chagas' disease in the Amazon Basin: I. *Trypanosoma cruzi* infections in silvatic mammals, triatomine bugs and man in the state of Pará, north Brazil. Trans. R. Soc. Trop. Med. Hyg., 73: 193-204.

Lainson, R., J. J. Shaw, R. D. Ward, P. D. Ready, and R. D. Naiff.
1979. Leishmaniasis in Brazil: XIII. Isolation of *Leishmania* from armadillos (*Dasypus novemcinctus*) and observations on the epidemiology of cutaneous leishmaniasis in North Pará State. Trans. R. Soc. Trop. Med. Hyg., 73: 239-242.

Lainson, R., J. J. Shaw, P. D. Ready, M. A. Miles, and M. Póvoa.
1981. Leishmaniasis in Brazil: XVI. Isolation and identification of *Leishmania* from sandflies, wild mammals and man in north Pará, with particular reference to *Leishmania braziliensis guyanensis*, caustive agent of "pian bois." Trans. R. Soc. Trop. Med. Hyg., 75(4): 530-536.

Miles, M. A.
1979. Transmission cycles and the heterogeneity of *Trypanosoma cruzi*. Pp. 117-196 *in* Biology of the Kinetoplastida. Vol. II (W. H. R. Lumsden and D. A. Evans, eds.). Academic Press, London, New York and San Francisco, 738 pp.

Miles, M. A., T. V. Barrett, R. H. Hoff, and K. E. Mott.
1977. Epidemiological studies of Chagas' disease in Brazil. Trans. R. Soc. Trop. Med. Hyg., 73: 135-136.

Miles, M.A., P.J. Toye, S.C. Oswald, and D.G. Godfrey.
1979. The identification by isoenzyme patterns of two distinct strain-groups of *Trypanosoma cruzii* circulating independently in a rural area of Brazil. Trans. R. Soc. Trop. Med. Hyg., 71: 217-255.

Miles, M. A., A. Souza, M. Povoa, J. J. Shaw, and R. Lainson.
1978. Isozymic heterogeneity of *Trypanosoma cruzi* in the first autochthonous patients with Chagas' disease in the Amazonian Basin. Nature, Lond., 272: 819-821.

Miles, M. A., S. M. Lanham, A. S. de Souza, and M. Póvoa.
1980. Further enzymic characters of *Trypanosoma cruzi* and their evaluation for strain identification. Trans. R. Soc. Trop. Med. Hyg., 74: 221-237.

Miles, M. A., M. Povoa, A. A. Souza, R. Lainson, and J. J. Shaw.
1981. Chagas' disease in the Amazon Basin: II. The distribution of the *Trypanosoma cruzi* zymodemes 1 and 3 in Pará state, north Brazil. Trans. R. Soc. Trop. Med. Hyg., 75: 667-674.

Montgomery, G. G.
1985. Movements, Foraging and Food Habits of the Four Extant Species of Neotropical Vermilinguas (Mammalia; Myrmecophagidea) (this volume).

Montgomery, G. G. and M. E. Sunquist.
1978. Habitat selection and use by two-toed and three-toed sloths. Pp. 329-359 *in* The Ecology of Arboreal Folivores (G. G. Montgomery, ed.). Smithsonian Institution Press, Washington, D.C. 574 p.

Newton, B. A., and J. K. Burnett.
1972. DNA of kinetoplast: a comparative study. Pp. 127-138 *in* Comparative Biochemistry of Parasites (H. van den Bossche, ed.). Academic Press, New York and London. 516 p.

Pipkin, A. C.
1968. Domicillary reduviid bugs and the epidemiology of Chagas' disease in Panama (Hemiptera: Reduviidae: Triatominae). J. Med. Entomol., 5: 107-124.

Rey-Matiz, H.
1941. Observaciones sobre trypanosomas en Colombia. I. Observaciones sobre *Trypanosoma cruzi*. Rev. Fac. Med. Bogotá, 10: 25-49.

Rodrigues, B. A., and G. B. Mello.
1942. Contribuição ão estudo do tripanosomiase americana. Mem. Inst. Oswaldo Cruz, 37: 77-90.

Sarich, V.M.
1985. Xenarthran systematics: albumin immunological evidence. (this volume).

Shaw, J. J.
1964. A possible vector of *Endotrypanum schaudinni* of the sloth *Choloepus hoffmanni* in Panama. Nature, Lond., 210: 107-108.

1966. The relationship of *Endotrypanum* to other members of the Trypanosomatidae and its possible bearing upon the evolution of certain haemoflagellates of the New World. Proc. First. Internat. Congress Parasit., Rome 1964., 1: 332-333.

1969. The Haemoflagellates of Sloths. London School of Hygiene and Tropical Medicine Memoir No.13., H.K. Lewis and Co. Ltd., London. 132 p.

1981. The behaviour of *Endotrypanum schaudinni* (Kinetoplastida Trypanosomatidae) in three species of laboratory-bred Neotropical sandflies (Diptera: Psychodidae) and its influence on the classification of the genus *Leishmania*. Pp. 232-241 *in* Parasitological topics, a presentation volume to P.C.C. Garnham F.R.S. on the occasion of his 80th birthday, 1981 (E.V. Canning, ed.), Soc. Protozol., Special Publ. 1.

Shaw, J. J., and R. G. Bird.
1969. The endoerthrocytic habitat of a member of the trypanosomatidae, *Endotrypanum schaudinni* Mesnil and Brimont, 1908. Z. Tropenmed. Parasit., 20: 144-150.

Shaw, J. J., R. Lainson, and H. Fraiha.
1969. Considerações sobre a epidemiologia dos primeiros casos autoctones de Doenca de Chagas registradas em Belèm, Parà, Brasil. Rev. Sàude Pùbl., São Paulo, 3: 153-157.

Sleigh, M. A.
1979. Radiation of the Eukaryote Protista. Pp. 23-53 *in* The Systematics Association Special Volume No. 12, The Origin of Major Invertebrate Groups (M. R. House, ed.). Academic Press, London, New York and San Francisco. 515 p.

Sunquist, M. E. and G. G. Montgomery.
1973. Activity patterns and rates of movement of two-toed and three-toed sloths (*Choloepus hoffmanni* and *Bradypus infuscatus*). Journal of Mammalogy, 54: 946-954.

Walton, B. C., and O. E. Sousa.
1967. Trypanosomes of the lesser anteater, *Tamandua tetradactyla*, from Panama. J. Parasit., 53: 956-961.

Wilcock, C., and P. E. C. Manson-Bahr.
1974. Manson's Tropical Diseases. Bailliere Tindall, London. 110 p.

Zeledon, R., C. Ponce, and E. de Ponce.
1975a The isolation of *Leishmania herreri* sp. n. from sloths in Costa Rica. Am. J. Trop. Med. Hyg., 24: 706-707.

1975b. Isolation of *Trypanosome rangeli* from Costa Rican sloths. Am. Soc. Parasit. Program and Abstracts, 60.

Zeledon, R., C. Ponce, and J. Murillo.
1979. *Leishmania herreri* sp. n. from sloths and sandflies of Costa Rica. J. Parasit., 65: 275-279.

ANTHROPOD ASSOCIATES OF SLOTHS

ARTHROPOD ASSOCIATES OF SLOTHS

J.K. Waage[1] and R.C. Best[2]

[1]Imperial College at Silwood Park, Ascot, Berks, SL5 7PY, U.K. [2] Instituto Nacional de Pesquisas da Amazônia, C.P. 478, 69.000 Manaus, Amazonas, Brazil; and Research Associate, Vancouver Public Aquarium, Vancouver, Canada.

Abstract

Sloths are host to a variety of arthropods which may be divided into two distinct guilds. The hematophagous guild is dominated by biting flies and ticks, and is not unlike that of other similarly sized mammals, while the coprophagous guild is a unique assemblage of moths, beetles and mites which appear dependent upon the dung of the host for development and utilize the sloth primarily for phoresis. We review the literature on arthropod associates of sloths and present new information on the biology of sloth ticks, moths, beetles, and mites. Recent studies on the arthropod fauna of sloths near Manaus, Brazil, reveal a highly diverse and dynamic community, with some species undergoing great seasonal population fluctuations while others remain relatively constant in numbers. No less than nine species of the coprophagous guild were found in this study area, suggesting intense competition among their larval, dung-feeding stages. The life histories of these phoretic associates are discussed, and hypotheses are presented to explain how their unusual host associations have evolved and why they should be so well represented on sloths.

Abstracto

Los perezosos sirven de hospederos a una variedad de artrópodos que pueden dividirse en dos grandes gremios. El gremio hematófago es dominado por moscas picadoras y por garrapatas, y en ésto los miembros del gremio no difieren de otros mamíferos de tamaño similar; por otra parte, el gremio de los coprófagos constituye una asociación singular de polillas, escarabajos y ácaros, que parecen depender para su desarrollo del estiércol de hospedero y que utilizan al perezoso, primordialmente, como medio de transporte. En este artículo revisamos la literatura sobre artrópodos asociados con perezosos y presentamos nueva información sobre la biología de las garrapatas, las polillas, los escarabajos y los ácaros de los perezosos. Estudios recientes sobre la fauna de artrópodos de los perezosos en las cercanías de Manaus, en Brasil, revelaron la existencia de una comunidad altamente diversificada y dinámica en las que unas especies mostraron una gran fluctuación estacional de población, mientras que otras permanecieron constantes. Por lo menos nueve especies del gremio coprófago fueron localizadas en el área de estudios, sugiriendo la existencia de una intensa competencia entre los estados larvarios que se alimentan de estiércol. El ciclo de vida de estos insectos asociados en su medio de transportación es discutido, y se presentan varias hipótesis para explicar cómo ha evolucionado esta insólita asociación y la razón por su presencia en perezosos.

Resumo

As preguiças são hospedeiras de uma variedade de artrópodos que podem ser divididos em dois grupos distintos. O grupo hematófago é dominado por moscas picadoras e carrapatos e não é diferente do de outros mamíferos de tamanho similar, enquanto que o grupo coprófago é uma associação única de mariposas, besouros e ácaros que paracem depender das fezes do hospedeiro para se desenvolverem e utilizam a preguiça principalmente para forese. Revisamos a literatura em artrópodos associados às preguiças e apresentamos novas informações sobre a biologia dos carrapatos, mariposas, besouros e ácaros de preguiças. Estudos recentes da fauna de artrópodos de preguiças de uma região próxima de Manaus, Brasil, revelam uma comunidade altamente diversa e dinâmica, com algumas espécies sendo submetidas a grandes flutuações sazonais de população enquanto outras permanecem relativamente constantes em número. Nove espécies do grupo coprófago foram encontrados nessa área de estudo, sugerindo intensa competição entre seus estágios larvais, que se alimentam das fezes. O ciclo de vida desses associados foréticos são discutidos e hipóteses apresentadas para explicar como suas associações hospedeiras únicas evoluiram e porque elas seriam tão bem representadas nas preguiças.

Like most terrestrial vertebrates, tree sloths (Bradypodidae and Megalonychidae) play host to a variety of arthropod associates. The community of bloodsucking arthropods associated with sloths is typical for a medium-sized tropical mammal, and includes biting flies (Culicidae, Psychodidae, Tabanidae), triatomine bugs, lice, ticks and mites. The arboreal habit of sloths accounts for the general absence of some hematophagous groups, such as the Siphonaptera, which usually require some form of nest habitat for larval development. In addition to this bloodsucking guild, sloths carry a highly specific community of commensal arthropods, including moths, beetles, and mites. The relationship between these arthropods and their host is rather unique and its study may reveal a great deal about the evolution of insect-vertebrate associations. In this paper we review literature on the arthropod associates of sloths, and report new information from ongoing studies in Manaus, Brazil, on sloth ticks, moths, and beetles.

These studies have been made at two principal collecting sites. One is in second-growth *terra firme* forest on the outskirts of Manaus at the National Amazonian Research Institute (INPA). The second site is a small *varzea* island, Ilha Curari, in the Rio Solimões near Manaus (described by Irmler, 1975). The small island is subject to seasonal flooding, and is inundated for approximately four months each year. During this time, sloths continue to live in the trees.

The study began in May 1977 and data collected until July 1979 are included here. During that period we captured and examined 102 *Bradypus tridactylus* L. and 17 *Choloepus didactylus* at INPA in Manaus and an additional 86 *Bradypus variegatus* Schinz from Ihla Curari. The three-toed sloths were identified after Wetzel and Avila-Fires (1980) and Wetzel (1985) and the existence of two species has been confirmed by karyotyping (Jorge et al., 1985).

The sloths were captured by hand and care was taken not to disturb the moths, which tend to fly off if the sloth is handled roughly. Some sloths lost moths due to rough handling during capture. These were not included in the analysis of moth numbers. All of the larger arthropods encountered were collected and representative samples of mites were also taken for identification. These collections were made by carefully preening each animal shortly after it was captured. Arthropods collected were preserved in alcohol or, in the case of moths, were pinned and dry-mounted.

The arthropod data from the terra firme sloth population at Manaus have been related to the mean monthly precipitation patterns of this area (Ministry of Agriculture Meteorology Station, Manaus) where there is a short, 3-month dry season from June through August. We considered the annual flooding of Ilha Curari (data from the Port Authority, Manaus; rainfall data are not available for Ilha Curari) to be the most important environmental factor in governing sloth arthropod populations.

The findings of the present study are combined below with information from other sources. In general, arthropod associates of sloths are very poorly known, and further research in this area will undoubtedly add new species and information to the survey presented here. Three-toed sloths have been more intensely studied than two-toed sloths, probably as a consequence of their greater abundance and gentler disposition (Montgomery and Sunquist, 1975). The confused systematics of the genus *Bradypus*, and the poor host records given in the literature make it very difficult to assign arthropod species to particular *Bradypus* species and therefore, as yet, little can be said about host specificity.

THE BLOODSUCKING GUILD

Insects.-- Most information on hematophagous associates of sloths has come indirectly through studies of sloths as reservoirs for human diseases. Thus for Diptera, the presence of *Leishmania brasiliensis* and other protozoan parasites in sloths (Herrer and Telford, 1969; Christensen and Herrer, 1976) implicates their vectors, psychodid flies, as sloth parasites. In Panama, *Phlebotomus sanguinarius* Fairchild and Hertig and *P. trapidoi* Fairchild and Hertig (Diptera, Psychodidae) have been recorded feeding on both *Bradypus* and *Choloepus* (Thatcher and Hertig, 1966). A virus isolated from the mosquito *Aedes serratus* (Theob.) (Diptera, Culicidae) in Brazil has also been found there in *Bradypus tridactylus*, suggesting the mosquitoes also commonly attack sloths (Pinheiro et al., 1962). Tempelis and Galindo (1970) found that several species of the mosquito genus *Deinocerites* will attack *Bradypus* in Panama, and Shaw (1964; 1969) observed mosquitoes and Tabanidae attacking *Choloepus* and *Bradypus*. Seymour (1985) summarizes the results of field and laboratory studies which show that sloths are maintenance hosts of several important mosquito-borne virus diseases, including Saint Louis encephalitis and Venezuelan encephalitis (found particularly in two-toed sloths, *Choloepus*). He also reports several arbovirus diseases found in sloths which are vectored by *Culicoides* midges (Diptera, Ceratopogonidae). Sloths are unusual virus hosts, in that they show a high incidence of intense infections with remarkably high amounts of virus circulating in their blood for prolonged periods, but they show no symptomatic indications of many of the diseases (Seymour, 1985).

Sloths may also serve as hosts to bloodsucking Hemiptera. A variety of triatomine bugs and the lygaeid *Clerada apicicornis* Signoret have been fed in the laboratory on *Choloepus didactylus* (see Christensen and Herrer, 1976, for review). The presence of *Trypanosoma cruzi* in *Bradypus* from Panama (Pipkin, 1968) suggests that its triatomine vectors are parasites of sloths in the field. Miller (1971) mentions the triatomine *Belminus rugulosus* Stål as a possible sloth specialist, though very little is known of its biology. Specimens of *Belminus* have been collected from tree trunks in Panama (P. Galindo, pers. comm.).

To the best of our knowledge, no sucking lice (Anoplura) have been recorded from sloths. The mallophagan louse *Lymeon gastroides* (Cummings) has been found attached to the hair of *Choloepus* in Guiana and Brazil (Emerson and Price, 1975; Werneck, 1950; Ward, 1953). Sloths seem to have no closely associated Siphonaptera; however, the flea *Polygenis atopus*

Guimarães, which feeds primarily on rodents and marsupials, has been recorded from *Bradypus infuscatus* in Venezuela (Tipton and Machado-Allison, 1972).

Ticks.-- Sloths are hosts to a number of tick species (Ixodidae). Up to six species of ticks of the genus *Amblyomma* have been reported from two-toed and three-toed sloths in Panama, Venezuela, Guiana and Brazil (Aragão, 1936; Floch and Fauran, 1958; Aragão and Fonseca, 1961; Fairchild et al., 1966; Jones et al., 1972; this study). Only two of these, *Amblyomma geayi* Neumann and *A. varium* Koch may be considered true sloth specialists as they are invariably found on sloths of both genera and rarely found on other hosts. The other amblyommid ticks recorded from sloths, *A. auriolatum* Pallas, *A. cajennense* (Fab.), *A. calcaratum* Neumann and *A. oblongoguttatum* Koch, appear to infest sloths only accidentally, as they are rare in collections and known to have other primary hosts in the adult stages.

In contrast to other studies (Aragão, 1936; Fairchild et al., 1966), we found larval and nymphal stages of *A. geayi* in relatively large numbers on sloths from forests in the vicinity of Manaus (Fig. 1). This may indicate that this tick utilises only a single host species during its life cycle, in contrast to the more common three-host cycle of amblyommid ticks (Hoogstrall, 1973). The unequal frequencies of larvae, nymphs and adults may indicate the use of alternate hosts in some stages, but more likely reflects different "frequencies of encounter" for different stages, a function of the relative abundance of each stage (larvae > nymphs > adults) and the time spent by each stage on the host (adults >> nymphs ⩾ larvae). Nothing is yet known about the lengths of feeding bouts of larval or nymphal *A. geayi*. However, if they are similar to other amblyommids (for example, *A. cajennense*), the larvae feed for about 3-6 days and the nymphs 5-7 days before they become engorged and leave the host (Flechtmann, 1973). Similar feeding times have been recorded for *A. striatum* by Fonseca (1935). We found that adult male *A. geayi* may stay for over three weeks on a sloth. In contrast, females must leave the sloths after engorging to oviposit, and this may be why they are found in much lower numbers relative to males (females/males = 0.28). Sloth ticks are among the largest of the ixodids; a single female of *A. varium* was found to contain 12,954 eggs (Rohr, 1909), and engorged females may weigh up to 7.5 g (Floch and Fauran, 1958; Fairchild et al., 1966).

At INPA in Manaus, the frequency of infestation of *Bradypus* by *Amblyomma* spp. was 99.0 percent (102 sloths), while for *Choloepus* it was 86.7 percent (15 sloths). No ticks were found on sloths collected at Ilha Curari; tick life cycles may have been interrupted by the seasonal inundation there. The tick load of *Choloepus* was about 50 percent that of *Bradypus*, with mean

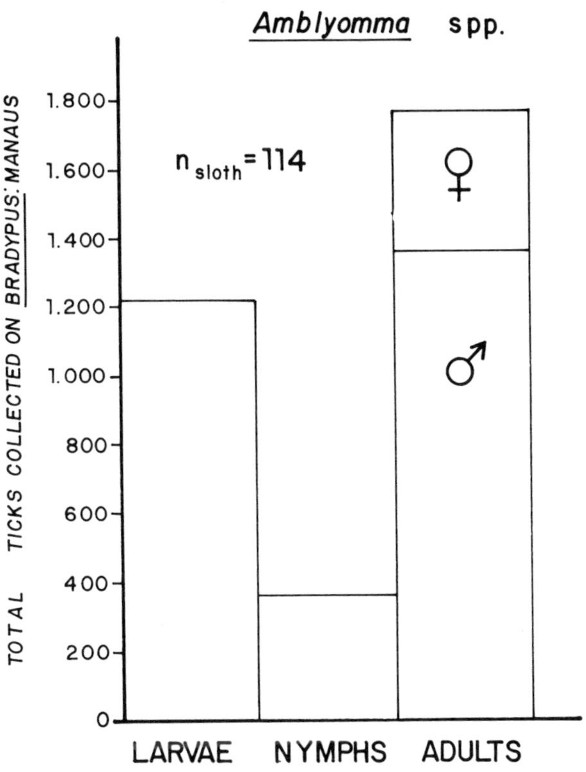

Figure 1. The relative abundance of the three life stages of ixodid ticks (*Amblyomma geayi* and *A. varium*: combined data) found on three-toed sloths *Bradypus tridactylus* at Manaus. The large number of larvae probably reflects their strongly clumped distribution shortly after hatching, while the greater proportion of adult males to females in the sample is probably due to the females leaving the sloths to oviposit after engorging and being fertilized by the less dispersive males.

numbers of ticks per sloth being 16 and 33 respectively; ticks were primarily *A. geayi*, with only occasional individuals of *A. varium* appearing in the collections. There were no significant differences in the number of ticks on male and female *Bradypus* (mean ticks/male sloth = 34.7 ± 33.7; mean ticks/female sloth = 32.9 ± 36.1), indicating that ticks are impartial to the sex of their host.

There was no apparent relationship between numbers of ticks of any life-stage per sloth and the seasonal difference in rainfall at Manaus (Fig. 2).

Nothing is known of the host-finding behavior of either *A. geayi* or *A. varium*. As they are rarely encountered other than on sloths, it must be assumed that their life cycle is intricately linked with the habits of their hosts. Home ranges of sloths are relatively restricted, being about 1.6 ha for *Bradypus* and 2.0 ha for *Choloepus* in Panamanian forests (Montgomery and Sunquist, 1975). Within these home ranges, eight or less trees may account for over 50 percent of the sloth's usage of the

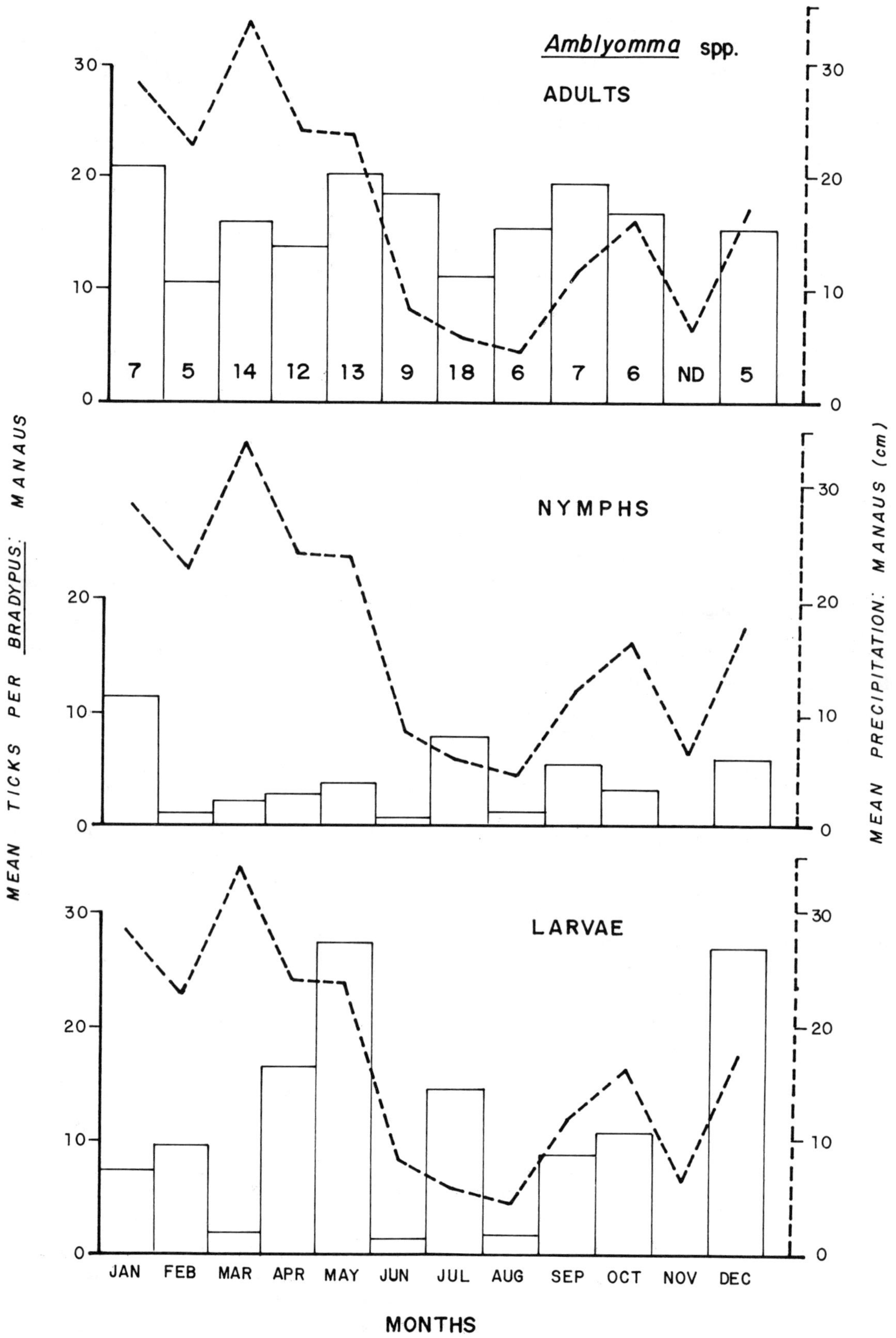

Figure 2. The mean numbers of ticks collected per month (**histograms**) showed no apparent correlation with the mean monthly precipitation in the Manaus area (**dashed line**). The numbers within each histogram represent the number of infested sloths used in calculating the monthly mean in this and the following figures. **ND** signifies that no data are available for that month.

area. Under these conditions, any life stage of the tick dropping off the host after engorging would have a relatively high chance of encountering the same host again. As sloths have overlapping home ranges (Montgomery and Sunquist, 1975), a potential exists for the ticks to change hosts. Engorged ticks dropping off the host from rarely used trees would likely be lost from the population, as would ticks dropping to the water in inundated forests such as those on Ilha Curari.

THE COPROPHAGOUS GUILD

The assemblage of coprophagous moths, beetles and mites is more properly called commensal than parasitic, because there is no evidence that any of the species feed on blood or other living tissues. Thus they do not have the potential to affect host fitness as might the blood-sucking guild, though the effect of either guild on sloth fitness has yet to be examined. All of these commensal associates exhibit an association with sloth excrement, and it is therefore appropriate to describe first the unusual defecatory habits of the host sloths.

While sloths spend the great majority of their time in the forest canopy, there is strong evidence that most defecation occurs on the forest floor, three-toed sloths descending from the trees about once each week apparently for this sole purpose (Montgomery and Sunquist, 1975) and two-toed sloths at invervals of about 4 days (Eisenberg and Maliniak, 1985; Merritt, 1985). Sloth dung takes the form of small, dense, angular or oval pellets, about 8-12 mm in diameter, roughly a cupful of these pellets is deposited in a pile with every defecation; the average dry weight of the feces passed each defecation by a three-toed sloth is 56 gm (Montgomery and Sunquist, 1975). Pellets of the dung of two-toed sloths are larger and more ovoid in shape than pellets of three-toed sloth dung.

Piles of dung are commonly found beneath leaf litter at the base of trees, particularly where large vines from the canopy curve sharply as they meet the ground. Sometimes several dung piles of varying age are found in such localities, suggesting repeated or common use of defecation sites. While defecation on the ground is thought to be the rule for sloths, there are known instances of sloths defecating from trees (Hoke, 1976). Montgomery (unpublished data) observed and photographed a three-toed sloth while it descended to the forest floor, dug a shallow depression with its stubby tail, defecated and urinated into the depression, then covered the feces with leaves by moving its hind limbs in a sweeping motion as it climbed the tree. Montgomery and Sunquist (1975) used miniature radio transmitters to measure the intervals between defecations by individual three-toed sloths, and to estimate the amounts of feces passed per defecation. A transmitter was inserted into a sloth's rectum after each defecation, and passed with the feces when the sloth next defecated. Piles of dung were then radio-located on the forest floor. In all 13 instances during the study when a radio-marked sloth defecated, the sloth descended from the forest canopy, made a small depression with its tail in the leaf litter and defecated in the depression. None of the five sloths used in the study defecated in a place where old sloth dung was present.

The contribution, if any, of defecation on the ground to sloth fitness is difficult to understand. A sanitary function does not seem likely, as defecation from the treetops would be just as effective. By burying dung, sometimes repeatedly in the same place, sloths may prevent ground predators from using the distribution of dung to determine the distribution of sloths overhead. On the other hand, while descending and defecating on the ground, sloths expose themselves to high risks of predation from those very same predators. Montgomery and Sunquist (1975) speculated that sloths may fertilize the trees they commonly use, and thus pay back part of the cost to the tree of having its leaves cropped. However, much more needs to be known about this habit before its importance to sloth biology can be determined. Whatever its function, defecation on the ground is crucial to the life histories of the commensal associates discussed below, and probably has played an important role in their evolution.

The Sloth Moths.-- Certainly the most extraordinary commensal associates of sloths are the sloth moths. This name has been given to a number of species of pyralids, in the subfamily Chrysauginae, which live as adults in the fur of tree sloths. *Bradypus* spp. appear to be the major hosts for these insects, and individual sloths sometimes carry in excess of 120 moths. Low infestations of moths are occasionally found on *Choloepus*, but no moth species has yet been found which is more common on this host than on *Bradypus*.

The first sloth moths may have been collected by Henry Bates during his famous travels on the Amazon (Westwood, 1877), but the species he obtained from the fur of *Bradypus* was not described until 1906. The description of this moth, *Bradypodicola hahneli* (Spuler, 1906), from the Guianas and later Brazil (Spuler, 1906; see also Costa Lima, 1950; Viette, 1954) was followed by a description of another species, *Cryptoses choloepi*, from *Choloepus* (and later *Bradypus*) in Panama (Dyar, 1908a; 1908b; 1912). Ihering (1914) described a third species, *Bradypophila garbei*, from *Bradypus marmoratus* in southern Brazil. Wetzel (1985) considers *B. marmoratus* to be *B. variegatus*. Recent collections in and near Manaus, Brazil, have turned up two new species of *Cryptoses*, *C. waagei* Bradley, which resembles *C. choloepi* and the other, *C. rufipictus* Bradley possessing a striking sexual dimorphism in wing venation and pattern (Bradley, 1982). While the early species descriptions suggested that different moth genera might occupy different parts of the range of *Bradypus*, the recent Brazilian studies reveal considerable sympatry among species and even the coexistence of several species locally on one individual host sloth. We have not found moths on *Choloepus* in Brazil.

From early collections, nothing was learned about sloth moths save that the adults could be found in the dense fur of hosts shot from trees. It is surprising, therefore, that by 1926 we can read in Hering's *Biologie der Schmetterlinge* of the "peculiar pyralids, *Bradypodicola hahneli* Spuler, which live as larvae and adults in the fur of sloths in South America" (our translation). Variations on this entirely fanciful larval lifestyle persist to the present day (Attenborough, 1979). Doubt concerning the habits of the larval stages arose when Casebeer and Hogue (1966), examining *Cryptoses choloepi* on *Bradypus infuscatus* in Costa Rica, found no traces of larval or pupal stage in the fur of sloths heavily infested with adult moths. The true life history of *Cryptoses choloepi* was elucidated by Waage and Montgomery (1976), who found that the larval stage is spent in the sloth dung.

Gravid female moths presumably leave the sloth during defecation and oviposit on the fresh dung, laying their eggs singly on the pellets. The early instar larvae feed on the surface of these pellets, covered by a light silken web. In later stages these webs are modified into long silken tubes, encrusted with frass, in which the larvae feed and, finally, pupate. Newly emerged moths are often caught at light-traps (Wolda, 1985), presumably on their way to the forest canopy to locate a sloth. While the sex ratio of *Cryptoses choloepi* at emergence is about 1:1, adult males on *Bradypus* outnumber females by about 3:1. This suggests that females are less tenacious in their sloth association than males, possibly being lost from the sloth-bound population during oviposition, when the host sloth ascends to the forest canopy.

For female *Cryptoses choloepi*, association with sloths appears to facilitate the location of fresh, unexploited, oviposition sites. Hence, a phoretic relationship is implied. There is no evidence that *Cryptoses* feeds on the sloth, though it is possible that it might take some nourishment from nasal secretions or simply from rainwater trapped in the host's fur. Protection from predation might be another benefit of the relationship, a hypothesis supported by the cryptically striped wing pattern of *Cryptoses choloepi* which renders it difficult to see against a background of coarse hairs.

The advantage to male *Cryptoses* of association with sloths must be a somewhat different one. Their high proportions on the host may reflect its use as a form of lekking ground, where males accumulate to compete for arriving females. Females with the highest potential fitness are clearly those on sloths, as they are the individuals most likely to secure resources (sloth dung) for their larvae. Males would therefore be maximizing their own fitness if they remain on the sloth and mate with as many of these females as possible.

The life history of other sloth moths is less well known. In Brazil, *Bradypodicola hahneli* and *Cryptoses rufipictus* collected near Manaus have been reared on sloth dung and exhibit larval habits very much like those of *Cryptoses choloepi*. The larval stages of *Cryptoses waagei* from Manaus are unknown, but are likely to be similar to *Cryptoses choloepi*. Larvae of *Bradypophila garbei* have not been collected, although a closely related species, probably belonging to the same genus (J.D. Bradley, pers. comm.), is known to develop under bark and in soil in Costa Rica (Becker, 1974).

The adult behavior of the two new sloth moths collected from Manaus, *Cryptoses waagei* and *C. rufipictus*, appears to be similar to that of *Cryptoses choloepi*. Both species live on the surface of the sloth's fur, running or flying when disturbed. Occasionally they will burrow into the fur of the host. In contrast to these species, *Bradypodicola hahneli* is characteristically found deep within the fur of the sloth, and is capable of extremely rapid movement in this environment. Most adults of *Bradypodicola hahneli* have truncated wings and possess neither the ability nor the disposition to fly. Moths eclose from dung piles, however, with fully developed wings and fly readily. Apparently, once the moths have settled on a sloth their wings are broken by abrasion, such that only the basal half of the forewing remains. This change bears interesting parallels to the blood-sucking hippoboscid fly *Lipoptena cervis* L., which sheds its wings soon after settling on its deer host (Edwards et al., 1939). The sex ratio of *B. hahneli* from *Bradypus* (1.4:1, n = 19) appears to be closer to unity than that of *C. choloepi*, as does that of *Cryptoses rufipictus* (0.78:1, n = 69), while the sex ratio of *Cryptoses waagei* is skewed towards males (2.9:1, n = 27). Specimens of *B. hahneli* are often collected *in copula* from the host, unlike the other sloth moth species which, to our knowledge, have never been found mating.

Feeding by sloth moths other than *C. choloepi* has not been observed, and it is not known whether these species feed at all as adults. The tongue and labial palpi of *Cryptoses* spp. are well developed and appear functional, whereas in *Bradypodicola* these structures are reduced; the tongue is shorter and apparently weaker, but possibly still functional (J.D. Bradley, pers. comm.). Both males and females of *Bradypodicola* display a marked physogastry (swelling of the abdomen) which may reflect the intake of some nutrient; a similar physogastry is found in the fly *Carnus hemapterus* Nitzsch., which lives on birds and feeds on quill gland secretions (Bequaert, 1942; Capelle and Whitworth, 1973).

Great variation in local population size has been

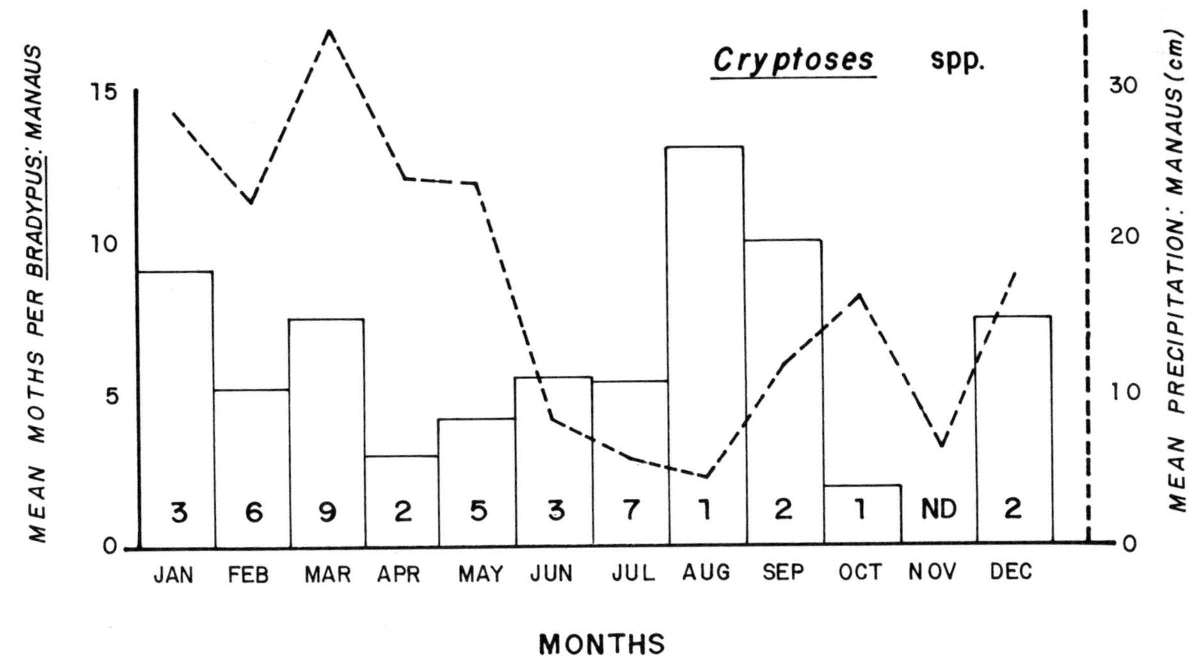

Figure 3. The mean numbers of *Cryptoses* spp., captured from three-toed sloths *Bradypus tridactylus* at Manaus were variable and showed no apparent correlation with amount of precipitation. This may indicate that the sloth feces, buried in the litter at the forest floor, may remain moist enough to allow the survival of larvae in the dung even during the short dry season.

observed for sloth moths. In Panama, (Waage and Montgomery, 1976) the numbers of *C. choloepi* collected on sloths in July and August 1974 ranged from 19 to 132 individuals with a mean of 49 per sloth (n = 7). In collections made in Manaus from 1977 to 1979, the numbers of *Cryptoses* spp. ranged from 0 to 24, with a mean of 6.1 per sloth (n = 42). Specimens of *B. hahneli* were rarely encountered on sloths in Manaus, but on the nearby Ilha Curari they were encountered at an average frequency of 1.9 per sloth with a maximum of 7 on any individual. Two sloths collected on terra firme near the Rio Negro had 8 and 11 *B. hahneli* each.

The population sizes and dynamics of sloth moths are most certainly dependent on the availability of sloth dung and, implicity, on sloth density. Seasonal factors such as rainfall may also have secondary effects on sloth moth populations; however, our data from Manaus (Fig. 3) show no clear trends. The apparent low seasonal variation in the numbers of *Cryptoses rufipictus* and *C. waagei* may indicate that the sloth feces, buried in the litter of the forest floor, may maintain a stable microclimate that allows the larval stages of the moths to survive even in dry season. In the very different environment of Ilha Curari, where the forest floor is continuously flooded during a four-month period, the populations of *Bradypodicola* seem to rise slightly during the nonflooded season (Fig. 4), and then slowly decline while the forest is inundated. These apparent seasonal changes must be viewed with caution, as there are relatively few data available on which to base the interpretations.

The most plausible hypothesis for the evolution of the unusual life history of the sloth moths is that the phoretic association of the adult is secondary. A variety of Lepidoptera have dung-feeding larval stages (Waage, 1979) and, indeed, larvae of a common tineid moth of stored products, *Tinea setomorpha rutella* Zeller, have been recovered from sloth dung at Ilha Curari, where they form frass tubes very similar to those of the pyralid sloth moths. Occasional phoretic behavior in a coprophagous insect would increase the speed of finding fresh, unexploited, larval resources, and thereby increase the number of offspring it leaves. Thus, under conditions where phoresis is possible, we would expect selection for it. Following the establishment of phoretic behavior, the close association of adult insect and vertebrate host might facilitate the evolution of further adaptations, such as feeding on the host. The flightless adults of *Bradypodicola* might represent a stage in the development of more intimate moth/sloth associations.

Sloth Beetles.-- The adults of a number of species of scarab beetles are commonly found in the fur of *Bradypus* spp. The small (3 mm) coprine *Trichillum bradyporum* was described by Boucomont (1928) from specimens collected from inside the anus of a sloth in Costa Rica. A member of this genus, *Trichillum adisi* (Ratcliffe, 1980), has been found in abundance on

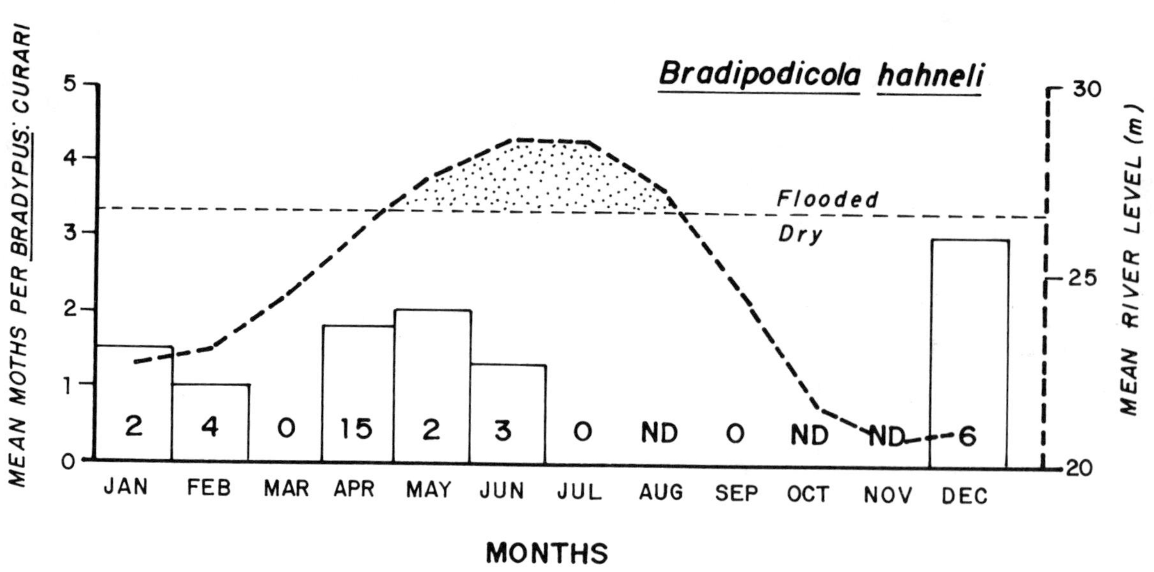

Figure 4. The monthly collections of the sloth moth *Bradipodicola hahnlei* from three-toed sloths *Bradypus variegatus* at Ilha Curari showing variations in numbers throughout the year. **Dashed lines** indicate the mean river level **(m)**.

sloths in the central Amazon during our studies. We have collected more than 980 individuals from a single sloth from Ilha Curari. These small beetles are usually found near the elbows or on the flanks behind the knees of the sloth, buried deep within the fur. They do not appear to be very active on the host, and we have never observed them on the surface of the fur except when they have been disturbed. Balthasar (1939) confirms the presence of *T. bradyporum* in the anal region of sloths from Costa Rica, while Montgomery (pers. comm.) has found an unidentified beetle in the feces taken directly from the rectum of sloths in Panama.

Another genus of sloth beetles, *Uroxys*, has been recorded from sloths in Panama, Brazil, Bolivia, and Colombia. The first described species was *Uroxys gorgon* from the fur of the head and neck of a *Bradypus* shot in Colombia (Arrow, 1933). We found an additional two species of this genus, *Uroxys* sp. nr. *batesi* Harald and *Uroxys besti* (Ratcliffe, 1980), the latter occuring relatively frequently in the thick fur of the shoulders and flanks of *Bradypus* in the Manaus area. Halffter and Matthews (1966) refer to a species of *Uroxys* recorded from *Bradypus* in Bolivia and G.G. Montgomery recently collected a *Uroxys*-like beetle from a *B. variegatus* in Panama. No beetles have been recorded from *Choloepus*.

Many coprine scarabs feed as larvae in animal dung, which suggests that sloth beetles, like their moth counterparts, might be phoretic coprophages. Balthasar (1939) suggests that *Trichillum* feed on dung about the anus of the sloth as adults, but does not suggest a larval diet for these beetles. Our studies with *Trichillum adisi* in Manaus have demonstrated that the larval stages of this beetle are spent inside the dung pellets of the sloth, usually one or two per pellet, and pupation occurs there also. *Uroxys besti* from Manaus, too large to develop in a single dung pellet, has been observed to tear up and bury pellets in burrows. Adults can be maintained for weeks on such dung, upon which they presumably feed. The burrows may be only for feeding and do not necessarily indicate the suitability of sloth feces as a larval medium for *Uroxys*.

The specialized host association of sloth beetles has been substantiated by dung-trap studies in the Manaus area, in which sloth beetles were never caught in traps baited with rotten fish and human feces (B. Ratcliffe, pers. comm.), and by studies in Panama with traps baited with sloth dung, which caught no dung beetles of any species (O. Young, pers. comm.).

Interestingly, on Barro Colorado Island in Panama, sloth dung has been found which contained very high densities of the larvae of a *Trichillum* and of another beetle, *Aphodius* sp. (identified by B. Ratcliffe), but adults of the latter species have not been collected on the sloths (J.K. Waage, pers. obs.). Adults of several other scarabaeids have been collected in sloth dung on Ilha Curari. This may be an abnormal situation as the lack of leaf litter on the island means that the feces are exposed and more accessable to other species of dung beetles.

Our studies represent the first attempts to understand the biology and demography of the sloth beetles. The variation in the numbers of *Trichillum adisi* on sloth populations at Ilha Curari and Manaus were considerable, with an average of 59.2 per sloth and a maximum of over 978 per sloth on Ilha Curari (n = 49 sloths), and an average of 4.3 per sloth and a maximum of 49 per sloth in Manaus (n = 26 sloths). In contrast, *Uroxys besti* was not encountered on Ilha Curari, but occurred

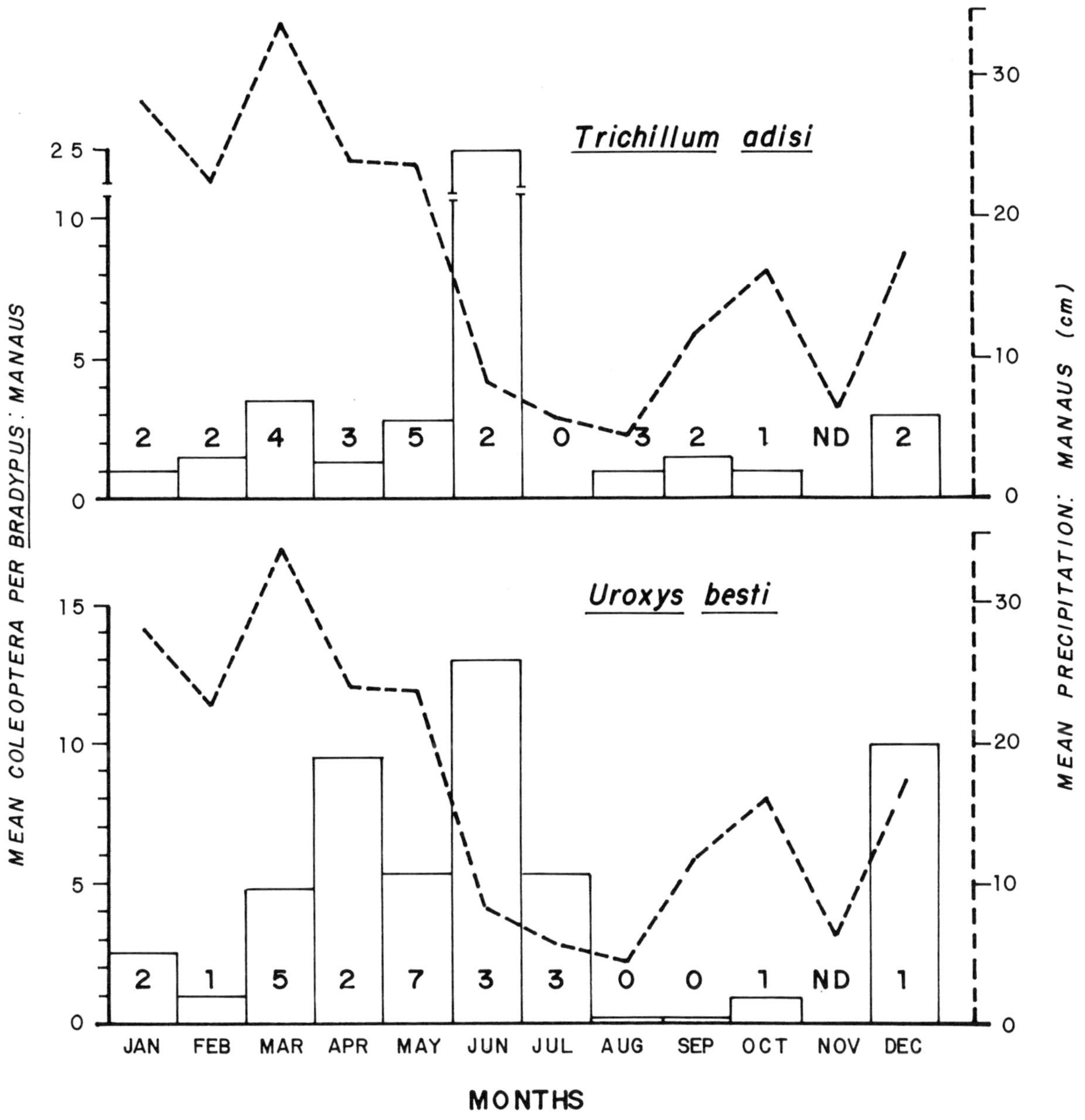

Figure 5. Monthly numbers of the sloth beetles *Trichillum adisi* and *Uroxys besti* collected at Manaus.

on sloths in Manaus at an average density of 6.1 beetles per sloth, with a maximum of 37 per sloth. *Uroxys* sp. nr. *batesi* was not recorded in Manaus, but two individuals were found on a *Bradypus* caught 54 km north from Manaus on highway BR 174, and another collected from a *Bradypus* on the Rio Negro near Tapuruquara, 700 km northwest from Manaus. These are the first records of this species associated with a sloth; the sloths in question also carried several specimens of *Uroxys besti*.

Seasonal factors affecting sloth beetle populations are unknown. The numbers of *Uroxys besti* and *Trichillum adisi* collected each month showed no evidence of correlation with rainfall, although both species showed a similar, but unexplained, peak in numbers during June (Fig. 5). In contrast, the Ilha Curari population of *Trichillum adisi* showed a strongly defined seasonality in response to the annual flooding of the island (Fig. 6). Peak beetle densities (average up to 110 beetles per sloth) occurred during the nonflooded season, while beetles were rarely found during the season of inundation. This may indicate that the feces necessary to complete the beetles' life cycle are absent because they are deposited into the water and washed away in the current. Apparently at least some of the adult beetles survive the season of inundation, or there is recruitment

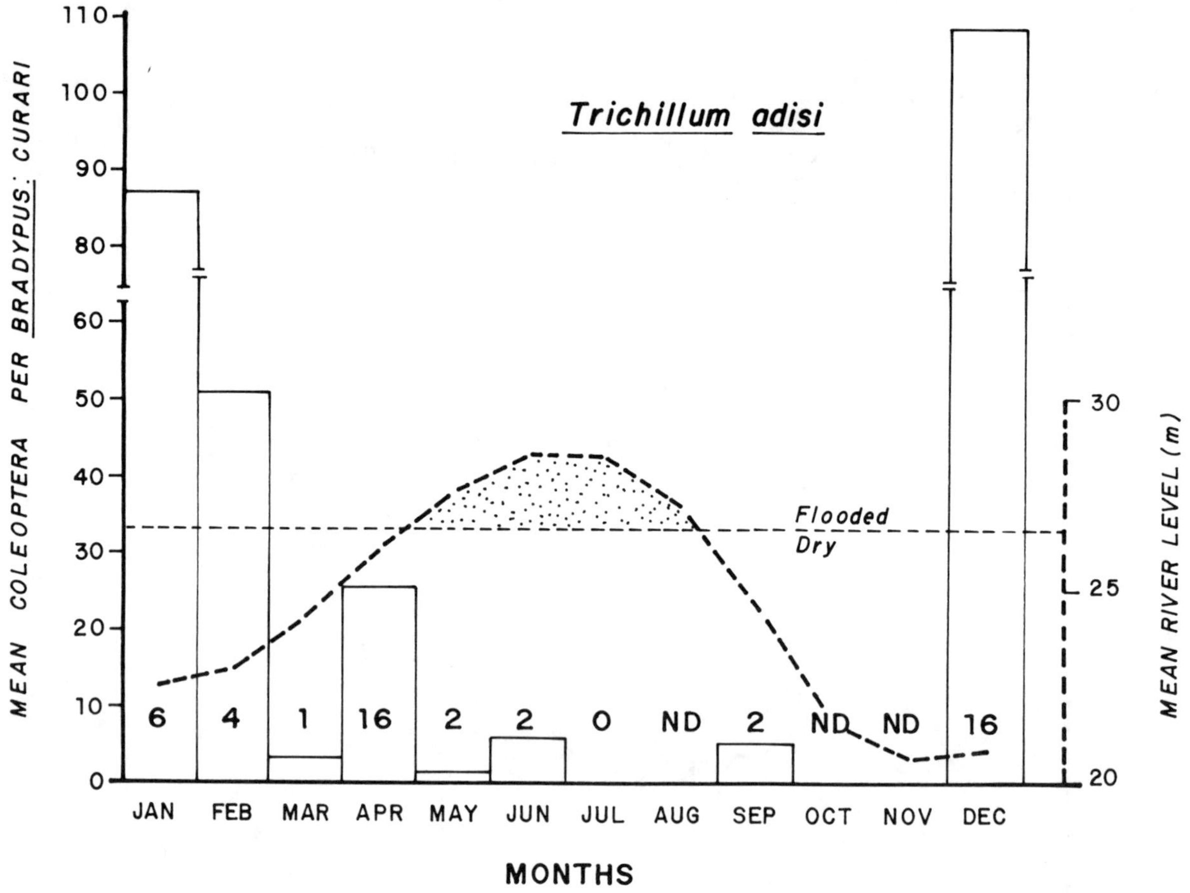

Figure 6. The monthly collections of *Trichillum adisi* beetles from *B. variegatus* sloths at Ilha Curari. Because the sloths must defecate into the water during the season of flooding, the food resource of the beetles becomes unavailable, and recruitment into the beetle population is interrupted.

from *terra firme* beetle populations. Thus adults are available to sustain the population in this seasonally unfavorable environment.

Sloth beetles are but one of a variety of coleopterous associates of mammals (Waage, 1979). Most beetles which live in mammal fur are associated with nests and nest detritus as larvae. However, an interesting parallel to the beetle/sloth association is found in species of *Onthophagus* (Scarabaeidae) which infest the fur of kangaroos, *Macropus,* in Australia, and develop as larvae in dung (Matthews, 1972). Beetles of these species aggregate in the anal region of their host, leaving immediately upon defecation to prepare and bury dung balls. It is likely that the wide dispersion of dung-producing kangaroos and the rapid dessication of dung in the arid habitats they frequent has created strong pressures for the evolution of this phoretic relationship. The claws of these beetles are highly specialized for clinging to the fur of their host species. It has also been suggested (Arndt, in Norris, 1970) that beetles might feed on the dermis of kangaroos, indicative of an even more intimate association. Comparable evidence of morphological or feeding specializations has not been found for sloth beetles. In the neotropics occasional records of associations between scarabaeids (e. g., *Glaphyrocanthon* spp.) and howler monkeys *Alouatta* might indicate yet another dung beetle/mammal association (Halffter and Matthews, 1966; O. Young, pers. comm.), but this remains to be investigated.

Sloth Mites.-- Apart from the original descriptions of a number of mites specific to sloths, little work has been done on the biology of these arthropods. Fonseca (1935) described the females and protonymphs of the bloodsucking *Liponissus inheringi* (Acarina, Liponissidae) from *Bradypus tridactylus* in southern Brazil and, later, *Lobalges trouessarti* (Sarcoptiformes, Epidermoptidae) (Fonseca, 1954a) and *Edentalges bradypus* (Sarcoptiformes, Psoroptidae) (Fonseca, 1954b) from the same host species. A second species of Psoroptidae *Edentalges choloepi* was described by Fain (1964) from *Choloepus didactylus* and has also been recorded from Surinam (Fain, 1965; Fain and Lukoschus, 1971). We have also found on *Bradypus* small parasitic mites of the suborder Acaridida which are as yet unidentified.

We commonly found mites of a complex of three species of macrochelid acarines (*Macrocheles*; Krantz, 1983) on three-toed sloths from both Ilha Curari and Manaus sloth populations. We found mites on 52.4 percent of a sample of 96 sloths examined; some sloths may have had mites which could have been found with a more detailed examination. These mites were most commonly found in the anal region of the sloth, running deep in the fur near the anus, or in large numbers at the anus and inside the rectum of the sloth. With the aid of an otalaryngoscope, the mites were observed inside the anus, where they appeared to remain primarily on the fecal pellets. They were not observed attached to the sloth as if feeding, and it seems likely that these mites are either predatory or coprophilous, as are other members of the genus (Evans and Hyatt, 1963; G.W. Krantz, pers. comm.).

Even more interesting is the apparent relationship between two of the three species of mites *Macrocheles impae* and *M. uroxys*, and the sloth beetles *Uroxys besti* and *Trichillum adisi* on which they were often found. A close association has been shown between certain scarab beetles and macrochelid mites (Evans and Hyatt, 1963; Halffter and Matthews, 1966), and the mites and beetles that live on sloths seem to have a similar relationship. We found only adult females of these two species of macrochelid mite on the beetles, principally on *Uroxys besti* but occasionally also on *Trichillum adisi*. In Panama, similar mites have also been found in apparent phoretic association with sloth moths, *Cryptoses choloepi* (Waage, pers. obs). The role of the beetles, and perhaps the moth, in the mite's life history would seem to be one of transporting adult females from a diminishing dung pile to another sloth, which in turn would produce a fresh dung pile when it defecated. As a result of the beetle being phoretic on the sloth and the sloth being the source of food (dung) for the mite, these two mite species could have become secondarily phoretic on the sloth, and thus able to utilize dung as it accumulates in the cloaca prior to periodic defecation. However, the finding of only adult female mites adhering to the beetles may indicate that only females which have already been fecundated are transported from a dung pile, and that the beetles may act solely as a means for fecund females to emigrate to a fresh dung source. Evans and Hyatt (1963) suggest that the phoretic females of this genus may be faculatively parthenogenic, and Costa (1967) reports that larviviparity and oviviparity are about equally common and depend on environmental conditions. The third species *Macrocheles lukoschusi* has only been found on the sloths, and not in association with beetles. In this species adult males, adult females, and deuteronymphs were all abundant in the collections, implying that this species is primarily or completely dependent on the sloths and that the phoresis on the beetles has probably been completely eliminated.

CONCLUSIONS

The most striking character of the arthropod community on sloths is the large proportion of coprophagous species which appear to associate with the host primarily for phoresis. In other respects, sloths seem to host a typical arthropod fauna, with a notable absence of fleas and anoplurans, which are themselves rather discontinuously distributed across mammalian groups. The phoretic coprophage lifestyle is not unique to sloth associates, it is found as well in other Coleoptera, e.g., *Onthophagus*, in the large group of muscid flies associated with ungulates, and in many insects which are dependent as larvae on feces and other substances in animal nests (Waage, 1979). That phoretic coprophages are found on other vertebrates indicates that the uniqueness of sloths is not in their possession of such associates, but in the possession of such a great diversity of them. We might therefore ask what it is about sloths which makes them "targets" for the evolution of phoretic associations.

The evolution of any sort of phoresis requires (1) selection pressure for phoresis acting on the insect and (2) characteristics in the host facilitating the phoretic behavior. As regards the first criterion, arthropods using sloth dung exploit a larval food resource which is very patchily distributed in the forest. The relative lack of odor of *Bradypus* dung (*Choloepus* dung is more odoriferous and apparently less important to arthropods) and the fact that it is buried beneath leaf litter may make this resource more difficult to locate through olfaction than other sorts of dung. It is as yet unknown how restricted sloth commensals are to sloth dung. But, if the larval associations of these insects are as specific as the adult associations appear to be, considerable selection pressure must exist to evolve and maintain a phoretic relationship.

Clearly, however, many specific coprophages might benefit from phoresis if it increased the efficiency of locating dung for larval development. The potential for the evolution of phoresis may differ between coprophagous species as a result of differences in the probability of encounter with the potential host, differences in host pelage as a substrate for attachment, and differences in the host's grooming behavior. Grooming, for instance, is known to greatly limit the growth of ectoparasite populations on vertebrates (Waage, 1979), and may be even more effective against incipient phoretic species which are not adapted for attachment to the surface of the host.

From this perspective, the three-toed sloth *Bradypus* makes an ideal substrate for phoresis. Accidental association between coprophage and host may occur during the long defecation periods (several minutes), which often occur near sites of previous defecations. Insects can easily lodge in the dense fur coat of *Bradypus*, and grooming of this species is slow and relatively ineffective. During grooming by *Bradypus*, for instance, *Cryptoses* may be seen to advance slowly in a wave in front of the moving claws of the forefoot, disturbed, but by no means dislodged from the host. Perhaps the relative absence of moths and beetles on *Choloepus* reflects the thinner fur and more active habits of this sloth and the consequent higher risk to phoretic insects.

It is tempting to speculate, therefore, that the great diversity of phoretic coprophages on sloths is the evolutionary consequence of (1) strong selection for phoresis and (2) ideal conditons for successful transport on the xenarthran hosts. Once established, phoresis may lead to the development of close adult interaction with the host, including, ultimately, true parasitism. The biology of sloth commensals might provide, therefore, a glimpse along one of the evolutionary pathways by which free-living organisms become parasitic.

The complex guild of phoretic coprophages on sloths poses interesting problems of community structure and competition. In the forests around Manaus, this guild consists of nine species; *Cryptoses rufipictus, C. waagei, Bradypodicola hahneli, Trichillum adisi, Uroxys* sp. nr. *batesi, Uroxys besti*, and 3 species of *Macrocheles*, not to mention the species of tineid and scarabaeids collected from dung but not from sloths. Clearly, while there is little potential for interaction of adults of these species on sloths, with the exception of the positive relationship between mites and beetles, the potential for competition between larval stages is considerable. Dung piles have been recovered on Barro Colorado Island in Panama with a large proportion of pellets converted to moth frass and many of the remaining pellets infested by *Trichillum* and *Aphodius* larvae. However, not all sloth dung is broken down rapidly, and it may remain intact and recognizable for several months after being deposited on the forest floor (Montgomery and Sunquist, 1975). Perhaps not all dung is suitable for moth or beetle larvae because of the presence of phytochemicals or other substances (G.G. Montgomery, pers. comm.). It would appear that the composition of local phoretic coprophage communities and the population densities of the species present may be largely determined by intra- and interspecific competition for limited larval resources. These resources are, in turn, related to the local density of the sloth hosts and perhaps to the kind of leaf on which the sloths have been feeding.

Such intense competition for dung would serve to increase the selective pressure for more efficient means to locate dung. The discovery of the beetle *Trichillum bradyporum* inside the anus of a sloth (Boucomont, 1928) suggests that oviposition in this species may occur before defecation, thus giving the larval stages at least several days of incubation before entering into competition with other coprophages. The sloth moth *Bradypodicola hahneli* appears to have an exceptionally rapid incubation period from egg to larva (possibly less than 24 hours), and may in this way secure a head start for its larvae over the slower-developing *Cryptoses*. The mite *Macrocheles lukoschusi* appears to have foregone the need to wait for sloth defecation and, through entering and leaving the cloaca freely, it uses the stored feces as a food and larval resource without having to leave the sloth.

Further studies on the reproductive behaviour and ecology of this guild of sloth moths, beetles, and mites are necessary before the role of competition in community structure and dynamics can be fully understood.

ACKNOWLEDGMENTS

We thank the many people who helped with the sloth project at INPA, particularly Megumi Yamakoshi, José Cladiston dos Santos Queiroz, Astrogildo da Silva, Gilberto de Assis Ribeiro, Lucivaldo Sampaio da Rocha, Vera Maria F. da Silva, and Moira Sucharov. We also acknowledge valuable collaboration of Joachim Adis during the initial stages of the work. We especially thank Gene Montgomery, whose enthusiasm for sloths led us into the work and whose comments on the manuscript have much improved it. G.W. Krantz, J.E. Keirans, J.D. Bradley and B.C. Ratcliffe identified the mites, ticks, moths, and beetles, respectively. Finally, we thank Hazel Mackenzie for typing the manuscript and Jorge Dacio for preparing the figures. The work was supported by the National Research Council of Brasil (CNPq) and by a Sigma Xi grant to J.K. Waage.

LITERATURE CITED

Aragão, H. de B.
1936. Ixodidas Brasileiros e de alguns paises limitrophes. Mem. Inst. Oswaldo Cruz, 31(4): 759-843.

Aragão, H. de B. and F. da Fonseca.
1961. Notas de Ixodologia IX: O complexo ovale de genero *Amblyomma*. Mem. Inst. Oswaldo Cruz, 59(2): 131-148.

Arrow, G. H.
1933. The genus *Uroxys* (Coleoptera, Copridae), with descriptions of some new species. Ann. Mag. Nat. Hist., 11: 389-399.

Attenborough, D.
1979. Life on earth. A natural history. Collins Sons and Company, Ltd., London. 319 pp.

Balthasar, V.
1939. Monographie der Gattung *Trichillum* Har. Vest. csl. zool. Spol. Praze, 6-7: 11-26.

Becker, V.
1974. Studies on the shoot borer, *Hypsipyla grendella* (Zeller) (Lep., Pyralidae). XXVI. A new genus and three new species of Microlepidoptera (Pyralidae and Gracillariidae) associated with *Carapa, Cedrela* and *Swetenia* in Costa Rica. Turrialba, 24: 332-335.

Bequaert, J.
1942. *Carnus hemapterus* Nitzsch, an ectoparasitic fly of birds, new to America. Bull. Brooklyn Ent. Soc., 37: 140-149.

Boucomont, A.
1928. Coprophages d'Amerique de Sud nouveaux ou peu connus. Bull. Soc. Ent. Fr., 1928: 186-194.

Bradley, J. D.
1982. Two new species of moths (Lepidoptera, Pyralidae, Chrysauginae) associated with the three-toed sloth (*Bradypus* spp.) in South America. Acta Amazonica, 12: 649-656.

Capelle, K.J. and T.L. Whitworth.
1973. The distribution and avian hosts of *Carnus hemapterus* (Diptera: Milichiidae) in North America. J. Med. Entomol., 10: 525-526.

Casebeer, R.S. and C.L. Hogue.
1966. Hideaway for moths. Natural History, 75: 40-41.

Christensen, H.A. and A. Herrer.
1976. Neotropical sand flies (Diptera: Psychodidae), invertebrate hosts of *Endotrypanum schaudinni* (Kinetoplastida: Trypanosomatidae). J. Med. Entomol., 13: 299-303.

Costa, M.
1967. Notes on Macrochelids associated with manure and coprid beetles in Israel. II. Three new species of the *Macrocheles pesentii* complex with notes on their biology. Acarologia, 9: 304-329.

Costa Lima, A. da.
1950. Insetos do Brasil. Vol. 6 Lepidopteros. 2nd part. Esc. Nat. Agron., Serie Didatica No. 8, 420 p.

Dyar, H.G.
1908a. A pyralid inhabiting the fur of the living sloth. Proc. Ent. Soc. Wash., 9: 142-144.

1908b. A further note on the sloth moth. Proc. Ent. Soc. Washington, 10: 81-82.

1912. More about the sloth moth (Lepidoptera: Pyralidae). Proc. Ent. Soc. Washington, 14: 169-170.

Edwards, F.W., H. Oldroyd and J. Smart.
1939. British blood-sucking flies. William Clowes and Sons, Ltd., London. viii + 156 pp.

Eisenberg, J.F. and E. Maliniak.
1985. Maintenance and reproduction of the two-toed sloth *Choloepus didactylus* in captivity. (this volume).

Emerson, K.C. and R.D. Price.
1975. Mallophaga of Venezuelan mammals. Brigham Young Univ. Science Bull., Biol. Series, 20(3): 1-77.

Evans, G.O. and K.H. Hyatt.
1963. Mites of the genus *Macrocheles* Latr. (Mesostigmata) associated with Coprid beetles in the collections of the British Museum. Bull. Brit. Mus. (Nat. Hist.) Zool., 9(9): 327-401.

Fain, A.
1964. *Edentalges choloepi sp. n.* acarien parasite cuticole du paresseux didactyla *Choloepus didactylus* (L.) (Psoroptidae: Sarcoptiformes). Z. Parasitenk., 25: 103-107.

1965. Les Acariens producteurs de gale chez les Edentés et les Marsupiax (Psoroptidae e Lobalgidae, Sarcoptiformes). Bull. Inst. Royal Sci. Nat. Belg., 41: 1-42.

Fain, A. and F. Lukoschus.
1971. Parasitic mites of Suriname II: skin and fur mites of the families Psoroptidae and Lobalgidae. Acta. Zool. Pathol. Antverp, 51: 49-59.

Fairchild, G.B., G.M. Kohls and V.J. Tipton.
1966. The ticks of Panama. Pp 167-219, *in* Ectoparasites of Panama (R.L. Wenzel and V.J. Tipton, eds.). Field Museum of Natural History, Chicago. xii + 861 pp.

Flechtmann, C.H.
1973. Acaros de importancia medico-veterinária. Livraria Nobel S. A., São Paulo. 192 pp.

Floch, H. and P. Fauran.
1958. Ixodides de la Guayane et des Antilles Francaises. Arch. Inst. Pasteur Guyane Francais, 19(446): 1-94.

Fonseca, F. da.
1935. Notas de acariologia XIII: Novas especies sul-americanas de parasitos do genero *Liponissus* Kolenati, 1858 (Acarina, Liponissidae). Mems. Inst. Butantan, 10: 3-17.

1954a. Notas de acarologia XXXVIII: *Sarcoptiformes* da preguica; *Lobalges trovessarti* gen. n., sp. n. (Acari, Epidermoptidae). Mem. Inst. Butantan, 26: 85-92.

1954b. Notas de acarologia XXXIX: Sistematica e filogenese de *Psoralgidae* Oudemans, Sarcoptiformes parafagistas de mamiferos com morfologia de Acari plumicolas. Mem. Inst. Butantan, 26: 93-167.

Halffter, G. and E.G. Matthews.
1966. The natural history of dung beetles of the subfamily Scarabaeinae (Coleoptera:Scarabaeidae). Folia Ent. Mex., 12-14: 1-322.

Hering, M.
1926. Biologie der Schmetterlinge. Springer Verlag, Berlin. 480 pp.

Herrer, A. and S.R. Telford, Jr.
1969. *Leishmania braziliensis* isolated from sloths in Panama. Science, 164: 1419-1420.

Hoke, J.
1976. Discovering the world of the three-toed sloth. Franklin Watts, New York. 87 pp.

Hoogstrall, H.
1973. Acarina (ticks). Ch. 5, Pp 91-103, *in* Viruses and Invertebrates (A.J. Gibbs, ed.). North-Holland Publ. Co., Amsterdam, Netherlands. xvi + 673 pp.

Ihering, R. von.
1914. As traças que vivem sobre a preguiça, *Bradypophila garbei* n. gen., n. sp. (Lepid. Fam. Pyralid.) Rev. Mus. Paulista, 9: 123-127.

Irmler, U.
1975. Ecological studies of the aquatic soil invertebrates in three inundation forests of Central Amazonia. Amazoniana, 5: 337-409.

Jones, E.K., C.M. Clifford, J.E. Keirans and G.K. Kohls.
1972. The ticks of Venezuela (Acarina, Ixodoidea) with a key to the species of *Amblyomma* in the western hemisphere. Brigham Young Univ. Sci. Bull., Biol. Ser., 17(4): 1-40.

Jorge, W., A.T. Orsi-Souza and R. Best.
1985. The somatic chromosomes of xenarthra. (this volume).

Krantz, G.W.
1983. Three new species of *Macrocheles* (Acari: Gamasida: Macrochilidae) associated with 3-toed sloths, *Bradypus* spp. (Edentata: Bradypodidae) in Brazil and Suriname. Acarologia, 24: 3-12.

Matthews, E.G.
1972. A revision of the Scarabaeinae dung beetles of Australia. Aust. J. Zool. Suppl., 9: 1-330.

Meritt, D.A., Jr.
1985. The two-toed Hoffman's sloth, *Choloepus hoffmanni* Peters. (this volume).

Miller, N.C.E.
1971. The biology of the Heteroptera. E.W. Classey, Hampton, U.K. xiii +206 pp.

Montgomery, G.G., and M.E. Sunquist.
1975. Impact of sloths on neotropical forest energy flow and nutrient cycling. Pp. 69-98, *in* Tropical ecological systems: trends in terrestrial and aquatic research. (F.B. Golley and E. Medina, editors). Springer Verlag, New York. xiii +398 pp.

Norris, K.R.
1970. General biology. Insects of Australia. Melbourne University Press, Melbourne. xiii + 1029 pp.

Pinheiro, F., M. Pinheiro, G. Bensabath, O.R. Causey and R. Shope.
1962. Epidemia de virus oropouche em Belem. Rev. Serv. esp. Saúde públ., 12: 15-23.

Pipkin, A.C.
1968. Domiciliary reduviid bugs and the epidemiology of Chaga's disease in Panama (Hemiptera; Reduviidae: Triatominae). J. Med. Entomol., 5: 107-124.

Ratcliffe, B.C.
1980. New species of Coprini (Coleoptera: Scarabaeidae: Scarabaeinae) taken from the pelage of three-toed sloths, with a brief commentary on scarab-sloth relationships. Coleop. Bull., 34: 337-350.

Rohr, C.J.
1909. Estudos sobre Ixodidas do Brasil. Thesis Faculty of Medicine, Rio de Janeiro. 204 pp.

Shaw, J.J.
1964. A possible vector of *Endotrypanum schaudinni* of the sloth *Choloepus hoffmanni* in Panama. Nature, 201: 417-408.

1969. The haemoflagellates of sloths. London School of Hygiene and Tropical Medicine, Memoir 13, H.K. Lewis and Co., London, 132 pp.

Seymour, C.
1985. Sloths as hosts of arboviruses. (this volume).

Spuler, A.
1906. Uber einen parasitisch lebenden Schmetterlinge, *Bradypodicola hahneli* Spuler. Biol. Centralbl., 26: 690-697.

Tempelis, C.H. and P. Galindo.
1970. Feeding habits of five species of *Deinocerites* mosquitoes collected in Panama. J. Med. Entomol., 7: 175-179.

Thatcher, V.E. and M. Hertig.
1966. Field studies on the feeding habits and diurnal shelters of some *Phlebotomus* sandflies (Diptera; Psychodidae) in Panama. Ann. Ent. Soc. Amer., 59: 46-52.

Tipton, V.J., and C.E. Machado-Allison.
1972. Fleas of Venezuela. Brigham Young University Science Bulletin., Biol. Series 17: 1-115.

Viette, P.
1954. A propos de *Bradypodicola hahnlei* Spuler, trouvé sur un parasseux (Lep. Pyralidae). Bull. Soc. zool. France, 79: 448-451.

Waage, J.K.
1979. The evolution of insect/vertebrate associations. Biol. J. Linn. Soc., 12: 187-224.

Waage, J.K. and G.G. Montgomery.
1976. *Cryptoses choloepi*: a coprophagous moth that lives on a sloth. Science, 193: 157-158.

Ward, R.A.
1953. Brief notes on the Mallophaga. I. 2. The host of *Lymeon gastroides* (Cummings) (Trichodectidae). Entomological News, 64(8): 203-204.

Werneck, F.L.
1950. Os malofagos de mamiferos. II. Ischnocera (continuação de Trichodectidae) e Rhyncophthirina. Edição Inst. Oswaldo Cruz (Rio de Janeiro), 207 pp.

Westwood, J.O.
1877. Entomological Notes. Trans. R. Ent. Soc. Lond., p. 436.

Wetzel, R.M., and F.D. de Avila-Pires.
1980. Identification and distribution of the recent sloths of Brazil. Rev. Brasil. Biol., 40: 831-836.

Wetzel, R.M.
1985. The identification and distribution of recent xenarthra (= edentata). (this volume).

Wolda, H.
1985. Seasonal distribution of sloth moths (*Cryptoses choloepi* Dyar) (Pyralidae: Chrysauginae). (this volume).

SEASONAL DISTRIBUTION OF SLOTH MOTHS *CRYPTOSES CHOLOEPI* DYAR (PYRALIDAE; CHRYSAUGINAE) IN LIGHT TRAPS IN PANAMA

Henk Wolda

Smithsonian Tropical Research Institute, P.O. Box 2072, Balboa, Panama.

Abstract

On Barro Colorado Island, Panama, seasonality of captures of the sloth moth *Cryptoses choloepi* Dyar in light-traps is assumed to reflect real seasonality in the population. The early wet season maximum followed a suitable season for increase during the dry season and was followed by unsuitable conditions later in the rainy season. In Las Cumbres, Panama, the dry season was much more severe and this may have caused an increased activity of the sloth moths during January and February, resulting in an extra peak in the numbers entering light-traps. In Fortuna, a relatively nonseasonal site, the sloth moth showed changes in abundance that were also nonseasonal.

Abstracto

La variación estacional de las capturas de la polilla de perezosos *Cryptoses choloepi* Dyar en trampas de luz en la isla de Barro Colorado en Panamá, reflejan una verdadera variación estacional en la población de estas polillas. El máximo registrado durante la primera parte de la temporada lluviosa, siguió una temporada favorable al incremento de la población durante la estación seca, pero luego fué seguido por condiciones desfavorables más adentrada la estación lluviosa. En el área de las Cumbres, en Panamá, la estación seca fue mucho más severa y ésto puede haber sido la causa de un incremento de actividad de las polillas durante enero y febrero, lo cual resultó en un pico adicional en el número de polillas capturadas en las trampas de luz. En la Fortuna, que constituye un sitio sin diferencias estacionales, las polillas de los perezosos mostraron cambios en abundancia que tampoco pueden tildarse como cambios estacionales.

Resumo

Na ilha de Barro Colorado, Panamá, a sazonalidade de mariposas de preguiça *Cryptoses choloepi* Dyar nas armadilhas de luz, reflete sazonalidade real na população. O início da estação úmida máxima segue uma estação conveniente para aumentar durante a estação seca e é seguida por condições inconvenientes mais tarde na estação chuvosa. Em Las Cumbres, Panamá, a estação seca é muito mais severa e isso pode causar uma aumentada atividade das mariposas de preguiças durante janeiro-fevereiro, causando um pico extra no números que entram nas armadilhas de luz. Em Fortuna, um local não sazonal, a mariposa de preguiça apresenta mudanças na abundância que também não são sazonais.

Throughout the animal kingdom many instances are found of two or more species, often of different phyla, being closely associated in some form of parasitism, symbiosis, or commensalism. In many cases the nature of the interaction remains largely a mystery. This is, to a large extent, still the case for the pyralid moth *Cryptoses choloepi* Dyar, which lives in the fur of the three-toed sloth *Bradypus infuscatus* and the two-toed sloth *Choloepus hoffmannii* in Panama. There are other forms of sloth moth of the genus *Cryptoses* and of a second genus *Bradypophila* (Waage and Best, 1985); two or more moth species can coexist on the same host sloth. The sloth moth was assumed to be an ectoparasite (Grassé, 1951) and it has been stated that both the adult and the immature stages live on the sloth (Dyar, 1908; Askew, 1971). Casebeer and Hogue (1966) examined numerous sloths in Costa Rica and found only adult moths, with no evidence of eggs, larvae or pupae. Waage and Montgomery (1976) discovered that the larvae are not found on the sloths themselves, but on their dung.

Sloths of both species come down to the forest floor to defecate about once per week, the two-toed sloth leaving the feces on top of the litter, the three-toed sloth digging a shallow hole with its tail and covering the feces with litter (Montgomery and Sunquist, 1975). Mated females of *Cryptoses* presumably leave the defecating sloth to lay eggs on the dung and they, as well as newly emerged adults, have to fly to find a sloth.

It is assumed (Waage and Montgomery, 1976) that the advantages for the moths are (1) the females are directed by their hosts to suitable oviposition sites, (2) they may receive some protection from avian predators, and (3) they may find nutrients in secretions of the sloth skin and the algae which abound in the fur. These hypotheses still await testing.

The advantage to the sloth, if any, is not clear. In fact, the situation is rather peculiar. The feces of sloths are very resistent to decomposition and this has been assumed to be advantageous to the trees the sloths feed on and to the sloths themselves (Montgomery and Sunquist, 1975). If this is true, it is difficult to understand why the sloths should increase the decomposition rate of their feces by helping one of the few agents which can attack them, the larvae of the sloth moth, to get there.

The adults do not spend all their lives on the sloths. Young moths, freshly emerged from the sloth droppings, have to take to wing to find a suitable host (Waage and Best, 1985). Ovipositing females have to fly up to find a sloth again and may even at times fly off the sloth to find oviposition sites. Adults of both sexes may occasionally leave their host for feeding or some other activity. These flying moths are active at night and are captured in light-traps. Data on sloth moth abundance in light-traps in three localities in the Republic of Panama are presented here, with emphasis on the seasonal changes in abundance of moths in those traps.

An examination of the light-trap records of *Cryptoses* in Panama may provide some clues about the ecology and behavior of that species, although more detailed study of the living moths in their natural environment will still be needed.

LIGHT TRAPS AND LOCALITIES

The light-traps were of the Pennsylvania type (Wolda, 1980a) and sloth moths were sorted out and counted in samples from three sites: Las Cumbres, Barro Colorado Island (BCI), and Fortuna.

Las Cumbres is a residential area 10 km north of Panama City, 120 m above sea level, where a trap was operated all night, every night for a number of years (Wolda, 1980a). Sloth moths were sorted out from the insects trapped from July 1974 until March, 1977.

Barro Colorado Island in Gatun Lake, Panama, is covered with tropical forest (Croat, 1978). A set of two light-traps was installed on a ridge near trailmarker Snyder-Molino 3 in June, 1973, one near ground level, and one high in the canopy overlooking a large area of forest. These traps operated every night, all night (Wolda, 1977, 1982). In early 1977 (after week 9), the procedure of trapping was changed in that insects were no longer caught in a dry cloth bag where they were killed with carbon tetracholoride, but fell directly into Kahle's solution, a mixture of alchohol, formaldehyde and glacial acetic acid.

Fortuna refers to the dam site on the Rio Chiriquí in very wet old forest, 1050 m above sea level, in the mountains of western Panama (Adames, 1977; Wolda, 1978b, 1980). In late September 1976, the trap was placed in the crown of a large tree in the middle of good forest. Late in June 1977, the trap was lowered and left some 3 m above the forest floor. The trap was in operation there until early July 1979.

The climate in Las Cumbres is characterized by a strong dry season of 4-5 months from January-April. Annual rainfall averages 1500 mm. On BCI the rainfall pattern is similar, but it is generally wetter, with an annual rainfall of 2500 mm. Fortuna does not have a dry season, and the rainfall is reasonably even throughout the year. Mean annual rainfall is 4500 mm (Caballero, 1978).

RESULTS

The Sloth Moth In Traps On BCI.--There were some notable differences between the years in the seasonal pattern, but in general numbers of moths were very low during the first ten weeks of the year (January to mid March), the early part of the dry season (Fig. 1). They increased during the later part of the dry season and reached a peak at the start of the rainy season, around week 18 (mid May). The larger numbers caught

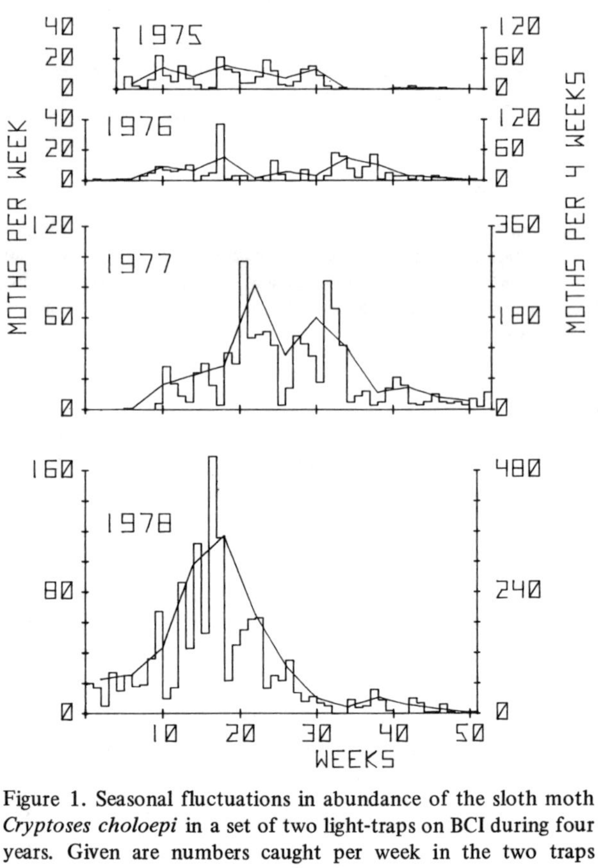

Figure 1. Seasonal fluctuations in abundance of the sloth moth *Cryptoses choloepi* in a set of two light-traps on BCI during four years. Given are numbers caught per week in the two traps (histogram, left ordinate) and per four weeks to eliminate effects of moon phase (continuous line, right ordinate).

in 1977 and 1978 could reflect a real increase in abundance as compared with the previous years, but may also have to do with the different repelling/attracting action of the chemicals used in the traps. In 1978 the peak was sharp and followed by a sudden decline; in the other years the peak was broad. In all years the numbers declined towards the end of the rainy season. The dry season of 1977 is the longest on record and the rains started late. In 1978 the rains started unusually early and this rainfall pattern undoubtedly caused the peak in *Cryptoses* abundance to be earlier in 1978 and later in 1977. Some mid dry-season rains which occurred in 1975 (Wolda, 1978a) may have affected the seasonal distribution of the sloth moth in that year.

In the dry season most moths were caught in the low trap, during the early rainy season most were caught in the canopy trap, and at the end of the year both traps caught about equal numbers (Table 1).

Table 1. Distribution of the numbers (N) of sloth moths *Cryptoses choloepi* caught in the two light-traps on BCI. The percentage of the moths caught in the canopy trap (C) is given for three seasons in four successive years. The differences among the seasons are highly significant.

Year	*Dry Season* *N*	*C*	*Early Wet Season* *N*	*C*	*Late Wet Season* *N*	*C*
1975	77	40.2	151	67.5	3	(33.3)
1976	65	40.0	107	72.0	57	47.4
1977	165	39.4	638	58.9	131	48.9
1978	291	37.8	899	51.4	72	52.8
Total	598	38.8	1795	56.7	263	49.4

The Las Cumbres Trap.--The seasonal pattern observed in Las Cumbres is very different from that on BCI (Fig. 2). There was a broad peak in abundance during the first half of the rainy season, which is similar to BCI, but then the numbers increased again, to go down only during the first half of the dry season (1976, 1977) or at the very end of the rainy season (1974). Maximum abundance seems to have occured in January rather than May.

The Fortuna Trap.--During the period when the trap was at canopy level numbers were much higher than later in the year (Fig. 3). This could be partly due to more moths flying at the level of the canopy, but considering the differences between canopy and ground level on BCI (Table 1), I doubt that this is the only explanation. I believe that at least part of the decrease is real, especially because the numbers were already declining at the canopy level, and that decrease continued after the level of the trap was changed. The numbers did not increase again until late in 1978 and reached high levels towards the end of the observation period in 1979. A clear

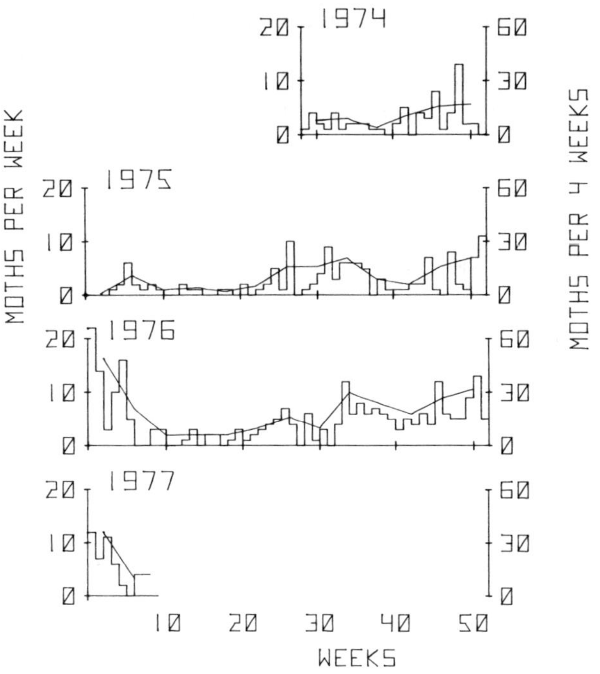

Figure 2. Seasonal fluctuations in abundance of the sloth moth *Cryptoses choloepi* in a light-trap in Las Cumbres. Legends as in Fig. 1.

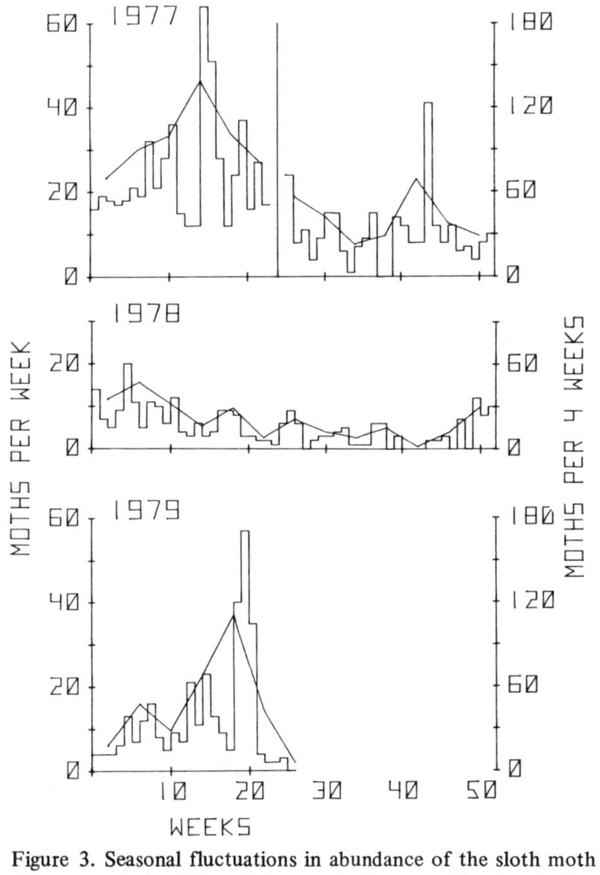

Figure 3. Seasonal fluctuations in abundance of the sloth moth *Cryptoses choloepi* in a light-trap in Fortuna. The data before the vertical line in 1977 refer to the trap at canopy level. After that the trap was at the same site but at a lower level. Legends as in Fig. 1.

pattern which repeats itself each year, such as was found in BCI and Las Cumbres, did not occur at Fortuna. A gradual increase in numbers reaching a peak around weeks 15-20 was observed in 1977 and 1979, but not in 1978.

DISCUSSION

There is little information which could help predict seasonal changes in abundance of *Cryptoses* caught by a light-trap. There are climatic changes, and in two of the three sites there is a clear alternation of wet and dry seasons. However, as far as is known, this has little effect on the behavior of the sloths. Published papers on sloths do not refer to presence or absence of seasonal changes in the ecology of the sloths except that mortality seems to be highest towards the end of the rainy season in the forest on BCI (Montgomery and Sunquist, 1978). This is possibly because of a combination of the increased cloudiness during that part of the year which may interfere with the thermoregulation of the sloths and the low rate of production of new leaves in the forest, probably an important part of the available food, towards the end of the rainy season (Wolda, 1978a; 1982). Reproduction in the sloths occurs throughout the year, with a broad peak centered around September (G. G. Montgomery, pers. comm.). A higher mortality of the sloths could mean an increased number of *Cryptoses* in the air looking for a new host. After a peak in reproduction of the sloths there are more sloths around, which means that the production of feces in the forest increases and that it is easier for the moths to find a host. Both factors could be significant, both for the actual population size and for the number of individuals of *Cryptoses* caught in the light traps.

Both species of sloths are common on BCI (Montgomery and Sunquist, 1975) but there is no evidence of seasonal changes in abundance, although one might expect these to occur when mortality and reproduction are seasonally variable. Similarly, no evidence was found that the sloths seasonally change habitat selection, their food, or that they disperse seasonally (Sunquist and Montgomery, 1973; Montgomery and Sunquist, 1978). The sloths themselves, at the present state of our knowledge, do not provide clues on the causation of the fluctuations in abundance in *Cryptoses*.

There are no data with which one could test the hypothesis that the changes in abundance of the moths in the traps reflect real changes in abundance of the moths on the sloths. Waage and Montgomery (1976) and Waage and Best (1985) give counts of the number of moths per sloth, but they found a large variance between individuals. The sloths used are not well distributed over the year, so that no conclusion can be drawn about the presence or absence of seasonal fluctuations in moth density.

The only available information on seasonal variation in abundance of the sloth moth in Panama is given in the present paper. Until evidence to the contrary is found, I assume the changes in moth abundance in traps reflect real changes in abundance in the population.

On BCI (Fig. 1) the weather pattern, especially the transition from dry to wet season, seems to play an important role. The late arrival of the rainy season in 1977 compared with the unusually early start of the rains in 1978 is reflected in the timing of the peaks in abundance of the sloth moth. It seems possible that the wet fur of the sloths in the rainy season, especially in October-November when the days can be very wet, is not a very suitable habitat for the moths and the numbers decrease, while during the dry season the numbers can increase again. If sloth dung is left to dry, the larvae in it will perish (Waage and Montgomery, 1976), but it seems doubtful that this will be an important factor in the forest of BCI, even during the dry season.

In Las Cumbres both species of sloths are commonly seen, but no systematic study of them has been made in this area. Sloth moths were observed on the sloths, but no counts are available.

The light-trap data from Las Cumbres (Fig. 2) present a picture which is very different from that of BCI (Fig. 1). Both sites are similar in that there is a peak in sloth moth numbers in the first part of the rainy season. The large peak at Las Cumbres in the early dry season, with low numbers during the rest of the dry season, is very different from the pattern on BCI. A similar difference exists between the two sites in the seasonality of some Cicadellidae (Homoptera) (Wolda, 1979; 1980a; 1982). They too have a large peak in abundance in the light-trap in the early dry season and it is clear here that this peak is not caused by a real increase in abundance in the population, but by an increased activity, a search for food and oviposition sites. It seems possible that the explanation for the sloth moth is similar. Within the forest on BCI no seasonal changes in habitat were found, but Las Cumbres has a dry season which is more severe than that on BCI. In the man-influenced environment in Las Cumbres, the sloths possibly move away from the gardens in the vicinity of the trap in the dry season to very young forest somewhat further away. Newly emerged moths then would have a hard time finding a host, would fly around more and end up in the trap in relatively large numbers. I hypothesize that the January-February peak is not a peak in abundance but a peak in activity. Here sloth dung may very well dry out during the dry season, keeping moth numbers low.

In Fortuna the climate is as nonseasonal as one may find anywhere in the tropics. The sloth moths vary in abundance, but the pattern varies between years (Fig. 3). The best explanation is that the variation is nonseasonal, thus supporting the view that the weather pattern is an important factor causing the seasonality in the sloth moths. Insects in general in Fortuna tend to be less seasonal than the ones in places like BCI or Las Cumbres. However, several species are highly seasonal, showing that there are seasonal factors insects can clue in on (Wolda, 1980b; Wolda and Fisk, 1981).

In Fortuna sloths may be rare. During an environmental impact study, no sloths were observed in the area around the trap (Adames, 1977) and according to people who frequent the area sloths were habitually shot during several years. I doubt that hunters have managed to exterminate the sloths, as finding them in a forest is no easy task and many parts of the forest are not easily accessible, but it seems that the density of the sloths is at best rather low.

The sloth moth, however, is not rare at all in the traps. Either the density of the sloths is grossly underestimated in the area or, with a lower density of the sloths, the moths have to fly more to find a host and have increased chances of being caught in the trap. With the information available it cannot be decided which situation applies.

ACKNOWLEDGMENTS

I gratefully acknowledge the help received in sorting the insect samples and in counting the *Cryptoses* in those samples by Saturnino Martínez and Miguel Estribí. The light-trap program on BCI is partly supported by the Environmental Sciences Program of the Smithsonian Institution. The Instituto de Recursos Hidráulicos y Electificación (IRHE) kindly permitted the operation of the light-trap in Fortuna and I am very much indebted to Ing. Cecilio Estribí and Mr. Lionel Quiros of the RENARE in Chiriquí Province for their help received with the Fortuna project. Gene Montgomery critically read the manuscript.

LITERATURE CITED

Adames, A. J.
1977. Evaluación ambiental y efectos del projecto hidroeléctrico Fortuna. Lotería 254, 255, 256: 1-538.

Askew, R. R.
1971. Parasitic Insects. London, Heinemann. 316 pp.

Casebeer, R. S. and C. L. Hogue
1966. Hideaway for moths. Nat. Hist. 75: 40-41.

Caballero, J. M.
1978. Situación fisica, meteorología, año 1976. Estadistica Panameña 121, clima.

Dyar, H. G.
1908. A pyralid inhabiting the fur of the living sloth. Proc. Entom. Soc. Washington, 9: 142-144.

Grassé, P.-P.
1951. Traité de Zoologie, Tome X, p.330. Paris, Masson et Cie. 1948 pp.

Montgomery, G. G. and M. E. Sunquist
1975. Impact of sloths on neotropical forest energy flow and nutrient cycling. Pp. 69 to 98, *In* Tropical Ecological Systems: Trends in terrestrial and aquatic research. (F. B. Golley and E. Medina, editors,) New York, Heidelberg, Berlin, Springer-Verlag. 398 pp.

1978. Habitat selection and use by two-toed and three-toed sloths. Pp. 329 to 359, *In* The ecology of arboreal folivores. (G. G. Montomery, ed.), Smithsonian Institution Press, Washington, D. C. 574 pp.

Sunquist, M. E. and G. G. Montgomery
1973. Activity patterns and rates of movement of two-toed and three-toed sloths (*Choloepus hoffmanni* and *Bradypus infuscatus*). J. Mamm. 54: 946-954.

Waage, J.K. and R.C. Best.
1985. Arthropod associates of sloths. (this volume).

Waage, J.K. and G. G. Montgomery.
1976. *Cryptoses choloepi*: A coprophagous moth that lives on a sloth. Science, 193: 157-158.

Wolda, H.
1977. Fluctuations in abundance of some Homoptera in a neotropical forest. Proc. III Int. Symp. Trop. Ecol., Lubumbashi, Zaire, 1975. Geo-Eco-Trop, 3: 229-257.

1978a. Seasonal fluctuations in rainfall, food and abundance of tropical insects. J. Anim. Ecol. 47: 369-381.

1978b. Estudios ecológicos sobre insectos en la región de Fortuna y su importancia para la conservación de la naturaleza. ConCiencia 5: 3-6.

1979. Fluctuaciones en abundancia de insectos en el bosque tropical. Pp. 519-539 *In* Actas del IV

Symposium de Ecología Tropical (Panama, 1977), 1038 pp.

1980a. Seasonality of tropical insects. I. Leafhoppers (Homoptera) in Las Cumbres, Panama. J. Anim. Ecol. 49:277-290.

1980b. Fluctuaciones estacionales de insectos en el trópico: Sphingidae. Memorias del VI Congreso de la Sociedad Colombiana de Entomología "SOCOLEN," julio 1979: 10-58.

1982. Seasonality of tropical insects. II. Leafhoppers (Homoptera) on Barro Colorado Island, Panama Canal Zone. Pp. 319-330 *In*: Ecology of a Tropical Forest: Seasonal Rythms and Long-term Changes. (E.G. Leigh, A.S. Rand and D.M. Windsor, eds). Smithsonian Institution Press, Washington, D.C., 488 pp.

Wolda, H. and F.W. Fisk.
1981. Seasonality of tropical insects II. Blattaria in Panama. J. Anim. Ecol. 50: 827-838.

SEASONAL DISTRIBUTION OF THE LARGE SLOTH BEETLE *UROXYS GORGON* ARROW (SCARABAEIDAE; SCARABAEINAE) IN LIGHT TRAPS IN PANAMA

Henk Wolda and Miguel Estribí

Smithsonian Tropical Research Institute, P.O. Box 2072, Balboa, Panama.

Abstract
In light traps on Barro Colorado Island, Panama, the large sloth beetle *Uroxys gorgon* occurs year round but is more abundant in the rainy season. In Fortuna, a relatively non-seasonal area, the seasonal distribution of the beetle tends to be bimodal. There are indications of dispersal flights both early and late in the rainy season.

Abstracto
En trampas de luz en la Isla de Barro Colorado, Panamá, el escarabajo grande del perezoso *Uroxys gorgon* se encontró a través del año, pero tenía una abundancia superior en la estación lluviosa. En Fortuna, un sitio relativamente no estacional, la distribucíon estacional del escarabajo fue bimodal. Existen indicaciones de vuelo de dispersión al principio como al final de la estación lluviosa.

Resumo
Em armadilhas leves na ilha de Barro Colorado, Panamá, o grande coleóptera se preguiças *Uroxys gorgon* ocorre durante todo o ano, embora seja mais abundante na estação de chuvas. Em Fortuna, uma área relativamente não-sazonal, a distribuição estacional do coleóptera tende a ser bimodal. Há indicações de vôos de dispersão tanto no princípio como no fim da estação de chuvas.

Among the various arthropods which live in the fur of the three-toed sloth *Bradypus infuscatus* are a number of species of dung beetles (Scarabainae) (Halffter and Matthews, 1966; Waage and Best, 1985). At least two of these, *Uroxys gorgon* Arrow and *U. metagorgon* Howden and Young, have been found on sloths in Panama (Howden and Young, 1981). The only information in the literature on seasonal variation in abundance of beetles on sloths is for *Uroxys besti* and *Trichilium adisi* in Brazil (Waage and Best, 1985). This paper describes the seasonal variation in abundance of *Uroxys gorgon* in light-traps in Panama.

LIGHT-TRAPS AND COLLECTION LOCALITITES

Sloth beetles were identified and counted in samples of insects caught with light-traps in five localities in Panama. These sites are Barro Colorado Island (BCI), Fortuna, Boquete, Miramar and Corriente Grande. BCI, 9° 9' N; 79° 45' W, is a forested island in the Panama Canal area, 120m above sea level (Leigh et al., 1982). Miramar, 9° 0' N; 82° 15' W, is an area of disturbed forest on the coast of Bocas del Toro Province, consisting of pastures at the foot of a forested slope with an undergrowth of cacao, at sea level. Corriente Grande, 9° 17' N; 82° 32' W, is in a lowland forest, some 65 km northwest of Miramar in Bocas del Toro, 100 m above sea level, on the Changuinola river. The Boquete site, 8° 48' N; 82° 26' W, is in a small forest remnant surrounded by coffee plantations at 1350 m elevation in Chiriquí Province. Fortuna, 8° 44' N; 82° 16' W, is in wet forest, 20 km East of Boquete, at 1050 m elevation.

Scarab beetles were sorted from 3 years of samples from BCI and Boquete, 2 years and 9 months at Fortuna, 1 year at Miramar and 4 months at Corriente Grande. Only one light-trap operated in all sites except BCI where there was one trap in the canopy and one at a low level in the forest. At Fortuna, the trap was in the canopy for 9 months and at lower level for the next 2 years. BCI and Boquete have a 4 month dry season from January through April (for BCI see Leigh et al., 1982). Fortuna does not have a dry season (McElravy et al., 1982). Miramar and Corriente Grande also have no dry season, but rainfall is reduced both in March and September (Wolda and Galindo, 1981).

RESULTS

Very few *Uroxys gorgon* were caught at the Bocas del Toro sites. At Miramar one sloth beetle was captured in January 1979 and in Corriente Grande 2 were captured, one in March and one in May 1980. The reasons for this low abundance in the traps are not clear. The species is known from Bocas del Toro (Howden and Young, 1981) and it seems unlikely that sloths have been hunted out, especially in Corriente Grande. The species has not been reported from Chiriquí Province (Howden and Young, 1981), and was absent in the samples from Boquete. However, a number of individuals were obtained in Fortuna, also in Chiriquí Province.

As in many other insects, the numbers of *Uroxys gorgon* in the light-traps is strongly depressed during full moon. In order to not let this effect interfere with a study of seasonal variation in abundance, the samples were lumped per period of four weeks. There are 13 such periods per year. The abundance of the beetle on BCI per four weeks is indicated in Fig.1. The series of data starts in March 1977, which is why the abscissae in Figs.1 and 2 begin with period 3. In both 1977 and 1978 there was a strong tendency for numbers of beetles to be lower during the dry season (periods 1-4). A quadratic polynomial described the seasonal changes in abundance significantly better than a linear one in both years. In 1979 the numbers also increased briefly at the end of the dry season, but returned to a low level immediately where they stayed until the end of the year. The rains in 1977 started much later than in 1978 or 1979 and the increase in the beetles also started later.

In both 1977 and 1978 there is a tendency toward bimodality in the seasonal distribution of the percentages of the beetles in each 4-week period which were caught in the canopy on BCI (Fig.2). In both years a bimodal polynomial describes the seasonality significantly better than a unimodal or linear one. A higher percentage of the individuals was caught in the canopy both at the beginning and at the end of the rainy season. In 1979 there was also a peak in the percentages of beetles early in the year but there was no indication of a second peak at the end of the year. The change in percentages in 1979 can best be described as a gradual decrease from March onwards.

The numbers of sloth beetles caught at Fortuna were rather low, 14 during the first 9 months when the trap was in the canopy, and 26 during the next 2 years with the trap at a lower level. The seasonal distribution of the beetles in the lower trap (Fig.3, both years combined) tends towards bimodality, with one peak in March - April and one in October - November.

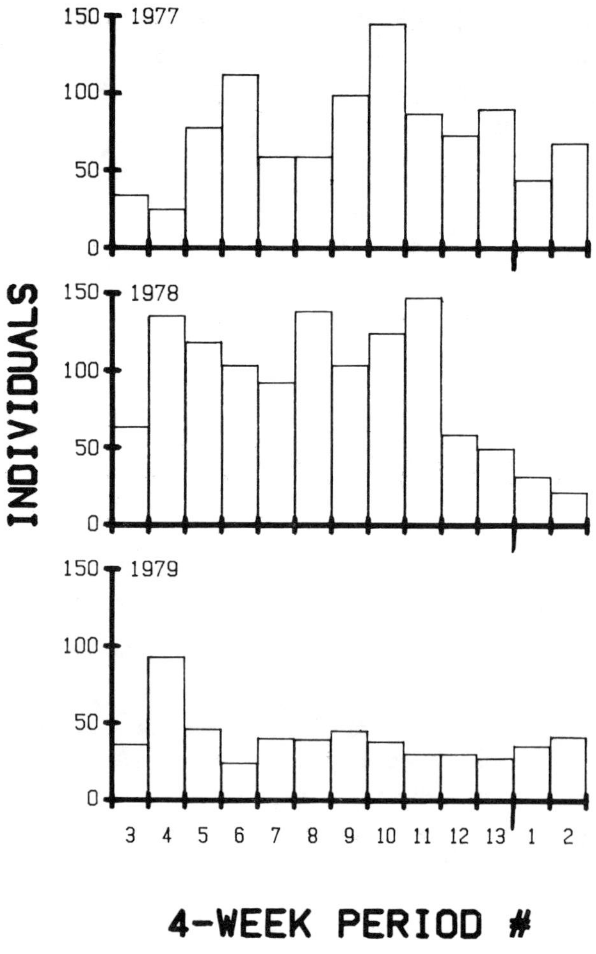

Figure 1. Seasonal distribution of the large sloth beetle *Uroxys gorgon* in light-traps on Barro Colorado Island, Panama. The numbers are given per four-week period. The data start in March 1977, period 3, and end in late February 1980, period 2. Data from two light-traps, one in the canopy and one at a lower level, are combined.

DISCUSSION

Waage and Best (1985) concluded that the number of the sloth beetles *Uroxys besti* and *Trichilium adisi* "showed no evidence of correlation with rainfall, although both species showed a similar, but unexplained, peak in numbers during June." However, in both species the numbers per sloth were lower in the dry season, which fits nicely with the data from light-traps reported here for BCI. The lower numbers observed during the dry season in the light-traps do not seem to be some artifact. During the rainy season, the numbers remained at a relatively steady level, unlike the sloth moth *Cryptoses choloepi* Dyar, which had a sharp seasonal peak on BCI (Wolda, 1985).

The peaks of the number of beetles in the light-trap in the forest canopy are interesting because they occurred at the times of the year when various insects in Panama disperse (Wolda, 1977). This dispersal presumably occurs mostly in and above the canopy of the forest, which suggests that *Uroxys* also has dispersal flights at

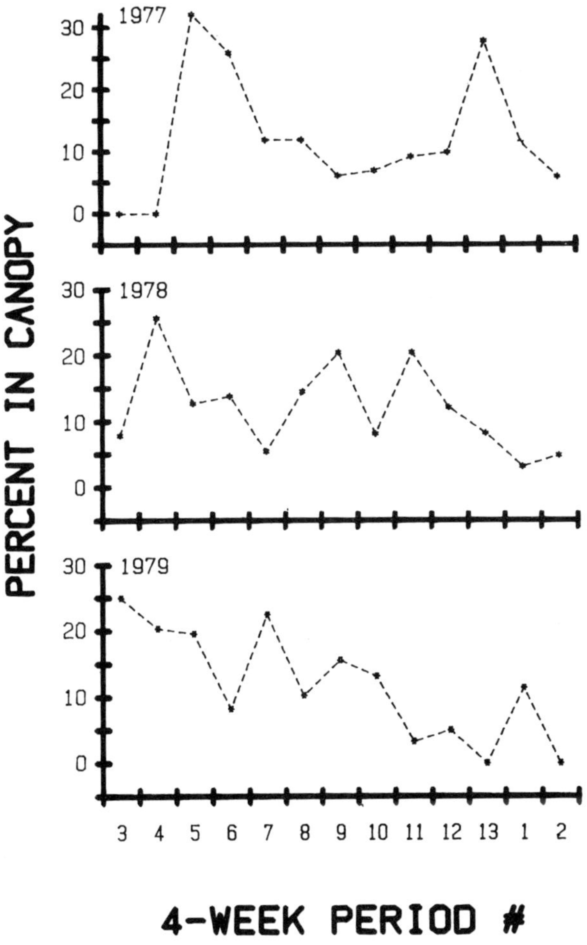

Figure 2. The percentage of the sloth beetles caught on Barro Colorado Island, Panama, which were in the light-trap in the canopy, per four-week period.

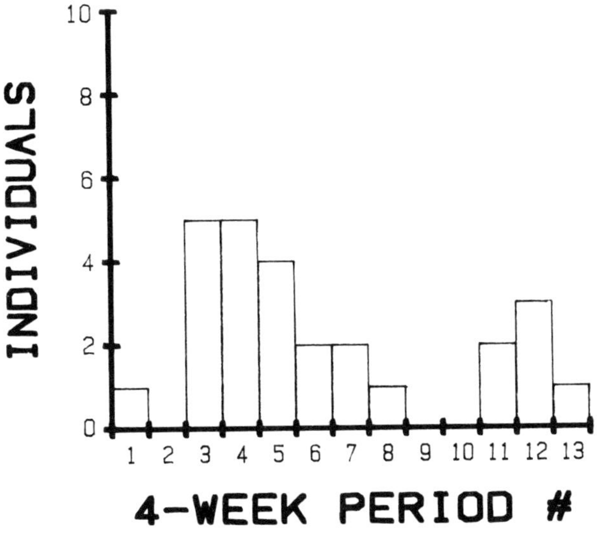

Figure 3. The number of sloth beetles per period of 4 weeks in a light-trap in Fortuna, Panama. Two years of data are combined.

the beginning and at the end of the rainy season. This taken together with the lower abundance during the dry season suggests that conditions are less then optimal for the immature stages of the beetle during the dry season. Part of the adult population possibly enters reproductive diapause at the end of the rainy season, disperses off the sloths, perhaps to sites with at least some moisture and resumes reproductive activities at the end of the dry season when they disperse back to sloths. This scenario is not unlikely considering the diapause known to occur in other tropical beetles (Paarman, 1977; Wolda and Denlinger, 1984). For many tropical insects diapause in adults is weak. It affects usually only part of the population, the insects remain active and the only sign of diapause is a halt of reproductive activities.

In Fortuna the distribution of the sloth moth *Cryptoses choloepi* was non-seasonal (Wolda, 1985), but the present data suggest that the sloth beetle there varies in abundance seasonally, at least in the light-trap. The amount of rain per month does not vary over the year in Fortuna, but during December through February it tends to fall in almost continuous drizzle. During the rest of the year it falls in tropical rainstorms of at most a few hours each. In December through February the weather is much wetter. This could cause reduced flight activity in the beetles, explaining the low numbers during periods 13 - 2. The low numbers during late July through September (periods 8 - 10) remain unexplained. An alternative explanation, which would cover both peaks, is that the species was really very rare in Fortuna, and that the specimens caught were really dispersing from elsewhere. The peaks do coincide with the dispersal peaks on BCI. Fortuna is only a few km away from the seasonal Pacific slopes in Chiriquí Province, from which such dispersal could occur. However, apart from Fortuna, the sloth beetle is not known from Chiriquí Province and was not observed in Boquete, only 20 km away in an area with a seasonal weather pattern. This makes this hypothesis unlikely.

ACKNOWLEDGEMENTS

We gratefully acknowledge help received from various persons and institutions during the collecting and analysis of the beetles samples, including Saturnino Martínez, Bonifacio De León, Ing. Cecilio Estribí, Nessim Yishui, the Instituto de Recursos Hidráulicos y Electrificación (IRHE) and the Dirección de Recursos Naturales Renovables (RENARE) in Chiriquí Province. Bruce Gill, Carleton University, Ottawa, Canada, kindly identified the beetles. The work on BCI was partly supported by the Environmental Sciences Program of the Smithsonian Institution, that in Miramar by a Smithsonian Research award.

LITERATURE CITED

Halffter, G. and E.G. Matthews.
1966. The natural history of dung beetles of the subfamily Scarabaeinae. Fol. Entomol. Mexic., 14-16: 1-281.

Howden, H.F. and O.P. Young.
1981. Panamanian Scarabaeinae: Taxonomy, distribution and habits (Coleoptera, Scarabaeidae). Contr. Am. Entomol. Inst. 18: 1-204.

Leigh, E.G., A.S. Rand and D.M. Windsor.
1982. Ecology of a tropical forest: Seasonal rhythms and long-term changes. Smithsonian Institution Press, Washington, D.C., 468 pp.

McElravy, E.P., H. Wolda and V.H. Resh.
1982. Seasonality and annual variability of caddisflies adults (Trichoptera) in a "non-seasonal" tropical environment. Archiv Hydrobiol. 94: 302-317.

Paarman, W.
1977. Propagation rhythms of subtropical and tropical Carabidae (Coleoptera) and its control by exogenous factors. Pp. 49 to 60, *in* Advances in invertebrate reproduction, Volume I (K.G. Adiyodi and R.G. Adiyodi, editors). Peralam-Kenoth, Karivellur, Kerala, India.

Waage, J.K. and R.C. Best.
1985. Arthropod associates of sloths. (this volume).

Wolda, H.
1977. Fluctuations in abundance of some Homoptera in a neotropical forest. Geo-Eco-Trop, 1: 229-257.

Wolda, H.
1985. Seasonal distribution of sloth moths (*Cryptoses choloepi* Dyar) (Pyralidae: Chrysauginae). (this volume).

Wolda, H. and D.L. Denlinger.
1984. Diapause in a large aggregation of a tropical beetle. Ecol. Entomol. (in press)

Wolda, H. and P. Galindo.
1981. Population fluctuations of mosquitos in the non-seasonal tropics. Ecol. Entomol. 6: 99-106.

ECOLOGY OF LIVING SLOTHS

MAINTENANCE AND REPRODUCTION OF THE TWO-TOED SLOTH *CHOLOEPUS DIDACTYLUS* IN CAPTIVITY

[1] **John F. Eisenberg and Eugene Maliniak**

National Zoological Park, Smithsonian Institution, Washington, D.C. 20008. [1] Present Address: Department of Zoology, University of Florida, Gainesville, Florida 32611.

Abstract
The maintenance and reproductive history of two-toed sloths *Choloepus didactylus* at the National Zoological Park are reviewed. Defecation rate varied according to age and reproductive state; intervals between defecations averaged 3.8 days for four individuals. Gestation is at least 44 weeks (309 days). There is no evidence for delayed implantation or sperm storage. The interbirth interval during which a female successfully weans her young is longer than an interbirth interval following the early death of a young. A lactation anestrus exceeding 3 months is implied. *Choleopus* exhibits the reproductive pattern of a "K" strategist with a long life span, reduced litter size, and long interbirth intervals.

Abstracto
Fue revisado en el Parque Zoológico Nacional el mantenimiento e historia reproductiva de monos perezosos de dos dedos *Choloepus didactylus.* La rata de defecacion variaba de acuerdo a edad y estado reproductivo; intervalos entre cada defecacíon tuvieron un promedio de 3.8 días para cuatro individuos estudiados. El tiempo de gestación fue de casi 44 semanas (309 días). No hubo evidencia de el implantación astrasada o carga de esperma. El intervalo entre el estado infantil-juvenil durante el cual una hembra exitosamente desteta su cría más largo que un intervalo infantil-juvenil seguido por muerte prematura de un juvenil. Se implica una lactación anestrus de más de 3 meses. El *Choloepus* exhibe un patrón reproductivo de una estrategia "K" con larga vida, período de alumbramiento reducido, e intervalos prolongados de nacimiento.

Resumo
A manutenção e a história reprodutiva de preguiças de dois dedos *Choloepus didactylus* no National Zoological Park são revisadas. O ritmo de defecação variou de acordo com a idade e estado reprodutivo. Iintervalos entre as defecaçaões foram em média 3.8 dias para 4 indivíduos. A gestação é de pelo menos 44 semanas (309 dias). Não há evidência de implantação retardada ou armazenamento de esperma. O intervalo entre partos durante o qual a fêmea desmama com sucesso a cria é maior do que o intervalo entre partos que se segue após a morte precoce de uma cria. Um anestro lactacional de mais de 3 meses está implicado. O *Choloepus* exibe o padrão reprodutivo de um estrategista "K", com uma vida média longa, tamanho de ninhada reduzido e intervalos entre partos prolongados.

The two-toed sloth, *Choloepus didactylus*, has been maintained and has bred at the National Zoological Park (NZP) since 1950. This report concerns studies since 1 September 1973, with special attention to the patterning of reproduction.

ANIMALS AND METHODS

In 1973, five two-toed sloths were on exhibit: one adult male, three adult females, and a male juvenile born 9 March 1972. The group was examined by G.G. Montgomery and Mitchell Bush on 27 August 1973, prior to moving the animals to new quarters; the abdomen of the female was X-rayed, and pregnancy was indicated by palpation and from the radiograph. The adult female and her juvenile were placed separately in the tropical animal house of the Office of Zoological Research. The female gave birth on 3 April, and the juvenile was removed on 4 April. The adult male was transferred to these same quarters on 20 June 1974, and the three sloths lived in the same room until the following July. A third young was born on 27 July 1975. On 5 August, the adult male was separated from the female, juvenile and her new infant. On 27 August, the juvenile was separated from the female and her infant.

The sloths were housed in a room which, after mid-September, contained arboreal feeding platforms and nest boxes (Fig. 1). During late fall and winter, temperature was maintained at 21°C to 30°C and relative humidity at 30 percent. Lighting in winter was 12L: 12D. In late spring and summer the animals were maintained at the ambient temperature (20°C to 38°C)

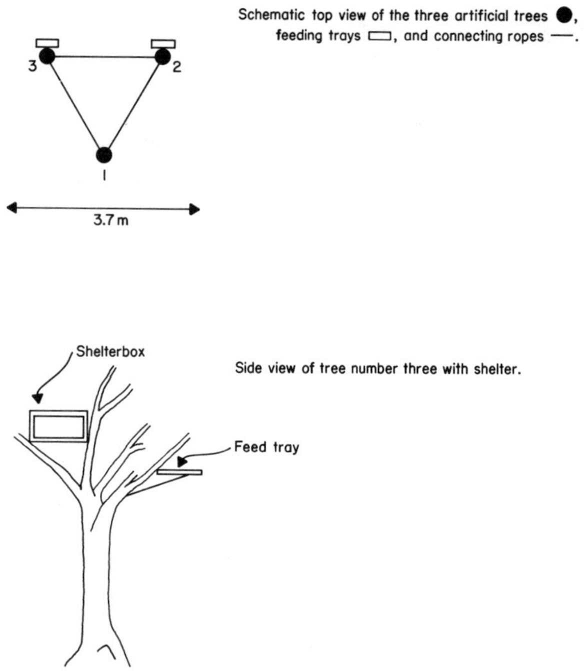

Figure 1. Arrangement of three artificial trees, nesting boxes, feeding trays and connecting ropes for the study of *Choloepus* in captivity. Trees 1 and 3 each contained two nest boxes, and each of the trees contained a feeding platform.

humidity (25 percent to 90 percent) and light. The sloths were fed kale, sweet potatoes, apples, carrots and a few pellets of Purina monkey chow daily. Water was provided at all times. Feeding was sporadic and, although the sloths ate some food almost every day, the amount varied greatly. Sleeping locations during the day, defecation intervals, and associations of sloths were noted daily throughout the study.

RESULTS AND DISCUSSION

Defecation Intervals.--From 1 September to 29 November 1973, the juvenile male defecated with an average interval of 3.38 days, as did his mother (Table 1). Meritt (1985) reports a defecation interval of 3 to 8 days in *Choloepus hoffmanni*, and points out that defecation and urination occur simultaneously. The fecal piles of the two sloths were easily distinguished by pellet size, shape, and relative amount of feces. From 29 November to 3 April 1974, the average interval between defecations for the male was 4 days; for the female, 4.6 days. The average for the entire period was 7.8 defecations per month per animal.

The two animals defecated on the same night from 2 to 7 times per month, thus indicating either a degree of metabolic synchrony or social facilitation of defecation.

The young male and his mother only occasionally defecated near the same point (range 0-4 times/month). The subadult male tended to defecate at the base of either tree 1 or tree 2 (Table 2), while the female seldom defecated at the base of tree 1 and tended to defecate at the base of tree 3 (Table 2). The female consistently fed more from food placed on the platform in tree 3, thus she defecated at the base of the tree in which she most frequently fed. After the adult male was introduced the adults tended to defecate at different sites when they defecated on the same night.

Fecal depositions by the infant were first noted at the beginning of June, when it was 2 months old, and were initially sporadic but became more regular by the end of June. The infant's feces were adjacent to those of the female. Presumably, when the female descended to defecate, the youngster was also stimulated to defecate; it rode on her almost continuously. The female and infant defecated at the same place and time until the end of September 1974, when the young began defecating at separate loci.

Defecation intervals for the female from April through June while she was lactating (3.4 days) were less than the intervals preceeding parturition (4.4 days). The maturing young showed an average defecation interval of

Table 2. Defecation and sleeping sites for a mother-young *Choloepus* dyad. The mean number of days per month that a site was used is shown.

	Adult Female Tree Number			*Subadult Male Tree Number*		
Site	*1*	*2*	*3*	*1*	*2*	*3*
Sleeping	9.9	11.0	7.1	11.3	11.4	5.9
Defecation	0.1	2.6	4.1	2.6	4.1	0.9

Table 1. Monthly average interval between defecations in days.

	1973 *S*	*O*	*N*	*D*	*1974* *J*	*F*	*M*	*A*	*M*	*J*	*J*	*A*	*S*	*O*	*N*	*Mean*
Juvenile Male	3.0	3.9	4.4	3.9	3.9	5.6	4.4	4.5	4.4							4.22
Adult Male										5.5	4.6	3.4	2.4	4.4	4.5	4.13
Adult Female																
(Prepartus)	3.3	4.4	4.4	4.4	5.2	4.6	4.4									4.39
(Postpartus)								3.1	2.6	3.5	3.8	3.5	2.3	4.4	4.5	3.46
Maturing young										2.5	2.2	2.6	2.3	2.5	2.5	2.43

less than 2.4 days. If the interval between defecations reflects rate of food intake, then the female's metabolic rate may have increased during lactation and afterward while she carried her young about.

Associations of sloths.--Except for several days in January, the juvenile male and his mother tended to sleep in the same nest box, often touching, between January and March.

During July 1974, after being placed with the female, the adult male rested near her and the infant during 75 percent of spot checks. In August he was in association with the female and infant during resting periods 100 percent of the time, but this association with them fell to slightly less than 50 percent during the subsequent months. During August he was often seen embracing the female and it is possible that mating occurred then.

Copulation--In July 1976, the young born on 27 July 1975 was removed from the mother. The adult male was again introduced on 28 July and a copulation was witnessed by Larry Newman. Approximately 20 minutes after the introduction of the male, the pair of sloths reached an open area of branches which facilitated their maneuvering into a copulatory position. The male, who was following the female from behind, reached up and grasped her, halting her slow forward movement. The female then held still, hanging from the branch by all four limbs, while the male moved up and interposed himself between the female and the branch. The male then moved around until they were face to face and his posterioventral surface was opposed to the female's vaginal region. The male made about ten thrusts before achieving penetration. After intromission the male thrusted three or four times in rapid succession before withdrawing. The male was left with the female until 11 August, but no further copulations were witnessed. The female gave birth 11 months later on 18 June 1977.

Gestation Period--Veselovsky (1966) reported a 5-month 20-day gestation period for *Choloepus didactylus*, based on observations of a presumed copulation with a subsequent captive birth. Stone (1957) reported a gestation of at least 8 months 23 days, the interval between the death of an adult male and a subsequent birth in captivity by the surviving female. In a similar case, Van Doorn (1971) reported a gestation exceeding 8.5 months and speculated that delayed implantation might have occurred.

Two-toed sloth gestation probably exceeds or equals 10.5 months. At the beginning of this study on 27 August 1973, the adult female was pregnant; an embryo was felt by palpation and seen on a radiograph. The female subsequently gave birth on 3 April 1974, without additional access to an adult male. A detectable fetus was thus present some seven months prior to birth. On 28 July 1976, a copulation was witnessed, and no other copulations were seen between 28 July and 11 August when the male was moved. A young was born on 18 June 1977. If conception occurred on 28 July, then the gestation period for that pregnancy was more than 11 months (325 days). Assuming subsequent copulations in the intervening two weeks, then gestation might be as little as 10 months (309 days).

It has been suggested (Van Doorn 1971) that two-toed sloth females can store sperm. The colony at NZP was established in 1950 and the founding male died 4 October 1967. A new male was acquired 27 February 1968 and introduced to the females about three months later. The last sloth that could have been fathered by the founding male was born 29 March 1967, six months prior to his death. There were no additional births in the colony for some five years, until 9 March 1972. This lag in births following the death of the founding male strongly suggests that sperm storage did not occur.

The reproduction records for two-toed sloths at the NZP since 1952 were analyzed. The basic group included two reproducing females and one adult male. Although individual females were not marked, we infer from an analysis of interbirth intervals the duration of lactation anestrus and the average interval between births in cases where the young died shortly after birth, and in instances where the young was raised to social weaning.

When a female gave birth and the infant died within two weeks, she was likely to give birth again after an average interval of 14 months (423 days, range 361 to 622 days, 14 cases). If a female gave birth and raised the young to the age of social weaning then the average interbirth interval was 26 months (800 days, range 472 to 1093, 6 cases). The shortest of these interbirth intervals was 472 days, but the newborn young died shortly thereafter and may have been premature. Impregnation is apparently delayed for 15 months until the previous young becomes independent (26-month interbirth interval minus 11-month gestation) or, in cases when the young dies soon after birth, for a lactation anestrous period of about 3 months (14-month interbirth interval minus 11-month gestation).

Development of Young -- Meritt (1985) describes the behavior of a mother two-toed sloth toward its newborn young and during the early weeks of its development. The last date on which the young born on 3 April 1974 at NZP was seen at the nipple was 22 June. On this date the young was also first seen away from the female and was moving back and forth on a branch about 30 cm from her. The infant began taking food from its mother's mouth while she fed during the first week of its life, and the quantity of vegetable matter ingested increased dramatically in June. Meritt (1985) noted that water was consumed by young at 7 days of age, solid food at 15 days for a male and 27 days for a female infant. By 5 months of age, *Choloepus hoffmanni* young were regularly feeding on solids and were away from the mother (Meritt, 1985). Feces from the infant became noticeable in early June and his defecation rate was about twice that of the mother in July (Table 1). The mother continued to carry the young on her excursions until the end of October and, although the infant and mother always slept together the mother's tendency to carry the young during sustained locomotion declined in November, 1974 when the young sloth was 7 months old.

At birth the infant's weight was about 4 percent of the mother's weight. At six months of age, the infant's weight was approximately 11 percent of the mother's weight; the infant reached about 21 percent of the female's weight by nine months of age. Infant primates generally begin to locomote independently when the weight of the young is about 15 percent-20 percent of the mother's weight (Kleiman, 1977). If the female was able to move away from the young, the pair might thus have separated in November. With the female incapable of moving away, the young persisted in the infantile behavior of riding on the mother well into its second year of life.

Montgomery and Sunquist (1975, 1978) and Montgomery (1985), based on a long-term field study, report that some wild female *Bradypus infuscatus* give birth at intervals of 12-13 months. Gestation appears to take six months, lactation less than four weeks. The young is carried for about five additional months, whereupon the female abandons the young in a portion of her home range. Presumably the young learns habitat utilization patterns from the mother prior to "social weaning" and ''inherits'' part of the mother's home range (Montgomery and Sunquist, 1978). The mean adult weight of *Bradypus infuscatus* is 2.7 kg. *Choloepus didactylus* has a mean adult weight of 3.5 kg. The gestation of *Bradypus* is slightly longer than predicted from a regression of gestation time against mean body weight for a series of primates, while *Choloepus* is even more deviant (Eisenberg, 1981).

Based on records from our captives, female two-toed sloths do not breed before 3 years of age. Males are not sexually mature before 4.5 years of age. Females may remain reproductively active into their twenty-fifth year and one male was sexually active over a period of some 16 years and was at least 21 years old at his death.

The two-toed sloth is an extreme "K" strategist with a long potential life span, long interbirth interval, and extended period of reproduction.

ACKNOWLEDGMENTS

The authors greatly acknowledge the advice and assistance provided by Dr. G.G. Montgomery in the initial formulation of this study. Dr. Mitchell Bush was extremely helpful in the X-raying of the animals and providing invaluable advice.

LITERATURE CITED

Eisenberg, J.F.
1981. The mammalian radiations: An analysis of trends in evolution, adaptation and behavior. University of Chicago Press, Chicago. 610 pp.

Goffart, M.
1971. Function and form in the sloth. Pergamon Press, New York. 225 pp

Kleiman, D.G.
1977. Monogamy in mammals. Quart. Rev. Biol., 52(1): 36-69.

McCrane, M.P.
1966. Birth, behaviour and development of a hand-reared two-toed sloth, *Choloepus didactylus*, at Washington Zoo. International Zoo Yearbook, 6:153-163.

Meritt, D.A., Jr.
1985. The two-toed Hoffman's sloth, *Choloepus hoffmanni* Peters. (this volume).

Montgomery, G. G.
1985. Movements, Foraging and Food Habits of the Four Extant Species of Neotropical Vermilinguas (Mammalia; Myrmecophagidea) (this volume).

Montgomery, G.G. and M.E. Sunquist
1975. Impact of sloths on neotropical forest energy flow and nutrient cycling. Pp. 69-111, *in* Tropical ecological systems: trends in terrestrial and aquatic research (Ecological Studies, Vol. II). (F.B. Golley and E. Medina, editors). Springer Verlag, New York, 398 pp.

1978. Habitat selection and use by two-toed and three-toed sloths. Pp. 329-359, *in* The ecology of arboreal folivores. (G.G. Montgomery, editor). Smithsonian Institution Press, Washington, D.C., 574 pp.

Stone, W.D.
1957. The gestation period of the two-toed sloth. J. Mamm., 38: 419.

Sunquist, M.E. and G.G. Montgomery.
1973. Activity patterns and rates of movement of two-toed and three-toed sloths (*Choloepus hoffmanni* and *Bradypus infuscatus*). J. Mamm., 54: 946-954.

Van Doorn, V.
1971. Verzögerter Zyklus der Fortpflanzung bei Faultieren. Zeit. Kolner Zoo, 14: 1-8.

Veselovsky, Z.
1966. A contribution to the knowledge of the reproduction of the two-toed sloth, *Choloepus didactylus* at Prague Zoo. International Zoo Yearbook, 6: 147-153.

THE TWO-TOED HOFFMANN'S SLOTH, *CHOLOEPUS HOFFMANNI* PETERS

Dennis A. Meritt, Jr.

Assistant Director, Zoological Gardens, Lincoln Park, Chicago, Illinois 60605.

Abstract
The natural history, behavior, nutrition, physiology, reproduction and development of the two-toed Hoffmann's sloth, *Choloepus hoffmanni* Peters is summarized, based on captive studies and field work.

Abstracto
Se hace resumen de la historia natural, el comportamiento, la nutrición, la fisiología, la reproducción y el desarrollo del perezoso de Hoffmann, *Choloepus hoffmanni* Peters que tiene dos dedos en los pies, en base a estudios en cautiverio e investigación en el terreno.

Resumo
A história natural, comportamento, nutrição, fisiologia, reprodução e desenvolvimento do preguiça de dois dedos de Hoffman *Choloepus hoffmanni* Peters são sumarizados, baseados em estudos em cativeiro e a campo.

This paper summarizes work carried out between 1968 and 1977 on Hoffmann's two-toed sloth *Choloepus hoffmanni* Peters. The study included captive animals at the Lincoln Park Zoological Gardens, Chicago, Illinois, and field work in the Republic of Panama beginning in 1974. Field studies have continued annually since then. At present there are two off-display colonies of *C. hoffmanni* at the Zoological Gardens under continuing study and observation. Some adults were born in captivity while others were captured in the wild in Colombia and Panama.

RANGE

C. hoffmanni ranges from Nicaragua, through Central America into Panama, and then southward into Colombia, Venezuela, Guiana, Ecuador, Peru, Brazil and Bolivia (Wetzel, 1985). It is unknown from Argentina, Chile, Paraguay or Uruguay (Cabrera, 1957). This arboreal mammal is found in areas of tropical rain forest, mixed deciduous forest, and secondary growth, but is not restricted to these habitats.

SIZE, COLOR AND EXTERNAL FEATURES

The adult weight varies from 2.7 to 8.1 kg (6 to 18 lb). The hair is typically lighter in color than the "common" two-toed sloth, *C. didactylus*, the general body hair coloration being near blonde, buff, tan or light brown in adults. There may be a shading of the hair from light to dark over the head and back, but not obviously so. The facial hair is characteristically lighter, being blonde, buff, or nearly white in some animals. *C. hoffmanni* lacks the dark shoulder or forearm markings found in *C. didactylus*. In the wild, the light-colored hair is often shaded pale green from algae which grows on the hairs.

In most mammals, the hair on the back is directed down and around to the abdomen. In *C. hoffmanni* the abdominal hair is parted in the middle, flowing toward the midback. For an upside down way of life, this allows for the efficient runoff of water from the abdomen.

Hair length, texture and density varies considerably among adults. *C. hoffmanni* from Peru have short, dense, coarse hair and, in addition, have short noses and rounded faces with a relatively larger head when compared with sloths from Panama or Colombia, which have elongated faces and smaller heads.

This arboreal folivore has forelegs equipped with two large curved claws (see Mendel, 1985), while the hind limbs have three claws. Most individuals have a palpable tail remnant which is not visible through the hair coat.

The ears are rounded and thickened but are almost always covered by hair. In older adults, the four large foremost teeth (caniniforms) may be visible while the sloth's mouth is closed and occasionally the upper two cause a groove or tract in the edge of the lower lip.

The eyes are quite mobile and can be partially retracted when the lids are tightly closed. They may make the sloth appear pop-eyed during periods of stress or prior to the onset of aggressive behavior, when the eyes appear to rise from their orbits, almost as if being forced forward. The pupil never fully dilates in proportion to the eyeball size, even at death. It always appears as a circular opening which varies in size with light intensity and frequency. The eyelids are muscular and can effectively protect the eyes against injury.

The external genitalia of *C. hoffmanni* are small, inconspicuous and close to the rectal oriface. It was once thought that the sloth had a cloaca as in reptiles and birds (Weber, 1928). Close examination reveals that the genital and rectal oriface are separated by a bridge of skin and tissue. This narrow bridge led some to call this area a psuedocloaca.

The testis are intra-abdominal. The male genitalia appears anterior to the rectal oriface, in adult males as a protuberance about the diameter of the tip of a human finger. At the end of this protrubrance is a small opening about pinhead in size. No penis can be expressed. The area between the rectum and the genitalia is covered with skin and sparsely haired (see McCrane, 1966, for *C. didactylus*).

The external genitalia of the female is closer to the rectal oriface than in the male and is without haired skin between the two. In the female, the same protruberance is present although it is smaller. At first glance, there appears to be a terminal opening but manipulation reveals that the prominence is cleft its entire length, terminating at the lip of the rectum.

Sexing of an intact, agitated sloth is hazardous and difficult. The usual procedure is to inspect the genitalia while the sloth is in a resting position on a perch. If this fails, the animal is removed from its cage, placed abdomen down on a smooth floor and the genitals visually inspected and probed if necessary.

SEX RATIO

C. hoffmanni received at Lincoln Park Zoological Gardens came from various sources, countries of origin and over a ten-year period. The sex ratio among these sloths may be somewhat biased by local collectors and survivability prior to arrival in the United States. A total of 80 individuals were received, including 36 males and 44 females; 45 percent male and 55 percent female. Of those born in the zoological gardens, five were male and six female.

Goffart (1971) did not present data on sex ratios in *C. hoffmanni*. Sex ratios of 72 *C. hoffmanni* were determined from three sources, all in the Republic of Panama (Meritt and Meritt, 1976). Comparison of these samples revealed populations that contained 94.1 percent females in one case; 93.5 percent females in a second sample and 91.6 percent females in a third. All of the sloths involved were from central Panama, giving a combined population of 93.1 percent females.

Preliminary work with the three-toed sloth, *Bradypus variegatus*, in Panama indicates a population with a proportion of 33 adult males to 35 adult females per 100 sloths. Young and juveniles make up the balance of this sample (total sample 113 sloths).

NUTRITION

In nature, *C. hoffmanni* eat a variety of plant material. including buds, leaves, flowers, fruit, twig tips and young stems. Young leaves are preferred to older mature leaves and in some cases overipe fruit is more acceptable than fruit in other stages of ripeness. The diversity of plant material suitable for these sloths is still being discovered. Studies by Montgomery and Sunquist (1975) and the author's observations indicate that numerous plant species are utilized. Animal material is probably also consumed when available, based on captive observations, perhaps including bird eggs, nestlings, a wide variety of insects, lizards and carrion.

In captivity, *C. hoffmanni* accepts a wide variety of food including celery, carrots, lettuce, escarole, spinach, green beans, apple, orange, grapes, cooked white potato, cooked sweet potato, banana, bread, peas, fruit cocktail and several types of meat or meat products. These items were offered and tested over time and were accepted by the majority of sloths (Meritt, 1970; 1973; 1975; 1977). Although there was individual variation with respect to food preferences, spinach, banana, and apple were preferred by the majority of acclimated animals over all other food items, in the order listed.

Sloths were routinely offered equal amounts of apple, sweet potato, lettuce or escarole, and spinach, with lesser amounts of orange, banana, and green beans. About 10 percent of the volume of food offered the sloths was a meat mixture also offered to small mammals at the Zoological Gardens. At the time of feeding, a vitamin-mineral powder with an added source of vitamin K was sprinkled over the food, along with ground hard-boiled eggs.

Over the past ten years, various sizes and shapes of food have been tried. Food was at times chunked, diced, sliced and offered as elongated thin pieces. In a comparison study of diced food versus chunked or sliced food, there was a 50 percent increase in consumption with the diced form and a dramatic decrease in spillage and wasted food. In diced form there was little or no wastage due to dropping and the animals had the least difficulty in eating. Large and small sloths were able to eat equally well and there was no need for the animals to pick up food prior to chewing and swallowing. They fed directly from the food containers with their mouths rather than reaching into the containers with their foreclaws.

The sloths clearly demonstrated an interest in meat or meat products. Meat was offered and accepted daily as a matter of routine. Various items were tried and accepted before using the meat mixture fed to carnivorous small mammals. The items included fresh ground horsemeat, canned dog food, moistened kibbled dog food, canned salmon and fresh ground smelt.

Food was offered in the middle of the reversed light period, between 1300 and 1400 hours daily. It was available to the sloths until 1000 hours the next day. Placing of the food containers in their respective areas of the cage usually resulted in feeding activity by the sloths. Although the time that food was available may seem excessively long, it allowed all individuals time to feed unimpeded, and reduced excessive food compeition among cage mates.

Water was available at all times and, despite the high water content of the food, the sloths drank frequently and regularly. Average daily water consumption for acclimated individuals was 170 cc (6 ounces) per animal per day.

There is no known vitamin deficiency that affects sloths in captivity if offered a balanced diet. There was a known need for a source of vitamin K in other xenarthrans and this additive was prophylactically given as part of the sloth diet. There was a known need for additional calcium during lactation in small mammals, and sloths received ground eggshell and calcium lactate powder prior to parturition and for the first few weeks of life of the developing young.

FOOD CONSUMPTION

During field work in 1974 and 1980 selected sloths were captured in Panama, acclimated and later sent to the zoological gardens. During the acclimation period of one to four weeks, sloths were housed in outdoor cages. They had access to water at all times and were fed diced apple, banana, carrot, mango and spinach. This basic diet was supplimented daily with a powdered vitamin, mineral and protein additive. Food was available for a minimum of 22 of each 24 hours. The sloths had access to sun or shade, and were exposed to outdoor temperature and relative humidity in central Panama during dry season.

Daily food consumption by individual sloths averaged 340 g per day per adult sloth (wet weight; range 150 to 507 g per day). Thus, an average adult *C. hoffmanni* consumed 85 g (wet weight; range 37.5 to 127 g) of food per day per kilogram of body weight.

Food consumption was also recorded for fully acclimated adult sloths in the Zoological Gardens. Food intake averaged 250 g per day per adult (wet weight; range 150 to 400 g). Food consumption in acclimated sloths decreased noticeably prior to elimination. The animals become lethargic and were reluctant to move about. The abdomen became full, rounded and distended.

ELIMINATION

The elimination habits of sloths are quite different from those of other mammals. The amount of stool and urine, the consistency and interval between eliminations is unique. Observers unfamiliar with the toilet habits of sloths have been prompted to report periodic bouts of constipation in their animals. The usual interval between eliminations in healthy juvenile and adult sloths is three to eight days. Eisenberg and Maliniak (1985) report periods of 3 to 5 days between eliminating by captive adult *C. didactylus.* Three-toed sloths in the wild descend from the trees and defecate on average about once each 6 days (Montgomery and Sunquist, 1975). Both urine and stool are passed at the same time and deposited in the same location by the sloth. Healthy sloths do not defecate or urinate daily under normal conditions of good health and physical condition, although sloths with a heavy parasite load may pass small amounts of loose stool daily.

Some arboreal mammals are at a decided disadvantage when they descend to the ground. This is particularly true of a mammal with extreme adaptations for arboreal life such as the sloth, which has difficulty moving across open ground. When sloths descend to the ground or to the base of a tree to eliminate, they are subjected to increased predator encounters because of their location and body attitude. Retention of both urine and feces for several days decreases the likelihood of predator encounters by reducing the frequency of descents to the ground. The longest interval between eliminations in confined sloths during this study period was six days.

Urination and defecation occurred at the same time in all cases, always on the ground. The feces were firm, formed into a compact mass made up of individual pellets. It was passed as single pellets or as a mass of compressed pellets. The usual volume per adult sloth was in excess of two cupfuls. The urine when voided, usually in copious amounts, was clear, yellow to brown and had a characteristic penetrating odor. The specific gravity was 1.023 (range: 1.020 to 1.031) and the pH 7.1 (range: 7.0 to 9.0). Measured amounts of urine averaged 808 cc (range: 340 to 1600 cc) per elimination. This amounts to about 178 cc per kg of body weight per urination per adult sloth. Weight loss due to elimination products has exceeded 30 percent of the sloth's body weight in exceptional cases. Laboratory testing for urine sugar, protein, urobilinogen, ketones, blood and bilirubin were negative in healthy, acclimated *C. hoffmanni.*

GROOMING

The incidence of mutual grooming in captive sloths is low, having been observed only during the course of an infant/mother relationship. The intentional, purposeful scratching of one sloth by another was not observed during a study period exceeding ten years. Field observations yielded similar results. On rare occasions, an adult sloth approached another and licked at its face and the hair over the nape of the neck. This usually prompted the sloth that was licked to smack its lips and sniff, nuzzle or lick the sloth that made the initial contact.

One form of self-grooming was centered about feeding and was usually in response to bits of food on the face, nose or wedged on the roof of the mouth. These food bits were dislodged with a foreclaw from the palate, using short flexing strokes of the forearm and claws. Bits of food clinging to the face and nose were clawed at in an attempt to loosen them or the sloth used its forelimb foot pad to brush the pieces off. These pieces were usually licked from the food pad after removal from the face.

Another form of self-grooming is best described as persistent scratching. The onset was spontaneous and was not obviously correlated with any other activity. A sloth hanging from a perch or resting with its back on a shelf would suddenly begin scratching with one rear leg. The areas scratched most frequently were the sides of the abdomen, lower back and neck region at the junction of the chest, in that order of frequency. This grooming was persistent and often lasted 1 to 6 minutes without interruption or change in speed or frequency of contact. It occurred more often during the sloth's night period and was most often seen in sloths that had just moved from a resting or sleeping position where they had been inactive for an extended period.

A similar form of self-grooming was performed with the foreclaws and usually occurred when the sloth was in an open resting position, with its back supported on a perch or shelf. The activity was spontaneous, the two areas most frequently scratched were the upper neck near the junction of the jaws and the external genitalia. While scratching near the jaw, a claw might be probed into the corner of the mouth and when this occurred, the scratching slowed, becoming more cautious. The same was true if the claw entered the rectal or genital oriface while scratching in these areas. The scratching was rhythmic and almost mechanical in nature, once started.

CIRCADIAN RHYTHM

Sloths in captivity are largely inactive. They rest or sleep about twenty hours per day (Goffart, 1968). In the wild they remain active through most of the night, and show about 11 hours of activity per diel (Sunquist and Montgomery, 1973). There has been disagreement regarding the nocturnal or crepuscular nature of this sloth (Lundy, 1952; Goldman, 1920). One study utilizing a movement-sensitive cage under controlled environmental conditions, concluded that *C. hoffmanni* was nocturnal (Howarth and Toole, 1973). Sloth activity did not start until about 1930 hours and remained at a heightened level until approximately 0530 hours. Peak activity started after the sloth's sunset and continued for about two to three hours. Sunquist and Montgomery (1973) reported the nocturnal nature of *C. hoffmanni*, in free-ranging, radio-tracked animals. They demonstrated essentially no crepuscular activity among individuals in the study group.

If adult sloths were maintained in an area illuminated by natural light or a combination of artificial and natural light, they invariably were inactive during the daylight hours. The only exception to this was influenced by feeding. If the animals were provisioned by their keepers during the day, it was not uncommon for the sloths to become active, moving to the food source and eating. Cage cleaning and servicing also caused some activity during the daylight hours.

When sloths were fed at night and serviced at the same time, daytime activity was almost nonexistant, except for minor shifts in resting or sleeping position without major location change. Sloths maintained under artificial illumination and reversed lighting responded accordingly. Red flourescent lighting as well as red spot-lamps, was used to illuminate cages housing sloths on a reversed schedule. The levels of illumination were such that the sloths could be easily seen. In this scheme, red light illuminated the animal area from 0700 hours to 1900 hours daily and white flourescent lighting from 1900 hours to 0700 hours. The animals were thus more likely to be active during daytime and asleep or inactive during the night. This allowed direct observation of active animals during a time convenient for visitors. Under this schedule, the sloths were serviced, manipulated and fed during their active period. The sloths adjusted to a reversed activity schedule without apparent difficulty. The bulk of active bouts occurred during the red light period. During the entire length of the white light period, most sloths remained stationary, resting or sleeping. An occasional individual would reposition itself or feed briefly.

BODY AND AMBIENT TEMPERATURE

Thermoregulation in *Choloepus* sp. has been reviewed by

Goffart (1971). The mean body temperature of *C. hoffmanni*, taken rectally, has been reported as 34.4°C with a range of 33.4°C to 36.2°C (Britton and Atkinson, 1938; Enders and Davis, 1936; Enger and Bullock, 1965; Goffart, 1967).

Body temperature was recorded intermittantly at the Zoological Gardens, and established a body temperature range of 32°C to 36°C. Body temperatures of captive sloths were recorded in Panama with a Yellow Springs model 42 SF Tele-thermometer, with the attached small-animal probe inserted a minimum of 10 cm into the rectum of the unrestrained sloth. The probe lead was not taped or otherwise secured to the animal's body. The sloths were free to move and climb about. Temperatures were recorded between 0730 and 1600 hours daily on two males and two females, for two weeks, under conditions of temperature and relative humidity in Panama during dry season. Sloths were subjected to similar temperature and humidity conditions as their free-ranging counterparts. The sloths had a mean rectal temperature of 32.9°C, range 30.5° to 34.2°C. Both the average and the range fall below those previously reported for this species, the minimum temperature by nearly 3°C (Meritt, 1974a).

During March and April of 1974, sloths were observed in nature and in captivity in Panama. Total observation time exceeded 150 hours. As ambient temperature increased, the sloths became inactive, assumed resting body positions and slept. Without exception, all perceptable activity ceased for a minimum of four hours, centered on midday (1200 hours). Air temperature during the study hours ranged from 24°C to 37°C. By the time the midpoint in the temperature range had been attained, the sloths were asleep in an open resting position with head extended, not tucked into the chest or abdomen. One or both forelimbs were folded on the chest or hung loosely away from the body. The rear legs grasped the tree trunk, limb or vine holding the sloth in position. No sloths were observed active between 0900 and 1500 hours on sunny days when the ambient temperature was at or above 30°C except for simple shifts in body position not involving a major change in body attitude.

RESPIRATORY RATE

The respiratory rate in *C. hoffmanni* is reported as 13 to 14 per minute. It decreases when rectal temperature exceeds 37°C (Irving, et al., 1942). Excited or agitated sloths breathe at a rate of 10 to 30 per minute (Britton, 1941).

Respiratory rates for captive sloths at 30°C ambient temperatures were variable depending upon the animal's state of alertness and activity. Sleeping adults of both sexes respired at a mean rate of 13 per minute (10-18); awake but inactive animals, 46 per minute (38-52) and awake, and active sloths, 73 per minute (65-78). Repiratory rates for both free-ranging and captive sloths in Panama were recorded. The resting rate for sloths at 30°C ambient temperature was 9 per minute with a range of 8 to 11 per minute (Meritt, 1974a). Differences between adults and juveniles, or between adult males and females were not seen. Montgomery and Sunquist (1972) noted a significant difference between the respiration rate of an adult male and pregnant female *Bradypus variegatus* in Panama.

CHEMICAL IMMOBILIZATION

C. hoffmanni of known weight were immobilized using Sernylan (phencylidine hydrochloride) at 1.65 mg per kg in combination with promazine hydrochloride at 0.85 mg per kg (Meritt, 1974b). The drugs were administered intramuscularly using a hand syringe. The onset of effect was marked by relaxation of the forelimbs and associated loss of coordination and grasping reflex in these limbs; loss in the rear limbs followed shortly thereafter. Total immobilization was reached with a relaxation of the head and neck, the jaw partially relaxed. Within an average of 7 minutes (range: 5 to 10 minutes) these animals were handled without fear of injury to the sloth or handler. Within 21 minutes (range: 15 to 26 minutes), the immobilization was complete and procedures carried out. Recovery times varied invidually, but were complete within five hours.

The use of these drugs in sloths was not contrindicated; no detrimental side effects were noted in any of the animals. Seizures or convulsions have been observed in felids and ursids using this combination but not in our experience with xenarthrans (Meritt, 1972).

BLOOD PARAMETERS

Venous blood was collected from four sloths, three female and one male, while under sernylan and promazine anesthesia. All animals were mature, acclimated individuals in good health and condition. The sloths had not been fed for a minimum of eight hours prior to sampling.

Hemoglobin values were 11.7, 13.6, 13.6 and 13.6 g per 100 ml, respectively. Hematocrit values ranged from 33 to 38.7 with a median of 35 percent. Red blood cells ranged from 3.85 X 10^6 to 5.04 X 10^6. White blood cells ranged from 13,400 to 21,000 per mm^3. Differential counts for the four sloths showed basophilic cells 1 to 3, eosinophils 0 to 5, myelocytes 0, juveniles 0, stab cells 0 to 3, seg cells 40 to 49, lymphocytes 40 to 49 and monocytes 0 to 3. These values fall within the range of variability of previous reports (Toole 1972; Marvins and Shook 1963).

MENSES

The onset of menstruation in *C. hoffmanni* was detectable as the presence of small amounts of blood on the external genitalia and in the vaginal opening. Blood spots were seen on perches or resting areas. Swelling or edema of the vaginal lips was noted in most cases. In some cases, there was obvious separation between the genitalia and the rectal oriface. In two cases, there was a hard, dark piece in the vaginal opening which was almost scablike in appearance and consistency. This scab was easily removed.

The menstral cycle spanned 5 to 7 days. Days 1 through 3 were marked by the presence of rusty colored fluid or blood in and on the vaginal opening. On day four, the external genitalia were dry and clean but slightly puffy. Insertion of a cotton swab into the vaginal opening revealed some blood-tinged fluid. Day 5 through 7 usually showed genital edema but no fluids or secretion, internally or externally. In only one case the cycle was prolonged for 10 days in a large, old adult maintained separately.

ESTRUS

The only reliable indicator of the onset of estrus is a behavioral change requiring careful and repeated observation to detect. Four females were observed in estrus during the study period. During one of these observation periods, two females were noted rubbing against one another in a near vent to vent position but they separated when the observer was noted. No change in genital appearance or secretion or discharge was noted in any of these females during this time.

When in estrus, the females actively solicited the male by pressing close to him, attempting to rub their ano-genital region against his. When the females were near or touching the male, he was not aggressive but, on the other hand, showed no real interest and no sexual behavior. His overall activity did not increase during these 10 to 14 day periods of apparent female interest and estrus. The male usually responded to the female's attention by shifting body position slightly and tightly curling up in a closed resting position. He did not leave his favored perch area during these periods.

SEXUAL MATURITY

The three-toed sloth *Bradypus tridactylus* may take up to six years to reach adult size and weight (Tirler 1966). Veselovsky (1966), described mature examples of *C. didactylus* that were about three years old but not reproducing in captivity until nearly ten years old. He deduced from a captive-born individual that this species would need 30 to 36 months to reach definitive size and weight. No estimate was given as to the age at sexual maturity.

All males arrived in our colony as adults of unknown age, and no infant nor juvenile males reached sexual maturity during the study period. There was, however, conclusive proof for sexual maturity at about 2 years in the female *C. hoffmanni*. A female born on 11 August 1972 died on 30 December 1974. Autopsy revealed a well-formed and developing fetus *in utero*. Thus, two years and four months after her own birth, she was sexually mature and nurturing her own baby. The onset of sexual maturity in a female *C. hoffmanni* is thus at or about two years of age. Eisenberg and Maliniak (1985) reported that captive *C. didactylus* do not breed in captivity before they are 3 years old, and that males are not sexually mature before they are 4.5 years old.

PARTURITION

Diagnosing pregnancy in Hoffmann's sloth was difficult at best. Because of fetal position, abdominal musculature and pelvic skeletal structure in the adult, physical palpation yielded negative results in nearly all cases. Full urinary bladders and colons substantially increased the girth in some sloths, even if they were not pregnant.

Postpartum, there was a total absence of blood, blood-tinged fluids or tissue debris in or around the genitalia. The external genitalia were not obviously swollen, torn or distended. Newborn sloths were alert, mobile and strong. In a captive situation, the baby could climb onto an adult sloth other than its mother and be accepted. Adult sloths were observed next to and physically touching females shortly after parturition without signs of hostility or aggression. If a transfer occurred during the first two days of life the survival of the newborn was in doubt. Suckling by the newborn apparently stimulates milk flow, for no obvious breast development or fluid was noted prior to parturition. Nuzzling and suckling stimulated the flow, usually within the first 48 hours.

None of several births that occurred in the colony were observed. Single offspring were always born during the sloth's daylight period when they were normally not under observation.

NEWBORN SLOTHS

Newborn *C. hoffmanni* weigh from 340 to 400 g (Phillippott et al., 1965). Newborns at the zoological gardens weighed from 350 to 454 g in seven cases. Sloths were born with their eyes open, teeth present and pigmented, the ear pinnae were patent, the claws were fully formed and hard. The sloths were fully haired except over the ventral surface where the hair was sparse and less dense. Newborn sloths vocalized loudly, were capable of coordinated movement, were able to hold their head erect and the grasping reflex was strong. They were also capable of biting if so motivated. They were able to cling or crawl about the mother's abdomen and successfully transfered from one adult sloth to another without assistance. The usual position for the newborn was spread on the mother's abdomen, head to her chest, grasping the body hair of the adult's abdominal sides. The claws were sharp and very effective in keeping this hold. Sex determination in the newborn was as difficult as in the adult and required careful observation.

MOTHER-INFANT RELATIONSHIP

Mother sloths were solicitous of their single offspring in all cases. During the first several days after parturition, a mother spent considerable time vigorously licking the head and face of the baby. The ano-genital region was also cleaned daily. Urine and fecal material passed by the infant were consumed by the mother. Elimination was stimulated by the licking of the female, at least during the first few weeks of life.

Individual mothers showed varying degrees of agitation towards other adult females in the colony postpartum. In some cases, other females came to and licked the newborn, particularly its head, face and dorsal surface. This was usually noted within twelve hours of birth and the mother usually tolerated their assistance and attention. In two cases, the presence or proximity of another sloth caused the mother to hiss, strike out weakly with a foreclaw and shift body position or change cage location. When the new position had been reached, the mother curled over the baby, using her limbs as drapes over the infant's body and shielding it.

The newborn sloth spends the first few weeks of its life in intimate contact with its mother, clinging to her chest and abdomen. Because of this association and proximity, the mother attended to the infant's needs almost spontaneously. Mutual lickings of the face and head was observed. Mothers were slow and deliberate while moving about the colony and appeared conscious of the safety of the chosen route. Areas with a high concentration of sloths or those particularly tangled with perches were avoided initially.

As the infant grew and started first attempts at exploration or play, the mother was always nearby. Usually the infant left the mother and explored in the immediate vicinity of her body. By the age of two months, explorations lasted longer and the distances away from the mother were greater. On several occasions, an infant strayed too far or encountered another sloth. If the direct route to mother was not blocked, the infant rapidly returned to her and to a clinging position. If the route was blocked or if the way back was too complicated to maneuver, the infant sloth vocalized almost immediately. These vocalizations were loud and characteristically were bleats lasting from 30 to 90 seconds (Montgomery and Sunquist, 1974) repeated until there was an obvious response from the mother. In one case, the number of bleats was ten, by the tenth the mother had reached the infant and it crawled onto her abdomen. The mother responded by moving from her position directly to the infant. Mothers responded to these distress calls beginning at about sixty days and until the offspring was nearly six months old. Specifically, responses were noted at sixty, sixty three, seventy two, one hundred and fifty and one hundred eighty days in at least three offspring.

DEVELOPMENT OF YOUNG

Based on the intensive observation of developing sloths of known age and parentage, a tentative development schedule in captivity was established. Newborns nursed for periods of from 6 to 15 minutes at a time, several times daily. Water was consumed independantly at 7 days of age. The first observed solid food consumption was at 15 days for a male and 27 days for a female infant. Play behavior was noted at 15 and 19 days of age. The baby first assumed a hanging or upside down position between 20 and 25 days of age. The developing sloth actively sought solid food with its mother between 30 and 42 days after birth. The first foods were meat mixture, banana, cooked sweet potato and hard boiled egg. Sleeping in an adult position off the mother was noted at 50 days. The first apparent independant exploration was observed at 41 days for a female and 54 days for a male. Exploration away from the mother and in close proximity to other sloths was noted at 59 and 65 days of age.

By five months of age, the sloth was feeding on solids regularly and away from its mother. By six months the sloth was eliminating in typical adult fashion, on the cage floor, head up and eyes partially closed. At this age, the sloth was still in behavioral association with its mother, but was capable of independent existance.

Although infants were capable of independent exploration and existence much sooner, one offspring was still tolerated in a carrying position on its mother's abdomen at eight months of age. Despite its large size and considerable weight, the mother carried it without apparent difficulty.

There was no sudden nor dramatic break in ties between mothers and their offspring. At no time was a developing sloth obviously chased by its mother, nor were any aggressive behaviors noted between the two. The relationship gradually changed from maternal dependance to one of companionship and association, as the sloth matured. Even two years after birth, a female offspring was most often found in close proximity to its mother, often being active, feeding or resting at the same time as the adult and in association with her. The bond between the two was strong and without apparent hostility.

ACKNOWLEDGMENTS

Dr. Lester E. Fisher, Director, allowed and encouraged these studies. The keepers in the Small Mammal and Lion Houses contributed time, care and behavioral observations. The American Philosophical Society, Philadelphia, through its Penrose Research Fund supported portions of this work. I am grateful for the assistance and cooperation of all. Gail Meritt assisted in many facets of these studies and was especially supportive of this work.

LITERATURE CITED

Britton, S.W.
1941. Form and function in the sloth. Quart. Rev. Biol., 16: 13-34; 190-207.

Britton, S.W.. and W.E. Atkinson.
1938. Poikilothermism in the sloth. J. Mammal., 19: 94-99.

Cabrera, A.
1957. Catalogo de los Mammiferos de American del Sur. Rev. Mus. Argentino Cienc. Nat. "Bernardino Rivadavia", Cienc. Zool., 4: 1-308.

Eisenberg, J.F. and E. Maliniak.
1985. Maintenance and reproduction of the two-toed sloth *Choloepus didactylus* in captivity. (this volume).

Enders, R.K. and D.E. Davis.
1936. Body temperatures of some Central American mammals. J. Mammal., 17: 165-166.

Enger, P.S. and T.H. Bullock.
1965. Physiological basis of slothfulness in the sloth. Hvalradets Skriftet, 48: 143-160.

Goffart, M.
1967. Variation circadienne de la temperature chez un Paresseux, 1' Unau, *Choloepus hoffmanni*. J. de Physiol., 59: 415-416.

1968. The problem of slothfulness in the didactyl sloth, *Choloepus hoffmanni* Peters. Electromyography, 8: 245-251.

1971. Function and form in the sloth. Pergamon Press, Oxford. 225 pp.

Goldman, E.A.
1920. Mammals of Panama. Smith. Misc. Coll., 69. (5), Public. No. 2498: 1-309

Horwarth, S.T. and J.F. Toole.
1973. Some observations on the circadian rhythm of *Choloepus hoffmanni*, the two-toed sloth, Lab. An. Sci., (3): 377-379.

Irving, L., P.F. Scholander, and S.W. Grinnel.
1942. Experimental studies of the respiration of sloths. J. Cell. Comp. Physiol., 20: 189-210.

Lundy, W.E.
1952. The upside-down animal. Nat. Hist., 61: 114-119.

Marvins, H.N. and B.R. Shook.
1963. Hematological studies on the two-toed sloth, *Choloepus didactylus*. Comp. Biochm. Physiol., 8: 187-189.

McCrane, M.P.
1966. Birth, behavior and development of a hand reared two-toed sloth, *Choloepus didactylus*. International Zoo Yearbook, 6: 187-189.

Mendel, F.C.
1985. Adaptations for suspensory behavior in the limbs of two-toed sloths. (this volume).

Meritt, D.A.
1970. Edentate diets currently in use at Lincoln Park Zoo, Chicago. International Zoo Yearbook, 10: 136-138.

1972. Edentate immobilization at Lincoln Park Zoo, Chicago. International Zoo Yearbook, 12: 218-220.

1973. Edentate diets II. Two-toed sloths. Lab. An. Sci., 23: 543-545.

1974a. Grant Report, (Penrose Fund). Amer. Phil. Soc. Yearbook No. 6782: 358-359.

1974b. A further note on the immobilization of sloths. International Zoo Yearbook, 14: 160-161.

1975. Edentates. The Ark. Vol. II. 4: 2-5; 12-15.

1976. A simplified method for diagnosing parturition

in Hoffmann's sloth, *Choloepus hoffmanni*. International Zoo Yearbook, 16: 152-153.

Meritt, D.A. and G.F. Meritt.
1976. Sex ratios of Hoffmann's sloths, *Choloepus hoffmanni* Peters and Three-toed sloth, *Bradypus infuscatus* Wagler in Panama. Amer. Midl. Nat., 96 (2): 472-473.

1977. Edentate nutrition. *in* CRC Handbook Series in Nutrition and Food. Vol. I: 541-548. (M. Rechcigl, Jr. editor). CRC Press Inc., Cleveland.

Montgomery, G.G., and M.E. Sunquist.
1972. Resting respiration rates of free-living three-toed sloths. Mammalia, 36(3): 530-532.

1974. Contact-distress calls of young sloths. J. Mammal., 55(1): 211-213.

1975. Impact of sloths on neotropical forest energy flow and nutrient cycling. Pp. 69-98. *in*: Trends in tropical ecology: ecological studies IV (E. Medina and F. Golly, editors). Springer-Verlag, New York.

Philipott, E., M. Goffart and A. Dresse.
1965. Le systeme surrenalo-sympathique ches la parreseax, *Choloepus hoffmanni* Peters, Arch. Int. Physiol. Bioch., 73: 476-504.

Sunquist M.E. and G.G. Montgomery.
1973. Activity patterns and rates of movement of two-toed and three-toed sloths *(Choloepus hoffmanni* and *Bradypus infuscatus*). J. Mammal., 54(4): 946-954.

Tirler, H.
1966. A sloth in the family. The Harvill Press, London, 47 pp.

Toole, J.F.
1972. Blood chemistry of the sloth. Lab. An. Sci. 22 (1): 118-121.

Veselovsky, Z.
1966. A contribution to the knowledge of the reproduction and growth of the two-toed sloth, *Choloepus didactylus* at the Prague Zoo. International Zoo Yearbook. VI: 147-153.

Weber, M.
1928. Die Saugetiere, 2nd Ed. Gust. Fisher Verlag, Jena.

Wetzel, R.M.
1985. The identification and distribution of recent xenarthra (= edentata). (this volume).

SLOTHS AND OTHER MAMMALIAN PREY OF THE HARPY EAGLE

Robert J. Izor

Field Museum of Natural History, Chicago, Illinois 60605

Abstract

Skeletal remains of about 83 prey items were gathered from a Harpy Eagle nest site. All remains were mammalian, consisting of about one-third sloths, one-third primates, and smaller numbers of opossums, porcupines, agoutis, and kinkajous.

Abstracto

Los restos de los esqueletos de alrededor de 83 elementos de sus presas, fueron rescatados de un sitio de nidaje del Aguila Harpía. Todos los restos procedían de mamíferos y consistían de como un tercio de perezosos, un tercio de primates, y de números menores de zarigüeyas, puercoespines, agutíes y "cusumbíes" (= kinkajous).

Resumo

Restos esqueléticos de cerca de 83 presas foram reunidos de um ninho de uma águia harpia. Todos os restos eram de mamíferos, sendo constituído de um terço de preguiças, um terço de primatas e um número menor de gambás, porcos-espinhos, cotias e juparás.

Because it is the largest and most powerful of the birds of prey, the Harpy Eagle *Harpia harpyja* has a significant influence on the neotropical fauna of which it is a part. This study deals with the food habits of a pair of Harpy Eagles while they incubated and raised a chick.

MATERIALS AND METHODS

Data were gathered during two periods of field observation in southwestern Guyana, from May through December 1974 and March through April 1975. Observations were made on many aspects of the eagles' biology, some reported elsewhere (Rettig, 1977; 1978). Skeletal remains of food items were gathered from near the base of the tree and from the nest. This material was brought to the Field Museum of Natural History, Chicago, Illinois, for identification. Additionally, photographic and sight records were analyzed for agreement with the results of the physical data.

RESULTS

The original material comprised 106 samples, representing collections from the ground under the nest tree, from the nest itself, and some odd lots of recovered bones. Some lots contained more than one individual, since carcasses remained in the nest for many days and there were often several skeletons "shedding" pieces simultaneously. Conversely, some collections made many days apart contained parts recognizably of the same animal. At an observed hunting rate of once or twice each week, the approximately 83 individuals found represent a large majority of the kills made during the year of study.

Three items found under the nest were not considered; a partial ilium of the toad *Bufo*, and a tibia and a fibula of a very young brocket deer fawn *Mazama*. The toad bone was almost certainly not from a prey item, and could have been under the tree for reasons unassociated with the eagles. The brocket deer fawn, while certainly eaten by the eagles, represents an imponderable since including it would necessitate consideration of all other species in which only the young are in the vulnerable 1 to 5 kg weight class. This could only be accomplished with extensive and unavailable data on reproduction and growth characteristics.

Including the *Mazama*, at least eleven species of mammals, belonging to six orders, were represented in the skeletal remains (Table 1). All but the most fragmentary and featureless pieces were allocated at least to a restricted part of a family.

Cebid monkeys constituted about one-third of all

Table 1. Relative numbers and biomass of mammals in samples collected from the vicinity of the nest of the Harpy Eagle. For comparison, estimates of faunal composition from other neotropical areas are summarized.[1]

	Average Weight(kg)	*N*	*This Study* *%N*	*% Biomass*	*Surinam* *%N*	*% Biomass*	*Barro Colorado Island* *%N*	*% Biomass*
Marsupialia								
Didelphidae								
Didelphis	1.8	3	3.6	2.2	2.2	1.2	5.5	3.0
Metachirus	1.8	0			1.2	0.7	0.1	0.1
Philander	1.4	0			0.1	0.1	2.1	0.9
Caluromys	1.3	0			0.4	0.2	<0.1	<0.1
Xenarthra								
Bradypodidae								
Bradypus	3.2	7	8.4	9.3	33.2	32.7	60.2	59.0
Choloepus	4.3	19	22.9	34.1	12.1	16.0	14.8	19.5
Myrmecophagidae								
Tamandua	4.0	0			3.7	4.6	<0.1	<0.1
Dasypodidae								
Dasypus	4.0	0			15.2	18.7	<0.1	<0.1
Euphractus	4.9	0			<0.1	<0.1	–	–
Primates								
Cebidae		29[2]	34.9	31.4				
Alouatta	5.5	1	1.2	2.3	6.8	11.5	5.3	8.9
Ateles	5.0	0			<0.1	<0.1	0.1	0.2
Cebus	2.6				<0.1	<0.1	1.3	1.0
Rodentia								
Erethizontidae								
Coendou (2 species)	1.8	11	13.3	8.3	20.9	11.6	<0.1	<0.1
Dasyproctidae								
Dasyprocta	2.0	5	6.0	4.2	3.2	2.0	7.4	4.5
Carnivora								
Procyonidae								
Nasua	3.0	0			<0.1	<0.1	3.2	2.9
Potos	2.5	8	9.6	8.3	1.0	0.8	<0.1	<0.1
Mustelidae								
Eira	4.0	0			<0.1	<0.1	<0.1	<0.1
Felidae								
Felis	5.0	0			<0.1	<0.1	<0.1	<0.1

[1] Estimates from Eisenberg and Thorington (1973); includes only species with 1.0-5.5 kg adult weight.

[2] May include up to four species (see text).

individuals eaten (Table 1). Four species may be included under Cebidae: *Cebus apella, C. nigrivittatus, Chiropotes satanus*, and *Pithecia pithecia*. No means was found to satisfactorily distinguish all femori of these species, and the bulk of the cebid remains were posterior vertebrae, pelves, and femori. All of these individuals appeared to have been mature. Compared to their estimated abundance in the general mammal fauna, the cebids are greatly overrepresented in prey remains.

Sloths made up another one-third of the individuals identified and were predominately the two-toed sloth *Choloepus*. This is in line with their probable abundance as a family, although the two genera were disproportionately represented. In the censuses from other areas *Bradypus* is 2 to 3 times as numerous as *Choloepus* (Montgomery and Sunquist, 1974), whereas it is less than half as numerous in the prey remains. Skulls and forelimbs predominated in the material, and there were few young.

Two species of prehensile-tailed porcupine, *Coendou*, were identifiable in the remains. They are easily separable by size, and it appears that the larger species, *C. prehensilis*, was represented in the eagle's diet less frequently than the figures from Surinam might suggest (Table 1). The other rodent found, *Dasyprocta*, was the primary terrestrial prey.

Small numbers of kinkajou *Potos flavus* and oppossum *Didelphis* remains were also found. No conclusions on eagle preference for these can be drawn because of the small sample. Notably absent from the remains were armadillos *Dasypus* and tamanduas *Tamandua*.

DISCUSSION

Several aspects of selectivity are indicated in this sample of Harpy Eagle prey. It is self-evident that the eagle prefers mammals as food. Despite the wide variety of reptiles and birds in the local fauna, none were apparently taken. The reason for this is unclear, but the behavioral aversion to nonmammalian prey is persistent even in captivity (Foerster, 1972).

Another conclusion may be that the eagles prefer arboreal hunting, although the data are actually not conclusive on this point. The apparent predominance of arboreal prey is more parsimoniously explained as a preference for monkeys, since the other arboreal species (sloths, porcupines, and kinkajous) are not consistently taken as prey out of proportion to their availability.

A complex interaction of physiological and behavioral traits probably causes the apparent preference for two-toed sloths *Choloepus* and three-toed sloths *Bradypus*. Sunquist and Montgomery (1973) found in Panama that both genera minimized their activities around sunrise, the usual hunting time for the Harpy Eagle. At this time the ambient temperature is at its minimum and the sloths, slowed by thermal stress, are high in the trees seeking early sunshine to warm themselves (Montgomery and Sunquist, 1978). This, of course, exposes them to attack. Combinations of metabolic limitations may be forcing the sloths to risk this exposure.

One of the most remarkable features of the Harpy Eagle are the enormously developed feet and claws. Difficulties in preying on the robust and tenacious sloths necessitate such weaponry. Since the bird must strike and dislodge the sloth from its tree without losing flight speed, impact and penetration of the talons is critical (described in Retting, 1978). Even after sustaining an attack and being carried some distance back to the nest, some sloths were still alive.

Part of the sloth's resistance to injury is due to its broad and sturdy ribs which enclose most of the torso and are well connected to the spine. This should not be taken, however, as a specific defensive adaptation against eagle predation because it is widespread in the xenarthrans and may be an ancestral trait.

Voous (1969), discussing predatory potential in 30 neotropical raptors including *Harpia*, showed a nearly linear relationship between body weight and claw size. The slope of this line is such that a fivefold increase in body weight only doubles claw length. If prey size is proportional to claw size, obviously a huge increase in the bird's body weight is necessary before a substantially different array of prey species becomes available.

Cebid monkeys are the group which is most preferentially taken, relative to their presumed abundance. Despite their agility, cooperative defenses, and visual abilities, they are unable to escape the eagle's attack. Fowler and Cope (1964) and Foerster (1972) noted that the Harpy Eagle has excellent hearing and can easily locate animals by sound. Fowler also described their astonishing speed and maneuverability while flying through the forest canopy and middle stories. It appears that a monkey troop would have no warning of the eagle's approach until the bird emerged from the surrounding foliage.

All of the primate remains found were adult, a result contrary to expectations in most predator-prey systems. This may simply be due to disintegration or consumption of fragile juvenile skeletons, or may reflect selection of peripheral adult members of the troop.

It is evident that the Harpy Eagle is well adapted for exploiting its environment. Its prey are primary consumers in the most productive part of the tropical ecosystem, the forest canopy. With only a single intermediary, the Harpy Eagle utilizes the foliage (howler monkeys, sloths), fruit (other cebids, kinkajous), and twigs and bark (porcupines). Few other species can tap such a rich and broadly based resource. Even apart from its notoriety as the only avian predator of man, this eagle is a remarkably successful carnivore.

ACKNOWLEDGEMENTS

I thank the members of the field party, especially Neil Rettig, for the opportunity to identify and analyze their material. Assistance was also given by Sarah Derr, Philip Hershkovitz, and Ron Pine. During part of this project I was employed under a grant to the Division of Mammals, Field Museum of Natural History, from the National Science Foundation.

LITERATURE CITED

Eisenberg, J.F. and R.W. Thorington, Jr.
1973. A preliminary analysis of a neotropical mammal fauna. Biotropica, 5: 150-161.

Foerster, J.
1972. Notas biologicas sobre *Harpia harpyia* (Linne). Neotropica, 18: 146-148.

Fowler, J.M. and J.B. Cope.
1964. Notes on the harpy eagle in British Guiana. The Auk, 81: 257-273.

Montgomery, G.G. and M.E. Sunquist.
1974. Impact of sloths on neotropical forest energy flow and nutrient cycling. Pp. 69 to 95, *in* Tropical Ecological Systems (E. Medina and F. Golley, editors). Ecological Studies Vol. II. Springer-Verlag, New York, N.Y. 398 pp.

1978. Habitat selection by two-toed and three-toed sloths. Pp. 329-359, *in* The ecology of arboreal folivores (G. G. Montgomery, editor). Smithsonian Institution Press, Washington, D.C. 574 pp.

Rettig, N.
1977. In quest of the snatcher. Audubon, 79(6): 26-49.

1978. Breeding behavior of the harpy eagle (*Harpia harpyja*). The Auk, 95: 629-643.

Sunquist, M.E. and G.G. Montgomery.
1973. Activity patterns and rates of movement of two-toed and three-toed sloths (*Choloepus hoffmanni* and *Bradypus infuscatus*). J. Mamm., 54: 946-954.

Voous, K.H.
1969. Predation potential in birds of prey from Surinam. Ardea, 57: 117-148.

ECOLOGY OF LIVING VERMILINGUAS

IMPACT OF VERMILINGUAS (*CYCLOPES, TAMANDUA* : XENARTHRA = EDENTATA) ON ARBOREAL ANT POPULATIONS

G. Gene Montgomery

Smithsonian Tropical Research Institute, Apartado 2072, Balboa, Republic of Panama.

Abstract

Two species of vermilinguas (neotropical mammalian anteaters) *Cyclopes didactylus* and *Tamandua mexicana*, eat arboreal ants on Barro Colorado Island, in Panama. Biomass and diversity of arboreal ants were estimated by fogging insecticide into the crowns of trees, then catching and identifying samples of ants which fell. Diets of vermilinguas were studied by direct observation and by analyses of stomach contents and feces. Habits of vermilinguas were studied by radio-marking and following selected individuals. *Cyclopes* were arboreal and nocturnal, while *Tamandua* spent about 40 percent of their activity and feeding time in trees and were sometimes nocturnal, depending on the individual. *Cyclopes* fed entirely on ants. One-third of the food items identified from *Tamandua* were ants, the remainder termites. There was about 20 percent overlap in the kinds of ants eaten by *Tamandua* and *Cyclopes*, but they tended not to eat high proportions of ants of the same morpho-species. *Cyclopes* concentrated its feeding on three morpho-species of ants. *Crematogaster* sp. and *Solenopsis* sp. (Myrmicinae) and *Pseudomyrmex* sp. (Pseudomyrmecinae) ranked highest and accounted for about one-half the ants identified. Feeding by *Tamandua* was concentrated on one ant morpho-species each feeding period, but not the same morpho-species from day to day or from individual to individual. *Procryptocerus belti* and *Crematogaster* sp. (Myrmicinae), and an unidentified ant morpho-species ranked highest, and together accounted for about 45 percent of the total ants in *Tamandua* diets. Each *Cyclopes* ate 700 to 5,000 ants per day depending on its age and sex, and each *Tamandua* about 9,000 ants; both species of vermilingua ate 1 larvae or pupae for each 9 adult ants. *Cyclopes* cropped about 2,200 and *Tamandua* about 300 arboreal ants per ha per day, a total of about 2,500 arboreal ants. The daily cropping rate was less than 0.8 percent of the seasonal minimum standing crop of arboreal ants.

Abstracto

Dos especies de mamíferos hormigueros, *Cyclopes didactylus* y *Tamandua mexicana* segaren las poblaciones de hormigas arbóreas de la isla de Barro Colorado en Panamá. Se estudió la biomasa y diversidad de hormigas arbóreas cubriendo la copa de los árboles con una niebla de insecticida y luego capturando e identificando una muestra de las hormigas que caían al suelo. Las dietas de los hormigueros fue estudiada por observación directa y el análisis del contenido estomacal y las heces. Los hábitos de los hormigueros fueron estudiados marcándo los con transmisores de radio y siguiendo el movimiento de individuos escogidos. Los hormigueros *Cyclopes* demostraron ser arbóreos y nocturnos, mientras que los *Tamandua* pasaron como el 40 por ciento de sus períodos de actividad y de alimentación en los árboles y a veces fueron nocturnos, dependiendo, en parte, de variaciones entre individuos. *Cyclopes* se alimentó exclusivamente de hormigas, mientras que sólo 1/3 de los elementos alimenticios identificados de *Tamandua* fueron hormigas, y el resto eran termitas. Hubo como un 20 por ciento de coincidencia en las especies morfológicas de hormigas ingeridas por *Tamandua* y *Cyclopes* , pero ambas especies mostraron la tendencia a ingerir una proporción alta de una sola especie morfológica. *Cyclopes* concentró su alimentación en unas tres especies morfológicas de hormigas, con las especies *Crematogaster* sp. y *Solenopsis* sp. (de la subfamilia Myrmecinae) y *Pseudomyrmex* sp. (de la subfamilia Pseudomyrmecinae) figurando como las más comunes y respondiendo por más o menos la mitad de las hormigas identificadas. *Tamandua* concentró su alimentación sobre una especie de hormigas durante cada período de alimentación, no siendo esta especie de hormigas la misma de día a día o de individuo a individuo. Las hormigas *Procryptocerus belti* y *Crematogaster* sp. (Myrmicinae) y un especies no identificada figuraron como las más comunes, y entre ellas sumaron como el 45 por ciento del total de hormigas en la dieta de *Tamandua.* Cada *Cyclopes* ingirió de 700 a 5,000 hormigas por día, dependiendo de su edad y de su sexo, y cada *Tamandua* ingirió como 9,000 hormigas; ambas especies de hormigueros se comieron como l empolladura (de larvas o de pupas) por cada 9 hormigas adultas. *Cyclopes* cosecharon como 2,200 y *Tamandua* como 300 hormigas arbóreas por hectárea y por día, o sea un total de unas 2,500 hormigas arbóreas. La tasa de cosechas fue de menos del 0.8 por ciento del mínimo estacional de la cosecha en pie de hormigas arbóreas.

Resumo

Duas espécies de tamanduás *Cyclopes didactylus* e *Tamandua mexicana* colher populações de formigas arbóreas na ilha de Barro Colorado, no Panamá. A biomassa e diversidade de formigas arbóreas foram estudadas pulverizando-se inseticida nas copas das árvores, coletando-se e identifacando-se, então, uma amostra de formigas que caem. Dietas de tamanduás foram estudadas por observação direta e análises de conteúdo estomacal e fecal. Os hábitos dos tamanduás foram estudados por rádio-rastreamento e seguindo-se indivíduos selecionados. *Cyclopes* são arbóreos e noturnos, enquanto *Tamandua* passam cerca de 40 percentagem de seu tempo de atividade e alementação nas árvores e sendo, ás vezes, noturnos, dependendo em parte do indivíduo. *Cyclopes* se alimentam inteiramente de formigas. um tercô da alimentação identificada de *Tamandua* foi de formigas, o restante cupins. Houve cerca de 20 percentagem de superposição das morfo-espécies de formigas comidas por *Tamandua* e *Cyclopes*, mas eles apresentaram tendências a não comer altas proporções de formigas da mesma morfo-espécie. *Cyclopes* concentrou sua alimentaçõa em cerca des três morfo-espécies de formagas, com *Crematogaster* sp. e *Solenopsis* sp. (Myrmicinae) e *Pseudomyrmex* sp. (Pseudomyrmicinae) apresentando-se em maior quantidade e contando cerca da metade das formigas identificadas. A alimentação de *Tamandua* estava concentrada em uma espécie de formiga em cada período alimentar, não o mesmo do dia-a-dia ou de indivíduo a indivíduo. *Procryptocerus belti* e *Crematogaster* sp. (Myrmicinae) apresentarme-se em maior quantidade e juntas contavam cerca de 45 percentagem do total de formigas nas dietas de *Tamandua.* Cada *Cyclopes* comeu cerca de 700 a 5.000 formigas por dia, e cada *Tamandua* cerca de 9.000 formigas, ambas as espécies de tamanduãs comeram cerca de 1 jovem (larva ou pupa) por cada 9 formigas adultas. *Cyclopes* consumiu cerca de 2.200 e *Tamandua* cerca de 300 formigas arbóreas por ha por dia, num total de cerca de 2.500 formigas arbóreas. A taxa de consumo diária foi de menos que 0,8 percentagem da população sazonal, mínima, de formigas arbóreas.

There are four species of vermilinguas (neotropical mammalian anteaters), including: *Cyclopes didactylus*, the silky anteater; *Tamandua mexicana*, the banded anteater; *Tamandua tetradactyla*, the golden anteater; and *Myrmecophaga tridactyla*, the giant anteater. These vermilingua genera form a three-member feeding guild which includes a small arborial ant specialist *Cyclopes*, a medium-sized scansorial (arboreal and terrestrial) predator of ants and termites *Tamandua*, and a large terrestrial ant specialist *Myrmecophaga* (Montgomery, 1979). Two genera of armadillos, *Priodontes*, the giant armadillo, and *Cabassous*, the naked-tailed armadillos, may specialize on colonies of social insects as well, and thus add to the feeding guild; their food habits have not yet been studied in detail (Redford, 1985).

The ranges of the two *Tamandua* species do not overlap (Wetzel, 1975), but in many parts of the neotropics members of all three genera live in the same areas (Wetzel, 1985) and indeed may have overlapping home range spaces (Montgomery, 1979).

Members of two of the three genera of vermilinguas feed on arborial ants. There is thus a potential for competition for arboreal food and living space between *Cyclopes* and *Tamandua*. The impact of vermilinguas on arborial ant populations is a result of the combined feeding activities of members of these two genera.

METHODS

Capture, Radio-tracking and Sampling Diets.-- Data reported here are mainly from *Tamandua mexicana* and *Cyclopes didactylus* captured and radio-marked on Barro Colorado Island, Panama (BCI) then tracked as they moved through the forest during periods when they were active. A total of 16 *Tamandua mexicana* were captured and marked during 1973 and 1974, and 12 *Cyclopes didactylus* were captured and marked during 1977. The moist tropical forest on BCI has been described elsewhere (Croat, 1978), as has the secondary forest near the laboratory clearing where the radio-marked vermilinguas lived (Montgomery and Sunquist, 1978).

Tamandua were captured near the laboratory clearing, usually by grasping the tail by hand, but in some instances with a noose attached to a long pole. *Cyclopes* were captured with the noose, by using a long pole from the ground, or by climbing the tree and working from adjacent branches. After capture, a drug CI 744 was injected to subdue the animals and they were weighed and measured. Drug dosages were about 5 mg CI 744 per kg of *Cyclopes* and 10 mg CI 744 per kg of *Tamandua*.

A sample of the stomach contents of *Tamandua* was taken by intubation, and feces were collected when vermilinguas defecated in the laboratory. *Tamandua* feces were also collected from the forest on BCI, as were contents of the stomachs of *Cyclopes* and *Tamandua* found dead on roads near the Panama Canal.

The vermilinguas were radio-marked with 151 mHz miniaturized radio transmitters (Montgomery, 1985). Each marked animal was radio-located daily (Montgomery et al., 1973) and selected individuals were followed during major parts of their activity periods on some dates (Montgomery and Lubin, 1977). It was necessary to recapture the animals periodically to renew transmitter batteries or to change the adhesive tape which held the transmitter on the animal's tail. At those times, samples of stomach contents were taken from each *Tamandua*. Each *Cyclopes* was held overnight in a small cage and all feces was collected from the cage before the animal was released.

Each sample of stomach contents or feces was subsampled to provide about 10-15 percent of the material by volume, and then all arthropod heads in the subsample were identified to morpho-species (a set of individuals that looked alike and looked different from all other such sets). In some cases species identification of ants and termites was made by appropriate specialists, while some morpho-species remain to be identified.

In stomach samples from *Tamandua* some ants and termites were relatively undigested and whole, while in other samples of stomach contents and in all the samples of feces arthropods had been digested and the parts of their exoskeletons separated. The remains of each arthropod could be identified on the basis of the head capsule, which was not digested and which remained relatively whole as it passed through the vermilingua's digestive tract.

Individual radio-marked *Tamandua* were followed as they moved in the trees or on the ground during daylight hours. It was usually possible to see the animals as they moved and fed, but occasionally it was necessary to radio-locate an individual when visual contact was temporarily lost. Samples of prey taken by *Tamandua* during those periods of observation were collected whenever possible, including prey that fell from arborial sites where the vermilinguas fed.

It was not possible to observe *Cyclopes* as they moved and fed during their activity periods at night, and no samples of their food were collected in the field.

Estimating Diversity and Density of Arboreal Ants.-- An estimate of the diversity and density of ants potentially available to vermilinguas foraging in the tree crowns was made by fogging pyrethrum insecticide into the crowns of 18 *Luehea seemannii* trees (Table 1). A quantified sample of the ants and other arthropods which fell from the trees was then collected on sheets spread beneath the trees. Collections of this kind are not permitted on BCI, thus the study was done along Pipeline Road near Gamboa, on the mainland, approximately 8 km from BCI. The forest where the sampling was done was superficially similar to the forest near the laboratory clearing on BCI, where the radio-marked vermilinguas lived. However, quantitative measures of similarities between the two forests or the ant faunas in them are not available. For this paper, I assume that samples of ants from trees along Pipeline Road are representative of ant density and diversity in the portion of the forest on BCI where the radio-marked vermilinguas lived.

RESULTS

Activity, Movements and Home Ranges.-- *Cyclopes* radio-tracked in this study only moved and fed in arboreal situations at night. A translocated individual that we had previously radio-tracked was also completely nocturnal (Sunquist and Montgomery, 1973). During daylight hours *Cyclopes* rested, curled in a tight ball usually within 20 m of the ground on small lianas (Montgomery, 1983). *Cyclopes* were observed feeding during daylight, immediately after they were released after having been held overnight.

When a *Cyclopes* fed, it stood with its body perpendicular to the direction of a small stem. It grasped the liana stem or small branch with its hind feet, turning one foot in the opposite direction to the other. In this way it gripped the stem tightly and maintained torque and balance without using the front feet. It then bent over and inserted its large, sharp foreclaws into the stem to open a slit in it. This slit was held open with the fore-

Table 1. The diversity and population density of arboreal ants in *Luhea seemannii* trees along Pipeline Road was highest in early wet season, and lowest in late wet season.

Season	*m^2 in Sample*	*Number of Trees*	*Thousands of Ants/ha (S.D.)*	*Morpho-species Per Tree (S.D.)*
Dry	195	5	707 (734)	29.4 (6.10)
Early Wet	217	5	2,404 (4,112)	34.8 (10.71)
Late Wet	211	8	351 (190)	22.5 (5.58)

claws and the long, mobile tongue was inserted into the slit. Ants were withdrawn into the mouth with the tongue. Once it had taken ants from a location it moved down the stem or branch and repeated the procedure. These small vermilinguas presumably took ants from nests inside the small limbs and lianas. Thus, the small stems provided both movement and feeding substrate. Taylor (1985) noted that the hind foot of *Cyclopes* is adapted to provide increased ability to grasp limbs of small diameter.

Wilson (1959) reported that many ant species were to be found in hollow, dead twigs in lowland rain forest in Mexico. Carroll (1979) reported an average of 1 to 5 stem-nesting ant species per tree in Costa Rica, and a maximum species density for such ants as about 1 per m^2. Although most of the trees sampled there were small, there were 40 to 70 dead stems in each tree that were occupied by ants, representing 20 percent to 50 percent of the dead stems in each tree crown; pseudomyrmecines were the most common ants.

Cyclopes tended to move through trees with large masses of liana stems in their crowns, and to use such trees for daytime resting sites. A *Cyclopes* was either within a large mass of lianas or the tree it was in had large numbers of liana stems in the crown in 280 of the 324 nocturnal radio-locations (87 percent) when presence or absence of lianas was noted. There were masses of lianas in the crowns of 212 of 250 trees used as daytime resting sites by *Cyclopes* (85 percent). Only about 20 percent of trees in portions of the forest on BCI near where the radio-marked *Cyclopes* lived had large amounts of lianas in their crowns, while about half the trees examined had no associated lianas (Montgomery and Sunquist, 1978).

Tamandua were much more individualistic with regard to timing of their activity, the movement and feeding substrate that each used, (Montgomery and Lubin, 1977), and apparent food preferences. For this paper it was useful to combine all data from radio-marked *Tamandua*, realising the possible errors that might result from the relatively small sample sizes and from the large individual-to-individual variation.

An individual *Tamandua* might have a daily period of activity that began at any time of day or night, but then this activity period continued for about 8 hours. During the period of activity there were usually one or two times when the *Tamandua* rested for 30 min or more before continuing. Except for these brief rest periods and for a rare long feeding bout, *Tamandua*, like *Cyclopes* moved almost continuously. They usually fed for less than 1 min at each ant or termite colony that they attacked (Montgomery and Lubin, 1977).

Tamandua generally used larger limbs and lianas more than 5 cm in diameter as their substrate for both feeding and moving in arboreal situations. Adult *Tamandua* are more than 10 times as heavy as adult *Cyclopes* (Table 2), and need larger stems and branches to support them; their hind feet are not modified for grasping small stems (Taylor, 1985). Many of the nests of ants and termites that were attacked by *Tamandua* were large ones on the surface of tree trunks or large branches, or were nests attached to limbs and large stems of lianas. *Tamandua* occasionally fed from nest openings that penetrated the wood, and sometimes broke away portions of the wood to reach ants or termites inside. Usually only the claw of one forefoot was used, and the wood or nest material was broken away with a lateral twisting motion, quite different from the motion used by *Cyclopes* to penetrate wood. This twisting motion was predicted by Taylor (1978; 1985) on the basis of anatomical studies. Alternately, *Tamandua* opened the softer nests of *Azteca* ants or the surface of nests of *Nasutitermes* termites with delicate touches of the nest surface with the tip of the foreclaw, or ripped away the harder portions of the nests by using the foreclaw as a rigid hook and pulling it downward through the nest material.

The movements of *Tamandua* during arboreal feeding took them through many different tree crowns in each activity period. As an example, observation of *Tamandua* number 807 on 8 dates showed that she spent an average of 3.5 min in each tree that she entered (SD = 5.26; n = 479 trees). She spent 3 min or less in 73 percent of those 479 trees, entering a different tree on average with each 8.0 m of travel (SD = 2.65). During 23 periods of observation, each of which was continuous for 20 min or more (total 1,541 min; maximum period 203 min; mean = 67 min; SD = 49), she entered an average of 18.6 trees per h (SD = 5.99 trees). During these 23 periods of observation, the *Tamandua* moved a total of 3,397 m (mean per period = 148, SD = 96) at an average rate of 132 m per h. She was observed feeding or attempting to feed once per 6.9 min (SD = 3.37), and stopped to feed on ants or termites once each 16.7 m of travel (SD = 10.83). Projecting this feeding rate to an 8-h long activity period would indicate that the arboreal *Tamandua* might attempt to feed from 50 to 80 different colonies of ants and termites daily.

Cyclopes were not observed as they moved and fed at night, and they were radio-located at less frequent intervals (so that more individuals could be followed simultaneously), thus no such measure of the number of trees used per unit time is available for them. The best estimate of number of trees used is the rate of movement of *Cyclopes* (Montgomery, 1985) times the number of trees per m of travel by arboreal *Tamandua*. Adult male *Cyclopes* thus may have passed through an estimated 38 trees, nulliparous females through 27 trees, and parous females through 19 trees each night.

Table 2. Population structure, density and biomass estimates for 2 species of vermilinguas on Barro Colorado Island, Panama. Numbers of individuals captured during the study are shown in parentheses.

	Cyclopes didactylus	Tamandua mexicana
Mean Weight (kg)		
Adult Male	0.240 (4)	–
Adult Female	0.214 (7)	–
Adult, Both Sexes	–	3.73 (11)
Juvenile	0.124 (2)	3.16 (5)
Young	0.067 (5)	0.60 (1)
Vermilinguas per ha	0.77	0.13
Biomass, kg per ha	0.05	0.18

Home range spaces of *Tamandua* and *Cyclopes* overlapped on BCI (Montgomery, 1979; 1985) and indeed they fed from some of the same trees. Home ranges of *Tamandua* were approximately 5 times as large as those of *Cyclopes* females and twice as large as those of *Cyclopes* males. Home range spaces of female *Cyclopes* were exclusive of those of other adult females, while the home range space of a male *Cyclopes* was exclusive from those of other males but overlapped approximately three female home range spaces (Montgomery, 1979; 1985). Home range spaces of *Tamandua* generally did not completely overlap each other, although some overlap near the edges of the home ranges might have occurred (Lubin et al., 1977).

Thus while *Cyclopes* and *Tamandua* fed in the same areas and in some of the same trees, the actual places where they obtained their food were quite different. *Cyclopes* fed from places which *Tamandua* could not reach because they were too heavy, while *Tamandua* fed from colonies of ants and termites which *Cyclopes* were too small to reach or from nests which the smaller *Cyclopes* lacked the strength to open.

Vermilingua Population Density and Biomass.-- The population density and biomass of both *Cyclopes* and *Tamandua* is low on BCI, relative to such primary consumers as sloths and howler monkeys (Montgomery, 1979). Home range spaces of *Tamandua* were about 25 ha in area (erroneously reported as 65 ha in Montgomery, 1979). Population densities of *Tamandua* were estimated as about 0.05 per ha (Lubin et al., 1977), primarily on the basis of the rate at which they attacked nests of termites and *Azteca* ants. That estimate seems reasonable in terms of the numbers of *Tamandua* seen in this study during field work on BCI, and by other workers (Eisenberg and Thorington, 1973). Biomass of *Tamandua* is here estimated at about 0.2 kg per ha, based on the estimate of population density and the population structure and average weights of individuals captured (Table 2).

The population density and biomass of *Cyclopes* is here estimated on the basis of home range sizes, degrees of home range overlap, population structure, and weights of animals captured during the study (Table 2). One adult male *Cyclopes* had a relatively large home range space of about 11 ha that overlapped the home range spaces of 2 adult females (leaving space for a third female), but did not overlap the home range space of an adjacent adult male (Montgomery, 1979; 1985). Home range spaces of adult females were smaller, averaging 2.8 ha (n = 4; SD = 0.49), and three females could live on the home range space of each adult male. The home range spaces of adjacent adult females did not overlap each other.

A young *Cyclopes* is dependent on its mother for movement from tree-to-tree until it is about 3/4 her size although as it grows it tends more and more to move about in the tree where the mother leaves it while she forages at night. While the larger young moves independently of the mother to some extent soon before dispersing, it lives entirely within a restricted portion of its mother's home range space. Dispersal by the young is a relatively straight-line movement away from the natal home range space, apparently continuing until an empty territory is encountered (Montgomery, 1983; 1985).

I estimate from *Cyclopes* captured on BCI (Table 2) that there are about one-half as many young and pre-dispersal juveniles, and about one-fifth as many dispersing young as adult females, based in part on the gestation period and the length of time it takes to produce an independent young (Montgomery, 1983; 1985).

To estimate *Cyclopes* population density and biomass, I assumed that there are 3 adult female home range spaces of 2.5 ha each within each 11 ha male home range space. I assumed no overlap of male home range spaces. I assumed that one-half of the females carry dependent young that weigh about 0.07 kg, and one-fifth of the females have independent dispersing young that weigh about 0.12 kg. Adult males are assumed to weigh 0.24 kg, and adult females 0.21 kg.

Using these numbers, the population density of adult males is 1 per 11 ha (0.09 per ha), that of adult females 1 per 2.5 ha (0.40 per ha), that of dependent young 0.20 per ha, and that of independent young, 0.08 per ha. Biomass of adult males in kg per ha is thus 0.022, that of adult females 0.084, that of dependent young is 0.014, and that of independent juveniles is 0.010. The total estimated *Cyclopes* biomass is thus 0.130 kg per ha, and there are an estimated 0.77 *Cyclopes* per ha of forest on BCI (Table 2).

Table 3. Overlap in the diets of *Cyclopes didactylus* and *Tamandua mexicana*. Twelve percent of 135 ant morpho-species were shared by the two kinds of vermilinguas. About one-fifth of the ant morpho-species and one-third of the individual ants eaten by each species of vermilingua were involved in the dietary overlap.

	Cyclopes didactylus	*Tamandua mexicana*
Vermilinguas		
Individuals Sampled	12	17
Number of Samples	42	29
Total Arthropods	15566	18056
Ants		
Percent of Total[1]	100	33 (8.8 to 96.4)
Percent Arboreal	100	76
Ant Morpho-species		
Total	76	77
Mean Number per Sample (SD)	18.3 (6.3)	8.6 (3.3)
Percent Arboreal	100	57
Percent Shared	21	21
Ants of Shared morpho-species[2]	5566 (36 percent)	1642 (28 percent)
Ant Brood		
Percent of Total[3]	9 (0 to 16.8)	3 (0 to 16.1)
Termite Morpho-species[4]	0	13

[1] The remaining arthropods in samples from *Tamandua mexicana* were termites.
[2] In *Tamandua mexicana*, these represented 9 percent of all dietary items, including termites.
[3] Based on complete analysis of stomach contents of 3 *Cyclopes didactylus* and 29 stomach samples from *Tamandua mexicana*.
[4] Based on complete analysis of 5 stomach samples from *Tamandua mexicana* in which the prey was relatively undigested.

Identity of Arthropods in Vermilingua Diets.-- A total of 39 individual vermilinguas provided samples for analysis of dietary preferences and overlap of the diets of the two species (Table 3). Some were sampled repeatedly; 12 *Cyclopes* provided 42 fecal samples, while 17 *Tamandua* provided a total of 29 stomach samples. The 61 samples were sub-sampled for analysis, and the sub-samples included a total of about 32,000 arthropods.

Cyclopes fed exclusively on ants, while *Tamandua* included about 67 percent termites along with the ants in their diets (Table 3). On average, each sample from *Cyclopes* contained about 18 ant morpho-species, twice as many as were found in each sample from *Tamandua*. The only termite remains found in *Cyclopes* feces were of one *Nasutitermes* which was probably ingested by accident. The samples from the two vermilingua species contained no other arthropods except for a homopteran which appeared to be a mimic of one of the ant species (myrmicinae: *Zacryptocerus*) which was used as food by *Tamandua*. The mimic was probably living in association with the ants that represented its model, and was ingested accidentally by the vermilingua, along with the ants. Similarly, 1 of the 10 samples of stomach contents from *Tamandua longicaudata* in Venezuela contained a homopteran ant mimic and its model.

Ant brood (larvae and pupae) represented 3 percent of all dietary items (including termites) identified from *Tamandua* stomachs (Table 3). There were thus about one-tenth as many ant brood as adult ants. *Cyclopes* feces contained almost no brood, with larvae or pupae being found in only 2 of the 42 samples. However, this may have been because brood was broken up and made unrecognizable during digestion. Three samples of *Cyclopes* stomach contents taken from dead individuals contained the following proportions of brood; 16.8 percent, 8.2 percent, and 0.0 percent (total 9.1 percent of 3,865 dietary items).

Both *Cyclopes* and *Tamandua* consumed more than 75 morpho-species of ants (Table 3). In addition, *Tamandua* used at least 13 morpho-species of termites as food. A total of 135 ant morpho-species were identified from both vermilingua species combined, and 16 morpho-species, about 12 percent of these, were used by both species of vermilingua. Those 16 ant morpho-species represented about 20 percent of the ant morpho-species in the diets of both species of vermilinguas. The individual ants of those shared morpho-species represented about 36 percent of the diet of *Cyclopes* and 9 percent of the dietary items from *Tamandua* (28 percent of the ants in *Tamandua* diets).

Diversity, Density and Consumption of Arboreal Ants.-- The samples of arthropods from tree crowns near Pipeline Road came from a floristically much more restricted portion of the habitat than that used for feeding by arboreal vermilinguas. All samples were from one species of canopy tree *Luehea seemannii* and from associated lianas and epiphytes which grew in the crowns of the 18 trees. The diversity of the ant fauna in those trees was probably less than that in the forest canopy as a whole, but the density of ants living in the crowns of those trees may have more closely reflected that in other parts of the forest. Sampling was done at 3 times of the year and possibly was adequate to measure seasonal faunal variation.

The average *Luhea* tree crown contained more than 20 morpho-species of ants, even in late wet season when diversity and biomass were lowest (Table 1). Trees sampled in early wet season tended to have more ant morpho-species if there were more species of lianas growing in the tree crown. There was a significant positive correlation between the number of ant morpho-species and the number of liana species in the tree crown ($r^2 = 0.85$, $p < 0.01$, 4df). This correlation was not significant in late wet season ($r^2 = 0.27$, 3df) or in dry season ($r^2 = 0.05$, 4df). Tree crowns with more liana species had a significantly higher biomass of ants than those with fewer species of lianas in early wet season ($r^2 = 0.40$, $p < 0.05$, 4df) and dry season ($r^2 = 0.34$, $p < 0.05$, 4df), but not in late wet season.

About 40 percent of ants identified from *Cyclopes* feces were of morpho-species commonly represented in the samples of arthropods from tree crowns along Pipeline Road (Table 4). Of the 76 morpho-species found in *Cyclopes* feces, 34 morpho-species (45 percent) were also collected from the crowns of the 18 *Luehea* trees. Ants of those 34 morpho-species represented 83 percent of all ants identified from *Cyclopes*, while those 34 morpho-species included 74 percent of all the ants identified from the tree crowns.

About 25 percent of the ants eaten by *Tamandua* were of morpho-species commonly found in the *Luehea* tree crowns (Table 4). Of the 77 ant morpho-species identified from *Tamandua*, 23 morpho-species (30 percent) were also found in samples from the crowns of the 18 trees. Ants of thòse 23 morpho-species represented 50 percent of all ants identified from *Tamandua* and 93 percent of all ants collected by sampling the 18 trees.

Cyclopes were usually consistent among themselves in choosing the same ant morpho-species for the main part of their diet. High-ranking ant morpho-species were found in most of the fecal samples, and from most of the individual *Cyclopes* (Table 5).

Those ant morpho-species which formed most of *Cyclopes* diets were relatively unimportant in the diets of *Tamandua*, and there was little rank correlation between morpho-species eaten by *Cyclopes* and those eaten by *Tamandua*. The 5 highest ranking ant morpho-species in *Cyclopes* diets totalled more than 60 percent of the ants they ate, while those same 5 morpho-species contributed less than 6 percent of the ants used by *Tamandua* and less than 2 percent of the total dietary items, including termites, identified from *Tamandua*.

The ant morpho-species which formed most of *Cyclopes* diets were not those that ranked highest in the collections from the forest canopy (Table 5). The 5 highest ranking morpho-species in the *Cyclopes* diet formed less than 6 percent of all ants collected by sampling the canopy trees. However, those morpho-species used most by *Cyclopes* were apparently widely distributed in the forest, and perhaps usually available regardless of the season, being found in a high proportion of the trees sampled and at about the same relative frequency in all seasons. The second and fourth ranking morpho-species were exceptional, in that *Solenopsis* B ranked second in the overall diet of *Cyclopes* but was used by only one-fifth of the individuals, and was collected from only one-third of the trees in the canopy samples, while *Camponotus abdominalis* was collected from fewer than half the trees and ranked relatively low in the canopy samples, but was

Table 4. **The 10 highest ranking ant morpho-species in the samples taken by fogging insecticide into the crowns of 18 *Luhea seemannii* trees included only some of the highest ranking ant morpho-species in the vermilingua diets. They accounted for only 42.6 percent of the ants in the diet of *Cyclopes didactylus* and 25.6 percent of the ants in the diet of *Tamandua mexicana*. A dash (-) indicates that individuals of a morpho-species were not found. Samples from the trees were taken in dry season (D), early wet season (EW) and late wet season (LW). Twenty-nine samples of stomach contents from 19 *Tamandua mexicana* were analyzed, and contained 1,642 ants. There were 42 samples of feces from 12 *Cyclopes didactylus*, and these contained 5,566 ants. There were 72,258 ants in the samples from the 18 *Luehea* trees. Morpho-species identification numbers are in parentheses.**

	Forest Canopy					Cyclopes didactylus				Tamandua mexicana			
	Rank												
Ant Species	*D*	*EW*	*LW*	*% of Total*	*% of Trees*	*Rank*	*% of Total*	*% of Samples*	% of Cyclopes	*Rank*	*% of Total*	*% of Samples*	*% of* Tamandua
Azteca sp. (6)	1	1	2	62.7	100	13	1.6	52	58	33	0.3	10	16
Dolochoderus bispinosis (101)	4	2	1	21.0	55	-	-	-	-	9	3.1	17	21
Camponotus auricomis (21)	2	3	4	3.0	89	5	5.9	93	100	<61	>0.05	3	5
Crematogaster B (8)	3	5	5	2.0	94	1	22.1	98	92	7	4.0	28	24
Zacryptocerus minutus (1)	6	9	7	1.1	83	6	4.7	83	92	-	-	-	-
Pseudomyrmex A (9)	5	10	9	1.0	89	3	7.6	98	100	-	-	-	-
(106)	-	-	3	0.9	11	-	-	-	-	47	0.05	3	5
(55)	10	4	11	0.8	50	-	-	-	-	-	-	-	-
Procryptocerus belti (24)	8	5	12	0.7	94	23	0.7	43	58	1	18.1	21	26
(43)	22	23	6	0.5	72	-	-	-	-	-	-	-	-

found in most of the samples of *Cyclopes* feces.

Many of the most important ant morpho-species in *Tamandua* diets did not occur in the diet of *Cyclopes* or in the samples from the forest canopy (Table 6). The 5 highest ranking morpho-species, which accounted for more than 55 percent of the ants eaten by *Tamandua*, included less than 2 percent of the ants eaten by *Cyclopes* and less than 1 percent of the ants collected from the 18 *Luehea* trees. The ant morpho-species most commonly eaten by *Tamandua* were not of genera found in most samples of litter from the forest floor on BCI, which included *Pachycondyla, Odontomachus, Solenopsis*, and *Pheidole* (Williams, 1941).

It is impossible to know which of the ants found in *Tamandua* stomachs and feces were actually consumed in an arboreal context, because these vermilinguas fed in both arboreal and terrestrial situations, and because individuals of at least some of the ant species are found in trees and on the ground. Of the individual ants in *Tamandua* stomachs, 76 percent were assumed to have been arboreal by the following criteria; 28 percent were of morpho-species eaten by *Cyclopes* (and thus arboreal), 24 percent were of additional morpho-species on which arboreal *Tamandua* were seen feeding, and 24 percent were of additional morpho-species collected by sampling the crowns of *Luehea* trees. Of the 77 morpho-species identified from *Tamandua* stomachs, 44 morpho-species (57 percent) were thus considered to be arboreal.

Unlike *Cyclopes, Tamandua* were not consistent among themselves in the ant morpho-species which they ate.

Table 5. **The top 10 ranking ant morpho-species eaten by *Cyclopes didactylus* showed little rank correlation with ant morpho-species eaten by *Tamandua mexicana*. *Cyclopes* tended to feed on ants that were in the 18 *Luhea seemannii* tree crowns sampled with pyrethrum insecticide, but there was little rank correlation between the occurrence of ants in the diet of *Cyclopes* and the occurrence of those morpho-species in the tree crowns. Numbers of vermilinguas and tree crowns sampled, number of samples of stomach contents and feces, and total numbers of ants identified are given in Table 4. Morpho-species identification numbers are in parentheses.**

	Cyclopes didactylus				Tamandua mexicana				Luhea *tree crowns*				
									Rank				
Ant Species	*Rank*	*% of Total*	*% of Samples*	*% of* Cyclopes	*Rank*	*% of Total*	*% of Samples*	*% of* Tamandua	*D*	*EW*	*LW*	*% of Total*	*% of Trees*
Crematogaster B (8)	1	22.1	98	92	7	4.0	28	24	3	5	5	2.0	94
Solenopsis B (15)	2	18.8	60	21	19	0.5	14	21	13	-	14	0.2	33
Pseudomyrmex A (9)	3	7.6	98	100	-	-	-	-	5	10	9	1.0	89
Camponotus													
abdominalis (5)	4	7.2	98	100	<17	1.1	28	26	9	17	<38	>0.05	44
auricomis (21)	5	5.9	93	100	<61	>0.05	3	5	2	3	4	3.0	89
Zacryptocerus minutus (1)	6	4.7	83	92	-	-	-	-	6	9	7	1.1	83
(63)	7	3.3	21	42	-	-	-	-	-	-	-	-	-
(19)	8	2.8	43	58	-	-	-	-	-	-	-	-	-
Zacryptocerus belti (35)	9	2.5	43	58	8	3.8	14	16	-	-	-	-	-
(29)	10	2.5	33	67	-	-	-	-	-	-	-	-	-

Table 6. **The 10 highest ranking ant morpho-species in the diet of *Tamandua mexicana* were not the highest ranking ones in either the samples of feces from *Cyclopes didactylus* or the samples from the crowns of *Luehea seemannii* trees. Numbers of vermilinguas and tree crowns sampled, number of samples of stomach contents and feces, and total numbers of ants identified are given in Table 4. Morpho-species identification numbers are in parentheses.**

	Tamandua mexicana				Cyclopes didactylus				Luhea *Tree Crowns*				
Ant Species	*Rank*	*% of Total*	*% of Samples*	*% of* Tamandua	*Rank*	*% of Total*	*% of Samples*	*% of* Cyclopes	*D*	*Rank EW*	*Rank LW*	*% of Total*	*% of Trees*
Procryptocerus belti (24)	1	18.1	21	26	23	0.7	43	58	8	5	12	0.7	94
(139)	2	17.0	55	63	-	-	-	-	-	-	-	-	-
Crematogaster A (12)	3	10.6	41	53	<37	0.1	5	17	<46	-	-	>0.05	5
(168)	4	5.9	7	11	-	-	-	-	-	-	-	-	-
(152)	5	5.7	21	26	-	-	-	-	-	-	-	-	-
(134)	6	4.1	41	42	-	-	-	-	-	-	-	-	-
Crematogaster B (8)	7	4.0	28	24	1	22.1	98	92	3	5	5	2.0	94
Zacryptocerus belti (35)	8	3.8	14	16	10	2.5	43	58	-	-	-	-	-
Dolichoderus bispinosus (101)	9	3.1	17	21	-	-	-	-	4	2	1	21.0	55
(104)	10	3.0	31	42	-	-	-	-	18	<48	<32	0.1	44

Only ants of the second-ranked morpho-species were found in more than one-half of the samples, and only the second- and third-ranked morpho-species were found in samples from more than half the individual *Tamandua* (Table 6). This tendency toward individuality in food habits reflects the individuality shown by *Tamandua* in the timing of their activity and in the different degree to which each was arboreal (Montgomery and Lubin, 1977).

Ant Cropping rates.-- Rates at which *Cyclopes* and *Tamandua* cropped ants were estimated by counting ant heads in weighed samples of feces. After each capture, *Cyclopes* were caged individually in the laboratory overnight and almost invariably defecated (43 of 45 times). The feces passed by a *Cyclopes* each night were assumed to represent the normal nightly feces output in the wild, and thus to represent the amount of food eaten per night (*Cyclopes* were inactive during daytime). Dry weight of feces from adult males averaged 1.37 gm (n = 7; SD = 0.26), from adult females 0.97 gm (n = 19; SD = 0.21), from independent young 0.61 gm (n = 5; SD = 0.26), and from dependent young 0.24 gm (n = 4; SD = 0.09). All ant heads from a weighed sub-sample representing about 15 percent of the feces (mean = 17.5 percent; range 8 to 100 percent; n = 35) were identified to morpho-species and counted. The samples contained an average of 408 ant heads each (range 103 to 1199 heads). Based on stomach contents of *Cyclopes*, ant brood was ingested in an approximate ratio of 1 larvae or pupae for each 9 adult ants (Table 3). I assume that each fecal sample represented ingestion of 10 percent brood, even though brood was not separated and counted from the samples.

Individual *Cyclopes* cropped 700 to 4700 ants on average daily, depending on their age, and on the dependency of young on their mothers for milk and mobility (Table 7). Among adults, the males tended to consume more ants each day than the smaller females. Independent young ate fewer ants than adults, although adults and independent young ate about the same amount per unit body weight, about 20,000 ants and 2,000 ant brood per kg of vermilingua.

Table 7. **Estimated numbers of ants consumed daily by *Cyclopes didactylus*.**

Age and Sex	*Total Samples*	*Mean Ants per Day ± SD*	*Ants per kg* Cyclopes *per Day*
Dependent Young	4	713 ± 310	7550
Independent Young	5	2605 ± 999	21510
Adult Female	19	3642 ± 2404	17920
Adult Male	7	4710 ± 1243	16590

The total number of ants which *Cyclopes* consumed daily was estimated from the population density and the age and sex structure of the vermilingua population (Table 2), and from rates at which individuals of the various sex and age classes consumed ants (Table 7). For example, there were 0.09 adult males per ha which fed on ants at a daily rate of 4,710 ants and 471 ant brood per individual, equivilant to 424 ants and 42 ant brood per ha per day. Similarly, adult females ate 1,457 ants and 146 brood, independent young 209 ants and 21 brood, and dependent young 143 ants and 14 brood per ha per day. Total cropping by *Cyclopes* is thus estimated as 2,233 ants and 233 brood per ha per day.

Estimates of daily rates at which *Tamandua* cropped ants were less precise because they were based on counts of ant heads in sub-samples of feces passed by only two individuals captured for marking, and because of the individual-individual variability in the species of ants which *Tamandua* ate and the proportion of ants in their daily diet.

The number of ants per gm of feces was first adjusted

by calculating the number of ants which would have been in each gm had the *Tamandua* eaten only ants:

adjusted ants/gm = (ants/gm)/(percent ants X 100).

The mean adjusted estimate was 531 ants/gm of feces (SD = 63.3 gm). The average dry weight of 7 *Tamandua* fecal masses was 80.0 gm (SD = 15.9 gm). For each stomach sample from *Tamandua* the estimated number of ants which had been consumed was thus:

number of ants = 80 x [531 x (ants/total arthropods)].

For the 29 stomach samples, the mean estimated number of ants consumed per defecation was 18,114 (SD = 12,575).

In captivity, *Tamandua* defecate at 2 to 6 day intervals (Meritt, 1975); there are no data concerning the interval between successive defecations by wild *Tamandua*. For this paper, I assume that each *Tamandua* fecal mass represented the feces output for 2 days, and thus the ants and termites eaten during 2 daily activity periods. Daily intake was thus about 9,000 ants. Based on analyses of *Tamandua* stomach contents (Table 3), I assume that brood was consumed in a ratio of 1 brood item per 9 adult ants, or 10 percent of the total food items. Under those assumptions, the daily cropping rate by *Tamandua* was equivilant to about 2,500 ants and 250 ant brood per kg of vermilingua.

Tamandua cropped fewer arboreal ants than *Cyclopes* because only a part of their feeding was arboreal, and only a part of their diet was ants. For the 0.05 *Tamandua* per ha (Table 2), the estimated individual cropping rate of 9,057 ants per day represents 453 ants and 45 brood per ha. If 76 percent of those ants were arboreal (Table 3), then *Tamandua* ate an estimated 344 arboreal ants per ha per day.

Total cropping of arboreal ants by the two vermilingua species combined is thus estimated as about 2,500 ants and 250 ant brood per ha per day. This daily cropping rate is about 0.11 percent of the standing crop of arboreal ants when population densities of arboreal ants are highest in early wet season (density of ant brood was not estimated), and 0.76 percent of the standing crop of arboreal ants in late wet season when ant population densities are lowest (Table 1).

DISCUSSION

Vermilinguas of both species moved much further during each daily activity period than would appear to have been necessary for them to obtain their daily ration of ants and termites (Montgomery, 1985). Why did the vermilinguas expend so much energy in travel, when there was apparently sufficient food available in much smaller areas of the forest canopy than they used each day? Each arboreal vermilingua passed near many ant colonies of many different species during each activity period, and fed from a large number of those colonies, as indicated by the diversity of ants in each of the samples of their diets. The arboreal vermilinguas ate relatively few ants from each colony, as was reported elsewhere for the terrestrial giant anteater *Myrmecophaga* (Montgomery and Lubin, 1977).

A mosiac of the territories of dominant ant species, each with a number of associated non-dominants, provides a potentially diverse diet for a vermilingua in any small portion of the forest canopy. Moreover, the packing-in of many ant species which differ in their ecological requirements allows for a greater total ant biomass than would be possible if only the dominant species were present.

When the vermilinguas moved through the trees, they passed through a mosiac of ant territories. Territories of the dominant ant species (in both the behavioral sense and in terms of colony sizes) form a mosiac in the forest canopy, with the dominant species spaced through aggressive interactions such that the foraging territories of dominants usually do not overlap (Brown, 1959; Room, 1971; Leston, 1973; Majer, 1976a). Within the foraging territory of a colony of a dominant species, other ant species may occur (Room, 1971) and the overlap of colony foraging territories may include co-existing dominants (Majer, 1976a) as well as a large number of other species which generally occur with dominants, but in much lower numbers. These latter species have been termed non-dominants (Majer, 1972). Room (1971) found that the 5 common dominants on his study area had distinct communities of as many as 11 species of non-dominants associated with them. Colonies of the non-dominants (minor ants) may be more easily eliminated by predation by other ants (Greenslade, 1971), and perhaps more susceptible to depredations by vermilinguas.

Majer (1976b) showed that about 73 percent of trees on a cocoa plantation supported one or more species of dominant ants. All 18 *Luehea seemannii* trees sampled with pyrethrum insecticide along Pipeline Road in Panama had a dominant ant species in their crown. These included ants of the genera *Azteca, Monacis,* and *Crematogaster*. Of 9 canopy trees sampled near Manaus, Brazil, each of 8 had 1 dominant ant species in their crown, while one of the trees had co-dominant ants. Dominant ants were of the genera *Crematogaster, Hypoclinea*, and *Azteca* (unpublished data).

Samples from the crowns of the 18 *Luehea seemannii* trees in Panama (Table 1) yielded an average of 22 to 35 ant morpho-species per tree, depending in part on season of the year, and 35 to 240 ants per m^2, while samples from canopy trees near Manaus, Brazil, indicated averages of 19 ant morpho-species per tree and about 45 ants per m^2 (unpublished data). Majer (1972) identified

44 arboreal ant species in a 2.25 ha plot, and Room (1971) 58 arboreal ant species in 250 m^2 of cocoa plantation in Ghana. Leston (1973) reported an average of 171 ants per m^2 in seven 25 m^2 samples taken from the canopy of cocoa farms with a pyrethrum insecticide fog, and those samples included an average of 13 ant species each (range 10 to 15 species). In 6 similar samples from primary tropical forest canopy, there was an average 356 ants per m^2 (Leston, 1973). Elsewhere he reported (Leston 1971) collecting an average 623 *Oecophylla longinoda* per m^2; ants of many other species were collected as well. A total of 67 ant species were collected by Room (1971) in 168 samples taken by hand from the canopy in 30 cocoa farms; an average of 3.5 to 6.8 species were caught by hand per 5 minutes from any one place, depending in part on which species of ant was the dominant where the sample was taken.

During their daily activity periods, the vermilinguas passed through many trees, certainly more than would have been necessary to satisfy either their qualitative or quantitative dietary needs for arboreal ants. The forest on BCI contains some 380 trees over 15 cm diameter at breast height per ha (Montgomery and Sunquist, 1978). Thus the home range space of each adult *Cyclopes* could have contained at least 1,000 and the home range space of each *Tamandua* at least 9,000 such trees. *Cyclopes* tended to move through, and presumably feed in, trees with large numbers of liana stems in their crowns. Some 21 percent of trees examined by Montgomery and Sunquist (1978) had many liana stems in their crowns, thus perhaps 500 trees or more well suited for feeding by *Cyclopes* were present on each of their home range spaces.

Each *Cyclopes* could probably have met its daily need for ants, both in terms of numbers and in terms of species diversity, from the crown of 1 large canopy tree, while each *Tamandua* could have satisfied its daily needs from 2 or 3 large trees at most. Radio-marked *Tamandua* entered an average of about 18 trees per hour, or perhaps 140 trees per 8-hour long activity period, while each *Cyclopes* moved through 20 or more trees as it fed each night. Certainly any one of the larger trees contained enough ants to provide the 5,000 ants eaten daily by an adult *Cyclopes*. Larger trees on BCI have crown surface areas on the order of 175 m^2 (from data in Montgomery and Sunquist, 1978), and there were an estimated 35 to 240 ants per m^2 of canopy surface (Table 1). The average sample of feces from *Cyclopes* contained 18 ant morpho-species (Table 3). *Luehea seemannii* trees had an average of from 22 to 35 morpho-species in their crowns, and there is no reason to suspect that large canopy trees of other species had fewer ants of fewer species.

There are at least 4 possible reasons for the seeming excessive expenditure of energy during foraging. The first is that the relatively long-distance daily movements had little to do with foraging per se, but were a necessary component of the social behavior of these solitary animals. The long-distance movements perhaps provided for dispersion of scent marks by which the vermilinguas communicated their presence to neighboring conspecifics.

The dispersion pattern of the home range spaces of adult *Cyclopes*, which corresponded to a harem-territorial social system (Montgomery, 1979; 1983; 1985) would require signalling among neighbors, and scent is the most likely communication mode. Visual signalling is precluded by the nocturnal habits of the animals and the poor lateral visibility through the forest canopy. Adult *Cyclopes* were not vocal while being handled, while they were caged in the laboratory, or while they were being radio-tracked in the wild. However, young and juvenile *Cyclopes* emitted loud distress calls when handled or separated from their mothers in the laboratory.

The spatial correlates of the social system of *Tamandua* are less obvious, but *Tamandua* are solitary and perhaps territorial (Lubin, et al., 1977). Individuals of both vermilingua species were seen drag-marking, apparently with anal scent glands, and *Tamandua* (but not *Cyclopes*) has a strong, offensive odor detectable from several meters away. It seems likely that some of the long-distance movement is a component of the vermilingua's social behavior, and thus that a part of the apparent high cost of foraging goes toward communication and maintenance of exclusive foraging areas.

A second possibility is that vermilinguas take prey of a large number of species during each activity period to intake a mix of some chemicals, and reduce the cost of handling noxious defensive chemicals. They surely ingest such chemicals with almost every ant and termite they eat. The reasons for ingesting a mixture of ant and termite species would thus be analagous to those suggested by Freeland and Janzen (1974) for herbivores which ingest a mixture of leaves to balance the toxic effects of any one of the leaf phytochemicals.

There is little information about the particular chemicals used for defence by most of the ant species eaten by *Tamandua* and *Cyclopes*. However, the defensive chemicals of close relatives of some of the species of ants have been studied (Roth and Eisner, 1962), and the few species reported on show a wide range of different noxious chemicals (Citrol, Citronellal, Limonene, etc.). Beard (1963) discussed the effects of some ant venoms on vertebrates, and pointed out that the venoms injected into the vertebrate skin can have severe effects. There are no data on the effects of ingesting large quantities of ant venoms, but the vermilinguas surely must have to detoxify some of the chemicals while digesting the ants, in order to avoid being poisoned. *Tamandua* ingest large numbers of *Nasutitermes* termite soldiers (Lubin and Montgomery, 1981). Some of the defensive chemicals used by these soldiers, particularly α -pinene, can do considerable physiological damage when digested or inhaled.

While the amount of a noxious chemical contained in a single ant may be small, the cumulative dosage resulting from a feeding bout or feeding period when several thousand ants are ingested can be considerable. If, for example, all of the estimated 4,700 ants consumed by an adult male *Cyclopes* each night were *Solenopsis*, then the 240 g vermilingua might consume (4,700 X 50 μ g) of venom (based on data reviewed by Beard, 1963), or more than 900 mg of venom per kg of vermilingua. In dogs, 2 mg of dried bee venom per kg of dog is a lethal dose.

A third possible explanation for long-distance movements of foraging vermilinguas is that they continuously crop the ant colonies on their feeding space to such a low level that there are relatively few susceptable ants available from each colony. It seems unlikely that the level of cropping of arboreal ants by these two species of vermilinguas, in the absence of other mortality factors acting on the ants, would be sufficient to cause permanent damage to either the ant community as a whole, or to the individual ant colonies. Many of the colonies would usually contain many times the number of ants of a species consumed by an individual vermilingua for its entire daily feeding (cf Baroni-Urbani, 1978) and many colonies were fed from during each feeding period. Even at a time of year when ant populations were estimated to be at their lowest, the vermilinguas took fewer than 0.8 percent of the total ants per day.

However, there are few published indications of the amount of depradation which an ant colony can sustain, or the rate at which damaged colonies recover. Way (1953) reported that recovery of colonies of the dominant ant *Anoplolepsis custodiens* after treatment with one of three kinds of insecticide (DDT, BHC, or Dieldren) occurred within 15 to 75 days depending on which insecticide was used. Hanna, et al. (1956) also failed to reduce ant population levels *(Crematogaster)* with application of DDT and removal of nests; colonies apparently recovered within 6 weeks. Not all ant species are equally resistant to such depradation. Most colonies of *Oecophylla* sp. were either eradicated or colony members changed the location of their nest after bi-weekly or monthly application of DDT or Lindane (Vanderplank, 1960). Areas where nests of *Oecophylla* were completely removed required more than 1 year for recovery of populations of this species, meanwhile other ant species recovered more quickly, or invaded the area (Leston, 1971).

It seems unlikely that the vermilinguas would cause as much damage to the colonies by eating ants as was inflicted by application of insecticide. However vermilinguas, particularly *Tamandua*, can cause considerable structural damage to the nest itself or to the supporting substrate for the nest, in the process of finding ants on which to feed. This damage, and the energy required to repair it, may have a depressive effect on the colony which is greater than that of actual numbers of ants removed by feeding. Therein may lie a major difference between the impact of mammalian predators on ant populations, and the damage to ant populations caused by predatory ants and other invertebrates; only the mammals cause significant damage to the nest and nest substrate.

Vermilinguas cannot possibly reach all arboreal ant colonies or all parts of most colonies, even with their strong forelimbs and foreclaws and their long, mobile tongues. There are many nests inside tree limbs and branches, protected by wood that is too strong for even *Tamandua* to penetrate. Generally, only the portions of the nests which are on the surface or which are in soft or decaying wood are susceptable to attack by *Tamandua*. *Cyclopes* can probably penetrate most small stems (about 1 cm diameter or smaller), but cannot reach all of the places where these small stems grow. *Cyclopes* are not strong enough to penetrate larger stems unless the wood has been softened by decomposition. Thus there are ant colonies or portions of those colonies which are unaffected by the vermilinguas, and which can repopulate areas where the vermilinguas may have reduced local population densities.

A fourth possible explanation for the long distances moved by the vermilinguas is that colony defences are usually so effective that only a short period of feeding is possible at each feeding stop. Data supporting this view have been presented elsewhere (Montgomery and Lubin, 1977; Lubin et al., 1977; Lubin and Montgomery, 1981). Observations of feeding *Tamandua* and *Myrmecophaga* show that they usually stay for 1 min or less at each place they stop to feed. The movement patterns of *Cyclopes* (Montgomery, 1985) suggest a similar pattern of feeding, although actual feeding was rarely observed. The length of time the vermilinguas stay to feed after disturbing a colony of ants or termites is approximately equal to the time required to mobilize members of defensive castes to the place where the vermilingua attacks the colony (Montgomery and Lubin, 1977; Lubin and Montgomery, 1981).

As members of the defensive castes are mobilized, the potential of that colony as a food source rapidly decreases. This is because the vermilingua must spend increasing amounts of time and energy grooming ants or termites from its body, because of the increasing pain and discomfort inflicted by members of the defensive caste, and perhaps because of the cumulative effects of chemicals which are injected into or squirted onto the vermilingua.

Not all ants and termites have defensive castes which react to an attack on the colony by mobilizing to the point of attack. In these species, colony members flee from the point of disturbance, going deeper into the wood or nest structure, and thus effectively temporarily remove that nest or colony as a food source. The rate at which colony members leave the point of attack determines the length of time which a vermilingua can efficiently spend feeding at that place before moving on.

These two kinds of restrictions which individual colony members place on the length of time which a vermilingua can feed from any one colony help to explain why most feeding stops are of short duration, but do not explain why a vermilingua does not simply move to an adjacent colony of ants within the same tree. Usually, *Tamandua* left a tree after feeding from only 1 or 2 colonies within it, although there were probably many more colonies, at least in the larger tree crowns. It seems possible that the adjacent colonies were alerted by the attack of a vermilingua, even though the colonies were of different species, and were thus alarmed before the vermilingua arrived. A general alert of all or most colonies within a tree would further reduce the potential feeding time at subsequent colonies, and perhaps make it inefficient for the vermilingua to attack other ants in the same area.

I feel that the low rate of cropping, selective feeding on only some species of ants or termites, and high investment in energy for foraging result mainly from chemical, structural and behavioral defences which the social insects use against vermilinguas. These defences prevent destruction of arboreal ant colonies, and probably limit population densities of arboreal vermilinguas below levels which the apparently rich food source represented by arborial ants might allow.

LITERATURE CITED

Baroni-Urbani, C.
1978. Appendix 1: Adult populations in ant colonies. Pp. 334-335, *In* Production ecology of ants and termites (M.V. Brian, ed.). Cambridge Univ. Press, Cambridge. 409 pp.

Beard, R.L.
1963. Insect toxins and venoms. Ann. Rev. Entomology 8: 1-18.

Brown, E.S.
1959. Immature nutfall of coconuts in the Solomon Islands. II. Changes in ant populations and their relation to vegetation. Bull. Ent. Res., 50: 523-558.

Carroll, C.R.
1979. A comparative study of two ant faunas: the stem-nesting ant communities of Liberia, West Africa and Costa Rica, Central America. Am. Naturalist, 113(4): 551-561.

Croat, T.
1978. Flora of Barro Colorado Island. Stanford Univ. Press, Stanford. 943 pp.

Eisenberg, J.F. and R.W. Thorington.
1973. A preliminary analysis of a neotropical mammal fauna. Biotropica, 5: 150-161.

Freeland, W.J. and D.H. Janzen.
1974. Strategies in herbivory by mammals: the role of plant secondary compounds. Am. Naturalist 108: 269-289.

Greenslade, P.J.M.
1971. Interspecific competition and frequency changes among ants in Solomon islands coconut plantations. J. Appl. Ecol., 8 (2): 323-352.

Hanna, A.D., E. Judenko, and W. Heatherington.
1956. The control of *Crematogaster* ants as a means of controlling the mealybugs transmitting the swollen-shoot virus disease of cacao in the gold coast. Bull. Ent. Res., 47: 219-226.

Leston, D.
1971. Ants, capsids and swollen shoot in Ghana: interactions and the implications for pest control. Proc. 3rd International Cocoa Research Conference, Accra, 1969: 205-221.

1973. The ant mosiac-tropical tree crops and the limiting of pests and diseases. PANS, 19: 311-341.

Lubin, Y.D. and G.G. Montgomery.
1981. Defences of *Nasutitermes* termites (Isoptera, Termitidae) against *Tamandua* anteaters (Edentata, Myrmecophagidae). Biotropica 13 (1): 66-76.

Lubin, Y.D., G.G. Montgomery, and O.P. Young.
1977. Food resources of anteaters (Edentata: Myrmecophagidae) I. A year's census of arboreal nests of ants and termites on Barro Colorado Island, Panama Canal Zone. Biotropica, 9(1): 26-34.

Majer, J.D.
1972. The ant mosiac in Ghana cocoa farms. Bull. Ent. Res., 62: 151-160.

1976a. The maintenance of the ant mosiac in Ghana cocoa farms. J. Appl. Ecol., 13(1): 123-144.

1976b. The ant mosiac in Ghana cocoa farms: further structural considerations. J. Appl. Ecol., 13(1): 145-156.

Meritt, D.A.
1975. The lesser anteater *Tamandua tetradactyla* in captivity. International Zoo Yearbook 15: 41-45.

Montgomery, G.G.
1979. El grupo alimentico (feeding guild) del oso

hormiguero: convivencia y especialización de las presas de sustento de los osos hormigueros neotropicales (Edentata, Myrmecophagidae). Conciencia, 6(1):3-6.

1983. *Cyclopes didactylus* (Tapacara, Serafin de Platanar, Silky anteater). Pp. 461-463, *In* Costa Rican natural history (D.H. Janzen, ed.). Univ. Chicago Press, 816 pp.

Montgomery, G.G.
1985. Movements, foraging and food habits of the four extant species of neotropical vermilinguas (Mammalia; Myrmecophagidae). (this volume).

Montgomery, G.G. and Y.D. Lubin.
1977. Prey influences on movements of neotropical anteaters. Pp. 103 -131. *In* Proceedings of the 1975 Predator Symposium (R.L. Phillips and C. Jonkel, eds.). Montana Forest and Conservation Experiment Station, University of Montana, Missoula, 268 pp.

Montgomery, G.G. and M.E. Sunquist.
1978. Habitat selection and use by two-toed and three-toed sloths. Pp. 329-359, *In* The ecology of arboreal folivores (G.G. Montgomery, ed.). Smithsonian Insitution Press, Washington, D.C. 574 pp.

Montgomery, G.G., W.W. Cochran and M.E. Sunquist.
1973. Radiolocating arboreal vertebrates in tropical forest. J. Wildl. Magmt., 37(3): 426-428.

Redford, K.H.
1985. Food habits of armadillos (Xenarthra: Dasypodidae). (this volume).

Roth, L.M. and T. Eisner.
1962. Chemical defences of arthropods. Ann. Rev. Entomol. 7: 107-136.

Room, P.J.
1971. The relative distribution of ant species in Ghana's cocoa farms. J. Anim. Ecol., 40: 735-751.

Sunquist M.E. and G.G. Montgomery.
1973. Activity pattern of a translocated silky anteater (*Cyclopes didactylus*). J. Mammal., 54(3):782.

Taylor, B.K.
1978. The anatomy of the forelimb of the anteater (*Tamandua*) and its functional implications. J. Morph., 157(3): 347-367.

Taylor, B.K.
1985. Functional anatomy of the forelimb in vermilinguas (anteaters). (this volume).

Vanderplank, F.L.
1960. The bionomics and ecology of the red tree ant, *Occophylla* sp., and its relationship to the coconut bug *Pseudotheraptus wayi* Brown (Coreidae). J. Anim. Ecol., 29: 15-33.

Way, M.J.
1953. The relationship between certain ant species with particular reference to biological control of the coreid, *Tharaptus* sp. Bull. Ent. Res., 44: 669-691.

Wetzel, R.M.
1975. The species of *Tamandua* Gray (Edentata, Myrmecophagidae). Proc. Biol. Soc. Washington, 86: 25-34.

Wetzel, R.M.
1985. The identification and distribution of recent xenarthra (= edentata). (this volume).

Williams, E.C.
1941. An ecological study of the floor fauna of the Panama rain forest. Bulletin of the Chicago Academy of Science, 6 (4): 63-124.

Wilson, O.E.
1959. Some ecological characteristics of ants in New Guinea rain forests. Ecology, 40 (3): 437-447.

MOVEMENTS, FORAGING AND FOOD HABITS OF THE FOUR EXTANT SPECIES OF NEOTROPICAL VERMILINGUAS (MAMMALIA; MYRMECOPHAGIDAE)

G. Gene Montgomery

Smithsonian Tropical Research Institute, Apartado 2072, Balboa, Republic of Panama.

Abstract
Individuals of two species of vermilinguas (anteaters), *Cyclopes didactylus* and *Tamandua mexicana* were studied in a tropical forest in Panama, and the two other extant species, *Tamandua tetradaetyla* and *Myrmecophaga tridactyla*, in a seasonally flooded savanna in Venezuela. Their movement patterns, foraging habits and prey species were determined. *Cyclopes* ate only ants, but the other species, especially *Tamandua*, included termites as a significant part of their diet. There was very little dietary overlap in the ants taken by the two Panamanian species, but the two Venezuelan species were very similar in the selection of their prey. Seasonal changes in composition of the diet were studied in all four species. The vermilinguas rarely cause major damage to the prey colonies they attack and thus preserve their foood sources.

Abstracto
Se estudiaron dos especies de vermilinguas (osos hormigueros), *Cyclopes didactylus y Tamandua mexicana* en una selva tropical en Panamá, y dos otras especies, *Tamandua tetradactyla y Myrmecophaga tridactyla*, en una sabana estacionalmente inundada en Venezuela. Se determinaron sus movimientos, su comportamiento alimenticio, y sus especies de presas. *Cyclopes* se alimentó sólo de hormigas, pero las demás especies, especialmente *Tamandua*, incluyeron termites como un componente significativo de su dieta. Hubo muy poca similitude en las hormigas capturadas por las dos especies panameñas, pero las dos especies venezolanas fueron muy similares en las selección de sus presas. Se estudiaron los cambios estacionales en la composición de la dieta de las cuatros species. Los hormigueros raramente causaron mayor daño a las colonias de las presas que atacaron, preservando así sus fuentes de alimentación.

Resumo
Se estudaram duas espécies de vermilínguas (formigueros), *Cyclopes didactylus* e *Tamandua mexicana* num bosque tropical de Panamá, e de outras espécies *Tamandua tetradactyla* e *Myrmecophaga tridactyla* numa savana com inundações estacionais. Se determinaram seus padrões de movimentos, seus hábitos alimentares e suas presas. *Cyclopes* se alimentava sòmente de formigas, mas as outras espécies, especialmente *Tamandua*, incluiram térmitas como parte siignificativa de sua dieta. Houve pouca similitude nas espécies de formigas as quais se alimentavam das especies panamenhas mas as espécies venezuelanas selecionaram especies similares. Se estuadaram as mudanças de estaçoes na composição da dieta das qua tro espécies. Os formigueiros, em raras ocasiões, causam maior prejuízo às colônias de suas presas e deste modo conservam sua fonte de alimento.

Ants and termites form one of the most abundant, diverse, and ubiquitous potential food resources in the tropics. Ants are probably numerically the commonest of free-living insects (Leston, 1973). Ground-nesting ants *Lasius flavus* in British grassland reach population densities of 4,000 to 6,000 per m^2 (Waloff and Blackith, 1962). There are on average more than 680 ants per m^2 in the litter layer on Barro Colorado Island, in Panama (BCI) (Willis, 1976). Single species of arboreal ants may reach 600 per m^2 (Leston, 1971), and there are 35 to 240 ants per m^2 in the canopy of Panamanian forest (Montgomery, 1985). There are 45 to 50 species per tenth ha in savanna habitat (Baroni-Urbani et al. 1978), and even greater species diversity in tropical forest canopies (Majer, 1976). Species diversity of termites is lower, but their population densities may be considerably higher (cf Wood and Sands, 1978).

Colonies of ants and termites are relatively long-lived (Lubin, et al., 1977; Baroni-Urbani et al., 1978), with colonies living more than 4 years but less than 10 years in some species (Greenslade, 1971), and as much as 19 years in others (Waloff and Blackith, 1962). New workers can be recruited rapidly if the nest is damaged (cf. Vanderplank, 1960), and turn-over of workers is relatively rapid even without catastrophic loss (Nielson and Josens, 1978).

Ant population levels remain relatively constant throughout the year in spite of environmental changes and fluctuations in population levels of other arthropods (Leston, 1973). This is probably not because they are unable to adjust their populations to seasonal change (Janzen and Schoener, 1968), but because the colonial nests of ants and termites serve to buffer the colony against depredation by predators and against climatic fluctuations (Wilson, 1971; Culver, 1974). Thus individuals of a given colony will usually be potentially available for a predatory ant-or termite-eater.

In view of the apparent availability of ants and termites as food, surpisingly few mammalian forms have evolved to exploit this food resource, suggesting that it is a difficult kind of food to utilize (Montgomery, 1979). World-wide, in no part of the tropics are there more than three or four species of mammalian ant- and termite- eaters which have specialised anatomically, behaviorally and physiologically for cropping colonies of social insects. Those few mammalian forms which have evolved to feed on colonies of social insects live at relatively low population densities and crop small proportions of the available food. In this paper and elsewhere (Montgomery, 1985), I present data which indicate that social insects are well able to defend their colonies against attack by vermilinguas (neotropical mammalian anteaters) (Montgomery and Lubin, 1977; Lubin and Montgomery, 1981), and thus probably limit the population density and shape the movements and activity of these mammalian forms.

The range of habitats used by vermilinguas is from seasonally very dry savanna to very moist forests at low to moderate altitudes (Handley, 1976; Wetzel, 1985). In at least some of those habitats and in many parts of their species ranges, members of the three vermilingua genera live in sympatry, sharing the same parts of the habitat.

The three extant genera of vermilinguas have widely disparate sizes. They range in body weight from the 0.25 kg *Cyclopes didactylus*, through the 3.5 kg *Tamandua mexicana* and *T. tetradactyla*, to the 30 kg *Myrmecophaga tridactyla*. The three genera are widely separated ecologically, with *Cyclopes* being a specialized arboreal, nocturnal anteater, *Tamandua* a relative generalist which is active during both day and night, feeding on both ants and termites from terrestrial and arboreal sites, and *Myrmecophaga* a terrestrial specialist, feeding mainly on ants, usually at night.

In this paper, I explore the bases for niche separation among sympatric neotropical vermilinguas. I give details of the foraging behavior and movement patterns, and report data on food habits. I then offer further thoughts on why vermilinguas appear to be limited in their current population densities and in their evolutionary radiation by defences of the colonial arthropods they use as food.

STUDY AREAS AND METHODS

Study Areas.-- *Cyclopes* and *T. mexicana* were studied on Barro Colorado Island (BCI), in the lowlands of central Panama (Croat, 1978). The secondary forest where the study animals lived has been described elsewhere (Montgomery and Sunquist, 1978). That portion of Panama has a seasonal pattern of rainfall, with an approximately 4-month long dry season beginning in January, and an 8-month long wet season beginning in May. Ánnual rainfall is about 26 cm, 90 percent of it falling during the wet season (Dietrich et al., 1982).

T. tetradactyla and *Myrmecophaga* were studied on a privately-owned cattle ranch (Hato Masaguaral) in the llanos of central Venezuela (Montgomery and Lubin, 1977). The study animals lived mainly in seasonally flooded palm savanna (Troth, 1979), but used higher areas of sand dunes as well, particularly in wet season. That part of Venezuela is seasonally very dry, with almost no rain falling from December through March. Average annual rainfall for a period of 25 years was about 15 cm, most of it falling during the period May through October (from data in Troth, 1979). Flooding is severe in wet season, and parts of the study area which are ankle-deep in dust in March are 1 to 2 m deep in flood water in June.

Radio-location and Observation.-- Most data in this paper are from radio-marked, wild vermilinguas. Individuals were captured in the wild, anesthetized, radio-marked, and returned to where they were captured. *Cyclopes* and arboreal *Tamandua* were returned to the base of the tree from which the individual was captured. Each animal was captured as often as necessary to renew batteries or transmitter attachment, or to take samples of stomach contents or other data.

Selection and Capture of Study Animals.-- There was never a surplus of study animals of any of the 4 species. Almost all individuals found in suitable portions of the study areas were captured and radio-marked, except those which escaped during attempts to capture them.

All *Cyclopes* used in the study were initially seen by Bonifacio de León while he conducted surveys in the forest on BCI. After he reported sighting an individual, it was captured with a noose attached to a 2- m long pole; I climbed the tree if necessary and used extension poles to reach the animal. This procedure was also used to recapture radio-marked *Cyclopes*. After capture, the *Cyclopes* was carried to the laboratory building in a small holding cage. There it was anesthetized with 1 or 2.5 mg of CI-744 (4-9 mg per kg), weighed and measured, and a new transmitter was taped to its tail. It was left in the cage overnight before being returned to the forest. The *Cyclopes* almost invariably defecated during the night (43 of 45 opportunities), and all feces were carefully collected from the cage. The feces were dried in an oven at 60°C, weighed, and a sub-sample was retained so that an estimate of the number of ants in the feces could be made.

T. mexicana in Panama were usually picked up from the ground by the tail, held by the tail until 20-25 mg CI-744 (48 mg per kg) could be injected into the tail, then placed in a bag and transported to the laboratory. *T. mexicana* are quite aggressive and capable of inflicting serious wounds with their powerful foreclaws, and this capture method was dangerous. A slightly safer method was to noose the vermilingua around the chest, then hold it by the tail and the noose while injecting CI-744 into the tail. The noose-pole was not usually carried during field observations, but a syringe and CI-744 were, thus allowing capture of vermilinguas encountered in the forest.

T. tetradactyla in Venezuela were nocturnal, and were usually captured during daytime from the crowns of palm trees (*Copernicia tectorum*) or other trees in which they slept. They were much less alert and aggressive than *T. mexicana* and they were usually drugged with CI-744 (48 mg per kg) by climbing adjacent trees until the vermilingua could be reached, then injecting the drug into their hip or tail before they moved.

Myrmecophaga were usually drugged where they slept on the ground during daytime, by injecting 100-250 mg CI-744 (4-11 mg per kg), 150-500 mg ketamine (8-24 mg per kg), or 250-400 mg Sernylyn (12-13 mg per kg) with a syringe-dart fired from a Cap-chur pistol. CI-744 was the most effective of the capture drugs used, but was not always available. Sernylyn was an effective capture drug, but the animals required a longer recovery period, while ketamine was only marginally satisfactory, and was used only when the other drugs were not available. Rarely, a moving *Myrmecophaga* was shot with the syringe-dart by running ahead of it and waiting for it to run or walk past. A sleeping *Myrmecophaga* could be approached to within 1 m or less, and the dart was aimed at the animal's hip. After being struck by the dart, the *Myrmecophaga* usually awoke and ran away. If the animal was not radio-marked, it was necessary for other persons to surround the capture site, so as to know the direction the animal ran, and to attempt to follow it until the drug took effect. *Myrmecophaga* ran as much as 200 m before the drug took effect. If the animal was radio-marked, or sleeping in a relatively open area, I could capture it by myself, using radio-location if necessary to find the drugged individual after it stopped moving. *Myrmecophaga* were not taken to the laboratory for radio-marking. Instead, measurements and samples of stomach contents and feces were taken, and the transmitter was attached in the field. The animals remained immobile for about 30 minutes after the drug took effect, and were left asleep in the shade to recover. They were always able to travel and feed that same night.

Transmitter Attachment.-- Transmitters (AVM Instrument Co., Champaign, Ill.) were taped to the animal's tail with adhesive tape (usually Elastikon; Johnson and Johnson, New Brunswick, N.J.) with hairs of the tail incorporated in the wraps of the tape to help hold the transmitter in place. All transmitters except those used on young *Cyclopes* being carried by their mothers had whip antennas, and the weight of the transmitter and length of the antenna varied with the size of the animal. Transmitters for adult *Cyclopes* weighed 6-11 gm, and had antennas 7-10 cm long. Transmitters used on young *Cyclopes* weighed 3 gm. Transmitters used on *Tamandua* weighed about 30 gm, and had a 15-cm long antenna, while transmitters used on *Myrmecophaga* weighed about 50 gm and had an antenna 30 cm long. Depending on the weather, the transmitters could be left in place on *Cyclopes* for about 30 days, and on *Tamandua* and *Myrmecophaga* for at least 75 days before it was necessary to change the batteries or the tape. Some necrosis of tissue under the transmitters occurred during rainy season, and the animals were recaptured more often then to change the location of the transmitter on the tail.

Radio-location and Observation.-- On many dates during the study, all radio-marked individuals were located once, by walking to the place where the animal was sleeping or less commonly to where it was active (only *T. mexicana* were usually active during daylight). Triangulation was used only to give an approximation of the animal's location, to make approaching it more efficient. All radio-locations were precise, and the radio-marked individual was usually seen. These daily locations were marked with colored plastic flags with the date and time written on them.

Individuals were radio-tracked and followed during large portions of their activity cycles on selected dates. When possible, they were observed as they moved through the trees (*Tamandua*) or on the ground (*Tamandua* and *Myrmecophaga*). It was not possible to see *Cyclopes* because they traveled in tree crowns at night, and usually moved through thick masses of lianas (Montgomery, 1985). Plastic flags were used to mark the places where the vermilingua was radio-located or seen, or a place directly below an arboreal animal. Flags were placed at 2 to 5 minute intervals when the animal was moving. The time and date were written on each flag, and it was tied to vegetation.

The flagged locations were surved and mapped with a compass and surveyors measuring tape. These data were used to compute the X-Y coordinates of each of the locations. The coordinates were stored, with the date and time of the location, and used for mapping movement paths and home range spaces, and for computing rates of travel and other movement parameters (Montgomery and Lubin, 1977). A HP-9830 computer system was used for data reduction and analysis.

Field notes were recorded on a tape recorder, and

included activity of the animal, time spent feeding, the kind of prey being eaten if this could be determined, whether or not specimens of the prey were collected from the feeding site or from pieces of wood or debris which fell from a site where an arboreal animal fed, and the approximate direction the animal moved. When vermilinguas were followed at night (*Myrmecophaga* and *T. tetradactyla* in Venezuela; *Cyclopes* in Panama) headlamps were used and, for *Myrmecophaga*, horses or mules were used for mobility when the animal moved rapidly between places where it fed at a more leisurely pace. *Myrmecophaga* were followed on foot whenever possible. Arboreal *Cyclopes* presented a particularly difficult observational problem, because they moved at night through the forest canopy in areas where there were many lianas. When a *Tamandua* moved through trees in daytime, it was possible to keep the animal in view most of the time, and to record some of its activities on video-tape using a portable video-recorder (Sony AV-3400). Radio-location was used to track *Cyclopes* at night, and to relocate individuals of the other species when they were temporarily lost to view during periods when they were being followed and observed.

Sampling for Food Habits Analyses.-- Samples of prey were taken from places where vermilinguas fed, carried in plastic bags labeled with the time of the feeding observation, and identified or placed in preservative in the laboratory after a period of observation. Samples of stomach contents were taken by intubation from anesthetized vermilinguas when they were captured for radio-marking or recaptured to change batteries or transmitter attachment. Samples of stomach contents and feces were also taken from vermilinguas found dead on roads near the study areas. Complete fecal masses from *Cyclopes* were collected from the cage in which the animal was held overnight. Complete fecal masses, each representing a defecation by *Tamandua* or *Myrmecophaga*, were collected in the field, then dried and weighed. Vermilingua feces were easily recognized and distinguished from the feces of other mammals by their size and characteristic odor. All insect remains in subsamples of these samples were then identified and counted, to give estimates of daily food intake.

Insect remains in stomach contents and fecal samples were sorted, counted, and identified to morpho-species (all individuals that looked alike). Except in some samples of stomach contents which contained very fresh insects, identifications and counts were based on head capsules, which usually survived the digestive process with little damage. A reference collection of ants and termites collected from the study areas was made, and head capsules and whole individuals from the stomach contents and fecal material were identified by comparison with it. In some cases, the morpho-species have been identified and named. A reference collection containing samples of all the morpho-species is kept in Panama.

The total length and the length of the head capsule of a sample of whole ants was measured to the nearest 0.1 mm. The lengths of head capsules from some of the samples of ants from vermilingua stomach contents were measured as well.

RESULTS AND DISCUSSION

Activity, Foraging, and Movement Patterns.-- *Cyclopes* were strictly nocturnal, arboreal ant-eaters (Table 1). With respect to their food habits (Montgomery, 1985), they showed considerable agreement from day-to-day, season-to-season, and individual-to-individual.

Tamandua were scansorial, moving, feeding and resting both on the ground and in trees. They fed on a combination of ants and termites. In trees, they used mainly branches rather than lianas as a substrate for moving and feeding, but sometimes fed from ants or termites living inside liana stems. They sometimes fed from large nests of ants or termites, both while foraging on the ground and in trees, but more commonly took a few individuals from each of many small nests or covered trails of ants or termites (Montgomery and Lubin, 1977). *Tamandua*, particularly *T. mexicana* were individualists with respect to timing of activity, and in their food habits (Montgomery, 1985). An adult *Tamandua* tended to be active at about the same times each diel, but different individuals in the population were active at different times of day and night. Some *Tamandua* in both Panama and Venezuela were mainly arboreal, while others tended to move and feed mainly on the ground (Montgomery and Lubin, 1977).

Myrmecophaga were terrestrial, primarily nocturnal, and primarily ant-eaters. They appeared to specialize by feeding mainly on ants of the genus *Camponotus* (see below), and to show general dietary agreement from day-to-day and from individual-to-individual. There was an indication that their diet seasonally included more or less termites.

Vermilinguas of all three genera spent considerable amounts of energy and a large portion of each day in finding and consuming social insects. Certainly most of the time they spent foraging was spent in searching for prey, not in consuming it. The vermilinguas found their prey by scent, exposed it by using strong foreclaws, then captured and consumed the prey with a long, highly mobile tongue. During observations of feeding vermilinguas, it became obvious that they fed only from nests or covered feeding trails of ants, termites, or both, and not on scattered individuals. Their foraging pattern can best be described as a long series of short stops to feed, interspersed with relatively rapid, almost straight-line movement through the habitat. This movement pattern was as typical of arobreal *T. mexicana* while they moved through the trees, as it was of terrestrial *Tamandua* and *Myrmecophaga*. Most feeding stops which could be observed (*Tamandua, Myrmecophaga*) were of short duration, usually less than 1 min, and an individual vermilingua fed from 80 or more different colonies of social insects during its daily activity period (cf. Montgomery and Lubin, 1977; Montgomery, 1985). Although it was not possible to observe *Cyclopes* while they moved and fed in the forest canopy at night, their pattern of movement suggests that they too moved almost constantly through all the hours of darkness, and probably fed from very many colonies of ants each night (Montgomery, 1983; 1985). The high species diversity of ants in each day's ration of food certainly supports the idea that *Cyclopes*' pattern of feeding was similar to that of the other vermilinguas (see below).

The vermilinguas apparently monitored some colonies and cropped them repeatedly during the course of several weeks (Montgomery and Lubin, 1977), taking relatively few individuals from the colony each time they fed there. Extensive damage to nests rarely occurred, then usually only when members of large, fat-rich reproductive castes were present (Montgomery and Lubin, 1977).

Potential Competition Between Vermilinguas .-- In both Panama and Venezuela, *Tamandua* had activity patterns, movement substrates, and food habits which overlapped in part with the sympatric vermilinguas, *Cyclopes* in Panama and *Myrmecophaga* in Venezuela. Direct comparisons of the movement patterns and food habits of the sympatric vermilinguas on both study areas show some ways in which competition for food and living space might have been reduced (Table 1). These included differences in the timing of activity, differences in the degree to which the animals were arboreal or terrestrial, differences in the actual places where the animals fed, and differential selection of prey, primarily termites.

In both Panama and Venezuela, sympatric radio-marked individuals of different vermilingua species had overlapping home range spaces. They thus were microsympatric, and in a general way had access to the same colonies of ants and termites. Sizes of the home range spaces were related to the body sizes of the animals (Table 1), and approximately 2 adult male and 5 adult female *Cyclopes* could fit on the home range space of one *T. mexicana* (Montgomery, 1979; 1985). Approximately 20 *T. tetradactyla* home range spaces would fit on the home range space of one *Myrmecophaga*.

The two species of *Tamandua* differed considerably in the size of their home range spaces (Table 1). *T. tetradactyla* living in the palm savanna had home range spaces more than 15 times as large as those of *T. mexicana* living in moist lowland forest. The palm savanna has little vertical structure (Troth, 1979) and lower plant species diversity, in comparison with moist lowland forest (Croat, 1978), and probably has lower species diversity of both ants and termites. The larger home range spaces in Venezuela may have been related to the need for the animals to travel further to gain each day's ration of diverse kinds food.

The two species of vermilinguas in Panama used very similar portions of the habitat, and arboreal *T. mexicana* were observed feeding in some trees where *Cyclopes* were radio-located. However, *Cyclopes* were completely arboreal, and those *T. mexicana* which were terrestrial had access to different colonies of ants and termites.

There were additional ways in which the two species of vermilinguas gained access to different sources of food (Montgomery, 1985). *Cyclopes* were entirely nocturnal (Sunquist and Montgomery, 1973), while at least some *T. mexicana* were active during daytime (Montgomery and Lubin, 1977). Day-active *T. mexicana* might thus have had differential access to day-active social insects, although nests of the colonies would certainly be available (but perhaps less attractive or vulnerable) throughout the day and night. *Cyclopes* are able to reach and feed from lianas and small stems in the forest canopy (Montgomery, 1983; 1985) which are too small and weak to support the heavier *T. mexicana*. On the other hand, *Tamandua* are larger, and have a more strongly developed forearm and foreclaw (Taylor, 1985). They are thus able to open durable carton nests, or nests in wood, which are inaccessable to *Cyclopes*.

In Venezuela, radio-marked *Myrmecophaga* and *T. tetradactyla* also had overlapping home range spaces. Except for times when *T. tetradactyla* fed or slept in trees, vermilinguas of both species used very similar parts of the habitat. A possible exception was during wet

Table 1. Comparison of body sizes, home range spaces, movements, and food habits of the four extant species of vermilinguas.

Species	*Weight (kg)*	*Home Range (ha)*	*Daily Movement (m)*	*Time of Activity*	*Percent Ants*
Cyclopes didactylus	0.2	5	300	Night	100
Tamandua mexicana	3.8	25	1000	Night and Day	33
Tamandua tetradactyla	4.2	375	3000	Night and Day	49
Myrmecophaga tridactyla	26.2	2500	11000	Mainly Night	96

season when the llanos was flooded. *Myrmecophaga* were able to walk through flooded portions of the study area on which the water was too deep for *T. tetradactyla* to reach without swimming. *T. tetradactyla* were not radio-tracked during times when the study area was flooded, thus there are no data to indicate whether they changed their movement and feeding patterns in wet season. There are no indications from the radio-tracking data that *Myrmecophaga* responded to flooding by changing their patterns of movement or use of their home range spaces.

Diets of Vermilinguas.-- Vermilinguas find their prey by scent, expose it by breaking open nests or wood, or by digging with a strong foreclaw, then capture and ingest prey with a long, mobile tongue. All four species of vermilinguas were selective in choosing their diet, feeding entirely on social insects (Table 2). The only other food items in the samples of stomach contents and feces were mimics of social insects. Stomach contents and feces from all four species did not contain large amounts of soil, wood chips, or pieces of carton nest (certainly less than 1 percent of the volume), in spite of the obvious difficulty of ingesting ants or termites without taking in the substrate on which the insects were found. I thus cannot agree with McNab (1974; 1980) who felt that a possible reason for the low metabolic rate of vermilinguas might relate to ingestion of large amounts of indigestable detritus with the food.

The mechanism for the extreme selectivity in intake of food is not at all clear. All food enters the mouth on the vermilingua's tongue (Naples, 1985), either held there by the ridged structure of the tongue and saliva, or because the ant or termite has bitten the tongue. Arthropods other than ants and termites are probably excluded from the diet because the vermilinguas feed only where ants and termites are concentrated in nests or covered feeding trails. Other arthropods, and ants or termites of other species would be excluded from the nest or feeding trail of any colony by the members of the defensive caste of that colony. However, it is not at all apparent how the vermilinguas avoid ingesting soil, leaves, and pieces of the nest material. The possibility that the structure of the tongue, or the viscosity or some other propery of the saliva, is optimized to selectively hold objects of the size and configuration of the usual prey should be investigated.

The smallest of the vermilinguas *Cyclopes* was truly an ant-eater, with no indication that other arthropods, including termites, were ingested, even by accident (Table 2). There were probably as many or more termites as ants in the forest canopy on BCI, even though there were probably fewer species. Given that the selection of food was done at night, probably by finding a concentration of ants inside a small stem on the basis of scent, from outside the stem (Montgomery, 1983; 1985), with the final selection probably on the basis of taste of the food items, it seems obvious that *Cyclopes* selected against feeding on termites. How this was done so precisely is not at all clear, because there were surely times when termites and ants mingled in the confusion of having their nests or feeding trails disturbed by the vermilingua. I suspect that some final selection of food items was done after the insects were in the mouth of the *Cyclopes,* and that termites were deliberately rejected. If that was the case, then it perhaps indicates that termites are unsuitable as food, perhaps even poisonous to *Cyclopes*, even though termites are eaten by *Tamandua*, and to a much lesser extent by *Myrmecophaga* (Table 3).

Table 2. Ants in the diets of the four species of vermilinguas. All arthropods found in the samples of stomach contents and feces were either ants or termites, except for 2 individuals of a homopteran mimic of one of the ants commonly found in *Tamandua* stomachs. The numbers of vermilinguas sampled are shown in parentheses. Data from animals found dead off the study areas are included.

	Number Of Samples	*Number Of Arthropods*	*Percent Ants In Samples*	*Number Of Ant Morpho-species*
Panama (Moist Lowland Forest)				
Cyclopes	42(12)	15,584	100	76
T. mexicana	29(17)	18,056	33	77
Venezuela (Seasonally Flooded Palm Savanna)				
T. tetradactyla	19(16)	10,595	49	32
Myrmecophaga	28(10)	16,627	96	23

The diet of *Cyclopes* was diverse, with a total of 76 morpho-species found in the samples (Table 2). There was a great deal of dietary agreement among the individual *Cyclopes,* and five of the 6 highest ranking morpho-species (which included 66 percent of the ants) were found in samples from 90 percent or more of the *Cyclopes* whose feces was sampled (Montgomery, 1985). Nearly half the diet of *Cyclopes* was composed of ants of the 3 highest ranking genera (Table 3), with *Crematogaster* the highest ranking genus. That genus was represented in the samples by only 1 morpho-species (Montgomery, 1985).

Both species of *Tamandua* ate termites, as well as ants (Table 2). Identification of termites in the stomach contents and feces to morpho-species was more difficult than identification of ants, but the diets of *Tamandua* included at least 10 species of 6 genera (*Nasutitermes, Microcerotermes, Armitermes, Leucotermes, Coptotermes* and *Calcaritermes*).

Most of the termites which could be identified were *Nasutitermes*. As examples, a total of 1,527 termites were identified in 6 samples of stomach contents taken from *T. mexicana* on BCI during 1974. More than 92 percent of those were *Nasutitermes*, 5 percent were *Coptotermes*, 2 percent were *Microcerotermes*, and the remainder were *Leucotermes* and *Armitermes*. Termites which were identified from a total of 63 feeding stops when termites were eaten by *T. longicaudata* in Venezuela while 3 individuals were being radio-tracked and observed on 10 dates, included 71 percent stops to feed on *Nasutitermes*, 27 percent stops to feed on *Microcerotermes*, and 2 percent stops to feed on *Coptotermes*.

Termites formed a much smaller part of the diet of *Myrmecophaga* (Table 3). More than 90 percent of the 708 termites identified from 25 samples of stomach contents taken from 7 *Myrmecophaga* on the study area were *Nasutitermes*.

While a large part of the diet of both *T. mexicana* and *T. tetradactyla* was termites, these vermilinguas also ate many ants of many morpho-species (Table 2). The morpho-species diversity of ants in the diet of *T. mexicana* in Panama was considerably higher than that of *T. tetradactyla* in Venezuela. There were more than twice as many morpho-species in the samples from the former species, with about the same number of individual ants. In Panama, *T. mexicana* and *Cyclopes* diets included almost the same number of ant morpho-species. The species diversity among ants in the moist lowland forest was probably greater than that in the seasonally flooded palm savanna, and the vermilinguas studied in Panama thus may have had more ant species available as food.

Table 3. The three highest ranking ant genera in the diets of the four species of vermilinguas. The diet of *Cyclopes* included only ants (15,584 food items), that of *Tamandua mexicana* included 71 percent termites (14,056 food items), that of *T. longicaudata* included 35 percent termites (7,413 food items), and that of *Myrmecophaga* included 5 percent termites (13,992 food items). Only data from animals captured on the study areas are included.

Study Area / *Vermilingua*	*Ant Genus*	*Percent Of Diet*	*Percent Of Ants*
Panama (Moist Lowland Forest)			
Cyclopes	*Crematogaster*	22	22
	Solenopsis	19	19
	Pseudomyrmex	8	8
T. mexicana	*Camponotus*	12	41
	Azteca	7	24
	Crematogaster	3	10
Venezuela (Seasonally Flooded Palm Savanna)			
T. tetradactyla	*Solenopsis*	18	28
	Formicine, sp. 213	17	26
	Camponotus	12	19
Myrmecophaga	*Camponotus*	61	65
	Solenopsis	23	24
	Pheidole	6	6

Unlike *Cyclopes*, individual *T. mexicana* tended not to eat the same ant morpho-species as other individuals living in the same area, or to be consistent in the morpho-species of ants which each individual ate from day-to-day (Montgomery, 1985). None of the 10 highest ranking morpho-species were found in samples from more than 65 percent of the individuals sampled or in more than 55 percent of the samples. Most high ranking morpho-species were found in fewer than half the total samples and in samples from fewer than half of the individuals. Similarly, *T. tetradactyla* in Venezuela were not consistent among themselves in the ant morpho-species which they ate, or in the morpho-species eaten from day-to-day. Of the four highest ranking morpho-species, only the fourth-ranked was in more than 40 percent of the samples (73 percent), or was used by more than 6 of the 9 individuals.

Myrmecophaga diets included the fewest ant morpho-species, and had the lowest morpho-species diversity among diets of the four kinds of vermilinguas (Table 2). Like *Cyclopes*, *Myrmecophaga* showed considerable dietary agreement among individuals, and from sample to sample for each individual. More than 65 percent of the ants were of the genus *Camponotus* (Table 3), and more than 40 percent were of the highest-ranking morpho-species (*Camponotus abdominalis*). Ants of that species were in samples from all of the 10 *Myrmecophaga* sampled on and off the study area, and in 90 percent of the 29 stomach samples.

Size of Ants in the Diets.-- The sizes of ants eaten by the vermilinguas were studied by measuring the length of head capsules in samples of stomach contents and feces (Table 4). The relationship between ant head size and body size was estimated by measuring ants of several morpho-species collected whole in the field. There was a highly significant positive correlation between the length of an ant's head capsule and the total length of the ant ($r^2 = 0.84$, $N = 74$, $p < 0.01$).

Most of the diet of *Cyclopes* consisted of small ants (length < 4 mm), and there were significantly fewer

Table 4. Sizes of ants in the diets of vermilinguas. Small ants were less than 4 mm in length, and large ants were longer than 8 mm.

Study Area / *Vermilingua*	*Samples*	*Average Percent (Range)* *Small Ants*	*Large Ants*
Panama (Moist Lowland Forest)			
Cyclopes	4	86 (56-90)	1 (0-1)
T. mexicana	21	65 (13-89)	2 (0-12)
Venezuela (Seasonally Flooded Palm Savanna)			
T. tetradactyla	16	46 (16-92)	1 (0-11)
Myrmecophaga	8	24 (2-44)	18 (1-35)

large ants (length > 8 mm) in the diet of *Cyclopes* than in the diet of *T. mexicana* ($t = 3.14$, 22 df, $p < 0.01$). The diet of *Myrmecophaga* included mainly medium-sized and large ants, and there were significantly fewer small ants in samples from *Myrmecophaga* than in samples from *T. tetradactyla* ($t = 3.39$, 49 df, $p < 0.01$). This result is somewhat surprising, in view of the large amount of dietary overlap between the two species of vermilinguas in Venezuela (Table 5).

Dietary Overlap.-- There was relatively little overlap in that portion of the diets of *Cyclopes* and *T. mexicana* formed by ants. Many of the most important ant morpho-species in the diet of *T. mexicana* did not occur at all in the samples from *Cyclopes*, and ant morpho-species which formed most of the diet of *Cyclopes* were relatively rare in the diet of *T. mexicana* (Montgomery, 1985). Of the 112 total morpho-species of ants used by one or the other species of vermilingua, only 11 percent were found in samples from both *Cyclopes* and *T. mexicana* (Table 5), and more than 60 percent of the individual ants in those samples were not of shared morpho-species. When it is considered that the *T. mexicana* diet included many more termites than ants, and that no termites were found in samples from *Cyclopes* (Table 2), it is apparent that the diets of these sympatric vermilinguas were quite different.

By contrast, the portions of the diets of the sympatric *T. tetradactyla* and *Myrmecophaga* formed by ants were quite similar (Table 5). All except 1 of the ants identified from *Myrmecophaga* were of morpho-species also used by *T. tetradactyla*, and 92 percent of the ants in samples from *T. tetradactyla* were of ant morpho-species used by *Myrmecophaga*. The only major difference in the food habits of the vermilinguas in the palm savanna in Venezuela was that half the diet of *T. tetradactyla* consisted of termites, while *Myrmecophaga* ate relatively few termites. Even that difference tended to decrease in wet season, when *Myrmecophaga* ate increasing numbers of termites (see below).

Table 5. The numbers of ant morpho-species identified in the diets of the four species of vermilinguas, and overlap in the ants eaten by sympatric vermilinguas in Panama and in Venezuela. Only data from animals captured on the study areas are included. Percent shared species and percent shared ants are the respective proportions of the total species or ants eaten by one kind of vermilingua which were also in the diet of the sympatric vermilingua.

Study Area / *Vermilingua*	*Total Ants*	*Morpho-species*	*Percent Shared Species*	*Percent Shared Ants*
Panama (Moist Lowland Forest)				
Cyclopes	15,584	76	16	42
T. mexicana	4,068	60	20	26
Both Species	19,652	112	11	38
Venezuela (Seasonally Flooded Palm Savanna)				
T. tetradactyla	4,818	23	74	92
Myrmecophaga	13,284	18	94	100
Both Species	17,726	23	74	99

Table 6. Rainfall on the part of the llanos of Venezuela where *Myrmecophaga* and *T. tetradactyla* were studied during 1978. From about day 150 onward, an increasing depth of surface water accumulated where the animals moved and fed.

Season	*Cumulative Rainfall (mm)*
Dry, day 0 to day 90	2.1
Transition, to day 150	97.9
Wet, to day 210	541.5

Seasonal Changes in Food Habits.-- All 4 species of vermilinguas live in a seasonal environment, in which changes are less in ambient temperature than in humidity, rainfall and accumulation of soil moisture. Because of flooding, the contrast between wet and dry seasons in the llanos of Venezuela was perhaps more striking than in central Panama. However, changes in the forest on BCI during the 4-month dry season are also considerable (Leigh and Windsor, 1982).

In the llanos, rain water accumulates because of poor drainage, and covers much of the study area to a depth of 1 m or more. There are areas which are not seasonally covered with water, but the vermilinguas did not restrict their movements and feeding to those higher portions of the habitat. Very little rain fell during the dry season of 1978, about 10 cm fell during the transition from dry to wet season in April and May, and about 44 cm fell during June and July, the first 2 months of the wet season (Table 6).

There was a marked change in the diet of *Myrmecophaga* with the beginning of the 1978 wet season and subsequent flooding of the llanos (Table 7). During dry season, the repeated samples of stomach contents from radio-marked *Myrmecophaga*, as well as 21 samples of prey collected while the radio-marked individuals fed, showed their diet to consist almost entirely of *Camponotus* ants (94 percent of 5,636 food items in stomachs and 86 percent of 21 feeding observations). No *Solenopsis* ants, and very few termites were found in the stomachs of *Myrmecophaga* during dry season.

Heavy rains began in mid-May, on day 138, although there were a few light rains prior to day 100. After mid-May, *Myrmecophaga* increasingly began to include *Solenopsis* ants in thier diet (Table 7). As surface water accumulated among the grass clumps and scattered palms, the proportion of *Solenopsis* ants in the diet continued to increase, the proportion formed by *Cam-*

Table 7. Change in food habits of *Myrmecophaga* with change from dry to wet season in Venezuela during 1978. A total of 12,386 arthropods were identified to morpho-species in 21 samples of stomach contents from 5 radio-marked individuals. Numbers of ants in all morpho-species of identified genera were added to compute percentages of all ants formed by all the genera. These percentages are shown, with the rank in parentheses.

	Dry Season	*Transition*	*Wet Season*
Genus			
Camponotus	95.1 (1)	61.1 (1)	24.7 (2)
Solenopsis	0.0	23.7 (2)	67.3 (1)
Pheidole	3.6 (2)	9.2 (3)	3.0 (4)
Doryline, sp. 210	1.3 (3)	5.8 (4)	5.0 (3)
Total			
Samples	10	5	6
Myrmecophaga	3	4	5
Termites	58	0	650
Ants	5,578	2,109	3,991
Percent Shared			
Dry-Wet		71.7	
Transitions	92.7		99.3
Morpho-species	7	10	8
Percent Shared			
Dry-Wet		60	
Transitions	70		80

ponotus decreased, and a significant portion of the diet (14 percent) was formed by *Nasutitermes* termites.

There was considerable agreement in the kinds of ants which were eaten in both dry season and during the transitional period, as well as in those which were eaten both during the transitional period and in wet season (Table 7). About 72 percent of all ants in samples from the two seasons (transition excluded) were of morpho-species found in samples from both seasons.

There are no adequate data to indicate changes in relative population densities of the two ant species and of the termites with the coming of the rains, or to indicate whether flooding of the llanos might have made *Camponotus* colonies less available to the vermilinguas. There was no indication that the movement patterns of the radio-marked vermilinguas changed, although the kind of detailed observations which could only be made by following them on foot while they fed were very difficult when the llanos was flooded and the new grass grew. Detailed data from the season of flooding are scanty.

Ants of the genus *Camponotus* might be the preferred food, but become less available to *Myrmecophaga* during the rainy season, and both *Solenopsis* and *Nasutitermes* might be foods of lesser quality or preference that are used to make up the remainder of the diet. This idea is supported by data from *T. tetradactyla* living in the same areas, which fed on *Solenopsis* and *Nasutitermes* during both wet and dry seasons (Table 8). Certainly colonies of both kinds of arthropods were present year-round. The proportion of the *T. tetradactyla* diet formed by both *Solenopsis* ants and *Nasutitermes* termites decreased during the wet season (Table 8), while the proportions of *Camponotus* in the diet remained relatively constant.

An alternative idea is that *Solenopsis* and *Nasutitermes* became more available to *Myrmecophaga* during periods of inundation by being flooded out of their normal nests. Ants and termites living in flooded areas were perhaps less available to *T. tetradactyla* because the smaller *Tamandua* might have been less able to forage in deep water than *Myrmecophaga*, which have much longer legs. This view is supported by finding large numbers of *Solenopsis* on the grass which grew up through the water. Bites and stings from *Solenopsis* and other species of ants added to the difficulty of moving across the flooded llanos.

The major seasonal change in the diet of *T. tetradactyla* was inclusion of fewer termites during wet season (Table 8). All of the 21 morpho-species of ants found in samples from *T. tetradactyla* in wet season, were also found in samples from the dry season. More than 90 percent of the ants in samples from dry season were of morpho-species also found in wet season. The diversity of ants in the diet of *T. tetradactyla* was almost the same during both seasons.

Table 8. Seasonal differences in food habits of *T. tetradactyla* in Venezuela. Data from 1975, 1977 and 1978 were combined for animals captured on the study area. Numbers of ants in all morpho-species of identified genera were added to compute percentages of all ants formed by the highest ranking genera. These percentages are shown, with the ranks in parentheses.

	Total	*Dry Season*	*Wet Season*
Genus			
Solenopsis	30.6	45.9 (1)	14.4 (2)
Formicine, sp. 213	19.0	36.9 (2)	0.0
Camponotus	20.2	18.8 (3)	21.7 (1)
Paracryptocerus	5.5	10.6 (4)	0.9 (12)
Pheidole	8.0	10.6 (5)	5.2 (3)
Total			
Samples	15	8	7
Termites	2,595	1,805	790
Percent of Diet	35.0	44.4	23.6
Ants	7,413	4,818	2,257
Percent Shared		92.5	51.2
Morpho-species	23	21	23
Percent Shared		100	91

The seasonal differences between dry and wet seasons on BCI in Panama were far less obvious than those in the llanos of Venezuela. However, during 1977 there was a period of 75 days, from March 1 to April 15, when less

Table 9. Cumulative rainfall on Barro Colorado island in central Panama, where *Cyclopes* was studied during 1977. Wet season began on about day 137, when moderate amounts of rain fell after a long period of dry weather. Data are from Windsor (1982).

Season	*Cumulative Rainfall (mm)*
Dry	
Day 0- Day 60	119.6
Day 61- Day 121	144.6
Day 122- Day 136	149.6
Wet	
Day 137- Day 153	310.4
Day 154- Day 214	714.5

than 1 cm of rain fell (Table 9). In the following 78 days, about 56 cm of rain fell.

There were no appreciable changes in the diet of *Cyclopes* with the change in season. Ants of many of the same species ranked high in the diet in both seasons (Table 10), and termites were not eaten in either season. The proportion of *Solenopsis* ants decreased in wet season, while the proportion of *Pseudomyrmex* ants increased. However, even those changes were slight.

The morpho-species diversity of ants in the diet of *Cyclopes* was higher during wet season, but in fact most of the species which were added to the diet were represented by very few individuals. Thus there was more than 95 percent agreement between the kinds of ants which were eaten in the two seasons (Table 10).

Table 10. Seasonal differences in food habits of *Cyclopes* during the 1977 dry season and wet season on Barro Colorado island, in central Panama. No arthropods other than ants were found in the samples of feces. Percentages of all food items formed by the highest ranking morpho-species are shown, with the ranks in parentheses.

	Total	*Dry Season*	*Wet Season*
Morpho-species			
Crematogaster sp. 18	22.0	23.8 (1)	17.9 (1)
Solenopsis sp. 15	18.6	22.6 (2)	9.7 (4)
Camponotus abdominalis	7.2	7.4 (3)	6.6 (6)
Unknown sp. 19	2.7	5.0 (4)	8.1 (5)
Pseudomyrmex sp. 9	7.6	4.5 (5)	14.9 (2)
Zacryptocerus minutus	4.6	3.9 (6)	6.4 (7)
Unknown alate, sp. 63	3.3	0.5 (17)	9.8 (3)
Total			
Samples	42	25	17
Cyclopes	11	8	9
Termites	0	0	0
Ants	15,584	10,907	4,677
Percent Shared		96.9	98.8
Morpho-species	76	51	71
Percent Shared		90.2	64.8

Table 11. Seasonal differences in food habits of *T. mexicana* on Barro Colorado island, in central Panama during dry season and wet season 1973 and 1974. Numbers of ants in all morpho-species were added to compute percentages of all ants formed by the highest ranking genera. These percentages are shown, with the ranks in parentheses.

	Total	*Dry Season*	*Wet Season*
Genus			
Camponotus	41.2	34.3 (1)	45.5 (1)
Azteca	23.7	31.1 (2)	19.1 (2)
Zacryptocerus	5.8	10.5 (3)	2.8 (4)
Solenopsis	4.6	9.6 (4)	1.4 (6)
Pheidole	1.6	1.9 (5)	1.5 (5)
Crematogaster	10.4	0.7 (6)	16.6 (3)
Total			
Samples	19	9	10
Tamandua	9	4	5
Termites	9,988	2,950	7,038
Percent of Diet	71	54	82
Ants	4,068	1,576	2,492
Percent Shared		92.5	51.2
Morpho-species	60	37	48
Percent Shared		64.9	50.0

The major seasonal change in the diet of *T. mexicana*, like that of *T. tetradactyla* in Venezuela, was inclusion of substantially more termites in wet season (Table 11). Most of the ant genera retained their exact rank or a similar one. However, the seasonal differences in the kinds of ants in the diet were far greater than in *Cyclopes*. There were more ant morpho-species in wet season, and only about half the ants eaten in wet season were of morpho-species which had been eaten in dry season. The dry season of 1974 on BCI began earlier, and was more severe than that of 1977 (Table 12).

Table 12. Cumulative rainfall on Barro Colorado island in central Panama, where *T. mexicana* was studied during 1973 and 1974. Dry season began on about 15 December, 1973, and wet season began on about 20 May, 1974. Data are from Rubinoff (1974) and Windsor (1975).

Season	*Cumulative Rainfall (mm)*
Wet	
Day 272- Day 350 (1973)	250.3
Dry	
Day 350 (1973)- Day 140 (1974)	10.6
Wet	
Day 141- Day 212 (1974)	61.5

I have no ready explanation for the increase in importance of termites in the diets of *Myrmecophaga* and both species of *Tamandua* with the coming of the wet season. One possibility for the increase is that the vermilinguas selectively fed on nests which contained reproductives (alates and pre-alates) of the termites. Only a small part of the termites in the stomach samples were alates, but soldiers might have been ingested as well, in order to reach alates. Reproductives of many species of ants and termites are differentially produced in the weeks before the rains begin. These forms are considerably larger than members of other castes, and have a higher caloric value per unit weight (Montgomery and Lubin, 1977). *Tamandua* apparently seek out termite nests which contain reproductives, and feed more often and for longer periods on those nests (Montgomery and Lubin, 1977).

The major conclusion from seasonal comparison of vermilingua diets is that most of the species of ants, and probably of termites as well, are available and are used as food during both seasons. This is in spite of obvious depredations by the mammalian predators (as indicated by the presence of ants and termites in the diets), and the major seasonal environmental fluctuations.

Summary Remarks.-- The vermilinguas probably take a very small part of the total standing crop of ants (Montgomery, 1985) and termites (Lubin, et al., 1977; Lubin and Montgomery, 1981). They apparently have feeding patterns which involve regular or repeated cropping of a large number of colonies of social insects (Montgomery and Lubin, 1977), rather than one of eating many individuals of any given colony during a feeding period. I feel that this is primarily because the insects are able to defend their colonies against the mammalian predators (Montgomery and Lubin, 1977; Lubin and Montgomery, 1981), but have presented alternative explanations elsewhere (Montgomery, 1985).

Nest structures of many species of colonial arthropods are very difficult to penetrate (Montgomery and Lubin, 1977), even for large mammals such as *Myrmecophaga* and *Tamandua*, genera which are specialized for tearing open durable materials (Taylor, 1985). These structures, and mechanical and chemical attack by the ants and termites, protect the reproductives of the colonies, and provide the colony with the means to recoup the losses and grow to survive major depredations. Other species of colonial insects accomplish the same end by nesting in holes in the ground or in durable wood, and are thus relatively inaccessible.

Elsewhere in this paper I pointed out that colonial nests of ants and termites buffer the colonies against depredations by predators, arthropod and mammalian alike, and against environmental changes. Thus a colony can survive losses of many of its members, and still remain as a potential food resource which is relatively stable in both time and space. I suspect that the vermilinguas, and probably other mammalian ant- and-termite eaters, have evolved in response to the availability of this relatively stable and diverse food supply. However, the defences remain effective, and probably influence the movements and home range sizes of vermilinguas of all 4 species, and limit their population densities far short of what one might suspect the apparently available food supply might allow. Those defences have undoubtedly also influenced the evolution of vermilinguas, probably limiting the radiation of vermilinguas to the few species which are known, whereas the diversity and abundance of ants and termites might have promoted the evolution of many diverse forms of mammalian predators.

LITERATURE CITED

Baroni-Urbani, C.
1978. Appendix 1: adult populations in ant colonies. Pp. 334-335, *In* Production ecology of ants and termites (M.V. Brian, ed.). Cambridge Univ. Press, Cambridge, 409 pp.

Croat, T.
1978. Flora of Barro Colorado island. Stanford Univ. Press, Stanford. 943 pp.

Culver, D.C.
1974. Species packing in Caribbean and north temperate ant communities. Ecology, 55: 974-988.

Dietrich, W.E., D.M. Windsor, and T. Dunne.
1982. Geology, climate, and hydrology of Barro Colorado island. Pp. 21-46, *In* The ecology of a tropical forest (E.G. Leigh, A.S. Rand, and D.M. Windsor, eds.). Smithsonian Institution Press, Washington, D.C. 468 pp.

Greenslade, P.J.M.
1971. Interspecific competition and frequency changes among ants in Solomon islands coconut plantations. J. Appl. Ecol., 8(2): 323-352.

Handley, C.O., Jr.
1976. Mammals of the smithsonian Venezuelian project. Brigham Young Univ. Science Bull., Biological Series, 20(5) 89 pp.

Leigh, E.G., Jr. and D.M. Windsor.
1982. Forest production and regulation of primary consumers on Barro Colorado island. Pp. 111-122, *In* The ecology of a tropical forest (E.G. Leigh, A.S. Rand, and D.M. Windsor, eds.). Smithsonian Institution Press, Washington, D.C. 468 pp.

Leston, D.
1971. Ants, capsids and swollen shoot in Ghana: interactions and the implications for pest control. Proc. third Int. Cocoa Res. Conf., Accra, 1969: 205-221.

1973. The ant mosiac-tropical tree crops and the limiting of pests and diseases. PANS, 19: 311-341.

Majer, J.D.
1976. The ant mosiac in Ghana cocoa farms: further structural considerations. J. Appl. Ecol., 13(1): 145-156.

McNab, B.K.
1974. The energetics of endotherms. Ohio J. of Science 74(6): 370-380.

1980. Food habits, energetics, and the population biology of mammals. Am. Nat. 116(1): 106-124.

Lubin, Y.D. and G.G. Montgomery.
1981. Defences of *Nasutitermes* termites (Isoptera, Termitidae) against *Tamandua* anteaters (Edentata, Myrmecophagidae). Biotropica 13: 66-76.

Lubin, Y.D., G.G. Montgomery, and O.P. Young.
1977. Food resources of anteaters (Edentata: Myrmecophagidae) I. A year's census of arboreal ants and termites on Barro Colorado island, Panama Canal Zone. Biotropica 9(1): 26-34.

Montgomery, G.G.
1979. El grupo alimentico (Feeding guild) del oso hormiguero. Convivencia y especializacion de las presas de sustento de los osos hormigueros neotropicales (Edentata, Myrmecophagidae). ConCiencia 6(1): 3-6.

1983. *Cyclopes didactylus* (Tapacara, Serafin de Platanar, Silky Anteater). Pp. 461-463, *In* Costa Rican natural history (D.H. Janzen, ed.). Univ. Chicago Press, Chicago, 816 pp.

1985. Impact of vermilinguas (*Cyclopes, Tamandua*: Xenarthra = Edentata) on arboreal ant populations. (this volume).

Montgomery, G.G. and Y.D. Lubin.
1977. Prey influences on movements of neotropical anteaters. Pp. 103-131, *In* Proceedings of the 1975 predator symposium (R.L. Phillips and C. Jonkel, eds). Montana forest and conservation experiment station, Univ. of Montana, Missoula, 268 pp.

Naples, V.L.
1985. The superficial facial musculature in sloths and vermilinguas (anteaters). (this volume).

Nielsen, M.G. and G. Josens.
1978. Production by ants and termites. Pp. 45-53, *In* Production ecology of ants and termites (M.V. Brian, ed.). Cambridge Univ. Press, Cambridge, 409 pp.

Rubinoff, R.W.
1974. 1973 environmental monitoring and baseline data. (R.W. Rubinoff, ed.). Smithsonian Institution, Washington, D.C. 465 pp.

Sunquist, M.E. and G.G. Montgomery.
1973. Activity pattern of a translocated silky anteater (*Cyclopes didactylus*). J. Mammal. 54(3): 782.

Taylor, B.K.
1985. Functional anatomy of the forelimb in vermilinguas (anteaters). (this volume).

Troth, R.G.
1979. Vegetational types on a ranch in the central llanos of Venezuela. Pp. 17-30, *In* Vertebrate ecology in the northern neotropics (J.F. Eisenberg, ed). Smithsonian Institution Press, Washington, D.C. 271 pp.

Vanderplank, F.L.
1960. The bionomics and ecology of the red tree ant, *Oecophylla* sp., and its relationship to the coconut bug *Pseudotheraptus wayi* Brown (Coreidae). J. Anim. Ecol., 29: 15-33.

Waloff, N. and R.E. Blackith.
1962. The growth and distribution of the mounds of *Lasius flavus* (Fabricius) (Hym: Formicidae) in Silwood Park, Berkshire. J. Anim. Ecol., 31: 421-437.

Wetzel, R.M.
1985. The identification and distribution of recent xenarthra (= Edentata). (this volume).

Wilson, O.E.
1971. The insect societies. Belknap Press, Cambridge, Mass., 548 pp.

Windsor, D.M.
1975. 1974 environmental monitoring and baseline data. (D.M. Windsor, ed). Smithsonian Institution, Washington, D.C. 252 pp.

Wood, T.G. and W.A. Sands.
1978. The role of termites in ecosystems. Pp. 245-292, *In* Production ecology of ants and termites (M.V. Brian, ed.). Cambridge Univ. Press, Cambridge, 409 pp.

ECOLOGY OF THE GIANT ANTEATER *MYRMECOPHAGA TRIDACTYLA* IN SERRA DA CANASTRA, MINAS GERAIS, BRAZIL: A PILOT STUDY

J.H. Shaw[1], T.S. Carter[2,4], and J.C. Machado-Neto[3]

[1]Department of Zoology, Oklahoma State University, Stillwater, Oklahoma 74078, USA. [2]Escola Superior de Florestas, Universidade Federal de Vicosa, 36.570 Vicosa, Minas Gerais, Brazil. [3]Instituto Brasileiro de Desenvolvimento Florestal, Avenida Contorno 8.121, 30.000 Belo Horizonte, Minas Gerais, Brazil. ([4]Present address: Department of Zoology, Oklahoma State University, Stillwater, Oklahoma 74078, USA.)

Abstract

Foraging patterns, prey abundance, and population densities of giant anteaters *Myrmecophaga tridactyla* were studied in a new national park in Brazil's eastern highlands. Frequency and duration of feeding bouts were determined through direct observations. Stratified random plots were used to estimate prey abundance. Giant anteater population densities were estimated through road counts and modified time:area counts. Mean feeding bout length was 30.8 seconds, with 69 percent of the bouts being less than 35 seconds, at an average rate of 0.6 bouts/minute. Nest density averaged 2.46 x 10^5/km^2 for ants and 1.58 x 10^5km^2 for termites. Anteater population density was estimated at 1-2/km^2. Anteaters were not spatially exclusive but one ritualized agnostic encounter may have been related to defense of space. Results of this study are compared with another done in Venezuela. The role of claws in predator defense is discussed.

Abstracto

Se estudiaron los patrones forrajeros, la abundancia de la presa y la densidad de población de los hormigueros gigantes *Myrmecophaga tridactyla*, en un nuevo parque nacional en las tierras altas orientales de Brasil. Por medio de observaciones directas se determinaron la frecuencia y la duración de los encuentros con la presa. Se utilizaron parcelas estratificadas escogidas al azar para determinar la abundancia de la presa. La población de los hormigueros gigantes fue estimada por medio de recuentos en caminos y por recuentos modificados de tiempo por área. El tiempo promedio por encuentro fue de 30.8 segundos, mientras que el 69 porciento de los encuentros duraron menos de 35 segundos, con una tasa promedio de 0.6 encuentros por minuto. La densidad de los nidos de la presa fue de 2.46 x 10^5/km^2 para las hormigas, y de 1.58 x 10^5/km^2 para las termitas. La densidad de población de los hormigueros fue estimada en 1-2/km^2. Los hormigueros no son espacialmente excluyentes, aunque un encuentro agnóstico ritualizado que fue observado pudo estar relacionado con mecanismos de defensa del espacio vital. Se comparan los resultados de este estudio con otro realizado en Venezuela. El papel de las garras en la defensa del depredador es discutida.

Resumo

Os padrões de alimentação, as presas básicas e a densidade populacional do tamanduá-bandeira *Myrmecophaga tridactyla* foram estudados em um novo parque nacional situado no Brasil Central. A frequência e a duração das atividades alimentares foram determinadas através de observações diretas. Circulos selecionados ao acaso nos vários habitats, foram usados para estimar populações das presas básicas. As estimativas da densidade populacional do tamanduá-bandeira foram obtidas através de contagens de estrada e contagens temp/area modificadas. A duraçaõ média das atividades alimentares foi 30,8 segundos, sendo que 69 por ciento delas não ultrapassou 35 segundos, com uma taxa média de 0,6 atividades/minuto. A densidade média de ninhos foi 2,46 x 10^5km^2 para formigas e 1,58 x 10^5 por km^2 para termitas. A densidade populacional do tamanduá-bandeira foi estimada em 1-2 por km^2. Os tamanduás não são especialmente exclusivos porém um encontro agonístico ritualizado pode estar relacionado á territorialidade. Os resultados deste estudo são comparados com outro realizado na Venezuela. O papel defensivo das garras é discutido.

The objectives of this investigation were to make preliminary estimates of giant anteater *Myrmecophaga tridactyla* population densities, available prey base, and foraging patterns. These estimates were then to furnish the basis for further, more intensive field studies planned for 1979-80. The pilot study reported here was carried out in February and March, 1978.

THE STUDY AREA

Serra da Canastra National Park is in western Minas Gerais at 20-21°S latitude and 46-47°W longitude. The 760 km^2 park is atop a plateau that averages about 1,400 m in elevation. The park's topography results from pediplantation (Cole, 1960). Climate is mesothermal tropical of altitude (CWb Köppen) with dry winters and rainy, mild summers.

Prior to expropriation in 1977, Serra da Canastra was privately owned and, despite its limited accessibility, used for cattle grazing. The combined effects of intensive grazing and periodic burning probably helped retard vegetative succession and keep most of Serra da

Canastra as a grassland. Although the relative importance of human activity in maintaining the grassland is unknown, Serra da Canastra has been predominantly grassland since as early as 1819 (Saint-Hiliare, 1847). Gallery forests persist, generally on lowland or riparian sites too wet to burn. Now that cattle have been removed from the park and a policy of fire protection implemented, forests may expand wherever soil depths and fertility permit. The successful establishment of plantation pines (*Pinus* sp.) just beyond the park's west boundary indicates that at least some sites within the park may someday support upland forests.

The park contains a variety of xenarthrans. Besides the giant anteater, these include the lesser anteater *Tamandua tetradactyla* and several species of armadillo: *Euphractus sexcinctus, Dasypus novemcinctus, D. septemcinctus, Cabassous unicinctus squamicaudis, C. tatouay* and *Priodontes maximus*.

Xenarthrans seem clearly to be the dominant group of wild mammals in Serra da Canastra. Their ecological importance has been documented in a neotropical forest (Eisenberg and Thorington, 1973), but, to our knowledge, this is the first indication of their dominance in a neotropical grassland.

MATERIALS AND METHODS

Prey abundance was estimated by counting the number of ant and termite nests on randomly located 10-m^2 sample plots. An ant or termite "nest" was defined as any sheltered aggregation of these insects that constituted a potential feeding site for giant anteaters. We could not determine how many of these nests were breeding colonies, nor did we attempt any biomass estimates. Sampling was stratified by the main habitat types including open grasslands, rocky grasslands, and both abandoned and fresh termite mounds. No sampling was done in the gallery forests. Insects were collected from the nests and identified to subfamily for ants, and family for termites, by criteria from Borror, et al. (1976).

Population densities for giant anteaters were estimated by road counts and modified time-area counts. Road counts were conducted from 1.5 to 0.5 hours before sunset at a constant speed of 35 kph. When an anteater was sighted, its perpendicular distance from the road was estimated by at least two independent factors. The rolling, mountainous terrain caused considerable variation in the distances at which anteaters could be seen. Thus, the density estimates were made by assuming that the area censused equalled the distance driven times twice the mean sighting distance.

Modifications of Goodrum's (1940) time:area counts were carried out on foot from between 1.5 hours and 0.5 hour before sunset. High hills and ridges were climbed and the surrounding areas scanned systematically through 7 x 35 binoculars. The distance to each anteater was estimated, as were the maximum distances over which anteaters could be spotted. Areas surveyed were delineated on topographic maps and their sizes determined with a compensating polar planimeter. Population density was estimated by dividing the maximum number of anteaters seen during an observation period by the area surveyed.

Giant anteater feeding behavior and prey selection were determined through direct observation. By remaining downwind, we could easily and consistently approach to within 10-15 meters of active anteaters with no apparent disturbance. Feeding bouts (i.e., the time an anteater actually spent feeding on a single nest) were timed with a stopwatch. We could tell that an anteater was feeding when its snout was kept on the ground and muscular action was visible in the gular region.

RESULTS

More sample plots were taken in open grassland habitat because of its greater abundance. Throughout the four habitat types, the overall ratio of ant nests to termite nests was about 1.5:1. By far the greatest abundance of both ant and termite nests occurred among old, abandoned termite mounds (Table 1). Among fresh (and presumably active) termite mounds, the ratio of ant nests to termite nests was 0.54:1.

Five counts were conducted from the main road in the eastern half of Serra da Canastra between 10 February and 8 March, resulting in calculated densities from 0.17 anteaters/km^2 to 1.31 anteaters/km^2 (Table 2). The higher counts were obtained on 10 and 13 February, and may have been due to a subsequent shift in anteater

Table 1. Estimated densities of ant and termite nests by habitat type.

Habitat Type	*No. of 10^2 m Plots*	*Nests per km^2 Ant*	*Termite*	*Total*
Open Grassland	20	205,000	5,000	210,000
Rocky Sites	10	205,000	35,000	240,000
Active Mounds	10	130,000	240,000	370,000
Old Mounds	10	485,000	505,000	990,000
Overall Mean		**246,000**	**158,000**	**404,000**

Table 2. Results of counts of giant anteaters from roads, Serra da Canastra National Park.

Date	*km Driven*	*Number Sighted*	*Mean Distance m*	*Estimated Anteaters Per km^2*
Feb. 10	14.7	9	234	1.31
Feb. 13	25.9	11	165	1.29
Feb. 26	20.7	6	325	0.46
Mar. 6	15.9	4	750	0.17
Mar. 8	18.8	11	1266	0.23

Table 3. **Estimated density of giant anteaters by the modified time:area count method.**

Area km²	*Anteaters Observed*	*Anteaters Per km²*
6.13	7	1.1
2.50	5	2.0
4.39	5	1.1

activity. Throughout most of February, anteaters were seen only after about 1500 hours. By early March, sightings of active anteaters were common after about 1200 hours. Reduced rainfall during this period may have been responsible for the activity change, since we rarely saw anteaters foraging during rainy periods. This evident shift toward earlier onset of activities may have contributed to the lower density figures obtained in March.

The modified time:area count was applied in three areas, yielding density estimates of from 1.1 anteaters/km² to 2.0 anteaters/km² (Table 3). The slightly higher and more consistent results of this technique, compared with counts from roads, are probably largely due to the increased visibility.

Five hundred forty-six feeding bouts were timed on 17 different occasions (Table 4, Fig. 1). Mean bout length was 30.8 seconds at an average rate of 0.6 bouts per minute. Forty-three percent of the bouts were of less than 20 seconds duration and 69 percent were less than 35 seconds (Fig. 1). While active, anteaters spent an average of 30.8 percent of their time actually feeding. The time between bouts was devoted to moving between feeding sites and to grooming. Grooming consisted

Table 4. Anteater feeding patterns by observation session.

Session	*Secs Observed*	*Feeding Bouts*	*Length in secs ± SD*	*Bouts Per min*	*% Time Feeding*
1	960	7	19.3± 8.14	0.44	14.0
2	1200	9	65.0± 33.75	0.45	48.8
3	3420	19	23.1± 21.44	0.33	12.8
4	3420	34	51.3± 33.92	0.60	51.0
5	1020	9	27.6± 11.65	0.53	24.4
6	3120	40	20.0± 19.37	0.77	25.6
7	2640	21	36.0± 27.08	0.48	28.7
8	3165	20	45.2± 35.86	0.38	28.6
9	2460	38	22.9± 21.90	0.93	35.3
10	2400	39	33.5± 27.05	0.98	54.5
11	2700	35	21.7± 14.02	0.78	28.1
12	7020	46	37.1± 22.48	0.39	24.3
13	4080	54	25.1± 19.42	0.79	33.2
14	2520	28	36.4± 19.89	0.67	40.4
15	5460	30	32.5± 20.11	0.33	17.9
16	5340	77	18.5± 10.73	0.87	26.6
17	3600	40	44.5± 30.22	0.67	49.5
Mean		**30.8**		**0.60**	**30.8**

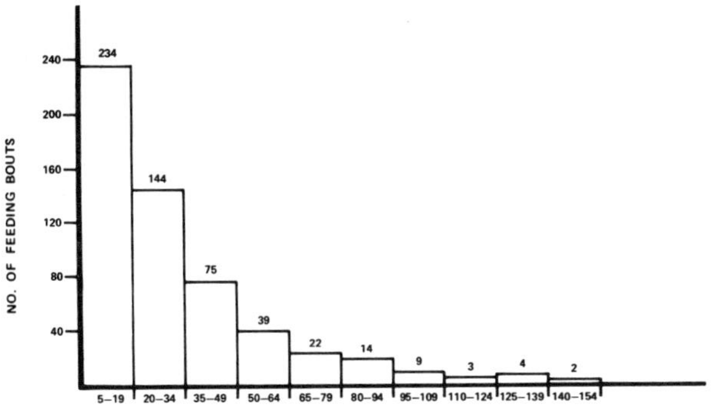

Figure 1. Frequency distribution of the length of giant anteater feeding bouts.

simply of 2 to 3 swipes of the forepaw along the snout. On two occasions, anteaters interrupted their feeding activities to roll in the sand. Movements between feeding sites tended to be linear, with animals seldom altering their course by more than a few degrees. One interesting exception to this movement pattern was an erratic course taken by an anteater that had suffered injury or infection to its right eye, and seemed incapable of sight on that side. This optical impairment may have interfered with its ability to maintain a more constant heading. Foraging anteaters moved at rates of about 100 to 250 m per hr.

Samples of insects were taken from 58 feeding sites and compared with overall availability of ants and termites (Table 5). Myrmicine ants were found at 72 percent of the feeding sites, followed by formicine ants (12 percent) and dolichoderine ants (3 percent). Termites were found at 12 percent of the feeding sites. Most of the ants were less than 5 mm in length.

With only two exceptions, all of our observations were of solitary anteaters. Two anteaters, apparently adults of about the same size, were observed together. They sniffed and touched one another several times before one of them moved away to forage. The second followed within about 25 meters.

The second observation involving two anteaters was an agonistic encounter on 24 February. At 1205 hours,

Table 5. A comparison of estimated prey availability with prey selection.

Prey	*% of Nests on 50 Sample Plots*	*% of 58 Feeding Bouts*
Ants		
Myrmicines	43	72
Formicines	4	12
Dolichoderines	0	3
Unidentified	14	0
Termites	39	12

after a foggy morning, an anteater was seen approaching woody vegetation along a creek. Upon reaching the woody vegetation, the animal curled up under some shrubs and appeared to sleep. Periodically, the anteater extended its nose vertically and sniffed. It changed locations only once, a distance of 2 m, between 1237 and 1730. At 1735 hours, it left the creek and wandered into an area of old termite mounds in which a second giant anteater was feeding. This feeding anteater appeared to scent the approaching one and ran directly toward it. The two anteaters circled one another with tails raised. Several times they pushed at each other's shoulders with their forepaws, keeping claws tucked. The movements were not swift, nor did they appear to be delivered with much force. After a few minutes of circling and pushing the two began "foraging" about 10 m apart with heads pointed away from each other and tails raised while "feeding" rapidly in what was probably redirected aggression. Gradually they separated, each going in the direction from which it had come and maintaining the tails in a raised posture.

Giant anteaters, unlike their smaller relatives, are usually regarded as completely terrestrial. On 26 February a small (and presumably immature) giant anteater interrupted its feeding activities and climbed some 3 m up the vertical trunk of a small, broad-leafed tree with a diameter at breast height (dbh) of about 30 cm. After remaining at 3 m for about 1 minute, the anteater slowly climbed down, posterior end first, and resumed foraging. The tree contained no insect nests or other obvious attractants, but its bark showed clearly the set of single claw marks. The anteater showed no signs of alarm so its climb was apparently not an escape response.

Additional evidence of arboreal activities was obtained while following another giant anteater on 1 March. The foraging animal paused to scratch the bark of a small (3 m) broad-leafed tree that had two nearly horizontal branches at a height of 1 m. Bark had earlier been scraped from these limbs and claw marks identical to those seen on 26 February were abundant.

DISCUSSION

The relative importance of termites in the diet of giant anteaters has been debated in the scientific literature. Carvalho and Kloss (1951), Carvalho (1966), and Preble (1924) reported that the species actually prefers termites over ants. In their Venezuelan study area, Montgomery and Lubin (1977) reported that giant anteaters subsisted primarily on large formicine ants of the genus *Camponotus*, and found termites only rarely in stomach and fecal samples. Only 12 percent of the feeding sites examined in our study contained termites. Montgomery (1985) summarizes analyses of 28 stomach samples from a total of 10 giant anteaters in the llanos of Venezuela. In those samples, there was a total of 16,627 arthropods, 96 percent of them ants and 4 percent termites. There were seasonal differences in feeding on termites, with an increase in the proportion of the diet formed by termites as the wet season progressed and the llanos was flooded; after about 550 mm of rain had accumulated, 14 percent of the food items were termites.

In 15 hours of direct observations of giant anteaters, we witnessed three defections, all near surface water. All of the 15 additional scats we found were located within about 25 m of surface water. Although we lack sufficient data to test this spatial relationship statistically, we suspect that surface water or related habitat conditions stimulate defecation. Insofar as we could determine, all the scats contained exclusively ants, with only one scat containing remains of several millipedes (Diplopoda) and large ant heads.

Foraging patterns of giant anteaters in Serra da Canastra differed from those in Hato Masaguaral (Venezuela) studied by Montgomery and Lubin (1977). The anteaters in Venezuela were nocturnal and ranged over long distances, averaging about 11 km per nightly foraging period. They fed primarily on *Camponotus abdominalis*, a nocturnal ant. In Serra da Canastra, the anteaters were at least partially diurnal and during daylight hours moved at a far slower rate. Feeding, insofar as we could determine, was exclusively on ant or termite nests, rather than on foraging insects. Most of the nests were those of small myrmicines taken at approximately the same frequency as they occurred. We have no data on nocturnal movements or foraging of anteaters in Serra da Canastra. However, Carvalho (1966) reported that Dr. Otto Schubart found some 900 specimens of *Camponotus*, predominantly *C. abdominalis*, in the stomach of a giant anteater from Alto São Francisco (in or near the present Serra da Canastra). Giant anteaters may shift their feeding habits between daylight and nighttime periods as the activities and availabilities of ants change.

Montgomery and Lubin (1977) observed feeding bouts by the lesser anteater *Tamandua tetradactyla* and concluded that individual bouts were limited by the insects' defenses, similarly (Lubin and Montgomery, 1981, Montgomery, 1985) *Tamandua mexicana* and the silky anteater *Cyclopes didactylus* may show patterns of movement and feeding largely influenced by defenses of the potential prey. We are not certain what terminates feeding bouts by giant anteaters in Serra da Canastra. Ants did not swarm over giant anteaters as much as Montgomery's films show that they do the lesser anteaters, nor did the giant anteaters spend much time and effort trying to remove swarming ants from their faces. Furthermore, the ants being fed upon in Serra da Canastra were entirely ground-dwellers, presumably with tunnels extending outside the reach of anteater tongues. Our own efforts to collect ants were hampered by the fact that ants often quickly disappeared underground following disturbance. Thus, we suspect that reduced availability of ants, rather than their defenses, terminated many of the feeding bouts we observed.

Both our study and that of Montgomery and Lubin (1977) found that giant anteaters maintained a fairly constant direction when foraging. This movement pattern presumably results in more optimal foraging since it minimizes the chances of revisiting sites depleted of food (Krebs, 1978).

We can make preliminary estimates of cropping rates in Serra da Canastra. If the anteater population averages $1.5/km^2$, and the foraging rate is 36 nests/hr for 8 hr each day, then 432 $nests/km^2$ are raided per day. At this rate, it would take about 569 days to crop the estimated 246,000 $nests/km^2$. Understanding the dynamics of this prey-predatory relationship will require not only more data on the variables mentioned above, but also estimates of (1) differential survival of cropped versus uncropped nests, (2) preferences of anteaters for different types of ants, and (3) establishment rate for new nests. Only with additional study will we be able to make conclusions regarding the role of food in regulating giant anteater populations.

We believe that the giant anteater population density in Serra da Canastra is either exceptionally high, the animals more diurnal than in other parts of the species' range, or both. Throughout the literature, authors report the species as being nowhere abundant. Serra da Canastra's remoteness, its habitat conditions, and perhaps its lack of large predators, may all contribute to giant anteater abundance.

Territoriality in giant anteaters has not yet been confirmed. In Serra da Canastra, it was fairly common to see one anteater foraging in an area where another anteater with different pelage characteristics was seen earlier. However, the one agonistic encounter we observed may have been spatially related, and it occurred in the most productive habitat type. Further study is planned to test the extent to which temporo-spatial separation is maintained, especially in areas of high prey density.

Numerous references (cf. Morris, 1965; Taylor, 1978; Walker, 1975) attribute the large claws and powerful forelimbs of anteaters as adaptations for breaking into insect nests. We never observed giant anteaters ripping into termite mounds or tearing into ant nests. Furthermore, a review of published literature has led us to conclude that the primary function of the claws in *Myrmecophaga* is predator defense, a conclusion discussed in greater detail elsewhere (Shaw and Carter, 1980). This view is supported by Taylor (1985), who argues for it primarily on the basis of anatomical evidence. Osgood (1939) reported that giant anteaters disemboweled dogs. In provoked attacks on humans, anteaters have used claws (Cutright, 1940; Tate, 1931). A young anteater in captivity defended itself by positioning its claws to protect it head (Bickel et al., 1976). The Myrmecophagidae all have expanded ribs which can increase the stability of the thorax and, in turn, increase the stability of the vertebral column (Jenkins, 1970). This arrangement enhances the striking force that anteaters can muster, especially from a bipedal posture.

Without an effective predator defense, such a slow-moving and generally unwary species as the giant anteater would have vanished long ago. For some 2 million years it has shared its range with both the jaguar *Felis onca* and the cougar *Felis concolor*. Prior to the arrival of the big cats, there were large, marsupial predators such as *Thylacosmilus* (Colbert, 1966).

Finally, it is probably no mere coincidence that the most terrestrial of the Myrmecophagidae is also by far the largest. Its size, combined with effective defenses against predators, made possible its survival in more exposed, terrestrial habitats.

ACKNOWLEDGMENTS

This research was supported by the Brazilian Institute for Forest Development (IBDF) in Minas Gerais. The director, the late Dr. Ivens Pinto Franqueira, graciously made support available, and his staff furnished generous assistance.

LITERATURE CITED

Bickel, C.L., G.K. Murdock and M.L. Smith.
1976. Hand-rearing a giant anteater (*Myrmecophaga tridactyla*) at Denver Zoo. International Zoo Yearbook 16: 195-198.

Borror, D., D. DeLong and C. Triplehorn.
1976. An introduction to the study of insects (4th edition). Holt, Rinehart, and Winston. New York. 852 pp.

Carvalho, J.C.M.
1966. Novos dados sobre a alimentação do tamanduá bandeira, *Myrmecophaga tridactyla* (Lineaus, 1758). Anais da Academia Brasileira de Ciencia. 38 (Supl.): 341-346.

Carvalho, J.C.M., and G.R. Kloss.
1951. Sobre a alimentação do tamanduá-bandeira (*Myrmecophaga tridactyla* L. 1758). Rev. Brasil Biol. 11(1): 37-42.

Colbert, E.W.
1966. Evolution of the vertebrates: a history of the backboned animals through time. John Wiley and Sons, Inc., New York. 479 pp.

Cole, M.M.
1960. Cerrado, caatinga and pantanal: the distribution and origin of the savannah vegetation of Brazil. Geogr. J. 126(2): 168-179.

Cutright, P.R.
1940. The great naturalists explore South America. The MacMillan Company. New York. 340 pp.

Eisenberg, J.F., and R.W. Thorington, Jr.
1973. A preliminary analysis of a neotropical mammal fauna. Biotropica, 5(3): 150-161.

Goodrum, P.D.
1940. A population study of the gray squirrel in eastern Texas. Texas Agric. Experi. Sta. Bull. 591. 34 pp.

Jenkins, F.A., Jr.
1970. Anatomy and function of expanded ribs in certain edentates and primates. J. Mammal. 51(2): 288-301.

Krebs, J.R.
1978. Optimal foraging: decision rules for predators. Chapter 2, Pp. 23-63, *in* Behavioral ecology: an evolutionary approach (J.R. Krebs and N.B. Davies editors). Sinauer Associates, Inc., Sunderland, Massachusetts. 494 pp.

Lubin, Y.D., and G.G. Montgomery.
1981. Defenses of *Nasutitermes* termies (Isoptera, Termitidae) against *Tamandua* anteaters (Edentata, Myrmecophagidae). Biotropica, 13(1): 66-76.

Montgomery, G. G.
1985. Movements; Foraging and Food Habits of the Four Extant Species of Neotropical Vermilinguas (Mammalia; Myrmecophagidea) (this volume).

Montgomery, G.G., and Y.D. Lubin.
1977. Prey influences on movements of neotropical anteaters. Pp. 103-131, *in* Proceedings of the 1975 Predator Symposium. (R.L. Phillips and C.J. Jonkel editors). Montana Forest and Conservation Experiment Station, School of Forestry, University of Montana, Missoula. 268 pp.

Morris, D.
1965. The mammals, a guide to the living species. Harper and Row, New York and Evanston. 448 pp.

Osgood, W.H.
1939. The great anteater group at the Field Museum. The Scientific Monthly, 48: 484-485.

Preble, E.A.
1924. The great anteater. Nature 4(3): 158.

Saint-Hiliare, A.
1847. Viagem as nascentes do Rio São Francisco e pela provencia de Goyaz. Companhia editora nacional. São Paulo. 341 pp.

Shaw, J.H. and T.S. Carter.
1980. Giant anteaters. Natural History 89(10): 62-67.

Tate, G.H.H.
1931. Random observations on habits of South American mammals. J. Mammal. 12(3): 248-256.

Taylor, B.K.
1978. The anatomy of the forelimb in the anteater (*Tamandua*) and its functional implications. J. Morphology 157:347-368.

Taylor, B.K.
1985. Functional anatomy of the forelimb in vermilinguas (anteaters). (this volume).

Walker, E.P.
1975. Mammals of the World (3rd ed.). Johns Hopkins University Press. 1,500 pp.

ECOLOGY OF LIVING ARMADILLOS

NAKED-TAILED ARMADILLOS, *CABASSOUS* SP.

Dennis A. Meritt, Jr.

Lincoln Park Zoological Gardens, Chicago, Illinois 60614

Abstract

Cabassous is a widely distributed Central and South American armadillo that is nocturnal, found singly, produces a single offspring, and is behaviorally distinct from other armadillo types. Body weights, core temperature, activity cycle, and reproductive biology are reported, as are behavior and captive husbandry techniques.

Abstracto

Los armadillos del género *Cabassous* están extensamente distribuídos en Centro y Sur América. Es un armadillo nocturno y solitario que produce una sola cría y es distinto en su comportamiento a otros tipos de armadillos. En este artículo se informa sobre su peso corporal, temperatura, ciclos de actividad y biología de reproducción, como también su comportamiento y técnicas de manejo en cautiverio.

Resumo

Cabassous é um tatu da América Central e do Sul, amplamente distribuído, noturno, encontrado solitariamente, produz um único filhote e apresenta um comportamento distinto de outros tipos de tatus. Pesos do corpo, temperatura interna, ciclo de atividade e biologia reprodutiva são pormenorizados, bem como comportamento e técnicas de criação em cativeiro.

Members of the genus *Cabassous* are distributed from Central America through South America (Wetzel, 1985). Living examples are known from Honduras, Costa Rica, and Panama in Central America. The extreme southern portion of its range is southwestern Paraguay and adjacent areas of northern Argentina. Various authors have identified five species, although the genus appears to be in need of a revision based upon more recent work (Wetzel, 1985). Walker (1968) recognizes five species while Cabrera and Yepes (1960) list four, as does Cabrera (1957). Yepes (1935) lists two species, *C. unicinctus* and *C. loricatus*, from Argentina in the provinces of Chaco, Formosa, and Misiones, as well as Santa Fe and Santiago del Estero.

Cabassous sp. are usually called the naked-tailed armadillos because of the absence of scutes or plates on the tail. Characteristically, the tail is unprotected, soft to the touch and covered with skin; shields are absent. In many ways, *Cabassous* sp. resemble a smaller version of the giant armadillo, *Priodontes*; some observers confuse *Cabassous* sp. for baby or young giant armadillos. A campesino along the Rio Paraguay was convinced the *Cabassous* sp. he was caring for was to grow up in a few weeks to become a *Priodontes* (Meritt, unpublished observation).

Cabassous has been reported to inhabit grassland and upland plateaus. In Paraguay, these armadillos were captured in semiarid lowland and moist palmettos, while in Panama this armadillo is found along major waterways (Rio Chepo), highland areas and along embankments adjacent to roadways (La Hagua), and in moist lowland forest (Bayano River basin and Barro Colorado Island; Montgomery, unpublished data). In northern Argentina, burrows, excavations, and live animals were found along the Rio Pilcomayo (Meritt, unpublished observation).

Burrows of *Cabassous* sp. are easily identified. The entrances are usually in open ground or at the base of an embankment. The shape of the opening is nearly identical with that of the armadillo's carapace and is rarely much larger in overall dimensions. Typically, the area adjacent to the entrance is free of debris, excrement, and living vegetation. The soil is usually hard packed and not mounded at the entrance to the burrow. Active burrows have a strong musky odor. *Cabassous* sp. in captivity does not construct a nest even when provided with nesting material and a nest retreat (barrel, nest box, hollow log).

All field observations of active naked-tailed armadillos indicate nocturnal activity which begins shortly after sunset; the animals returned to their burrows before sunrise in all cases. On occasion, individuals remained intermittantly active after sunrise, but always in the immediate area of the burrow entrance. Captive *Cabassous* also show a strong nocturnal tendency. Wild-caught naked-tailed armadillos from Panama *C. centralis* weighed from 2.0 to 3.5 kg. Those from Guyana *C. unicinctus*, 2.2 to 4.8 kg. Two *C. tatouay* from the eastern margin of the Chaco, Paraguay, weighed over 5 kg each.

Characteristically and typically, when handled, startled, or picked up, *Cabassous* sp. urinate in copious quantity. Simultaneously, the armadillo vocalizes, sounding like a grunting, snorting pig. At times, small amounts of loose feces are passed and spattered about by the tail. While sleeping abdomen up, *Cabassous* often urinate on themselves. In males, the penis is enlarged, mobile, and serpentine in its movements during REM sleep. In females, the tail is laid across the genitalia and moves up and down during sleeping urination. Robinson and Lyon (1901) noted an unusual propensity for urination in a *C. lugubris* captured in Venezuela.

The normal stool of *Cabassous* sp. is formed and pelleted. Small particles of soil and insect remains may be found in the feces. Wild *Cabassous* probably feed largely on ants and termites (Redford, 1985). Wild-caught *Cabassous* frequently are heavily parasitised with ascarids, hookworms, and strongyloides. Captive specimens also have pelleted stool and show a preference for eliminating in a specific area of their enclosure or habitat.

Field measurements of core body temperature of several examples yield an average of 33°C with a range of 32° to 34°C at ambient temperatures averaging 29° C. All of the approximately 100 measurements from 16 individuals were recorded with the rectal probe inserted 5 cm (Yellow Springs Instruments Tele-thermometer). These low body temperatures are typical of xenarthrans, and have also been reported by McNab (1985).

Two separate births afforded the opportunity to observe and weigh the single offspring. The weights at birth of the one male and one female were 100 and 113 g, respectively. Each baby weighed about 3 percent of its mother's body weight. At birth, newborn *C. centralis* from Panama were naked, hairless, the ear pinnae closed, eyes and eyelids sealed and in a primative state. The claws were hardened, the foreclaw greatly enlarged, as in the adults. The banding pattern and scutes were formed, but the skin was soft, with no firmness. The skin was pink, glossy, and warm to the touch. The mouth and nares were closed, nearly touching; the nose tip and lower jaw were soft, wrinkled, and appeared slightly distorted.

Captive examples fared well on modified small mammal meat mix (Meritt, 1970; 1973; 1977) and were housed in similar fashion to other armadillos under study at the Lincoln Park Zoological Gardens (Meritt, 1976).

Although usually found singly in nature, compatable breeding groups containing up to two males and four females have been established at the Lincoln Park Zoological Gardens. Unlike other species of armadillo, individual *Cabassous* rarely interacted and, at times, appeared to ignore the presence of others in the same habitat or enclosure. Animals were active or inactive without respect to others in the same enclosure. Feeding behavior was not stimulated by another individual eating, as is the case in other armadillo genera.

As a zoological exhibit, *Cabassous* sp. are rather uninteresting, lethargic, and unattractive. As subjects of behavioral, physiological and natural history studies, they promise to be unique, intriguing, and an enigma among the living armadillos.

ACKNOWLEDGEMENTS

Dr. Lester E. Fisher, director of the Lincoln Park Zoological Gardens, allowed these studies and thanks are due for support and encouragement. Animal keepers in the Small Mammal and Lion houses assisted in daily care and observations. The American Philosophical Society, Philadelphia, partially funded some of the field research. Gail Meritt shared difficult field conditions.

LITERATURE CITED

Cabrera, A.
1957. Catalogo de los mamiferos de america del sur. I. Imprenta y Casa Editora Coni., Buenos Aires, Argentina, 307 pp.

Cabrera, A. and J. Yepes.
1960. Mamiferos sud americanos. Ediar S.A. Editores, Buenos Aires, Argentina, 160 pp.

McNab, B.K.
1985. Energetics, population biology, and distribution of xenarthrans, living and extinct. (this volume).

Meritt, D. A.
1970. Edentate diets currently in use at Lincoln Park Zoo, Chicago. Int. Zoo Yb., 10: 136-138.

1973. Edentate diets. I. Armadillos. Lab. Anim., Sci., 23: 540-542.

1976. The nutrition of edentates. Int. Zoo Yb., 16: 38-46.

1977. Edentate nutrition. Pp 541 to 547 *in*: CRC Handbook Series in Nutrition and Food. Vol. 1. (Milsolav Rechcigl, Jr. ed.) CRC Press, Cleveland, Ohio, 645 pp.

Redford, K.H.
1985. Food habits of armadillos (Xenarthra: Dasypodidae). (this volume).

Robinson W. R. and M. W. Lyon.
1901. An annotated list of mammals collected in the vicinity of La Guaira, Venezuela. Proc. U.S. Nat. Mus. Vol. XXIV: 135-162.

Walker, E. P.
1968. Mammals of the world, I. John Hopkins Press. Baltimore, Maryland, 644 pp.

Wetzel, R.M.
1985. Taxonomy and distribution of armadillos, Dasypodidae. (this volume).

Yepes, J.
1935. Las especies argentinas del genero (*Cabassous*). Physis, XI: 438-444.

THE FAIRY ARMADILLO, *CHLAMYPHORUS TRUNCATUS* HARLAN

Dennis A. Meritt, Jr.

Zoological Gardens, Lincoln Park, Chicago, Illinois 60614

Abstract

Chlamyphorus truncatus Harlan, the fairy armadillo, is nocturnal, fossorial, and rarely encountered in nature. Its life history still needs substantial study. Descriptions of natural habitat, probable food sources, and body structure are presented. Behavioral notes and body measurements for known animals are detailed.

Abstracto

Chlamyphorus truncatus Harlan, el armadillo duende, es nocturno y cavador, y en raras ocasiones se encuentra en la naturaleza. Su historia natural aún necesita estudios substanciales. Aquí se presenta una descripción de su habitat natural, las probables fuentes de su alimentación y su estructura corporal. También se detallan notas sobre su comportamiento y medidas corporales de animales conocidos.

Resumo

Chlamyphorus truncatus Harlan, o tatu pichiego menor, é noturno, cavador e raramente encontrado na natureza. Sua história de vida ainda necessita de estudos substanciais. Uma descrição do habitat natural, prováveis fontes de alimento e estrutura do corpo são apresentados. Notas de comportamento e medidas do corpo para animais conhecidos são minudenciados.

The fairy armadillo *Chlamyphorus truncatus* Harlan is known from Argentina, in the provinces of Mendoza, San Luis, and La Pampa. Additionally, it may be found in portions of the provinces of Catamarca, La Rioja, Buenos Aires, and Cordoba (Cabrera, 1957; Wetzel, 1985). Its common name in Spanish is "Pichi ciego". The current range of this diminutive armadillo is unknown, its population numbers unrecorded, and essentially nothing is known about its natural history.

During seven weeks in the spring of 1972 a field study was done in Argentina and Paraguay. The bulk of this time was spent in Argentina, in the provinces of Mendoza, San Juan, Cordoba, and Buenos Aires. This report is based upon that field effort supplemented by observations in captivity, primarily of a captive male *Chlamyphorus truncatus*.

Hershkovitz (1969) described the fairy armadillo as fossorial and "rat-sized". This is the smallest of the armadillos (Table 1). Head-body length is given as 125-150 mm, with a tail length of 25 mm. (Walker, 1968). Rood (1970) gives a total length of about 150 mm and a weight of 120 g. Cei (1967) gives an overall length of about 5 in. and likens its size to that of a mole. Minoprio (1945) gives an excellent series of measurements on several specimens. Morris (1965) describes the fairy armadillo as the smallest of the xenarthrans, about 127 mm long when adult. Wetzel (1985) gives measurements of two unsexed adult specimens from Argentina (La Pampa? and Mendoza).

Table 1: Measurements of a recently deceased adult male fairy armadillo *Chlamyphorus truncatus* provided by the Chicago Zoological Park. This is in all probability the armadillo referred to by Rood (1970). All measurements in mm.

Total body length	171
Head length	44
Body length	89
Tail length	38
Neck girth	19
Chest girth (at pectoral level)	101
Abdominal girth (at navel)	95
Tail base girth	19
Distance between eyes	25
Length of mouth opening	32
Distance between external nares	2
Front claw length	19
Hind claw length	3

Cabrera and Yepes (1960) describe taxonomically three forms of *Chlamyphorus*: *C. truncatus* , the Pichi ciego menor; *C. t. minor* , the Pichi de las pampas; and *C. t. patquiensis*, the Pichi ciego de las provincias centrales. Cabrera (1957) describes one form of *Chlamyphorus*, the species *truncatus*. No further subspecies are delineated in this work. Minoprio (1945), Morris (1965), Cei (1967), Walker (1968, 1975), and Rood (1970) all refer to the fairy armadillo as *Chlamyphorus truncatus* with no subspecific designation (see also Wetzel, 1985).

HABITAT AND FOOD HABITS

The habitat of the fairy armadillo in Mendoza and San Juan provinces is harsh, dry, without appreciable rainfall, sparsely populated, and with distinct flora and fauna. Soil type is variable but, in the areas where sign of this armadillo was most prevalent, sandy to mixed sand and poor topsoil were predominant. Vegetation is limited, reduced in size, and characteristic of this environment. Shrubs, bushes, and trees all show reduced leaf size and many smaller plants are thorny with tough bark covering the branches. Cei (1967) listed and illustrated some of the shrubs, trees, and flowers of the region. Predominent in this habitat in Mendoza and San Juan provinces is *Prosopis* sp., a type of mesquite. The Pichi armadillo, *Zaedyus pichi caurinus*, is known to feed on seeds and pods of this shrub (Meritt, 1973), as does the little hairy armadillo *Chaetophractus vellerosus* (Greegor, 1985). Cattle, horses, and goats roam within large tracts of this habitat. Mammalian fauna includes the Pichi armadillo, *Zaedyus pichi*, and mara, *Dolichotis patagonium*. Avian forms include several psittacines and tinamou, *Tinamus* sp. Both mammals and the latter bird are used regularly as food by the indigenous peoples.

Food for the fairy armadillo in nature has been variously described as ants, insects, roots, and underground stems, also worms and plant material; in addition, available items in the Argentine habitat include spiders, soft-bodied grubs, seeds, pods, flowers, and occasional carrion. The eggs of ground-dwelling birds, small lizards, and snakes should be considered as other possible sources of food as well, as these are readily available.

Rood (1970), in describing the food habits of a captive individual, lists bread and milk, whole oats, with occasional beetles or beetle larvae, type not specified. He found no evidence to suggest that *Chlamyphorus* eats roots or underground stems or tubers. A captive male maintained at the Chicago Zoological Park from February 1970 until December 1971 ate crickets, mealworms, white bread soaked in milk, and grains of barley or oats. When offered appropriate sized pieces of banana, apple, or sweet potato, no interest was shown.

Small, pelleted fecal material is characteristic of the fairy armadillo. Living animals and active burrows do not have any noticeable odor, as in *Cabassous* (see Meritt, 1985).

EXTERNAL ANATOMY

Sexing of a male was easily accomplished as the penis is expressable from the sheath or skin covering. There are no external testicles.

External ear orifices are nearly impossible to locate in living fairy armadillos and there is no sign of an ear per se. The carapace appears to act as a tentlike cover over the body, with fine white hair all the way up and under the carapace. The underside of the body is haired and is soft without toughened skin. The attachment of the carapace appears to be along the spine but is loose and flexible. The head shield and movable bands over the body and pelvic or rear shield are flexible and not distinct from one another. The butt plate is firm and more solid/dense than the shielding on the balance of the carapace. The tail is the hardest, firmest piece of the entire carapace assembly. It is spoon-shaped, rather stiff, and hard to the touch with little actual flexibility, and the posterior portion is spoonlike. The tail is firmly attached to the butt plate but does not appear to be flexible at the junction with this shield.

ACTIVITY PATTERN

Despite being described as nocturnal or crepuscular, a captive male was intermittantly active during the day under dim white light. When higher levels of white light were turned on, the armadillo burrowed down and out of sight into the sandy loam provided in its enclosure.

A captive animal was observed to sleep and rest with its head and nose tucked under its lower abdomen. The armadillo was most often seen in this position on the soil surface but occasionally slightly under the soil with only a small portion of its carapace exposed.

When active, the armadillo was methodical in its movements about the enclosure. It showed clear direction and its actions were not frantic or hasty. The fairy armadillo never tried to dig through the barrier walls and rarely raised its body off of the soil surface. If presented with a barrier on the soil surface that was too large to walk over, it dug under it rather than investigating its length. The large nails on the forelimbs were efficient scoops for moving soil slightly to the side of or under the armadillo.

BURROWS

Minoprio (1945) described the burrow systems of the fairy armadillo and there is no need to attempt to expand on this fine work. Three types of burrows or excavation were identified in the field in Argentina. The entrances to the excavations were typically shaped and conformed to the carapace outline of the living animal. Frequent inhabitants of abandoned burrows were tarantula spiders (family Theraphosidae), one per burrow. At dusk, on overcast cool days, and during light rain, these arachnids came to the burrow openings and at times were confused for retreating armadillos. Coral snakes *Elaps* sp. were also found in abandoned deep burrows of *Chlamyphorus truncatus*. Their presence presented certain hazards in exploring fairy armadillo burrows by hand.

RARITY

During over 200 hours of sunup to sundown field exploration, no active fairy armadillos were observed in primary habitat. During the same time, dozens of other armadillos were encountered and observed, including *Zaedyus pichi caurinus*, seven-banded *Dasypus septemcinctus* and La Plata three-banded *Tolypeutes matacus*. Discussions with Argentine campesinos and ranch owners as well as desert hermits, nearly all born and raised in fairy armadillo habitat, revealed little about this elusive and little-known xenarthran. The armadillo was known but no patterns of activity, no time of day, no particular season was clearly identified as the best or most appropriate to locate the animal. To give one a prospective on field difficulties, more fossilized glyptodont remains were located, with more ease, than were *Chlamyphorus*.

It is clear that a full-scale, in-residence, in-nature field effort is necessary to further define the characteristics, habits, and natural history of this living enigma, the fairy armadillo.

ACKNOWLEDGMENTS

Dr. Lester E. Fisher allowed and encouraged these studies. The American Philosophical Society (Penrose Fund) partially funded a portion of this work. Fito provided invaluable field assistance. Don Alejandro and Don Fructoso shared their home and talents in the field. Gail Meritt provided encouragement and technical assistance, and shared unusual field conditions.

LITERATURE CITED

Cabrera, A.
1957. Catalogo de los Mamiferos De America Del Sur. I. Imprenta y Casa Editora Coni, Buenos Aires, Argentina, 307 pp.

Cabrera, A. and J. Yepes.
1960. Mamiferos Sud Americanos. II Ediar S.A. Editores, Buenos Aires, 160 pp.

Cei, J. M.
1967. Pichiciego and Portulaca. Animals. London, 10:176-177

Greegor, D.H.
1985. Ecology of the little hairy armadillo *Chaetophractus vellerosus*. (this volume).

Hershkovitz, P.
1969. The Evolution of Mammals on Southern Continents. VI The Recent Mammals of the Neotropical Region: A zoogeographic and ecological Review. Quart. Rev. Bio., 44: 1-70.

Meritt, D. A.
1973. In nature the behavior of armadillos. Pp. 383-384. *in*: Yearbook, 1972. The American Philosophical Society. Philadelphia, 839 pp.

Meritt, D.A., Jr.
1985. Naked-tailed armadillos, *Cabassous* sp. (this volume).

Minoprio, J. D. L.
1945. Sobre el *Chlamyphorus truncatus* HARLAN. Acta Zoologica Lilloana, 3: 5-58.

Morris, D.
1965. The Mammals, A guide to the living species. Harper and Row. New York, 173 pp.

Rood, J. P.
1970. Notes on the behavior of the pygmy armadillo. J. Mamm., 51: 179.

Walker, E. P.
1968. Mammals of the World. I. John Hopkins Press., Baltimore, Maryland, 501 pp.

Wetzel, R.M.
1985. Taxonomy and distribution of armadillos, Dasypodidae. (this volume).

ECOLOGY OF THE LITTLE HAIRY ARMADILLO *CHAETOPHRACTUS VELLEROSUS*

David H. Greegor, Jr.

Alaska Pacific University, Anchorage, Alaska 99504.

Abstract

The little hairy armadillo *Chaetophractus vellerosus* is one of several species of armadillos inhabiting the Monte Desert of Argentina. *Chaetophractus* is omnivorous, which is important for living in deserts that have relatively few insects. *Chaetophractus* avoids summer heat and water problems by being fossorial and nocturnal. In winter, they are diurnal. Their burrows are unaffected by daily and possibly seasonal air temperature variations due to their depth. Burrow temperature is below armadillo body temperature even during the hottest summer days. Because of their fossorial habits, *Chaetophractus* distribution is limited by the substrate. They are uncommon on rocky slopes, but common on sand dunes and apparently prefer to burrow on sloping rather than flat surfaces. The relative medullary thickness (RMT) indicates that the *Chaetophractus* kidney has a higher ability to concentrate urine than expected for a xeric mammal of its weight. *Chaetophractus* remains in water balance on a reduced water diet.

Abstracto

El pequeño armadillo peludo *Chaetophractus vellerosus* es una de varias especies de armadillos que habitan el Desierto Monte de Argentina. *Chaetophractus* es omnívoro, lo cual es importante para una vida en desiertos que tienen relativamente pocos insectos. Los *Chaetophractus* evitan el calor del verano y los problemas por escasez de agua, por sus hábitos cavadores y nocturnos. Durante el invierno, ellos se vuelven diurnos. Sus madrigueras no son afectadas, debido a su profundidad, por variaciones diarias, y posiblemente estacionales, en la temperatura del aire. Las temperaturas dentro de las madrigueras siempre están por debajo de la temperatura corporal de los armadillos, aún durante los días más calurosos del verano. Debido a sus hábitos cavadores, la distribución de *Chaetophractus* es limitada por la naturaleza del substrato de la tierra. Son poco communes en laderas rocosas, pero se encuentran comúnmente en áreas de dunas arenosas y, aparentemente, prefieren cavar sus madrigueras en superficies de laderas, más que en tierras planas. El espesor medular relativo (RMT) indica que el riñón de *Chaetophractus* tiene una mayor habilidad para concentrar la orina, que la esperada de un mamífero desértico de su peso. Los *Chaetophractus* permanecen en balance hídrico hasta en un régimen reducido de ingestión de agua.

Resumo

O pequeno tatu peludo *Chaetophractus vellerosus* é uma das várias espécies de tatus que habitam o Monte Desert, na Argentina. *Chaetophractus* é onívoro, fato de importância por viver em desertos, onde há relativamente poucos insetos. *Chaetophractus* evita o calor do verão e os problemas de água sendo cavador e noturno. No inverno, eles são diurnos. Suas tocas não são atingidas pelas variações diárias da temperatura do ar, e possivelmente sazonal, dada a sua profundidade. A temperatura da toca é inferior à temperatura corpórea do tatu, mesmo durante os dias mais quentes de verão. Face ao seu hábito cavador, a distribuição de *Chaetophractus* é limitada pelo substrato. Eles são incomuns em declives rochosos mas comuns em dunas de areia e aparentemente preferem entocar-se em declives do que em superfícies planas. A espessura medular relativa (RMT) indica que o rim de *Chaetophractus* tem uma alta capacidade para concentrar urina, mais do que o esperado para um mamífero xeromórfico de seu peso. *Chaetophractus* conserva um balanço hídrico mesmo numa dieta reduzida em água.

The Monte Desert of Argentina is located in the pre-Andean region occupying the intermountain valleys where moist air currents rarely penetrate, and extends from 24° 35' S to 44° 20' S and 62° 54' W on the Atlantic coast to 69° 50' W (Morello, 1958). Precipitation, which occurs mostly in summer, is generally less than 200 mm per year with much of the area being described by Meigs (1951) as arid. Much of the Monte has a cover of resinous shrubs, largely represented by the family Zygophyllaceae. Trees, where present, are on the river borders where they form gallery forests.

The Monte Desert is physically and physiognomically quite similar to the Sonoran Desert of North America. From 1967 to 1974, investigations were conducted by the International Biological Program-Structure of the Ecosystem Subprogram (IBP-SES) to determine the structural and functional similarity of these two desert ecosystems which have evolved separately but under similar climatic regimes. In 1971, while an IBP-SES participant, I undertook a 3-year field and laboratory study of the little hairy armadillo *Chaetophractus vellerosus* a common mammal of the northern Monte.

The literature on armadillo ecology consists largely of studies of the common long-nosed armadillo *Dasypus novemcinctus* in North America (Kalmbach, 1943; Taber, 1945; Clark, 1951; Fitch et al., 1952; Breece and Dusi, 1985; Galbreath, 1985; Layne and Glover, 1985). Talmage and Buchanan (1954) summarized research efforts on *Dasypus* until 1954 in a comprehensive monograph. Studies of the natural history of any of the South American armadillos are virtually nonexistent, but field research on some species has begun (Meritt, 1985a, 1985b). Crespo (1944) contributed a monograph on the burrow characteristics of several species; others (Hudson, 1922; Krumbiegel, 1940; Redford, 1985; Wirtz et al., 1985) have commented on armadillo food habits.

In addition to an interest in the natural history of *Chaetophractus* it was my objective to determine how this desert armadillo confronts the problems of high temperature and aridity. I studied their movements, activity periods, temperature environments, burrows, diet, and renal capabilities.

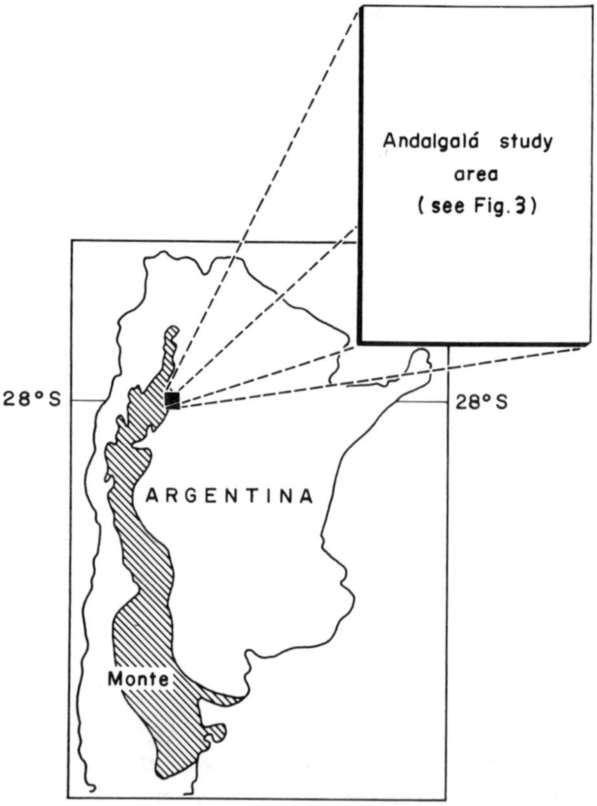

Figure 2. The Monte phytogeographic province of Argentina, as described by Morello (1958).

SPECIES DESCRIPTION

Two subspecies of *Chaetophractus vellerosus* have been described by Yepes (1958); *C. v. vellerosus* is distributed in western Bolivia and the northwestern provinces of Argentina whereas *C. v. pannosus* is found south of *C. v. vellerosus*.

Chaetophractus v. pannosus were not on my study area. Individual *C. v. vellerosus* collected during this study had from six to eight moveable bands and the mean weight of 147 individuals was 840 grams (Fig. 1). The male was the heavier and larger of the sexes, but both were heavier in winter than in summer (males 3.9 percent, females 8.6 percent) due to the addition of a dense, subcutaneous fat layer.

Figure 1. The little hairy armadillo, *Chaetophractus vellerosus*, of northwestern Argentina.

THE STUDY AREA

The village of Andalgalá (elevation 1,080 m) is located in the Bolsón (basin) de Pipanaco, a large valley in the northern Monte Desert (Figs. 2, 3). Andalgalá, situated on the northern edge of the basin, receives an average rainfall of 293 mm; the settlement of Pilciao, 20 km south of Andalgalá receives 130 mm (Morello, 1958). The single rainy season occurs during the summer months of December through February. *Chaetophractus* is common throughout the area but its abundance varies with soil condition and proximity to civilization.

My main study area was in sand dunes 15 km south of Andalgalá (Fig. 4). The dunes support a good *Chaetophractus* population relating to the excellent burrowing conditions. The vegetation on the dunes is predominantly *Bulnesia retama* and *Larrea cuneifolia*, shrubs of the Zygophyllaceae. Along the ephemeral stream courses, *Cercidium praecox, Acacia aroma, A. furcatispina*, and *Prosopis* sp. are common. The rivers of the area, Río Pozo and Río Vírgin, are seasonal; the nearest permanent water is the Río Andalgalá, 20 km north. Rainfall is comparable to that at Pilciao.

To a lesser extent, *Chaetophractus* were investigated in other habitats. Between the dunes and Andalgalá, the terrain flattens to support monotypic stands of *Larrea*

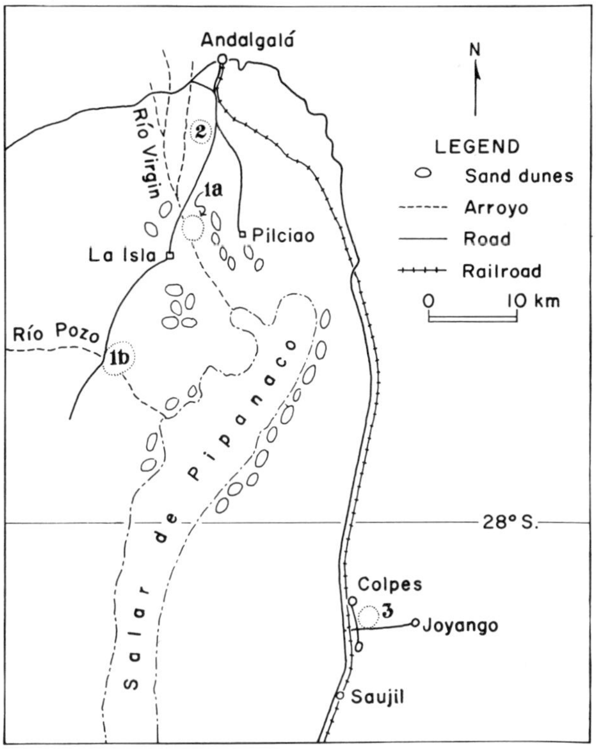

Figure 3. Andalgalá, Argentina and surrounding area.

Figure 4. Optimal habitat for *Chaetophractus vellerosus*. Shown are sand dunes several kilometers south of Andalgalá, Argentina.

cuneifolia. Populations of *Chaetophractus* are less dense, probably because burrowing into the hard-packed soil is more difficult. *Chaetophractus* are also less common on the rocky sides of the basin, owing to the lack of soil.

METHODS

Movements and Home Range.-- I monitored movements of seven individuals for periods up to five consecutive days, with a trailing device designed for box turtles (Breder, 1927) and later modified (Stickel, 1950). Following the capture of an armadillo, I attached a spool of polyester thread (455 m) to the tail with epoxy. The spool did not impede movements or noticeably affect behavior of captive individuals. The armadillo was released where captured and left undisturbed until the following day whereupon I mapped its activities by following the thread trail. In this manner, I obtained information about distance and route of travel, home range, burrows, and foraging behavior.

Diet.-- During December (summer) 1971, I collected 48 *Chaetophractus* on the sand dunes, and preserved the stomach contents in formalin. I collected 36 additional animals during the winter (May to July) of 1972. I initially placed stomach contents to be analyzed on ordinary window screen (mesh diameter 1.5 mm) supported over a 4.0-l beaker and washed. I refiltered the filtrate (< 1.5 mm) through cheesecloth and classified it as unidentifiable. It is entirely possible that fragments of arthropods were overlooked by this procedure, but examination of the filtrate indicated that it was predominantly sand. I placed the unfiltered material (> 1.5 mm) on a large aluminium pan, then separated and identified it with a dissecting microscope. I then determined volumetric displacement of water for each category, including the unidentifiable filtrate.

Temperature and Activity Periods.-- A vegetation transect approximately 10 km long was established by the IBP on the Joyango bajada (slope) near the village of Colpes, 60 km south of Andalgalá (Fig. 3). At the base of the transect (elevation 840 m), I monitored summer and winter ambient and burrow temperatures. Temperature differences between the Joyango transect and the Rio Vírgin dunes were not significant.

From November through January 1971-1972 (summer), I recorded hourly temperatures for a 3-day period (72 hours) each month. I placed thermistors in shade and in full sunlight at ground surface, 15 cm and 2 m above the surface, and in one burrow in continuous shade (15 cm deep) and one in sunlight (50 cm deep). During June 1972 (winter), I repeated the procedure at the Joyango site for a 24-hour period. In July, I moved the recording equipment to the dunes to record winter soil temperatures with thermistors buried 100, 50, 30, 20, and 10 cm, and armadillo body temperatures.

I determined activity periods of *Chaetophractus* by observing captive animals in outdoor enclosures and individuals in the field, and recorded ambient air temperatures at the time of observation.

Burrow Analysis.-- I excavated and measured a total of 27 burrows. I also measured burrow densities and locations in various vegetation and soil types.

Renal Capabilities: RM and Water Balance.-- The details

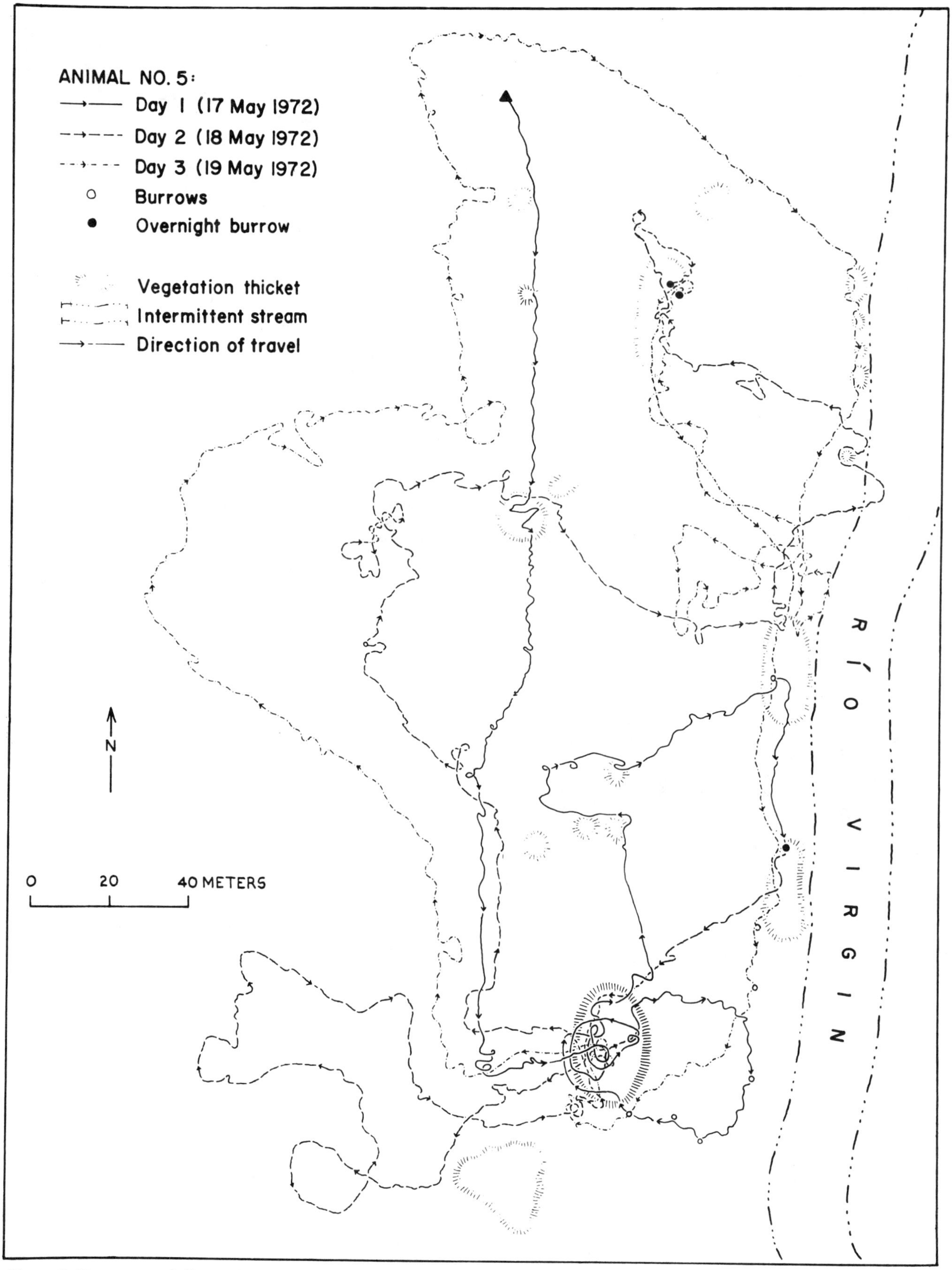

Figure 5. Movements of *Chaetophractus vellerosus* number 5 for three consecutive days, 17-19 May 1972, along the Rio Vírgin near Andalgalá, Argentina (Greegor, 1980b).

of the relative medullary thickness and water balance procedures have previously been reported (Greegor, 1975). To determine the renal capabilities of *Chaetophractus*, live individuals were transported to the University of Arizona in 1972. *Chaetophractus* was compared with *Dasypus novemcinctus* from Louisiana obtained through the Gulf South Research Institute, New Iberia, Louisiana. Captive armadillos were initially fed freshly ground horse meat (67.6 percent water) to determine whether they could maintain their weight. They were then fed partially dehydrated meat (40.6 percent water).

The concentrating potential of the kidneys was determined using Sperber's (1944) equation for relative medullary thickness,

$$\text{RMT} = \frac{\text{Medulla Thickness X 10}}{\sqrt[3]{\text{Kidney Size}}}$$

whereupon a value was determined for each kidney. Ten kidneys representing nine adult individuals were used for each of the two species.

RESULTS AND DISCUSSION

I monitored the movements of seven individuals during the study period (1971 to 1972) and followed two individuals (nos. 5, 6) for longer than one day (Greegor, 1980b). During a 5-day period, number 6 traveled 1,077 m (range 2 to 908 m per night), probed 222 times for food, and either investigated or dug 19 burrows. Number 5 traveled the maximum distance recorded during a single activity period, averaging 1,032 m for three consecutive nights (650 m, 1045 m, 1400 m, respectively). During that period, the armadillo was active within an area of 3.4 ha, and it periodically retraced portions of the previous evening's route and revisited the same burrow (Fig. 5). Layne and Glover (1977) reported the longest movement for *Dasypus novemcinctus* over a 24-hour period was 488 m. They estimated a home range size of 5.7 ± 1.7 ha for *Dasypus* in Florida. Other estimates for *Dasypus* home ranges have been 3.4 ha (Clark, 1951) and 19.8 ha (Fitch et al., 1952). *D. novemcinctus* adults had home ranges averaging 3.5 ha in Alabama, while juveniles had home ranges of about 1.5 ha (Breece and Dusi, 1985).

Chaetophractus did not necessarily return to the same "home" burrow following activity. Number 5 returned to the same burrow 2 of 3 nights and number 6 on 3 of 5 nights.

The places where the marked animals probed for food indicated that the armadillos were foraging for plant tubers of *Heliotropium mendocinum* and ants on the open dunes and for *Acacia* and *Prosopis* pods within the vegetation thickets. The stomach content analyses indicated that insects were their major food in summer; 46 percent insects by volume (Greegor, 1980a). In winter, the insect volume dropped to 25.7 percent, being replaced by plant material (50.7 percent) as the main food (Fig. 6).

In winter, the diet of *Chaetophractus* consisted of almost 51 percent plant material, compared to 22 percent in summer. Most ground insects disappear by winter and *Chaetophractus* fed heavily on *Prosopis* pods (30.9 percent by volume) and tubers and roots of other plants. It is apparent that *Chaetophractus* rely heavily on plant material, particularly *Prosopis* pods. Several workers (Kalmbach, 1943; Baker, 1943; Taber, 1945; Fitch et al., 1952) reported low percentages of plant material in *Dasypus* stomachs. However, Hamilton (1946) found 80 percent of the stomach contents of northern Texas armadillos for August was black persimmon fruit, which indicates they relish succulent fruits when available.

Rodents comprised 20.2 percent of the diet by volume in summer and 9.2 percent in winter. *Eligmodontia typus* and *Phyllotis griseoflavus* were identified. One stomach contained four nestling *P. griseoflavus*. Lizards and amphisbaenids were the next most frequently encountered vertebrates in the stomachs; amphisbaenids were in three stomachs and in one stomach was a

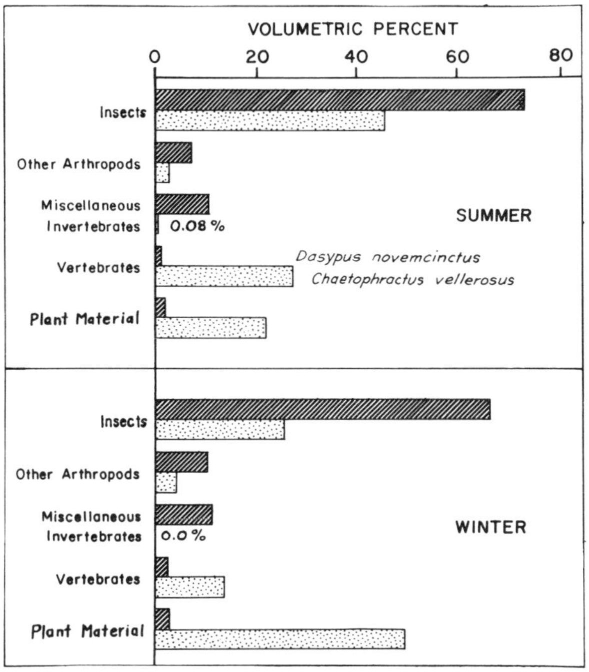

Figure 6. Summer and winter diets of *Chaetophractus vellerosus* and *Dasypus novemcinctus* (Greegor, 1980a). Data for *Dasypus* represent the summer (June and July) and winter (December and January) averages from Kalmbach (1943).

specimen of the lizard, *Teius*. More vertebrates were consumed in summer than in winter, probably owing to decreased prey activity in winter and reduced numbers of vulnerable individuals then. The proportion of vertebrates consumed by *Chaetophractus* (percent volume--13.9 winter, 27.7 summer) is higher than previously reported for an armadillo (Kalmbach, 1943; Fitch et al., 1952; Hudson, 1922; Krumbiegel, 1940).

Throughout the year, ground-dwelling insects were never abundant in the sand dunes. Those insects most frequently seen were also those most often eaten by *Chaetophractus*. In contrast, the abundance of ground insects in the *D. novemcinctus* habitat is greater owing to wetter conditions and the presence of leaf litter. Kalmbach (1943) found that 77.6 percent (volume) of the *Dasypus* diet was insects. Krumbiegel (1940) found primarily insects in the stomachs of several South American armadillo genera. The food habits of *Chaetophractus* implicate them as an opportunistic eurytrophic species, decidedly more so than any armadillo reported to date.

While not foraging, *Chaetophractus* remains underground in burrows. *Chaetophractus* burrow by scraping soil back to mid-body, lifting the body to draw the hind legs forward, and thrusting the dirt backward with the hind feet. Frequently, this results in a mound of soil at the entrance to the burrow. They are efficient burrowers; I watched one individual excavate at a rate of 100 cm/min in sandy soil. *Chaetophractus* digs in such a manner that it seals the burrow as it excavates. This certainly would help to maintain a more constant temperature and higher humidity in the burrow. Feniuk and Kazantzeva (1937) reported that the jerboa *Dipus sagitta* plugs its burrow in hot weather. The pygmy armadillo *Chlamyphorus truncatus*, using the posterior portion of its carapace, packs sand behind it when burrowing (Rood, 1970).

Many of the 27 burrows excavated were probably not permanent burrows, but temporary shelters. While *Dasypus* build nests inside their burrows (Clark, 1951; Talmage and Buchanan, 1954; Taber, 1945), I never found nesting material in a *Chaetophractus* burrow. In captivity, *Dasypus* constructs straw nests but *Chaetophractus* never did.

In the dune area, nearly 75 percent of the burrows were located on the sides of the dunes, while on the bajada, where much of the soil is impenetrable, burrows were clumped around vegetation and on arroyo banks where the only soft soil is found. The entrance tunnels of burrows on sloped ground were at lower angles (26°) than burrows on a horizontal surface (39°). Lower tunnel angles would require less effort to excavate and maintain, and would reduce the chance of flooding.

Burrows were clumped in the Río Vírgin dunes around the *Bulnesia retama* plants, the roots providing necessary support and protection in the loose sand. Crespo (1944) found that when the soil lacked natural cohesion, burrows were more numerous near vegetation.

Burrows excavated in sand dunes were frequently several meters long and 1 to 2 meters deep. The average depth of 27 burrows was 51 cm, which is adequate to provide protection from fluctuations in ambient temperature. Several burrows were deeper than 100 cm, one being almost 200 cm. My impression was that most permanent burrows in the dunes are at least 100 cm deep, which suggests that the annual temperature range of the burrow would be 18°C to 27°, using Sinclair's (1922) results from the Sonoran desert as a basis. Clark (1951) reported an average depth of 50 cm for *Dasypus* burrows, and Taber (1945) found a maximum burrow depth of 200 cm. Crespo (1944) reported very shallow burrows for *Chaetophractus* which seems unlikely, particularly because one of his study areas was at 4,000 m.

Soil temperatures measured in winter indicated that daily fluctuations are negligible in the dunes at 50 cm depth. Sinclair (1922) reported that fluctuations become negligible at 45 cm depths during summer in the Sonoran Desert. At 100 cm depth in the Monte Desert, the midwinter mean was 18.6°C (daily range 17.9°C to 19°C), which agrees with Sinclair's (1922) findings for the Sonoran Desert. Temperature of *Chaetophractus* burrows during winter ranged from 10°C to 20°C.

The temperatures of shallow burrows in full sunlight might easily exceed armadillo body temperatures (35° C), but *Chaetophractus* generally burrow in open spaces for food and not permanent shelter. A shallow burrow in the shade of Joyango bajada was cooler in summer and warmer in winter than a deeper burrow in the open, and the temperature was also more stable. During summer, the shaded burrow never exceeded 30°C, at least 4°C below *Chaetophractus* body temperature. Vorhies (1945) reported maximum burrow temperatures of 30°C for the banner-tailed kangaroo rat *Dipodomys spectabilis* in the Sonoran Desert.

Chaetophractus are nocturnal in summer and diurnal in winter, and their activity is probably limited by temperature. During summer, most activity occurs between 2200h and 0600h. If *Chaetophractus* emerges from its burrow prior to 2200h, it will experience ambient temperatures exceeding body temperature. By 0800h, the ambient temperature is 27°C and rising rapidly. Individuals were rarely observed above ground on a summer day. Summer afternoon air temperatures are frequently above 40°C, while surface temperatures approach 70°C. One individual was observed in winter foraging on sand having a surface temperature of 51°C.

In winter, armadillos were active during daytime and were not observed before 0900h or after 1800h. *Chaetophractus* probably does not leave or return to its burrow until the ambient temperature is near burrow temperature. One individual was observed above ground at 2300h at an elevation of 1,920 m with an air temperature of 12°C.

Body temperatures of two individuals inside burrows during the winter dropped 5.4°C and 6.0°C within one hour. Simultaneously, the burrow temperatures dropped several degrees but not to the same extent as did the body temperatures. Roig (unpubl.) reported that when the Patagonia armadillo *Zaedyus pichiy* experiences a rapid ambient temperature drop, the internal temperature drops suddenly. Thermoregulatory studies done by Johansen (1961) and Roig (1969; unpubl.) indicate that armadillo body temperatures are significantly lower during periods of inactivity.

The results of a comparison between *Chaetophractus vellerosus* and *Dasypus novemcinctus* renal capabilities have been previously published (Greegor, 1975). *Dasypus* was used as a contrast because it is definitely restricted in its distribution to mesic habitats, while *Chaetophractus* survives in the most arid portions of the Monte Desert.

Measurements of the kidney relative medullary thickness (RMT) indicate that the *Chaetophractus* kidney has a higher concentrating ability than expected for a xeric mammal of its weight (RMT = 7.5 ± 0.5). The RMT of the *Dasypus* kidney is lower than expected for a xeric mammal, yet slightly higher than a mesic mammal (RMT = 4.9 ± 0.4). When placed on a reduced water diet of dehydrated meat, *Chaetophractus* is able to remain in water balance whereas *Dasypus* is not. These results confirm that *Chaetophractus* has a kidney that is well adapted for an arid existence.

In conclusion, it seems quite apparent why *Chaetophractus* has so successfully flourished in a desert that has dramatic fluctuations in temperature, precipitation, and food supply. In marked contrast to the mesic *Dasypus* this armadillo has a highly concentrating kidney for water conservation and is a trophic generalist, which permits it to be flexible in an environment with a seasonally changing food supply. *Chaetophractus* avoids temperature extremes by seasonally shifting its activity periods and thereby utilizing only optimal temperatures when above ground. The burrows are sufficiently deep that the armadillo can always retreat to a temperature environment that is physiologically undemanding.

ACKNOWLEDGMENTS

I am grateful to Dr. G. Gene Montgomery for his critical editing of the manuscript. This project was supported, in part, by NSF (IBP-SES) Grant No. GB 27161.

LITERATURE CITED

Baker, R.H.
1943. May food habits of armadillos in eastern Texas. Am. Mid. Nat., 29: 379-380.

Breece, G.A., and J.L. Dusi.
1985. Food habits and home ranges of the common long-nosed armadillo *Dasypus novemcinctus* in Alabama. (this volume).

Breder, R.H.
1927. Turtle trailing: a new technique for studying the life habits of certain Testudinata. Zoologica, 9: 231-243.

Clark, W.K.
1951. Ecological life history of the armadillo in the eastern Edwards plateau region. Am. Mid. Nat., 46: 337-358.

Crespo, J.A.
1944. Contributión al conocemiento de la ecologiá de algunos dasipodidos (Edentata) argentinos. Rev. Argent. Zoogeogr., 4: 7-39.

Feniuk, B.K. and J.M. Kasantzeva.
1937. The ecology of *Dipus sagitta*. J. Mamm., 18: 409-426.

Fitch, H.S., P. Goodrum, and C. Newman.
1952. The armadillo in the southeastern United States. J. Mamm., 33: 21-37.

Greegor, D.H.
1975. Renal capabilities of an Argentine desert armadillo. J. Mamm., 56: 626-632.

1980a. Diet of the omnivorous armadillo, *Chaetophractus vellerosus* of northwestern Argentina. J. Mamm., 61: 331-334.

1980b. Preliminary study of movements and home range of the armadillo, *Chaetophractus vellerosus*. J. Mamm.61: 334-335

Hamilton, W.J.
1946. The black persimmon as a summer food of the Texas armadillo. J. Mamm., 27: 175.

Hudson, W.H.
1922. The Naturalist in La Plata. (6th ed.) E. P. Dutton and Co., N.Y., 394 pp.

Jacobs, J.
1985. Social behavior and space usage of the common long-nosed armadillo *Dasypus novemcinctus* in Mississippi. (this volume).

Johansen, K.
1961. Temperature regulation in the nine-banded armadillo (*Dasypus novemcinctus mexicanus*). Physiol. Zool. 35: 126-144.

Kalmbach, E.R.
1943. The armadillo: its relation to agriculture and game. Game, Fish, and Oyster Commission, Austin, Texas. Illus. 61 pp. + IV.

Krumbiegel, I.
1940. Die Saugetiere der Sudamerida-Expeditionen Prof. Dr. Kriegs. I. Gurteltiere. Zool. Anz., 131: 49-73.

Layne, J.N. and D. Glover.
1977. Home range of the armadillo in Florida. J. Mamm., 58: 411-413.

Layne, J.N. and D. Glover.
1985. Activity patterns of the common long-nosed armadillo *Dasypus novemcinctus* in south-central Florida. (this volume).

Meigs, P.
1951. World Distribution of Arid and Semi-Arid Homo-climates. UNESCO/NS/AZ/37.

Meritt, D.A., Jr.
1985. The fairy armadillo, *Chlamyphorus truncatus* Harlan. (this volume).

Meritt, D.A., Jr.
1985. Naked-tailed armadillos, *Cabassous* sp. (this volume).

Morello, J.
1958. La Provincia Fitogeographica del Monte. Tucuman, Argentina. Illus. 155 pp. + LVII.

Redford, K.H.
1985. Food habits of armadillos (Xenarthra: Dasypodidae). (this volume).

Roig, V.G.
1969. Termorregulacion en *Euphractus sexcinctus* (Mammalia, Dasypodidae). Physis., 39(78): 27-32.

(Unpublished). Observaciones sobre la termorregulacion en *Zaedyus pichiy*. University of Cuyo, Mendoza, Argentina, 5 pp.

Rood, J.P.
1970. Notes on the behavior of the pygmy armadillo. J. Mamm., 51: 179.

Sinclair, J.G.
1922. Temperatures of the soil and air in a desert. Mon. Weath. Rev., 50: 142-144.

Sperber, I.
1944. Studies on the mammalian kidney. Zoologiska Bidrag fran Uppsala., 22: 249-432.

Stickel, L.F.
1950. Populations and home range relationships of the box turtle, *Terrapene c. carolinea* (Linnaeus). Ecol. Mono., 20: 352-378.

Taber, F.W.
1945. Contribution on the life history and ecology of the nine-banded armadillo. J. Mamm., 26: 211-266.

Talmage, R.V. and G.D. Buchanan.
1954. The armadillo (*Dasypus novemcinctus*). The Rice Institute Pamphlet., 41: 135 pp.

Vorhies, C.T.
1945. Water requirements of desert animals in the Southwest. Univ. Arizona., Agric. Exp. Sta., Tech. Bull. No. 107: 487-525.

Wirtz, W.O., D.H. Austin and G.W. Dekle.
1985. Food habits of the common long-nosed armadillo *Dasypus novemcinctus* in Florida, 1960-61. (this volume).

Yepes, J.
1928. Los Edentata argentinos sistemática y distribución. Rev. Univ. Buenos Aires., 24: 461-515.

ACTIVITY PATTERNS OF THE COMMON LONG-NOSED ARMADILLO *DASYPUS NOVEMCINCTUS* IN SOUTH-CENTRAL FLORIDA

James N. Layne and Debbie Glover

Archbold Biological Station, Route 2, Box 180, Lake Placid, Florida 33852.

Abstract

Activity patterns of armadillos in south-central Florida were studied by track counts and direct observation during varying periods from 1972 to 1979. Armadillos were more active and relatively more nocturnal in summer than in winter. Temperature was the major factor influencing activity, with low temperatures in winter having a more pronounced effect than high summer temperatures. Peak activity tended to occur in the 20-25°C range throughout the year, with activity compressed into a shorter time span in winter. Air temperatures at times active armadillos were observed ranged from 6°C to 35°C with a mean of 26.8°C. Seventy percent of the observations of active armadillos were when temperatures were between 25°C and 32°C. Young individuals and adult females apparently were more active and more diurnal than adult males in cold weather. Estimates of metabolic rates during times of activity suggest that, by shifting to daytime activity in cold weather, an animal would expend about 50-60 percent less energy for thermoregulation than it would if it retained the typical nocturnal summer pattern. The metabolic advantage of nocturnal activity in summer is less obvious than for diurnality in winter. Thus, factors other than metabolic effects of temperature, such as relative humidity and food availability, may have a relatively more important influence on activity in warm weather. The labile activity pattern and well-developed nest-building habits allow the common long-nosed armadillo to tolerate a broader range of environmental conditions than would be permitted by its physiological capacities alone. These behavioral characteristics have undoubtedly contributed to the species' success in expanding its geographic and ecological range in North America.

Abstracto

Se estudiaron los patrones de actividad de los armadillos de la Florida sud-central, por medio de un recuento de huellas y de observaciones directas durante varios períodos comprendidos entre 1972 y 1979. Los armadillos tuvieron mayor actividad y fueron mayormente nocturnos durante los veranos que en los inviernos. La temperatura fue el factor de mayor influencia sobre la actividad, siendo más pronunciado el efecto de las bajas temperaturas de invierno que de las altas temperaturas de verano. El pico de actividad ocurrió entre los 20-25°C durante todo el año, mientras que la actividad se encerraba dentro de un lapso menor de tiempo durante el invierno. Las temperaturas del aire oscilaron desde 6°C hasta 35°C, con una mediana de 26.8°C, durante el período de actividad de los armadillos. Setenta porciento de las observaciones de armadillos activos ocurrieron cuando la temperatura estaba entre 25°C y 32°C. Durante épocas de frío, individuos jóvenes y las hembras adultas eran mayormente diurnos y más activos que los machos adultos. Estimados de la tasa de metabolismo durante períodos de actividad sugieren que el cambio a la actividad diurna durante épocas de frío, significa un gasto de alrededor de 50-60 perciento menos de energía en la termo-regulación, que si el animal retiene su típico patrón nocturno de actividad veraniega. La ventaja metabólica de la actividad nocturna del verano es menos obvia que la representada por la actividad diurna durante el invierno. Es así como factores distintos a los efectos metabólicos de la temperatura, tales como la humedad relativa y la disponibilidad de alimentos, pueden ejercer una influencia relativamente más importante sobre la actividad durante épocas de calor. La inestabilidad como patrón de actividad, acompañada de un bien desarrollado hábito de construir sus nidos, le permiten al armadillo de nariz larga tolerar un rango mayor de condiciones ambientales de lo que le permitiera su capacidad fisiológica por sí sola. Estas características de comportamiento han contribuído, sin duda alguna, al éxito que ha tenido la especie en la expansión de su distribución geográfica y ecológica en América del Norte.

Resumo

Os padrões de atividade de tatus ao centro-sul da Flórida foram estudados pela contagem de pegadas e observação direta durante períodos varíaveis, de 1972 a 1979. Os tatus eram mais ativos e relativamente mais noturnos no verão do que no inverno. A temperatura foi o maior fator que influenciava na atividade, com as baixas temperaturas no inverno tendo um efeito mais pronunciado do que altas temperaturas do verão. O pico de atividade tendeu a ocorrer num âmbito de 20° -25°C por todo o ano, com a atividade resumida a um período de tempo mais curto no inverno. Temperaturas do ar, quando tatus ativos foram observados, variavam entre 6°C e 35°C, com uma média de 26,8°C. Setenta por cento dos registros cairam entre 25°C e 32°C. Indivíduos jovens e fêmeas adultas aparentemente eram mais ativos e mais diurnos do que machos adultos em épocas frias. Estimativas de taxas metabólicas, durante períodos de atividade, sugerem que pelo deslocamento da atividade do período diurno, na época fria, um animal gastaria cerca de 50-60 por cento menos de energia para termorregulação do que gastaria se mantivesse o padrão noturno típico de verão. A vantagem metabólica da atividade noturna no verão é menos óbvia do que para a diurnalidade no inverno. Assim, fatores além dos efeitos metabólicos de temperatura, tais como umidade relativa e disponibilidade de alimento, podem relativamente ter uma influência mais importante na atividade, em épocas quentes. O padrão de atividade labîl e hãbitos de construção de ninhos bem desenvolvidos permitem ao tatu comum de focinho longo tolerar um âmbito mais vasto de condições ambientais do que seria permitido pelas suas capacidades fisiológicas somente. Essas características comportamentais têm contribuído para o êxito da espécie, na expansão de seu alcance geográfico e ecológico na América do Norte.

One of the most effective and widespread means by which animals minimize the influences of unfavorable environmental factors and maximize access to food and other essential resources is through the timing of their activity. Thus, a full understanding of the ecology of a species and its role in the biotic community is dependent upon knowledge of its activity patterns and the factors influencing them. The pronounced sensitivity of the metabolic rate of the common long-nosed armadillo *Dasypus novemcinctus* to environmental temperature (Johansen, 1961; McNab, 1980) suggests that activity behavior may be of particular significance in the ecology and geographic distribution of this species. Observations on activity of armadillos in the wild or in captivity have been reported by Anderson and Benirschke (1966), Breece and Dusi (1985), Burns and Waldrip (1971), Clark (1951), Fitch et al. (1952), Johansen (1961), Kalmbach (1943), Lehmann (1934, cited by Kalmbach, 1943), McCusker (1985), McNab (1980), Moore (1968), Prudom and Klemm (1973), Taber (1945), and Yaksh et al. (1967). However, only the studies of Moore (1968) on animals in an outdoor enclosure and Yaksh et al. (1967) on armadillos under laboratory conditions have dealt primarily with activity behavior.

We studied activity patterns of armadillos in south-central peninsular Florida. The armadillo was introduced into peninsular Florida during the 1920s and 1930s and rapidly spread thoughout most of the state (Bailey, 1924; Fitch et al., 1952; Neill, 1952; Sherman, 1943). In recent years, the species has appeared in the western part of the Florida panhandle (Stevenson and Crawford, 1974; Wolfe, 1968), presumably mainly through natural range expansion from the west. Although the armadillo is now one of the most ubiquitous and conspicuous mammals in Florida, its ecology and behavior in the state are poorly known.

DESCRIPTION OF THE STUDY AREA

The study was conducted on the Archbold Biological Station located 13 km south of Lake Placid (27°11'N, 81°21'W) in Highlands County, Florida.

The climate of the region is characterized by wet, hot summers and dry, cooler winters. Mean annual temperature is 22.2°C. Average maximum and minimum daily temperatures for the four warmest months (June-September) are 33.9° and 20.6°C, respectively. Corresponding values for the four coldest months (November-February) are 24.4° and 9.4°C. Lowest temperatures in winter are usually above freezing. Cold snaps with temperatures dropping below freezing during the night usually last only a few days and seldom occur more than two or three times each winter. Even during such cold spells, temperatures generally reach 15-20°C during the day. Minimum daily temperatures usually occur shortly before dawn and maximum temperatures between approximately 1300 hrs and 1500 hrs. Throughout the year relative humidity (RH) averages approximately 80 percent during the night and 30 percent to 40 percent at midday. Mean annual rainfall is 1,368 mm. The four wettest months are June, July, August, and September, with about 60 percent of the annual total being received during this period. Rain is of almost daily occurrence in summer, usually in the form of brief thunderstorms. Summer rains tend to occur later in the afternoon as the season progresses. Late winter and early spring are the driest parts of the year, with rather severe drought conditions sometimes occurring in spring.

Five major habitat types present in the study area include scrubby flatwoods, low flatwoods, dry ponds, brushy railroad right-of-way, and pasture. Scrubby flatwoods consist of scattered slash pines *Pinus elliottii* and sand pines *P. clausa* and a dense shrub layer. Sand live oak *Quercus geminata*, scrub oak *Q. inopina*, Chapman's oak *Q. chapmanii*, rusty lyonia *Lyonia ferruginea*, fetterbush *L. lucida*, and saw palmetto *Serenoa repens* are characteristic components of the shrub layer. Herbaceous ground cover is scarce, and patches of bare sand often occur between clumps of shrubs. The soil is sandy and well drained. Low flatwoods have soils with an organic hardpan and tend to be moister than scrubby flatwoods. They usually have rather dense stands of slash pine and either a heavy shrub layer of gallberry *Ilex glabra*, saw palmetto, or fetterbush, or a more open shrub layer with well developed herbaceous ground cover. Dry ponds are grassy depressions, usually without large trees, that may contain water during wet periods but often remain dry for months or years. They are typically surrounded by a zone of dense saw palmetto. The railroad right-of-way contains a shallow ditch with a rank growth of shrubs and ferns, grasses, and other herbaceous vegetation and drier, more sparsely vegetated margins. Common shrub species of this habitat are wax myrtle *Myrica cerifera* and elderberry *Sambucus simpsonii*. The pasture area has low grass with scattered pines and shrubs. Open water is scarce on the study area, being limited to several ditches and scattered transient ponds, most of which were dry during the periods of study. Although armadillos are common in the above vegetation types, these are not optimum habitats for the species. In the general region of the study locality, highest populations of armadillos occur in the moister forest and grassland habitats bordering the drier Lake Wales Ridge.

METHODS

The number of tracks made on a smoothed, bare, sand strip 4.5 km long during a given time interval was used as an index of activity. The census strip was approximately 2 m wide and was located along one edge of a sparsely vegetated sandy firelane averaging about 12 m in width (Fig. 1A). The census trail was marked off at 61-m intervals with numbered stakes to allow accurate recording of the location of tracks. The distinctive tracks of the armadillo were easily detected on the census strip even when the sand was dry and loose (Fig. 1B). Observations of foraging armadillos supported the use of track counts as an index of activity. The animals showed no hesitation in crossing firelanes but did not seem particularly attracted to them nor did they spend a disproportionate amount of time foraging in them.

The trackway traversed the five habitat types described above. Thirty-two percent of the route had scrubby flatwoods on both sides; 9 percent, scrubby flatwoods on one side and low flatwoods on the other; 6 percent, low flatwoods on both sides; 1 percent, dry pond on both sides; 22 percent, scrubby flatwoods on one side and railroad right-of-way on the other; 21 percent, low flatwoods on one side and railroad right-of-way on the other; and 2 percent, scrubby flatwoods on one side and pasture on the other.

The census strip was surveyed for tracks by driving slowly along the edge. A complete circuit of the route took about 30 minutes. The trackway was resmoothed by dragging a weighted board over it with a tractor either immediately after a circuit or at intervals of several days if the surface remained undisturbed enough to detect fresh tracks. In the latter case, tracks were brushed out or walked on after being recorded to avoid duplication on subsequent counts before the strip was dragged again.

Track counts were made on 8 diels (24-hr period from sunrise one day to sunrise the next) during July and August 1972, 147 diels from January to August 1973, and 10 diels each month in January and July from 1974 through 1979. Total number of tracks per diel and numbers at 2.5-hr intervals from 1600 hrs to 0700 hrs were recorded in 1972. In the January-August period during 1973, single counts, representing total activity during a 24-hr period, were made shortly after sunrise on 36 diels. On the remaining 111 diels, counts were made

Figure 1. **A.** View of track census strip. **B.** Armadillo tracks on census strip.

at sunrise and sunset to provide a measure of nocturnal and diurnal activity as well as total activity. On 89 of these diels, additional counts were made at 2-to 5-hr intervals to give a more detailed picture of the activity pattern. The 120 midsummer and midwinter counts from 1974 through 1979 were of total number of tracks per diel.

In addition to the track surveys, numerous direct observations were made on armadillo activity at all times of the year.

Temperature data for diels on which track surveys were conducted or armadillos observed were obtained from a thermograph housed in a standard weather instrument shelter at a height of 1.2 m near the beginning of the census strip. Mean day and night temperatures were calculated from values at 2.5-hr intervals from sunrise to sunset and sunset to sunrise, respectively. Because of differences between air temperature at the height of the recorder and at ground level where armadillos are active, as well as microclimatic variation between habitat types, the temperature data derived from the thermograph should be regarded as only an approximation of the actual temperature regime encountered by the animals. Different temperature measures, including mean night,

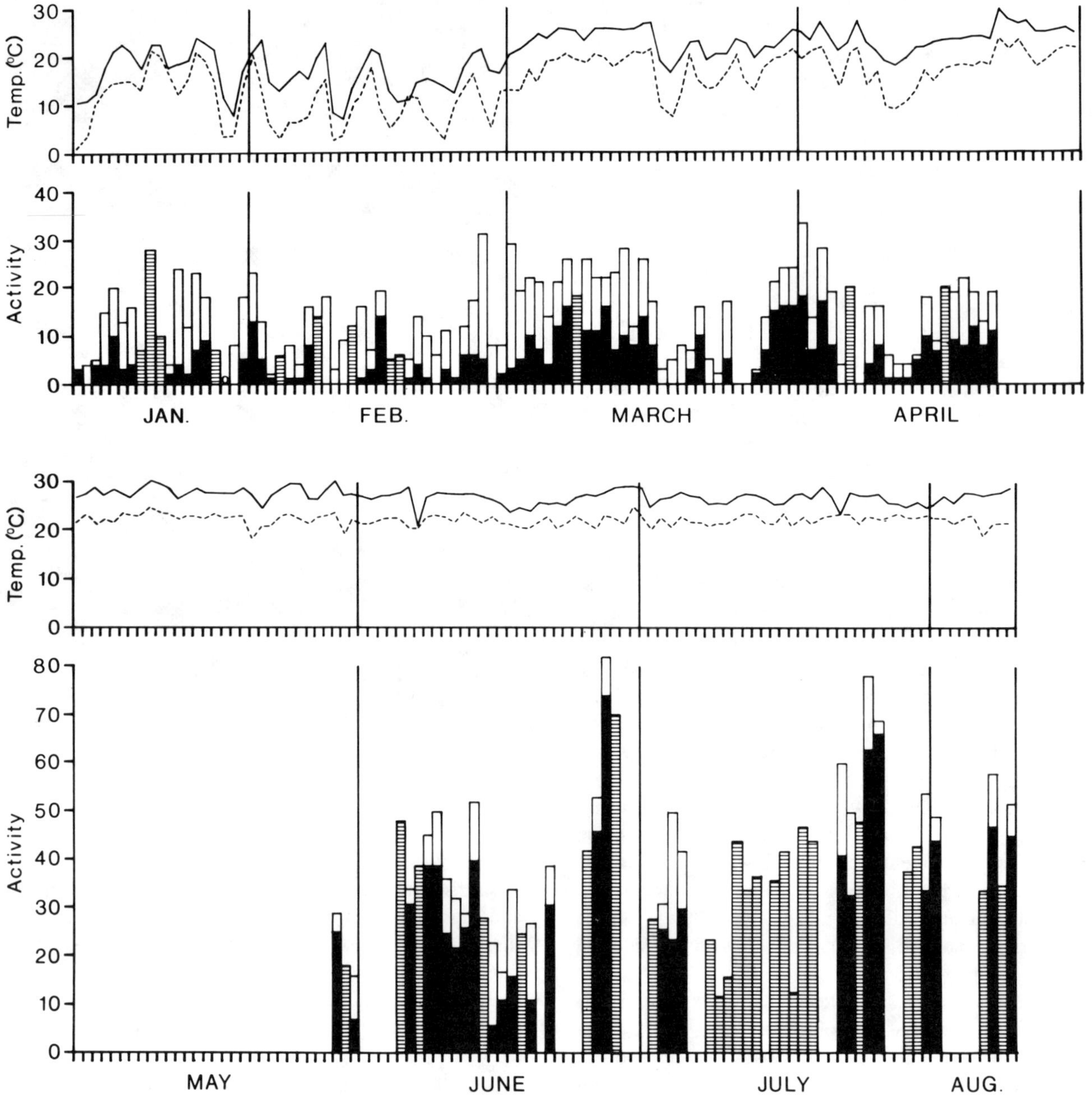

Figure 2. Daily activity (number of tracks per diel) of armadillos in south-central Florida from mid January to early August 1973. Solid and open portions of bars represent number of tracks made at night and during daylight, respectively. Shaded bars are total counts for 24-hour period (diel) only. Daily temperatures (solid line = mean daytime temperature, dashed line = mean night temperature) for the corresponding period are shown above the activity graph.

mean day, mean daily, minimum daily, and maximum daily temperature, were highly inter-correlated. Therefore, mean night temperature was arbitrarily selected for comparisons of effects of temperature on activity.

RESULTS

Track Surveys.-The average number of tracks per diel during each month in the period January-August 1973 ranged from 10.9 to 21.0 from January to May and from 40.0 to 45.6 in the June to August interval (Table 1). The number of tracks per diel was relatively more variable in winter and spring than in summer. Armadillos were predominantly diurnal (53 percent to 70 percent of the total tracks were made during the day) in winter and early spring and largely nocturnal (65 percent to 86 percent of the tracks were made at night) in late spring and summer (Fig. 2). As in the case of total

Table 1. Monthly means, standard deviations (SD), and coefficients of variation (CV) of total number of armadillo tracks crossing the census strip per diel, percentage of tracks made at night, and nighttime temperatures on diels when counts were made for the period January-August 1973.

	Total Number of Tracks per Diel				*Percent of Total Tracks at Night*				*Temperature (°C)*		
Month	*Diels*	*Mean*	*SD*	*CV*	*Diels*	*Mean*	*SD*	*CV*	*Mean*	*SD*	*CV*
January	19	12.6	8.2	65.0	14	35.2	27.8	79.0	13.1	6.2	47.3
February	28	10.9	6.8	62.4	22	29.7	26.5	89.2	9.4	4.7	50.0
March	29	17.0	8.3	48.8	28	41.9	24.7	58.9	16.6	3.8	22.9
April	20	15.4	8.0	51.9	18	46.3	21.4	46.2	16.0	3.6	22.5
May	3	21.0	7.0	33.3	2	65.0	29.7	45.7	21.6	2.0	9.3
June	20	40.0	16.0	40.0	11	67.7	20.4	30.1	21.7	1.0	4.6
July	23	40.8	16.3	40.0	8	72.1	14.7	20.4	22.0	1.0	4.5
August	5	45.6	10.6	23.2	3	86.0	4.6	5.3	20.9	1.2	5.7

Table 2. Means, standard deviations (SD), and coefficients of variation (CV) of numbers of tracks in ten censuses each month in January and July from 1974 to 1979.

	January			*July*		
Year	*Mean*	*SD*	*CV*	*Mean*	*SD*	*CV*
1974	17.5	4.0	22.9	44.1	12.6	28.6
1975	14.3	10.2	71.3	52.8	14.5	27.5
1976	14.5	8.7	60.0	58.2	25.1	43.1
1977	21.4	9.1	42.5	56.9	16.4	28.8
1978	13.0	7.2	18.2	64.6	9.3	14.4
1979	9.6	5.3	13.3	56.6	16.1	28.4

activity, the proportion of diurnal to nocturnal activity was less variable in late spring and summer than during the cooler part of the year. The track censuses in January and July over the six-year period 1974-1979 (Table 2) consistently showed higher and less variable counts in summer compared to winter, with January counts averaging only about 25 percent of the July counts.

Variation in daily and seasonal activity patterns is largely related to temperature. Comparison of track counts and corresponding temperatures (Fig. 2) shows a clear tendency for reduced activity at lower temperatures and increased activity at warmer temperatures during the winter-early spring period. There was a general increase in activity level, with no obvious relationship between track counts and the slight temperature fluctuations, during the summer months when temperatures are uniformly high and stable. Total track counts per diel and mean night temperatures for the entire period from January to August 1973 (Fig. 3) were positively correlated ($r = 0.69$, $p < 0.001$). The correlation was higher ($r = 0.63$) during the cooler months from January to April than for the warmer May-August interval ($r = 0.30$). The difference is significant ($p < 0.05$). The low but statistically significant correlation ($p < 0.05$) between activity and temperature in summer may be spurious, as at least part of the sharp increase in track counts in summer, which strongly influences the r

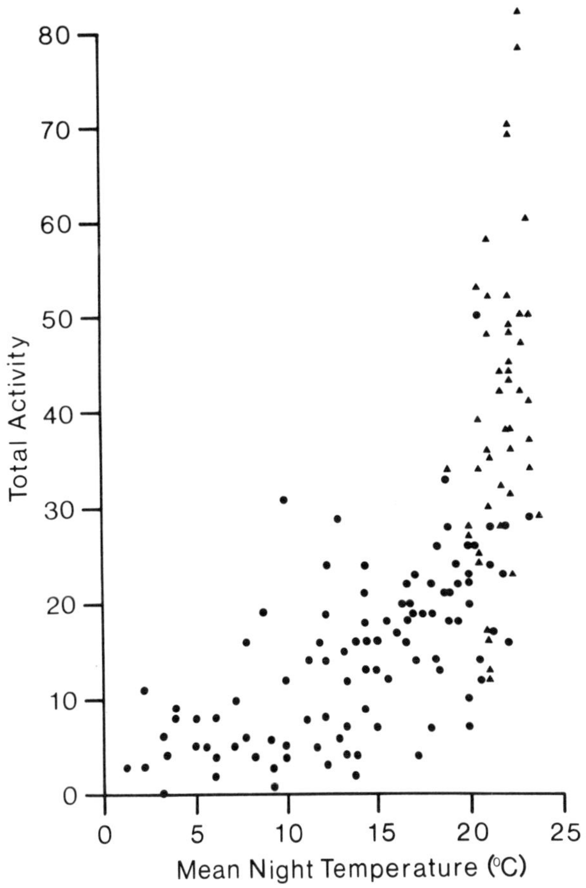

Figure 3. Relation of total activity (total track counts per diel) to mean night temperature. Dots are counts from January to May; triangles, counts from June to August.

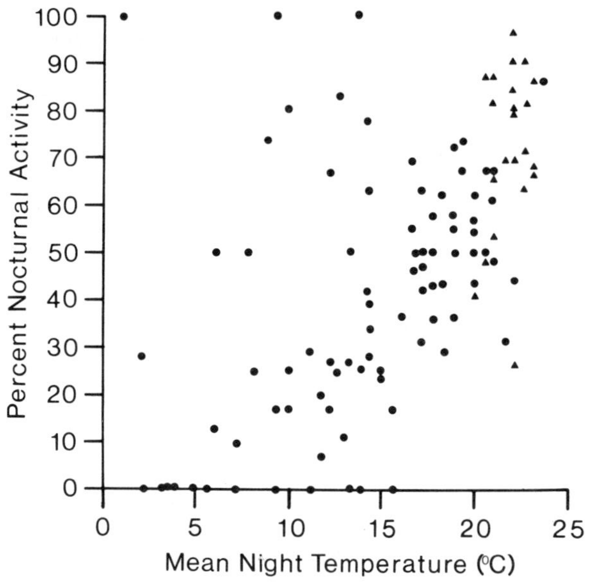

Figure 4. Relation of percentage of nocturnal activity (percent of total tracks made at night) to mean night temperature. Dots are counts from January to May; triangles, counts from June to August.

value, appeared to be due to an increased number of animals rather than higher individual activity levels (see Discussion).

There was a tendency for the relative amount of diurnal activity to be higher on diels with lower temperatures during winter and spring (Fig. 2). Until early April there were some diels each month when armadillos were active exclusively during the daylight hours and such greatly increased diurnal activity usually corresponded to sharp drops in temperature. After early April, no diels of entirely diurnal activity were recorded and the relative amount of nocturnal activity increased. The correlation coefficient between percentage of nocturnal activity and temperature during the January-August interval (Fig. 4) was 0.57, ($p < 0.001$), compared with 0.41 ($p < 0.001$) for the cooler months (January-April). There was no significant correlation between percentage of nocturnal activity and temperature during the warmer (May-August) interval. Within the winter-spring period, nocturnal activity and temperature were less strongly correlated in the colder months of January and February ($r = 0.14$) than in the milder March-April interval ($r = 0.61$). The difference is significant ($p < 0.01$).

Throughout the year, peak activity tended to occur in the 20°-25°C range, with low temperatures being avoided in winter and high temperatures in summer (Fig. 5). With the shorter daily interval with temperatures above 20° in winter (5-6 hr) as compared to summer (24 hr), activity was compressed into a shorter period of the day during the cooler part of the year. On colder days in winter and early spring, armadillos were frequently abroad in late morning, with peak activity occurring in the afternoon from 1430 hrs to 1700 hrs. On milder days during this period, activity was shifted to later in the day. These contrasting patterns are clearly evident in March and April (Fig. 5).

Temperature Range of Observed Activity.- Data from direct observation of armadillos agree with those from the track counts in indicating that most activity occurred at ambient temperatures above 20°C. Ambient air temperatures at times armadillos were observed moving about or foraging were obtained for 273 sightings of animals. The number of such values each month ranged from 3 to 45 (mean = 23). Although

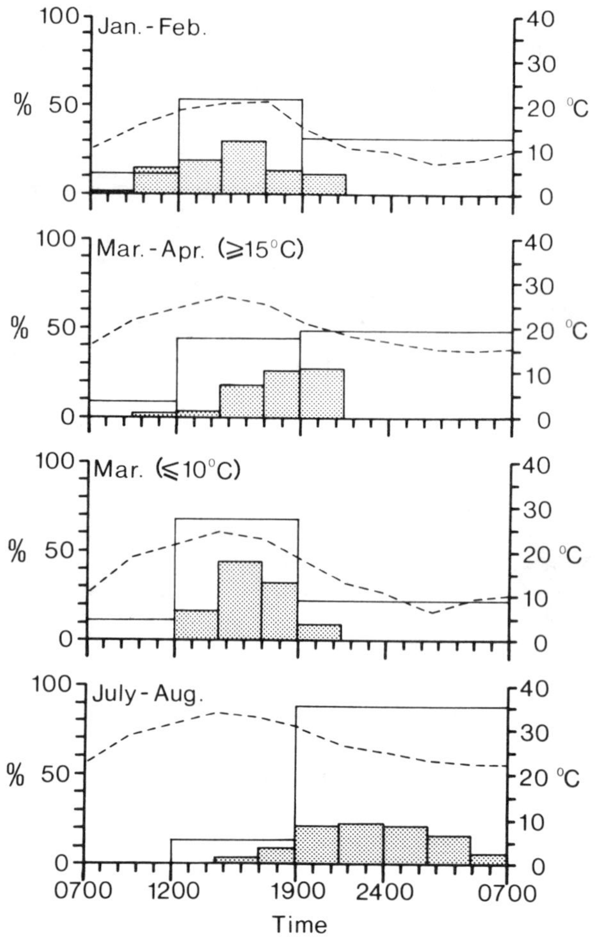

Figure 5. Patterns of diel activity at different seasons. Open and stippled bars represent the percentages of total numbers of tracks made during the indicated time periods. Counts for the different time intervals were made on different days. The dashed line is the mean daily temperature curve for the days on which track counts were made. Upper and lower graphs represent average winter and summer patterns over the typical temperature range encountered during these seasons. The two middle graphs show activity patterns on cooler (minimum temperature ≤ 10°C) and warmer (minimum temperature ≥ 15°C) days in early spring.

some observations were made during the night, most were made during the daylight hours. The lowest temperature at which an armadillo was seen foraging was 6°C at 2045 hrs on a day in January with minimum and maximum temperatures of 4° and 13°C, respectively. Four other individuals were active at temperatures of 7°, 8°, and 9°C between 1200 hrs and 1615 hrs during a two-day period in January with minimum temperatures of -5°C and maximums of 8°C and 16°C. The highest temperature at which an active armadillo was observed was 35°C at 1405 hrs on a July day with minimum and maximum temperatures of 25° and 37°C, respectively. The average temperature at time of observed activity was 26.8°C (SD = 5.5), with 70 percent of the records falling within the range of 25° to 32°C and only 11 percent below 20°C.

In summer, armadillos were most often observed during the day when the sky was overcast or after rain, particularly in late afternoon. Individuals abroad on hot, sunny days were usually in dense brush or shady wooded areas. In contrast, individuals active during the day in cold weather were often encountered in direct sun in open areas.

Of 67 active individuals captured during the cooler period of the year (December through March), 25 (37 percent) were males and 42 (63 percent), females. The difference from an expected 50:50 sex ratio is significant ($p < 0.05$, Chi-square test). However, the proportions of males (46 percent) and females (54 percent) among 35 animals captured during the warm season (May through September) did not differ significantly from equality. Seasonal differences in the proportion of large individuals of each sex (body weight $\geqslant 3$ kg) captured during activity are also suggested by the records. Relatively more large females (42 percent, N = 26) than males (23 percent, N = 13) were captured while foraging from December through March; whereas, the proportion of large males (90 percent, N = 10) exceeded that of females (77 percent, N = 13) from May to September. In both periods of the year, the majority of the animals was captured during the daylight hours, with relatively more being taken at night in the warmer months. Although differences in the actual age structure of the population at different times of the year complicates seasonal comparisons of activity patterns of young and adults, we gained the impression that there was a greater tendency for young armadillos to be more active and diurnal in cold weather than adults.

DISCUSSION

Even under the comparatively mild climate of southern Florida, armadillo activity patterns exhibit considerable daily and seasonal variation. Overall level of activity is reduced and the animals tend to be more diurnal in the cooler months, whereas total activity increases and the animals are more strongly nocturnal in the warmer part of the year. Temperature appears to be the major environmental factor influencing activity, the animals adjusting their activity to avoid low temperatures in winter and high temperatures in summer.

Low temperatures appear to have a more pronounced effect on activity than high temperatures. During the winter-early spring interval in 1973, total number of tracks per diel and the percentage of tracks made during the day clearly varied with temperature. Direct observations also showed that armadillos were less active and largely diurnal during the colder periods. However, despite the obvious general tendency for reduction in total activity and increase in the relative extent of diurnal foraging at low temperatures during this period, only about 40 percent of the variability in total activity level and 17 percent of the increased diurnality is accounted for by temperature effects. In addition, these activity parameters tend to be less strongly correlated with temperature during the coldest winter months (January and February) than in the warmer early spring months (March and April). These seemingly contradictory results are explained by the fact that, as a result of their inability to store food or become torpid to conserve energy, armadillos are often forced to forage during prolonged cold periods irregardless of environmental temperature; whereas, if intervals of mild weath-weather are frequent enough during the cooler time of the year, they can restrict their activity to periods of preferred temperatures.

Armadillos forced to be active in cold weather not only have the problem of high energetic costs of thermoregulation and muscular activity for foraging but may face food scarcity as well. Thus, under conditions of prolonged low temperatures, individuals may expend more energy in activity than they gain through their feeding. On several occasions in late winter and early spring (February and March), we observed armadillos feeding on the pulp of rotting oranges lying on the ground, suggesting a shortage of their usual animal foods. The effect of low temperatures in limiting availability of insects and other arthropods in winter and early spring may be exacerbated by the dry conditions that often prevail during that time of year.

The finding of weakened and dead armadillos during and following cold spells in Texas and Louisiana has been reported by several authors (Clark, 1951; Fitch et al., 1952; Kalmbach, 1943; Taber, 1945). Extensive mortality may be caused by cold weather. Fitch et al. (1952) estimated that up to 80 percent of the armadillo population was killed by an unusually severe freeze in Louisiana in 1948. Although large-scale deaths from cold weather have not been reported in Florida, we have observed emaciated and weakened animals in our study area after prolonged cold spells. These are most frequently young individuals. Peter Cone (pers. comm.)

noted fewer signs of armadillos in an area in west Florida following the severe winter of 1976-77, suggesting that the population may have been significantly reduced by the cold.

The sharp increase in average number of tracks per day on the census strip between late May and early June 1973 does not appear to be temperature related, as it occurred long after the advent of high, stable, summer temperatures. It probably is due, in part, to an increase in the population from recruitment of young rather than to an increase in activity level *per se*. The fact that all juveniles (body weight < 1 kg) captured during the study were taken from May to August supports this suggestion. However, an actual increase in activity as a result of breeding behavior of adults might be partly responsible for the high track counts in summer. Presumed breeding activity in this population has been observed in June and August. Moore (1968) also recorded a sharp increase in activity between late April and June in armadillos in an outdoor enclosure and suggested breeding activity as a possible cause.

We are unable to offer an explanation for the persistence of relatively pronounced day-to-day fluctuations in number of tracks in warmer months when there is very little variation in the daily temperature cycle. Variability in total diel activity in summer was not clearly related to the minor day-to-day temperature differences, occurrence or amount of rainfall, cloud cover, or other obvious environmental factors. Daily variation in the percentage of diurnal activity in summer appeared to be largely correlated with rainfall, with more daytime activity occurring on days with rain. Increased activity of armadillos after rain may be due to cooler temperatures, better foraging conditions, or both. The track counts in 1973 suggest a higher proportion of diurnal activity earlier in summer (June, 32 percent; July, 30 percent) than later (August, 14 percent). This may reflect the tendency for rain to come so late in the afternoon in late summer that armadillos are not stimulated to become active until near dark. It is also possible that greater food abundance in late summer as a result of a buildup of arthropod populations allows armadillos to more closely restrict their activity to the nighttime hours during that season.

Our data suggest that there may be sex and age differences in activity responses of armadillos at different seasons, with adult females and younger animals tending to be more active and perhaps more diurnal in cold weather than adult males. Young animals may be forced to forage more in unfavorable weather than adults because of lower fat stores (McNab, 1980) which provide both energy reserves and insulation. Adult females may have to forage more in cold weather than adult males because of the additional metabolic demands of gestation. Gause (1980, cited by McNab, 1980) found that pregnant females and immatures were the first individuals to begin foraging during prolonged cold periods in northern Florida. However, Burns and Waldrip (1972) captured more males than females during winter. Lehmann (1934, in Kalmbach, 1943) noted that females and young were more active during the day than males and that almost all individuals collected in late afternoon and early evening were females, while those collected later were mostly males. The time of year was not stated.

Daily and seasonal activity patterns of armadillos in this study were generally similar to those reported in other parts of the range in North America. Observations on wild armadillos in Texas and Louisiana showed them to have a readily modified activity pattern, with activity being reduced and becoming relatively more diurnal at low temperatures (Burns and Waldrip, 1971; Clark, 1951; Fitch et al., 1952; Kalmbach, 1943; Taber, 1945). In some regions, aridity as well as temperature may affect activity. Clark (1951) found that during a dry period in the eastern Edwards Plateau region of Texas armadillos emerged much earlier than under normal summer conditions, presumably as a result of food scarcity. In a radiotelemetric study of activity of armadillos in a large outdoor enclosure in Texas (Moore, 1968), the animals foraged in the afternoon and early evening in winter and spring and were strictly nocturnal in summer (June). They were also less active in winter, the time spent foraging increasing from 1 hr per animal-day in February to 5.06 hr per animal-day in June.

Seasonal changes in physiological parameters may occur in association with shifts in activity patterns in armadillos. Weiss and Wislocki (1956) demonstrated seasonal variation in hematopoeisis in the marrow of the dermal bones of the carapace, possibly under the influence of temperature; and Burns and Waldrip (1971) suggested that body temperature and heart rate may show seasonal differences reflecting changes in activity and movements.

Common long-nosed armadillos are reported to be primarily nocturnal in captivity (Anderson and Bernischke, 1966; Johansen, 1961; Prudom and Klemm, 1973; Yaksh et al., 1967). However, the study of Yaksh et al. (1967) is apparently the only investigation of activity of this species in captivity under controlled conditions. Activity of a group of five individuals housed together in a room with natural lighting and an ambient temperature range of 25.5° to 27.2°C was largely confined to the dusk and night period (1600 hrs to 0500 hrs), with a single, pronounced peak about 2200 hrs. Prudom and Klemm (1973) reported that feeding usually occurred between 1800 hrs and 2000 hrs.

In view of the common long-nosed armadillo's obvious preference for nocturnal activity given a suitable temperature regime and adequate food supply, it is of interest that a related species, *Dasypus sabanicola*, in Venezuela is reported to be diurnal in flood savanna habitats with a mild average annual temperature of 27°C and only slight seasonal fluctuation (Pacheco and Naranjo, 1978).

Few data on ambient temperatures at times of armadillo activity are available. Except for three instances of activity at temperatures below 0°C, armadillos in Moore's (1968) study avoided activity below 5°C. From mid February to early March, foraging occurred from -5° to 25°C, with no observed tendency for any preference within that range. From late March to late June, activity occurred from 10° to 40°C. Although the extreme temperatures at which animals were active in his study generally agree with those observed in the Florida population, Moore's data suggest a greater tendency for his subjects to avoid high rather than low temperatures, which is the reverse of our results. This difference may reflect physiological differences between the two populations, the environmental differences between the two study areas, or some effect of the enclosure or other experimental conditions on the behavior of the animals in Moore's study.

Because of the lack of a well-developed coat of hair, the common long-nosed armadillo must depend largely on changes in metabolic rate to maintain normal body temperature. It also is unable to avoid unfavorable environmental conditions by becoming torpid. Therefore, the ability to readily adjust its activity schedule to take advantage of the most favorable available temperatures is an important behavioral adaptation. According to Johansen (1961) and McNab (1980), *Dasypus novemcinctus* has a minimum resting metabolic rate of approximately 250 ccO_2/kg-hr occurring at a thermoneutral point of 30°C (Johansen, 1961) or over a thermoneutral range of 28° to 38°C (McNab, 1980). Rough estimates of the metabolic consequences of seasonal modification of activity patterns of armadillos in our study population can be derived from these resting metabolism-temperature data, times of activity as determined by track counts, and ambient temperatures during activity periods. In January, an armadillo confining its foraging to the peak activity period indicated by track counts (1400-1700 hrs, mean temperature 23°C) would expend only about 14 to 20 percent more energy (depending upon which metabolic data set is used) than it would during the peak activity period in July (2100 hrs to 0000 hrs, mean temperature 25°C). However, if it continued to adhere to the July activity schedule in January, the energetic cost of activity during the 2100 hr to 0000 hr period (mean temperature 15°C) would be 50 - 60 percent higher than during the 1400 hr to 1700 hr period. In addition to the higher ambient temperature, a further advantage of diurnal activity in cold weather may be absorption of radiant energy, which would further reduce metabolic costs of heat production. Our observations suggest that armadillos active on cold winter days often deliberately forage in open areas in direct sun. The metabolic advantage of being nocturnal in summer compared to that of being diurnal in winter is less obvious. In fact, an armadillo foraging during the hottest part of the day in July (1400 hrs to 1700 hrs, mean temperature 32°C) would theoretically expend about 24-68 percent less energy than by being active during the observed peak activity period (2100 hrs to 0000 hrs). However, considering the potential problem of dissipating excess heat produced by muscular effort during activity at high summer daytime temperatures, actual oxygen consumption of day-active animals might be significantly higher than the estimated value.

The fact that energy expenditure for nighttime activity in summer may be about the same or possibly even higher than the daytime level suggests that factors other than direct effects of temperature on metabolic rate may be responsible for increased nocturnal activity during this season. Relative humidity may be an important factor. Avoidance of low daytime RH in summer by being active at night would contribute to water conservation. Considering the high summer rainfall in the study area, the increased evaporative water loss experienced by armadillos active during the hot part of the day would probably have no appreciable effect on their water balance during this season. However, water conservation resulting from nocturnal activity may be of greater importance in spring, when the weather is typically warm and dry and the animals are thus more likely to be stressed by lack of moisture.

Nocturnal activity in summer may also be related to prey availability. The abundance of invertebrates, particularly soil-inhabiting arthropods, at or near the surface where armadillos forage, may be greater at night than during the day because of the inhibiting effect of high daytime soil temperatures. Some indication of the magnitude of daily and seasonal soil temperature variation in the study area is given by data from an open area with sparse grass cover located near the beginning of the track census strip. In July, average maximum and minimum soil temperatures at 5 cm, the approximate maximum depth of armadillo foraging, were 35.6° and 25.9°C, respectively. Greater food availability, in addition to energy conservation, may also be an advantage of diurnal activity in winter. Higher daytime soil temperatures during cold periods in winter may be more favorable for invertebrates and small lizards and snakes than low temperatures at night. For example, average maximum and minimum soil temperatures at 5 cm depth in January at the site described above were 22.6° and 14°C, respectively. Investigation of the influence of temperature on activity of a variety of arthropods in the same area as this study has shown that the general low temperature threshold for activity of many species is about 18°C (James Carrel, pers. comm.).

In his study of armadillos in an outdoor enclosure,

Moore (1968) estimated actual energetic costs of foraging at different seasons. His calculations were based on Johansen's (1961) resting metabolic data, a constant added to the resting rate at a given ambient temperature to represent increased metabolic rate due to activity, number of hours spent foraging as determined by automatic radiotelemetric monitoring of the movements of the animals, and air temperatures recorded at times of activity. The energetic cost of foraging varied from a maximum of 15.2 Cal/animal-hr in late February and early March over an ambient temperature range of -5° to 40°C to a minimum of 9.1 Cal/animal-hr in June over a temperature range of 20° to 55°C. The difference is relatively small and reflects the fact that the armadillos modified their activity schedules to avoid extreme temperatures.

The common long-nosed armadillo has well-developed nest-building habits (Clark, 1951; Kalmbach, 1943; Lehmann, 1934, cited in Kalmbach, 1943; Taber, 1945). This behavior and the labile activity pattern constitute important thermoregulatory adaptations. By utilizing well-insulated nests during periods of inactivity and selecting a temperature range that minimizes the metabolic cost of thermoregulation during activity, the animal is able to exist under a broader range of climatic conditions than would be permitted by its physiological mechanisms alone. Thus, these behavioral characteristics have undoubtedly played a key role in the success of the common long-nosed armadillo in expanding its geographic and ecological range in North America.

ACKNOWLEDGMENTS

We thank W. E. Haas, F. E. Lohrer, and C. E. Winegarner for help in making track counts, capturing armadillos, and providing observations on armadillo activity; Jane Thomason for preparation of the figures; and Dorothy E. Carter for typing the manuscript.

LITERATURE CITED

Anderson, J. M., and K. Benirschke.
1966. The armadillo, *Dasypus novemcinctus*, in experimental biology. Lab. Animal Care, 16: 202-216.

Bailey, H. H.
1924. The armadillo in Florida and how it reached there. J. Mammal., 5:264-265.

Breece, G.A., and J.L. Dusi.
1985. Food habits and home ranges of the common long-nosed armadillo *Dasypus novemcinctus* in Alabama. (this volume).

Burns, T. A., and E. B. Waldrip.
1971. Body temperature and electro-cardiographic data for the nine-banded armadillo *(Dasypus novemcinctus)*. J. Mammal., 52:472-473.

Clark, W. K.
1951. Ecological life history of the armadillo in the eastern Edwards Plateau Region. Amer. Midland Nat., 46: 337-358.

Fitch, H. S., P. Goodrum, and C. Newman.
1952. The armadillo in the southeastern United States. J. Mammal., 33:21-37.

Gause, G. E.
1980. Physiological and morphometric responses of the nine-banded armadillo, *Dasypus novemcinctus,* to environmental factors. Ph. D. Thesis, Univ. Florida, Gainesville. (Cited in McNab, 1980).

Johansen, K.
1961. Temperature regulation in the nine-banded armadillo *(Dasypus novemcinctus mexicanus)*. Physiol. Zool., 34:126-144.

Kalmbach, E. R.
1943. The armadillo: its relation to agriculture and game. Game, Fish and Oyster Comm., Austin, Texas, 61 pp.

Lehmann, V.
1934. Armadillo investigations, 1934. MS report filed in Fish and Wildlife Serv. (cited by Kalmbach, 1943).

McCusker, J.S.
1985. Testicular cycles of the common long-nosed armadillo *Dasypus novemcinctus*, in north central Texas. (this volume).

McNab, B. K.
1980. Energetics and the limits to a temperate distribution in armadillos. J. Mammal., 61: 606-627.

Moore, A. M.
1968. A radiolocation study of armadillo foraging with respect to environmental variables. Ph.D. thesis, Univ. Texas at Austin, 106 pp.

Neill, W. T.
1952. The spread of the armadillo in Florida. Ecology, 33:282-284.

Pacheco, J., and C. J. Naranjo.
1978. Field ecology of *Dasypus sabanicola* in the flood savanna of Venezuela. Pp. 13-15 *in* The armadillo as an experimental model in biomedical research. Sci. Publ. 366 Pan Amer. Health Org., Washington, D.C.

Prudom, A. E., and W. R. Klemm.
1973. Electrographic correlates of sleep behavior in a primitive mammal, the armadillo *Dasypus novemcinctus*. Physiol. and Behavior, 10:275-282.

Sherman, H. B.
1943. The armadillo in Florida. Fla. Entomol., 26:54-59.

Stevenson, H. M., and R. L. Crawford.
1974. Spread of the armadillo into the Tallahassee - Thomasville area. Fla. Field Nat., 2: 8-10.

Taber, F. W.
1945. Contribution on the life history and ecology of the nine-banded armadillo. J. Mammal., 26:211-226.

Weiss, L. P., and G. B. Wislocki.
1956. Seasonal variations in hematopoieses in the dermal bones of the nine-banded armadillo. Anat. Rec., 126:143-162.

Wolfe, J. L.
1968. Armadillo distribution in Alabama and northwest Florida. Quart. J. Fla. Acad. Sci., 31:209-212.

Yaksh, T., L. J. Peacock, and C. L. Darby.
1967. Group circadian activity cycles in the nine-banded armadillo. Paper presented at the Annual Meeting of the Georgia Academy of Sciences, April 1967.

FOOD HABITS AND HOME RANGES OF THE COMMON LONG-NOSED ARMADILLO *DASYPUS NOVEMCINCTUS* IN ALABAMA

[1]**Gary A. Breece and Julian L. Dusi**

Department of Zoology-Entomology, Auburn University, Auburn, Alabama. [1]Present address: Georgia Power Company Environmental, 791 DeKalb-Industrial Way, Decatur, Georgia 30030.

Abstract
Volumetric analyses of 50 armadillo stomachs collected during a 12-month period in southwestern Alabama revealed that animal material formed approximately 98 percent of the diet during fall, winter, and spring, and 89 percent in summer. Insects formed approximately 66 percent by volume in spring, 54 percent in summer, 55 percent in fall, and 29 percent in winter; percentage of vertebrates, consisting almost entirely of reptiles, increased in winter and the amount of plant material increased in summer. To estimate home range size, 26 adult and 26 juvenile armadillos were marked and observed for a 13.5-month period on a 83 ha portion of the study area. The average minimum home range was 1.47 ha (0.25 to 2.85) for juveniles and 3.51 ha (0.41 to 10.68) for adults.

Abstracto
Los análisis volumétricos del contenido estomacal de 50 armadillos, colectados durante un período de 12 meses en la parte suroccidental de Alabama, revelaron que el 98 por ciento de la dieta estuvo constituída por material de origen animal durante el otoño, el invierno y la primavera y el 89 por ciento durante el verano. Los insectos formaron, aproximadamente, el 66 por ciento, por volumen, de la dieta durante la primavera, 54 por ciento en el verano, 55 por ciento en el otoño y 29 por ciento en el invierno. El porcentaje de restos de vertebrados, consistentes casi por completo de reptiles, aumentó durante el invierno, mientras que el material vegetal se incrementó en el verano.

Resumo
Análises volumétricas de 50 estômagos de tatus coletados durante um período de 12 meses a sudoeste do Alabama, revelaram que material animal formou aproximadamente 98 percentagem da dieta durante o outono, inverno e primavera e 89 percentagem no verão. Insetos formaram aproximadamente 66 percentagem do volume na primavera, 54 percentagem no verão, 55 percentagem no outono e 29 percentagem no inverno. Á percentagem de vertebrados, consistindo quase que inteiramente de répteis, aumentou no inverno e a quantidade de material vegetal aumentou no verão. Para estimar o tamanho da área de dispersão "home range", 26 tatus adultos e 26 juvenís foram marcados e observados por um período de 13,5 meses, numa porção de 83 ha da área de estudo. A área média de dispersão mínima foi de 1,47 ha (0,25 a 2,85) para juvenís e 3,51 ha (0,41 a 10,68) para adultos.

The common long-nosed armadillo *Dasypus novemcinctus mexicanus* has spread from Mexico across the southeastern United States in the past 100 years and is now a part of the fauna of all the Gulf states, Oklahoma, Arkansas, and Georgia (Humphrey, 1974). This rapid dispersion, along with its unique appearance and burrowing activities, has nearly always created persecution by residents of the areas invaded. Home owners complain of burrows damaging lawns and gardens. Hunters commonly claim that armadillos destroy the nests of game birds, primarily the Bobwhite Quail *Colinus virginianus* and Wild Turkey *Meleagris gallopavo*, and are a distraction and nuisance during the hunting season. Biologists are concerned about possible detrimental effects of armadillos on ground nests of birds, on populations of other small vertebrates, and possible competition with native species. Although considerable research has been done with the embryology, physiology, and histology of the armadillo, few natural history studies have been published.

This study is based upon the examination of the contents of stomachs from 50 adult armadillos collected on the Fred T. Stimpson Sanctuary in southwestern Alabama between July 1972 and June 1973. It documents the food habits of the armadillo in a relatively new portion of its range. Additionally, home range and movement observations were made from July 1972 through August 1973 on an 83 ha portion of the sanctuary.

Kalmbach's (1943) volumetric analysis of 169 armadillo stomachs, which revealed that the diet of the armadillo in southeastern Texas was 93 percent animal matter, was the initial work on armadillo food habits and one of the most thorough analyses to date. In that study, insects accounted for 78 percent, arachnids 2 percent, millipedes and centipedes 6 percent, miscellaneous invertebrates 6 percent, vertebrates 2 percent, and vegetable matter 7 percent of the diet. Fitch et al. (1952) found similar results in Louisiana, except for an indication of somewhat greater reliance on vegetable material. Wirtz et al. (1985) found that ants predominated among food items in the stomach contents of 186 armadillos collected in Florida; insects accounted for 97 percent of the food items, other invertebrates 1 percent, and vertebrates less than 0.1 percent. Redford (1985) in a review of literature on armadillo food habits, suggested that *Dasypus* in central and South America might rely even more heavily on ants and termites than they do in North America. Baker (1943) analyzed 25 stomachs collected in Texas during May in an attempt to determine if armadillos preyed on the eggs of the Bobwhite Quail. Insects made up 77 percent of the total volume; annelida 14 percent, vertebrates 1 percent, and vegetable matter 3 percent. No sign of bird eggs was found. Studies by Kennamer and Lunceford (1973) in Mississippi and Breece and Causey (1973) in Alabama also indicated that armadillos were of little significance as predators on bird eggs. Newman and Baker (1942) observed an armadillo feeding on newborn cottontail rabbits *Sylvilagus floridanus* in Texas. Hamilton (1946) found that black persimmon *Diospyros taxana* was a favorite summer food of armadillos in Texas.

The initial work on armadillo home range size was done in southern Texas by Clark (1951), who estimated minimum home range sizes for 3 animals of 3.4 ha (8.38 acres), 4.8 ha (11.94 acres), and 2.1 ha (5.28 acres), an average of 3.4 ha (8.52 acres) based on a 28-day study. A juvenile on the same study area had a range of 0.3 ha (0.69 acres). The ranges for all four armadillos studied overlapped to some extent.

Fitch et al. (1952) suggested a range of 20.2 ha (50 acres) after using the greatest distance between observations as diameters of circular ranges. This figure is based on data from 13 marked animals, 7 of which were seen more than once. The greatest number of observations of one individual was 7, over a 4.5-month period.

Taber (1945) stated that a decided movement to higher and drier terrain occurred in the rainy winter and the opposite occurred during the dry season. He also noted movement from thickets to open areas after dark.

STUDY AREA

The 2,240 ha Fred T. Stimpson Sanctuary in southern Clarke County, Alabama, is in the Lower Coastal Plain Physiographic Province. The sanctuary was intensively managed for the eastern wild turkey *Meleagris gallopavo silvestris* and provided turkeys and white-tailed deer *Odocoileus virginianus* for restocking programs. The primary management practices were controlled burning, planting summer and winter food plots, and maintaining permanent openings. There was no commercial logging until the summer of 1972, when minimal logging was allowed to remove trees damaged by the southern pine beetle *Dendroctonus frontalis*.

The topography is generally steep hills forming alternating ridges and ravines. Elevation varies from 5 to 100 m above sea level. Approximately 200 ha of the Tombigbee River floodplain are within the sanctuary. Portions of the floodplain are bordered by steep bluffs that rise abruptly up to 100 m above the floodplain. The soils in the area are various types of loams. Several salt springs and the remnants of salt lakes are located on the sanctuary (Wheeler, 1948).

Approximately 45 percent (1,005 ha) of the sanctuary is pine forest, 20 percent (441 ha) mixed pine-hardwood forest, 30 percent (684 ha) hardwoods, and 5 percent (114 ha) grassed clearings (Kirk 1973). Ninety km (54 miles) of access roads are present.

METHODS

Food Habits.-- The contents of stomachs from 50 adult armadillos collected from the upland ridges and valley areas of the sanctuary were analyzed. Twelve stomachs were collected in the spring, 18 during summer, 14 during fall, and 6 during winter; stomachs were removed in the field and preserved in 10 percent formalin. Armadillos usually had fed for at least one hour before being killed. Stomachs that contained less than 5 cc of food material were not used. No stomachs from juveniles were used in the study, although brief examinations of several stomachs from juveniles revealed the same food items as the adults ate. Juvenile feeding activity began as much as an hour before that of adults during the late summer and fall months. Most armadillos collected during spring, summer, and early fall were shot at dusk; during the cooler months, armadillos were collected earlier in the afternoon because armadillos were not active in the late evening.

The contents of each stomach were sorted under a 10-X dissecting microscope and each food item was identified when possible. Most food items were intact enough so that identification to family was possible. Volumes of food items from each stomach were measured by the amount of water they displaced. Sand and fragments of leaves and twigs occurred in all stomachs, but were not considered food since they probably were ingested incidentally.

Home Range and Movements.-- The movement and home range study was conducted on a 83 ha portion of the sanctuary. On this portion, there were deep hollows alternating with ridges running from north to south. Each of the valleys between the ridges contains a creek bed. The southern part of the area is a wide bottom within the floodplain of the Tombigbee River which is covered with 15 to 60 cm of water in spring during continued periods of heavy rainfall.

Fifty-one armadillos were captured and marked during the study (26 adults and 25 juveniles). Twenty-one adults and 18 juveniles were observed on enough occasions to derive a home range area for each.

Armadillos were captured by hand or netted. Each was marked with a fluorescent orange letter painted on the dorsal portion of the pelvic plate. Date, capture location, sex and weight were recorded, and each animal was returned to its capture site and released after the paint dried. The painted symbols lasted for approximately 6 months. Greater wear of the symbols occurred during the winter when the armadillos spent most of their time in burrows. Any individual with a symbol difficult to recognize was recaptured and repainted. During the summer armadillos frequently wallowed in mudholes. The resulting film of mud over the painted symbols made recognition difficult. Four adults (numbers 2, 7, 15 and 32) had scars or abnormal band formations that made field identification possible without the painted symbols.

The observation of marked armadillos began in July of 1972 and continued to mid-August of 1973. Ten to 14 days per month were spent on the study area from April through November. No observations were made in December and 5 to 8 days per month were spent on the area from January through March. During spring, summer, and fall, observations began 3 hours before dark and lasted until poor visibility made identification difficult. Designated routes were followed so that all parts of the study area received approximately equal amounts of observation.

Grid points were established at 30 m intervals, on compass lines that were 90 m apart. Each grid point was flagged and marked with an identification symbol. The location of each armadillo observed was mapped by taking a compass bearing and pacing a distance from the nearest grid point. For each observation, the time, the armadillo's behavior and the weather conditions were recorded. Although some armadillos could be approached to within 3 m without alarming them, most observations which lasted more than a few minutes were made from 15 to 25 m to avoid disturbing the animal. Several armadillos were followed for 3 hours without being alarmed. The presence of unmarked armadillos was noted. Temperature on the study area was recorded continuously with a hygrothermograph placed on the ground in the study area.

Locations of all armadillos seen 3 or more times were plotted on a map of the study area. Outside locations for each animal were connected by straight lines to form polygons representing the minimum home range area. The distance between the two extreme points of the home range was used as the diameter of a circle to calculate the circular home range area (Cockrum, 1962).

RESULTS AND DISCUSSION

Food Habits Study.-- Armadillos forage nearly all of the time they are outside their burrows. Feeding activity was primarily observed on moist or upland areas in leaf litter or around the base of trees or fallen logs. On several occasions, armadillos were seen tearing bark from fallen pine trees and feeding on insects from under the bark. Little feeding activity was seen in wet areas or the floodplain. Such areas are seasonally inundated, resulting in a very sparse leaf litter that would contain little food material.

Approximately 91 percent of the volume of the diet for all seasons was invertebrate material; vertebrate and plant material each accounted for slightly less than 5 percent (Table 1).

Adult and immature beetles combined accounted for 24 percent of the annual diet (Fig. 1). Both forms of beetles were found in each stomach (Table 1), the adult forms being more abundant in spring and summer and the immature forms in the fall and winter (Fig. 2). In Alabama, most beetle eggs are deposited in the spring and summer. The larva develop in the summer and fall, and the beetles overwinter as larvae, to pupate and change to adults in the spring. Fitch et al. (1951) considered beetles to be the principal food of armadillos, accounting for over 56 percent of the diet in summer and fall, 25 percent in winter and 34 percent in spring. Kalmbach's study (1943) indicated a similar value, 42 percent of the total diet. Wirtz et al. (1985) found beetles in most (96 percent to 100 percent) of the armadillo stomachs collected during all seasons in Florida; click beetle larvae (Elateridae), and scarab beetle larvae (Scarabaeidae) were the most important in the diet, but 25 families of beetles, including both adults and larvae, were found. This study indicates that beetles were utilized to a lesser extent in Alabama, but were still the principal food item. Terrestrial forms of Scarabidae and Carabidae comprised the bulk of the adult beetles. Immature beetles were primarily Scarabidae and Carabidae, but also included Elateridae, Tenebrionidae, and Lampyridae.

The armadillos fed by digging and probing in the soil and leaf litter and around the base of stumps and logs, as well as by tearing apart decayed logs. Slow-moving adult

Table 1. Percent occurrence and volume of food items found in 50 adult armadillo stomachs collected on the Fred T. Stimpson Sanctuary in southwestern Alabama between July 1972 and June 1973. T = trace; — = not found; numbers in parentheses indicate the number of stomachs examined.

Food item	*Spring (12)* % Occurrence	% Volume	*Summer (18)* % Occurrence	% Volume	*Fall (14)* % Occurrence	% Volume	*Winter (6)* % Occurrence	% Volume	*Total (50)* % Occurrence	% Volume
Insect:										
Orthoptera										
Grasshopper	25.0	0.8	44.4	0.6	57.1	1.1	—	—	38.0	0.7
Katydid	8.3	T	5.6	T	14.3	0.7	—	—	8.0	0.2
Camel Cricket	—	—	44.4	1.9	21.4	0.6	—	—	22.0	1.0
Mole Cricket	8.3	0.6	—	—	—	—	—	—	2.0	0.1
Field Cricket	16.7	0.2	11.1	0.1	35.7	0.2	—	—	18.0	0.2
Roach	8.3	T	11.1	T	14.3	T	—	—	10.0	T
Isoptera										
Termite	41.7	6.9	50.0	0.3	50.0	0.1	50.0	0.9	48.0	1.3
Dermaptera										
Earwig	—	—	5.6	T	—	—	—	—	2.0	T
Hemiptera										
Stink bugs	—	—	11.1	T	28.6	0.6	33.3	0.3	20.0	0.3
Homoptera										
Cicada	—	—	5.6	T	7.1	0.1	16.7	0.1	6.0	T
Coleoptera										
Beetle Adults	100.0	21.8	100.0	11.8	100.0	9.5	100.0	4.1	100.0	12.1
Beetle Larvae	100.0	6.4	100.0	7.7	100.0	19.2	100.0	8.1	100.0	11.6
Lepidoptera										
Caterpillars	58.3	3.8	77.8	2.2	92.9	6.0	66.7	2.8	76.0	3.7
Diptera										
Diptera Larvae	33.3	1.9	72.2	1.5	50.0	0.4	83.3	4.2	58.0	1.4
Hymenoptera										
Ants	91.7	20.9	94.4	28.2	100.0	15.4	100.0	4.8	96.0	21.2
Other	16.7	0.4	16.7	0.1	14.3	0.1	16.7	0.1	8.0	0.1
Other Invertebrates:										
Millipedes	91.7	9.9	100.0	15.2	100.0	16.0	100.0	23.3	98.0	15.2
Centipedes	83.3	1.2	88.9	2.4	100.0	5.6	100.0	2.1	92.0	3.2
Spiders	50.0	2.1	72.2	1.0	78.6	0.8	100.0	2.5	72.0	1.2
Ticks	—	—	—	—	—	—	16.7	0.1	2.0	T
Harvestmen	16.7	0.1	44.4	0.3	28.6	0.1	—	—	28.0	0.2
Scorpions	33.3	0.8	61.1	0.7	57.1	0.6	16.7	0.1	48.0	0.6
Snails	41.7	1.1	55.6	0.8	25.7	0.5	50.0	0.4	46.0	0.7
Earthworms	25.0	0.8	44.4	0.9	57.1	9.1	66.6	8.9	46.0	4.1
Unidentified	100.0	18.0	100.0	13.2	78.6	5.4	100.0	13.3	94.0	11.4
Total Invertebrate		**97.8**		**88.9**		**92.7**		**76.3**		**90.6**
Vertebrate										
Frog	8.3	0.4	—	—	7.1	1.2	—	—	4.0	0.4
Salamander	—	—	5.6	0.5	14.3	0.2	—	—	6.0	0.3
Green Anole	8.3	T	—	—	42.9	4.9	50.0	23.7	20.0	3.3
Green Anole Egg	50.0	1.7	55.6	0.6	—	—	—	—	32.0	0.6
Skink	—	—	—	—	14.3	0.4	—	—	4.0	0.1
Snake	—	—	—	—	14.3	0.6	—	—	4.0	0.2
Unidentified	8.3	T	—	—	—	—	—	—	2.0	T
Total Vertebrate		**2.2**		**1.1**		**7.3**		**23.7**		**4.9**
Plant										
Wild grape	—	—	61.1	10.0	—	—	—	—	22.0	4.5

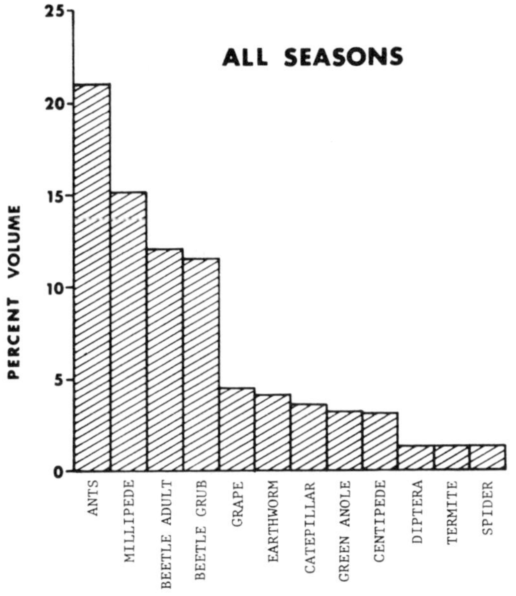

Figure 1. Percentage volume of the 12 most common food items occurring in stomachs of 50 armadillos collected in Alabama during all seasons, 1972 and 1973 combined.

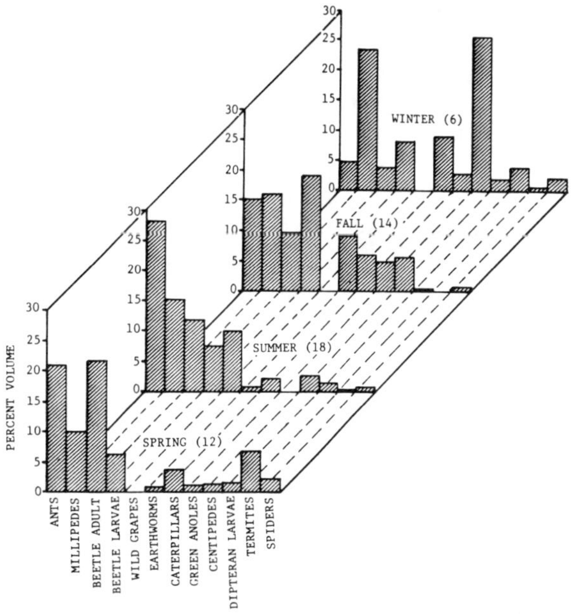

Figure 2. Seasonal breakdown of percentage volumes of the 12 most common food items in stomachs of 50 armadillos collected in southwest Alabama, 1972 and 1973 combined. Numbers in parentheses indicate the number of stomachs examined.

scarabs and ground beetles, commonly found in leaf litter, were probably easy prey. The immature beetle forms were most likely encountered in the soil, as well as in decaying wood of logs and stumps.

Ants occurred in 96 percent of the stomachs analyzed (Table 1), and accounted for 21 percent of the total annual diet (Fig. 1). Ants (Fig.2) were the most important in summer (28 percent) and least important in winter (5 percent). Previous studies (Kalmbach 1943; Fitch et al., 1952) show the same general seasonal trends. Kalmbach's (1943) study indicated that ants formed 14 percent of the diet for all seasons combined. Fitch et al. (1952) reported much lower values. In Florida, ants were found in most of the armadillo stomachs examined during all seasons, and were the most common type of food eaten (Wirtz et al., 1985). The local abundance of imported fire ants *Solenopsis invicta* may explain the greater importance of ants to the armadillo's diet in Alabama, although we did not identify ants to species. Fire ant mounds were abundant on the study area and nearly every mound observed during this study showed evidence of having been disturbed by armadillos. On two occasions, armadillos were observed tearing open mounds and feeding on fire ants.

Millipedes ranked third in volumetric contribution to the diet (Fig. 1), and although they were more important food items in the winter (23 percent), they were taken throughout the year, accounting for 10 percent of the diet in the spring, 15 percent in the summer and 16 percent in the fall (Fig. 2). Only one stomach examined did not contain millipedes (Table 1). Centipedes were found in 92 percent of the stomachs analysed, but were not as important as the frequency of occurrence would indicate, accounting for only 3 percent of the annual diet. Kalmbach (1943) found that millipedes and centipedes (Myriopoda) constituted 6 percent of the annual diet and were present in 151 of the 169 stomachs examined. Fitch et al. (1952) found that millipedes and centipedes were more important food items in the fall and winter; our data support this finding (Table 1).

Earthworms accounted for less that 1 percent of the diet in spring and summer, but increased in importance to 9 percent in both fall and winter (Fig. 2). The increase may relate to increased rainfall during the cooler months. The moist soil would draw the earthworms closer to the soil surface, making them more accessible to armadillos. Fitch et al. (1952) state that earthworms comprised 4 percent of the annual diet, but were more important in spring, accounting for 10 percent of the diet then. Kalmbach (1943) combined earthworms with miscellaneous invertebrates and thus reported no data specific to earthworms, although he mentioned that earthworms accounted for nearly one-fourth of the food in stomachs collected in January. He felt that soil fertility and moisture might account for conditions under which earthworms would be more easily obtained for foraging armadillos.

Caterpillars (lepidoptera larvae) occurred in 76 percent of the stomachs and accounted for 4 percent of the total volume (Table 1). During the fall, caterpillars were more important, accounting for 6 percent of the diet and occurring in 93 percent of the stomachs. No adult lepidopterans were found in the stomachs. Previous studies indicate caterpillars to be of greater importance, especially during the spring. Kalmbach (1943) found they comprised 8 percent of the total diet and Fitch et al. (1952) reported 5 percent.

Termites occurred in less than one-half of the stomachs analyzed, and can be considered important food items only during the spring months, when they comprised 7 percent of the diet (Fig. 2). During the spring, the stomachs contained large volumes of adult termites and eggs; only adults were found in the other seasons. Fitch et al. (1952) recorded low values throughout the year, but Kalmbach (1943) recorded 8 percent termites in May and 30 percent in October.

Spiders were frequent food items throughout the year but did not constitute an important food category (Table 1). Harvestmen, scorpions, and snails occurred with moderate frequency, but did not form a significant portion of the diet.

Reptiles and amphibians accounted for all the vertebrate food material. Green anoles *Anolis carolinensis* were found in 6 of 14 stomachs collected during the fall and 3 of 6 stomachs from the winter months. This lizard accounted for 24 percent by volume of the total winter diet and 3 percent of the annual diet. Only one green anole was found in a stomach collected in the spring and none in summer. These lizards are inactive during the cooler months and are probably unable to escape when an armadillo uncovers them. Lizard eggs, probably those of the green anole, were found in 6 of the 12 stomachs collected during spring and 10 of 18 stomachs collected in summer. None were found in fall or winter. Other vertebrates were found in the stomachs mainly in fall, and included the ground skink *Scincella laterale*, rough earth snake *Virginia striatula*, an unidentified tree frog *Hyla* sp., and a slimy salamander *Plethodon glutinosus*. Kalmbach (1943) and Fitch et al. (1952) found that reptiles and amphibians accounted for all the vertebrate material taken by armadillos; lizards were the most common vertebrate and were preyed upon most during colder weather. Lizard eggs occurred in stomachs collected during spring and summer in both studies.

Wild grapes *Vitis* spp. were the only plant food material found in the stomachs. Grapes occurred in 61 percent of the stomachs collected in the summer and 5 percent of the diet for all seasons (Table 1). Grapes occur only in the stomachs collected in summer because that is the only time of the year when grapes are on the ground and accessible to armadillos. Fitch et al. (1952) found plant material in nearly half of the stomachs examined, but at least some of this was incidentally ingested fragments of leaves and stems; these also occurred in all of the stomachs collected for this study but were not considered as food. Hamilton (1946) stated that the fruit of the black persimmon accounted for more than 80 percent of the stomach contents of armadillos collected in south Texas in August. Kalmbach (1943) reported a much lower value for plant material (2 percent), but mentioned that some fruits, especially blackberries *Rubus* spp., appeared to be sought after.

The food habits of the armadillos from Alabama show similar trends to studies from Texas, Louisiana and Florida. The diet consists primarily of animal material, mostly insects, of which beetles were most important. Ants, centipedes, and millipedes were relatively important invertebrate items in all studies; while vertebrates, primarily lizards, were important during the colder months. Small mammals and birds were of little or no significance. Vegetable material was limited to succulent fruits available during the summer months.

Based upon this and earlier studies, armadillos may best be described as nonselective opportunistic feeders that will ingest any food item that can be captured and which will fit into their mouths. The relative abundance of a food item is probably more important than the armadillo's preference for its taste in determining its importance in the diet.

The absence of bird eggs in the stomach analysis is significant in that it reinforces past studies that concluded that the armadillo is not a serious predator of ground nests. However, Breece and Causey (1973) found that predation on "dummy" quail nests may have been restricted to specific individuals and that the armadillos fed by breaking the egg and licking out the contents, rather than ingesting the entire egg. Thus, the absence of egg shell does not mean that the armadillo has not been eating eggs, and may invalidate the conclusion that the armadillo is not important as a predator on ground nests of birds.

Predation by armadillos on terrestrial vertebrate populations occurs in the cooler months and appears to primarily affect lizards, especially green anoles, more than any other group. The selectivity may be due to the lizard's poorly developed dormancy habits. During the winter, dormant green anoles can be found under pieces of bark, wood chips and other relatively exposed situations (Mount, 1975), rendering them vulnerable to armadillo predation. Most of these individuals would probably not survive the winter, so it is unlikely that the population would be adversely affected by the armadillos.

The similarities of armadillo food habits in different regions is expected, since its manner of feeding restricts it to the leaf litter and upper 5 to 10 cm of soil; a habitat characteristically occupied by arthropods and small vertebrates. Several mammals in the southern United States feed on the same organisms, but none have the specialized physical adaptations to be as efficient as the armadillo in exploiting this habitat zone. The two kinds of mammals which perhaps offer the most competition for food in this zone in the neotropics, the coatimundi *Nasua narica* and the vermilingua *Tamandua* are both absent from most of the armadillo's range in the United States. The long, stout front claws, pointed rostrum, and long sticky tongue, enable the armadillo to feed in a habitat that maintains an abundance of food items. This adaptation to occupy a relatively unexploited niche may partially explain the armadillos rapid rate of dispersion in the United States during the past several decades (Humphrey 1974).

Movements and Home Ranges.-- Average minimum home ranges for the 21 adult armadillos seen 3 or more times each were 3.5 ha (Table 2), while those of 15 juveniles were 1.5 ha (Table 3). There was a wide range of home range sizes (adults 0.4 to 10.7 ha; juveniles 0.2 to 2.8 ha). A large part of this variation was because the marked armadillos were neither seen an equal number of times nor observed for equal periods of time.

Home ranges estimated for juveniles were much smaller than comparable estimates for adults (Tables 2 and 3). The average number of observations was also distinctly lower. There was a trend for the home ranges of individuals seen most frequently to be larger.

Clark's (1951) 3.5 ha (8.5 acre) average home range for 3 adults was near the average value of 3.5 ha for this study. The 20.24 ha (50 acre) circular home range reported by Fitch et al. (1952) was twice as large as the 10.16 ha average determined in this study. This 20.24 ha (50 acre) figure was supported by only one individual with a range 505 m (1650 feet) long, or 1.21 to 8.50 ha (3 to 10 acres).

Adult home ranges overlapped extensively and there was no indication of territorial or aggressive interactions among the adult armadillos. Adults were often seen feeding within 3 m of each other, and once 3 adults were seen leaving one den.

On five occasions, adult armadillos were seen chasing juveniles. They made no attempts to harm the juveniles and the chases only lasted for 35 m. It could not be determined if the action was hostility or curiosity on the adult's part.

Population Dynamics.-- Fewer than half of the armadillos marked during 1972 were seen after the winter of 1972-73. Sixteen of the 26 adults marked between July and November 1972 were seen on the area the following spring and summer. Only 11 of the 25 juveniles marked in 1972 were observed on the area in 1973, after the winter. The other individuals either moved off the area after they were marked or died during the winter. Taber (1945) noted that many excavated burrows contained dead armadillos which had starved during a long period of below-freezing weather. This indicated that sick or weak armadillos may usually die in their

Table 2. Home range (HR) sizes for 21 adult armadillos marked in Alabama during 1972.

Months Observed	*Number of Observations*	*Minimum HR Area (ha)*	*Diameter of HR (m)*	*Circular HR Area (ha)*
Male				
July-June	10	2.83	304	7.29
Aug-June	25	8.90	408	15.33
Aug-June	10	4.57	403	12.78
Aug-June	7	1.32	182	2.63
Aug-June	6	2.82	472	17.52
Aug-June	6	3.24	287	6.22
Aug-Nov 72	6	2.61	274	5.91
Female				
July-Aug 73	61	10.68	457	16.47
July-Aug 73	59	10.66	411	13.29
July-Aug 73	10	2.09	465	16.93
Aug-July	16	2.14	381	11.39
Aug-June	13	0.81	296	4.37
Aug-June	9	2.82	457	16.41
Sep-June	7	3.08	502	19.85
Aug-April	9	1.25	213	3.57
Nov-April	5	0.82	79	2.99
Aug-Nov 72	10	1.44	274	5.90
Aug-Nov 72	5	8.00	579	26.32
Aug-Oct 72	4	1.40	152	1.82
Aug-Nov 72	4	1.98	274	5.90
Aug 72	3	0.41	190	2.84
Average	**14**	**3.51**	**1004**	**10.16**

Table 3. Home range (HR) sizes for 18 juvenile armadillos marked in Alabama during 1972 and 1973.

Months Observed	*Number of Observations*	*Minimum HR Area (ha)*	*Diameter of HR (m)*	*Circular HR Area (ha)*
Male				
Aug-April 73	18	2.16	183	2.63
Aug-Jan 73	4	–	175	2.40
Sep-May 73	6	2.55	404	12.78
Nov-June 73	10	1.35	233	4.28
Nov-June 73	5	1.06	312	7.65
July-Oct 72	3	0.25	128	1.28
Aug-Oct 72	5	2.13	206	3.31
Aug-Oct 72	5	0.91	91	0.66
Aug-Nov 72	3	–	213	3.57
Mar-June 73	10	2.03	259	5.27
Female				
Aug-June 73	3	2.26	206	3.31
Aug-June 73	9	2.85	381	11.39
Oct-May 73	4	–	383	11.51
Nov-June 73	9	1.07	160	2.00
Nov-June 73	7	0.57	137	1.48
Nov-June 73	5	1.06	312	7.65
July-Sep 72	5	0.30	122	1.17
Sep-Nov 72	3	0.55	328	7.29
Average	**6**	**1.47**	**687**	**4.78**

burrows, so that no trace of their bodies would be seen above ground. Clark (1951) reported a decline of similar magnitude in an armadillo population in Texas during the fall and winter of 1947.

One adult female was observed feeding with 4 juveniles in mid-June before the study began. Independent juveniles were observed in early July. The study was purposely continued through mid-August during the second year to gather information on reproduction. No juveniles were seen on the study area during the second year and only 2 lone juveniles were seen in other parts of the sanctuary. None of the female reproductive tracts examined contained embryos. No reason can be given for the low reproductive rate during the second year.

Activity Patterns.-- Juveniles were often seen feeding earlier in the day than adults. Several juveniles were sighted during late morning hours, while no adults were ever seen before noon during the summer. Both juveniles and adults were active during late afternoon. Activity was greatest from 2 hours before dark until 2 hours after dark. The majority of the sightings were made during the late evening hours, until darkness made identification impossible. Night observations were attempted, but were unproductive due to the difficulty of recognizing marked armadillos and in making accurate locations.

Armadillo activity decreased abruptly each day when the ambient temperature fell below 10°C. This was especially noticeable in late fall and early spring during periods of warm days and cool nights when abrupt temperature changes occurred. During a field study in Texas, no armadillos were seen above ground when temperatures were below 10°C; foraging was restricted to daylight hours and the number of days when armadillos were active was greatly reduced during the period October to March (McCusker, 1985). In Florida, most armadillo activity occurred at ambient temperatures above 20°C; armadillo activity was observed with temperatures as low as 6°C and as high as 35°C (Layne and Glover, 1985). In their study, peak activity throughout the year occurred when ambient temperatures were in the 20° to 25°C range, with consequent shortening of the daily period of activity in winter and more nocturnal and crepuscular activity in summer. In our study in Alabama, hot weather appeared to alter daily activities, especially during dry weather, and feeding activity usually began after sunset during periods when the peak daily temperature was above 32°C. During other times, activity began approximately 1 hour before sunset. In warm weather, strong winds tended to reduce activity and caused feeding armadillos to appear nervous and more wary than usual, while cold winter winds halted activity, except in protected areas. Rainfall had no noticeable effect on above-ground activities. Armadillos were observed on several occasions feeding during severe lightning and thunder storms and appeared to have no reaction to the disturbances caused by the storm.

Importance of Burrows.-- The burrows are probably an important home range component. Burrows were usually dug in the ground, but were also made under logs and in the bases of hollow trees. Usually, an armadillo would be within 60 m of a burrow and run straight to it when frightened. This was not always the case and, on several occasions, frightened armadillos ran past a burrow they had entered before in order to reach a burrow located further away. On many occasions, a frightened armadillo ran past several active burrows before finally taking shelter in another burrow. This possibly indicated that armadillos utilize only certain burrows within their home range, although many more are available. Usually, a frightened armadillo retreated to the burrow it had last occupied, although this was not always the case.

ACKNOWLEDGMENTS

The data collected in this study were used as part of a M.S. Thesis at Auburn University (Breece 1974). We acknowledge the aid and advice given by Messrs. Jim Davis, Bill Hamrick and Jim Keeler of the Alabama Department of Natural Resources, Division of Game and Fish and personnel of the Fred T. Stimpson Sanctuary, especially Messrs. Huey Dykes and Fred Pringle. We are grateful to Dr. Floyd Scott for help in identifying insect food items.

LITERATURE CITED

Breece, G.A.
1974. The ecology of the nine-banded armadillo in south-west Alabama. M.S. Thesis, Auburn University, 40 pp.

Breece, G.A.and M.K. Causey
1973. Armadillo depredation on "dummy" bobwhite quail nests in southwest Alabama. Proc. 27th Annual Conf. Southeastern Game and Fish Commissioners: 18-20.

Baker, R.H.
1943. May food habits of armadillos in eastern Texas. Amer. Midland Nat., 29: 379.

Clark, W.K.
1951. Ecological life history of the armadillo in eastern Edwards Plateau Region. Amer. Midland Nat., 46: 337-358.

Cockrum, E.L.
1962. Introduction to Mammalogy. The Ronald Press, New York, 463 pp.

Fitch, H.S., P. Goodrum and C. Newman
1952. The armadillo in the south-eastern United States. J. Mamm., 33: 21-37.

Hamilton, W.J.
1946. The black persimmon as a summer food of the Texas armadillo. J. Mamm., 27: 175.

Humphrey, S.R.
1974. Zoogeography of the nine-banded armadillo in the United States. Bioscience, 24(8): 457-462.

Kalmbach, E.R.
1943. The armadillo: its relation to agriculture and game. Texas Game, Fish and Oyster Comm. Bull., 61 pp.

Kennamer, J.E. and W.H. Lunceford, Jr.
1973. Armadillos tested as potential egg predators in the Mississippi Delta. Pp. 175 to 177 *in* Wild Turkey Management. (G.C. Sanderson and H.C. Schultz, editors). Univ. of Missouri Press, Columbia, 355 pp.

Kirk, J.P.
1973. Food availability and utilization by eastern wild turkey polts (*Meleagris gallopavo silvestris*) on the Fred T. Stimpson Sanctuary. M.S.Thesis. Auburn University, 76 pp.

Layne, J.N. and D. Glover.
1985. Activity patterns of the common long-nosed armadillo *Dasypus novemcinctus* in south-central Florida. (this volume).

Mount
1975. The reptiles and amphibians of Alabama. Auburn University Agricultural Experiment Station, 347 pp.

Newman, C.C. and R.H. Baker
1942. Armadillo eats young rabbits. J. Mamm., 23: 450.

McCusker, J.S.
1985. Testicular cycles of the common long-nosed armadillo *Dasypus novemcinctus*, in north central Texas. (this volume).

Redford, K.H.
1985. Food habits of armadillos (Xenarthra: Dasypodidae). (this volume).

Taber, F.W.
1945. Contributions on the life history and ecology of the nine-banded armadillo. J. Mamm., 26(3): 211-266.

Wheeler, R.J.
1948. Wild turkey in Alabama. M.S. Thesis, Auburn University, 92 pp.

Wirtz, W.O., D.H. Austin and G.W. Dekle.
1985. Food habits of the common long-nosed armadillo *Dasypus novemcinctus* in Florida, 1960-61. (this volume).

FOOD HABITS OF ARMADILLOS (XENARTHRA: DASYPODIDAE)

Kent H. Redford

Museum of comparative Zoology, Harvard University, Cambridge, Massachusetts USA 02138. (Present address: Florida State Museum, University of Florida, Gainesville, Florida USA 32611)

Abstract

The nine genera of armadillos can be divided into four groups according to modal feeding specializations. The carnivore-omnivores, which include *Zaedyus, Euphractus* and *Chaetophractus*, all generalist feeders, consume virtually any type of animal matter as well as many different types of fruit and tubers. The generalist insectivore (fossorial) group contains one genus, *Chlamyphorus*, with two species which consume insects and possibly some plant material. The generalist insectivore (terrestrial) group contains only *Dasypus*, a genus well studied in the southern United States. *D. novemcinctus* and the other species of *Dasypus* may feed more on termites than has been previously considered. Unlike the myrmecophagous armadillos, the long-nosed armadillos take larger animal prey and occasionally plant material. The ant- and termite-eater group consists of *Cabassous, Tolypeutes*, and *Priodontes*, which all eat almost exclusively ants and termites though *Cabassous* and *Tolypeutes* appear to take other arthropods more frequently than *Priodontes*.

Abstracto

Los nueve géneros de armadillos pueden ser divididos en cuatro grupos, de acuerdo a las especializaciones modales de alimentación. Los carnívoros-omnívoros incluyen los géneros *Zaedyus, Euphractus* y *Chaetophractus*, todos los cuales tienen alimentación generalizada, se comen virtualmente cualquier tipo de material animal, como también muchas clases diferentes de frutas y tubérculos. El grupo de insectívoros generalizados de cavadores, contiene un género, *Clamyphorus*, de dos especies que comen insectos y posiblemente algún material de origen vegetal. El grupo de insectívoros generalizados de hábitos terrestres contiene sólo al género *Dasypus*, un género que ha sido bien estudiado en el sur de los Estados Unidos. *D. novemcinctus* y las otras especies de *Dasypus*, tal vez se alimenten más de termitas de lo que se ha considerado previamente. A diferencia de los armadillos mirmecófagos, los armadillos de nariz larga capturan presas más grandes y, ocasionalmente, también ingieren material de origen vegetal. El grupo de comedores de hormigas y termitas consiste de los géneros *Cabassous, Tolypeutes* y *Priodontes*, todos los cuales comen casi exclusivamente hormigas y termitas, aunque miembros de los géneros *Cabassous* y *Tolypeutes* parecen tomar otros artrópodos con mayor frecuencia que los *Priodontes*.

Resumo

Os nove gêneros de tatus podem ser divididos em quatro grupos, de acordo com as especializações no modo de alimentação. Os carnívoro-onívoros, incluindo *Zaedyus, Euphractus* e *Chaetophractus*, todos generalistas quanto à alimentação, consomem virtualmente qualquer tipo de materia animal, bem como muitos diferentes tipos de frutos e tubérculos. O grupo generalista insetívoro (cavador) contém um gênero, *Chlamyphorus*, com duas espécies, que consomem insetos e possivelmente algum material vegetal. O grupo generalista insetívoro (terrestre) contém somente *Dasypus*, um gênero bem estudado no sul dos Estados Unidos. *D. novemcinctus* e a outra espécie de *Dasypus* podem alimentar-se mais de cupim do que têm sido considerado ate agora. Distintamente dos tatus mirmecófagos, os tatus de focinho longo caçam presa animal de maior tamanho e ocasionalmente material vegetal. O grupo que se alimenta de formigas e cupins é composto de *Cabassous, Tolypeutes* e *Priodontes* os guais todos se alimentam quase exclusivamente de formigas e cupins, embora *Cabassous* e *Tolypeutes* parece comer outros artrópodos mais freqüentemente do que *Priodontes*.

The armadillos, family Dasypodidae, include 18 extant species in 9 genera (Wetzel, 1985). They demonstrate the widest range in size, geographic distribution, and food habits of any of the three families of xenarthrans. Very little systematic work has been done on the food habits of armadillos; the bulk of information available consists of casual observations or examinations of the stomach contents of only a few animals. Still, it is instructive to gather what information is available in order to examine overall patterns in armadillo trophic specialization.

The armadillo genera show a range of trophic specialization (Table 1) from the generalized carnivore-omnivore through the generalist insectivore to the specialist insectivore (ants and termites). The placement of each genus within a food speciality group was done using information from published reports on food habits of wild animals, food preferences of captive animals,

Table 1. A summary of the food habits of armadillos. *Dasypus* may tend more toward being a specialist on social insects in Central and South America, where termites are more abundant during all seasons. Specialist insectivores feed mainly on ants, termites, or both.

Carnivore-omnivore	*Insectivore (generalist)*	*Insectivore (specialist)*
	Dasypus............................(*Dasypus*?)	
Euphractus		*Priodontes*
Chaetophractus		*Cabassous*
Zaedyus		*Tolypeutes*
	Chlamyphorus	

functional morphology and unpublished data from Paraguay and Brazil. Because of the paucity of information, this should be considered only as a broad outline of food habits within the dasypodids.

It is important to realize that what is being considered are "modal" feeding patterns (Eisenberg, 1978). There is probably a large degree of both temporal and geographical variation in the diet of most species. This variation is probably more pronounced among the carnivore-omnivores than in the other groups. Thus, the groupings of armadillo genera (Table 1) reflects not only the types of food taken, but also the variation in prey type.

CARNIVORE-OMNIVORES

The first group to be examined are the carnivore-omnivores, consiting of four or five species in three genera that comprise the tribe Euphractini; *Chaetophractus nationi* may be a high-altitude subspecies of *C. vellerosus* (Wetzel, 1985). These are the hogs of the armadillo world, eating whatever is available.

The only species in this group to have been well studied is *Chaetophractus vellerosus* (Greegor, 1985). The omnivory typical of the Euphractines was documented for the little hairy armadillo in northern Argentina (Greegor, 1974). In the study area in the Monte Desert, the major food item was insects (46 percent in summer and 26 percent in winter by volume), followed by a surprisingly large amount of plant material (51 percent in winter and 22 percent in summer). The plant material consumed during the winter was about 60 percent *Prosopis* pods, and the rest tubers and roots of other plants. The percentages of vertebrates identified in the diet of *C. vellerosus* (14 percent in winter, 28 percent in summer) are higher than any previously recorded for an armadillo. Rodents comprised 20 percent of the diet (by volume) in summer and 9 percent in winter (Greegor, 1985). This penchant for vertebrates is reflected in the fact that the stomach of one *Chaetophractus* collected in Paraguay contained the major portion of a small bird. In Paraguay *Chaetophractus* of both species were regularly attracted to and caught in large box traps baited with meat.

Though not nearly as well studied as *Chaetophractus vellerosus, Chaetophractus villosus* appears to be similarly catholic in its food habits; "it is omnivorous, and will thrive on anything from grass to flesh, found dead and in all stages of decay or captured by means of its own strategy" (Hudson, 1922). Specifically, Krumbiegel (1940) and Darwin (1933) recorded *C. villosus* feeding on "worms," carcasses, nestling birds and eggs, amphibians and smaller reptiles, beetles, larvae, and roots. Aplin (1908) comments on its preference for carrion, and Roth (1923) states that this species of armadillo, because of its habit of necrophagy, is not eaten by humans.

In most naturalists' accounts, distinction is not made between an animal being at a carcass to consume the flesh and being there to consume the insects attracted to the carcass. The jaws of members of the Euphractini are massive for an armadillo, so it is possible that they eat both flesh and insects at a carcass.

Hudson (1922) reports on the adroitness of a pet *C. villosus* in catching mice; sniffing them out in the grass and pouncing on them like a cat. He also reports, third hand, an episode in which a hairy armadillo killed a snake by squatting on it and sawing the serrated edges of its shell back and forth. When killed, the snake was partially eaten.

The second genus in the carnivore-omnivore group is *Zaedyus*, represented by the single species *Z. pichiy*. Simpson (1965) reports keeping an individual of this species to which putrefied meat was "nectar and ambrosia." Such a fondness for rotten flesh is reiterated by Walker (1975) and Kreig and Rahm (1961). They found *Z. pichiy* to be carnivorous first and omnivorous second, displaying a preference for the larvae and cocoons of ants. This fondness for insects is mentioned by Krumbiegel (1940) and documented by Meritt (1973), who found *Z. pichiy* to consume tarantula spiders and a species of soft-bodied grub. These armadillos also ate a large proportion of vegetable matter, particularly the pods of *Prosopis*.

Euphractus, represented by the single species *E. sexcinctus*, is the largest member of the carnivore-omnivore group, has the largest, most powerful teeth and uses these teeth to masticate more than any other armadillo (Moeller, 1975). *Euphractus* fights viciously with teeth and claws (Allen, 1916), and is the only armadillo that tries to bite when handled (D. A. Meritt, pers. comm.; pers. obs.).

Aplin (1894) accused *E. sexcinctus* of being less particular than any other armadillo in its diet, giving it the name "hog-armadillo." It feeds on carrion, small vertebrates, snails, worms and insects, particularly the adults, larvae and cocoons of ants (Walker, 1975; Krieg and Rahm, 1961; Moeller, 1975; Table 2). Schade

Table 2. Stomach contents of *Euphractus* and *Tolypeutes*, from Schaller (in press) and Redford (unpubl.).

Species	*Location*	*Stomach Contents*
Euphractus sexcinctus	Laguna Pora, Paraguay	2 termites; 4 Coleoptera larvae; fruit husk.
	Pocone, Brazil	95 percent ants, ant eggs, larvae and dirt; Cerambycidae beetle; Cuculionidae beetle; fruit pulp and seeds; masticated frog.
	Fazenda Acurizal, Mato Grosso, Brazil	95 percent palm fruit; 5 percent insects (cricket, spiders).
		100 percent turnip-like tuber.
		100 percent seed husks and fibers.
		95 percent fruit and fibers; 5 percent leeches and insects (beetle, grasshopper).
		33 percent fibers; 66 percent insects (mostly larvae).
		90 percent fibers and fruit rinds; 10 percent insects (ant, beetle, grubs).
		40 percent fibers; 60 percent insects.
Tolypeutes matacus	Chaco, Paraguay	Ants, ant pupae and eggs.
		Ants, ant eggs and larvae; termites.
		Ants; termites; beetle larvae.
		Ants; spider.
		Termites.
Tolypeutes sp.	Fazenda Acurizal, Mato Grosso, Brazil	Grubs; spider; ants; termites.
		Termites; spider.
		Ants; termites.

(1973) thought that *Euphractus* dug into *Atta* (leaf-cutter ant) nests and consumed adults and brood. In Paraguay, holes are often visible in the sides of *Atta* colonies, though what species of animal digs them is unknown. *E. sexcinctus* also eats plant material, including roots and various fruits (Moeller, 1975; Walker, 1975; Table 2). Schaller (pers. comm.) reports bromeliad fruits to be a common item in the diet of this species, and it will unearth and consume the hazelnut-sized nuts of a low palm (pers. obs.).

Many armadillos, particularly the members of the carnivore-omnivore group, have been reported to eat plant roots and tubers. Several species of termites specialize on roots and tubers, and it is possible that in some cases the armadillo may be eating termites that inhabit the root and not the plant material itself. An example of this has been suggested for *Chlamyphorus truncatus* by Minoprio (1945), who was unable to determine whether this species was eating a plant root or the insects parasitic on the root. However in many areas of Brazil, farmers say that both *Dasypus* and *Euphractus* dig up and consume manioc tubers.

The species in this carnivore-omnivore group are catholic in their food habits. Animal matter ranging from ants and carrion to birds and mice is taken in combination with many types of plant material. As Greegor (1974) has well documented for *C. vellerosus* the occurrence and abundance of certain items in the diets of these carnivore-omnivore species varies greatly with season. When more work has been done on these species throughout their ranges, it will probably become evident that there are not only seasonal, but also geographic shifts in diets. *Chaetophractus, Zaedyus*, and *Euphractus* probably eat whatever they can whenever they can.

GENERALIST INSECTIVORES (FOSSORIAL)

The second group of trophic specialists within the Dasypodidae is the fossorial-insectivore group. This group contains the one genus *Chlamyphorus* represented by two species (*Burmeisteria* is subsumed under *Chlamyphorus*, see Wetzel, 1985). They are the least studied of the armadillos. Both species are small, fossorial animals found in the southern half of South America and were originally reported to eat roots and stems (Cabrera and Yepes, 1960). However, further work on *Chlamyphorus truncatus* demonstrated that this species eats beetles, ants, insect eggs and larvae, worms, snails, and perhaps some plant material (Minioprio, 1945; Walker, 1975; Moeller, 1975). In captivity they eat beetles, earthworms, putrefied meat, *Tenebrio* larvae and cracked corn soaked in milk (Rood, 1970; Minioprio, 1945; Meritt, 1976). Nothing has been published on the food habits of *Chlamyphorus retusa* though Walker (1975) suggests that its food in the wild must be close to that of *C. truncatus*. Certainly the similarity in morphologies would suggest a convergence in food habits. Their fossorial habit, small size and apparent preference for sandy soil certainly limits the types of food available to these fossorial insectivores.

GENERALIST INSECTIVORES (TERRESTRIAL)

The third group of armadillos contains only the genus *Dasypus*, with six species (Wetzel and Mondolfi, 1979;

Wetzel, 1985). Extensive work on the food habits of the common long-nosed armadillo *D. novemcinctus* in the southern United States was largely prompted by its alleged consumption of the eggs of game birds (Breece and Dusi, 1985; Wirtz et al., 1985).

The results of much of the early work were summarized by Talmage and Buchanan (1954) who relied heavily on the work of Kalmbach (1944). In a study of the stomach contents of 232 *D. novemcinctus* from Texas, Kalmbach was able to identify 488 different food items. By volume, 93 percent of the food was of animal origin; 78 percent of the total volume was insect material, 6 percent was other arthropods, 2 percent other vertebrates and 7 percent was plant matter. Comparing three different studies, Talmage and Buchanan found 90 percent animal matter in *D. novemcinctus* stomachs; 75 percent of the total diet was insects, over half of which were Coleoptera, chiefly Scarabidae.

Wirtz et al. (1985) reported a detailed analysis of the stomach contents of 186 *D. novemcinctus* collected in Florida. Plant material was found in 8 percent to 18 percent of the stomachs, depending on the season; vegetation was most often individual leaves, stems, or rootlets and may have been ingested incidentally along with animal food material. Arthropods, primarily insects, formed the bulk of the diet. Vertebrates, primarily frogs, toads, and lizards were found in more than 20 percent of the armadillo stomachs. Breece and Dusi (1985) did a volumetric analysis of the contents of 50 *D. novemcinctus* stomachs collected in Alabama, and found that animal material formed 98 percent of the diet during fall, winter, and spring and 89 percent in summer (fragments of leaves and twigs were not considered as food). Wild grapes occurred in 61 percent of the stomachs in summer and accounted for 5 percent of the diet for all seasons. Their study shows the diet to consist primarily of animal material, mostly insects, of which beetles were most important. Work by other investigators on North American populations of *D. novemcinctus* further documents this species' largely insectivorous habits (Moore, 1968; Layne, 1976; Clark, 1951; Baker, 1943).

As Talmage and Buchanan pointed out in 1954 "there is a decided difference between what an armadillo (*D. novemcinctus*) is capable of eating and what it does eat in its natural habitat." In fact, the common long-nosed armadillo seems to be willing to eat nearly anything on occasion. Hamilton (1946), Fitch et al. (1952), Duges (1902), Kalmbach (1944), Taber (1945), and Greegor (1974) all discuss the amount and importance of plant matter in the diet of *D. novemcinctus*. In a much-quoted report, Newmann and Baker (1942) report this species to have eaten baby rabbits. Other vertebrates reportedly eaten include dead birds (offered to captive animals, Taber 1945; Clark 1951), reptiles and amphibians (Kalmbach, 1944; Layne, 1976; Wirtz et al., 1985; Breece and Dusi, 1985) and young mice offered to captive animals, (Redford, pers. obs.).

Recent studies confirm that reptiles and amphibians formed a significant and seasonally consistent portion of the diet of *D. novemcinctus*. In Alabama they (primarily lizards) formed about 5 percent of the annual diet (Breece and Dusi, 1985), whereas in Florida lizards were found in the stomachs less frequently, but the remains of frogs and toads were found in about 20 percent of the stomachs examined (Wirtz et al., 1985).

Newmann (1913) and Nelson (1918; in Talmage and Buchanan, 1954) described *D. novemcinctus* as being chiefly, if not exclusively, myrmecophagous. In fact, Audubon and Bachman (1854) report them being kept as house pets in order to eat ants. Kalmbach (1944), found ants and/or termites in more than 126 of 169 stomachs and states "where ants are plentifully abundant, the armadillo's food preferences seem to agree closely with those of . . .the true anteaters. Under such conditions, these insects may be ingested by the thousands." He found 40,000 ants of three species in one *D. novemcinctus* stomach and 13,000 termites in another stomach. Common long-nosed armadillos appear to prefer some species of ants over others (see Talmage and Buchanan, 1954; Kalmbach, 1944; Maller and Kare, 1967).

Despite the occasional large number of ant and termite individuals found in *D. novemcinctus* stomachs and the reported fondness of these armadillos for them, ants and termites remain a fairly minor part of the overall diet of North American *Dasypus*. In Kalmbach's study (Kalmbach, 1944), Hymenoptera and Isoptera made up only 19 percent of the total diet. This could be explained by the relatively low biomass of ants and termites in the northerly extent of *D. novemcinctus*' range.

However, beetles were the principle food item in Alabama (Breece and Dusi, 1985), even on a study area where there was a local abundance of fire ants (*Solenopsis*) in summer when ant populations might have been most accessible. Termites were important there as food only in spring, when they comprised 7 percent of the diet, while ants were found in 96 percent of the stomachs examined, but accounted for only 21 percent of the diet (by volume). In Florida, where ants accounted for about 53 percent of all food items identified, the average numbers of ants (adult, larvae, or eggs) per stomach varied from 379 in winter to 600 in spring (Wirtz et al., 1985).

Further study of *D. novemcinctus* and other species of *Dasypus* in South and Central America may show that members of this genus feed largely on a combination of ants and termites where termites are abundant. Hints of the greater dependence on social insects that may be found in Central and South American *Dasypus* come from several sources. Stomachs of two *D. novemcinctus* examined by Anthony-Mathews (1977) in Brazil contained large numbers of termites. In a field study of *D. sabanicola* in Venezuela, Pacheco and Naranjo (1978) found stomach contents to be 45 percent termites, 22 percent ants, 18 percent coleopterans (including larvae) and 15 percent "hymenopterans and others." Preliminary examination of fifteen *D. novemcinctus* stomachs from central Brazil shows them to be largely filled with ants and termites. Termites are less common and less accessable to foraging armadillos in the southern United States than in Central and South America. Schade (1973) reports many holes dug into the sides of *Atta* mounds, many of which he states were made by foraging *Dasypus* (foraging by *Dasypus* was not observed).

Patterson (1975), examining the morphology and biology of *Dasypus* said that this genus stood "poised on the road to myrmecophagy." In a comprehensive review of this genus, Wetzel and Mondolfi (1979) reiterate this, commenting that the long-nosed armadillos are probably fairly committed to myrmecophagy. Neither author distinguishes feeding on ants from feeding on termites, however. This conclusion is based on the cranial and dental morphology of members of this genus: they have a mandible proportionally smaller than any other armadillo; they have fewer teeth than any other armadillo; their teeth are smaller in proportion to the skull than any genus other than *Priodontes*; there are frequent gaps in adult toothrows and they have a poorly defined mandibular fossae. All of these factors are correlated with mammalian forms feeding on ants, termites, or both, and lead Wetzel and Mondolfi (1979) to speculate that "*D. novemcinctus* and other members of the genus are probably more dependent upon food requiring less mastication, such as softer and/or smaller items, than are most armadillos." However, in addition to ants and termites and their brood, many of the food items reported by Wirtz et al. (1985), Breece and Dusi (1985) and others are soft-bodied and do not require much mastication.

It seems probable that armadillos of the genus *Dasypus* are primarily insectivorous, with a large proportion of these insects being ants and termites. The low proportion of termites in the diet of the northerly populations of *D. novemcinctus* may be due to the relative scarcity of these social insects in the north.

ANT AND/OR TERMITE SPECIALISTS

The fourth trophic category of armadillos is composed of three genera, *Cabassous, Tolypeutes*, and *Priodontes*, which probably specialize in feeding on ants and termites.Very little work has been done on any of the three species of the genus *Cabassous*. Using data from field work and descriptive morphology, Patterson (1975) and Griffiths (1968) both concluded that the genus *Cabassous* is largely, if not completely myrmecophagous. Ingles' (1953) observations on an individual of *C. centralis* were expanded upon by Meritt in a field study in Panama (Meritt, 1985). This species uses its strong forelegs to dig into ant and termite nests, then extracts the prey with its long tongue. Heck (1920), Krumbiegel (1940), and Anthony-Mathews (1977) reporting on various species of *Cabassous* note the virtual lack of anything other than ants and termites in the stomachs of those individuals examined. Analysis of feces of several *C. unicinctus* from central Brazil revealed only ant and termite fragments. An interesting exception to this pattern of myrmecophagy is an unsubstantiated report of *Cabassous* feeding on small mice (Heck, 1940).

Anthony-Mathews (1977), commenting on other arthropods found in the stomach of a *Cabassous* that had been feeding almost exclusively on one species of termite, points out that the very low numbers of other species of beetles, ants, and termites were probably ingested when the animal was feeding on the termites whose mound it had opened. The occurrence of many species of ants, termites, and other arthropods in association with the nest of a mound-building species of termite should be kept in mind when interpreting the stomach contents of myrmecophagous mammals.

The second genus in this group of ant and termite feeders is *Tolypeutes*, the three-banded armadillo. Examination of stomach contents by several field workers (Eisentraut, 1952; Meritt, 1973; Table 2) confirms that they feed primarily on ants and termites. Prey other than ants and termites found in *Tolypeutes* stomachs includes spiders, beetle larvae, and worms. Meritt (1973) reports *T. matacus* to eat available carrion.

One of the prominent features of this genus is the enlarged third claw on the forefeet. In an exhaustive monograph on the morphology of *Tolypeutes*, Murie (1875) states that he doubts if this animal uses its forefeet for digging, as they appear to be very weak. This speculation is borne out by observations of animals foraging in the wild. *Tolypeutes* appears to dig only shallow foraging holes and, according to reports, does not dig its own burrow, using instead holes dug by other armadillos. Further work should resolve this apparent paradox between possession of forefeet morphologically suited for powerful digging and the absence of the necessary muscles and behaviors to use these specialized forefeet.

The third genus of ant- and termite-eaters is the giant armadillo, *Priodontes maximus*, the most myrmecophagous of the armadillos. It is an extremely good digger (Pocock, 1924; Morris, 1965), capable of destroying large termitaria (Moeller, 1975, pers. obs.). Ants, termites, and their larva and eggs are virtually all it is reported to eat (Krumbiegel, 1940; Krieg and Rahm, 1961; Heck, 1920; Krieg and Rahm, 1962; Walker,

1975; Kuhlhorn, 1939; Beebe, 1922; Moeller, 1975). As in *Cabassous* and *Tolypeutes, Priodontes* is reported to take other arthropods (Gijzen, 1965), although it is possible that these other arthropods were taken while the armadillo was foraging on ants and termites.

Walker (1975) reports *Priodontes* as eating snakes and carrion, while Hanif and Poonai (1968) report it to eat carrion and aquatic fauna. Azara (1801) even states that *Priodontes* frequently digs up graves in order to consume the corpses.

Captive *Priodontes* refused to eat a standard armadillo diet, eating only a similar but more souplike mixture, and refused mice, chicks, young rats, and cooked sweet potato (Meritt, 1976). This refusal to eat food that requires much chewing is another indication of the degree to which the giant armadillo is committed to myrmecophagy. Patterson (1975) reviewed and expanded upon Griffiths' (1968) summary of the characteristics of mammalian myrmecophages. *Priodontes*, according to Patterson, demonstrates the extent of its myrmecophagy by possessing certain of these morphological characteristics. It is a powerful digger, possesses a vermiform tongue and large salivary glands, has the glenoid articulation low on the skull with the posterior portion of the mandible having undergone reduction,and has small, highly variable teeth. Wetzel (1985) and Patterson (1975) comment on the highly variable number of teeth in *Priodontes* and Lönnberg (1942) lists a range in tooth number from various individuals of 65 to 98. Apparently there is little or no selection operating on the tooth row, consistent with their high level of myrmecophagy. Kuhlhorn (1938) provides a more detailed examination of the adaptations associated with myrmecophagy in the giant armadillo.

Cabassous, Tolypeutes, and *Priodontes* eat mostly ants and termites. Patterson (1975) felt that the first two genera are less specialized as myrmecophages than *Priodontes*, largely based on morphological evidence. *Tolypeutes* and *Cabassous* probably eat a wider range of foods than *Priodontes*, though all three genera confine themselves to small, soft foods that require little or no chewing.

From what little is known, the three genera of myrmecophages may obtain their ant and termite prey in different ways. *Cabassous* digs holes into ant or termite mounds or places where social insects are foraging (Ingles, 1953). *Tolypeutes*, on the other hand, digs shallow holes in search of its prey, while the giant armadillo either digs very large shafts into nests or completely destroys them. These different foraging styles may reflect not only the size and morphology of the armadillo, but also different species of prey (Redford, in press).

ACKNOWLEDGMENTS

I thank the following for making possible my research in Paraguay and Brazil: the Anderson, Richmond and Barbour funds of Harvard University, the Harris Foundation, Ralph Wetzel of the University of Connecticut, the Organization of American States, the National Geographic Society, and Sigma Xi. Valuable help was received from Frank and Barbara Rivera, Karen Anderson, Ralph Wetzel and George Schaller. I thank in particular George Schaller for his hospitality and permission to use unpublished data. Grateful thanks are extended to the following for encouragement and critical reading of the manuscript: D. Caldwell Hahn, John F. Eisenberg, Ralph M. Wetzel, G. Gene Montgomery and Kenneth P. Sebens.

LITERATURE CITED

Allen, J.A.
1916. Mammals collected on the Roosevelt Brazilian Expedition, with field notes by L.E. Miller. Bull. Am. Mus. Nat. Hist., 35: 559-610.

Anthony-Mathews, A.G.
1977. Studies on termites from the Mato Grosso State, Brazil. Academia Brasileira de Ciências, Rio de Janeiro. 267 pp.

Aplin, O.V.
1894. Field notes on the mammals of Uruguay. Proc. Zool. Soc. London, 1894: 297-315.

1908. Notes on the Tuco-tuco and the hairy armadillo. Zoologist, London, 12: 392.

Audubon, J. J. and T. Bachman.
1854. Quadrupeds of North America, III. V. G. Audubon, New York. 348 pp.

Azara, F. d'
1801. Essais sur l' histoire naturelle des quadrupeds de la province du Paraguay. Vol. 2. Paris. 499 pp.

Baker, R.H.
1943. May food habits of armadillos in eastern Texas. Am. Midl. Nat., 19: 379-380.

Beebe, W.
1922. The giant armadillo. Zool. Soc. Bull., 25: 117-120.

Breece, G.A., and J.L. Dusi.
1985. Food habits and home ranges of the common long-nosed armadillo *Dasypus novemcinctus* in Alabama. (this volume).

Cabrera, A. and J. Yepes.
1960. Mamiferos Sud Americanos. Carolos C. Wiedner, Buenos Aires. 238 pp.

Clark, W.K.
1951. Ecological life history of the armadillo in eastern Edwards Plateau region. Am. Midl. Nat., 46: 337-358.

Darwin, C.
1933. Journal of researches into the natural history and geology of the countries visited during the voyage round the world of H.M.S. "Beagle" under command of Captain FitzRoy, R. N. Edited by Nora Barlow, Macmillan, London. 451 pp.

Duges, A.
1902. Sur le tatou, *Cachicama novemcincta*. Mem. Soc. Cient. "Antonio Alzate" Mexico, 17: 35-42.

Eisenberg, J.F.
1978. The evolution of arboreal herbivores in the Class Mammalia. Pp 135 to 152 *in* The Ecology of Arboreal Folivores (G. G. Montgomery, editor). Smithsonian Institution Press, Washington, D.C. 574 pp.

Eisentraut, M.
1952. Vom Kugel-Gurteltier (*Tolypeutes conurus*) Natur. und Volk., 82: 43-48.

Fitch, H.S., P. Goodrum and C. Newmann.
1952. The armadillo in the south-eastern United States. J. Mamm., 33: 21-37.

Gijzen, A.
1965. Les tatous geants *Priodontes giganteus* Geoffroy et leurs apparentes. Zoo, (Antwerp), 31: 48-51.

Greegor, D.H.
1974. Comparative ecology and distribution of two species of armadillos, *Chaetophractus vellerosus* and *Dasypus novemcinctus*. PhD. thesis. Univ. of Arizona. 150 pp.

Greegor, D.H.
1985. Ecology of the little hairy armadillo *Chaetophractus vellerosus*. (this volume).

Griffiths, M.
1968. Echidnas. Pergamon Press. Oxford. 282 pp.

Hanif, M. and N.O. Poonai.
1968. Wildlife and conservation in Guyana. Man and Nature Ser., No. 8. 43 pp.

Hamilton, W.J.
1946. The black persimmon as a summer food source of the Texas armadillo. J. Mamm., 27: 175.

Heck, L.
1920. Brehm's Tierleben. Die Saugetiere. Erste Bande. Leipzig. 580 pp.

Hudson, W.H.
1922. The Naturalist in La Plata. E. P. Dutton and Co. New York. 394 pp.

Ingles, L.G.
1953. Observations on Barro Colorado mammals. J. Mamm., 34: 266-268.

Kalmbach, E.R.
1944. The armadillo: its relation to agriculture and game. Austin: Texas Game, Fish and Oyster Commission, Austin. 60 pp.

Krieg, H. and U. Rahm.
1961. Das Verhalten der Gurteltiere (Dasypodidae). Handb. Zool., 10(12c); 24-31.

Krumbiegel, I.
1940. Die saugetiere der Sudamerika - expeditionen Prof. Dr. Krieg. I. Gürteltiere. Zool. Anz., 131: 49-73.

Kuhlhorn, F.
1938. Das Fiesenguerteltier (*Priodontes giganteus* E. Geoffr.) als Anpassungsform. Zool Garten, 10: 107-114.

1939. Beziehunger zwischen Ernahrungsweise und Bam des Kauapparates bei einigen Gurteltier - und Ameisenbarenarten. Morph. Jahrb., 84: 55-85.

Layne, J.N.
1976. The armadillo. Fla. Nat., (June): 8-12.

Lönnberg, E.
1942. Notes on xenarthra from Brasil and Bolivia. Arkiv for Zoologi 34A, No.9: 1-58.

Maller, O. and M. Kare.
1967. Observations on the sense of taste in armadillos (*D. novemcinctus*). Anim. Behav., 15: 8-10.

Meritt, D.A.
1973. Note. *in* Yearbook Am. Phil. Soc. 1972: 383-384.

1976. The nutrition of edentates. International Zoo Yearbook., 16: 38-46.

Meritt, D.A., Jr.
1985. Naked-tailed armadillos, *Cabassous* sp. (this volume).

Minoprio, J.D.L.
1945. Sobre el *Chlamyphorus truncatus* Harlan. Acta zoologica Lilloana, 3: 5-58.

Moeller, E.
1975. Present-day edentates. Pp. 149 to 181, *In* Animal life encyclopedia. Mammals II. (B. Grizmek, editor) Van Nostrand Co. 635 pp.

Moore, A.M.
1968. A radiolocation study of armadillo foraging with respect to environmental variables. PhD thesis. Univ. Texas at Austin. 116 pp.

Morris, D.
1965. The mammals. Harper and Row, New York. 448 pp.

Murie, J.
1875. On the habits structure, and relations of the three-banded armadillo (*Tolypeutes conurus*, Is. Geoff.) Trans. Linn. Soc. Lond., 30: 71-132.

Newman, H.H.
1913. The natural history of the nine-banded armadillo of Texas. Am. Nat. 47: 513-539.

Newmann, C.C. and R.H. Baker.
1942. Armadillo eats young rabbits. J. Mamm., 23: 450.

Pacheco, J. and C.J. Naranjo.
1978. Field ecology of *D. sabanicola* in Venezuela. Pp. 13 to 17, *in* The armadillo as an experimental model in Biomedical research. Pan. Amer. Health Org. Sci. Publ., 366. 140 pp.

Patterson, B.
1975. The fossil aardvarks (Mammalia: Tubulidentata). Bull. Mus. Comp. Zoology, 147 No. 5. 67 pp.

Pocock, R.I.
1924. The external characters of the South American edentates. Proc. Zool. Soc. London, 1924: 983-1031.

Redford, K.H.
(in press). Patterns of mammalian myrmecophagy: ants and termites as food for mammals. Ann. Rev. Ecol. Syst.

Rood, J.P.
1970. Notes on the behavior of the pygmy armadillo. J. Mamm., 51: 179.

Roth, W.E.
1923. Richard Schomburgk's Travels in British Guiana 1840-1844. Daily Chronicle 1923, Georgetown. vol. 1, 402 pp; vol 2, 443 pp.

Schade, F.H.
1973. The ecology and control of leaf-cutting ants of Paraguay. Pp. 77-98, *in* Paraguay: Ecological Essays. (J. R. Gorham, ed.) Academy of Arts and Sciences of the Americas. Publisher, 269 pp.

Schaller, G.
(in press) Mammals and their biomass on a Brazilian ranch. Papeis avulsos de zoologia.

Simpson, G.G.
1965. Attending marvels. Time Inc. New York. 289 pp.

Taber, F.W.
1945. Contribution on the life history and ecology of the nine-banded armadillo. J. Mamm., 26: 211-226.

Talmage, R.V. and G.D. Buchanan.
1954. The armadillo (*D. novemcinctus*). The Rice Inst. Pamphlet col XLI No. 2, 135 pp.

Walker, E.P.
1975. Mammals of the World. Johns Hopkins Press, Baltimore. 1500 pp.

Wirtz, W.O., D.H. Austin and G.W. Dekle.
1985. Food habits of the common long-nosed armadillo *Dasypus novemcinctus* in Florida, 1960-61. (this volume).

Wetzel, R.M. and E. Mondolfi.
1979. The subgenera and species of long-nosed armadillos, genus *Dasypus*. Pp 43 to 63 *in* Vertebrate Ecology in the Northern Neotropics. (J. F. Eisenberg, editor). Smithsonian Institution Press, Washington D.C. 271 pp.

Wetzel, R.M.
1985. Taxonomy and distribution of armadillos, Dasypodidae. (this volume).

FOOD HABITS OF THE COMMON LONG-NOSED ARMADILLO *DASYPUS NOVEMCINCTUS* IN FLORIDA, 1960-61

W. O. Wirtz, David H. Austin and George W. Dekle

Department of Biology, Pomona College, Claremont, California 91711; Florida Game and Fresh Water Fish Commission, Avon Park, Florida 33825; Division of Plant Industry, Entomology Bureau, Florida Department of Agriculture and Consumer Services, Gainesville, Florida 32601.

Abstract

The armadillo in Florida is an opportunistic insectivore, taking other invertebrates and small vertebrates when available. Food habits of common long-nosed armadillos in Florida were determined on the basis of detailed analysis of contents of 186 stomachs, collected in all months of the year, chiefly from live oak hammock or long-leaf pine flatwoods. Over 167,000 food items were sorted, including 89,105 ants, 75,340 insects other than ants, 2,457 other invertebrates, and 147 vertebrates. The insects represented 11 orders, 75 families, and 90 genera; the other invertebrates represented 10 orders. Major items in the diet included Orthoptera (crickets, short-horned grasshoppers, cockroaches), Coleoptera (click and scarab beetle larvae, adult and larval ground beetles), Lepidoptera (larvae of common skippers, noctuid and acrolophid larvae), Diptera (crane fly, March fly, darkwinged fungus gnat, horse- and deerfly larvae), Hymenoptera (ants), Oligochaeta (earthworms), Diplopoda (millipedes), Chilopoda (centipedes), Araneida (chiefly lycosid spiders), and Anura (frogs). The greatest variety of food items was taken in October-December (74), but average number of prey per stomach was highest in April-June (1,535). Seasonal reproductive patterns of insects and prey immobility during cooler months probably influenced the type of prey taken at different times of year.

Abstracto

En Florida, el armadillo es un animal insectívoro oportunista, que toma otros invertebrados y pequeños vertebrados cuando están disponibles. Los hábitos alimenticios de los armadillos comunes de nariz larga de la Florida fueron determinados sobre la base de un análisis detallado del contenido de 186 estómagos colectados durante todos los meses del otoño, principalmente en las hamacas de encinas (live oak hammocks) o bosques de llanuras de pinos de hojillas largas. Se separaron más de 167,000 artículos individuales de alimentos, incluyendo 89,105 hormigas, 75,340 insectos de otras clases, 2,457 invertebrados de otros tipos y 147 vertebrados. Los insectos separados representaban a 11 órdenes, 75 familias y 90 géneros distintos; los otros invertebrados pertenecían a 10 órdenes. Los elementos principales de la dieta incluyeron a miembros de las Ordenes Orthoptera, como grillos, saltamontes de cuernos cortos y cucarachas; Coleoptera, como larvas de escarabajos de elatéridos y de escarabajos de tierra; Lepidoptera, como larvas de mariposas "skippers", de noctuídas y de acrolófidas; Diptera, como larvas de tipúlidos, de moscas de Marzo, de mosquitas de hongos y de tábanos; Himenoptera, como hormigas; Oligochasta, como lombrices de tierra; Diplopoda, como los milipies; Chilopoda, como los cenpies; Araneida, principalmente arañas licósidas y Anura, o ranas. La mayor variedad de elementos alimenticios aparecieron de octubre a diciembre (74), pero el número promedio de presas por estómago fue más alto durante los meses de abril a junio (1,535). Los patrones estacionales de reproducción en los insectos y la inmovilidad de la presa durante los meses de temperaturas más bajas, probablemente influyeron sobre el tipo de presa presente durante distintas épocas del año.

Resumo

O tatu, na Flórida, é um insetívoro oportunista, alimentando-se de outros invertebrados e pequenos vertebrados, quando disponíveis. Os hábitos alimentares dos tatus comuns de focinho comprido, na Flórida, foram determinados em base á análises minuciosas de 186 estômagos coletados em todos os meses do ano, principalmente em lombadas cobertas de carvalhos ou em planícies de pinheiros de folha comprida. Mais de 167.000 ítens de alimentos foram separados, incluindo 89.105 formigas, 75.340 insetos excetuando-se formigas, 2.457 outros invertebrados e 147 vertebrados. Os insetos representaram 11 ordens, 75 famílias e 90 gêneros, os outros invertebrados representaram 10 ordens. Os maiores ítens na dieta incluiram Orthoptera (grilos, gafanhotos, baratas), Coleoptera (larvas de besouros escarabeídeos e elaterídeos, larvas e adultos de besouros terrestres), Lepidoptera (larvas de diabinhos comuns, larvas de noctuídeos e acrofilídeos), Diptera (tipulídeos, bibionídeos, sciarídeos e larvas de mutucas), Hymenoptera (formigas), Oligochaeta (minhocas), Diplopoda (miriápodos), Chilopoda (centopéias), Aracnida (principalmente aranhas licósidas) e Anura (sapos). A maior variedade de ítens de alimentos foi consumida em outubro-dezembro (74) mas or número médio de presas, por estômago, mais alto foi em abril-junho (1.535). Padrões de reprodução sazonal de insetos e imobilidade de presas durante os meses mais frios provavelmente influenciaram no tipo de presa consumida em diferentes épocas do ano.

The common long-nosed armadillo, *Dasypus novemcinctus*, is distributed naturally from southcentral North America through Central America into South America (Wetzel, 1985). Within the last hundred years it has extended its range in Texas and neighboring states to the north and east (Bailey, 1905; Strecker, 1926; Hall and Kelson, 1959). Armadillos from Texas were introduced into Florida near Hialeah in Dade County in 1922 (Bailey, 1924), at Coca in Brevard County in 1924 (Fitch et al., 1952), and near Titusville in Brevard County in 1936 (Hardburger, 1950). By 1948, armadillos were present in all parts of nine Florida counties and portions of five others, with isolated records for 13 additional counties (Fitch et al., 1952). Expansion and northward extension of their range from both the Texas and Florida populations continues (Humphrey, 1974), and common long-nosed armadillos are currently found virtually throughout the southeastern United States and northward into Kansas and Missouri.

In 1959, concern about possible effects of armadillo depredations on ground-nesting game birds and agricultural crops (cf. Neill, 1952) led the Florida Game and Fresh Water Fish Commission to employ one of us (Wirtz) to conduct a study of the food habits of armadillos in Florida.

Table 1. Collection data for the 186 armadillos used in the study, expressed as percent of samples, except for the number of armadillos.

	Jan-Mar	*Apr-June*	*July-Sept*	*Oct-Dec*	*Total*
Number of Armadillos	52	49	35	50	186
County:					
Alachua				2	0.5
Hernando				8	2.2
Highlands	48	31	47	20	35.5
Marion	8		17	2	5.9
Osceola	4				1.1
Polk	19	59	11	10	25.8
Putnam	2				0.5
St. Johns	6		14	26	11.3
Volusia	13	10	11	32	17.2
Habitat:					
live oak hammock	62	47	66	60	58.2
long-leaf pine flatwoods	12	22	26	26	21.2
sandpine scrub	8			4	3.3
long-leaf pine turkey oak	12		8		1.6
river marsh/hammock	6	25		10	10.9
wiregrass-palmetto prairie		6			4.8
Time of Collection:					
0900-1259	4		3	20	11.1
1300-1659	54	16	12	51	49.1
1700-2059	42	68	76	29	38.0
2100-2459		16	9		1.8
Sex:					
male	46	62	31	47	47.5
female	54	38	69	53	52.5

METHODS AND MATERIALS

Between December 1959 and August 1962, 186 armadillos were collected, primarily by Austin and Wirtz, for analysis of stomach contents (Table 1); 86 percent of the sample was collected in 1960 or 1961. The sample was divided into 3-month periods for analysis; all periods are equally represented, except that only 35 animals were available for the July to September period. Specimens came primarily from Highlands (35.5 percent) or Polk (25.8 percent) counties in central Florida. Armadillos were most frequently taken in live oak hammock (58.2 percent) or long-leaf pine flatwoods (21.2 percent). Armadillos were more commonly observed in the afternoon (49.1 percent) or evening (38.0 percent); activity in the morning was most prevalent in cool months, while activity later at night occurred in warm months (Layne and Glover, 1985). Only adults were collected and males and females occurred approximately equally in the total sample, though males predominated in the April-June sample and females predominated in the July-September sample.

Stomach contents were sorted, counted, and identified; volumetric analysis proved impractical with such quantities of tiny organisms. Wherever possible, expert opinion was obtained in identification; Dekle did most of the insect identifications and Wirtz did all vertebrate identifications. We were unable to find persons interested in identifying members of numerous invertebrate orders, such as millipedes, centipedes, and arachnids. Preserved voucher specimens were retained by Wirtz.

RESULTS

A total of 167,039 food items was sorted from the 186 stomachs, or an average of 898 items per stomach. These items included 89,105 ants of 10 genera; 75,340 insects other than ants, representing 11 orders, 74 families and 80 genera; 2,475 other invertebrates representing 10 orders; and 147 vertebrates representing three orders (Table 2).

Centipedes (Chilopoda) occurred in over 50 percent of each seasonal sample, and were most prevalent in the October-December sample. The prevalence of millipedes (Diplopoda) was high in July-September and October-December samples (80 percent) and lowest in April-June (Fig. 1).

Spiders (Araneida), especially wolf spiders (Lycosidae) and some jumping spiders (Salticidae), were an important food item, especially in the April-June sample where they were present in 88 percent of the stomachs (Fig. 1). Spiders were frequently found singly, but some stomachs contained 10 to 20 individuals and one stomach held 71. Harvestmen (Phalangida) were found in eight stomachs; none were found in the July-September sample.

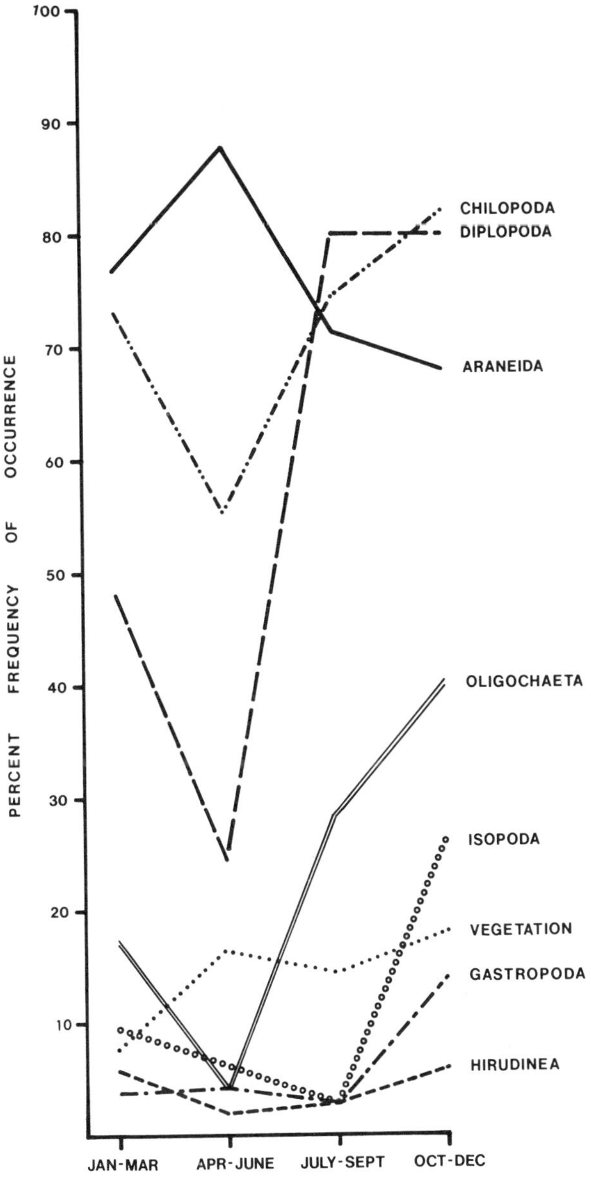

Figure 1. Seasonal occurrence of several invertebrate orders and vegetation found in stomachs of 186 armadillos in Florida.

Earthworms (Oligochaeta; Lumbriculidae) were found in 4 percent to 40 percent of the seasonal samples (Fig. 1) and were most prevalent in July-September and October-December. Maximum numbers of individuals per stomach were 72 in July-September and 25 in October-December.

Isopods (Porcellionidae) occurred in 3 percent to 26 percent of the samples (Fig. 1), being most prevalent in October-December. Isopods were usually found singly, but one stomach in the October-December sample contained 19 individuals. Only *Porcellio virgatus* was identified from the material collected.

Vegetation was present in 8 percent to 18 percent of the seasonal samples (Fig. 1), but most often as individual leaves, stems, or rootlets. Such vegetation may be incidentally ingested along with animal food material, and perhaps should not be considered as food (Breece and Dusi, 1985). The incidence of vegetation was highest in the October-December sample.

Small land snails (Gastropoda) were found in 12 stomachs and were most prevalent in October-December (Fig. 1). Leeches (Hirudinea) occurred singly in 8 stomachs, being more prevalent in January-March and October-December. Roundworms (Nematoda) were found in 6 stomachs; it is not known whether they were parasites or food items. Crayfish, *Cambarus*, occurred singly in 6 stomachs; none were found in July-September.

Among the vertebrates, anurans were most frequently taken (Table 3). None was found in the July-September sample, but they occurred in 24 percent to 29 percent of the remaining samples. Cricket frogs, *Acris*, and tree frogs, *Hyla*, predominated among the anurans taken. Toads, *Bufo*, were found in 2 stomachs; narrowmouthed toads, *Gastrophryne carolinensis*, in 3 stomachs; and true frogs, *Rana*, in 5 stomachs. Lacertilians were found in 4 percent to 26 percent of the seasonal samples and were most prevalent in October-December. Ground skinks, *Lagosoma laterale*, were found in 11 stomachs, 6 of them in October-December. The green anole, *Anolis carolinensis*, occurred in 9 stomachs, 8 of them in October-December. Two glass lizards, *Ophisaurus attentuatus*, were found in October-December, and single worm lizards, *Rhineura floridana*, were present in April-June and July-September. Hatchling water snakes, *Natrix*, and garter snakes, *Thamnophis* , were found in 4 stomachs. A complete fish skeleton 2 cm long was found in the October-December sample.

At least 24 reptile eggs were present in 14 stomachs; the seasonal frequency of occurrence varied from 2 to 16. Bird egg shells were present in two stomachs from April-June and one from the July-September sample. The absence of egg shell in an armadillo's stomach does not necessarily indicate that eggs have not been fed upon (Breece and Dusi, 1985), because they may break eggs and lick out the contents.

Among the insects, those orders most important as armadillo food were Coleoptera, Hymenoptera, Lepidoptera, Orthoptera, and Diptera (Fig. 2). Coleoptera were the single most important group, occurring in 96 percent to 100 percent of the seasonal samples. Hymenoptera, mainly ants, were next most important, occurring in 83 percent to 96 percent of seasonal samples. Lepidoptera were found in 77 percent to 94 percent of seasonal samples, Orthoptera in 68 percent to 92 percent, and Diptera in 48 percent to 92 percent. Four other insect orders (Homoptera, Hemiptera, Isoptera, and Dermaptera) were found regularly, but in seasonal frequencies below 14 percent.

Twenty-five families of beetles (Coleoptera) were

Table 2. Percentage frequency of occurrence of a total of 167,039 food items in 186 stomachs of the common long-nosed armadillo in Florida 1960-61.

Chilopoda, centipedes	71.0
Coleoptera, beetles	97.8
Alleculidae, comb-clawed beetles	12.9
Capnochra	
Buprestidae, metallic wood-boring beetles	0.5
Cantharidae, soldier beetles	1.6
Cantharis	
Chauliognathus	
Carabidae, ground beetles	37.1
Calosoma	
Harpalus	
Pterostichus	
Cerambycidae, long-horned beetles	10.8
Prionus	
Chrysomelidae, leaf beetles	4.3
Chlamys	
Curculionidae, snout beetles	21.5
Dermestidae, dermestid beetles	0.5
Dermestes	
Dytiscidae, predaceous diving beetles	1.1
Cybister fimbriolatus crotchi	
Elateridae, click beetles	84.4
Aeolus	
Alaus oculatus	
Ampedus	
Colaulon rectangularis	
Conoderus	
Ludius	
Melanotus communis	
Parallelostethus	
Endomychidae, handsome fungus beetles	0.5
Helodidae, marsh beetles	0.5
Prionocyphon	
Histeridae, hister beetles	1.1
Holsepta	
Saprinus	
Hydrophilidae, water scavenger beetles	2.2
Lagriidae, long-jointed bark beetles	3.8
Arthromacra	
Lampyridae, fireflies	4.8
Photinus	
Photuris	
Lucanidae, stag beetles	0.5
Meloidae, blister beetles	0.5
Phengodidae, glow-worms	3.2
Phengodes laticollis	
Ptilodactylidae, ptilodactylid beetles	0.5
Rhipiceridae, cedar beetles	0.5
Scaphidiidae, shining fungus beetles	1.1
Scarabaeidae, scarab beetles	89.2
Anomala	
Diplotaxis	
Phyllophaga glaberrina	
Strategus	
Staphylinidae, rove beetles	11.8
Philonthus	
Tenebrionidae, darkling beetles	4.8
Decapoda, crayfish	3.2
Dermaptera, earwigs	3.8
Labiduridae, long-horned earwigs	0.5
Labiidae, little earwigs	0.5
Diplopoda, millipedes	56.4
Diptera, flies	76.9
Bibionidae, March flies	28.5
Bibio	
Bombyliidae, bee flies	1.1
Calliphoridae, blow flies	1.6
Lucilia illustris	
Phormia regina	
Ceratopogonidae, biting midges	0.5
Ephydridae, shore flies	1.1
Lonchaeidae, lonchaeid flies	1.1
Mydaidae, mydas flies	1.1
Mydas clavatus	
Muscidae, muscid flies	1.6
Musca domestica	
Otitidae, picture-winged flies	1.1
Pseudotephritis	
Sarcophagidae, flesh flies	1.6
Sarcophaga	
Scatopsidae, minute black scavenger flies	1.1
Sciaridae, darkwinged fungus gnats	32.8
Sciara	
Stratiomyidae, solder flies	1.6
Tabanidae, horse and deer flies	22.6
Chrysops	
Tabanus	
Therevidae, stiletto flies	3.8
Psilocephala	
Tipulidae, crane flies	54.3
Tipula hermannia	
Gastropoda, slugs, snails	6.4
Hemiptera, bugs	6.4
Coreidae, leaf-footed bugs	1.1
Acanthocephala femorata	
Gelastocoridae, toad bugs	2.7
Gelastocoris oculatus	
Nerthra stygica	
Naucoridae, creeping water bugs	0.5
Pelocoris carolinensis	
Pentatomidae, stink bugs	2.2
Banasa packardii	
Mormidea lugens	
Nezara viridula	
Tetyra bipunctata	
Reduviidae, assassin bugs	2.2
Rhigidia cruciata	
Hirudinea, leeches	4.3

Table 2. Percentage frequency (continued).

Homoptera, cicadas, hoppers, scale insects	8.1
Cicadidae, cicadas	3.8
Margarodidae, giant coccids	0.5
Hymenoptera, bees, wasps, ants and allies	92.5
Apidae, bees	0.5
Diprionidae, conifer sawflies	1.6
Formicidae, ants	91.9
Aphaenogaster	
Camponotus abdominalis floridanus	
Camponotus socius	
Crematogaster ashmeadi	
Formica	
Neivamyrmex nigrescens	
Odontomachus brunneus	
Paratrechina longicornis	
Pheidole dentata	
Solenopsis	
Trachymyrmex septentrionalis	
Megachilidae, leafcutting bees	0.5
Sphecidae, sphecid wasps	0.5
Tenthredinidae, common sawflies	0.5
Isopoda, sow bugs	11.8
Isoptera, termites	3.2
Rhinotermitidae, subterranean and damp-wood termites	3.2
Lepidoptera, moths and butterflies	83.9
Acrolophidae, burrowing webworms	54.8
Acrolophus	
Arctiidae, tiger moths	3.8
Hesperiidae, common skippers	22.6
Hylephila phyleus	
Nepticulidae, nepticulid moths	0.5
Noctuidae, noctuid moths	34.4
Agrotis ypsilon	
Bleptina	
Epizeuxis	
Feltia ducens	
Feltia subterranea	
Litoprosopus futilis	
Mocis latipes	
Renia	
Spodoptera spp.	
Spodoptera eridania	
Notodontidae, prominents	6.4
Datana ministra	
Heterocampa manteo	
Lophodonta angulosa	
Schizura ipomeae	
Nymphalidae, brush-footed butterflies	0.5
Pyralidae, pyralid moths	15.6
Crambus	
Herpetogramma phaeopteralis	
Saturniidae, royal moths	0.5
Eacles imperialis	
Sphingidae, sphinx moths	1.6
Paonias excaecatus	
Tortricidae, tortricid moths	0.5
Nematoda, roundworms (endoparasites ?)	3.2
Neuroptera, fishflies, snakeflies, lacewings	1.1
Sialidae, alderflies	0.5
Chauliodes	
Odonata, dragonflies and damselflies	0.5
Oligochaeta, earthworms	22.0
Orthoptera, grasshoppers, crickets, cockroaches	80.6
Acrididae, short-horned grasshoppers	39.8
Melanoplus	
Romalea microptera	
Schistocerca	
Blattidae, cockroaches	32.8
Gryllidae, crickets	68.8
Gryllus spp.	
Nemobius ambitiosus	
Scapteriscus	
Gryllotalpidae	
Gryllotalpa hexadactyla	
Phasmatidae, walkingsticks	2.7
Tetrigidae, pygmy grasshoppers	4.8
Phalangida, harvestmen	4.3
Vegetation	14.0
Vertebrata	
Anura, frogs and toads	21.5
Bufonidae, toads	1.1
Bufo	
Hylidae, treefrogs	5.9
Acris	
Hyla	
Microhylidae, narrow-mouthed toads	1.6
Gastrophryne carolinensis	
Ranidae, true frogs	2.7
Rana	
Lacertilia, lizards	11.3
Amphisbaenidae, ringed lizards	1.1
Rhineura floridana	
Anguidae, glass lizards	1.1
Ophisaurus attentuatus	
Iguanidae, iguanid lizards	4.8
Anolis carolinensis	
Scincidae, skinks	5.9
Lagosoma laterale	
Serpentes, snakes,	2.2
Colubridae, colubrid snakes	2.2
Natrix	
Thamnophis	
Pisces, fish	0.5

Table 3. Seasonal percent frequency of occurrence/number of prey for vertebrate genera found in 186 armadillo stomachs from Florida.

	Jan-Mar	*Apr-June*	*July-Sept*	*Oct-Dec*	*Total*
Toads and Frogs					
Acris, Hyla	9.6/8	8.2/53		4.0/14	75
Bufo	3.8/3				3
Gastrophryne	2.0/1			4.0/2	3
Rana	3.8/2	4.1/2		2.0/1	5
Lizards					
Anolis		2.0/1		16.0/11	12
Lagosoma	3.8/2	4.1/2	2.9/1	12.0/7	12
Ophisaurus				4.0/2	2
Rhineura	2.0/1	2.9/1			2
Snakes					
Natrix, Thamnophis		4.1/2		4.0/2	4
Eggs					
Bird		4.1/4	2.9/1		5
Reptile	2.0/1	16.3/18	8.6/3	4.0/2	24

Table 4. Seasonal percent frequency of occurrence/number of prey taken of Coleoptera families found in 186 armadillo stomachs from Florida.

	Jan-Mar	*Apr-June*	*July-Sept*	*Oct-Dec*	*Total*
Alleculidae	15.4/52	4.1/3	14.3/22	18.0/178	255
Buprestidae	2.0/2				2
Cantharidae			5.7/2	2.0/1	3
Carabidae	32.7/147	36.7/97	82.8/299	10.1/8	551
Cerambycidae	13.5/11	12.2/10	11.4/12	6.0/3	36
Chrysomelidae	3.8/10			12.0/54	64
Curculionidae	34.6/329	28.6/40	11.4/4	8.0/185	558
Dermestidae				2.0/2	2
Dytiscidae		2.0/1		2.0/1	2
Elateridae	86.5/471	79.6/210	91.4/309	82.0/622	1612
Endomychidae			2.9/1		1
Helodidae				2.0/1	1
Histeridae		2.0/1	2.9/2		3
Hydrophilidae	1.9/1	2.0/1	2.9/1	2.0/1	4
Lagriidae	3.8/11		2.9/3	8.0/22	36
Lampyridae	3.8/2			14.0/21	23
Lucanidae			2.9/1		1
Meloidae	2.0/11				11
Phengodidae	1.9/5	2.0/1	2.9/3	6.0/3	12
Ptilodactylidae				2.0/3	3
Rhipiceridae		2.0/2			2
Scaphidiidae			5.7/3		3
Scarabaeidae	94.2/ 843	91.8/ 4221	94.3/ 2014	78.0/ 561	7639
Staphylinidae	13.5/35	6.1/5	20.0/13	10.0/15	68
Tenebrionidae		16.3/24	2.9/1		25

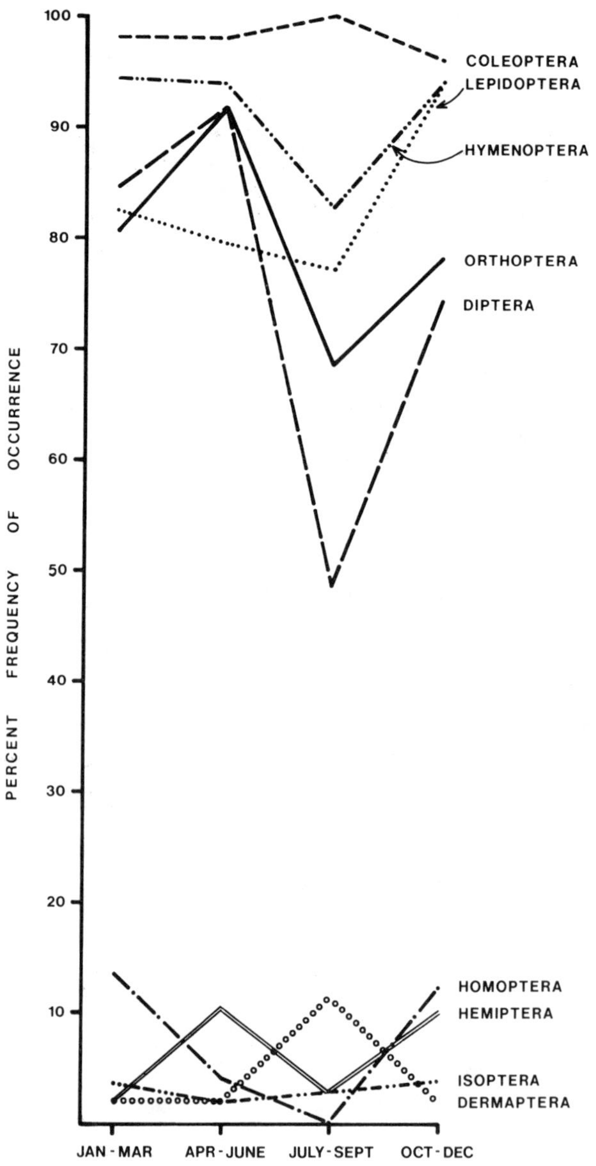

Figure 2. Seasonal occurrence of major insect orders found in stomachs of 186 armadillos in Florida.

represented in the stomachs (Table 4), but only seven families were regularly taken (Fig. 3). Click beetle larvae (Elateridae), known as wireworms, were found in 80 percent (April-June) to 91 percent (July-September) of the seasonal samples. Average number of larvae per stomach varied from 5.4 in the April-June sample to 15.2 in the October-December sample; four stomachs contained over 50 individuals, and one from the October-December sample held 150 larvae. Eight genera were identified (Table 2), *Melanotus* being most common and *Conoderus, Alaus,* and *Ampedus* occurring in lesser numbers. Scarab beetle larvae (Scarabaeidae) were found in 78 percent (October-December) to 94 percent (January-March, July-September) of the

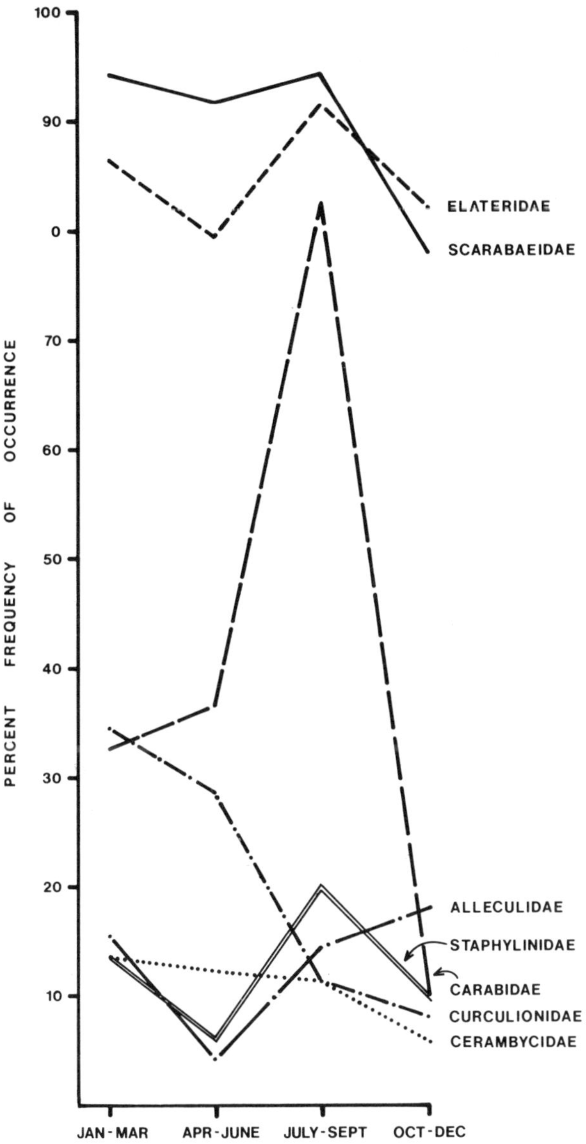

Figure 3. Seasonal occurrence of 7 major families of Coleoptera found in stomachs of 186 armadillos in Florida.

seasonal samples (Fig. 3). Average number of larvae per stomach varied from 14.4 in the October-December sample to 93.8 in the April-June sample; 31 stomachs contained more than 50 individuals, and 5 stomachs had more than 300 larvae. Only four genera were identified (Table 2), and of these the genus *Phyllophaga* was most frequently noted. Numbers of adult and larval ground beetles (Carabidae) varied among seasonal samples, with frequencies of 10 percent to 37 percent from October through June and 83 percent in the July-September sample (Fig. 3). Average number of individuals was 6.3 per stomach during the period of low incidence and 10.3 per stomach during July-September when large numbers of adults were taken in addition to larvae. Only three genera were identified (Table 2) due to extensive mastication of adults. The larvae of snout beetles (Curculionidae) occurred in 35 percent (January-March) to 8 percent (October-December) of the seasonal samples (Fig. 3). Average number of larvae per stomach varied from 1.0 in the July-September sample to 46 in the October-December sample. Most larvae were found singly, but a stomach in January-March held 227 larvae and two stomachs in October-December had 119 and 65 larvae, respectively, accounting for the high average number in the latter period. No larvae were identified to genus.

The remaining three families which regularly occurred were present in less than 20 percent of each seasonal sample. Comb-clawed beetle larvae (Alleculidae) were least abundant in the April-June sample and most abundant in October-December (Fig. 3). The average number per stomach varied from 1.5 in April-June to 19.7 in October-December, and they occurred singly or in numbers as high as 72 in one stomach from the October-December period. One genus was identified (Table 2).

Long-horned beetle larvae (Cerambycidae) were most abundant in the January-March sample and least abundant in the October-December sample (Fig. 3). Only 1 to 3 larvae, frequently 3 cm in length, were found per stomach. One genus was identified (Table 2). Rove beetle larvae (Staphylinidae) were most abundant in the July-September sample and least abundant in the April-June sample (Fig. 3). Two to five individuals were found per stomach and only one genus was identified (Table 2).

The remaining 18 families of beetles occurred in very low numbers (Table 4). Seven families (Dermestidae, Rhipiceridae, Buprestidae, Ptilodactylidae, Helodidae, Endomychidae, Lucanidae) were represented by one or two larvae found in one stomach (see Table 2 for genera identified). Eleven blister beetle larvae (Meloidae) were found in one stomach. Five families (Dytiscidae, Histeridae, Hydrophilidae, Scaphidiidae, Cantharidae) were represented by one or two larvae found in two to four stomachs. Firefly larvae (Lampyridae) were found in nine stomachs, maximum three per stomach, chiefly in the October-December sample. Glowworm larvae (Phengodidae) were found in six stomachs, maximum five per stomach, chiefly in the October-December sample. Darkling beetle larvae (Tenebrionidae) were found singly in nine stomachs, chiefly in the April-June sample. Long-jointed bark beetle larvae (Lagriidae) were found in seven stomachs, usually 5 to 6 per stomach. Case-bearing leaf beetle larvae (Chrysomelidae) were found in eight stomachs, 5 to 9 larvae per stomach.

Six families of ants, bees and their allies (Hymenoptera) were represented in the sample (Table 5), but only ants (Formicidae) composed a significant portion of the diet, occurring in 83 percent to 94 percent of the

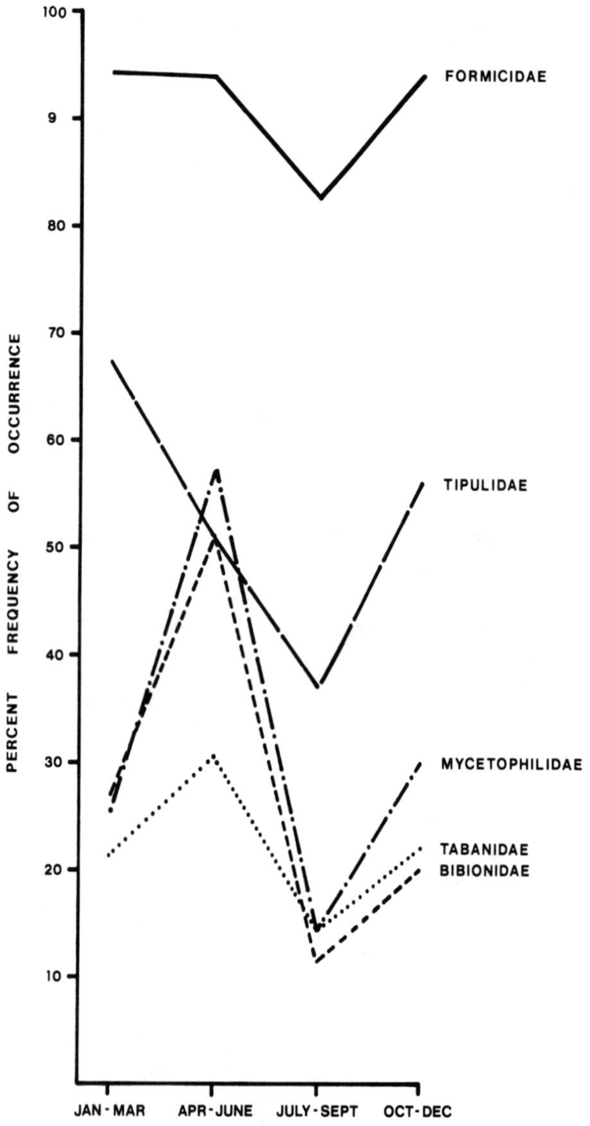

Figure 4. Seasonal occurrence of Hymenoptera: Formicidae, and of 4 major families of Diptera found in stomachs of 186 armadillos in Florida.

Table 5. Seasonal percent frequency of occurrence/number of prey taken of Hymenoptera families in 186 armadillo stomachs from Florida.

	Jan-Mar	*Apr-June*	*July-Sept*	*Oct-Dec*	*Total*
Apidae	2.0/1				1
Diprionidae		2.0/1		4.0/2	3
Formicidae	94.2/ 18591	93.9/ 27608	82.9/ 17193	94.0/ 25713	89105
Megachilidae			2.9/1		1
Sphecidae			2.9/2		2
Tenthredinidae				2.0/1	1

Table 6. Seasonal percent frequency of occurrence/number of prey taken of Lepidoptera families found in 186 armadillo stomachs from Florida.

	Jan-Mar	*Apr-June*	*July-Sept*	*Oct-Dec*	*Total*
Acrolophidae	65.4/793	36.7/53	57.1/91	60.0/394	1331
Arctiidae		4.1/2	2.9/8	8.0/5	15
Hesperiidae	19.2/21	46.9/251	14.3/5	8.0/10	287
Nepticulidae		2.0/1			1
Noctuidae	32.7/465	24.5/15	34.3/53	46.0/180	713
Notodontidae	2.0/1	4.1/3	5.7/2	14.0/7	13
Nymphalidae				2.9/1	1
Pyralidae	17.3/22	20.4/13	5.7/19	16.0/11	65
Saturniidae				2.0/1	1
Sphingidae				6.0/7	7
Torticidae				2.0/1	1

seasonal samples (Fig. 4). Average number of individuals, either adults, larvae, or eggs, per stomach varied from 379 in the January-March sample to 600 in the April-June sample; 10 stomachs contained over 2,000 individuals, and a stomach in the October-December sample held over 5,000 ants. Ten genera were identified (Table 2). Conifer sawfly larvae (Diprionidae) were found singly in three stomachs, while the remaining four families (Tenthredinidae, Sphecidae, Megachilidae, Apidae) were represented by single individuals; the honeybee (Apidae) was an adult (Table 5.)

Eleven families of moths and butterflies (Lepidoptera) were found in the stomachs (Table 6), but only five families were regularly taken by armadillos (Fig. 5). Most prevalent were burrowing webworms, the larvae of Acrolophidae, which occurred in from 37 percent (April-June) to 65 percent (January-March) of the seasonal samples. Average number of larvae per stomach varied from 2.9 to 4.6 in the April to September samples to 13.1 to 23.3 in the samples from October to March, seven stomachs contained over 50 individuals, and one stomach from January-March held 150 larvae. Only the genus *Acrolophus* was identified, and one stomach in the October-December sample contained 60 individuals. Noctuid moth larvae (Noctuidae), known as loopers or cutworms, occurred in from 24 percent (April-June) to 46 percent (October-December) of the seasonal samples (Fig. 5). Average number of larvae per stomach varied from 1.25 (April-June) to 27.4 (January-March); noctuids occurred singly in many stomachs, but three stomachs in the January-March sample contained 120 larvae or more. Ten genera of noctuids were identified (Table 2); the genera *Agrotis* and *Feltia* were most commonly found. Common skipper larvae (Hesperiidae) occurred in 47 percent of the April-June sample, but

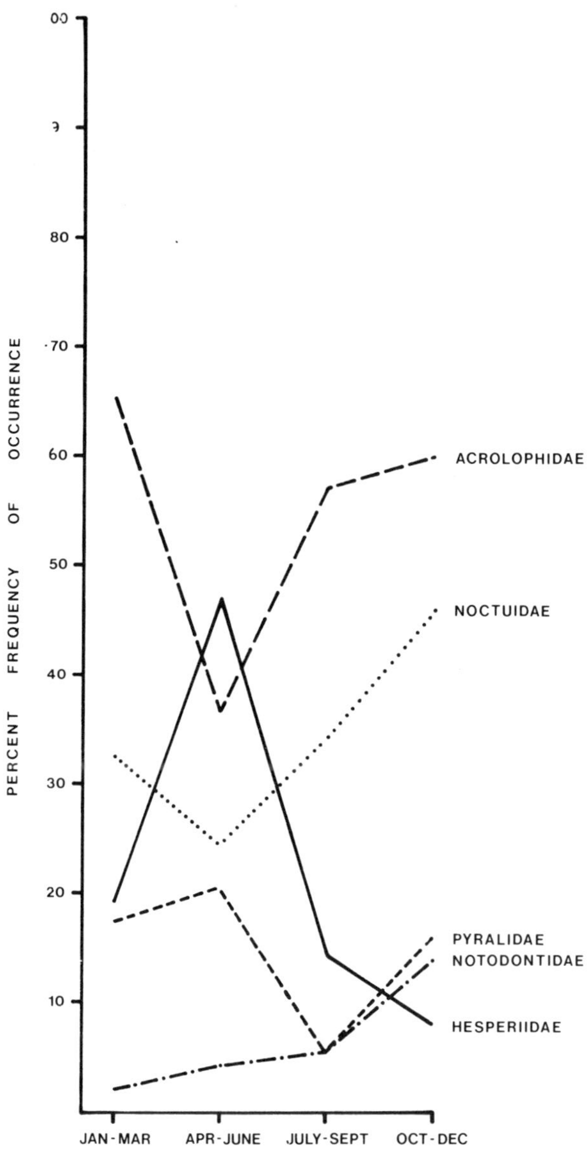

Figure 5. Seasonal occurrence of 5 major families of Lepidoptera (all larvae) found in stomachs of 186 armadillos in Florida.

were present in much lower numbers the remainder of the year (Fig. 5). Average number of larvae per stomach was 10.9 in April-June, but 1 to 3 during the remainder of the year. Larvae were usually found singly, but two stomachs in the April-June sample contained 72 individuals. Only the genus *Hylephila* was identified (Table 2). Pyralid moth larvae (Pyralidae) were found in from 6 percent (July-September) to 20 percent (April-June) of the seasonal samples (Fig. 5). Average number of larvae per stomach varied from 1.3 (April-June) to 9.5 (July-September), and one stomach from the latter period contained 18 larvae. The genera *Herpetogramma* and *Crambus* (Closewings) were identified (Table 2). Prominent moth larvae (Notodontidae) occurred in 2 percent to 14 percent of the seasonal samples (Fig. 5). Four

Table 7. Seasonal percentage frequency of occurrence/number of prey taken of Orthoptera families found in 186 armadillo stomachs from Florida.

	Jan-Mar	*Apr-June*	*July-Sept*	*Oct-Dec*	*Total*
Acrididae	28.8/29	63.3/330	37.1/20	34.0/40	419
Blattidae	38.5/61	28.6/46	17.1/17	42.0/85	209
Gryllidae	73.1/362	85.7/391	57.1/75	56.0/201	1029
Phasmatidae		2.0/2	5.7/2	4.0/5	9
Tetrigidae	3.8/2	8.2/7		6.0/13	22

genera were identified (Table 2), and only one or two larvae were found per stomach.

Tiger moth larvae (Arctiidae) were found in seven stomachs (Table 6), found in all samples except in January-March, and usually were found singly, except that one stomach in the July-September sample contained eight larvae. Sphinx moth larvae (Sphingidae) were found in three stomachs in the October-December sample. The remaining four families (Nymphalidae, Saturniidae, Tortricidae, Nepticulidae) were represented by single specimens from single stomachs.

Five families of Orthoptera were recorded from armadillos in Florida (Table 7). Crickets (Gryllidae) were most important, varying from 86 percent (April-June) to 56 percent (October-December) of the seasonal samples. Average number of individuals per stomach varied from 9.4 (January-June) to 5.8 for the remainder of the year. The genus *Scapteriscus* occurred in 21 percent of the January-March sample and 41 percent of the April-June sample. *Nemobius* and *Acheta* were found in the October-December sample, and single *Gryllotalpa* were present in the October-December and January-March samples. Short-horned grasshoppers (Acrididae) were present in 29 percent (January-March) to 63 percent (April-June) of the seasonal samples (Table 7). Average number of individuals per stomach was 10.6 in the April-June sample. In addition to adults, the October-December sample contained 539 eggs of this family. The genera *Melanoplus, Rhomaleum,* and *Schistocerca* were identified (Table 2). Cockroach adults or eggs were present in 17 percent (July-September) to 42 percent (October-December) of the seasonal samples. Maximum number of individuals per stomach was 17 in the latter period. Pygmy grasshoppers (Tetrigidae) and walkingsticks (Phasmatidae) were present in low numbers, the grasshoppers being most prevalent in the April-June sample and the walkingsticks in the July-September sample (Table 7).

Sixteen families of flies (Diptera) were represented in the stomachs analyzed (Table 8), but only four families were regularly taken (Fig 4). Crane fly larvae (Tipulidae) were most abundant, occurring in 37 percent

Table 8. Seasonal percent frequency of occurrence/number of prey taken of Diptera families found in 186 armadillo stomachs from Florida.

	Jan-Mar	*Apr-June*	*July-Sept*	*Oct-Dec*	*Total*
Bibionidae	26.9/ 8477	51.0/ 13474	11.4/ 59	20.0/ 955	22965
Bombyliidae			5.7/3		3
Calliphoridae	2.0/2			4.0/3	5
Ceratopogonidae	2.0/1				1
Ephydridae		4.1/5			5
Lonchaeidae		2.0/1		2.0/3	4
Mydidae		2.0/1		2.0/1	2
Muscidae		4.1/22		2.0/6	28
Otitidae		2.0/62		2.0/2	64
Sarcophagidae		4.1/129		2.0/3	132
Scatopsidae			2.9/1	2.0/9	10
Sciaridae	25.0/ 2186	57.1/ 27167	14.3/ 472	30.0/ 2599	32424
Stratiomyidae	2.0/1			4.0/10	11
Tabanidae	21.2/87	30.6/23	14.3/6	22.0/38	154
Therevidae	2.0/1		2.9/2	10.0/6	9
Tipulidae	67.3/ 2660	51.0/ 228	37.1/ 96	56.0/ 439	3423

(July-September) to 67 percent (January-March) of the seasonal samples. Average number of larvae per stomach varied from 7.4 (July-September) to 76 (January-March), 10 stomachs contained over 100 larvae, and one stomach from the January-March sample held 359. Only the genus *Tipula* was identified (Table 2). The occurrence of dark-winged fungus gnat (Sciaridae) and March fly (Bibionidae) larvae in seasonal samples was virtually identical (Fig. 4). Fungus gnat larvae varied in occurrence from 14 percent (July-September) to 57 percent (April-June). Average number of larvae per stomach varied from 94 in the July-September sample to 970 in the April-June sample, 10 stomachs contained over 500 larvae, and a stomach in the April-June sample held 15,400 individuals. Only the dark-winged fungus gnat, *Sciara*, was identified (Table 2). March fly larvae varied in occurrence from 11 percent (July-September) to 51 percent (April-June). Average number of larvae per stomach varied from 15 in the July-September sample to 605 in the January-March sample, 11 stomachs contained over 500 larvae, and one stomach in the January-March sample held 2,948 individuals. Only the genus *Bibio* was identified (Table 2), and one stomach in the October-December sample contained 408 individuals of this genus. Horsefly and deerfly larvae (Tabanidae) occurred in 14 percent (July-September) to 31 percent (April-June) of the seasonal samples (Fig. 4). Average number of larvae per stomach varied from 1.2 (July-September) to 7.9 (January-March), and tabanid larvae usually were found singly, though three stomachs contained over 10 larvae and one held 18 individuals. Two genera were identified (Table 2).

A single biting-midge larva (Ceratopogonidae) was identified (Table 8). Six families (Scatopsidae, Mydaidae, Bombyliidae, Otitidae, Lonchaeidae, Ephydridae) were represented by larvae found only in two stomachs, though 62 otitids were present in a single stomach from the April-June sample. Four families (Stratiomyidae, Muscidae, Calliphoridae, Sarcophagidae) were represented by larvae found in three stomachs; 125 sarcophagid larvae were present in one stomach in the April-June sample. One or two stiletto fly larvae (Therevidae) were found in seven stomachs, principally in the October-December sample.

Cicadas and their allies (Homoptera) were present in seasonal samples from October through June (Fig. 2), and occurred in low numbers per stomach. Only two families were identified (Table 9). Five families (Table 9) and nine genera (Table 2) of adult true bugs (Hemiptera) were found in 1.9 percent to 8.2 percent of the seasonal samples (Fig. 2). Hemipterans usually occurred singly in stomachs.

Termites (Isoptera) were present in very low frequencies in all seasonal samples (Fig. 2). Average number per stomach varied from 3.5 in the October-December sample to 317 in the January-March sample. Only one family was identified (Table 2). Two families of earwigs (Dermaptera) occurred in very low frequencies in seasonal samples (Fig 2, Table 9). Dragonfly (Odonata) wings were found in one stomach, and lacewing (Neuroptera) fragments were present in three stomachs (Table 9).

Crickets (Gryllidae) and short-horned grasshoppers

Table 9. Seasonal percent frequency of occurrence/number of prey taken of miscellaneous insect orders and families found in 186 armadillo stomachs from Florida.

	Jan-Mar	*Apr-June*	*July-Sept*	*Oct-Dec*	*Total*
Dermaptera	2.0/1	2.0/1	11.4/4	2.0/11	17
Labiduridae	1.9/11				11
Labiidae			2.9/5		5
Hemiptera	2.0/1	10.2/5	2.9/1	10.0/5	12
Coreidae				4.0/2	2
Gelastocoridae	1.9/1	8.2/4			5
Naucoridae		2.0/1			1
Pentatomidae	1.9/1	2.0/1		4.0/2	4
Reduviidae	1.9/1	2.0/2	2.9/1	2.0/3	7
Homoptera	13.5/18	4.1/2		12.0/6	26
Cicadidae	5.8/6	2.0/2		6.0/9	17
Margarodidae		2.0/7			7
Isoptera	3.8/635	2.0/249	2.9/44	4.0/7	935
Neuroptera		2.0/1		2.0/1	2
Sialiadae		2.0/1			2
Odonata				2.0/1	1

Table 10. Major components of the diet of the common long-nosed armadillo in Florida. Percent frequency in 186 stomachs, and the season when each component was important, are shown.

	Percent Freq-uency	*Season of Importance* Jan-Mar	Apr-June	July-Sept	Oct-Dec
Anura	21.5	X	X		
Araneida	76.3	X	X	X	X
Chilopoda	71.0				X
Coleoptera	97.8				
Carabidae (ad. and larv.)	37.1				X
Curculionidae (larvae)	21.5	X			
Elateridae (larvae)	84.4	X	X	X	X
Scarabaeidae (larvae)	98.2	X	X	X	X
Diploda	56.4			X	X
Diptera	76.9				
Bibionidae (larvae)	28.5		X		
Sciaridae (larvae)	32.8		X		
Tabanidae (larvae)	22.6		X		
Tipulidae (larvae)	54.3	X			
Lepidoptera	83.9				
Acrolophidae (larvae)	54.8	X			
Hesperiidae (larvae)	22.6		X		
Noctuidae (larvae)	34.4				X
Hymenoptera	92.5				
Formicidae	91.9	X	X	X	X
Oligochaeta	22.0				X
Orthoptera	80.6				
Acrididae	39.8		X		
Blattidae	32.8				
Gryllidae	68.8		X		X

(Acrididae) were prevalent in the April-June sample, and cockroaches (Blattidae) were common in the October-December sample (Table 10). Click beetle (Elateridae) and scarab beetle (Scarabaeidae) larvae formed a significant part of the diet throughout the year, while adult and larval ground beetles (Carabidae) were most important in the July-September sample, and snout beetle (Curculionidae) larvae were most prevalent in the January-March sample. Burrowing webworms (Acrolophidae) were most prevalent in the January-March sample, while common skipper (Hesperiidae) larvae were most important in the April-June sample, and noctuid moth (Noctuidae) larvae occurred most frequently in the October-December sample. Crane fly (Tipulidae) larvae were fed on heavily during January-March, while March fly (Bibionidae), dark-winged fungus gnat (Sciaridae) and horse- and deerfly (Tabanidae) larvae were taken predominantly during April-June. Ants (Formicidae) formed a significant part of the diet year-round. Earthworms (Oligochaeta) were taken most frequently in October-December, as were centipedes (Chilopoda). Millipedes (Diplopoda) formed a significant part of the diet from July through December. Spiders (Araneida, esp. Lycosidae) were an important food item throughout the year, while frogs (Anura) were taken chiefly from January through June.

A greater variety of food items was taken in the October-December period (74), but the average number of prey per stomach was low (679). Average number of prey items per stomach was highest for the April-June period (1,535) and variety was still high (63). The remaining periods are lower in both variety (53 to 56) and average number of prey items per stomach (610 to 712).

DISCUSSION

Although field data are needed concerning the foraging and prey-locating behavior of armadillos before definitive statements can be made, two factors probably influenced variety and numbers of prey consumed: seasonal reproductive patterns of arthropods, and increased vulnerability of invertebrates and vertebrates during cooler winter months. Increased consumption of centipedes, millipedes, earthworms, snails, and isopods during cooler months may have reflected the effect of lower temperatures on prey mobility. High consumption of lizards during cooler months probably was related to prey torpidity. Consumption of anurans in winter may be facilitated by their torpidity, while spring and summer consumption, when chiefly small anurans were taken, may have been related to reproduction.

In a warm temperate climate such as Florida, ants were apparently available year-round, as reflected by their importance in the diet. Orthoptera is the only insect order from which adults were regularly taken, and this consumption peaked in late spring and early summer when grasshoppers, crickets and cockroaches would be expected to be plentiful, and decreased noticably in late summer and early fall.

Dependence on Coleoptera was high throughout the year, but there was seasonal variation in the relative importance of families. Elaterid and scarabaeid larvae were important throughout the year, but less so in winter, while adult and larval carabid beetles increased dramatically in importance in the July-September period. High consumption of acrolophid and noctuid larvae during cooler months was on overwintering phases of the life cycle, while high consumption of hesperiid larvae, which are active leaf feeders, in the April-June period probably reflected their above-ground availability. Dipteran larvae were apparently most available in spring and early summer in Florida.

In Florida, armadillos prey largely on insect larvae. The larvae of 11 insect families which are heavily utilized by armadillos are found on or in the ground, in decaying plant material, or in carrion or dung (Carabidae, Elateridae, Scarabaeidae, Alleculidae, Cerambycidae, Noctuidae, Pyralidae, Acrolophidae, Tipulidae, Bibionidae, Sciaridae), and the larvae of 21 other families taken by armadillos in Florida are found in the ground, in moist soil, or in decaying plant material. The larvae of three families taken regularly by armadillos (Hesperiidae, Notodontidae, Curculionidae) are plant feeders; the notodontids are unique among these in that the larvae are gregarious and thus might be more available. Fourteen additional insect families taken by armadillos are foliage feeders; none were taken in significant numbers. The tabanids are unique among armadillo prey items in that the larvae are chiefly aquatic. Five families whose larvae are included in the diet are found chiefly in carrion or dung, and eight families, including the Carabidae, have larvae or adults which are predaceous.

Based on the kinds of organisms found in the stomachs, the picture is of a diet obtained primarily on or in the ground, frequently in moist places rich in decaying organic matter. That feeding frequently occurs in swampy habitats in Florida is indicated by the presence in the diet of hylid frogs, crayfish, leeches, snails, and at least 10 families of insects whose adults or larvae are closely associated with water (Naucoridae, Gelastocoridae, Dytiscidae, Hydrophilidae, Helodidae, Tipulidae, Ceratopogonidae, Stratiomyidae, Tabanidae, Ephydridae). Without information on the availability of commonly taken insect groups, we can only speculate as to whether armadillo feeding is selective, but the diversity of food items suggests that it is not.

The food habits of *Dasypus novemcinctus* have been reported by several authors, notably Baker (1943) and Kalmbach (1943) for Texas, Bailey (1931) for New Mexico, Fitch et al. (1951) for Louisiana, Bushnell (1952) for Florida, and Breece and Dusi (1985) for Alabama. An extensive summary of early data is presented by Talmage and Buchanan (1954), and food habits of *Dasypus* are reviewed by Redford (1985). All authors agree that beetles comprise a major portion of the diet, especially Scarabaeidae and Carabidae. Elateridae are apparently more important to armadillos in Florida than elsewhere. Ants, likewise, are an important dietary item in Florida and elsewhere, but not as important as early authors indicated (c.f. Audubon and Bachman, 1854; Bailey, 1905; Nelson, 1918; Newman, 1913). Lepidoptera and Diptera larvae and Orthoptera appear more important to armadillos in Florida than elsewhere, probably because the Florida sample is from more moist habitats. Spiders, centipedes and millipedes were taken frequently in Florida, as in Texas (Kalmbach, 1943) and Louisiana (Fitch et al., 1952). Frogs were more prevalent in the Florida sample than in Louisiana, while the lizard genera *Lagosoma* and *Anolis* were taken in both states (Fitch et al., 1952); *Anolis* were also taken in Alabama (Breece and Dusi, 1985). Plant foods were less important to armadillos in Florida than studies in Texas (Baker, 1943; Hamilton, 1946; Kalmbach, 1943) or Louisiana (Fitch et al., 1952) would indicate; such foods may not be preferred, but only utilized when insect food is not available in adequate quantity.

This study confirms, for a new portion of the range of *Dasypus novemcinctus*, that the species is an opportunistic insectivore, taking other invertebrates and small vertebrates when available. Its ecological relationships with insectivorous birds, reptiles, and other mammals is almost entirely unknown and would be a fruitful area for further study.

ACKNOWLEDGMENTS

Field aspects of this project were conducted under a grant from the Florida Game and Fresh Water Fish Commission and a grant-in-aid-of-research from the Society of the Sigma Xi to Wirtz while he was a graduate student at the University of Florida, Gainesville. A major portion of the insect identifications were done by the Entomology Bureau, Division of Plant Industry, Florida Department of Agriculture and Consumer Services, Gainesville, especially by G. W. Dekle. We are also indebted to the following for aid in identification of food items: Lewis Berner, University of Florida, Homoptera; Harold A. Denmark, Division of Plant Industry, Hymenoptera; Peter Drummond, University of Florida, Isopoda; Roland F. Hussey, University of Florida, Hemiptera; H. K. Wallace, University of Florida, Araneida; Robert E. Woodruff, Division of Plant Industry, Coleoptera. Completion of the project was accomplished under a grant to Wirtz from the Research Committee of Pomona College. Louis M. Luttrell served as research assistant at Pomona, and all figures were prepared by Helen J. Wirtz.

LITERATURE CITED

Audubon, J. J., and J. Bachman.
1846-1854. Quadrupeds of North America. 3 vols. V. G. Audubon, New York, 1107 pp.

Bailey, H. H.
1924. The armadillo in Florida and how it reached there. J. Mamm., 5:264-265.

Bailey, V.
1905. Biological Survey of Texas. N. Amer. Fauna, 25:1-221.

1931. Mammals of New Mexico. N. Amer. Fauna, 53:1-412.

Baker, R. II.
1943. May food habits of armadillos in eastern Texas. Amer. Mid. Nat., 29:379-380.

Breece, G.A., and J.L. Dusi.
1985. Food habits and home ranges of the common long-nosed armadillo *Dasypus novemcinctus* in Alabama. (this volume).

Bushnell, R.
1952. The place of the armadillo in Florida wildlife communities. Unpubl. M.S. thesis, Stetson Univ., 55 pp.

Fitch, H. S., P. Goodrum, and C. Newman.
1952. The armadillo in the southeastern United States. J. Mamm., 33:21-37.

Hall, E. R., and K. R. Kelson.
1959. The Mammals of North America. Ronald Press, New York, 1083 pp.

Hamilton, W. J., Jr.
1946. The black persimmon as a summer food of the Texas armadillo. J. Mamm., 27:175.

Hardburger, F. M.
1950. The nine-banded armadillo and the eastward expansion of its range. Turtox News, 28:174-177.

Humphrey, S. R.
1974. Zoogeography of the nine-banded armadillo in the southeastern United States. Bioscience, 24:457-462.

Kalmbach, E. R.
1943. The armadillo: its relation to agriculture and game. Texas Game, Fish and Oyster Comm., Austin: 1-61.

Layne, J.N. and D. Glover.
1985. Activity patterns of the common long-nosed armadillo *Dasypus novemcinctus* in south-central Florida. (this volume).

Neill, W. T.
1952. The spread of the armadillo in Florida. Ecology, 33:282-284.

Nelson, E. W.
1918. Smaller mammals of North America. Nat. Geo. Mag., 33:371-493.

Newman, H. H.
1913. The natural history of the nine-banded armadillo of Texas. Amer. Nat., 47:513-539.

Redford, K.H.
1985. Food habits of armadillos (Xenarthra: Dasypodidae). (this volume).

Strecker, J. K.
1926. The extension of the range of the nine-banded armadillo. J. Mamm., 7:206-210.

Talmage, R. V., and G. D. Buchanan.
1954. The armadillo. A review of its natural history, ecology, anatomy and reproductive physiology. Rice Inst. Pamphlet, Monog. Biol., 41:1-135.

Wetzel, R.M.
1985. Taxonomy and distribution of armadillos, Dasypodidae. (this volume).